The oil palm

TROPICAL AGRICULTURE SERIES

The Tropical Agriculture Series, of which this volume forms part, is published under the editorship of Gordon Wrigley.

The oil palm

(*Elaeis guineensis* Jacq.)

Third edition

C W S Hartley

CBE, MA, Dip. Agr. Sci., AICTA, FIBiol

Longman
Scientific &
Technical

Copublished in the United States with
John Wiley & Sons, Inc., New York

Longman Scientific & Technical,
Longman Group UK Limited,
Longman House, Burnt Mill, Harlow,
Essex CM20 2JE, England
and Associated Companies throughout the world.

Copublished in the United States with
John Wiley & Sons, Inc., 605 Third Avenue, New York,
NY 10158

© Longman Group Limited 1967, 1977

Third edition © Longman Group UK Limited 1988

First published 1967
Second edition 1977
Third edition 1988

British Library Cataloguing in Publication Data
Hartley, C. W. S.
    The oil palm (Elaeis guineensis Jacq.). – 3rd. ed. –
(Tropical agriculture series).
    1. Oil-palm
    I. Title   II. Series
    633.8'51   SB299.P3

ISBN 0-582-40400-2

Library of Congress Cataloging-in-Publication Data
Hartley, C. W. S. (Charles William Stewart), 1911–
    The oil palm (Elaeis guineensis Jacq.)/C. W. S.
Hartley. – 3rd ed.
        p.   cm. – (Tropical agriculture series)
    Bibliography: p.
    Includes index.
    1. Oil palm.   I. Title.   II. Series.
    SB299.P3H3   1988
    633.8'51 – dc19

ISBN 0-470-21070-2 (USA only)

Set in Linotron 202 10/11pt Times

Produced by Longman Singapore Publishers (Pte) Ltd.
Printed in Singapore

Contents

Disclaimer

NOTE:
The author has endeavoured to ascertain the accuracy of the state-
ments in this book. However, facilities for determining such accuracy with absolute certainty in relation to every particular statement
have not necessarily been available. Therefore, the reader is always
advised to seek an independent expert opinion before implementing
in practice any particular pesticide technique or method described
herein.

Preface to the third edition

The great expansion of oil palm planting in many parts of the world, but especially in South-east Asia, has continued during the last decade, and the availability of oil palm products on the world market has tripled during the period while at the same time consumption has increased dramatically in the producing countries themselves. In revising this book for a third edition opportunity has been taken to reduce its length by shortening the parts which deal with subjects, such as the palm groves, which are now of less agricultural or commercial importance, and to introduce the important new work on prospection, selection and breeding, tissue culture, pollination, pests and diseases, nutrition and mill sludge disposal.

As before, I am indebted to many organizations and individuals for helpful discussion and the provision of data, and I am especially grateful to Mr A. D. Marter of the Tropical Development and Research Institute for helping me update the statistics in Chapter 1, Dr J. A. Cornelius of the same Institute for assistance with Chapter 14, Dr R. H. V. Corley for the provision of photographs and information on tissue culture, and Dr van Heel for electron microscopy photographs of inflorescence development.

1987   C. W. S. Hartley

Preface to the second edition

For this new edition the text has been extensively revised. It is now ten years since the manuscript of the first edition was completed and since that time there has been great progress in research and an unprecedented enlargement of planted areas in Asia, Africa and America. Chapter 1 has been brought up to date and the widely different development methods being employed are discussed in this chapter and in Chapter 8.

In the revision of Chapters 3 and 4 particular account has been taken of the progress made in relating climate and soil to yield and of the work on growth analysis which has given a better understanding of the palm's performance under varying environmental and cultural circumstances. The part of Chapter 3 which deals with the African palm groves has been severely reduced as these areas are of dwindling importance in the total supply of oil palm products.

In Chapter 5 the recent work on heritability and on interspecific hybridization with the American oil palm is now incorporated. Methods of breeding currently employed are compared, while the latest prospections for new material and the research on growth factors in relation to selection and breeding are described.

Chapters 6 to 10 have been revised to take account of the progress made in cultural practices while Chapter 11 has been extensively rewritten and rearranged to allow for the substantial body of new data on many aspects of the nutrition of the palm. Chapter 13, on diseases and pests, has been brought up to date and expanded. In Chapter 14 the new work on oil quality is discussed and information is given on the oils of the American oil palm and the interspecific hybrid; reference is also made to new developments in both large and small mills.

Once again I am indebted to many research workers and organizations for assistance. In particular I wish to thank Mr A. H. Green and the Unilever Plantations Group for permission to quote from their Annual Reviews of Research, Messrs Harrisons and Crosfield for permission to make use of material from their Oil Palm Research Station Annual Reports, the Director of NIFOR, Nigeria, for supplying data for updating some experimental results, Dr B. S. Gray for information on Indonesian developments, the Department of Botany of the University of Birmingham for permission to quote from the thesis of Mr N. Rajanaidu, Dr J. A. Cornelius of the Tropical Products Institute, London, for oil analysis data, Mr J. J. McNerney and the Commonwealth Secretariat, and Oil World Publications, Hamburg, for export and other statistical data, and Dr R. H. V. Corley of the Oil Palm Physiology Unit, MARDI, and Unipamol Malaysia Ltd, for discussion and correspondence on physiology. I have also again been much helped by discussion with individual planters and members of research organizations in the many countries I have visited over the last ten years, and special mention should be made of the Institut de Recherches pour les Huiles et Oléagineux, Paris, whose publications, *Oléagineux* and Rapports Annuels, continue to be invaluable sources of information.

For this new edition, all data have been converted to the metric system. A conversion table has been provided at the end of Chapter 10 on p. 490. Fifteen new plates and eleven new text figures have been provided.

C. W. S. Hartley

5 October 1976

Preface to the first edition

During recent years there has been a considerable expansion of oil palm acreages in tropical Asia, Africa and America, and interest in the crop has been steadily increasing. At the same time there has been much improvement in the cultivation of the palm following research carried out by research institutes and plantation companies. Of particular value has been the interchange of information between the great producing regions of Africa and Asia and the realization that work done in one continent is often of great import to producers in another. Nevertheless, much of the work carried out has not been adequately published and original papers are sometimes difficult to obtain. It has therefore been my aim to provide in this book a comprehensive account of the oil palm as a plant, of the industry from its early beginnings to its present stage of development, and of the work carried out in all regions to improve cultivation, production and the extraction of the products. In so doing I have tried to interpret the difficulties that have been encountered in various parts of the world, to trace, historically and critically, the reasons underlying certain practices, and to draw attention to the experimental bases, where such exist, for present procedures.

I have been greatly assisted in the compilation of this book by the ready assistance I have received from many quarters. In the first place I have to thank the Managing Committee of the West African Institute for Oil Palm Research (now NIFOR) for assistance given to me and for permission to make use of material being the property of the Institute. Members of the research staff of the Institute, past and present, have contributed much to this book through their work. In particular I would like to acknowledge the help I have had during compilation from Mr G. Blaak, Mr T. Menendez, Mr S. C. Nwanze, Mr A. R. Rees, Mr J. S. Robertson, Mr R. D. Sheldrick, Mr J. M. A. Sly, Dr L. D. Sparnaaij, Dr P. B. H. Tinker and Mr A. C. Zeven; and I am especially grateful to Mr Robertson and Dr Tinker for reading and commenting upon parts of the text and to

Mr Rees for answering many queries on problems of germination and physiology generally.

Much assistance and data have also been generously given to me by the principal oil palm plantation companies and their research organizations, and for these I am very glad to be able to thank Mr D. L. Martin, Mr S. de Blank and Mr A. H. Green of Unilever Plantations Group, Mr B. S. Gray, Director of Research, and the headquarters staff of Messrs Harrisons and Crosfield Ltd, and Mr R. A, Bull, Director of Research (Oil Palms), Chemara Plantations Ltd. Discussions over the years with these veterans and stalwarts of the oil palm industry, and with many of their colleagues, have been of inestimable value to me. Dr J. J. Hardon, Oil Palm Geneticist, was kind enough to read and comment upon part of the text, and Mr B. J. Wood provided me with information on, and photographs of, Malaysian insect pests. My thanks are also due to Dunlop Plantations Ltd and Dunlop Malayan Estates Ltd for assistance in many ways, and to managers of oil palm estates in Malaysia, Africa and America, too numerous for separate mention but whose observations have often been of particular moment.

To the Department of Agriculture, Malaya, which first introduced me to the oil palm, and to Dr Ng Siew Kee, my thanks are due for the Malayan soils data included in the tables in Chapter 3 and for the data in Chapters 5 and 11 of certain field experiments. I would also like to thank the Director of Agriculture and his staff for many helpful discussions in Malaya in recent years.

In dealing with the oil palm in Sumatra my work was much facilitated by discussions and correspondence with workers conversant with the industry in that island. In particular I wish to thank Dr J. J. Duyverman and Mr J. Werkhoven of the Royal Tropical Institute, Amsterdam, Mr A. Kortleve of H.V.A. International, N.V., Mr F. Pronk, previously of AVROS, and Mr J. J. Olie and Mr M. J. van der Linde of Gebr. Stork and Co.; the latter kindly provided me with drawings and photographs and much information on processing plants.

My task has also been assisted by helpful discussion with research workers of the Institut de Recherches pour les Huiles et Oléagineux, Paris, and I have to thank M. Carrière de Belgarric, Director-General, Dr P. Prevot and M. M. Ollagnier for their friendly cooperation and for putting me in touch with their staff, both in Africa and America.

I have to thank the Ministry of Overseas Development for arrangements made for me to visit areas of oil palm development in a number of countries in South and Central America, and I am also grateful to the British Embassies in these countries for the very real assistance which they gave me. To Dr V. M. Patiño of Cali,

Colombia, my thanks are due for the supply of information on planting material and on introductions into Latin America, and on the American oil palm. Useful information from the American continent was also supplied to me by the United Fruit Company and, on insect pests, by Mr F. P. Arens of the FAO, Ecuador.

I should like particularly to thank Mr D. Rhind, CMG, for the many helpful comments he made during the final preparation of the chapters, and Mr E. O. Pearson, OBE, and his staff at the Commonwealth Institute of Entomology for checking the names of insect pests and supplying information and references.

The writing of this book has been made possible by the warm hospitality I have received from the Commonwealth Forestry Institute, Oxford, and I am especially grateful to Dr T. W. Tinsley, who welcomed me into his Section, to Professor M. V. Laurie for permission to work at the Institute, and to the Librarian, Mr E. F. Hemmings, and his staff for their unfailing help. Lastly, I have to thank my wife and children for some tedious work willingly done on data which I have used in this book.

C. W. S. Hartley

Three Gables,
5 October 1966

Acknowledgements

We are indebted to the following for permission to reproduce copyright material:

Bailey Hortorium Mann Library, Cornell University for figs. 2.1 (Rees 1960) & 2.5 (Rees 1963); the Chief Librarian of the Bibliothèque des Archives Africaines, Brussels for figs. 5.6 (Beirnaert 1933a), 10.6, 10.8 & 10.9 (Dufrane & Berger 1957); the Editor of *Biologist* for fig. 5.17 (Jones 1983); Cambridge University Press for figs. 4.4 & 4.5 (Corley 1973b); Centre for Agricultural Publishing & Documentation for tables 5.11 & 5.12 (van der Vossen 1974); Clarendon Press for figs. 4.2, 4.3 (Rees 1961) & 11.5 (Tinker 1964); Ghana Journal of Agricultural Science for table 11.19 (van der Vossen 1970); the author, Dr. van Heel for fig. 2.10 (van Heel *et al.* 1987); Incorporated Society of Planters, Malaysia for figs. 5.8 (Yeow *et al.* 1981), 9.3 & 9.4 (Corley *et al.*), 11.4 & 11.9 (Ollagnier & Ochs 1981), 11.7 (Warriar & Piggott 1973) and tables 4.10 (Syed *et al.* 1982), 4.11 (Donough & Law 1987) & 11.11 (Warriar & Piggott 1973); the Director of the Institut National pour l'Etude Agronomique du Congo for figs. 2.10 & 2.11 (Beirnaert 1935); the Director of *Oléagineux* for figs. 8.2 (Huguenot 1963) & 9.2 (Prevot & Duchesne 1955) and tables 5.14 (Meunier 1969) & 11.12 (Bachy 1969); Palm Oil Research Institute of Malaysia for fig. 5.16 (Rajanaidu *et al.* 1985a); the Editor, *Plant and Soil* for fig. 11.1 (Tinker & Smilde 1963a); Rubber Research Institute of Malaysia for figs. 10.1 to 10.5 (RRIM); Stork Amsterdam B. V. for figs. 14.10, 14.11 (Stork); Tropical Products Institute for fig. 11.3 (Ochs 1965); de Wecker Press for fig. 14.6 (De Wecker).

Whilst every effort has been made to trace the owners of copyright material, in a few cases this has proved impossible and we take this opportunity to offer our apologies to any copyright holders whose rights we may have unwittingly infringed.

We are grateful to the following for permission to reproduce copyright material upon which tables have been based:

The Director of Agriculture, Malaysia, for data from *Malayan Agricultural Journals*, **35**, 12, 1952: **41**, 131, 1958, and **20**, 16, 1932; Direction Générale de l'Administration Bibliotheque africaine for data by R. Pichel published in *Bulletin Agricole du Congo Belge*, **48**, 67; Butterworth and Co. (Publishers) Ltd for data from *Tropical Agriculture*, **39**, 271, 1962; The Clarendon Press for data from *Annals of Botany*, Vols 26 and 27, *Journal of Experimental Botany*, Vols 7, 9 and 12 and *Empire Journal of Experimental Agriculture*, Vol. 24; the Controller of Her Majesty's Stationery Office for data from *Soils of the Semporna Peninsula* by T. R. Paton; Institut National pour l'Etude Agronomique du Congo for data from *Publs. de l'INEAC Serie Tech.* **66**, 1961 and *Bulletin Agricole du Congo Belge*, **24**, 1933 and *Bulletin d'Information de l'INEAC*, **6**, 351; the Director of *Oléagineux* for data by C. de Berchoux and J. P. Gascon published in issues **20**, 1, 1965 and **18**, 713, 1963; the Editor of *The Planter* for data by B. S. Gray published in issue **42**, 16, 1966; the Tropical Products Institute for data on the 'Colour of samples of palm kernels received in the United Kingdom' by J. A. Cornelius, and the Managing Committee of the West African Institute for Oil Palm Research, Nigeria.

We are also indebted to the Commonwealth Economic Committee for supplying acreage and export statistics ahead of the annual *Vegetable Oils and Oilseeds*, and to the University of Birmingham for access to Mr N. H. Stilliard's unpublished thesis *The rise and development of the legitimate trade in palm oil with West Africa*. Many of the photographs have been supplied from the publications or collection of the West African (now Nigerian) Institute for Oil Palm Research. For the supply of certain other photographs we are indebted to Mr J. M. A. Sly, Mr T. Menendez, Mr B. J. Wood, Mr P. F. Arens, Gebr. Stork and Co. of Amsterdam, Société pour l'Equipement des Industries Chimiques of Paris, the Four Wheel Drive Auto Company, Clintonville, USA, Unilever Ltd, London and Ing. Sanchez Pates of Colombia.

Chapter 1

The origin and development of the oil palm industry

At the present time, the oil palm (*Elaeis guineensis* Jacq.) exists in a wild, semi-wild and cultivated state in the three land areas of the equatorial tropics: in Africa, in South-east Asia and in America. Of all oil-bearing plants it is the highest yielding, even the poorer plantations of Africa outyielding the best fields of coconuts, a crop which the oil palm has overtaken in the export field. Until recent centuries the palm has been confined to West and Central Africa, a region inhospitable even to the most adventurous of early traders, and it had to await the slave trade, starting after the early Portuguese voyages of the fifteenth century, before it could escape to another tropical land mass. It was not suited, like the coconut, to the foreshore, but it soon established itself meagrely behind the coastline in Brazil where it was little used except by the transported Africans who knew its value, and who, no doubt, had brought it with them. In Africa it remained a domestic plant, supplying a need for oil and vitamin A in the diet, and it was not until the end of the eighteenth and the beginning of the nineteenth centuries that it entered world trade.

The fruit of the oil palm is a drupe, the outer pulp of which provides the palm oil of commerce. Within the pulp or mesocarp lies a hard-shelled nut containing the palm kernel, later to provide two further commercial products, palm kernel oil (rather similar in composition to coconut oil) and the residual livestock food, palm kernel cake.

The origin of the oil palm

There is fossil, historical and linguistic evidence for an African origin of the oil palm. It has also been suggested on the evidence of an analysis by Friedel that fat found in a jar in a tomb at Abydos (*c.* 3000 BC) may have been palm oil (Raymond, 1961). Seward (1924) found no oil palm remains in the upper Cretaceous layers in eastern Nigeria but Zeven (1964) has reported fossil pollen from

Miocene and younger layers in the Niger delta as being similar to pollen of the oil palm as it grows today. That such pollen is found throughout these layers is strong evidence that the palm has been maintaining an existence in West Africa from very early times. Cook (1942) suggested a Brazilian origin for the oil palm; two of his grounds for this contention were that the palm grew 'spontaneously' in the coastal areas of that country and that all allied genera have an American origin. Zeven suggested that both *Elaeis guineensis* and another Cocoid palm, *Jubaeopsis caffra*, originated on the African side of the Tertiary land bridges which are believed to have lain between Africa and America, and that they became separated from other members of the tribe by the ocean. Corner (1966), however, considers that there are botanical and distributional grounds for discarding this theory and suggested that pre-Columbean transportation of *Elaeis* to Africa is probable.

The historical record of the oil palm is meagre and only recently have efforts been made to relate such records as exist to the main landmarks of exploration (Mauny, 1953; Rees, 1965; Zeven, 1965a). Colombus discovered South America in 1498 and Brazil was discovered by both the Portuguese and Spanish in 1500, but no real interest was taken in the country until the middle of the sixteenth century. If well-authenticated records of the oil palm in West Africa before or around these times can be established, then a Brazilian origin becomes less likely though the possibility of pre-Columbean transportation cannot be ignored.

Portuguese exploration of the Guinea coast started in 1434, Dutch and English exploration some 150 years later. There is no mention of the oil palm by the earlier Arab explorers or by Marco Polo. 'Palm trees' in abundance are mentioned in the account of Diogo Gomes's voyage of 1456 or 1457, but the account of the voyages of Ca' da Mosto (1435–60) (Crone, 1937) gives the first mention of a palm which strongly suggests the oil palm, namely:

> There is to be found in this country a species of tree bearing red nuts with black eyes, in great quantities, but they are small.

Of an oil used with food it was recorded that this

> has three properties, the scent of violets, the taste of our olive oil, and a colour which tinges the food like saffron, but is more attractive.

In the description give by Duarte Pacheco Pereira of his voyage of 1506–8 (*Esmeraldo de Situ Orbis*) mention is made of palm groves north of, and on an island off, the coast of Liberia, and of trade in palm oil (*azeite de palma*) near the Forcados river in Nigeria. Later Portuguese, Dutch and English accounts of voyages refer to palm wine as well as to the oil. These earlier accounts of palm groves and trade in palm oil, and the pre-Columbean description of the oil, make it certain that the palm could not have been brought from America by the Portuguese as claimed by Cook. In addition, the descriptions in the earlier herbals, although written after the discovery of America, are sufficiently early to make it likely that the plant described was native to West Africa and it is clear that the writers thought they were describing a plant indigenous to the Guinea coast. In certain cases specific reference is made to importation or to practices implying importation. For instance, Clusius (1605) stated that the palm was found on the Guinea coast and that the 'fruit, after addition of some flour of a certain root, was used by the Portuguese from San Thomé to feed their slaves during the whole journey' to America. Sloane (1696) reported that oil palms in Jamaica came from Guinea (Opsomer, 1956). Miller, in his *Gardeners' Dictionary* of 1768, stated that oil palm fruits had been carried from Africa to America by the Negroes. Finally, there are no early descriptions of the oil palm from Brazil.

Linguistic evidence also strongly supports a West African origin for the oil palm (Zeven, 1965a). All the West African vernacular names are short and can only be directly translated to mean oil palm while the Negro names for the oil palm in Surinam suggest that they are a corruption of the Yoruba, Fanti-Twi and Kikongo names of Africa; and the Brazilian name *dende* may be derived from the Kimbundu word *ndende* of Angola.

Habitat

Well before much of the above evidence was available, Chevalier (1934) was a strong champion of the African origin of the palm, and he suggested that its natural habitats were in *galeries forestières* or forest outliers either near *Raphia* or in associations of *Elaeis* and *Raphia*. The difficulty in determining the natural habitat of the oil palm lies in the fact that while it does not grow in the primeval forest it begins to flourish wherever man has cleared a part of the forest. It requires a relatively open area to grow and reproduce itself, and it thrives best when soil moisture is well maintained. It is natural to suppose therefore that the palm would find a place on the forest fringe near to rivers. Apparently wild associations of the oil palm and the *Raphia* palm have been noted bordering rivers in Zaire and on uncultivated islands, subject to annual flooding, in the Zaire and Ubangi rivers (Briey, 1922). Chevalier claimed that river habitats of this kind could be found right across Africa from Senegal to Angola and even to Mozambique. Similar habitats have now been assumed by 'escapes' in Sumatra and Malaysia.

The freshwater swamp has also been suggested as a natural habitat for the oil palm (Waterston, 1953). Palms growing in such swamps adjoining rivers have been seen in several parts of southern Nigeria. The palms are often to be found growing where there is standing, though not stagnant, water for many months of the year. However, these areas are always near to man's habitation, and it is not in any case possible to differentiate clearly between swamp habitats and the riverside habitats already mentioned. The differences are probably only ones of depth of penetration from the riverside, this depending to a large degree on the extent of the river flooding in the rainy season. As a producer, the oil palm will not tolerate permanently high water-tables in impervious soils, but it appears in its natural home to be tolerant of fluctuating water-tables and moving water in sandy or silty riverine soils. In short, it is most probable that the oil palm grows naturally near to rivers where the palms will be subject to less competition from the forest flora, where more light will therefore penetrate and where moisture is plentiful but not yet excessive.

The oil palm in its African setting

Some authors have attempted to distinguish between *palmeraies* (groves) *naturelles* or *spontanées* and *palmeraies subspontanées*, the supposition being that a 'spontaneous' grove or group of palms is one which has grown up without the intervention of man, while the 'subspontaneous' or semi-wild grove is one which owes its existence to the alteration of the natural vegetation by man in such a way that conditions suitable for unaided germination and development of oil palms have arisen. Chevalier (1934) even gave '*Elaeis bien spontanés*' a separate specific name, *E. ubanghensis*, claiming that it was distinguishable from the domesticated grove palm by the small bunches and fruit and the very thin mesocarp. While it is not difficult, in populated areas, to attribute the presence of palm groves to man's activities, in many isolated areas where oil palms are found it is virtually impossible to say with certainty whether the palms are truly wild or have found their position through the intervention, at some time in the past, of man. Even in an area which is under water for a considerable period of the year and which therefore gives the appearance of a natural habitat, the flooding may be of recent origin due to the progressive silting up or alteration in course of rivers during a comparatively short historical period of time.

The spread of the oil palm by seed may be through the agency of gravity and water, of animals, or of man. Movement by gravity and water must be of limited occurrence especially as the fruit does not float, but some groups of palms on the banks of rivers or streams may have arisen through the transport of fruit, perhaps on floating material, which has rolled down a slope into a stream. Animals considered to be responsible for the spread of seed are the common mammals, rodents and a few birds of tropical Africa. The main agency is thought to be man who has, for the most part, spent a wandering life in Africa and has been in the habit of carrying bunches from palm to homestead, dropping or throwing away odd fruit on the way.

Geographical distribution in Africa

The distribution before the First World War was summarized by Schad (1914) and this has been brought up to date by Zeven (1967). On the west side of the continent the oil palm occurs at 16° N near St Louis in Senegal. Inland it has been recorded on the upper Niger at 13° N near Bamako. Though there are planted palms around Dakar and in the Gambia, no extensive areas of palms are found in the countries of the coast until the Fouta Djallon district of Guinea is reached at the 10–11° N parallel. From here the real palm belt of Africa runs through the southern latitudes of Sierra Leone, Liberia, the Ivory Coast, Ghana, Togoland, Benin (Dahomey), Nigeria, Cameroon and into the equatorial regions of the Republics of Congo and Zaire. The northern limit of this belt varies with the isohyets and with topography. On the coast there is a small area surrounding Accra where rainfall falls to 65 mm per annum and the presence of grass savannah excludes the oil palm.

Except in the heavy summer rain areas of Guinea and Sierra Leone the oil palm only spreads beyond about 7° N into favoured river valleys under escarpments (e.g. west of the Jos plateau in Nigeria) or where underground water supplies are available. Beyond this latitude the oil palm gradually gives place to the palmyra palm (*Borassus aethiopicum*). However, small groups of palms occur in relatively dry areas in the Central African Republic and southern Sudan.

In Central Africa the main area of spread of the oil palm lies between 3° N and 7° S and extends through Zaire and parts of the Republic of Congo and Angola. The groves are most concentrated in the Kwilu and Kwango districts between 4° and 7° S and in Cabinda in Angola. In the coastal districts of Angola the climate is too dry though palms are found as far south as Mocamedes (15° S). Groves in the north are contiguous with those of southern Zaire, and plantations have been established south of the Rio Cuanza.

The spread of the oil palm into East Africa is believed to have been from Zaire. Dense stands of palms are found in the Semliki valley on the Zaire–Uganda border, in the Ruzizi plain between Lakes Kivu and Tanganyika and in a belt about 19 km wide on the

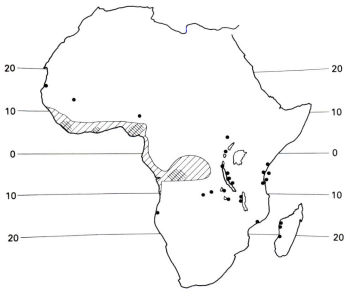

Fig. 1.1   An impression of the distribution of the oil palm in Africa. The shaded areas represent the 'palm belt' where the main groves are situated and where the palm is ubiquitous. The black dots represent the position or boundaries of isolated areas of colonization. The areas of heaviest population and production are double hatched. (Adapted from Zeven (1967) and earlier authors.)

eastern shore of Lake Tanganyika. Most of East and South-east Africa is far too dry for the oil palm and it appears only at altitudes below 1,000 m near lakes or watercourses where a reasonable rain-fall is assured. It exists on the west shore of Lake Malawi as far as 13° S and is also to be found on Lake Bangweulu and Lake Mweru in Zambia and scattered about in the south-eastern regions of Zaire at about 10° S wherever moisture conditions are favourable.

These areas are separated from the coast by belts of very dry territory and it seems not unlikely that the oil palm was brought to the coast during the long period of the Arab slave trade. Today, oil palms are to be found sporadically from the Tana river in Kenya to Dar es Salaam in Tanzania and on the islands of Zanzibar and Pemba (between 3° and 7° S).

The oil palm is also found on the island of Madagascar (Malagasy Republic) on a stretch of the west coast from south of Cap St André to the basins of the Manambolo and Tsiribihina rivers and as far south as 21° of latitude where it is known as *tsingilo* (Jumelle and Perrier, 1911). Climatically, the east coast of Madagascar seems more suited to the oil palm and although at one time it was thought to be indigenous on the west coast it seems more probable that it was brought across the sea by Africans of continental Africa when they entered the island and that its distribution is largely due to chance introduction.

The oil palm is a plant of the lowlands and in the great areas of productive groves it is found from sea-level to about 300 m. However, when mountainous country adjoins these areas or where the oil palm has been transported to high tablelands having sufficient rainfall, it has been able to survive at much greater heights. The most interesting mountainous habitat is the Cameroon mountain which is not far removed from the greatest grove area of Africa in eastern Nigeria. Here palms occur at over 1,300 m. In Guinea, the palm grows at 1,000 m in the mountains of the Fouta Djallon, and in East Cameroon it is found in abundance at 1,000 m in two favoured localities (Portères, 1947). Most of the East African inland localities are at a height of around 1,000 m.

Within 7° of the equator and in some exceptional areas at higher latitudes, wherever a large population is to be found, the typical palm groves of West and West Central Africa have arisen. Dense groves are found most continuously and extensively in eastern Nigeria, but the palm is so widespread that it is true to say that it would be difficult to stand anywhere in the palm belt except in the heart of a primary forest and not be able to see a few palms. In areas of highest population the groves are found at their densest, but in areas of relatively low population, such as in parts of western Nigeria, 'palm bush' country has developed where the density is rarely more than thirty adult palms to the hectare. In Zaire the groves are to be found mainly in the south of the country. In some areas dense groves comparable to those of eastern Nigeria exist, but in general the groves are less uniform and occur in long narrow belts or in patches alternating with forest or derived savannah.

Before the establishment of plantations in the Far East and in Zaire the groves of Africa were the sole source of palm oil and kernels on the world market. Large quantities of palm oil have always been consumed locally in West and Central Africa, and in the last quarter-century the rapidly increasing populations have removed from the export field the bulk of the surplus which used to be shipped to Europe or elsewhere; the Ivory Coast is now the only country in Africa with a significant export.

In areas where the fruit quality, as judged by the thickness of the mesocarp layer, is poor, kernels may sometimes be extracted from the fruit without the prior extraction of the mesocarp oil. This, together with the local consumption of palm oil, accounts for the very high ratio of kernels to palm oil which has always been characteristic of the African trade.

Besides providing the main source of palm kernels for export and an important source of palm oil for local consumption, the grove palms can produce large quantities of palm wine for consumption in urban areas and villages. In many parts of West Africa the palms surrounding large towns are regularly used for tapping for wine; the industry is an important one and the wine has dietetic value because of its vitamin B content. The value of an oil palm for wine production depends largely on its position and there is no doubt that where there is a steady market for wine a higher return can be obtained from tapping than from the harvesting of bunches.

The grove palms also provide a number of other products which are of importance in the domestic life of the African peoples. Many of these products, such as leaves for thatching, petioles and rachises for fencing or for protecting the tops of mud walls, may decrease in importance with the increasing use of more durable materials in the building of houses. Until recently little use was made in Africa of the palm kernel, although occasionally the kernel oil was expressed for soap-making and a very crude method of extracting the oil by frying the kernels in a large pan was to be found in Togoland and elsewhere. In Zaire, Nigeria and Benin (Dahomey) large kernel crushing industries have arisen using kernels from both plantations and groves; by 1972 African countries were exporting around 100,000 tonnes of palm kernel oil annually, and a fluctuating trade has continued.

The ash obtained by burning the refuse which remains after stripping the fruit from the bunches is rich in potash and is sometimes used in soap-making.

Early trade

West Africa was isolated from Europe throughout the Middle Ages; but when the Ottoman Empire blocked the way to the Orient, Portuguese seamen pushed their way round the west coast in a quest for spices, establishing a fort at Elmina in 1482. The discovery of the Americas soon diverted attention from this trade, however, and when interest was revived it was in a trade for slaves to sustain the development of the new countries of the Western world. The slave trade proper survived from 1562 to 1807, and during the whole of that time, though palm oil is mentioned as food for slaves, there was virtually no other commerce on the west coast. Vessels followed the 'triangular passage'; European goods were exchanged for slaves on the coast, slaves were exchanged for West Indian or American goods on the other side of the Atlantic and the ships then returned to Europe.

At the beginning of the nineteenth century the illegal slave trade had to be replaced by what came to be known as the 'legitimate trade' between Europe and the coast, but palm oil hardly figured among the commodities which, at that time and in very small quantities, supplemented the traffic in slaves. Ivory and timber were the most important of these, though gold dust, pepper, rice and gum copal also played a part. Although very small amounts of palm oil had been traded in England in 1588 and 1590, the first import records show that in 1790 less than 130 tonnes were imported into the United Kingdom and in the year of the abolition of the slave trade (1807) the quantity was no more.

The dominance of the slave trade and the methods of trading afford an explanation for the almost complete lack of knowledge of palm oil in Europe at that time. Until 1804 there was not even an accepted name for the oil in English, the Latin form *oleum palmae* often being used in documents. Some quaint ideas were held by traders about the origin of the oil, some believing that it was drawn off from the roots. Trade restrictions, navigation laws, the small number of traders, lack of access to the interior, local disturbances, disease and the continuance of illegal traffic in slaves all helped to prevent a rapid increase in trade. Imports into the United Kingdom from West Africa reached 1,000 tonnes in 1810, but thereafter a fluctuating trade continued well into the 1820s. It has been said that the trade was almost established in the legitimate commerce of Africa by 1830, while the developments which took place after that year made its future certain (Stilliard, 1938). These developments were the taking of more active measures for the suppression of the slave trade and the active encouragement of the palm oil trade by the British Government.

The main port was Bonny in the Niger delta, of which the missionary, Hope Waddell, wrote in 1846, 'By its safe and extensive anchorage, its proximity to the sea and connection with the great rivers of Central Africa, Bonny is now the principal seat of the palm oil trade as it was formerly of the slave trade.' By 1830 the oil trade had become larger in the delta than the trade in men and more than half the oil was handled at Bonny (Waddell, 1863).

The merchants' difficulties were not over, however. During the 1840s the expected increase in the trade did not take place. This stagnation was caused by an unexpected increase in the illegal slave trade induced by a demand from Brazil and Cuba. The increase in the number of slaves shipped was considerable and the measures which had to be taken against the trade were temporarily detrimental to the palm oil trade itself. Vessels engaged in the legitimate trade were held on suspicion of being slavers, often on quite trivial grounds. Matters were not brought under control until the 1850s, soon after a British Consul for the Bights of Biafra and Benin had been appointed at Fernando Po. This was followed by the capture of Lagos and the winning over of the traffic in the Bight of Benin to the legitimate trade. Henceforward palm oil was to be exported from the Benin river as well as from Bonny and Calabar in the area known as the Oil Rivers (Stilliard, 1983; Dike, 1956).

The trade was carried forward in an unusual manner. The European trader did not leave his ship; this was through justified fear of disease (as much as one-quarter of the crew often died even when remaining aboard in the rivers) and because the coastal tribes controlled the trade, buying oil from the interior where it was produced and, as middlemen, selling to the traders. The use of vast quantities of cowries, taking the place of modern currency, hindered the completion of transactions in the interior, and poor extraction methods led to arguments over quality. The regulation of the trade and the prevention of a resurgence of slave trading were assisted in the 1840s by the presence of a cruiser squadron in West African waters. The advantages and disadvantages of this protection were much argued at the time and the squadron gave place in the 1850s to the consular system, which was generally considered an improvement. The inauguration of this system gave rise, in its turn, to the idea of having settlements ashore for the trade. The value of British settlements, as opposed to the merchants' own commercial treaties with the chiefs of the coast, was much debated during the 1960s and there were many merchants who were not in favour of British penetration.

The establishment of the legitimate trade in palm oil was made possible by the industrial revolution in Europe which led to a new type of demand for tropical commodities. These were not only required for immediate consumption, but were needed as raw material for industries. 'People began to take washing seriously', and the demand for vegetable oils for soap-making increased rapidly as did the use of both animal and vegetable oils as lubricants for machinery. During the 1830s between 11,000 and 14,000 tonnes of palm oil per annum were exported from West Africa. A fluctuating trade continued during the 1840s and 1850s but by the end of the 1860s 25,000–30,000 tonnes per annum were being exported from the delta area alone, with prices varying between £34 and £44 per tonne (Dike, 1956). From then onwards until the First World War, exports steadily increased, over 87,000 tonnes, valued at nearly £1,900,000, being exported from British colonial territories in 1911 (Billows and Beckwith, 1913).

This large increase in importation was caused by the oil's wider uses. In the late eighteenth century it was imported for use by wool combers and soap boilers. It continued in use for soap-making throughout the nineteenth century; mixed with tallow and suet to make yellow soaps and, later, bleached for the manufacture of toilet soaps. In the middle of the nineteenth century a London firm patented a process for the treatment of fatty acids as a result of which palm oil became widely used in the manufacture of stearic candles. It was also used as a lubricant in the early days of the railways. Margarine manufacture started in the 1870s and the best-quality palm oil was employed for this, though palm kernel oil was, at that time, generally more suitable. In the 1890s a new use for palm oil was found in the tin-plate industry; oxidation of the iron was prevented by a coating of palm oil before the actual plating was carried out.

About 1850 both copra and palm kernels began to take a place in the provision of oil for soap-making. The oil from these two products is very similar. Palm kernel oil is almost colourless and odourless and came to be regarded as better adapted for soap-making than palm oil. Apart from this factor the kernels, after oil expression, left a valuable cattle cake which was increasingly used by farmers in Europe. Exports of palm kernels started in 1832 but production, carried out entirely by African women who extracted the kernels by cracking individual nuts, did not increase rapidly until the latter half of the nineteenth century; by 1905 exports of kernels from British colonial territories were over 157,000 tonnes, more than two and a half times the export of palm oil, and by 1911 over 232,000 tonnes valued at more than £3,400,000 were exported. The rise of this village industry in the British colonial territories was even more spectacular than that of palm oil; about 75 per cent of the world production came from Nigeria and about 18 per cent from Sierra Leone.

In spite of the increase in the trade in palm oil, palm kernels and other oils and fats, the demand at the beginning of the twentieth century proved greater than the supply. The soap and margarine industries were in competition. In 1913 it was stated that as a result of this competition the price of both coconut and palm kernel oil had almost doubled in the previous 10 years.

It was this situation which began to focus attention on the sources of supply and methods of production. Little was known of the native methods of palm oil extraction and many early descriptions were inaccurate or incomplete. Hand cracking of nuts for kernel extraction was slow and tedious and there had been some opposition to the trade by chiefs who dealt in palm oil. There was a general realization that a much larger quantity of palm products would be forthcoming if more efficient methods were employed.

Thus the second decade of the twentieth century saw the beginning of two developments which were to change, during the inter-war period, the pattern of oil palm exports. These were the Lever concessions in the Belgian Congo (now Zaire) and the establishing of the Deli oil palm as a plantation crop in Sumatra and, somewhat later, in Malaysia.

Early in the century, Sir William Lever (Lord Leverhulme), in his search for a further supply of oils and fats, approached the British Colonial Office for rights in Sierra Leone to introduce mills for treatment of palm fruit and to acquire land on which to cultivate the crop. A second application, in 1908, was confined to the first object. After negotiations entailing counter-proposals by the Government, a concession in Sierra Leone was given in 1911. In the Gold Coast (now Ghana) and Nigeria, however, the Governors were opposed to such concessions on the grounds that the land belonged to the native peoples and that the Crown made no claim to be able to dispose of it in any way. Some fears were aroused among the peoples themselves and these were voiced in meetings and deputations. Resistance to concessions was also supported by a section of public opinion in England and by a number of prominent merchants trading with West Africa.

The concession system failed in Sierra Leone. The cause of failure was said to be that the price offered compared unfavourably with what could be earned by extracting and selling oil and kernels by native methods. Whether this was so or not, insufficient fruit was brought to the mill for it to be operated successfully, and this might have been partly due to the long distances of carry needed in the relatively sparsely covered Sierra Leone groves. The great grove areas of West Africa therefore continued to be exploited throughout the next 30–40 years by crude native methods. Production and exports rose steadily, though not by any means spectacularly, throughout the inter-war period and the quality of the palm oil remained low. The British West African territories continued to dominate the trade in palm kernels, but events elsewhere reduced the relative importance of West Africa in the trade in palm oil.

Having failed to obtain the concessions he required in West Africa, Sir William Lever turned his attention to the great areas of largely unexploited groves in the Congo basin. In 1911 he obtained large concessions of land for the Huileries du Congo Belge (HCB) and entered into an agreement with the Belgian Minister for the Colonies by which, in certain defined regions, the company agreed to erect oil mills in return for a guarantee that all fruit harvested would be processed in them. The establishment of this company not only led to a rapid increase of oil and kernel exports, which in 1912 amounted in total to less than 8,000 tonnes, but also encouraged the exploitation of groves outside the Lever concessions. In 1933, new arrangements were made for concessions of palm grove areas by zones, each concession being limited to 15 years and subject to definite rules regarding methods of harvesting, purchase price, etc. and including an undertaking to plant small estates of selected oil palms.

By 1935, exports from Zaire had reached over 56,000 tonnes of palm oil and over 64,000 tonnes of kernels (Anon., 1949). Half the oil exported and a sixth of the kernels were produced by HCB. Outside this company the production contained a lower percentage of oil than of kernels, the amount of oil produced by village methods being relatively small and to a great extent locally consumed. The cracking of nuts for the sale of kernels for cash soon became an important village occupation, however, though Nigerian production was never rivalled.

The West African trade in palm kernels took a peculiar course in the early part of the present century. At the outbreak of the First World War, although British colonies were responsible for most of the production, Germany was buying about three-quarters of the kernels exported, and a considerable seed-crushing industry had been set up. During the war West African production had to be directed to the United Kingdom and, indeed, was required for the large quantities of margarine to be manufactured. New seed-crushing mills were erected and, to maintain the industry after the war, a protective export duty, to be remitted on exports to Commonwealth countries, was imposed in West Africa. This duty soon became unpopular, however, and was abolished in 1925.

The advance, retreat and improvement of the palm groves

Until after the First World War palm oil and kernels were supplied entirely from the groves of Africa and, to a much lesser degree, of Brazil. The phenomenon of the semi-wild or subspontaneous grove was first studied by Father Hyacinthe Vanderyst (1919) in southern Zaire, but it was not until after the Second World War that the West African groves were studied with any thoroughness. Vanderyst postulated that, in areas where the palm was already part of the natural vegetation, the factor of greatest importance in the development of a grove was the growth of a dense population. The cutting down of forest areas for annual crop cultivation removed dense shade and created conditions suitable for the rapid establishment of the palm. Moreover, the sooner the farmer returned to the area to cultivate it again, the better the conditions for palm establishment, since long periods of secondary forest growth would re-impose shade and restrict seedling development. With an increasing population, therefore, the secondary forest gradually gave place to palm grove, while where areas were abandoned, owing in some parts of Zaire to sleeping sickness, the groves decayed and slowly gave place to secondary forest.

Father Vanderyst also examined the occurrence of groves in the less favourable plateau country of the Kwilu district and showed how high grassland, under-populated and subject to grass fires, was unsuitable for grove establishment, whereas in lower areas, where villages had been established in the past and fields had been cultivated, groves of varying density had grown up.

The greatest areas of palm groves, supporting large populations and providing the highest quantities of oil and kernels, are to be found in West Africa. All along the coast from Sierra Leone to

West Cameroon, the oil palm has multiplied wherever population has increased, except in those areas, e.g. Ghana and western Nigeria, where cocoa has become the dominant economic crop or, as on the Accra plain, where rainfall has been well below 1,000 mm per annum. Population has, in fact, proved of more importance in the spread of the oil palm than has rainfall; in Benin (Dahomey) where the rainfall is often less than 1,200 mm per annum, the oil palm dominates the scene because there is no other suitable cash crop.

South-eastern Nigeria has been the most important area for grove exploitation although mid-west Nigeria has always produced as high an export of palm kernels. Studies in these regions provided information on the history and composition of groves and made it possible to piece together the factors giving rise to groves of different types, to determine their production and to consider their improvement and likely future (Waterston, 1953; Hartley, 1954; Zeven, 1965b). Comprehensive descriptions of these and other groves, their development, exploitation and decline have been provided by Zeven (1967, 1972), and for a fuller understanding of the historical sequence the reader is referred to his works or to earlier editions of this book. There is now a rapid decline in the world importance of the grove palms partly because conditions for their natural development and exploitation are disappearing through the unrelenting growth of the human population, and partly because of the rapid advances made by new plantations, both large and small. The circumstances which gave rise to the heyday of the groves and to their decline will therefore only be summarized briefly here.

Areas with semi-wild palms may be classified as follows.

1. *Secondary forest with oil palms*. These are areas where food-crop farming is occasional, the human population less than 20 per km^2, and the number of stemmed palms not more than about 70 per hectare. A typical block of 40.5 hectares in mid-western Nigeria gave a mean annual bunch yield over 12 years of 1,446 kg per hectare. With the cessation of farming in the area the palm population declined through deaths from *Ganoderma* Trunk Rot and the failure of self-sown seedlings to survive. Stemmed palms are classed as 'smooth-stemmed' or 'rough-stemmed', the latter so called because they have not yet dropped their leaf bases. In all types of grove the rough-stemmed palms contribute very little, and the seedlings nothing, to yield.

2. *Palm bush*. This is a transitional stage between secondary forest with palms and the more complicated structure of heavily populated regions. The human population is between 60 and 120 per km^2,

stemmed palms number between 75 and 150 per hectare, and bunch yields are usually over 2,000 kg per hectare.

3. *Dense groves and farmland palms*. These groves are typical of south-eastern Nigeria where population pressure is high and the groves have reached their highest development, with the palms in some areas in almost pure stand with only small shrubs or shade-tolerant food crops growing between them. Waterston (1953) classified them into subtypes according to their stage of development or recession (Table 1.1). Land with E-type palms can become dense groves (B-type) again when a household moves. A subtype in Sierra Leone, described as *farmland with palms* (Hartley, 1961), covers wide areas of high population where 'bush-rotation' farming is carried on among palms which suffer regularly from scorching when the bush is being burnt off. As a result of both more frequent burning and increasing population, both the number of palms and their yield are declining.

Table 1.1   *The development of dense groves and farmland palms*

Grove sub-type	Origin and characteristics
A. Primary compound palms	Palms growing around homesteads set up in areas of original forest
B. Dense grove palms	Dense groves have developed around the old homestead following the movement of its inhabitants into a new forest clearing or into farmland
C. Farmland palms in degraded groves	These areas have become largely farmland because of the reduction of the stand in the dense groves through disease or cutting out
D. Open farmland palms	Palms often grow up in small numbers in areas which have been selected as farmland
E. Compound palms near homesteads set up in farmland	When population pressure has become very high new homesteads must, perforce, be constructed in farmland

Table 1.2 gives the composition and yield in 1949–51 of plots of different grove subtypes in an area surveyed at Asuten Ekpe in south-eastern Nigeria. This area was originally thought to be one of *dense grove and farmland with palms* with several plots of the A subtype, i.e. palms around homesteads set up in original forest. Zeven (1967) showed, however, that in fact this area belonged to a fourth type.

4. *Thinned grove*. Groves of this type, though having similar sub-types to original *dense grove and farmland with palms*, belong to a second or subsequent generation or cycle of groves, and they show

Table 1.2 *The mean composition and yield of plots of different grove sub-types of thinned grove at Asuten Ekpe, south-eastern Nigeria (1949–51)*

Grove type	Number of palms per hectare			Yield of fruit bunches (tonnes per hectare per annum)
	Smooth-stemmed	Rough-stemmed	Total	
B. Dense groves	94	106	200	3.07
C. Degraded groves with farmland palms	86	62	148	2.55
D. Farmland	57	35	92	1.55
E. Groves around compounds in farmland	52	62	114	1.91

evidence of thinning to accommodate the increased farming needs of a population now grown to 250–300 per km^2. A resurvey of the palms at Asuten Ekpe showed that thinning was progressive in all subtypes, yield over a 6-year period of compound palms or dense plots (subtypes E and B) was still 2.9 tonnes of bunches per hectare per annum, while yields of the subtypes C and D were 2.1 and 1.6 tonnes respectively. Thus yield per palm had increased considerably to compensate for the decrease in the number of palms.

Yields of original dense grove can be much higher. Table 1.3 shows the relation between density and yield in one such grove at Abak in south-eastern Nigeria. The densest plots were found to be the highest yielding even though the density of stemmed palms per hectare was well above the optimum density for plantations. Zeven (1967) showed that though the optimum density for this grove was over 300 stemmed palms per hectare, calculated optimum densities varied greatly from grove to grove according, he suggested, to age, soil fertility and grove size. Annual yields of B subtype dense groves

Table 1.3 *The relation between density and bunch yield in a dense palm grove at Abak, Nigeria*

Density group (stemmed palms per hectare)	No. of plots	Mean density per hectare	Bunch yield per stemmed palm (kg)	Bunch yield per hectare (tonnes)
1–50	13	26.8	35.5	0.95
51–100	27	73.4	25.9	1.91
101–150	19	131.1	24.4	3.20
151–200	21	170.3	19.3	3.29
201–250	14	221.8	18.2	4.04
251–300	7	271.4	18.9	5.13
Over 300	4	318.7	14.3	4.57

in Nigeria may be expected to average less than 3 tonnes of bunches per hectare, though double this yield may be had in small areas under exceptional circumstances, and a few individual palms can be very high yielding.

The initial reason for the lower yield in groves than in plantations is low leaf production; this varies from 10 to 20 per annum with smooth-stemmed palms, and the leaves have an average life of 21 months (Zeven, 1965b). The inflorescence abortion rate of groves tends to be at least double that of mature plantations in West Africa. Fruit in the groves are predominantly *dura*, but the proportion of the desirable thin-shelled *tenera* form varies widely and is thought to be a measure of the amount of selection which has proceeded in the area. Though usually amounting to only 5–20 per cent, the *tenera* palms of one grove (Ufuma) in eastern Nigeria were found to constitute 43 per cent of the population. In the Ivory Coast *tenera* percentages varied from 0 to 41, shell-less *pisifera* palms 0 to 5 per cent, and the green-fruited *virescens* 0 to 3 per cent (Meunier, 1969).

The Brazilian groves, established naturally during the period of the slave trade on a belt 5–8 km wide from the Bay of Salvador to Maraú in the state of Bahia, are very similar in composition and yield to the groves of south-eastern Nigeria (Savin, 1966; Hartley, 1968a, 1977).

Whereas increasing pressure of population brought the groves to their land dominance and optimum yield, so further population pressure led first to *thinned groves* and then to their further decline. The demand for further farmland can have either of two effects. In some areas the thinning process will continue, leading to the *sparse grove* typical of the Porto Novo area of Benin (Dahomey) (Zeven, 1967). Here a stand as low as forty stemmed palms per hectare is maintained by natural regeneration and felling as needed, while intensive food-cropping is carried on between the palms. In other areas farmland tends to become separated altogether from the dense groves and, owing to frequent burning and cultivation, regeneration of palms does not take place. As the land becomes exhausted, such farmland may be succeeded by derived savannah with a few islands of palm grove remaining. This succession is visible in the more northern parts of south-eastern Nigeria and in the Ivory Coast and tends to be hastened by the practice of regularly burning the grassland. Depopulation follows, and redevelopment is a difficult problem; in the Ivory Coast the soil is intrinsically fertile enough for the re-establishment of the oil palm as a plantation crop (Surre *et al.*, 1961), but on the poor, sandy Nigerian 'bad lands' such usage is less likely to be successful.

Methods of improving the yield of groves without replanting have frequently been suggested, but the complexities of ownership and organization have usually precluded their widespread adoption (Sparnaaij, 1958; Desneux and Rots, 1959). The early trials, reviewed by Zeven (1967), usually entailed clearing the undergrowth, removing epiphytes, pruning and thinning to varying stands per unit area. In grove plots 'improved' in this way, yield increases have often been attributable to edge effects, i.e adjacent replanted or thinned and cleared plots have reduced competition for light. Some real effects have been recorded, however; in an experiment in south-eastern Nigeria, for instance, cultivation and the reduction of the smooth-stemmed palms by 57 per cent led to a small reduction in yield for the first 3 years followed by a 42 per cent increase over pretreatment yield during the next 8 years (Zeven, 1967). At the same time the control (untouched grove) showed a small increased yield owing to edge effects.

It is difficult to draw overall conclusions from such scanty results. Some dense groves are already suffering from too high a population of palms and thinning will improve their performance; others will not yet be in this condition and marked results cannot therefore be expected. All that can be said is that where replanting cannot be undertaken for the time being it may be worth while to seek out for thinning those groves choked with an excess of palms.

The application of fertilizers to grove palms is never practised and has rarely been tried. Experiments in both the Ivory Coast and south-eastern Nigeria have, however, shown that responses of 20 per cent or more can be obtained to potassium chloride or sulphate. In Nigeria, a single application of 3.4 kg of potassium sulphate produced a 20 per cent response which lasted over 6 years.

Zeven (1967) has described the successive schemes of grove rehabilitation by partial or complete replanting which have followed one another in Nigeria. The earliest scheme, in the 1920s and 1930s, was combined with the introduction of the curb press (p. 695) and failed through a combination of reasons, agricultural, economic and legal; and at this time, of course, *tenera* planting material was not available. The second scheme followed the introduction, in the 1940s, of the Pioneer oil mill (p. 731), and included subsidies for participating farmers; but it was no more successful even though selected planting material had by now become available. A third scheme, also subsidized, was instituted as part of the Nigerian Six Year Development Plan of 1962 (Menakaya, 1961), and was based on a calculation that a plantation would provide 2.8 tonnes more palm oil and 0.4 tonne more kernels than an 'average' grove, and that the subsidy would be recouped in less than 2 years of the palms coming into bearing. The scheme, which made a fair start, was overtaken by the Nigerian Civil War, and it was not until the late 1970s that modified schemes were introduced with international aid.

The main practical problem for the grove farmer has been how to retain a reasonable income while the planted palms are coming into bearing. Several methods were suggested for partially or gradually removing the old stand, and some of these were embodied in the thinning experiment in south-eastern Nigeria which has already been referred to. In this experiment the undisturbed dense grove was compared with complete replanting, complete replanting under a half stand (seventy-four palms per hectare) of old palms which were cut down after 4 years, and two treatments of thinning and partial replanting. In the latter treatments the yield after 13 years continued to be dominated by the grove palms and the young planted palms produced an insignificant crop. In the same period the plots which retained a half stand for 4 years gave as high a yield as those of the standard replanting treatment in terms of oil and kernels, and had the advantage of giving a return from the old palms during the interim; but Zeven (1967) has pointed out that there remains the temptation to retain the old palms for longer than 4 years and that these palms are costly to harvest (and nowadays it might be difficult to have them harvested at all). He therefore recommended either complete felling and planting or planting under a further reduced stand of only thirty-seven palms per hectare.

The establishment of a plantation industry

The growing of the oil palm in plantations started in the Far East, and, strangely, there was no direct connection between the African groves and the establishment of this new industry. The earliest record of the introduction of oil palms to the East Indies is of four seedlings, two from Bourbon (Réunion) or Mauritius and two from Amsterdam, which were planted in the Botanic Gardens at Buitenzorg, now Bogor, in Java in 1848.

The Deli palm

Unfortunately, the real origin of these Buitenzorg palms remains a mystery. In a report on the Buitenzorg gardens dated 23 March 1850, Teysmann records '*Elaeis guineensis* received from the Hortus Botanicus at Amsterdam and through Dr D. T. Price of Bourbon. This palm produces an oil which is of great trade interest on the Guinea coast.' In 1835 and 1856 Teysmann reported that the introductions of 1848 had flowered after 4 years. At first it was thought that the plants were either male or female, but their monoecious character was soon discovered and Teysmann's interest clearly lay in determining whether the palms would prove a more useful producer of oil than the ubiquitous coconut palm (Hunger, 1924).

It has not been possible to trace the origin of the Amsterdam plants and confusion over the other two specimens resulted from Teysmann's later mention of 'Bourbon or Mauritius'. As late as 1917 it was stated that the oil palm does not occur wild in Mauritius (nor, presumably, on Réunion), but it is known to have existed in the Botanical Gardens of Mauritius well before 1863. It can only be supposed that the Amsterdam plants came from Africa, and that the Mauritius or Réunion plants came from the same source but were routed through one of these islands.

There are two other puzzling features of these remarkable importations. Hunger reported that not one of the original palms existed in the early 1920s, yet it is widely believed, and more recent reports confirm (Rutgers, 1922; Terra, 1953), that two of the four originals still survive and are over 20 m in height. Secondly, the progeny of these four palms, transferred to Deli in Sumatra and eventually to give rise, under the term Deli palm, to the industry of the Far East, were uniform and their fruit were of a special constitution in certain respects. Though the fruit is *dura* (thick-shelled) in form, the spikelets of the bunches end in short spikes instead of long spines. The fruit are larger and contain a much higher proportion of mesocarp (around 60 per cent) than the general run of African *dura*; they are paler coloured and the oil percentage of the mesocarp is usually rather lower. Some observers claim to see differences in the colour of the leaves and the shape of the crown. What is most remarkable is that the population of palms descended from the Buitenzorg parents was relatively uniform.

It has often been stressed by me [wrote Hunger], that the most important point . . . has been the fortunate fact that the Deli oil palms grown for ornamental purposes and which have provided the seed for the first regular plantings have proved to be of such a peculiarly valuable variety. [And he continued.] As soon as *E. guineensis* was grown commercially at Deli, *a priori* selection gave no difficulties as the available ornamental palms were already of a very productive kind and furthermore, from the beginning and relatively speaking, they bred true to type.

These facts cannot but suggest that either all four palms were from the same parent or that only one of the palms was the progenitor of the Deli palms; the former seems the more likely.

In spite of the enthusiasm of Teysmann, the publicity given by the Government of Holland to the usefulness of the plant, and the establishment of plots at Banjar Mas in Java and at Palembang in Sumatra before 1860, no industrial plantation was to be fully established for 50 years. Rutgers (1922) attributed this to the lack of enthusiasm of the local authorities in Java and to the doubts of estates both about profitability and the milling methods to be employed. In 1878 a plot was established in the Economic Gardens from seed from the Bogor Botanic Gardens; plants from the same source had been established in the Deli district of Sumatra 3 years earlier. Ornamental avenues began to be established on tobacco estates in Sumatra's east coast district from 1884 and further planting from these avenues became widespread. Rutgers considered that the strongest evidence that the avenue palms had a common origin in the Bogor palms is that they were similar to those established in the Economic Gardens and that the source of the latter and of the first-recorded importation into Sumatra was the same.

Plantations in the Far East

The foundation of the industry is generally attributed to M. Adrien Hallet, a Belgian with some knowledge of oil palms in Africa, who planted palms of Deli origin in 1911 in the first large commercial plantations in Sumatra. Hallet's plantings on Sungei Liput (Atjeh) and Pulu Radja (Asahan) estates are recorded (van Heurn, 1948) as being contemporary with the establishment of 2,000 palms by a German, K. Schadt, on his Tanah Itam Ulu concession in Deli. Hallet recognized that the avenue palms growing in Deli were not only more productive than palms in Africa, but had a fruit composition superior to the ordinary *dura* palms of the west coast. A potential oil content of 30 per cent in the fruit was recognized right from the start (Leplae, 1939). Within 3 years, 2,600 hectares had been planted, but stagnation followed owing to the First World War and lack of information on likely profits and on extraction methods. In the meantime, however, M. H. Fauconnier, who had been associated with M. Hallet, had established during 1911 and 1912 some palms of Deli origin at Rantau Panjang in the Kuala Selangor district of Malaysia. These palms were in full bearing by 1917 and in that year the first seedlings were planted on an area later to be known as Tennamaram Estate.

The industry grew rapidly in Sumatra, but did not gain its full momentum in the Far East until the 1930s. In 1925 there were 31,600 hectares planted in Sumatra and only 3,348 in Malaysia (Jagoe, 1952), but by 1938 the areas had risen to 92,300 and 29,196 hectares respectively. With over 120,000 hectares, an industry of considerable importance capable of producing more oil than was being exported from Africa had, in the space of about 20 years, been established.

Plantations in Africa

Meanwhile some interest in planting the oil palm had emerged in Africa. In both the French territories and the Belgian Congo (Zaire) experimental plantings were started in the early 1920s. The

Germans had already planted areas of palms in Cameroon prior to the First World War and in Nigeria an interesting plot had been established near Calabar between 1912 and 1916.

By the early 1920s the Belgians had decided that plantations were likely to be more profitable than the exploitation of groves and that the rare thin-shelled *tenera* fruit, with its high percentage of oil-bearing mesocarp, would be more valuable than imported thicker-shelled Deli palms (Leplae, 1939). M. Ringoet, who had seen the beginnings of the plantation industry in Sumatra, established plantings at Yangambi, near Stanleyville (now Kisanganu), in 1922 and at a number of other places during the next few years. Open-pollinated selected *tenera* seed was used on commercial plantations as early as 1924 (Godding, 1930) and it was soon realized that the *tenera* was a hybrid which would nevertheless give a high proportion of its progeny in the same thin-shelled form. At the same time there seemed no reason to doubt that the *tenera* palm would provide as high a bunch and fruit yield as the *dura* palm. Improved exploitation of the groves, by clearing the undergrowth, providing paths, etc. continued alongside this planting, and by 1939 there were 14,038 hectares of African planted plots. About one-third of the European plantations were not yet in bearing.

In British West Africa a belated attempt was made to make the authorities aware of progress elsewhere and to induce them to encourage similar developments. In 1923 a committee recommended the Secretary of State for the Colonies to encourage commercial firms to introduce oil mills under similar conditions to those imposed in the Belgian Congo and at the same time to allow these firms to establish limited plantations. Recommendations were also made that the groves should be opened up for exploitation and that Africans should plant areas of palms (Ellis *et al.*, 1925). These recommendations went largely unheeded and no significant progress was made during the inter-war period when enthusiasm was high in Sumatra, Malaysia and Zaire. Twenty years were to pass by before mills were introduced and an appreciable improvement in oil quality could take place. Nigeria, the foremost exporter until 1934, took little or no part in the expanding world trade in palm oil. Attention was confined to improving the groves, the planting of small areas by the local population and the introduction of small hand-operated presses. Experimentation was inadequate and the idea of replacing groves by small plantations of unselected material was largely abandoned in Nigeria in the late 1930s in favour of planting small plots on forest land. This policy was slightly more successful: 3,727 hectares owned by 5,530 farmers were planted by the end of 1938, but land tenure systems impeded further progress (Buckley, 1938; Mackie, 1939). There was considerable extension in the use of hand presses, but this was not accompanied by an appreciable improvement in oil quality. Useful small plantings were established on agricultural stations and in 1939 an oil palm research station was started near Benin on an area of over 1,600 hectares.

In the mid-1930s Sumatran palm oil exports rapidly overtook those of Nigeria, but West Africa as a whole maintained her place as the predominant exporter of palm kernels. In 1936, Nigeria was providing nearly half and Sierra Leone nearly one-tenth of the world's exports. Exports from British territories in Africa in 1936 were over 410,000 tonnes. The territories under French rule had not developed plantations but, though only small exporters of oil, they provided substantial quantities of kernels for the European market, a peak export figure of over 170,000 tonnes being reached in 1936.

The development of the industry

The world production of oil palm products has always been impossible to assess accurately owing to the unrecorded quantities produced in the groves and outlying groups of semi-wild palms for domestic use and sale. Estimates made by the FAO have suggested a rise from about 2.2 million tonnes palm oil and 1.2 million tonnes kernels in 1972 to 6.2 million tonnes palm oil and 2 million tonnes kernels in 1982 (Anon., 1977, 1984). This spectacular rise in palm oil production is attributable largely to Malaysia and other Asian countries whose kernel production is only about one-fifth of their palm oil production. African countries on the other hand, with the partial exception of the Ivory Coast, have shown only a small rise in palm oil production and a fall in kernel production, though the latter is still more than half that of palm oil.

The Second World War put the whole of the Far Eastern industry temporarily out of the export market, though exports from Africa, particularly of kernels, were only slightly curtailed. The quantities of oil palm products exported from early in the century are shown in Table 1.4, and world prices of these products, in pounds sterling to 1973 and thereafter in US dollars, are given in Table 1.5 (Anon, 1948–72). Prices remained fairly steady after the war until 1972, though the palm kernel oil price showed a greater tendency to rise than those of the other products. During the last decade the supply of oil palm products rose rapidly, largely owing to Malaysia's expansion, and prices began to show larger fluctuations. This tendency was then accentuated by the marked increase, followed by a short-lived decrease, in production in Asia induced by the introduction in 1981 of the pollinating weevil, *Elaeidobius kamerunicus*. Prices first sank dramatically and then rose to levels never before reached, and they continued, until 1985, to be high enough to prolong the era of steady and widespread oil palm planting which had started

Table 1.4 Exports of oil palm products from producing countries *(thousand tonnes per annum)*

Countries	1909–13	1924–7	1928–31	1932–5	1936–9	1940–1	1942–5	1946–9	1950–3	1954–7	1958–61	1962–5	1966–9	1970–3	1974–7	1978–81	1982–5
Palm oil																	
Nigeria	83	121	131	126	138	132	133	136	176	189	179	134	48	8	9	0	0
Sierra Leone	9	4	?	?	1	1	—	?	?	†	—	—	—	—	—	—	—
Ivory Coast	6	7	7	17	19	9	8	5	13	17	1	1	1	37	97	71	57
Benin (Dahomey)	13	17	13	—	—	—	—	—	—	—	10	11	10	12	7	4	3
Zaire	2	17	33	48	68	63	85	102	136	148	168	125	123	96	46	7	8
Angola	2	4	3	4	3	4	6	12	11	10	11	16	13	8	3	—	—
Malaysia	—	1	3	15	47	52	—	40	51	57	88	122	254	618	1,157	2,042	2,996
Indonesia	—	12	44	117	205	164	—	36	117	128	115	118	161	217	370	366	376
Papua New Guinea	—	—	—	—	—	—	—	—	—	—	—	—	—	—	21	35	104
Others	8	12	11	11	21	9	9	11	10	8	14	15	11	9	35	29	163
Total	123	195	248	341	502	434	241	344	516	557	586	542	621	1,005	1,745	2,554	3,707
Palm kernels																	
Nigeria	175	262	257	297	339	312	326	329	390	446	432	400	226	194	203	90	41
Sierra Leone	48	72	61	73	75	46	39	64	74	60	57	55	48	50	19	10	10
Ivory Coast	6	13	11	70	77	49	48	56	77	71	15	12	10	20	29	9	7
Benin (Dahomey)	35	48	31	—	—	*	9	18	10	15	53	42	6	6	3	—	—
Liberia	—	—	—	—	—	—	—	—	—	—	8	8	16	9	—	—	—

Table 1.4 *(cont'd)*

Countries	1909–13	1924–7	1928–31	1932–5	1936–9	1940–1	1942–5	1946–9	1950–3	1954–7	1958–61	1962–5	1966–9	1970–3	1974–7	1978–81	1982–5
Zaire	7	73	71	56	90	38	54	65	88	50	28	6	2	—	—	—	—
Angola	6	7	6	6	8	6	8	12	11	10	9	15	14	8	4	—	—
Malaysia	—	*	*	3	8	6	—	5	11	14	22	19	32	20	23	26	45
Indonesia	46	80	93	76	83	55	62	60	77	73	99	84	74	70	40	42	66
Others	264	397	368	464	532	422	422	477	585	633	599	550	344	325	283	133	69
Total	323	557	538	605	721	527	546	619	772	780	757	674	466	423	350	201	180
Palm kernel oil																	
Nigeria	—	—	—	—	—	—	—	—	—	—	—	1	33	34	22	40	24
Benin (Dahomey)	—	—	—	—	—	—	—	—	—	—	—	4	19	23	8	8	8
Ivory Coast	—	—	—	—	—	—	—	—	—	—	—	—	—	—	3	9	14
Angola	—	—	—	—	—	—	—	—	—	—	—	2	1	1	1	—	—
Zaire	—	—	—	—	*	*	3	13	15	40	55	38	44	30	27	18	15
Malaysia	—	—	—	—	—	—	—	—	—	—	—	—	—	44	107	199	380
Indonesia	—	—	—	—	—	—	—	—	—	—	—	—	—	—	16	4	26
Others	—	—	—	—	—	—	—	—	—	—	1	1	4	3	4	29	21
Total	—	—	—	—	*	*	3	13	15	40	56	46	101	135	188	307	488

* very small.

Sources: Commodities Div., Commonwealth Secretariat; Oil World Publications; Tropical Development and Research Institute.

Table 1.5 *Prices of palm oil, palm kernels and palm kernel oil (per tonne to nearest £ or US $)*

	(1) Palm oil	(2) Palm kernels	(3) Palm kernel oil	Notes
	(Liverpool or London, £ per tonne)			
1911 (Dec.)	29	18	37	(1), (2) Liverpool landed. (3) Liverpool ex mill
1919 (Dec.)	85	38	91	
1923 (Dec.)	37	20	40	
1931 (Dec.)	20	12	23	
1939 (Av.)	13	9	18	
1945 (Dec.)	41	—	48	(1), (3) Govt. selling prices
1955 (Av.)	86	51	93	(1) 5% f.f.a.,* (3) crude, naked, ex mill
1959 (Av.)	89	69	133	
1963 (Av.)	81	54	99	
1964 (Av.)	86	54	105	
1965 (Av.)	97	64	127	
1966 (Av.)	85	56	95	
1967 (Av.)	81	58	107	
1968 (Av.)	71	74	153	
1969 (Av.)	77	64	128	
1970 (Av.)	108	70	154	
1971 (Av.)	108	60	139	
1972 (Av.)	87	46	102	
1973 (Av.)	155	107	187	

	(1) Palm oil[†]	(2) Palm olein[‡]	(3) Palm stearin[§]	(4) Palm kernels[¶]	(5) Palm kernel oil[‖]
	(United States $ per tonne)				
1974	710	—	—	482	926
1975	416	—	—	207	492
1976	415	—	—	228	523
1977	543	—	—	323	637
1978	620	—	—	366	699
1979	669	—	—	498	988
1980	586	—	—	344	671
1981	578	573	—	340	591
1982	439	476	—	265	458
1983	502	548	432	366	709
1984	729	736	537	532	1,037
1985	501	543	402	291	551

* Malaysian palm oil has normally commanded a premium of about £2 per tonne over West African oil

[†] Malaysian/Sumatran (resale) c.i.f. continent.

[‡] RBD, Malaysian c.i.f. Rotterdam.

[§] Ditto, F.o.b.

[¶] Nigerian, c.i.f. Europe.

[‖] Malaysian, c.i.f. UK.

Sources of recent data: Oil Palm News and Tropical Development and Research Institute.

at the beginning of the 1960s. From that time exports from certain African countries, notably Nigeria and Zaire, had been much reduced both by internal disorders and by rapidly increasing internal demand (Hartley, 1972). This latter demand led to widespread planting in Latin American countries where populations were increasing faster than their vegetable oil production, and when world prices at last slumped in late 1985 internal prices in several countries maintained their high level.

A short account is given below of postwar development in the more important producing countries.

Indonesia (Sumatra)

A peak of 109,755 hectares was reached in 1940 but following the neglect of the war years the Sumatran plantations were only slowly brought back into cultivation and in the late 1940s and early 1950s planting was only on a very moderate scale. The area under nominal cultivation had by 1956 increased by about 15 per cent over its 1938 level, but this was largely on account of plantings undertaken in the early war years. Thereafter there was little expansion, and even by 1962 the area in production was less than prewar and was yielding at a low rate.

In the late 1960s the treecrop plantations in Indonesia were largely reorganized into estate groups (the Perusahan Negara Perkebunan or PNPs); seven of these, situated in Sumatra, contained oil palm plantings covering a total area, in 1971, of about 90,000 hectares. The remaining area of about 36,000 hectares was still held by private companies. During the early 1970s there was considerable injection of capital by the World Bank and the Asian Development Bank for the rehabilitation of the PNPs and their expansion. The area under oil palms doubled during the decade and by 1985 exceeded half a million hectares producing more than 1.25 million tonnes of palm oil (Taniputra and Madjenu, 1980; Taniputra *et al.*, 1987). Smallholders' schemes were started and further large areas surveyed for expanding the industry. Private estates took little part in this expansion which was not confined to Sumatra but extended to Kalimantan, Sulawesi and West Irian.

An important recent development in Indonesia has been the conversion of large areas of *Imperata cylindrica* (lalang) to oil palms both by PNPs and private companies. These areas had established themselves following indiscriminate forest felling for food cropping, but by ploughing or herbicide treatment, and with suitable manuring, it has proved possible to establish productive fields of oil palms.

As in other producing countries, an increasing quantity of the palm oil produced is being consumed locally. The proportion exported has, however, tended to vary considerably, falling to 23 per cent in 1981 and rising again to 45 per cent in 1983 when production exceeded 900,000 tonnes (Anon., 1984).

Malaysia

Rehabilitation was faster in Malaysia after the war than in Sumatra and production was again in full swing by 1947. Acreages were slowly but steadily increased throughout the 1950s and thereafter planting proceeded at a fast rate. This was due not only to confidence in the future of the market for palm oil, but also to the need felt by plantation companies to diversify their interests and, in particular, to reduce their dependence on rubber. Thus much of the new planting in West Malaysia was at first on land hitherto carrying rubber or coconuts, but later huge areas of forest were allotted to oil palm planting.

The rate of planting in Malaysia accelerated in the 1960s and 1970s. In 1960 there were only 55,000 hectares, but by 1975 half a million hectares had been added and a total of a million hectares was reached by 1980 (Wood, 1981). A major part of the planting was done by federal and state development authorities, Government-sponsored settlement schemes accounting by the early 1980s for half the total planted area. About 52,000 hectares were planted by independent smallholders. In the well-known Jengka Triangle and other Government schemes, land was allotted at the rate of 4 hectares per settler, but it is worked by cooperative effort in a 'block-groups' system. In West Malaysia a high proportion of the available coastal alluvium has already been taken up, and the bulk of the more recent planting has therefore been on inland soils and, in some places, on steep land.

There were three especially important features of the industry's extraordinary expansion: firstly, there was a sudden swing towards local processing, from virtually none prior to 1975 to around 90 per cent in 1980, by which time 42 refineries were treating 9,000 tonnes of crude oil per day (Wood and Beattie, 1981); secondly, with 147 plantation mills operating by 1981, there was a major problem of river and land pollution by mill effluent. The industry had therefore to evolve, in a very short time, methods of effluent disposal (p. 721) which would satisfy increasingly stringent Government requirements. Lastly, through the sheer size of the expansion, there have been misgivings about rural labour shortages, about the effects of forest destruction on steep hillsides, and about the resulting high costs of development, particularly on the remoter planting sites.

Other Asian countries

Small oil palm industries were started during the last quarter-century in India, Sri Lanka, Philippines, China (Hainan) and Thailand. In the latter country the plantings are concentrated in the southern part of the Kra peninsula. Production had risen to 25,000 tonnes of palm oil by 1982 (Anon., 1982c).

Nigeria

Nigeria lost her foremost place in palm oil exports to Zaire in 1962 and regained it only temporarily in 1964–5. The decline of the oil palm industry through the early 1960s in terms of palm oil and kernel exports was the result of increased domestic consumption following population growth and the low producer prices allowed by the marketing boards which had a monopoly of purchase for export. The producer prices given were little more than half the f.o.b. prices and discouraged both planting and harvesting (Hartley, 1963).

The outstanding achievement in Nigeria in early postwar years was the improvement of the quality of palm oil as judged by its free fatty acid (f.f.a.) content. This was achieved by the imposition of a large price differential between oil with less than 4.5 per cent, and later with less than 3.5 per cent, f.f.a. (special or edible grade) and oils of higher f.f.a. contents (technical oil grades). The change-over was assisted by the establishment of Pioneer oil mills in the grove areas of eastern Nigeria. These found little difficulty in producing special grade oil and small producers using the curb press soon found it necessary and profitable to fall into line. Exports from Nigeria in 1950 contained 0.2 per cent of the special grade oil but by 1963 93.7 per cent of all purchases for export were of this grade. (Anon., 1963.)

In the 1950s Nigeria had discouraged large plantations even in under-populated areas, except for the limited number that could be financed by the Eastern and Western Nigeria Development Corporations. Some of the latter, particularly in eastern Nigeria, were well sited and made a small contribution to exports, but others were poorly sited and went out of production. By the end of the 1950s the work of the West African Institute for Oil Palm Research (set up with its Main Station near Benin City in 1952) was becoming better known, and regional governments began to see the need for, and the possibilities of, developing the industry on sounder lines. Development plans in the early 1960s included expansion of palm oil production by further Development Corporation plantations, by settlement schemes and by rehabilitation of palm grove areas, but the discouraging influence of the Governemnt's fiscal and marketing policies, as exercised by the marketing boards, had not been removed.

The Civil War which broke out in 1967 reduced the palm oil exports to a few thousand tonnes and palm kernel and kernel oil exports to half their former level. Following the war kernel and kernel oil exports recovered to some degree but palm oil exports, which came mainly from the areas of conflict, showed little sign of substantial recovery. With her large and increasing population it is unlikely that Nigeria will again play a large part in palm oil exporting until the large plantation and grove rehabilitation schemes started in the 1970s under World Bank auspices have developed much further. In 1950 nearly half the palm oil entering world trade was extracted by village methods from the African palm groves. The Nigerian Civil War removed most of this oil from the market and oil of grove origin is unlikely ever again to form any substantial part of the trade (Hartley 1972).

Zaire

The oil palm economy of Zaire was little affected by the war in comparison with other producing countries and exports of oil and kernels continued to expand. In 1949 a 10-year plan for the development of the country was published (Anon., 1949). This recommended increased production from plantations and from the groves for export, local consumption and, in the case of kernels, local crushing and export of palm kernel oil. A production of 300,000 tonnes of oil and 130,000 tonnes of kernels was envisaged by 1959. It was intended that a considerable portion of this production should also be used for (i) local manufacture of margarine, (ii) production of soap, and (iii) production of palm kernel cake for local use and for export. The plan allowed for a considerable increase in small-acreage planting in addition to expansion of plantations financed from overseas.

This plan was in large measure fulfilled. In 1949 large plantations covered 103,000 hectares; by 1958 plantations covered 147,000 hectares of which 107,000 were in bearing and producing 77,000 tonnes of palm oil and 31,000 tonnes of kernels (Anon., 1960). Yields from the best plantations were far above this low level brought about no doubt by poor production in the diseased areas of the south. Production of oil on certain estates reached 3 tonnes per hectare. By 1958 *small* plantations owned by Zaireans covered 98,000 hectares of which 57,000 hectares were in bearing.

By 1959 Zaire was exporting over 180,000 tonnes of palm oil, which was 75 per cent of her recorded commercial production. In the years of unrest which followed, commercial production sank to 160,000 tonnes and exports to 80,000 tonnes (1965) (Anon., 1972). During the late 1960s and early 1970s there was a fair recovery, but it became apparent that Zaire would consume an increasing proportion of her production. By 1974 commercial production was estimated at 190,000 tonnes and it continued at about the same level, but exports fell to 63,000 tonnes in 1974 and by the end of the decade they had become insignificant (Anon., 1976–9). Zaire did not take part in the expansion of planting that has been a feature of other producing countries, and established plantation companies were reluctant to increase their acreages. The great research station of the Institut National pour l'Etude Agronomique du Congo Belge, which had done so much for oil palm cultivation, ceased to operate on its previous scale.

The Ivory Coast

The Ivory Coast is climatically the most favoured of the smaller African countries, but no considerable plantation industry developed between the wars. However, important plans were put into operation in the early 1960s following successful plantings of Deli *dura* × *tenera* and *pisifera* material by the Institut de Recherches pour les Huiles et Oléagineux (IRHO) in areas of forest and of derived savannah. 'Le Plan Palmier à Huile' allowed as a first stage for the planting in five selected areas of a total of 4,700 hectares between 1962 and 1965 (Boyé, 1964). The scheme provided, in each area, for a central plantation block, a mill with a capacity of 12 tonnes of bunches per hour and the organization of 'satellite' holdings, owned individually or cooperatively, on the periphery of the block. The second stage, entrusted to the Société d'État pour le Développement du Palmier à Huile (SODEPALM), followed a search for suitable larger areas and allowed for the planting of 33,000 hectares between 1965 and 1968 with possible extensions to 75,000 or even 150,000 hectares by 1975. It was intended, through the system of plantations and satellite holdings, to immerse the local population in these ventures both as workers and owners. Considerable areas of selected progenies were established by the IRHO at La Mé to provide seed for the areas to be planted.

The scheme was remarkably successful. By 1969 out of 54,000 hectares planned 52,633 had been planted (Carrière de Belgarric, 1970) and by 1981 there were nine industrial plantations totalling 52,000 hectares and peripheral village plantings of 33,000 hectares (Anon., 1982a). Many of the village plantings are of a high standard but a serious problem from the early 1970s was the invasion of *Eupatorium odoratum*, a weed which was not controlled in the village plantings and soon invaded the plantation blocks. Other problems have been Vascular Wilt disease (*Fusarium oxysporum*) and the Oil Palm Leaf Miner (*Coelaenomenodera elaeidis*). Management and maintenance are now under the state organization 'Palmindustrie' which also manages the coconut industry. The 'Plan 1981–5' aims at extending the industrial plantations by 15,000 hectares and the village plantings and replantings by 19,000 hectares to give a total production for the country of over a million tonnes of bunches by 1990 (Anon., 1983).

The Ivory Coast was not a significant exporter of palm oil until 1970 but from that year exports steadily increased, reaching 114,000 tonnes by 1975. Thereafter there was a gradual fall in exports through local processing and consumption. Kernel exports have also declined for the same reasons.

Benin (Dahomey)

The climate of the country renamed Benin is too severe for high yields but the south has supported little else but sparse areas of palm grove. Palm kernels constituted nearly half the total exports in the early 1960s, but in later years an increasing quantity was crushed locally and since 1973 exports have been entirely in the form of kernel oil.

In spite of the poor yield expectations a great effort has been made by the Société Nationale pour le Développement Rural du Dahomey (SONADER) to increase producton of palm oil products. By 1970 this organization had planted around 22,000 hectares in small farmers' cooperative schemes and was aiming to extend its plantations to produce about 40,000 tonnes of palm oil by 1978 (Carrière de Belgarric, 1970). Published figures of production show, however, that bunch yields have been lower than expected and that only around 1 tonne of palm oil per hectare is being obtained from adult areas (Dissou, 1972). By 1982 there were over 28,000 hectares of plantations of which 900 were irrigated, but palm oil production was under 15,000 tonnes (Anon., 1982b).

Cameroon

Cameroon was only a small exporter of kernels in the early postwar period. However, with the very considerable expansion of planting by the Cameroons Development Corporation (CDC) and by Unilever in West Cameroon (Courade, 1978), and later, with the inauguration of the Société Camerounaise de Palmeraies (SOCA-PALM) in East Cameroon, the country was expected to become a considerable exporter of oil palm products. By 1973 the CDC had over 10,000 hectares and SOCAPALM over 7,000 hectares under oil palms, and an outgrowers' scheme was started with loans, transport and milling provided by SOCAPALM (van der Belt, 1981).

Other African countries

Most other countries in Africa entered world trade through exports of kernels, or, more recently, kernel oil. Sierra Leone used to export as much as 60,000 tonnes of kernels, and local crushing of kernels began in the early 1970s, but exports of both kernels and kernel oil gradually fell away and large quantities of palm oil are now imported. Development schemes including outgrowers and covering some 7,000 hectares were started in various parts of the country but the results were well below expectations (Hartley, 1979).

Liberia was estimated to have increased her palm oil production to 30,000 tonnes by 1982, when there was a small export. Plantations covering nearly 7,000 hectares were planted by the Liberian Palm Products Corporation (Anon., 1982c).

West and Central Africa as a whole failed to keep pace with its own needs for oil palm products and is rapidly disappearing from the export scene. Development schemes designed to rectify this situation, besides being of insufficient size to match the population increases, often failed to reach expectations owing to over-optimistic assessments of climatic data, the inclusion of unsuitable soils, destruction of plantings by dry-season fires, and generally poor management.

The Pacific

Several islands in or bordering the Pacific Ocean have started to plant oil palms during the last quarter-century. Both Papua New Guinea and the Solomon Islands have very suitable climates and exceptionally favourable soils. In the former country three projects, each of which comprised a nucleus estate with adjacent smallholdings, cover a total area of more than 25,000 hectares (Christensen and Densley, undated), and from an early date an oil palm experimental station established jointly by commercial and governmental interests has provided seed. Similar schemes have been started in the Solomon Islands.

South and Central America

The growing of the oil palm constitutes a new industry in Latin America. By the end of the 1950s adult producing plantations existed only in Venezuela, Colombia, Ecuador, Costa Rica, Nicaragua and Honduras. The areas in each case were small, and the palm oil and kernels were disposed of locally. About 1959 an increased interest began to be taken in the crop as it came to be realized that suitable growing conditions existed, and that both oils could replace the coconut and other oils which were being imported in increasing quantities.

Development has not been uniformly successful in Latin America and many difficulties have been encountered. Some unsuitable soils were initially used in Brazil while in Colombia, Panama and Peru development has been checked by disease. Elsewhere lack of capital and the remoteness of otherwise suitable areas have hindered expansion.

Colombia

Plantations have been established in Colombia in widely distributed areas but principally in the Magdalena valley and in areas of the Pacific coastal plain. By 1966 15,835 hectares had been planted but in the 7 years to 1973 this total increased only to 20,500 hectares. The failure of the industry to increase as rapidly as expected was due to serious outbreaks of Bud Rot and the Sudden Wither disease, *Marchitez sorpresiva*, which devastated the greater parts of two of the principal plantations. In the last decade, however, planting was resumed, and a plantation devastated by Bud Rot was replanted with the hybrid *E. guineensis* × *E. oleifera* which was found to be largely unaffected by the disease (Anon., 1974). By 1985 there were over 65,000 hectares planted and the production of palm oil was 120,000 tonnes which was 75 per cent of the country's total oils and fats production though less than half her needs. Large areas of suitable land are still available and proposals for planting a further 65,000 hectares by 1990 were accepted by the Government in 1984 (Vargas Tovar, 1984; Guerra de la Espriella, 1984).

Plantations in Colombia vary greatly in size. The largest covers 5,000 hectares (Anon., 1971), but medium-sized estates of 100–400 hectares have been successful and small mills, suitable for these acreages, have been manufactured locally.

Ecuador

Development in Ecuador was until recently confined almost entirely to a small region north-west and south of Santo Domingo de Los Colorados on the Pacific plain (Hartley, 1968b), but two large plantations were established in the 'Oriente' east of the Andes in the early 1980s. The first small plantation was planted in 1953 but further planting did not get under way until the beginning of the 1960s when an experiment station was set up in the region by the Instituto National de Investigaciones Agropecuarias. By 1975 planting was proceeding at the rate of about 3,000 hectares per year and in 1985 the area reached nearly 45,000 hectares of which about 15,000 hectares were in the Oriente. As in Colombia the size of holdings varies considerably. Many medium-sized estates of around 100 hectares were established without mills in the hope that the bunches produced would be sold to nearby larger estates with mills. In the event many small mills were erected in the 1970s and 1980s and so far milling capacity has kept pace with planting.

Brazil

Brazil has a tremendous potential for the crop but development has been slow for financial, physical and geographical reasons (Nascimento *et al*, 1981). South of Salvador in Bahia state lie the semi-wild palm groves established during the slave trade. These are estimated to cover 20,000 hectares and to produce some 6,000 tonnes palm oil and 1,500 tonnes palm kernels annually. Plantations in this area and in Pará and Amapá covered around 12,000 hectares by 1981 when the total palm oil production of the country was estimated to be 20,000 tonnes. Further plantings totalling 30,000 hectares were planned. Near Belém in Para, where climatic conditions are good, a 1968 plantation of 1,500 hectares was recently expanded to 5,000 hectares and a further area of 20,000 hectares on the Acara river is being planted. In the Amapá territory there is an interesting project to convert areas of ancient, annually fired, derived savannah to oil palm cultivation. But the largest potential areas are considered to be in the Amazonas state, west of Coari (Ooi *et al.*, 1981).

Costa Rica

On the Pacific coast large plantations were developed by the United Brands Company, the area planted reaching nearly 17,000 hectares by 1981. An experiment station maintained by the company does breeding, selection and seed production. On the Caribbean coastal plain, where the rainfall is heavier and more evenly distributed, there have been some small plantings and high yields have been reported (Matamoros and Gonzales, 1971). There are also over 600 hectares of *E. guineensis* × *E. oleifera* hybrids.

Honduras

The earliest of the American plantings was established in Honduras in 1943 by the United Fruit Company (now United Brands) and the areas planted on the northern coastal plain now exceed 20,000 hectares. A unique scheme of communal plantings (*cooperativas*) was started in 1971 in the Aguán valley, and by 1985 the planted area had reached over 10,000 hectares. The palms in Honduras were very little affected by the hurricane and floods which devastated habitations and other crops on the northern coast in 1974.

Other American countries

Plantations also exist in south-western Mexico, Panama, Venezuela, Surinam and Peru. For Latin America as a whole it can be said that oil palm areas have steadily expanded, though insufficiently to satisfy rising populations. In Peru, very suitable areas exist in the Huallaga valley, and one estate of over 10,000 hectares has given high yields (Martin, 1969). Other large plantings were started in 1983. In Surinam the first estate was started in 1969 and by 1982 over 8,000 hectares were planted (Renooy, 1982).

Development methods and problems

Of particular interest during the last two decades has been the diversity of methods by which oil palm industries have been developed and the controversy that this has engendered (Hartley, 1970, 1972). Everywhere there has been a desire for smallholders to participate to a greater extent, but in the more traditional plantation countries such as Malaysia a doubt has persisted of the ability of the smallholder regularly to provide bunches of the correct ripeness to fulfil the more stringent quality demands of palm oil purchasers. For this reason the Malaysian Federal Land Development Authority, in following the 'block-group method' of their rubber schemes, did not go so far as eventually to allocate individual oil palm lots to settlers (Taib, 1966), though recently there have been allocations of 40 hectares to be worked together by 10 settlers' families. In the Lower Aguán project of Honduras, the *cooperativas* have 200–350 hectares each, cultivated by about 80–120 heads of families.

　　The idea of nucleus estates with peripheral smallholdings from which bunches are brought to a central mill has gained popularity in other regions and has been most successful in the Ivory Coast (Phillips, 1965; Carrière de Belgarric, 1970). It must be admitted, however, that in the latter country it is the estates which dominate whereas in a predominately smallholders' scheme the nucleus estate would be expected to be relatively small and to provide processing facilities and an example of first-rate cultivation methods. The question of the proportion of the land to be occupied by the nucleus estate and the smallholdings is not an easy one, however, since those financing a scheme are unwilling to erect a large mill unless they feel assured of receiving a sufficient tonnage of bunches for it. In addition to providing a mill the nucleus estate is able to organize transport of bunches, and perhaps also to undertake the clearing and planting of the smallholders' areas and to provide planting material and advice. In the Mosa scheme in New Britain (Papua

New Guinea), where the planting and harvesting of oil palms by smallholders has been conspicuously successful, the proportion of estate to smallholdings was 3 : 4 (Manderson, 1971) and proportions of 2 : 3 or 1 : 2 have been maintained on more recent schemes (Christensen and Densley, undated).

Development in Benin (Dahomey) and in Cameroon stand in marked contrast: in the latter country the schemes undertaken entailed estate planting only, whereas in Benin (Dahomey) cooperative smallholder schemes were the rule. In East Cameroon it was argued that as there had been no tradition for oil palm planting by local farmers the initial developments should be on an estate scale, but when the plantation schemes grew to a certain stage outgrowers' plantings were started (van der Belt, 1981).

A feature of development in Colombia and Ecuador has been the planting of medium-sized holdings financed by local entrepreneurs or by loans from state development banks and having small locally designed mills. Some of the latter are unattached to plantations, depending entirely on bought-in bunches. In general, these small plantations have, with improving cultivation, been increasingly successful, and the widely held view that only large oil palm enterprises are likely to be viable has not been borne out; indeed, in some countries the exacting legislation on labour and housing makes costs on large estates often higher than on small ones, and attempts to reduce labour costs by mechanization have not usually been successful. However, the smaller holdings have had their own problems: non-resident ownership has often led to inefficient maintenance and harvesting, and the regular monitoring and control of pests and diseases are much more difficult in an area of scattered and variously managed holdings. In both Colombia and Ecuador oil palm planters' associations have been formed, but their influence has fluctuated; and in areas of this type, owing to the easy sale and immediate local usage of the palm produce, much less attention is paid, by planters and millers alike, to quality than where an export market is being supplied. As a result, the numerous small mills vary very greatly both in the efficiency of extraction and in the standard of their products.

The development of plantation crops as a whole is said to be constrained by an inelasticity of demand (Dwyer, 1982), but this applies more to the beverages (tea, coffee and cocoa) than to the coconut or oil palm, since growth in demand for vegetable oils depends primarily on population increases and secondarily on increasing consumption per capita. A new constraint in several countries is, however, shortage of labour, brought about by a drift of young people to manufacturing and service industries in large urban centres. This shortage can eventually be felt as much on smallholdings as on large plantations (Webster, 1983). Though economic circumstances or social preferences may reverse this trend, the oil palm industry is reacting to it by providing higher-yielding material for replanting, shorter palms, and mechanized methods of bringing in the crop.

The importation and usage of oil palm products

Imports

The palm oil trade from West Africa was traditionally, and to some degree remains, with Europe. The trade in kernels has followed a similar tradition. Only in three periods have countries outside Europe become large importers of oil palm products. The last of these periods started in the early 1970s and has continued.

The United States entered the market for palm oil prior to the First World War, and between 1909 and 1913 was estimated to have taken about 20 per cent of the world's supply. Demand built up after the war and the United States was purchasing half the world's exports by 1930. Purchases continued at a high level until 1937, reaching 183,000 tonnes, but thereafter a decline set in and by the end of the Second World War imports had levelled off at less than 30,000 tonnes. The withdrawal of the United States as a large purchaser was due firstly to increased domestic production of vegetable oils, principally soya bean oil, for cooking fat and soap manufacture, and, later, to technological advances which reduced palm oil requirements in the steel industry.

The entry of the United States into the palm kernel trade was small and short-lived, starting in 1926 and ending with the Second World War. However, both immediately before and after the war the United States was a substantial importer of palm kernel oil, taking its supplies firstly from Western Europe and later direct from Zaire. In 1934–8 she took 60 per cent of the total world imports of palm kernel oil, and in 1937 she took 90 per cent. When purchases then ceased the international commerce in palm kernel oil became negligible until production in Zaire started after the war. The United States then entered the field once more and by 1953 was purchasing two-fifths (22,500 tonnes) of the world's small export supply. This rose to nearly 40,000 tonnes in 1960. Considering that she was by far the largest exporter of vegetable oils this import seemed a curious anomaly. By Public Law 480 and various aid programmes, however, the United States so maintained the quantity and price of her own oil supplies (principally soya bean) that it often paid American manufacturers to import foreign oils instead of purchasing domestic supplies (Faure, 1961).

Within Europe, the United Kingdom was always the principal importer of palm oil, taking nearly half of Europe's supply before the Second World War and a like proportion in the late 1950s. During the early 1960s this proportion fell to about one-third and later to less than one-third of Europe's imports; this was largely due to the decline in Nigerian production, since the United Kingdom had traditionally taken a large share of her supply. Other substantial importers were, and continue to be, West Germany, Holland, Belgium, France, Italy and Portugal.

During the 1970s a new pattern of palm oil purchases emerged; Europe's imports tended to remain static and so by 1983 they constituted only 19 per cent of the total, though the United Kingdom continued to take more than any other European country. United States purchases again declined, and the Indian subcontinent, with more than 30 per cent, became the largest importing area, closely followed by the Middle East, the Soviet Union, East Africa, Japan, Korea and Nigeria; and there is now a substantial residual import, amounting to about 19 per cent, taken by a large number of other countries including Australia and several South American states. The products of Malaysian plantations are now very widely scattered over the globe and have invaded many erstwhile exporting areas.

The principal importer of palm kernels before the Second World War was Germany; she constantly took nearly half the world's supplies. After the war the United Kingdom became the principal purchaser, taking her supply almost exclusively from Nigeria and Sierra Leone. In the early 1950s she was taking well over half the world's supply, but from 1954 her imports tended to fall. Europe still dominates the kernel and kernel oil markets, taking about half the imports of both products. The United Kingdom remains the main importer of kernels, though with the enlarged market in kernel oil, on a much smaller scale than hitherto. Kernel imports as a whole are now little more than one-third those of kernel oil for which the United States is the only substantial buyer outside Western Europe.

Usage of palm products

Both palm oil and palm kernel oil are used in the manufacture of margarine, compound cooking fat and soap. The proportion in which the oils are used for these purposes depends on their quality and the supply of other vegetable oils. In manufacture, oils and fats are very largely interchangeable and if an oil used for a certain purpose is in short supply, another oil may be substituted and used for that purpose and the proportions in which the latter oil is used for various industrial processes will change.

Palm oil is used much more for the manufacture of edible products than previously because of the great improvement in its quality. Before the Second World War palm kernel oil was used almost entirely for soap-making, but after the war it became used to a much greater extent in a wide variety of edible products.

In recent years there has been a substantial change in the usage of palm oil both in the United Kingdom and elsewhere. Usage in the three 'traditional outlets' – soap, margarine and compound cooking fats – has declined. The main reason for this decline has been the substitution of animal fats, tallow in the case of soap and lard in the case of the edible products. The substitution of marine oils also contributed to the decline at the beginning of the period (Cheshire, 1965). In 1971, however, 36 per cent of the palm oil imported into the United Kingdom was still being used for margarine and compound cooking fats. Other and new outlets are absorbing the remaining supply without difficulty: among these are fatty acid manufacture, additives to animal feeding stuffs, the potato crisp industry and the baking, biscuit and ice-cream trades (Jacobsberg, 1976; Moolayil 1976).

Palm oil hydrogenation, refining and fractionation have now started in producing countries and a variety of products are being supplied both to the local and overseas market. These include margarine, *vanaspati* (vegetable ghee), shortenings, frying fats, cooking oil, salad oil, and oil products for use in ice-cream, creamers, coffee whitener, mellorine, confectionery, biscuits, bread and pastry (Wood and Beattie, 1981). Poorer-quality oil, and refining residues not used for edible products, are employed in the so-called 'technical uses', e.g. soaps, resins, candles, glycerol, fatty acids, ink, polishing liquids, cosmetics and the tin-plate industry (Schwitzer, 1980). In Malaysia the fractionation of palm oil into palm stearin and palm olein for separate sale leads to distinct usages of these products, e.g. shortenings, *vanaspati* and bakery fats from palm stearin, and cooking oils, margarine and salad oils from palm olein, the latter finding its application chiefly where oils liquid above 18–20 °C are required (Berger, 1981).

Malaysia now exports seven basic oil palm products: crude palm oil (CPO); neutralized or semi-refined palm oil (NPO); refined, bleached and deodorized palm oil (RBD); palm olein; palm stearin; palm kernel oil; palm kernel cake or meal (Pritchard, 1975; Wood and Beattie, 1981).

Lastly, the possibility of using palm oil as a fossil fuel substitute should be mentioned. Countries like Brazil which lack a sufficient supply of petroleum have already produced quantities of fuel alcohol from, for instance, sugar-cane. Plants chosen would generally produce either carbohydrates for conversion to alcohol, or a 'methanolizable' oil as a diesel substitute. This is technically possible with palm oil, but the overriding considerations are likely to be the cost of production and whether, in countries where fats and oils for food are already in short supply, the diversion of large areas of land from food farming would be acceptable (Martin, 1981).

Palm kernel oil usage has also undergone some change in recent years. Both palm kernel and coconut oils are required for their physical and chemical properties; they melt rapidly at a temperature just below that of the human body and when solid are hard and brittle. The decline in usage of these oils for margarine and compound cooking fats has been similar to that of palm oil. Usage in soap manufacture has, however, been maintained. Nut oils are particularly suited to toilet soaps and soap powders and as there is a tendency for this side of the trade to expand, the usage of palm kernel oil in soap manufacture may continue to be stable. Other outlets for palm kernel oil, particularly in the manufacture of detergents, have compensated for its reduced usage in the main edible products.

The place of oil palm products in world trade

The interchangeability of oils and fats in manufacture has already been mentioned. The oil palm has to compete in a market containing some thirteen principal vegetable oils and oil seeds, two types of marine oil, and three categories of animal fats. However, the versatility of palm oil and the very high oil yield per hectare should make it possible for the oil palm to compete very easily with other oils and fats within the world export market. The only substantial disadvantage the oil palm suffers is the relatively high extraction costs of oil and kernels on the plantations; but this disadvantage should be more than balanced by the high yields and by the fact that the main product, palm oil, does not need extraction off the plantations as is the case with oil from oil seeds.

When export statistics are studied, however, it is seen that the proportion of the world market held by oil palm products did not substantially alter until quite recently; indeed for half the postwar period their relative position in a gradually expanding market declined (Table 1.6). This decline can be accounted for firstly by the economic uncertainty, restrictions and unrest in the former principal producing countries of Africa, and secondly by the rise in shipments of competing oils and fats such as soya bean oil, rape seed and sunflower seed oils, and lard and tallow (Anon., 1972). However, the early 1970s saw a marked change, with oil palm oils rising steadily, owing to Asian production, from under 10 per cent to over 20 per cent of the total world exports of oils and fats. In the 10-year period 1972–5 to 1982–5, during which coconut oil

Table 1.6   *The position of oil palm products in world exports of vegetable and animal oils, net exports of oils, oil seeds and animal fats (thousand tonnes oil equivalent per annum)*

Oil	1934–8	1949	1954	1958–61	1962–5	1966–9	1970–3	1974–7	1978–81	1982–
Edible										
Groundnut	818	538	666	821	968	1,019	798	752	650	6
Soya bean	386	268	285	1,184	1,555	1,916	3,062	4,195	6,351	7,25
Cottonseed	193	119	349	269	342	206	322	341	415	39
Rapeseed	53	27	34	101	158	373	681	712	1,100	1,3
Sunflower	33	30	28	206	351	999	748	742	1,259	1,73
Sesame	69	36	41	56	73	86	100	105	116	12
Olive	136	42	113	151	162	181	288	222	238	29
Total	1,688	1,060	1,516	2,788	3,609	4,780	5,999	7,069	10,129	11,74
Edible-industrial										
Coconut	1,057	1,021	1,109	1,154	1,290	1,213	1,274	1,296	1,307	1,2
Palm kernel	320	353	400	414	364	319	329	341	296	5
Palm	444	495	579	577	539	627	1,015	1,745	2,554	3,7
Total	1,821	1,869	2,088	2,145	2,193	2,159	2,618	3,382	4,157	5,4
Industrial										
Linseed	620	621	637	437	444	400	438	280	417	3
Castor	86	83	104	152	189	207	223	183	210	1
Tung	79	61	47	58	42	51	54	48	37	
Total	785	405	788	647	675	658	715	511	664	6
Vegetable oil total	4,294	3,334	4,392	5,580	6,477	7,597	9,332	10,962	14,950	17,8
Animal and marine										
Whale and fish	552	416	482	559	621	689	620	556	692	8
Butter, lard, tallow	854	930	1,304	1,749	2,092	2,268	2,559	2,788	3,078	2,9
Animal and marine total	1,406	1,346	1,786	2,308	2,713	2,957	3,179	3,344	3,770	3,7
Grand total	5,700	4,680	6,178	7,888	9,190	10,554	12,511	14,306	18,720	21,5
Palm and palm kernel oils:										
as per cent of vegetable oils	17.8	25.4	22.3	17.8	13.9	12.5	14.4	19.0	19.1	23.0
as per cent of all oils and fats	13.4	18.1	15.8	12.6	9.8	9.0	10.7	14.6	15.2	19.2

* Average for 1982 and 1983 only.

Sources: Commodities Div., Commonwealth Secretariat; Oil World Publications; Tropical Development and Research Institute.

exports stood still, export availabilities of oil palm oils more than quadrupled and came to constitute nearly a quarter of all vegetable oils entering world trade.

References

Anon. (1948–72) *Vegetable oils and oilseeds. A review.* Compiled annually (until 1972) by the Commonwealth Economic Committee's Intelligence Branch and published by HMSO 1948 onwards.

Anon. (1949) *Plan décennal pour le développement economique et social du Congo Belge*, 2 volumes (1949), Brussels.

Anon. (1960) Volume jubilaire 1910–60. *Bull. agric. Congo belge*, 266 pp.

Anon. (1963) *Nigerian trade summary*, Dec. 1963, Chief Statistician, Lagos, Federal Republic of Nigeria.

Anon. (1971) Rendement du palmier à huile sélectionné en Colombie et potentiel de production. *Oléagineux*, **26**, 244.

Anon. (1972) Prospects for supplies of palm oil and palm kernels in 1980. *FAO Bull. Agric. Econ. Statist.*, **21**(4).

Anon. (1974) Replanting diseased oil palm areas with *Elaeis oleifera* × *E. guineensis* hybrids at La Arenosa estate in Colombia. *Oil Palm News*, No. 18, 1.

Anon. (1976–9) Oil palm market reviews. *Oil Palm News*, Nos 21–4.

Anon. (1977) Oil palm market review. *Oil Palm News*, No. 22, 19.

Anon. (1982a) Bilan 1980 de palmindustrie. *Afrique Agriculture*, No. 77, Jan. 1982, 13.

Anon. (1982b) Les huileries en Afrique noire francophone. *Europe Outremer (France)*, 1982 (629–30), 41–53.

Anon. (1982c) World forecast: record oilseed production, *J. Amer. Oil Chemists' Soc.*, **59**, 752A.

Anon. (1983) Le plan quinquennal de développement économique de la Côte d'Ivoire. *Afrique Agriculture*, No. 90, Jan. 1983, 41.

Anon. (1984) Oil palm market review. *Oil Palm News*, No. 28, 18.

Berger, K. G. (1981) Food uses of palm oil. *PORIM Occasional Paper*, No. 2, 1–12.

Billows, H. C. and **Beckwith, H.** (1913) *Palm oil and kernels, the consols of the west coast.* Charles Birchell, Liverpool.

Boyé, P. (1964) Le Plan 'Palmier à Huile' de Côte d'Ivoire. *Oléagineux*, **19**, 1.

Briey, J. (1922) Le palmier à huile au Mayumbe. *Mem. Rapp. Matièr grass.*, **2**, 112.

Buckley, F. E. (1938) The native oil palm industry and oil palm extension work in Owerri and Calabar provinces. *Third W. Afr. agric. Conf.*, **1**, *Papers*, 207.

Carrière de Belgarric, R. (1970) Deux exemples de développement du palmier à huile: La Côte d'Ivoire et Le Dahomey. *Coopér. Développement*, **30**, 15–28.

Cheshire, P. C. (1965) *The market for oil palm products with particular reference to the United Kingdom market.* Oil Palm Conference, London, 1965.

Chevalier, A. (1934) La patrie des divers *Elaeis*, les espèces et les variétés. *Revue Bot. appl. Agric. trop.*, **14**, 187.

Christensen, J. and **Densley, D. R. J.** (undated) *Oil palm agriculture in the economy – a series of review papers.* Dept of Primary Industry, Papua New Guinea, 17 pp.

Cook, O. F. (1942) A Brazilian origin for the commercial oil palm. *Sci. Monthly*, **54**, 577.

Corner, E. J. H. (1966) *The natural history of palms.* Weidenfeld and Nicolson, London.

Courade, G. (1978) Les plantations industrielles d'Unilever en Cameroun. *Cahiers d'ONAREST*, **1**(2), 91–159.

Crone, G. R. (1937) The voyages of Cadamosto and other documents on Western Africa in the second half of the fifteenth century. Hakluyt Society, Series II, 80.

Desneux, R. and **Rots, O.** (1959) Vers une exploitation plus intensive et plus rationelle des palmeraies subspontanées du Kwango. *Bull. agric. Congo belge*, **50**, 295.

Dike, K. O. (1956) *Trade and politics in the Niger Delta, 1830–85.* Oxford University Press, 250 pp.

Dissou, Machioudi (1972) Développement et mise en valeur des plantations de palmier à huile au Dahomey. *Cahiers d'Etudes Africaines*, **12**(47), 485.

Dwyer, G. D. (1982) The economic role of plantation in development: a personal viewpoint. *TAA Newsletter* (Seminar on plantations and development), Dec. 1982, 2.

Ellis, W. D. *et al.* (1925) *Palm oil and palm kernels.* Report of a committee appointed by the Secretary of State for the Colonies, September 1923, to consider the best means of securing improved and increased production. Colonial No. 10, HMSO.

Faure, J. C. A. (1961) Address given at the International Association of Seed Crushers' Congress, Stockholm, 20 pp.

Godding, R. (1930) Observation de la production de palmiers sélectionnés à Mongana (Equateur). *Bull. agric. Congo belge.*, **21**, 1263.

Guerra de la Espriella, A. (1984) Palma Africana: el nuevo cultivo Colombiano. *Palmas*, **5**(3), 9.

Hartley, C. W. S. (1954) The improvement of natural palm groves. *J. W. Afr. Inst. Oil Palm Res.*, **1**(2), 8.

Hartley, C. W. S. (1961) The oil palm in its Sierra Leone setting. Mimeographed paper, WAIFOR.

Hartley, C. W. S. (1963) The decline of the oil palm industry in Nigeria. NIFOR, Benin City, Nigeria. Mimeograph.

Hartley, C. W. S. (1968a) Report on oil palm research and development in Brazil. *Commun. Tecn. CEPLAC*, **17**, 1–32.

Hartley, C. W. S. (1968b) The oil palm in Ecuador. *Oil Palm News*, **5**, 9.

Hartley, C. W. S. (1970) Oil palm research in Africa and Malaysia. In *Change in Agriculture*, ed. Bunting, A. H., Duckworth, London.

Hartley, C. W. S. (1972) *The expansion of oil palm planting.* In advances in oil palm cultivation, eds Wastie, R. L. and Earp, D. A., Incorp. Soc. of Planters, Kuala Lumpur.

Hartley, C. W. S. (1977) The development of groves. In *The oil palm*, 2nd edn, pp. 77–96, Longman, London.

Hartley, C. W. S. (1979) The oil palm in Sierra Leone. *Oil Palm News*, No. 23, 2.

Heurn, F. C. van (1984) De Oliepalm. In *De Landbouw in den Indischen Archipel*, eds Hall, C. J. J. van, and Koppel, C., II A, pp. 526–98, Van Hoeve, The Hague.

Hunger, F. W. T. (1924) *De Oliepalm (Elaeis guineensis)* 2nd edn *Historisch onderzoek over den oliepalm in Nederlandsch-Indië.* Brill, Leiden, 383 pp.

Jacobsberg, B. (1976) Malaysian International Symposium on palm oil processing and marketing, 1976, review paper. *Oil Palm News*, No. 21.

Jagoe, R. B. (1952) 'Deli' oil palms and early introduction of *Elaeis guineensis* to Malaya. *Malay. agric. J.*, **35**, 3.

Jumelle, H. and Perrier de la Bathie, H. (1911) Le palmier à huile à Madagascar. *Matières grasses*, **4**, 2065.

Leplae, E. (1939) Le Palmier à huile en Afrique, son exploitation en Congo Belge et en Extrème-Orient. *Mem. Inst. r. colon. belge Sect. Sci. nat méd.*, **7**(3), 108 pp.

Mackie J. R. (1939) Annual report of the Agricultural Department for the year 1937, Nigeria.

Manderson, A. (1971) The Mosa experiment. Oil palm in Papua New Guinea. *Aust. External Territ.*, **11**, 4, 22.

Martin, G. (1969) Le programme de développement du palmier à huile du Huallaga au Pérou. *Oléagineux*, **24**, 259.

Martin, G. (1981) Le bilan énergétique de la culture du palmier à huile: une approche. *Oléagineux*, **36**, 273.

Matamoros Ramfrez, F. and **Gonzalez Soto, G.** (1971) Elsayo comparativo de hibridos y variedados de palma Africana. *Bol. Tecn. Minist. Agr. Ganderia*, Costa Rica, 59.

Mauny, R. (1953) Notes historiques autour des principales plantes cultivées d'Afrique occidentale. *Bull. Inst. fr. Afr. noire*, **15**, 684, and (1961) Tableau géographique de l'Ouest Africain au moyen âge. *Mem. Inst. fr. Afr. noire*, No. 61, Dakar, 587.

Menakaya, O. C. (1961) Eastern Nigerian oil palm rehabilitation scheme. *Agric. Bull.*, No. 3, Mimeographed.

Meunier, J. (1969) Étude des populations naturelles d'*Elaeis guineensis* en Côte d'Ivoire. *Oléagineux*, **24**, 195.

Moolayil, J. (1976) Uses of palm oil. *Malaysian Int. Symposium on palm oil processing and marketing 1976*. Preprint. 23 pp.

Nascimento, J. C. *et al.*, (1981) First Brazilian national oil palm research programme. In *Oil Palm in Agriculture in the Eighties*, ISP, Kuala Lumpur.

Ooi, S. C. *et al.* (1981) The oil palm industry in Brazil – current status and future potential. In *Oil Palm in Agriculture in the Eighties*, ISP, Kuala Lumpur.

Opsomer, J. E. (1956) Les premières descriptions de palmier à huile (*Elaeis guineensis* Jacq.). *Bull des Séanc. Acad. r. Sci. Colon.* (*outre Mer*), **2**, 253.

Phillips, T. A. (1965) Nucleus plantations and processing factories: their place in the development of organised smallholder production. *Trop. Science*, **12**, 3.

Portères, R. (1947) Aires altitudinales des Raphias, du Dattier sauvage et du Palmier à huiles au Cameroun français. *Revue Bot. appl. Agric. trop.*, **27**, 203.

Pritchard, J. L. R. (1975) Refining palm oil. *Oil Palm News*, **20**, 5.

Raymond W. D. (1961) The oil palm industry. *Trop. Sci.*, **3**, 69.

Rees, A. R. (1965) Evidence for the African origin of the oil palm. *Principes*, **9**, 30.

Renooy, J. (1982) De Ontwikkeling van de oliepalmcultuur in Suriname. *Suralio Magazine* (*Suriname*), **14**(1), 3.

Rutgers, A. A. L. (1922) *Investigation on oil palm at the General Experimental Station of AVROS*. Ruygrot & Co., Batavia.

Savin, G. (1966) Restauration de la palmeraie naturelle de Bahia. Résultats préliminaires. *Oléagineux*, **21**, 431.

Schad, H. (1914) Die geographische Verbreitung der Ölpalme (*Elaeis guineensis*). *Tropenpflanzer*, **18**, 359–91, 447–62.

Schwitzer, M. K. (1980) Non-food uses for palm oil. *Oléagineux*, **35**, 261.

Seward, A. C. (1924) A collection of fossil plants from south-eastern Nigeria. *Bull. Geol. Surv. Nigeria*, **6**, 66.

Sparnaaiji, L. D. (1958) The Palmeries of the Kwango district. In Notes on a visit to Research Stations, Oil palm plantations and palmeries in the Belgian Congo during April, 1958. Hartley, C. W. S. Mimeographed WAIFOR paper.

Stilliard, N. H. (1938) The rise and development of legitimate trade in palm oil with West Africa. Unpublished thesis, University of Birmingham.

Surre, Chr., Fraisse, A. and **Boyé, P.** (1961) Plantation de palmier à huile sur le sol de forêt et sur savane de *Imperata*. *Oléagineux*, **16**, 91.

Taib Bin Haji Andak (1966) Land development in Malaysia under the Federal Land Development Authority – description of programme and techniques of development implementation. *Proc. Seminar Malaysian Centre for Dev. Studies*, Kuala Lumpur.

Taniputra, B., Lubis, A. U., Pamin, K. and **Suleimi Syukus** (1987) Progress of oil palm industry in Indonesia in the last fifteen years (1971–1985). *Int. Oil Palm Conf.*, Kuala Lumpur, June 1987.

Taniputra B. and **Madjenu, M.** (1980) Balai penelitian perkebunaan Medan. *Indonesian Agr. Res. & Dev. J.*, **2**(3), 59.

Terra, G. J. A. (1953) private communication WAIFOR Ref. 420.

van der Belt, H. (1981) Socio-economic characteristics of 'outgrowers' linked to industrial oil palm plantations: the case of Dobombari, Cameroun. In *Essays in rural sociology*, pp. 1–23, Agric. Univ. Wageningen.

Vanderyst, Father Hyac. (1919) Contributions à l'étude du palmier à huile au Congo Belge. 4°. Origine des palmeraies du Moyen-Kwilu. *Bull. agric. Congo belge*, **10**, 70.

Vargas Tovar, E. (1984) Palma Africana: Motor de desarrollo. *Palmas*, **5**(3), 5.

Waddell, H. M. (1863) *Twenty-nine years in the West Indies and Central Africa.*

Waterston, J. M. (1953) Observations on the influence of some ecological factors on the incidence of oil palm diseases in Nigeria. *J. W. Afr. Inst. Oil Palm Res.*, **1**(1), 24.

Webster, C. C. (1983) Plantations and development: rubber and oil palms in South East Asia. *TAA Newsletter* (Seminar on plantations and development), **3**(1), 2.

Wood, B. J. (1981) Technical developments in oil palm production in Malaysia. *Planter, Kuala Lumpur*, **57**, 361.

Wood, B. J. and **Beattie, T. E.** (1981) Processing and marketing of palm oil. *Planter, Kuala Lumpur*, **57**, 379.

Zeven, A. C. (1964) On the origin of the oil palm. *Grana palynol.*, **5**, 50.

Zeven, A. C. (1965a) The origin of the oil palm (*Elaeis guineensis* Jacq.). *J. W. Afr. Inst. Oil Palm Res.*, **4**, 218.

Zeven, A. C. (1965b, 1968) Oil palm groves in Southern Nigeria. Part I. Types of groves in existence. *J. Nigerian Inst. Oil Palm Res.*, **4**, 226, and Part II. Development, deterioration and rehabilitation of groves. *Ibid.*, **5**, 21.

Zeven, A. C. (1967) The semi-wild oil palm and its industry in Africa. *Agr. Res. Rpts*, No. 689.

Zeven, A. C. (1972) The partial and complete domestication of the oil palm (*Elaeis guineensis*). *Econ. Bot.*, **26**, 274.

Chapter 2

The botany of the oil palm

Classification

The family of palms, the Palmae, has always formed a distinct group of plants among the monocotyledons. Although Bentham and Hooker's *Genera Plantarum* placed the palms with the Flagellari-aceae and Juncaceae under the series Calycinae, Engler and Prantl's system allowed them a place by themselves under the order Principes. In the comparatively recent classification of Hutchinson, the Palmae remain alone, though in the order Palmales. Here the oil palm, *Elaeis guineensis* Jacq., is grouped with *Cocos* and other genera under the tribe Cocoineae. The anatomical studies of Tomlinson (1961) support this grouping.

The genus *Elaeis* was founded on palms introduced into Martinique, the oil palm receiving its botanical name from Jacquin (1763, 1780) in an account of American plants (Bailey, 1933). *Elaeis* is derived from the Greek word *elaion*, oil, while the specific name *guineensis* shows that Jacquin attributed its origin to the Guinea coast. From time to time other specific names have been attached to supposed species of *Elaeis*, but none has shown any signs of permanency other than *E. melanococca*, now named *E. oleifera*, and *E. madagascariensis*, the legitimacy of which is doubtful. The Index Kewensis lists fourteen names, the majority of which have disappeared from the literature. Many of them either refer to quite different palms or are synonymous with *E. guineensis*. Of passing interest is the American palm *E. odora* Traill, also named *Barcella odora* and classified by some botanists with *Elaeis* in a subtribe Elaeideae.

Elaeis madagascariensis Becc. was described by Odoardo Beccari (1914a) as a separate species on the basis of material sent to him by Professor Jumelle. This material was distinguished from *E. guineensis* because in the male flower the fused filaments of the staminal tube were shorter and the anthers were erect, instead of spreading, at anthesis, while the fruits were smaller and rounded and surrounded by larger bracts. In view of the wide variation in many minor characters in the oil palm it is doubtful if these differences justify the naming of a separate species; Jumelle himself only considered this palm might have the status of a variety distinguished mainly by its red fruit (Jumelle and Perrier de la Bathie, 1911).

There remain therefore three species, *Elaeis guineensis, E. odora* and *E. oleifera*; the latter will be described and discussed later.

The early descriptions of the oil palm have been reviewed by Opsomer (1956) who claims for Mathias de Lobel (Lobelius, 1538–1616) the earliest botanical description and illustration of the fruit which he named *Nucula Indica*. Lobelius reported that the palm was found in Gunea. The brief descriptions were published in his *Plantarium seu Sirpium Historia* of 1570 (London) and 1576 (Antwerp) and his *Kruydtboeck* of 1581.

Mention of the oil palm is scattered throughout the works of de l'Escluse (Clusius, 1526–1609); the descriptions in the revised edition of the herbal (*Cruydt-boeck*, Leiden) of 1608 of de Dodoens (Dodonaeus, 1516–85) and in Bauhin's *Historie des Plantes* of 1650 are also attributed to him. There is some confusion over description and nomenclature in the earlier works but in the later ones the palm is called *Nucula Indica Secunda* and *Palma Guineenis*. The descriptions taken together are remarkably complete and it is recorded that thin-shelled nuts appear among the more numerous thick-shelled ones. A dried spikelet was illustrated in Clusius's last work and in Dodonaeus's *Cruydt-boeck* and this probably accounts for Clusius's erroneous belief that palm oil was extracted from the kernel, turning red on the journey from Africa (Plate 2.1).

Opsomer states that nothing of consequence was written on the oil palm for over 150 years between the time of Clusius and the first, and lasting, modern botanical description of Jacquin (1763) (Plate 2.2). However, Sloane's *Catalogus Plantarum* (London, 1696), besides making first mention of the spiny petioles, records the importation of the palm from the Guinea coast into Jamaica. It was from material from another West Indian island, Martinique, that Jacquin was to write his description and to name the plant. Jacquin's description is detailed, but he describes the flowers as either female or *hermaphroditi steriles* and seemed unaware that flowers of the two sexes were in separate inflorescences. The production of male and female bunches was first recorded by Miller in his *Gardener's Dictionary* (London, 1768). Before the end of the century Gaertner, in his *De Fructibus et Seminibus Plantarum* (Stuttgart, 1788), gave a more detailed description of the flower parts, recording that the male and female flowers are on separate inflorescences (Plate 2.3).

Elaeis guineensis is a large feather-palm having a solitary columnar stem with short internodes. It is unarmed except for short spines on the leaf base and within the fruit bunch. The irregular set of the leaflets on the leaf gives the palm its characteristic appear-

Pl. 2.1   A dried spikelet and some fruit with decomposing mesocarp, a nut and a kernel of the oil palm (R. Dodonaeus, 1608).

ance. The palm is normally monoecious with male or female, but sometimes hermaphrodite, inflorescences developing in the axils of the leaves. The fruit is a drupe which is borne on a large compact bunch. The fruit pulp which provides palm oil surrounds a nut the shell of which encloses the palm kernel.

The distinguishing of varieties of the oil palm has been attempted by many workers. These attempts have in most cases been unsatisfactory since in the wild state each palm is a hybrid in respect of certain of its characters. Most of the early attempts at classification are unworthy of mention since they were based on a very small acquaintance with the palm, and no knowledge of the inheritance

Pl. 2.2   Jacquin's drawings of an oil palm, a fruit and a nut (N. J. Jacquin 1763).

of the characters described. Of interest, however, is the first description by Preuss (1902) of the *lisombe* palm, a name used in Zaire, Cameroon and Nigeria for the thin-shelled *tenera* fruit form and employed to denote parental stock in quite recent times. Both Chevalier (1910) and Jumelle (1918) divided the species into subspecies according to the outer appearance of the fruit, while Beccari (1914b) extended Chevalier's classification. These classifications were unsatisfactory owing to their failure to attribute all the possible fruit variations to each subspecies, and it was left to Janssens (1927) and Smith (1935) to provide the first simple classifications which, in their essentials, have stood the test of time.

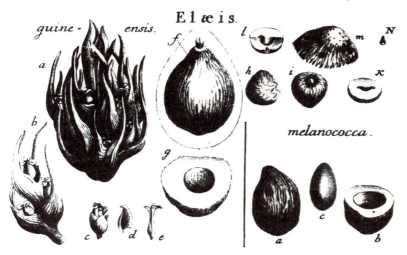

Pl. 2.3 Gaertner's drawing of a spikelet, flowers and fruit of *E. guineensis* together with the nut of his supposed *E. melanococca*, considered by Bailey, however, to be another *E. guineensis* nut (J. Gaertner, 1788).

Although nothing was known of the inheritance of the characters described, Janssens recognized that the fruit forms *dura* and *tenera*, distinguished by the thickness of shell, could be found among fruit types of different external appearance. Thus both the common fruit type *nigrescens* and the green-fruited *virescens* were divided by Janssens into three forms *dura*, *tenera* and *pisifera* (the latter called *gracilinux* – following Chevalier – when *virescens*). The white-fruited *albescens* was also recognized but only a *dura albescens* had been found. Similarly, although *dura* and *tenera* forms of the mantled-fruited *Poissoni* were found, no green-fruited mantled specimens were discovered. Smith, however, recognized both mantled and unmantled *nigrescens* and *virescens* fruit, called them 'types', and divided all four into thick-shelled and thin-shelled 'forms'. This simple procedure, described as the most complete and logical of the empirical classifications (Beirnaert and Vanderweyen 1941a), established the use, in English publications, of the fruit-type and fruit-form classification, thus eliminating the need of the term variety for material which might be heterozygous in many of its characters. That 'variety' was inappropriately used for the *tenera* form was recognized by Beirnaert (Beirnaert and Vanderweyen, 1941b), and in the Far East Schmöle (1930) used the term fruit form as early as 1929.

Morphology and growth

The seed

The embryo of palm seeds is always small and the cotyledon is never erected as a green photosynthetic organ. Instead, the cotyledon apex becomes enlarged and, as the haustorium, absorbs the food reserves of the endosperm (Tomlinson, 1961). Thus in the oil palm the seed is adapted to support a developing seedling for many weeks after germination.

The oil palm seed is the nut which remains after the soft oily mesocarp has been removed from the fruit. It consists of a shell, or endocarp, and one, two or three kernels. In the great majority of cases, however, the seed contains only one kernel since two of the three ovules in the tricarpellate ovary usually abort. Abnormal ovaries sometimes occur and four- or five-seeded nuts may, very rarely, arise from these. In botanical terms the kernel is the seed (Purseglove, 1972), but in common parlance the word 'seed' is used for the nut, comprising shell and kernel (Surre and Ziller, 1963), since it is the nut which is stored, germinated and then planted in a nursery (Fig. 2.1, A and B).

Nut size varies very greatly and depends both on the thickness of the shell and the size of the kernel. Typical African *dura* nuts may be 2–3 cm in length and average 4 g in weight (250 to the kilogram). Deli *dura* and some African *dura* nuts are larger, weighing up to 13 g. African *tenera* nuts are usually 2 cm or less in length and average 2 g (500 to the kilogram). Very small nuts weighing 1 g are not uncommon.

The shell has fibres passing longitudinally through it and adhering to it, and they are drawn into a tuft at the base. Each shell has three germ pores corresponding to three parts of the tricarpellate ovary, though the number of functional pores will of course correspond with the number of kernels developed. A plug of fibre is formed in each germ pore and these fibres are cemented together at the base to form a plate-like structure continuous with the inner surface of the shell (Hussey, 1958).

Inside the shell lies the kernel. This consists of layers of hard oily endosperm, greyish-white in colour, surrounded by a dark-brown testa covered with a network of fibres. Embedded in the endosperm and opposite one of the germ pores lies the embryo.

The embryo is straight and about 3 mm in length. Its distal end lies opposite the germ pore but is separated from it by a thin layer of endosperm cells, the testa and the plate-like structure referred to above. These three structures have been together called the operculum, but they are separate. In the quiescent state the bud is already well developed laterally within the distal end of the embryo.

In longitudinal section the apex with two differentiated leaves and the rudiments of a third can be distinguished, though the radicle is only poorly differentiated (Vallade and Lucien, 1966). Opposite the bud there is a longitudinal split in the wall of the embryo (the *fente cotylédonaire*). This part of the embryo is separated by a small constriction from the cotyledon which will develop into the haustorium. Within the cotyledon a system of procambial strands has developed.

The endosperm above the embryo is demarcated by a ring of cells of small size. When germination takes place the endosperm ruptures in this region and a disc consisting of endosperm, testa and the germ-pore plate is extruded from the germ pore together with the fibre plug (Plate 2.4).

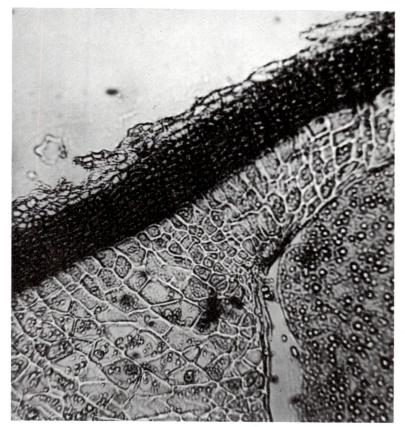

Pl. 2.4 Longitudinal section of a *tenera* kernel showing the distal end of the embryo enclosed by the endosperm and dark testa. Note that the endosperm is continuous above the embryo; rupture will take place through the smaller cells at the corner.

The process of seed germination is illustrated in Fig. 2.1. The emerging embryo forms a 'button' (commonly called the hypocotyl but considered by Henry (1951) and other botanists to represent the petiole of the cotyledon) which rapidly gains a plumular projection, while from the end of the embryo itself the persistent radicle emerges. The plumule and radicle both emerge through a cylindrical, persistent ligule close to the seed. It is of interest here to

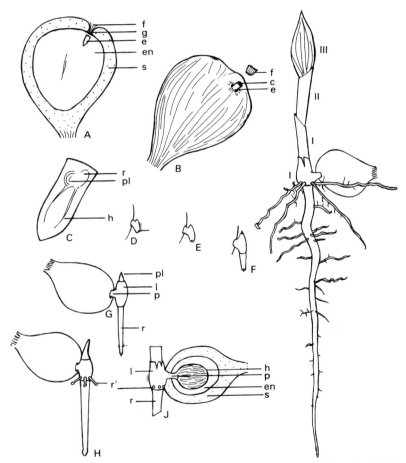

Fig. 2.1 Oil palm seed and early growth of seedling (Rees, 1960): **A**, longitudinal section of seed through embryo; **B**, just germinated seed; **C**, median longitudinal section of embryo; **D, E, F, G**, successive stages in the early growth of the embryo; **H**, production of adventitious roots; **I**, four-week-old seedling; **J**, section of seed at stage 7 to show haustorium; c, cap of testa; e, embryo; en, endosperm; f, fibre plug; g, germ pore; h, haustorium; l, ligule; p, petiole; pl, plumule; r, radicle; r', adventitious root; s, shell; I–III, plumular leaves.

note that palm seedlings have been classified by Gatin (1906) and by Tomlinson (1960) according to seed and seedling structure and mode of germination. In certain palms the embryo is exserted from the seed by the growth of a cotyledonary extension organ termed the apocole. The embryo may be carried some distance from the seed and buried as much as 60 cm below the soil surface. This appears to be an ecological adaptation to dry habitats and is exhibited by the *Borassus* palm whose range overlaps that of the oil palm. Such palms may have ligules or may not. Palms without a long apocole, however, usually have curved embryos and non-persistent radicles. The oil palm does not therefore fit conveniently into any of the groupings, although in structural respects it most resembles those palms having ligules and elongated apocoles. The palm is adapted to a seasonal climate, but it appears that its ecological adaptation is a physiological one (see p. 304) and that a strongly developed apocole would be of no value to it (Rees, 1960).

Within the seed the haustorium develops steadily. This organ has a yellow pigment and is convoluted along the long axis of the nut thus providing a greater surface area for absorption. After about 3 months the spongy haustorium has absorbed the endosperm and completely fills the nut cavity.

The seedling

The seedling has 3 months to establish itself as an organism capable of photosynthesis and absorption of nutrients from the soil (Anon. 1955, 1956).

The plumule does not emerge from the plumular projection until the radicle has reached 1 cm in length. The first adventitious roots are produced in a ring just above the radicle–hypocotyl junction and they give rise to secondary roots before the first foliage leaf has emerged (Fig. 2.1, H and I). The radicle continues to grow for about 6 months by which time it has reached about 15 cm in length. Thereafter the numerous primary roots develop in its place.

Two bladeless plumular sheaths are produced before a green leaf emerges. The latter is recognized by the presence of a lamina, and it emerges about 1 month after germination. Thereafter, one leaf per month is produced until the seedling is 6 months old. A 2-month-old seedling is shown in Fig. 2.2.

After 3–4 months the base of the stem becomes a swollen 'bulb' and the first true primary roots emerge from it. These are thicker than the radicle and grow at an angle of 45° from the vertical. Secondary roots grow out in all directions. During this second period in the seedling's life the leaves become successively larger and change in shape. The leaves of the adult palm are pinnate, but this form is only reached in stages. The first few leaves are lanceo-

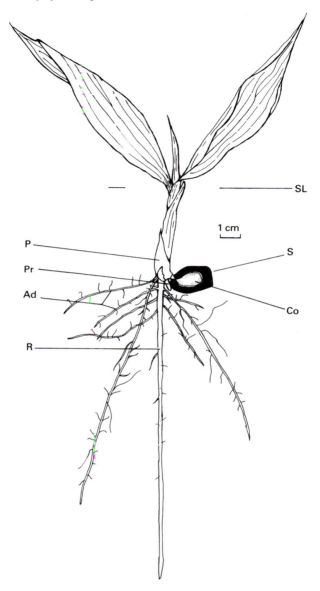

Fig. 2.2 Young seedling 2 months after planting: SL, soil level; S, shell; Co, cotyledon (haustorium); P, plumule; R, radicle; Pr, primary root; Ad, adventitious roots. The nut is shown in section (Anon., 1956).

late with a midrib to half their length; two veins proceed from the end of this midrib to the tip of the leaf. In later leaves a split appears between these veins and the leaf becomes bifurcate. This type of leaf is quickly followed by leaves in which splits divide the laminae between the veins into leaflets or pinnae, although the latter are still joined to one another at the apex. Later still the leaflets become entirely free, though when the leaf opens the tip of the leaflet is always the last part to become entirely unattached.

Young pinnate leaves differ from maturer leaves – to be described later – in the following respects: the leaflets are inserted directly on to the midrib, without pulvini; the lower leaflets do not degenerate into spines; they are less xeromorphic than mature leaves.

The development of the stem and stem apex

In common with other palms, early growth of the oil palm after the seedling stage involves the formation of a wide stem base without internodal elongation. A broad base is formed on which the stem column can rest firmly. Thus, although young palms in which stem base formation is not yet complete may be blown over by high winds, the mature palm's stem is rarely disturbed. Where palms are found leaning over, or are lying on the ground, this is due to soil shrinkage following drainage or loss of topsoil, or both.

The palm has one terminal growing point. Very occasionally branched palms develop from two or more growing points. It is believed that this branching is the result of damage to the apical cells resulting in the formation of two or more primordia with the independent power of growth and differentiation. The separate stems grow vertically, close together.

The apical meristem lies in a basin-like depression at the apex of the stem. In mature palms this depression is 10–12 cm in diameter and 2.5–4 cm deep. The apex itself is conical and is buried in the crown of the palm within a soft mass of young leaves and leaf bases commonly known as the 'cabbage', which is edible. The young leaves, which are yet to elongate, are largely composed of leaf bases with lateral extensions. The remainder of the leaf is reduced to small apical corrugations. The depression in which the apex lies is the result of the peculiar method of primary growth of palms which has been described by Tomlinson (1961) as follows:

> . . . the stem virtually completes its thickening growth before internodal elongation occurs. The apical meristem proper contributes little to the stem tissues but is largely a leaf-producing meristem. The tubular bases of the leaf primordia increase in diameter to keep pace with the increase in diameter of the nodes on which they are inserted. This thickening

growth is brought about by the activity of a meristem which is continuous beneath successive leaf bases and in which cell-division is largely in a tangential plane. Since it brings about increase only in diameter it is known as a primary thickening meristem and internodal elongation only begins where its activity has ended, that is below leaves the bases of which have widened to reach its outer margin and where the stem has almost achieved its maximum diameter.

There are as many as fifty leaves from the centre of the depression to the highest point of the rim (Plate 2.5).

During the early years, while the wide stem base is being formed, the base of the stem assumes the shape of the inverted cone. It is

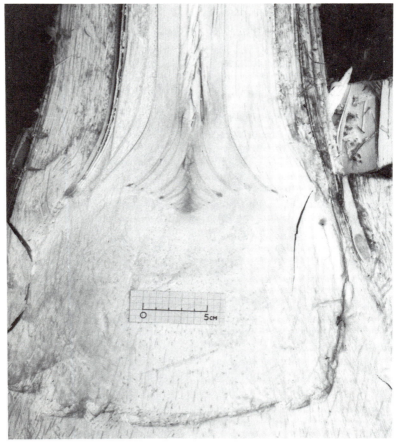

Pl. 2.5 Median longitudinal section through the apex of a young oil palm. Note the basin-like depression containing the apex and young unelongated leaves.

from this cone that the adventitious primary roots are continually being formed both below ground and above it. As soon as the internodes begin to elongate a columnar stem with adhering leaf bases is formed. Although each stem segment may be described as an internode plus leaf, it should be mentioned that the node is only indicated externally on old palms by the leaf scar; internally there is no boundary between adjacent internodes. The leaf bases adhere to the stem for at least 12 years, sometimes much longer. They fall away gradually, starting to fall from the base, the crown or the middle of the stem. When all leaf bases except a few near the crown have been lost, the palm is said to be 'smooth-stemmed' instead of 'rough-stemmed'. In a palm grove, a palm rarely becomes smooth-stemmed until it has grown, at least partially, above the surrounding vegetation and is in bearing. On the smooth-stemmed palms the scars of the leaf bases and those of the leaf sheaths (which encircle the stem) are clearly seen.

The manner in which the leaves are arranged with regard to the axis of the palm is known as its phyllotaxis. The leaves are produced at the apex in an orderly arrangement which, seen from above, is very roughly triangular. A fourth leaf in order of production does not, however, fall into place exactly above the first since the angle two successive leaves make with the axis (the divergence angle) varies about a mean of 137.5°. The arrangement therefore gives rise to sets of spirals or parastichies. This is illustrated in Fig. 2.3 (Henry, 1955a; Anon, 1961).

In well-grown plants two sets of spirals can be seen, eight running one way and thirteen the other. Such an arrangement is described as (8 + 13). If the leaf bases are numbered in the order of leaf formation (the 'genetic spiral') this becomes clear, since, one way, every eighth leaf is seen to be in the same spiral while, the other way, every thirteenth leaf appears in the same (more nearly vertical) spiral. The distance between two leaves in a thirteen spiral or parastichy is a measure of the time of production of thirteen leaves and is termed thirteen plastochrones. Other parastichies can be seen on the palm (see Fig. 2.3) – three for instance, but the larger the parastichy number the more nearly it approaches the vertical. For instance a near-vertical twenty-one parastichy can be distinguished on the diagram. The foliar spirals are in either direction, left-handed or right-handed; in two surveys in Malaysia nearly 53 per cent of the palms were left-handed, but there was evidence that this character was not genetically determined (Arasu, 1970).

More modern studies of the phyllotaxis of the oil palm have been based on Richards's *Phyllotaxis Index* which in turn is calculated from the plastochrone ratio, the latter being the ratio between the transverse distance of a primordium from the centre and that of the immediately preceding primordium. More simply, an equivalent

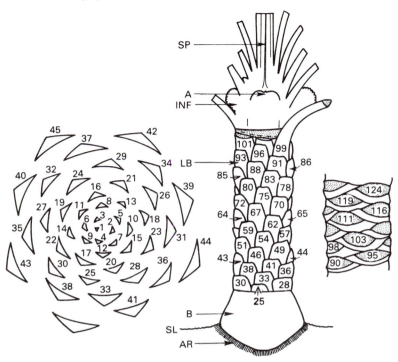

Fig. 2.3 (*Left*) Diagram of the phyllotaxy of the oil palm, adapted from Henry. (*Centre*) Diagrammatic representation of the stem; the sectioned upper portion shows the apex (A) surrounded by leaves, the spear (SP) above and the mature leaves with inflorescences (INF) laterally. The leaf bases are numbered in chronological order of formation from the base upwards and the 5, 8, 13 and other contact parastichies can be seen. Leaf bases have been omitted from the bulbous base of the palm (B). SL, soil level; AR, adventitious roots. (*Right*) Diagram of a portion of a stem from which the leaf bases have dropped. In contrast to the stem of the central diagram the 5 contact parastichy ascends from left to right, the 8 contact parastichy from right to left. The leaf sheath scars have been omitted for simplicity (*J. WAIFOR*, 1961).

phyllotaxis index can be calculated from the radius of the palm cylinder and the longitudinal distance separating two consecutive leaf insertions on the genetic spiral. Modifications of this index result from differing rates of longitudinal and radial growth during development (Rees, 1964) and may be attributable to physiological factors (Thomas *et al.* 1969).

The rate of extension of the stem is very variable and depends on both environmental and hereditary factors. Under extreme shade, growth of both leaves and stem is very slow indeed; in dense plantations or secondary bush, stem growth may be very rapid and the palm will assume an elongated appearance. Under normal plantation conditions, and particularly with heterogeneous planting material, there are often marked palm-to-palm differences, but the average increase in height will be from 0.3 to 0.6 m per year.

In high forest, palms may reach a height of 30 m but elsewhere they reach no more than 15 or 18 m. It is not possible to tell the age of individual grove palms since under heavy shade seedlings and young palms grow very slowly indeed. It is believed that many palms may be 200 years old or more. Of planted palms, the two surviving original Deli palms at Bogor, Indonesia, are more than 120 years old.

The width of the stem, unclothed by leaf bases, varies from 20 to 75 cm. In the Deli palm the diameter is said to vary from 45 to 60 cm (Jagoe, 1934a) but the stems of 'Dumpy' palm progeny are much wider. In plantations, the stem, after the initial bulge, is often remarkably uniform in width, but uneven stems are commonly seen in palm groves. This unevenness is due to alterations in the usage of the surrounding land, and probably also to the scorching of the crown during firing of the cut surrounding bush or to excessive wine tapping.

The stem functions as a supporting, vascular and storage organ. A wide central cylinder is separated from a very narrow cortex through which the leaf traces pass. The cylinder has a wide peripheral zone of congested vascular bundles with fibrous phloem sheaths, and the intervening parenchyma cells are sclerotic; thus this zone provides the main mechanical support of the stem. The vascular bundles are much less congested in the central zone. In common with other palms there is no cambium or callus formation and, although some palms show slight secondary thickening by cell division and expansion, this activity is not exhibited by the oil palm. Starch grains and silica-containing cells (stegmata) are abundant.

The courses taken by the vascular bundles within the stem are naturally of importance in the supply of nutrients to the crown by long-distance translocation. Early nineteenth-century workers made considerable progress in unravelling the vascular system, but a fuller understanding had to await the recent imaginative technique of Zimmerman and Tomlinson (1965) working with the small palm *Rhapis excelsa* while conjointly carrying out examinations of stems of larger palms. The general pattern is believed to be essentially similar in the oil palm and other large palms and may be briefly described from these authors' work. Firstly the continuity of the vascular bundles through the stem has been demonstrated and secondly the relationships of the vertical bundles with the leaf traces and with each other have been clarified. All bundles maintain their individuality and proceed indefinitely up the stem, giving off leaf traces at intervals. In proceeding up the stem the bundle slants towards the centre from the periphery and then bends sharply back towards the periphery and divides into several branches. One branch is the leaf trace which proceeds into the leaf. Others go into the inflorescence, others 'bridge' into neighbouring bundles while another bends vertically again as the vertical bundle and the process is repeated. It is this course, being followed by many thousands of bundles, which accounts for the crowding at the periphery and the even but sparser distribution in the centre. Finally, it has been shown that in the central uncrowded part of the stem the bundles do not remain on one side of the stem but take a spiral or helical course.

The leaf

In the crown of an adult palm a continuous succession of leaf buds or primordia are being separated laterally from the apical cone (Henry, 1955b, c). Development of the leaf is initially very slow. Some forty-five to fifty leaves are to be found in the crown; each remains enclosed for about 2 years and then rapidly develops into a central spear and finally opens (Broekmans, 1957). The base of the leaf completely encircles the stem apex and in the adult leaf the base is persistent as a strong fibrous sheet.

The mature leaf is simply pinnate, bearing linear leaflets or pinnae on each side of the leaf stalk. The latter may be divided into two zones, the rachis bearing the leaflets, and the petiole which is much shorter than the rachis and bears only short lateral spines (Anon., 1962). At the junction of petiole and rachis small leaflets with vestigial laminae (leaf blades) are found (Fig. 2.4). Petioles vary greatly in length and in the Deli palm may be as long as 1.2 m. Some petioles remain green for a considerable period.

The spines have been shown to be of two kinds which are named fibre spines and midrib spines (Fig. 2.5). The former are those on the petiole; they are very regular and are formed from the base of the fibres of the leaf sheath. The point at which these fibres break off is very regular, so the spines are nearly all the same length. Where the leaflets begin to occur they are poorly developed although they have the basal swellings of fully developed leaflets. The lamina of these poorly developed leaflets frequently becomes torn away, leaving a spine which was originally the leaflet midrib. These spines have the same irregularity of set as have the fully developed leaflets on the leaf (Rees 1963).

The leaf stalk is a hard fibrous body which may be as long as 8 m. At the tip it is almost circular in cross-section, but in the centre of the rachis it is asymmetrical with lateral faces where the leaflets are inserted. In the petiole the lateral faces disappear. The lower or abaxial face is much more strongly curved than the upper or adaxial face.

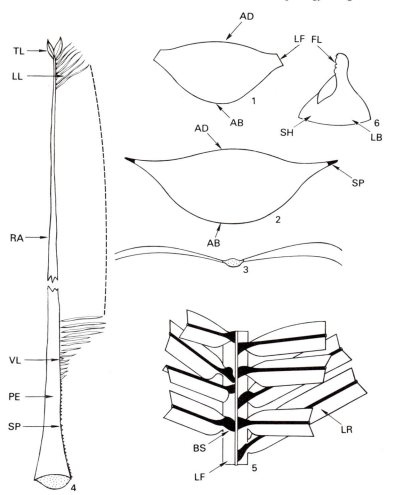

Fig. 2.4 The oil palm leaf: **1**, cross-section of rachis; **2**, cross-section of petiole; **3**, cross-section of leaf viewed end-on, showing two-ranked insertion of leaflets; **4**, diagram of oil palm leaf; **5**, central portion of rachis from above, showing irregular leaflet insertion; **6**, leaf apex of palm (*J. WAIFOR*, 1962). AD, adaxial face; AB, abaxial face; LF, lateral face; SP, spine; RA, rachis; PE, petiole; TL, terminal pair of ovate leaflets; LL, liner leaflets; VL, leaflets with vestigial laminae; LB, leaf base; FL, future green leaf; SH, leaf sheath completely encircling apex and through which younger leaves and stem grow; BS, basal swelling; LR, lower rank leaflet.

The leaflets are produced by the splitting of an entire leaf during the elongation of the leaf axis. Within the spear the leaflets are still attached to one another but are folded upwards and show clearly where the splitting is to occur. In an actively growing plant spears

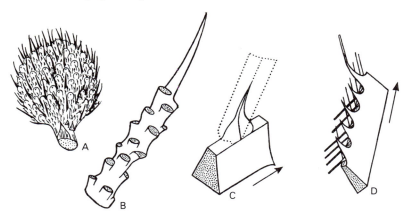

Fig. 2.5   Spines in the oil palm. Diagrams of **A**, fruit bunch with spiny branch tips; **B**, single branch of inflorescence with fruit removed; **C**, midrib spine, leaflet shown by dotted outline; **D**, fibre spines viewed from abaxial side. The arrows in **C** and **D** point to the leaf tip (Rees, 1963).

are produced one at a time and point vertically upwards. As the spear opens, another elongates rapidly to take its place. In severe dry seasons, however, and this is very noticeable in West Africa from January to March, many spears may elongate before the first of their number opens. In these circumstances it is not unusual to see half a dozen or more spears, many of them fully or nearly fully elongated, protruding from the centre of the crown. As soon as wet weather ensues the majority of these spears will open and the upper part of the crown takes on a light green appearance for a short period.

After the leaf has opened it is progressively displaced centrifugally as younger leaves emerge. Middle-aged leaves lie parallel to the ground with the tip bearing slightly downwards. Usually the adaxial face of the rachis faces upwards, but sometimes the leaf twists into a vertical plane or intermediate position.

Typically, the leaflets inserted on the lateral faces alternate in upper and lower ranks. There is no exact regularity, however, and two or more consecutive leaflets may appear in the same rank. Similarly within each rank the angle of insertion is often irregular, and very occasionally there is almost no 'ranking effect'. Generally, however, it is the provision of two ranks and the irregularity of leaflet insertion which gives the palm its shaggy appearance and distinguishes it, from a distance, from the coconut palm or *Elaeis oleifera*. Individual leaflets are linear in shape and each leaf has a terminal pair. Leaflets number some 250–300 per mature leaf and are up to 1.3 m long and 6 cm broad. The leaflet midrib is often very rigid and the laminae sometimes tear backwards from the tip. This increases the 'untidy' appearance of the leaf. There is a small basal swelling, resembling a pulvinus but with no motor function, at the insertion of the leaflet on the rachis. The stomata are on the abaxial surface of the leaflets, and their form and function are described in Ch. 4, p. 145.

Apart from the leaflet variation noted above, more striking leaf variations are to be found. In the *idolatrica* palm the leaflets do not separate normally and an entire or semi-entire leaf is formed. The midribs of all the unseparated 'leaflets' are in one plane. There is still some doubt concerning the inheritance of the *idolatrica* character, but self-pollinated *idolatrica* palms have bred true and the character is thought to be recessive (Zeven, 1964). Observations have suggested that the centre of distribution of the *idolatrica* palm lies between Ghana and the lower Niger. Many of these palms are

Pl. 2.6 The *idolatrica* palm, with fused leaflets.

found in Benin (Dahomey) and western Nigeria. Westwards and eastwards specimens are rarer and are often found only in botanic gardens or agricultural stations (Plate 2.6).

Other leaf peculiarities occur but have been insufficiently studied. They may be due to genetic, nutritional or pathogenic factors. Some will be described in other sections of this book.

The number of leaves produced annually by a plantation palm increases to between thirty and forty at 5 or 6 years of age. Thereafter the production declines to a level of twenty to twenty-five per annum. Leaf production of grove palms is much lower (see Ch. 1).

In the axil of each leaf there is a bud which may develop into a male, female or, occasionally, a hermaphrodite inflorescence. Very rare cases have been known, however, in which a vegetative bulbil-shoot is produced instead of an inflorescence. This has been termed 'vivipary' by Henry (1948a) who has described an original palm at Okeita, Benin, the shoots taken and developed from it, and similar palms in the Ivory Coast. While in some cases the shoots from 'viviparous' palms can be rooted and will produce similar bulbil-shoot-producing palms, in other cases no roots are formed and sexual buds are later produced either on the palm itself or at the extremity of the bulbil-shoot. In the latter cases there are also considerable malformations of both the vegetative and sexual parts of the shoots (Henry and Scheidecker, 1953; Davis, 1980).

The root system

The seedling radicle is soon replaced by adventitious primary roots emanating from the radicle–hypocotyl junction and then from the lower internodes of the stem which are formed into a massive basal cone or bole. The latter retains the capacity for producing roots well above ground level. Roots sometimes develop on the stem up to 1 m above ground but these normally dry out before reaching the soil.

In the mature palm thousands of primary roots spread rapidly from the bole and new primaries are continually replacing dead ones (Yampolsky, 1922). The vertical extent of the root system depends very largely on the presence or absence of a water-table. Two extremes may be cited. In Malaysia, Lambourne (1935) studied the roots of 11-year-old palms growing in soil where the water-table was as high as 1 m from the surface in dry weather. In these circumstances no primaries penetrated below this depth and the majority of roots were in the surface 45 cm. Individual primaries were found to a distance of 19 m from the stem and absorbing portions of roots were found at all intermediate distances. In contrast, Vine (1945) and Purvis (1956) examined root systems in free-draining sandy soils

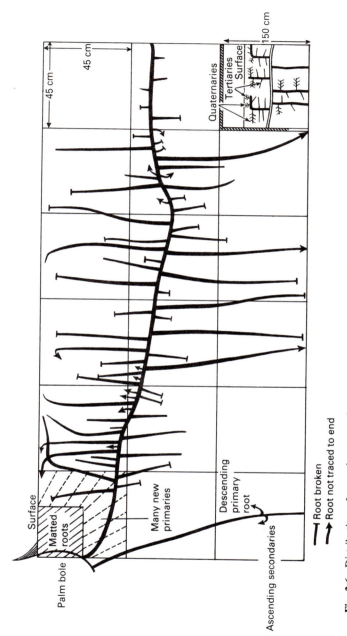

Fig. 2.6 Distribution of secondary roots from one primary root in a 3-year-old palm (Purvis, 1956). The drawing also shows the usual position of a descending primary root. *Inset* is a diagram of the development of tertiary and quaternary roots.

and found that primary roots may descend to great depths. It is this unimpeded root system which will now be described.

Primary roots extend either downwards from the base of the palm or radially in a more or less horizontal direction (Fig. 2.6). The descending primaries, which proceed directly from under the base of the palm are fewer in number than the radiating primaries and carry much fewer secondaries. Ruer (1969) has shown that these descending roots are for anchorage and play little or no part in the absorption of water.

The remaining primary roots appear from the base of the stem at all angles to the soil surface, but they tend to bend to the horizontal and few are found below 1 m. From these primary roots, 5–10 mm in diameter, secondary roots ascend and descend in approximately equal numbers, though with a slight preponderance of ascending roots. These secondaries are 1–4 mm in diameter and give rise in turn to horizontally growing tertiaries of 0.5–1.5 mm in diameter and up to 15 cm in length. From these are developed the mass of quaternaries of up to 3 cm in length and only 0.2–0.5 mm in diameter.

The ascending secondaries generally reach the surface of the soil while the descending ones may penetrate to a considerable depth.

The density of all classes of roots in the top 60 cm of soil usually decreases with distance from the palm, but with adult palms the total quantity of absorbing roots in successive surrounding circles increases at least to a radius of 3.5–4.5 m (Ruer, 1967a). The greatest quantity of roots is to be found between soil depths of 20 and 60 cm, and most of the absorption of nutrients has been shown to be through the quaternaries and absorbing tips of primaries, secondaries and tertiaries to this depth (Taillez, 1971). However, the exact depth of root concentration depends on the soil type (Chan, 1977).

Roots of all classes show a positive tropism towards superior conditions of water and nutrient supply and, with rotting felled vegetation or heaps of palm leaves, or under a good *Pueraria* cover, this may lead to a high density of quaternaries in the centre of the interline (Bachy, 1964). For instance, with a *Pueraria* cover and on good alluvial soil in Colombia tertiaries and quaternaries increased with distance from the palm, but where there was a grass cover the quantity of these roots declined with distance. Similarly, the quantity of roots is much reduced under the paths along the lines. Where the rooting volume is reduced by quantities of concretionary gravel the quanity of roots per palm is much reduced (Tan, 1977), and primary roots tend to become twisted and constricted, the root system lies nearer the surface and the tertiaries and quaternaries are coarser and more lignified (Taillez, 1971). Sub-aerial roots which grow up into loose decaying leaves are readily produced.

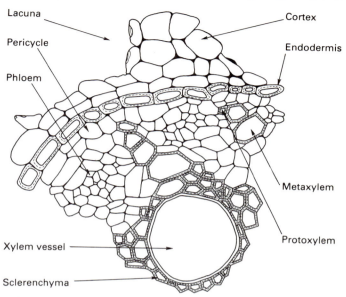

Lacuna

Pericycle

Phloem

Cortex

Endodermis

Metaxylem

Protoxylem

Xylem vessel

Sclerenchyma

Fig. 2.7 Transverse section of middle-aged seedling root to show stele (Purvis, 1956).

The anatomy of palm roots is described by Tomlinson (1961) and that of the oil palm in particular has been studied by Purvis (1956) and Ruer (1967b). The primary root consists of an outer epidermis and lignified hypodermis surrounding a cortex in which well-developed air lacunae are to be found. Within the cortex lies the central stele or vascular cylinder consisting of the surrounding lignified endodermis, the inner vascular strands of xylem and phloem and the pith or medulla which rapidly lignifies in old roots (Figs 2.7 and 2.8). The stele also contains lacunae. The secondary and tertiary roots have essentially the same structure as the primary roots. The unlignified tips of the growing primary, secondary and tertiary roots measure 3–4, 5–6 and 2–3 cm respectively. The quaternary roots are only 1–3 cm long, are produced in large numbers and are almost wholly unlignified. There are no root hairs and it is therefore reasonable to suppose that quaternary roots play the main part in the absorption of nutrients. Moreau and Moreau (1958) have studied in some detail the lignification of oil palm roots and the anatomy of production of the substituting roots which are formed when a seedling root is damaged or affected by disease. Such roots appear from just behind the point of damage or zone of disease and proceed in the same direction as the original root.

The roots of *Elaeis guineensis* (and other palms) are characterized

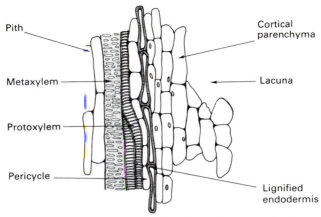

Fig. 2.8 Radial longitudinal section of old seedling primary root to show outer stele and inner cortex (Purvis, 1956).

by the presence of pneumatodes. These, although appearing on both underground and aerial roots, have been supposed to ventilate the underground roots: direct physiological evidence for this is lacking. Yampolsky (1924) found more pneumatodes on aerial than on underground roots in Sumatra, but the reverse is the case in West Africa. Moreover, they are commonest on seedlings grown in glasshouses or wherever the root system has been kept under water or in very moist conditions (Purvis, 1956).

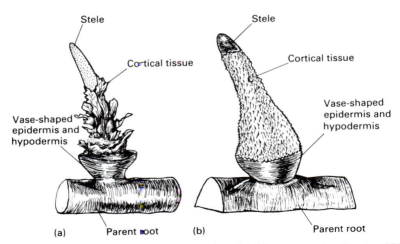

Fig. 2.9 Young, underground (a) and old, aerial (b) pneumatodes (Purvis, 1956).

In pneumatode-forming root shoots the epidermis and hypodermis rupture and the stele and cortex extrude. The latter then proliferates and its parenchymatous cells become suberized or, if the pneumatode is aerial or subjected to dry conditions, lignified (Fig. 2.9). If the growing point is unharmed after the rupture of the epidermis it remains attached as a cap and sometimes a normal root may develop again.

The firm anchorage of the adult palm is not only due to the descending primary roots. The old roots are strong and elastic and persist in the soil long after they have died. When death of a root occurs the cortex degenerates, leaving a tubular hypodermis with cortical fibres and the woody stele loose within.

The flowers and fruit

The oil palm is said to be *monoecious*, that is to say male and female flowers occur separately – and in this case usually in distinct male and female inflorescences – on the same plant. Detailed investigation of the flowers has shown, however, that each flower primordium is a potential producer of both male and female organs (Beirnaert, 1935a). In very rare cases both the androecium and the gynoecium develop fully to give a hermaphrodite flower. An inflorescence is initiated in the axil of every leaf but some inflorescences abort before emergence. Twin inflorescences in the leaf axil have been known. Each inflorescence is a compound spike or spadix carried on a stout peduncle 30–45 cm in length. Spikelets are arranged spirally around a central rachis in a manner which varies both with age and position on the rachis; however, equivalent phyllotaxis index measurements have shown little difference between male and female (Thomas *et al.*, 1970).

Van Heel *et al.* (1986) followed inflorescence and flower development by scanning electron microscopy from a very early stage and showed that, after the production of two lateral and anterior and posterior bracts, there follows a period during which parastichies of bracts are formed before the appearance of spikelets (Plate 2.7a). The development of spikelet primordia in the axils of the bracts then starts somewhat below the apex of the inflorescence. On each primordium two lateral and anterior and posterior bracts are first formed, but from this point development of the male and female spikelet is quite different, since the former produces, in spiral formation, a very much larger number of flower bracts, and the male flower, although developmentally regarded as a reduced triad (see p. 74), is solitary.

An inner and outer spathe tightly enclose the inflorescence until about 6 weeks before anthesis when the inner spathe begins to open. After a further 2 or 3 weeks the inner spathe splits; later both

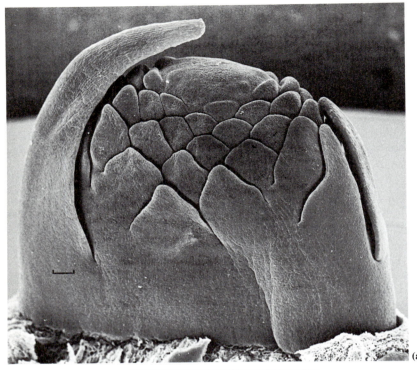

(a)

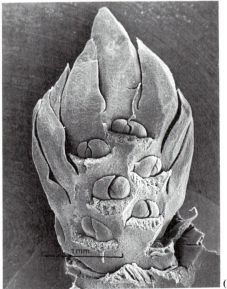

(b)

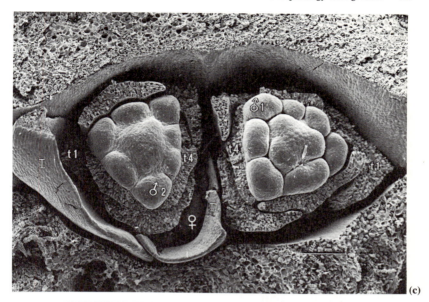

(c)

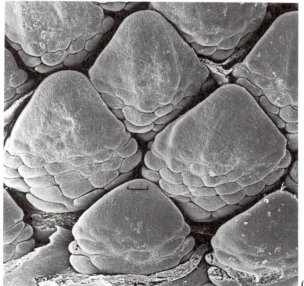

(d)

Pl. 2.7   Flower development. Electron microscopy photographs showing
(a) parastichies of bracts on the inflorescence central axis, with first lateral bracts;
(b) young female spikelet with some bracts removed exposing developing triads of
flowers; **(c)** the two male flowers of a triad with the female flower origin in the centre
(tepals removed); **(d)** male inflorescence showing the numerous bracts forming on
the developing spikelets (Van Heel *et al.*, 1987). I, II, III = bracteoles. t = tepal.
All bars except on (b) represent 0.1 mm.

spathes fray and disintegrate and the inflorescence pushes its way through. Six to ten long bracts are found below the lowest spikelet; two of them extend to the top of the inflorescence.

An inflorescence can be male, female or hermaphrodite and the order and proportions in which these are produced show little or no regularity. Occasionally there is a hermaphrodite inflorescence between the male and female series and these inflorescences are commoner in young palms. The number of spikelets per inflorescence varies greatly from palm to palm, but Beirnaert (1935b) showed that the variation in this respect between inflorescences of a given palm is very small and is independent of the sex of the inflorescence. In thirty-seven adult palms in Zaire the average number of spikelets per inflorescence was found to range from 100 to 283 and in almost all cases the coefficient of within-palm variation was very small indeed. In hermaphrodite inflorescences the sum of the male, female and mixed spikelets is always close to the average number of spikelets for male or female inflorescences of the palm concerned. An examination of a few inflorescences from a very much larger number (1,476) of Zaire palms showed that, though the range was great (85–285 spikelets), inflorescences with 125–165 spikelets were most frequent.

The female inflorescence and flower

The female inflorescence reaches a length of 30 cm or more before opening. Both male and female spikelets develop in the axils of the bracts of the inflorescence primordium (Plate 2.7b). These bracts become spinous in the female spikelet. The number of flower-subtending bracts on the latter are much fewer than on the male spikelet. The flowers are arranged spirally around the rachis of the spikelet; each is housed in a shallow cavity and subtended by a bract which is drawn up into a spine. At the end of the spikelet there is a spine of very variable length. The number of flowers in an inflorescence varies from palm to palm but in all cases there is a much larger number (twelve to thirty) on the central spikelets than on the lower or upper spikelets (twelve or less). The inflorescences will thus contain several thousand flowers. Those on spikelets at the base of the inflorescence open before those on spikelets at the top, and, within each spikelet, those at the base open first.

The functional female flower develops in a triad between two non-functional male flowers. This was first described by Beirnaert (1935b), though Van Heel *et al.* (1986) have shown that Beirnaert's positioning of the bracts and male flowers does not exactly correspond to the monochasial branching system which they consider the triad to be and which they term a cincinnus (Fig. 2.10 and Plate 2.7c). The tricarpellate ovary and rudimentary androecium of the female flower are enclosed by a double perianth of six sepaloid

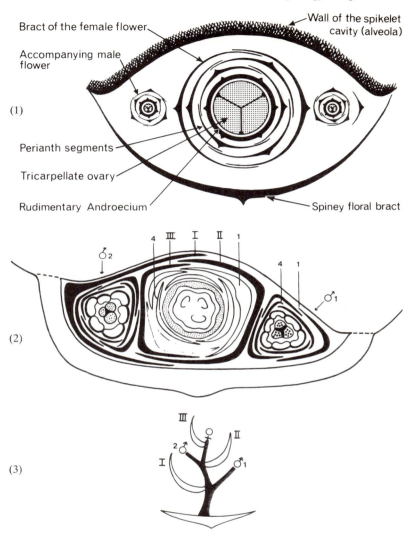

Bract of the female flower

Wall of the spikelet cavity (alveola)

Accompanying male flower

(1)

Perianth segments

Tricarpellate ovary

Rudimentary Androecium

Spiney floral bract

(2)

(3)

Fig. 2.10 (1) Vanderweyen's diagram of the female flower with accompanying rudimentary male flowers, in Beirnaert's *Introduction à la biologie florale du palmier à huile*, 1935; (2) Van Heel *et al.*'s (1987) developmental diagram of a triad of flowers on a female spikelet; and (3) diagram of cincinnal branching system.

segments in two whorls; these in turn lie within two bracteoles. The sessile stigma has three lobes; these are hairy, with a trichomatous crest forming the receptive surface (Lawton, 1981), and they exude moisture at the receptive stage. The sepals are about 2 cm long at

A   B

Pl. 2.8   Male and female inflorescences at two stages of development: **A**, before anthesis within (**1**) and freed from (**2**) the spathes; **B**, at anthesis. (*Below*) Some abnormal pollen-producing flowers at the end of female spikelets. (Unilever Ltd.)

anthesis. The rudimentary androecium has six to ten short projections; Beirnaert has described in detail the range of development of the rare hermaphrodite flowers which may occur both on the male and female inflorescences.

Occasionally two female flowers may develop, within a single pair of bracteoles, between the two accompanying abortive male flowers. Another unusual occurrence is the production of female flowers on the end of long peduncle-like stalks which are inserted directly on the rachis of the inflorescence. Inflorescence abnormalities are by no means uncommon in the oil palm and the tendency to abnormality must be taken into account by plant breeders (Plate 2.8). Anthesis in a normal inflorescence usually lasts for 36–48 hours but may be extended to a week.

The male inflorescence and flower

The male inflorescence is borne on a longer peduncle than that of the female inflorescence and contains long finger-like cylindrical spikelets. It is not spiny. The spikelet has short bracts and a short terminal projection. Spikelets measure between 10 and 20 cm in length and 0.8–1.5 cm in breadth.

Before opening, the sessile flower is completely enclosed in a triangular bract; it consists of a perianth of six minute segments, a tubular androecium with six, or rarely seven, anthers, and a rudimentary gynoecium with three projections corresponding to the trilobed stigma. It is considered to correspond to the first (non-functional) male flower of the triad. The three outer perianth segments are hard, the inner ones soft. The flower is 3–4 mm long and 1.5–2.0 mm wide. Spikelets of 8-year-old *tenera* palms in Malaysia were found to have an average of 785 flowers (Tan and Rao, 1979); there will thus be well over 100,000 flowers per inflorescence. Flowers begin to open from the base of the spikelet and all flowers on the spikelet have usually opened within 2 days, though during rainy weather opening may be prolonged to 4 days. Most pollen is shed during 2 or 3 days following the start of anthesis and production ceases within 5 days. Viability of late-produced pollen is low (Hardon and Turner, 1967). Inflorescences produce from 25 g up to 300 g of fresh pollen.

Hermaphrodite or mixed inflorescences

There is a great variety of hermaphrodite or mixed inflorescences formed. Usually male, female and mixed spikelets appear on the same bunch but in widely differing proportions and positions. Some palms are more disposed to the production of these inflorescences than others.

The mixed spikelets have, in varying proportions, female flowers at the base and male flowers at the summit. In between, 'geminiflorous' male flowers, corresponding to the accompanying male flowers, lie close together with no female flower between them. Further up the spikelet these give place to the normal single male flowers. Spikelets can contain either all three types of flower group, or male flowers and geminiflorous flowers alone, or they may have the superficial appearance of a female spikelet but contain geminiflorous flowers.

Young palms occasionally produce a peculiar type of inflorescence which has been called andromorphic. This has all the appearance and structure of a male inflorescence before it opens. Examination shows, however, that the male flowers have been replaced by small solitary female flowers arranged in the manner of flowers in a male inflorescence. Small fruit develop from the flowers, but the carpels are not firmly joined together and the resulting fruit have three lobes corresponding to the three partially separated carpels. Male flowers are also to be found in andromorphic inflorescences though some are deformed.

Pollen and pollination

Anthers of the male flowers dehisce by vertical slits. The pollen grains are at first oval, but at maturity they are two-celled and somewhat triangular in outline (Tan and Rao, 1979). For a long time it was thought that the oil palm was mainly wind pollinated and that pollination failures were caused by a low proportion of male inflorescences and unfavourable atmospheric conditions within the plantation. The unique researches of Syed (1978, 1979, 1981b) showed that, on the contrary, the palm is mainly insect pollinated. All the early authorities such as Lespesme (1946), Alibert (1945) and Beirnaert (1935a) believed that insects were not required for pollination, though Henry (1948b) thought they might play some part, and Jagoe (1934b) even postulated that the mild smell of aniseed emitted from the flowers owed its origin to a primitive ancestor. These views held firm for more than 40 years and led to much work being done on the wind movements of pollen and methods of assisted pollination.

It is now clear that the main pollinating agency in Africa is a subfamily – Derelominae – of weevils, among which species of *Elaeidobium* predominate, while in Malaysia it was established that *Thrips hawaiiensis* had adapted itself to the palm and could carry sufficient pollen to the female inflorescence if present in sufficient numbers. The species is polyphagous, but the oil palm strain of thrips appears to be absent from Sabah where pollination failures have been serious. In West Malaysia it is abundant in old palms but less frequent in new fields, and this may account for poor pollination of young palms (Syed, 1981a). *Elaeidobius kamerunicus* has recently been successfully transferred from West Africa to the Far East and

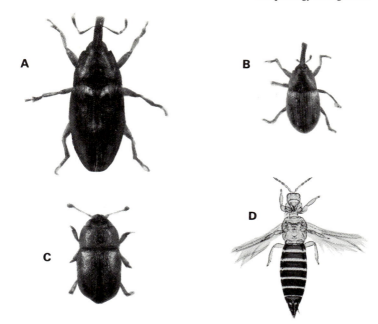

Pl. 2.9 The insect pollinators of the oil palm: **A**, *Elaeidobius kamerunicus* (Curculionidae, weevil), West and Central Africa, ♂ (× 10); **B**, *Elaeidobius subvittatus*, Africa and South America, ♀ (× 10); **C**, *Mystrops costaricensis* (Nitidulidae, sap beetle), South America, sex uncertain, (× 16); **D**, *Thrips hawaiiensis*, Asia, (× 24) (By courtesy of the Trustees of the British Museum (Natural History)).

become established on the oil palm (Syed *et al.*, 1982) with consequently improved fruit set (p. 176).

In Latin America the presence of *Elaeis oleifera* is obviously of significance in providing insect pollinators. Two pollinating beetles have been found on both *Elaeis* species. These are the sap beetle, *Mystrops costaricensis*, which is common in areas of high, even rainfall, and the weevil, *Elaeidobius subvittatus*, a native of Africa which is thought to have been fortuitously introduced perhaps in transported pollen samples (Wood, 1983). However, the exact status of these two insects, and the reasons for their uneven distribution, are not fully understood, and fruit set in America, though generally better than in Asia before the introduction of *E. kamerunicus*, is not always satisfactory. *Elaeidobius kamerunicus* has recently been introduced into Colombia (Syed, 1984; Corrado, 1984).

The fruit and fruit bunch

A short description of the morphology of the fruit is given here.

Fruit variation and the inheritance of fruit characters are discussed more fully in Chapter 5.

The fruit is a sessile drupe varying in shape from nearly spherical to ovoid or elongated and bulging somewhat at the top. In length it varies from about 2 cm to more than 5 cm, in weight from 3 g to over 30 g. The Deli fruit of the Far East are usually considerably larger than the fruit of Africa, though contrary to general belief fruit as large as Deli fruit are sometimes encountered in Africa.

The *pericarp* of the fruit consists of the outer *exocarp* or skin, the *mesocarp* or pulp (often incorrectly termed the pericarp – hence 'pericarp oil'), and the *endocarp* or shell. When measuring the pulp, the exocarp is included with the mesocarp. The endocarp together with the kernel forms the seed which has already been described.

There is one abnormal fruit type, variously known as *Poissoni*, mantled, and *diwakkawakka*, in which fleshy outgrowths or supplementary carpels surround the main part of the fruit. These have developed from the rudimentary lobes of the androecium of the female flower, but they are considered to be carpellary in character since they are often found to contain shell and kernel in the centre (Fig 2.11). Mantled fruit are rare; in one area of Nigeria only 33 mantled bunches were found from among 20,291 bunches harvested from grove plots over a 4-year period. In Angola the frequency was found to be 9 palms in 10,000. Similar figures have been quoted elsewhere (Zeven, 1973).

Fig. 2.11 Longitudinal sections of mantled fruit: (*left*) after Beirnaert (1935); (*right*) after Janssens (1927), in which one of the supplementary carpels shows complete development.

In external appearance the fruit varies considerably, particularly when ripening (Plates I and IV, between pp. 230 and 231). Moreover, the exocarp of the external fruit tends to be more pigmented than that of the internal fruit. By far the commonest type of fruit is deep violet to black at the apex and colourless at the base before ripening. Such fruit has been described as 'ordinary' or *nigrescens*. A relatively uncommon type is green before ripening, and this is called green-fruited or *virescens*. The latter changes at maturity to a light reddish-orange though the apex of the external fruit remains greenish. The frequency of the *virescens* type was found to be 50 in 10,000 bunches in a grove area in Nigeria and 72 in 10,000 in Angola.

The colour of the ordinary or *nigrescens* fruit varies to an appreciable extent on ripening and there is evidence that this is connected with carotene content. This colour difference in ripening was recognized at an early date by Chevalier (1910) who gave the names *communis* to 'fruit entirely red when ripe or with a small black or brown halo at the tip' and *semper-nigra* to 'fruit when ripe, black over the upper half but red at the base'. These differences are also recognized by some of the peoples in West Africa where different vernacular names are allotted to them (e.g. *abepa* and *abetuntum* in the Fanti and Twi languages of Ghana). The names *rubro-nigrescens* and *rutilo-nigrescens* have been proposed by Purvis (1957). The former fruit, when ripe, are defined as 'Cap – 00918, Garnet brown, sometimes tending to be darker, rarely extending over half the fruit; Base – 713, Indian orange, colour uniform to the base'. *Rutilo-nigrescens* is defined as 'Cap – Black, though it may show a brownish tinge at the edges, usually covering more than half the fruit; Base – colour not constant, tending to lighten towards base, the deepest being 13, Saturn red'. The colours and colour numbers refer to the Wilson Horticultural Colour Chart.* Nevertheless it is not a simple matter to allot fruit with any certainty to one or other of these subtypes and fruit of intermediate appearance can invariably be found. Moreover, the cap colour is deceptive; there are no real black caps since even the darkest when seen through transmitted light are reddish in colour.

The above description refers to the colour appearance of fruit commonly or occasionally found in palm groves and plantations. There is, however, a much more fundamental colour variation due to presence or absence of carotenoids. The *albescens* fruit, characterized by 'absence' of carotene in the mesocarp, is extremely rare. Actually, this fruit does contain a very small quantity of carotene. It was first noted in Ghana, under the name *abefita*, but was later named *albescens* by Beccari (1914b). It has been subsequently found in Zaire, Angola, Nigeria, the Ivory Coast and other parts of Africa. In Angola the frequency was found to be only 3 in 10,000 and it may be rarer in other parts of Africa.

Albescens fruit may be of *nigrescens* or *virescens* type; in Zaire the fruits are referred to as *albo-nigrescens* and *albo-virescens*. The difference is only in the cap of the fruit, the former's cap being

---

* The Wilson nomenclature may cause some confusion since the full hue 13, an orange colour, is called 'Saturn red' while its deeper shade 713, which appears much redder, is named 'Indian orange'.

dark-brown to black in appearance, the latter's green. The rest of the fruit is ivory coloured, ripening pale yellow. Only a very few *albo-virescens* palms have been found (Vanderweyen and Roels, 1949). *Nigrescens* and *virescens* fruit contain varying quantities of carotenoids in the mesocarp and this will be referred to again in Chapter 14. Exterior fruit may have as much as twice the carotene content of interior fruit.

In internal structure the most important differences are to be found in thickness of shell (Plate I). As shells of all thicknesses from less than 1 to 8 mm can be found it might be thought that a division of fruit into thick-shelled and thin-shelled forms would be somewhat arbitrary. However, a rare shell-less form was early noted in Africa and named *pisifera* owing to the pea-like shell-less kernels found in fertilę fruit. *Pisifera* palms always bear large quantities of female bunches. In many cases the majority of the bunches rot; these are known as infertile *pisifera* though the setting of a few fruit is of course necessary to identify them, as other abortive forms are encountered. Infertile *pisifera* tend to show strong vegetative growth. Fertile *pisifera* palms are less common.

Apart from the discovery of the *pisifera* it was also noted that in the majority of the thinner-shelled fruit there was a distinct ring of fibres embedded in the mesocarp but near to and encircling the nut. This can be clearly seen when the fruit is cross-sectioned. In 1935 Smith recorded that 'the present dividing line between thick- and thin-shelled forms is that the fruit of the latter contain the mesocarp fibre ring and the nuts can be readily cracked'. Subsequent genetical studies have shown that the thin-shelled form with a fibre ring – the *tenera* form – is a hybrid of the shell-less *pisifera* and the common thick-shelled *dura* form which has no fibre ring. Internal fruit form may therefore be described as being either:

(a) *Dura*: shell usually 2–8 mm thick though occasionally less, low to medium mesocarp content (35–55 per cent but sometimes, in the Deli *dura*, up to 65 per cent); no fibre ring;

(b) *Tenera*: shell 0.5–4 mm thick, medium to high mesocarp content (60–96 per cent, but occasionally as low as 55 per cent); fibre ring; or

(c) *Pisifera*: shell-less.

The term *macrocarya* has been used for *dura* palms with shell thickness of 6–8 mm, but the term has largely gone out of use as it has no genetical significance. It must be said, however, that in many parts of West Africa (e.g. Sierra Leone and western Nigeria) fruit which could be described as *macrocarya* forms a large proportion of the crop. In a grove survey in eastern Nigeria 27 per cent of the *dura* palms were classed as *macrocarya* and there are undoubtedly much higher proportions elsewhere.

The mesocarp of all fruit contains fibres which run longitudinally

through the oil-bearing tissue. This fibrous material usually constitutes about 16 per cent of the mesocarp but may vary from 11 to
21 per cent. The oil percentage varies from about 35 to 60 per cent.
The fibres of the fibre ring in *tenera* fruit are dark in colour; dark
fibres may also be distributed in other parts of the pulp though they
are usually in the central section. Light-coloured fibres are distributed regularly throughout the mesocarp.

The fruit bunch (Plate II, between pp. 230 and 231, and Plates
2.10 and 2.11) is ovoid and may reach 50 cm in length and 35 cm
in breadth. The bunch consists of outer and inner fruit, the latter
somewhat flattened and less pigmented; a few so-called parthenocarpic fruit which have developed even though fertilization has not
taken place (or possibly following partial abortion); some small
undeveloped non-oil-bearing 'infertile fruit'; and the bunch and
spikelet stalks and spines. In the parthenocarpic fruit endosperm
and embryo are absent and the centre is usually solid. Bunch weight
varies from a few kilograms to about 100 kg according to age and
situation but in adult plantations mean weights are 10–30 kg. Well-
set bunches carry from 500 to 4,000 fruit, a mean of about 1,500
being usual, with a fruit-to-bunch ratio of 60–70 per cent. Ripening
is usually from the apex downwards, the fruit becoming gradually
detached.

The female inflorescence at anthesis is in the axil of the seven-

Pl. 2.10   A bunch from a young Deli palm.

Pl. 2.11   A bunch from a young Nigerian *dura* palm.

teenth to twentieth leaf from the central spear. By the time the
bunch is ripe it is subtended by about the thirtieth to thirty-second
leaf, but the bunch leans out from its subtending leaf on to a leaf
in a lower whorl; it is not the leaf subtending the bunch which
supports it (Plate 2.12). The fruit develop steadily in size and weight

Pl. 2.12   Peduncle of a ripe bunch cut to show that the bunch was subtended by a
younger leaf than the one on which the bunch lay.

from about the fifteenth to nineteenth day after anthesis. Oil formation in the kernel and mesocarp takes place towards the end of a period of maturation during which the shell hardens and the embryo becomes viable (see Ch. 14, p. 675).

The American oil palm *Elaeis oleifera* (HBK) Cortés

The American oil palm was first described by Jacquin (1763) as growing in Colombia, and he illustrated the fruit with calyx. On the basis of this description and figure, Giseke, in editing Linnaeus's *Praelectiones in Ordines Naturales* (1792), adopted *Corozo* as the palm's generic name. However, the name *Elaeis melanococca* soon became widely used owing to its employment in a description of fruit and seeds of both *E. guineensis* and a supposed other species by Gaertner in his *De Fructibus et Semenibus Plantarum* (1788). *Melanococca* signifies 'black-berried' and Bailey (1933) believed that the seed figured by Gaertner was in fact a seed of *E. guineensis* (Plate 2.3). This has not been disputed, but Cook (1940) claimed that Giseke did not intend to adopt the name *Corozo*. For many years Bailey's nomenclature, *Corozo oleifera*, was generally accepted (Bailey, 1940; Salisbury, 1950), but it is now agreed (Wessels Boer, 1965; Corner, 1966) that both the ease of hybridization with *E. guineensis* and the degree of divergence between the two species justify separation on a specific rather than a generic level. The specific name *oleifera* was transferred by Bailey from the earlier *Alfonsia oleifera* of Humboldt, Bonpland and Kunth (1816).

   Elaeis oleifera is found in the tropical countries of South and Central America and has been described or collected from Brazil, Colombia, Venezuela, Panama, Costa Rica, Nicaragua, Honduras, French Guiana and Surinam (de Blank, 1952; Ferrand, 1960; Meunier, 1975). In Colombia, where it is called Corozo Noli, it is found in depressions between rolling areas of pasture land (Plate 2.13) and in damp or even swampy situations near or on the banks of rivers. In these situations it can be found in pure and dense stand, but in pasture land or in some river-bank habitats it is also found dispersed or in small groups. On the Magdalena river there are some farms where kernel oil is extracted or the fruit is fed to pigs and where palms are even selected, on bunch production and quality, for further planting. In Nicaragua, where the palm has both Spanish and English names – *coquito* and hone palm – it covers large areas on the banks of the San Juan and Escondido rivers. It tolerates both swampy and shady conditions, though under shade very few bunches are produced, while in low-lying terrain the palms on slightly raised areas are the healthier. In Costa Rica, where it

Pl. 2.13   A natural grove of *Elaeis oleifera* in the Sinú valley area of Cordoba Province, Colombia.

Pl. 2.14   Two *Elaeis oleifera* palms in the Sinú valley, Colombia. Note the coiling, recumbent trunks.

is called *palmiche*, some specimens are to be found on sloping or even steep pastures, but except in tracts of unoccupied riverside semi-swamp or in the few situations where use is being made of the bunches for oil extraction or pig feeding, the population is declining because cattle farmers consider the palm to be an encumbrance in their fields, giving no shade to their beasts. As land becomes more and more utilized for organized agriculture, so the population will decline, and the need now is to preserve as wide a gene pool as can be collected.

In Brazil, where it is called *caiaué* or *dende do Pará*, the palm is most common in the central region of Amazonas above Manaus, and here it is distinguished by its large fruit. In Surinam there are dense stands on poor, white, sandy soil, and the palms are of small size.

Apart from its domestic employment in many areas as a cooking oil, the mesocarp oil has been used locally for soap-making and as a lamp oil, while kernel oil has even been sold as a hair oil.

Morphology

The main feature of the palm, and one which distinguishes it from *E. guineensis*, is its procumbent trunk. An erect habit may be maintained for as long as 15 years, but thereafter a procumbent habit is generally assumed though the crown is in an erect position and the erect portion is usually 1.5–2.7 m high (Plate 2.14). Trunks lying on the soil for a distance of over 7.6 m have been measured. In certain areas there are types which appear to remain erect until the palm reaches a height of over 3 m. The leaf bases persist for only a short period.

The root development of the palm is similar to that of *E. guineensis*, but roots which may grow to 1 m in length are formed along the whole length of the procumbent trunk. It has been claimed that certain anatomical differences, namely greater lignification of the hypodermis and cortical parenchyma, less lacunae and the presence of tannins in the cells of the endoderm and phloem, account for resistance to certain diseases (see p. 621) (Arnaud and Rabéchault, 1972).

The leaf of *Corozo* also readily distinguishes it from *E. guineensis*. All the leaflets lie in one plane and have no basal swellings, and the spines on the petiole are short and thick. The number of leaves is sometimes larger, and may reach as many as forty-four and average thirty. In good specimens the leaflets are larger than those of *E. guineensis*, being up to 1.9 m long and 12 cm wide (Vallejo and Cassalett, 1974). There are usually more than 100 pairs).

The male inflorescence differs little from that of *E. guineensis*. The spikelets, of which there are between 100 and 200 varying in length from 5 to 15 cm, are pressed together until they burst through the spathe just before anthesis. The male flower is somewhat smaller with shorter anthers; the rudimentary gynoecium is more developed and has three marked stigmatic ridges.

The female inflorescence is distinguished by a spathe which persists after it has been ruptured by the developing bunch. The spikelets end in a short prong instead of a long spine. The number of flowers vary from 1,000 to 6,000 and they are sunk in the body of the spikelet; they are not subtended by a long bract as in the case of *E. guineensis*. As a result of these characteristics the bunch of *E. oleifera* is surrounded by the fibres of the spathe, and contains no long spines (Plate III, between pp. 230 and 231). The earliest inflorescences are usually male and are followed by hermaphrodite inflorescences having an increasing number of female spikelets. Andromorphic inflorescences are also not uncommon. The period of anthesis in the female inflorescence is much more erratic than in *E. guineensis* and may last for 3 or 4 weeks, or have two peaks, while other palms are completing anthesis in 3 or 4 days (Rao and Chang, 1982). Palms with a long anthesis duration show uneven ripening.

The bunches, being round and wide at their centre with a tendency to be pointed at the top, have a distinctly conical appearance. They usually weigh between 8 and 12 kg but occasionally reach 30 kg. The large number of small fruit, of which the normal ones alone may number more than 5,000, have been recorded in Colombia as weighing between 1.7 and 5.0 g. Parthenocarpic fruit, which are often present in even larger numbers and may constitute 90 per cent of all fruit, average 0.8 g. Higher fruit weights have been recorded though it is not certain that these were from pure *E. oleifera*. The mesocarp layer is thin and usually constitutes 29–42 per cent of normal fruit, though over 80 per cent in parthenocarpic fruit. Oil to mesocarp has been recorded as 23–38 per cent in Colombia, but in a more recent survey in four regions of that country the mean oil to fresh mesocarp was found to vary with locality from 16.7 to 22.6 per cent in normal fruit with lower oil percentages in parthenocarpic fruit (Vallejo and Cassalett, 1974). Shell thickness varies from 1 to 3 mm and the shell forms between 43 and 53 per cent of the fruit. There is between 13 and 22 per cent of kernel, and nuts with two kernels are fairly frequent and those with three kernels occasional.

Fruit-to-bunch ratios are often low; the mean percentage normal fruit to bunch in a survey in four regions of Colombia varied from 28.1 to 46.3 per cent with mean parthenocarpic fruit to bunch varying from 9.5 to 23 per cent. Within-region variations were very great and some bunches were found with normal fruit to bunch as low as 8.9 per cent or as high as 63.6 per cent. Normal plus parthenocarpic fruit usually constitute less than 60 per cent of the bunch, and as parthenocarpic fruit form such a high percentage of all fruit, kernel production is considerably lower than indicated by normal fruit analysis and total fruit-to-bunch percentages. The proportion of normal fertile fruit can be substantially raised by controlled pollination of the inflorescences in bags (Tam, 1981).

There are marked variations in fruit colour. About 90 per cent of palms have orange fruit at maturity, these having developed from immature fruit which were at first yellowish-green, then ivory coloured at the base and orange above. A less common type of fruit is yellow at maturity and has developed from immature fruit at first bright green, then turning olive-green and pale yellow (Hurtado and Nuñez, 1970). The small palms in Surinam are reported to have green immature fruit turning orange to red (Meunier, 1975). There is no evidence that fruit forms comparable to *dura*, *tenera* and *pisifera* exist in the populations of *E. oleifera*. Compared with *E. guineensis* the oil has a higher oleic content and iodine value (see p. 681). The carotene content is higher than that of the Deli palm but may be no higher than that of many *E. guineensis* palms in Africa.

The above description indicates that the palm has value in hybridization with *E. guineensis* on account of its slow growth in height and through the characteristics of its mesocarp oil. Interest in *E. oleifera* has since increased by the discovery of its resistance to Lethal Bud Rot in Colombia, a discovery which led to the establishment of the first commercial plantation of the hybrid of the two species of *Elaeis* (Anon., 1974).

The *E. guineensis* × *E. oleifera* hybrid

The two species have in the last few decades been widely hybridized on an experimental scale and the significance of the cross in selection and breeding will be discussed in Chapter 5 (p. 288). Botanically the hybrid is characterized by leaves which are considerably larger than those of either parent but retain the leaflet arrangement of *E. oleifera*. The characteristics of the latter palm as regards height increment, falling leaf bases, persistent spathes, parthenocarpy and fruit shape and colour are also retained in the hybrid. As both the flower-subtending bracts on the spikelets and the end-prong are only slightly longer than in *E. oleifera* the hybrid's bunches closely resemble those of that species. Andromorphism is not uncommon in young palms.

Under Malaysian conditions the early leaf production of the hybrid was found to be intermediate between that of the parent species, e.g. at 3–4 years *E. oleifera*, 22.5 leaves, *E. guineensis*, 36 leaves, hybrid, 29 leaves per annum; with leaf length *E. guineensis* was intermediate; e.g. at 4 years *E. oleifera*, 3.2 m, hybrid, 4.2 m and *E. guineensis*, 3.8 m (Tan Yap Pau, 1980). The number of leaflets per leaf is intermediate in the hybrid.

The internal fruit characters of the hybrid naturally depend on the fruit form (*dura, tenera* or *pisifera*) of the *E. guineensis* parent, but they are also influenced by the tendency of *E. oleifera* to produce both normal fruit and large quantities of parthenocarpic fruit of two types: those with a small nut with a liquid-filled cavity and those which are smaller and have only a lignified central core including a rudimentary cavity. With normal fruit, mesocarp to fruit in the *dura* cross varies from under 40 to over 50 per cent, but fruit from *tenera* and *pisifera* crosses has given mesocarp percentages of 58–74 per cent (Vallejo and Cassalett, 1974; Obasola, 1973). With parthenocarpic fruit mesocarp percentage depends on the degree of parthenocarpy. In a Malaysian trial large parthenocarpic fruit had a mean of 75 per cent mesocarp with 25 per cent shell, while the small type had 89 per cent mesocarp, the lignified core only accounting for 11 per cent of the fruit. Oil to mesocarp is intermediate between that of the parent species, and the distribution of fatty acids also appears to be intermediate (Hardon, 1969) (see Ch. 14, pp. 681–682).

Hybrid bunches either ripen normally or may show great irregularities; the latter may be caused by insufficient set of normal fruits, in which case higher fruit on the spikelets often fall out at the start of ripening, while in other cases a poor set may lead to full or partial abortion 1–3 months after anthesis. Normal fruit turn from black to yellow in the fifth month, then become orange-yellow and finally orange-red. The bunch may be considered 'ripe' when, per 10 kg of bunch, 1–2 fruit are loose and 10–20 are red and detachable. After this stage, all the fruit very rapidly become detachable (Anon., 1981).

Pollen grains of *E. oleifera* are mainly elliptical, not triangular like those of *E. guineensis*. Hybrid pollen is intermediate in shape and varies greatly in size. Its germination percentage is less than half that of the parent species (Arnaud, 1980).

References

Alibert, R. (1945) Pourquoi et comment on fait la fécondation artificelle sur le palmier à huile. *Farm and Forest*, **6**, 27.

Anon. (1955) Notes on the botany of the oil palm. 1. The seed and its germination. *J. W. Afr. Inst. Oil Palm Res.*, **1**(3), 73.

Anon. (1956) Notes on the botany of the oil palm. 2. The seedling. *J. W. Afr. Inst. Oil Palm Res.*, **2**, 92.

Anon. (1961) Notes on the botany of the oil palm. 3. The stem and the stem apex. *J. W. Afr. Inst. Oil Palm Res.*, **3**, 277.

Anon. (1962) Notes on the botany of the oil palm. 4. The leaf. *J. W. Afr. Inst. Oil Palm Res.*, **3**, 350.

Anon. (1974) Replanting diseased oil palm areas with *Elaeis oleifera* × *E. guineensis* hybrids at 'La Arenosa' estate in Colombia. *Oil Palm News*, No. 18, 1.

Anon. (1981) La maturité des régimes d'hybrides *Elaeis melanococca* × *E. guineensis.* *Oléagineux,* **36,** 127.

Arasu, N. T. (1970) Foliar spiral and yield in oil palms (*Elaeis guineensis* Jacq.). *Malay. agric. J.,* **47,** 409.

Arnaud, F. (1980) Fertilité pollinique de l'hybride *Elaeis melanococca* × *E. oleifera* et des espèces parentales. *Oléagineux,* **35,** 121.

Arnaud, F. and **Rabéchault, H.** (1972) Premiers observations sur les caractères cystohistochimiques de la résistance du palmier à huile au 'dépérissement brutal'. *Oléagineux,* **27,** 525.

Bachy, A. (1964) Tropisme racinaire du palmier à huile. *Oléagineux,* **19,** 684.

Bailey, L. H. (1933) Certain palms of Panama. *Gentes Herb.,* **3,** Fasc. II. 52.

Bailey, L. H. (1940) The generic name *Corozo.* *Gentes Herb.,* **4,** Fasc. X, 373.

Beccari, O. (1914a) *Palme del Madagascar.* Florence, p. 55.

Beccari, O. (1914b) Contribute alle conoscenza della palma a olio. *Agricoltura Colon.,* **8,** 255.

Beirnaert, A. (1935a) Introduction à la biologie florale du palmier à huile, *Elaeis guineensis* (Jacq.). *Publs INEAC,* Série Sci., No. 5.

Beirnaert, A. (1935b) Introduction à la biologie florale du palmier Elaeis. Organisation de l'inflorescence chez le palmier à huile. *Revue int. Bot. appl. Agric. Trop.,* **15,** 1091.

Beirnaert, A. and **Vanderweyen, R.** (1941a) Contribution à l'étude génétique et biométrique des variétés d'*Elaeis guineensis* Jacq. *Publs INEAC,* Série Sci., No. 27. Brussels.

Beirnaert, A. and **Vanderweyen, R.** (1941b) Influence de l'origine variétale sur les rendements. *Publs INEAC,* Com. No. 3 sur le Palmier à huile.

Blank, S. de (1952) A reconnaissance of the American oil palm. *Trop. Agric. Trin.,* **29,** 90–101.

Broekmans, A. F. M. (1957) Growth, flowering and yield of the oil palm in Nigeria. *J. W. Afr. Inst. Oil Palm Res.,* **2,** 187.

Chan, K. W. (1977) A rapid method for studying the root distribution of oil palm and its application. In *International developments in oil palm,* ISP, Kuala Lumpur.

Chevalier, A. (1910) *Les végéteux utiles de l'Afrique tropicale française.* Vol. 7, *Documents sur le palmier à huile.* Paris, 127 pp.

Cook, O. F. (1940) Oil palms in Florida, Haiti and Panama. *Natn. hort. Mag.,* 1940, 10.

Corner, E. J. H. (1966) *The natural history of palms.* Widenfeld and Nicolson, London.

Corrado, F. (1984) La conformacion de los racimos de la Palma Africana en las plantaciones de Colombia. *Palmas,* **5**(3), 66.

Davis, T. A. (1980) Double reversal of spadices in an African oil palm. *Planter, Kuala Lumpur,* **56,** 212.

Ferrand, M. (1960) Le Noli. *Oléagineux,* **15,** 823.

Gatin, C. L. (1906) Recherches anatomiques et chimiques sur la germination des palmiers. *Ann. Sci. Naturelles,* ser. 9, **3,** 191.

Hardon, J. J. (1969) Interspecific hybrids in the genus *Elaeis.* II. Vegetative growth and yield of F_1 hybrids *E. guineensis* × *E. oleifera.* *Euphytica,* **18,** 380.

Hardon, J. J. and **Turner, P. D.** (1967) Observations on natural pollination in commercial plantings of oil palm (*Elaeis guineensis*). *Expt. Agric.,* **3,** 105.

Henry, P. (1948a) Un *Elaeis* remarquable: le palmier à huile vivipare. *Revue int. Bot. appl. Agric. Trop.,* **28,** 422.

Henry, P. (1948b) Les facteurs de la pollinisation chez le palmier à huile. *Oléagineux,* **3,** 587.

Henry, P. (1951) La germination des graines d'*Elaeis.* *Revue int. Bot. appl. Agric. trop.,* **31,** 349.

Henry, P. (1955a) Note préliminaire sur l'organisation foliaire chez le palmier à huile. *Revue gén. Bot.,* **62,** 127.

Henry, P. (1955b) Sur le développement des feuilles chez le palmier à huile. *Revue gén. Bot.*, **62**, 231.

Henry, P. (1955c) Morphologie de la feuille d'*Elaeis* au cours de sa croissance. *Revue gén. Bot.*, **62**, 319.

Henry, P. and **Scheidecker, D.** (1953) Nouvelle contribution à l'étude des *Elaeis* vivipares. *Oléagineux*, **8**, 681.

Hurtado, J. R. and **Nuñez, G. R.** (1970) Estudio de la palmera Noli (*Elaeis melanococca*, Gaert.) y preliminares de su fitomejoramiento en Colombia. *Acta Agronica*, **20**, 9.

Hussey, G. (1958) An analysis of the factors controlling the germination of the seed of the oil palm, *Elaeis guineensis* (Jacq.). *Ann. Bot.*, NS, **22**, 259.

Jacquin, N. J. (1763, 1780) *Selectarum stirpium Americanarum historia.*

Jagoe, R. B. (1934a) Notes on the oil palm in Malaya with special reference to floral morphology. *Malay. agric. J.*, **22**, 541.

Jagoe, R. B. (1934b) Observations and experiments in connection with pollination of oil palms. *Malay. agric. J.*, **22**, 598.

Janssens, P. (1927) Le palmier à huile au Congo Portugais et dans l'enclave de Cabinda. Descriptions des principales Variétés de palmier (*Elaeis guineensis*). *Bull. agric. Congo belge*, **18**, 29–58, 59–92.

Jumelle, H. (1918) Les variétés de palmiers à huile. *Matières grasses*, **11**, 4923, 4883 and 5005.

Jumelle, H. and **Perrier de la Bathie, H.** (1911) Le Palmier à huile à Madagascar. *Matières grasses*, **4**, 6065.

Lambourne, J. (1935) Note on the root habit of oil palms. *Malay. agric. J.*, **23**, 582.

Lawton, D. M. (1981) Pollination and fruit set in the oil palm (*Elaeis guineensis* Jacq.). In *Oil Palm in Africa in the eighties* (Malaysian Oil Palm Conference), 241.

Lespesme, P. (1946) Les charancons floricoles des palmiers. *Agro. Trop.*, **1**(7–8), 400.

Meunier, J. (1975) Le 'palmier à huile' américain *Elaeis melanococca*. *Oléagineux*, **30**, 51.

Moreau, C. and **Moreau, M.** (1958) Lignification et réactions aux traumatismes de la racine du palmier à huile en pépinières. *Oléagineux*, **13**, 735.

Obasola, C. O. (1973) Breeding for short-stemmed oil palm in Nigeria. *J. Nig. Inst. Oil Palm Res.*, **5**(18), 43.

Opsomer, J. E. (1956) Les premières descriptions du palmier à huile (*Elaeis* guineensis. Jacq.). *Bull. des Séanc. Acad. r. Sci. Colon.* (*outre Mer*) **2**, 253.

Preuss, L. (1902) Die wirtschaftliche Bedeutung der Ölpalme. *Tropenpflanzer*, **6**, 450–76.

Purseglove, J. W. (1972) *Tropical crops. Monocotyledons 2.* Longman, London.

Purvis, C. (1956) The root system of the oil palm: its distribution, morphology and anatomy. *J. W. Afr. Inst. Oil Palm Res.*, **1**(4), 61.

Purvis, C. (1957) The colour of oil palm fruits. *J. W. Afr. Inst. Oil Palm Res.*, **2**, 142.

Rao, V. and **Chang, K. C.** (1982) Anthesis and fruit set in *Elaeis oleifera* (HBK) Cortés. *PORIM Bull.*, **4**, 27.

Rees, A. R. (1960) Early development of the oil palm seedling. *Principes*, **4**, 148.

Rees, A. R. (1963) A note on the spines of the oil palm. *Principes*, **7**, 30.

Rees, A. R. (1964) The apical organisation and phyllotaxis of the oil palm. *Ann. Bot.*, NS, **28**, 57.

Ruer, P. (1967a) Répartition en surface du système radiculaire du palmier à huile. *Oléagineux*, **22**, 535.

Ruer, P. (1967b) Morphologie et anatomie du système radiculaire du palmier à huile. *Oléagineux*, **22**, 595.

Ruer, P. (1969) Système racinaire du palmier à huile et alimentation hydrique. *Oléagineux*, **24**, 327.

Salisbury, Sir E. (August 1950) Private communication. WAIFOR 424/89.

Schmöle, J. F. (1930) The selection of oil palms (*Elaeis guineensis* Jacq.). *Proc 4th Pacif. Sci. Congress*, Java, 1929, Proc. 4, 185.

Smith, E. H. G. (1935) A note on recent research on empire products. (Extract from Botanical section Rep., S. Provinces, Nigeria, Jan.–June 1935.) *Bull. imp. Inst.*, Lond., **33**(3), 371.

Surre, C. and **Ziller, R.** (1963) *Le Palmier à huile*. G-P. Maisonneuve and Larose, Paris.

Syed, R. A. (1978) Studies on pollination of oil palm in West Africa and Malaysia. *Report, Commonwealth Institute of Biological Control*, 48 pp.

Syed, R. A. (1979) Studies on oil palm pollination by insects. *Bull. ent. Rev.*, **69**, 213.

Syed, R. A. (1981a) Pollinating thrips of oil palm in West Malaysia. *Planter, Kuala Lumpur*, **57**, 62.

Syed, R. A. (1981b) Insect pollination of oil palm: feasibility of introducing *Elaeidobius* spp. into Malaysia. In *The oil palm in agriculture in the eighties* (Malaysian oil palm conf.), 27 pp.

Syed, R. A. (1984) Los insectos polinizadores de la Palma Africana. *Palmas*, **5**(3), 19.

Syed, R. A., Law, I. H. and **Corley, R. H. V.** (1982) Insect pollination of the oil palm: introduction, establishment and pollinating efficiency of *Elaeidobius kamerunicus* in Malaysia. *Planter, Kuala Lumpur*, **58**, 547.

Taillez, B. (1971) La système racinaire du palmier à huile sur la plantation de San Alberto (Colombie). *Oléagineux*, **26**, 435.

Tam Tai Kin (1981) Investigations into fruit set capacities of the *Elaeis oleifera* under controlled conditions and germination requirements of the interspecific *E. oleifera* × *E. guineensis* (*pisifera*) hybrid seeds. *Planter, Kuala Lumpur*, **57**, 444.

Tan, K. S. (1977) Root development of oil palms on inland soils of West Malaysia. *Int. Conf. on role of soil physical properties*, Ibadan, Nigeria.

Tan, K. S. and **Rao, A. N.** (1979) Certain aspects of developmental morphology and anatomy of oil palm. *Histochemistry, developmental and structural anatomy of Angiosperms: a symposium*, pp. 267–85.

Tan Yap Pau (1980) Performance of *E. oleifera* and its hybrids. Typescript supplied to the author.

Thomas, R. L., Chan, K. W. and **Easu, P. T.** (1969) Phyllotaxis in the oil palm: arrangement of fronds on the trunk of mature palms. *Ann. Bot.*, NS, **33**, 1001.

Thomas, R. L., Chan, K. W. and **Ng, S. C.** (1970) Phyllotaxis in the oil palm: arrangement of the male/female spikelets on the inflorescence stalk. *Ann. Bot.*, NS, **34**, 93.

Tomlinson, P. B. (1960) Essays on the morphology of palms. 1. Germination and the seedling. *Principes*, **4**, 56.

Tomlinson, P. B. (1961) *Anatomy of the monocotyledons*. II. *Palmae*. Oxford.

Vallade, J. and **Lucien, P.** (1966) Aspect morphologique et cytologique de l'embryon quiescent d'*Elaeis guineensis* Jacq. *C. R. Acad. Sci.*, **262**, 856.

Vallejo Rosero, J. and **Cassalett, C.** (1974) Perspectivas del cultivo de los hibridos interespecificos de Noli (*Elaeis oleifera* (HBK Cortés) × Palma Africana de Aceite (*Elaeis guineensis*, Jacq.)) en Colombia. Instituto Colombiano Agropecuario, Colombia. Mimeograph.

Vanderweyen, R. and **Roles, O.** (1949) Les variétés d'*Elaeis guineensis* Jacquin du type 'Albescens'. *Publs INEAC*, Série Sci., **42**, 6.

Van Heel, W. A., Breure, C. J. and **Menendez, T.** (1986) The early devleopment of inflorescences and flowers of the oil palm (*Elaeis guineensis*, Jacq.) as seen through the scanning electron microscope. *Blumea*, **32**, 67.

Vine, H. (1945) Report of chemistry section, Agricultural Dept, Nigeria. Typescript.

Wessels Boer, J. G. (1965) *The indigenous palms of Surinam*. E. J. Brill, Leiden.

Wood, B. J. (1983) Note on insect pollination of oil palm in South and Central America. *Planter, Kuala Lumpur*, **59**, 167.

Yampolsky, C. (1922) A contribution to the study of the oil palm (*Elaeis guineensis*, Jacq.). *Bull. Jard. bot., Buitenz.*, **3**, 107.

Yampolsky, C. (1924) The pneumathodes on the roots of the oil palm (*Elaeis guineensis*, Jacq.). *Am. J. Bot.*, **11**, 502.

Zeven, A. C. (1964) The *Idolatrica* palm. *Baileya*, **12**, 11.

Zeven, A. C. (1973) The 'mantled' oil palm (*Elaeis guineensis*, Jacq.). *J. W. Afr. Inst. Oil Palm Res.*, **5**, 31.

Zimmerman, M. H. and **Tomlinson, P. B.** (1965) Anatomy of the palm *Rhapis excelsa*. I. Mature vegetative axis. *J. Arnold Arbor.*, **46**, 160.

Chapter 3

The climates and soils of the oil palm regions

The climatic features of the main areas of highest bunch production may be summarized as follows:
1. A rainfall of 2,000 mm (80 in.) or more distributed evenly through the year, i.e. no very marked dry seasons.
2. A mean maximum temperature of about 29–33 °C (85–90 °F) and a mean minimum temperature of about 22–24 °C (72–75 °F).
3. Sunshine amounting to about 5 hours per day in all months of the year and rising to 7 hours per day in some months; or solar radiation of around 350 cal per cm^2 per day.

It is in such a closely defined climate that the oil palm is cultivated in Indonesia and Malaysia. But in both these territories soil differences, and of course fertilizer practice, are responsible for quite substantial differences of yield. Moreover, in certain parts of these oil palm regions night temperatures may be a little lower than the above means and in some months of heavy rainfall sunshine hours may occasionally be below 4 per day; years of comparatively low rainfall are also not unknown.

While it is true that high yields will always be obtained on good soils under the climatic conditions stated above, the oil palm has been profitably cultivated in regions where rainfall is poorly distributed or very high, where temperatures are low in certain months or where sunshine hours are well below those of the Far East.

There are both botanical and economic reasons for the cultivation of the oil palm in areas where both climatic and soil conditions are much less favourable than those in the Far East. In the first place the plant is naturally adapted to areas of summer rainfall and winter drought. Though bunch production is reduced by drought conditions, 3 months without rain does not markedly reduce the health of the plant; growth of the bud continues, but spear leaves tend to remain unopened until the onset of wet weather and midday stomatal closure prevents excessive moisture loss.

Secondly, the oil palm is such a high producer of oil that even under the poorest plantation conditions, which may be exemplified by those parts of West Africa having 3 months without rain, few hours of sunshine during the wet season and poor, rapidly drying soils, production compares favourably with that of other oil-bearing crops. Thus plantation yields under very favourable Malaysian conditions have been 25–30 tonnes of bunches per hectare per annum while Nigerian estates have been producing only 8–11 tonnes. But the latter yield supplies, from *tenera* material, over 2 tonnes of palm oil with ½ tonne of kernels, a production exceeding in terms of oil that of the best coconut estates. Such production is rendered the more economic by the fact that labour in the low-production regions is often less expensive. Any definition of the limits of favourable climatic conditions for the oil palm is therefore extremely difficult.

Climate

In view of the difficulties of defining the limits of a 'suitable climate', as explained above, it is proposed to consider each factor in turn and to discuss its magnitude in relation to other factors. Meteorological data relevant to this discussion are given in Tables 3.1–3.3. Centres have been chosen to show contrasts within and between regions.

Rainfall and the water balance

The effect of prolonged drought on reducing the yield of the oil palm is well known and the physiological effects will be described in the next chapter. In areas where 2–4 months' drought are the rule, e.g. Benin (Dahomey), southern Zaire and parts of Nigeria, there is a tendency for big yield fluctuations from year to year, with a year of very low yield occurring at intervals of 4–6 years.

Benin presents an extreme example of a country with a large production of oil and kernels, but with a low and poorly distributed rainfall. Four months (November to February) are almost devoid of rain and the average annual rainfall is 1,232 mm at Pobé Experiment Station. However, on certain soils of a high water-holding capacity and which overlie underground water, production is at least double that of palms on soils not so favoured and, with good planting material, a yield of over 12 tonnes of bunches per hectare has been achieved (Ochs, 1963).

In Nigeria, water supply has been shown to be deficient through uneven rainfall distribution even in areas where the total annual rainfall exceeds 2,000 mm. Yields are correlated with dry-season rainfall and with measures of 'effective sunshine' which take into account the distribution of such rain as falls in the dry season (see p. 179). In areas adjoining Benin (Dahomey), where the rainfall is similar to that at Pobé, yields are very low, agreeing with those on the poorer Benin sites.

Southern Zaire (6–7° S) has a dry season of 2–4 months, with total precipitations of 1,400–1,900 mm. Yields have been very variable and disease has been the dominant feature of these plantings.

Dry seasons may be severe in several parts of tropical America where the oil palm is grown. In the north of Colombia at Aracataca (11° N) there are 5 months, December to April, which provide a mean of only 100 mm of rain; the remaining mean of 1,524 mm falls in the period May to November. In some years the total precipitation is below 1,000 mm. In many parts of the area, however, water-tables are high and irrigation water can be provided. Similarly, in the Palma-Sola area in Venezuela (10½° N) rainfall averages 1,450 mm, and 4 months, January to April, usually have less than 150 mm of rain. Although the soil is water-retentive and subsoil water is to be found at 180–360 cm below the surface in all seasons, the dry weather is sufficiently severe to inhibit the opening of spear leaves as in West Africa. Irrigation is being tried.

On the Pacific coast of Costa Rica (9½° N) the rainfall pattern is similar to that of northern Colombia, but the total annual rainfall is double, i.e. 3,300 mm. There are usually 3 months, January to March, with less than 100 mm, but occasionally the dry season can extend over the 5-month period December to April. Here clay soils overlie a water-table 120–180 cm below the surface, but the yield pattern is similar to that of West Africa, i.e. low yields coincide with the dry season and high yields with the beginning and middle of the wet season.

In recent years the concept of the water balance has played a much greater part both in assessing the suitability of areas for planting and in seeking the causes of yield fluctuations in already established regions (Hartley, 1973). Direct relationships have been found between the magnitude of water-deficit estimates and bunch yields (Surre, 1968; IRHO, 1969).

In calculating annual water deficits the year is divided into ten-daily or monthly periods and the following formula is applied:

$$B = Res + R - Etp$$

where B is the balance at the end of the period, Res is the soil water reserve at the beginning of the period and R and Etp are the rainfall and potential evapotranspiration during the period. The balance, B, is carried forward as the soil reserve at the beginning of the following period but this reserve has of course a maximum equal to the available water or field capacity.

The main difficulty with this method is determining the best estimates of available water stored in the soil and potential evapotranspiration. Available soil water will vary both with the physical

Table 3.1 Rainfall at centres of oil palm cultivation (mm)

Centre		Lat. and long.	No. of years	Jan.	Feb.	March	April	May	June	July	Aug.	Sept.	Oct.	Nov.	Dec.	Annual	
Asia																	
Malaysia	Telok Anson	4°2'N 101°1'E	34	185	191	217	242	171	96	124	122	79	284	289	246	2,348	Coastal plain
(West)	Paya Lang	2°35'N 102°40'E	20	106	138	146	136	145	95	117	90	142	155	217	191	1,678	Inland S
	Layang-Layang	1°15'N 103°30'E	18	135	150	158	230	167	149	164	169	170	210	230	248	2,180	Inland S
	Jerangau	4°59'N 103°9'E	25	297	168	132	144	196	206	214	256	287	343	569	839	3,651	E coast state
Malaysia	Mostyn	5°N 118°5'E	30	195	119	137	198	222	209	165	181	209	225	187	192	2,239	E coast
(East–Sabah)	Beluran	4°3'N 117°30'E	27	520	354	265	125	190	249	206	234	255	241	230	408	3,278	NE coast
Indonesia	Medan	3°35'N 98°41'E	58	114	91	104	132	175	132	135	178	211	259	246	229	2,487	E coast 20 km
(Sumatra)	Tindjowan	3°6'N 99°29'E	26	156	109	141	154	140	111	109	170	202	244	205	171	1,912	E coast 22 km
	Marihat Baris	2°58'N 99°6'E	21	311	223	287	305	296	214	201	277	358	452	411	292	3,627	60 km inland
Africa																	
Sierra Leone	Njala	8°6'N 12°6'E	39	12	22	79	127	251	364	418	517	437	338	180	38	2,822	Inland
Ivory Coast	La Mé	5°3'N 3°5'W	40	25	64	125	139	246	468	197	41	97	169	149	70	1,790	
Ghana	Aiyinasi	5°N 2°20'W	10	37	71	140	191	378	751	293	64	116	246	129	98	2,511	S West
Benin (Dahomey)	Pobé	6°6'N 2°4'E	40	14	38	96	136	189	190	114	50	132	159	35	9	1,162	
Nigeria	NIFOR, Benin	6°30'N 5°40'E	33	14	29	98	161	192	254	350	221	306	223	58	10	1,916	Mid-West
	Umudike	5°29'N 7°33'E	36	22	51	113	204	267	273	312	253	310	262	84	18	2,168	S East
	Abak	5°5'N 7°40'E	18	29	47	131	196	237	310	357	317	384	300	134	30	2,472	S East
	Idenau	4°30'N 9°10'E	27	27	72	165	199	194	345	627	820	609	302	127	21	3,508	Inland
Cameroon	Lobé	4°5'N 9°E	12	90	150	290	330	530	1,150	1,340	1,400	1,560	1,030	390	170	8,430	Coast
Zaire	Yangambi	0°49'N 24°29'E	30	85	99	148	150	176	127	146	169	181	235	183	123	1,822	N Congo basin
	Kiyaka	5°S 19°E	10	175	114	220	225	97	9	22	45	123	220	243	185	1,668	Kwilu

America																
Brazil	Paricatuba, Belém	1° 16'S 48° 8'W	17	359	405	494	410	398	202	184	179	203	139	113	219	3,306	Lower Amazon
	Taparoa, Bahia	13° 32'S 39° 6'W	5	118	135	305	215	183	193	131	152	86	116	123	142	1,899	Coast
Colombia	Aracataca	10° 35'N 74° 9'W	13	2	8	9	52	216	178	125	182	275	344	239	31	1,661	North
	San Alberto	7° 40'N 73° 30'W	17	47	55	123	295	374	228	187	213	322	405	338	117	2,704	Central
	Rio Mira, Tumaco	1° 33'N 78° 41'W	15	330	364	328	388	488	432	184	169	204	185	115	167	3,354	Pacific coast
	Casanare	6° 20'N 70° 30'W	7	3	62	107	340	359	313	283	272	218	273	117	27	2,374	Oriente
Ecuador	La Concordia	0° 05'N 79° 20'W	15	487	474	630	579	316	214	85	56	82	58	41	166	3,188	Pacific coast
Costa Rica	Quepos	9° 26'N 84° 9'W	28	57	25	39	110	340	394	420	422	436	606	312	146	3,307	Pacific coast
Honduras	San Alejo, Tela	15° 40'N 87° 40'W	19	214	234	116	67	59	166	183	253	235	322	412	405	2,666	North coast

Table 3.2 *Sunshine at centres of oil palm cultivation (hours per day)*

Centre	Lat. and long.	No. of years	Jan.	Feb.	March	April	May	June	July	Aug.	Sept.	Oct.	Nov.	Dec.	Annual Hrs per day	Annual Total
Asia																
Malaysia																
Kuala Lumpur	3° 7'N 101° 42'E	17	6.2	7.4	6.5	6.3	6.3	6.6	6.5	6.3	5.6	5.3	4.9	5.4	6.1	2,230
Chemara, Johore	1° 15'N 103° 30'E	11	5.5	5.9	6.3	5.8	5.9	5.7	5.7	5.5	4.5	4.6	4.0	5.3	5.3	1,940
Sumatra																
Medan	3° 35'N 98° 41'E	21	5.4	7.1	7.0	7.2	7.7	8.1	8.1	7.5	7.0	6.2	5.9	5.4	6.9	2,508
Sabah																
Mostyn	5°N 118° 5'E	24	5.9	6.5	7.0	6.9	6.1	5.8	6.1	6.2	5.8	5.6	6.3	5.7	6.2	2,245
Africa																
Sierra Leone																
Njala	8° 6'N 12° 6'W	32	7.2	7.3	6.7	6.1	6.0	5.1	2.9	2.0	3.5	5.6	6.1	6.4	5.4	1,971
Ivory Coast																
La Mé	5° 3'N 3° 5'W	28	5.6	6.4	6.3	6.5	5.5	3.5	3.3	2.6	3.0	4.9	6.0	5.8	5.0	1,819
Benin (Dahomey)																
Pobé	6° 6'N 2° 4'E	24	5.5	5.8	5.3	5.5	5.6	4.3	3.0	2.4	2.9	4.5	6.0	5.9	4.7	1,721

Nigeria																
NIFOR, Benin	6° 30'N 5° 40'E	15	5.6	6.0	4.9	5.3	5.4	4.2	2.6	2.4	2.6	4.2	6.0	6.4	4.6	1,692
W. Cameroon																
Idenau	4° 5'N 9° 10'E	9	5.2	6.7	4.7	5.0	4.7	2.6	1.6	1.0	1.3	2.4	3.5	4.5	3.6	1,306
Zaire																
Yangambi	0° 49'N 24° 29'E	10	6.6	6.8	6.0	6.1	6.0	5.5	5.0	4.4	5.2	5.1	5.5	5.7	5.6	2,054
Kiyaka	5° S 19° E	6	5.1	5.2	4.7	5.5	6.8	8.9	8.3	7.8	6.4	6.1	5.5	4.8	6.3	2,287
America																
Brazil Pará																
Paricatuba, Belém,	1° 16'S 48° 8'W	15	4.3	3.6	3.6	4.1	5.5	7.2	7.7	8.2	7.5	7.8	7.2	6.2	6.1	2,220
Iguape, Bahia	12° 30'S 39° W	4	7.1	8.2	6.9	6.3	4.6	4.5	4.6	6.4	7.1	7.6	7.7	5.6	6.4	2,323
Colombia																
Aracataca	10° 35'N 74° 9'W	13	8.6	8.6	8.8	7.7	7.2	6.8	7.8	6.9	7.1	7.2	6.9	8.4	7.7	2,792
San Alberto	7° 40'N 73° 30'W	4	7.3	5.4	3.8	4.6	5.3	5.3	6.4	6.0	6.0	6.0	5.1	7.0	5.7	2,070
Rio Mira, Tumaco	1° 30'N 78° 40'W	5	3.5	4.0	3.9	4.2	3.7	3.5	4.6	4.5	2.4	2.8	3.0	3.2	3.5	1,285
Ecuador																
La Concordia	0° 05'N 79° 20'W	10	2.4	2.8	3.3	3.7	2.7	1.8	2.2	1.7	1.4	1.4	1.3	1.9	2.2	808
Honduras																
San Alejo, Tela	15° 40'N 87° 40'W	10	6.1	6.5	8.0	8.2	7.9	7.0	7.3	7.8	6.7	6.5	5.7	5.5	6.9	2,533

Table 3.3 *Temperature at centres of oil palm cultivation (mean, mean maximum and mean minimum* °C)

Centre	Lat. and Long.	No. of years		Jan.	Feb.	March	April	May	June	July	Aug.	Sept.	Oct.	Nov.	Dec.	Av.
Asia																
Malaysia																
Telok Anson	4°2'N 101°1'E	20	Mean	27.6	28.0	28.4	28.7	28.7	28.5	28.1	28.0	28.0	27.9	27.7	27.4	28.1
			M. max.	32.6	33.1	33.5	33.7	33.6	33.4	33.1	32.9	32.7	32.4	32.1	32.0	32.9
			M. min.	22.6	22.9	23.2	23.6	23.8	23.5	23.1	23.1	23.2	23.3	23.2	22.8	23.3
Layang-Layang, Johore	1°15'N 103°30'E	18	Mean	25.6	26.0	26.6	27.1	27.3	27.0	26.6	26.7	26.7	26.9	26.6	25.9	26.6
			M. max.	30.6	31.4	32.5	32.7	32.7	32.4	31.7	31.8	31.9	32.1	31.5	30.3	31.8
			M. min.	20.5	20.6	20.6	21.4	21.8	21.6	21.6	21.5	21.4	21.7	21.6	21.4	21.3
Sumatra																
Medan	3°35'N 98°41'E	10	Mean	25.4	26.0	26.4	26.5	26.8	26.5	26.4	26.1	25.9	25.7	25.4	25.3	26.0
			M. max.	29.9	31.3	31.5	31.6	31.7	31.4	31.8	31.2	30.9	30.1	29.7	29.6	30.9
			M. min.	22.2	22.1	22.5	22.8	23.2	22.7	22.4	22.3	22.4	22.6	22.5	22.3	22.5
Africa																
Sierra Leone																
Njala	8°6'N 12°6'W	32	Mean	26.1	27.4	27.9	27.2	27.3	26.1	25.2	24.6	25.6	26.2	26.3	26.2	26.4
			M. max.	32.3	33.0	33.3	32.8	32.6	30.7	28.9	28.2	29.5	31.0	31.3	31.4	31.5
			M. min.	19.8	20.7	21.3	21.8	21.8	21.5	21.4	21.5	21.6	21.3	21.3	20.5	21.2
Ivory Coast 26.0																
La Mé	5°3'N 3°5'W	19	Mean	26.1	27.3	27.5	27.4	26.8	25.6	24.6	24.3	25.0	25.9	26.1	25.8	26.0
			M. max.	31.2	32.2	32.2	32.1	30.9	28.8	27.8	27.2	28.3	29.7	30.5	30.3	30.0
			M. min.	20.9	22.4	22.7	22.7	22.6	22.4	21.4	21.4	21.8	22.0	21.6	21.2	21.9
Nigeria																
NIFOR, Benin	6°30'N 5°40'E	12	Mean	26.3	27.5	27.4	26.9	26.5	25.6	24.6	24.3	25.0	25.6	26.2	25.8	26.0
			M. max.	30.9	32.7	32.4	31.5	30.9	29.4	27.6	27.4	28.3	29.6	30.7	31.2	30.2
			M. min.	21.6	22.3	22.4	22.2	22.0	21.7	21.5	21.3	21.8	21.6	21.6	21.4	21.8
Cameroon																
Lobé	4°30'N 9°10'E	15	Mean	26.7	27.4	27.4	27.1	26.8	26.0	24.9	24.5	25.2	25.8	26.6	26.5	26.2
			M. max.	31.4	32.1	31.8	31.4	31.0	29.6	27.8	26.9	28.2	29.4	30.7	30.9	30.1
			M. min.	22.0	22.7	22.9	22.8	22.6	22.4	22.0	22.1	22.1	22.1	22.5	22.0	22.3
Zaire																
Yangambi	0°49'N 24°29'E	10	Mean	24.1	24.1	24.3	24.4	24.1	23.6	23.1	23.0	23.2	23.3	23.5	23.4	23.7
			M. max.	30.2	30.8	30.6	30.3	30.1	29.5	28.5	28.4	29.2	29.1	29.3	29.0	29.6
			M. min.	19.6	19.4	19.9	20.3	20.0	19.8	19.3	19.5	19.4	19.5	19.7	19.5	19.7

America

Brazil

Location	Coordinates			J	F	M	A	M	J	J	A	S	O	N	D	Year
Iguape, Bahia	12° 31'S 39° W	8	Mean	25.5	26.8	26.6	26.4	25.2	23.6	22.5	21.9	22.1	23.7	24.3	25.5	24.5
			M. max.	30.6	32.3	31.9	31.1	29.2	27.1	26.2	25.8	26.4	28.6	29.1	31.0	29.1
			M. min.	20.3	21.4	21.2	21.6	21.2	20.1	18.7	17.9	17.7	18.8	19.5	20.4	19.9
Paricatuba, Belém, Pará	1° 16'S 48° 8'W	17	Mean	26.4	26.3	26.3	26.6	26.7	26.3	26.0	26.1	26.0	26.1	26.1	26.7	26.3
			M. max.	31.3	31.0	31.1	31.4	31.5	31.4	31.1	31.4	31.3	31.5	31.4	31.6	31.3
			M. min.	21.4	21.5	21.8	21.9	21.8	21.2	20.8	20.8	20.7	20.7	20.7	21.8	21.3

Colombia

Location	Coordinates			J	F	M	A	M	J	J	A	S	O	N	D	Year
Aracataca	10° 35'N 74° 9'W	13	Mean	27.3	27.4	27.5	28.9	28.4	27.8	28.0	27.9	28.1	27.5	27.5	27.8	27.8
			M. max.	33.3	33.2	32.8	34.6	33.8	33.0	33.7	33.9	33.9	33.7	33.0	34.0	33.5
			M. min.	21.4	21.7	22.3	23.2	22.9	22.7	22.4	22.3	22.4	22.2	22.1	21.7	22.3
Barrancabermeja	7° 4'N 73° 52'W	23	Mean	29.3	29.6	29.7	29.4	28.9	28.9	29.2	28.8	28.7	28.7	28.6	29.1	29.0
			M. max.	33.1	33.7	33.6	32.9	32.5	32.7	33.3	33.1	32.5	31.9	32.1	32.9	32.9
			M. min.	25.5	25.6	25.8	25.8	25.3	25.2	25.0	24.6	24.8	24.5	25.1	25.3	25.2
Rio Mira, Tumaco	1° 30'N 78° 40'W	5	Mean	26.7	27.6	28.2	28.2	27.8	28.4	28.4	28.1	28.2	27.7	27.9	26.6	27.8
			M. max.	28.7	29.7	30.6	30.6	30.2	30.7	30.9	30.4	30.5	30.0	30.5	28.5	30.1
			M. min.	24.7	25.4	25.7	25.8	25.7	26.0	25.9	25.6	25.8	25.4	25.3	24.6	25.5

Ecuador

Location	Coordinates			J	F	M	A	M	J	J	A	S	O	N	D	Year
La Concordia	0° 05'N 79° 20'W	8	Mean	25.2	25.2	25.5	25.8	25.1	24.3	23.6	23.5	23.8	23.6	23.3	24.1	24.4
			M. max.	29.0	29.5	30.0	30.2	29.0	27.8	27.5	27.5	27.6	27.5	27.2	27.9	28.4
			M. min.	21.3	20.9	20.9	21.3	21.2	20.5	19.6	19.4	20.0	19.7	19.3	20.3	20.4

Honduras

Location	Coordinates			J	F	M	A	M	J	J	A	S	O	N	D	Year
Tela	15° 43'N 87° 29'W	10	Mean	23.0	23.5	24.3	25.6	26.0	26.3	26.2	26.6	26.1	25.3	23.6	23.4	25.0
			M. max.	27.8	28.4	29.6	31.5	31.2	31.6	31.8	32.1	32.2	30.1	28.4	28.6	30.3
			M. min.	18.1	18.7	18.9	19.6	20.8	20.8	20.6	21.1	21.0	20.4	18.7	18.1	19.7

Costa Rica

Location	Coordinates			J	F	M	A	M	J	J	A	S	O	N	D	Year
Quepos	9° 26'N 84° 9'W	32	Mean	23.6	22.9	23.6	25.6	26.8	27.1	27.2	27.0	26.5	26.6	23.8	23.2	25.3
			M. max.	28.5	27.8	28.5	31.1	31.9	32.0	32.0	31.7	31.2	31.1	28.7	27.8	30.2
			M. min.	18.8	18.0	18.7	20.2	21.7	22.2	22.5	22.3	21.9	22.2	18.9	18.7	20.5

properties of the soil and the rooting depth of the plant. Thornthwaite has estimated for perennial plants that except in areas of shallow soil the water storage capacity available to mature plants with fully developed root systems varies around a mean that is equivalent to 10 cm of rainfall (Thornthwaite, 1948). Studies on Acid Sands soils in Nigeria indicated that 100 mm are available at field capacity in the top 122 cm of soil (Sparnaaij *et al.*, 1963), while in Ghana it is also considered that available water in the top 100–120 cm of the majority of forest soils is not more than 100 mm (Van der Vossen, 1969). However, 150 mm has been suggested in Malaysia (Nienwolt, 1965) while 200 mm in the top 200 cm of soil has been used by the IRHO for water-deficit calculations in West Africa and elsewhere (Surre, 1968).

Estimates of potential evapotranspiration by the method of Thornthwaite and Mather (1955, 1957) have been widely used. Figure 3.1 shows Thornthwaite's mean monthly *Etp* in five oil palm areas and illustrates the variation in rainfall and *Etp* patterns over a range of latitudes. Other *Etp* estimates have been employed. In Nigeria, the use of Penman's method and lysimeter determinations suggested that Thornthwaite's figures were too high (Sparnaaij *et al.*, 1963), while Chapas and Rees (1964), who compared four methods of calculation with measured evapotranspiration from watered grass, concluded that for Benin (Nigeria) the true annual value was around 120 cm. In Malaysia Nienwolt (1965) showed that evaporation is well correlated with saturation deficit and with sunshine hours and that computations based on these measures more nearly follow variations in observed evaporation from pans. Finally, IRHO, after comparing Thornthwaite's figures with direct determinations, and taking into account irrigation experiment results, have used values of 150 mm for months with less than 10 rainy days and 120 mm for months with 10 rainy days or more (Surre, 1968). For southern Nigeria this leads to an annual figure some 30 per cent higher than the estimate of Chapas and Rees (1964).

Van der Vossen (1969), who considered both available water and *Etp* to be overestimated in the IRHO system, compared water-deficit estimates obtained by that system with those based on the lower figures of other authors and found that although mean annual water deficits were much higher when calculated by IRHO methods there was a very exact linear regression between the two sets of values. For comparative purposes, therefore, IRHO values are useful and in combination with a soil suitability classification have been used for forecasts of the adult bunch yields to be obtained in thirty-five combinations of soil and climate (Table 3.4) (Olivin, 1968). It should be observed that the use of the soil classification, to be described later in this chapter, is to a large extent a method

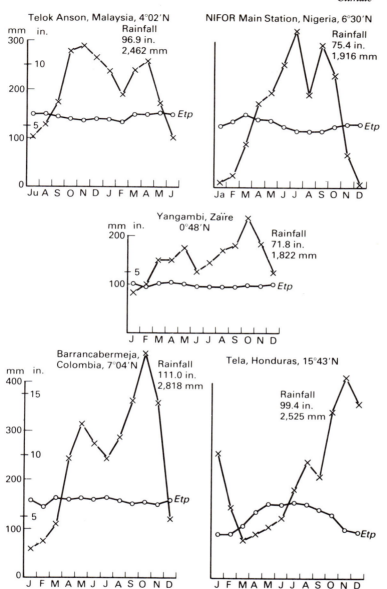

Fig. 3.1 Mean monthly rainfall and potential evapotranspiration (*Etp*) in five oil palm districts.

Table 3.4 *Bunch production estimates for various soil classes and levels of water deficit (tonnes bunches per hectare)*

Soil class	Water deficit (mm)						
	400	*350*	*300*	*250*	*200*	*100*	*Nil*
I	12	13	14	16	18	24	27
IIa	10	11	12.5	14	16	20	25
IIb	8	9.5	11	13	16	20	25
III	6	7.5	9	11	13	16	22
IV	4	5	6	8	9	12.5	16

Source: Olivin (1968).

Table 3.5 *Annual water deficit at a number of centres of oil palm cultivation, using IRHO constants for Etp and available soil water (mm)*

Centre	Mean annual rainfall	No. of years	Water deficit		
			Mean	Highest	Lowest
Pobé, Benin (Dahomey)	1,201	32	520	1,041	269
NIFOR, Nigeria	1,916	33	355	465	164
La Mé, Ivory Coast	1,993	22	254	703	28
Yangambi, Zaire	1,835	20	24	165	0
Bagan Datoh, W. Malaysia	1,837	10	169	375	0
Ulu Remis, W. Malaysia	2,300	14	5	67	0
Mostyn, E. Malaysia	2,322	18	12	158	0
San Alberto, Santander, Colombia	2,453	11	129	281	0

of correcting the constant 200 mm soil reserve maximum adopted. Table 3.5 shows the mean water deficit calculated by IRHO methods for a number of centres of oil palm cultivation. Figure 3.2 shows the relation between water deficit and bunch yield for IRHO Class I soils. In constructing curves the IRHO has used the deficit 28 months previous to the year of yield (IRHO, 1969) following the work of Sparnaaij *et al.* (1963). The latter authors evolved a parameter, 'effective sunshine', which was in effect a drought indicator. Effective sunshine measurement and its employment in predicting yield variations is described in Chapter 4 (p. 179) and its relation to nutrition is discussed in Chapter 11 (p. 511).

Oil palms are successfully cultivated in several areas of very heavy rainfall where the latter is always in excess of evapotranspiration. In coastal areas of the West Cameroon over 9,000 mm per annum are recorded and estates inland receive over 5,000 mm. Along the Pacific plain of tropical South America as far as Quevedo in Ecuador, and on the Caribbean side of Central America, areas

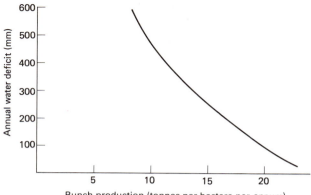

Fig. 3.2 Relation between the annual water deficit and bunch production on Class I soils (IRHO, 1969).

of very high rainfall are also encountered. In the former region this high rainfall is associated with constant cloudiness and the areas around Buenaventura in Colombia and Santo Domingo de los Colorados in Ecuador are among the dampest in the world. It is not thought that very high rainfall is in itself detrimental to yield, but it is often associated with low night temperatures, low solar radiation, waterlogging, constant soil leaching or erosion; so regions of very high rainfall rarely achieve the production of areas of more moderate, though even, rainfall found, for instance, in Malaysia.

Sunshine and solar radiation

The importance of a high level of solar radiation for the growth and bunch production of the oil palm has been inferred from several separate observations, but the exact requirements, either in terms of radiation or in hours of sunshine, for optimum yield are unknown. The importance of sunshine has been inferred from the following facts:

(a) Shading palms of all ages reduces growth and net assimilation rate.
(b) Shading adult palms reduces the production of female inflorescences.
(c) Pruning the leaves of adjacent palms increases the production of female inflorescences (Sparnaaij, 1960).
(d) A positive, though not high, correlation has been found between annual sunshine data and yield in the 12-month period 28 months later.
(e) In countries such as Nigeria with marked seasonal differences

in hours of sunshine, the differentiation of female inflorescences, as shown by the sex ratio at flowering 2 years later, is much higher during the months with many hours of sunshine than during the months with few (Broekmans, 1957).

It must, however, be noted with regard to (e) above that in other centres in the northern hemisphere such as Yangambi in Zaire and Aracataca in Colombia, where sunshine hours do not show such marked seasonal differences, the same pattern of female inflorescence differentiation obtains; the latter is therefore difficult to explain in terms of solar radiation alone. Again, sunshine hours at Yangambi are of a similar order to those of plantations in Malaysia and rainfall distribution is even. Yet bunch yields at Yangambi have been nearer to those of low-sunshine Nigeria than to those of the Far East (Vanderweyen, 1952). The only consistent climatic difference between Yangambi and the Malaysian plantations is one of temperature and this difference is quite marked (Table 3.3).

As might be expected, areas of very high rainfall have few hours of sunshine. At Calima in Colombia where rainfall averages over 6,000 mm per year, daily sunshine is recorded as varying from a mean of 2.1 hours in July to 4.1 hours in February with a total of only 1,243 hours in the year. Even lower figures have been recorded on the Pacific plain in Ecuador, but yields higher than those of the seasonal parts of West Africa are consistently obtained. From this it can be inferred that the effect of 3 months' drought is more serious than a reduction by about 50 per cent of the hours of sunshine and that, provided water is available to the plant throughout the year, the palm will tolerate a considerably lower level of radiation than obtains in the Far East.

It is unlikely that more will be learnt about this subject until more satisfactory methods than the recording of sunshine hours have been widely adopted for the measurement of radiation, and the effect of various light intensities on assimilation have been studied with the oil palm. It has been suggested that where sunshine hours are very low, the total sun and sky radiation may not always be correspondingly low. On the Pacific plain of Ecuador, where a comparatively thin layer of cloud covers the sky for long periods, figures of 900 hours sunshine per year and 264 cal per cm^2 per day have been given for Pichilingue (Olivin, 1980). The latter place, however, is south of the main oil palm areas, and it is probable that around Quinindé, where high yields are reported, the sunshine hours are similar but the radiation is higher. At Benin in Nigeria where radiation records were available for several years it was noticed that there is an inverse relationship between hours of sunshine and total sun and sky radiation over the period October to April. Whereas in the middle of this period the sunshine is sufficiently strong to burn the Campbell-Stokes recording card for

6–7 hours per day, the sky is hazy with dust blown from the Sahara, and although there is little or no rain the total radiation is not very high. However, at the beginning and end of the wet season, in April to May and October to November, there are less hours of sunshine recorded but, with clearer skies, the total radiation is higher. In general, at Benin, it was found that sunshine hours must drop below 5 per day before there is an appreciable fall in total radiation, but that radiation may also be low when the atmosphere is dust-filled in the dry season. These findings are shown graphically in Fig. 3.3 and it will be seen that while sunshine hours and radiation follow the same trends from April to November, this is not so from December to March.

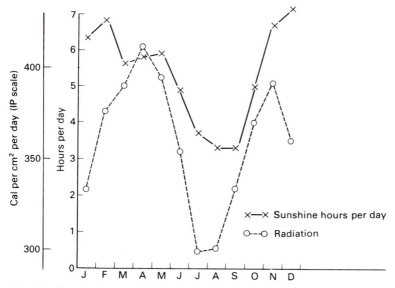

Fig. 3.3 Sunshine hours and total sun and sky radiation at NIFOR, Benin, Nigeria; mean monthly figures for the years 1958–63.

Temperature

Henry (1957) showed that growth of young seedlings is totally inhibited at 15 °C, and that growth at 25 °C is seven times, and at 20 °C three times, as rapid as at 17.5 °C. He estimated optimum temperature for growth to be 28 °C.

It is usual in countries of the humid tropics for mean minimum temperatures to be lower inland than near the coast or on a coastal plain. Low minimum temperatures are also found in areas of high latitude and, of course, where the elevation exceeds about 200 m.

Thus, while mean minimum temperatures in most oil palm-growing regions are over 21 °C (70 °F) in all months of the year, a high latitude area such as Tela in Honduras (15°43′ N) has 5 months with temperatures below 21 °C and 2 months around 19 °C (66.5 °F). This is accompanied by a shortened daylength and a mean maximum of only 27 °C (81 °F). In such a climate, which includes periods of even colder weather, the crop is reduced to very small proportions in the early months of the year and nearly 90 per cent of it is harvested in the 7 months from June to December. Investigations of the effect of temperature on the duration of bunch development and analogy with other crops suggest that this uneven yield distribution is due partly to delayed development of bunches during the colder months; but the combined effect of low temperature, shortened daylength and reduced rainfall in the winter months is mainly to increase abortion and to lower sex ratio (see Figs 3.1 and 4.13). In Bahia state, Brazil, even lower temperatures than in Honduras are experienced. The strongest fruiting period in the groves there is from November to March, but the effect of the low temperatures on the production of planted palms is not known.

The range of latitudes at which the oil palm can be grown differs from continent to continent owing to the effects on both temperature and rainfall of air and oceanic currents and of the great land masses (Ollagnier *et al.*, 1970). The range is wider in America than in West Africa where distance from the equator tends to be directly correlated with length of dry season. In Asia the oil palm areas are largely outside the influence of continental land masses, being separated from them by expanses of ocean. Thus the Asian oil palm belt is relatively narrow and climatic variations within it not great.

Comparatively low night temperatures are characteristic of the areas far inland in Zaire. At Yangambi all months of the year have a mean minimum between 19.3 and 20.3 °C (66.7–68.5 °F). This feature of the northern Zaire climate is the only one which distinguishes it markedly from that of the Far East and, as mentioned on p. 108, probably accounts for the lower yields. Mean maximum temperatures are also somewhat low, averaging 29.6 °C (85 °F).

Mention was made in Chapter 1 of the presence of the oil palm at relatively high elevation in Africa, where low night temperatures may also be expected to show their effect. It has been reported that in Sumatra palms at above 500 m come into bearing at least a year later than palms in the lowlands, and that their early yields are correspondingly reduced.

Soils

It was early realized that the oil palm can be grown on a wide range of tropical soils and that the natural regulation of a water supply to the roots was a more important factor than the intrinsic worth of the soil. Bunting *et al.* (1934), while suggesting that the most suitable soil was a 'loose alluvial loam overlying a friable clay subsoil', rightly confined their attention to four characteristics which in Malaysia would render the soil unsuitable. Later authors have drawn attention to the fact that in climatically marginal areas the characters of the soil become of greater relative importance. Naturally, in these areas it is a physical property of the soil, the available water within the rooting depth, which is of greatest importance, but the previous usage of the land and its nutrient reserves must also be taken into account.

Whenever possible, flat or gently undulating land should be chosen for the oil palm. Satisfactory yields can be obtained from hilly land in many countries, but costs of establishment and production will be increased. As will be seen in later chapters, plantations must be designed for easy transport to the oil mill of large quantities of bunches and loose fruit. The operations which will be much more expensive in hilly country will be road construction and maintenance, harvesting, and field maintenance, including maintenance of paths; additionally, inspection and supervision will be more difficult, transport vehicles will be subject to greater wear and tear and mechanization may prove impossible. Apart from these matters of expense there are the difficulties of preparing the land for planting without encouraging erosion, and the maintenance, in the later life of the planting, of a cover sufficient to prevent further erosion. A great deal can of course be done to prevent erosion and encourage covers, and on really steep land the palms themselves are normally planted on platforms or terraces, but all this work adds further to the expense. In some countries, e.g. Malaysia, a slope limit for cultivation is imposed by law, and oil palms can normally be planted on slopes below this limit.

The existence of palm groves or small groups of self-sown palms is not always an indication of soil suitability for plantations. The palm is capable of spreading into and maintaining itself in cleared areas near water very far from the centres of its normal climatic range. Apparently healthy escapes have been seen in a dry region at 18° S in Brazil. On the other hand, it may be possible, by control of the vegetation and the use of correct fertilizers, to renovate land by planting oil palms; and in certain savannah areas of West Africa, notably around Dabou in the Ivory Coast, the land, though degraded by food-cropping and now actually devoid of grove palms, has sufficiently good basic characteristics for such renovation. Lalang (*Imperata cylindrica*) areas have been similarly renovated in Indonesia and Malaysia and a large area of ancient derived savannah in the Amapá territory, Brazil, has recently been planted.

Unfavourable soil features

Soils which are unfavourable to the oil palm and which must be avoided are as follows:

1. *Poorly drained soils.* Soils may be poorly drained because of the high water-level of adjoining streams, rivers, etc. or they may be poorly drained because of their own structure. In the former case amelioration may be purely a matter of engineering while in the latter case, of which there are many examples, it may be impossible to drain the soil even when frequent and deep drains are provided. Soils of this kind are encountered in inland areas of undulating land both in Malaysia and Brazil. The effect of the lack of movement of water through the soil is most marked in young palms. If the soil can be sufficiently drained to bring the young palms through their early years, later growth and production may prove satisfactory. This is because the root system of the oil palm itself has a 'drying-out' effect through transpiration and through its effect on structure.

2. *Lateritic soils.* The term is here used to mean soils containing concretionary ironstone or plinthite (Mohr *et al.*, 1972) usually in the form of gravel, but sometimes in thick bands in the subsoil or, following erosion, becoming exposed at the surface. Soils with a small proportion of laterite may be quite satisfactory, but larger quantities of gravel or sheets of laterite near the surface lead to reduced rooting volume and rapid drying out of the soil in dry weather. Palms may, therefore, suffer drought conditions even in a climate where there is usually a satisfactory water balance.

3. *Very sandy coastal soils.* The oil palm does not grow satisfactorily on very sandy sea coasts where quite reasonable growth and yields are often obtained from the coconut. Occasionally very sandy soils are encountered inland and these are equally unsuitable.

4. *Deep peat.* Though palms can be successfully established where 90–120 cm (3–4 ft) of peat overlie a good clay subsoil, a satisfactory stand is difficult to obtain on deep peat of 250 cm (10 ft) or more. Contraction of the peat takes place on draining and the palms are unable to develop a root system sufficient for anchoring them firmly in the soil. Palms lean in all directions, some fall over, production is often low and access difficult. In spite of these disadvantages, however, much progress has been made during the last 20 years in the utilization of deep peat for oil palms, and a body of knowledge on planting techniques and manuring has been built up (Mohd Tayeb bin Hj Dolmat *et al.*, 1982; Rasmussen *et al.*, 1981).

5. *Other unsuitable soils.* There are a few other unsuitable soils, usually of limited area, which will be mentioned under individual regions.

The oil palm on the principal groups of tropical soils

Much progress has been made, since the first edition of this book was written, with the surveying and assessment of soils for oil palm planting, and much experience has been gained of growing the palm on widely varying terrain. The opening of very large new areas, often with the assistance of the international aid organizations, has made sound surveying and assessment imperative, and Charter's dictum (1957) that with 'areas of more or less undamaged forest . . . it is not possible to make accurate assessments of the value of the soil for any particular crop' is certainly no longer true for the oil palm. Nevertheless, Coulter's (1964) warning that surveys must be extended into fertility studies through field experiments still needs to be stressed. In South America, for instance, even where the fact that a soil can support oil palms may not be in doubt, its nutrient needs can remain baffling owing to a lack of controlled experiments.

Most provisional classifications of the soils of oil palm countries, though often following in broad outline the United States Department of Agriculture (USDA) system, are adapted to their own local circumstances. However, under the early USDA system the vast majority of soils bearing oil palms will be either latosols, alluvial soils or laterite soils (ground-water laterite) (Kellogg and Darol, 1949; Owen, 1951). The undesirable features of these latter soils have already been mentioned, but when laterization is not extensive they may be considered akin to latosols. The majority of the latter would be oxisols or ultisols under the USDA 7th Approximation and are also referred to in much recent literature as ferralitic soils.

Latosols

The latosols, which may be various shades of red, brown and yellow in colour, comprise all those soils in climates suited to the oil palm which are characterized by (a) low silica sesquioxide ratios of the kaolinitic clay fraction, (b) medium to low cation exchange capacities, (c) a relatively high degree of aggregate stability, (d) a usually low content of primary minerals except quartz. In other respects the latosols vary widely according to parent material and the amount of leaching; they are in fact the soils which develop over the great variety of parent material found in the wet tropics, and they cover vast areas. Their profiles are poorly differentiated though stone lines or some laterite concretions may sometimes appear. Colour may be indicative of parent material – soils from volcanic rocks are often reddish-brown – or may indicate the degree of leaching. Hence latosols used for oil palm cultivation are often classified by colour and derivation, e.g. 'reddish-brown latosols derived from basic igneous rocks' or 'yellow latosols, strongly weathered and derived from sandstones', etc.*

Alluvial soils

Alluvial soils form a very important group for the oil palm in spite of their great variations in fertility. Coastal and riverine alluvial soils are commonly planted with the oil palm in Asia and America. The marine clays when predominantly montmorillontic, not kaolinitic, are particularly productive. These soils may present considerable drainage problems, and where there has been an accumulation of organic materials they are often overlain by muck soils or by pure peat.

Alluvial soils in broad river systems, though in many ways similar to coastal alluvium, tend to have discernible sand fractions and to be kaolinitic and micaceous; in narrow inland valleys river alluvium tends to have the characteristics of the sedentary soils round about and may be very poor.

Classification of soils in production forecasting

With the opening up of large areas for oil palm cultivation in various parts of the world a useful method of classing soils has been developed by Olivin (1968). This is based to a large extent on an assessment of the unfavourable factors already mentioned. Soils encountered in prospection are classed according to four of their properties, viz. texture, quantity of gravel or stones, water permeability or lack of drainage, and chemical composition, the latter being considered of much the least importance. The textural classes are seven in number from sand (clay, 5 per cent) to heavy clay or excessively silty soils; four levels in the horizon are taken into account. The gravel classes are six in number and follow the percentage of gravel and stones in the four soil levels. For chemical composition classification pH, organic matter and exchangeable cations are taken into account. These four methods of classing the soil are drawn together into an 'agronomic classification' of which the main features are shown in broad outline in Table 3.6. For the detailed classification of each character Olivin's paper should be

---

* These distinctions of colour, particularly when associated with differences of rainfall, leaching and physical and chemical composition, find their place in certain nomenclatures in Africa, e.g. Charter's ochrosols and oxisols in Ghana (Charter, 1956), the ferrisols and ferralsols of Sys (1960, 1961) in Zaire, and the classification of ferralitic soils by French workers (Aubert, 1964; D'Hoore, 1965).

Table 3.6   *The characteristics of soils classified for oil palm cultivation (Olivin, 1968)*

Soil Class	Characteristics			
	Texture	Gravel and stones	Drainage	Chemical status
I	Sands to clays	None	Good	Organic – good Exch. cations – good
IIa	Sands to clayey sands	None or very little	Good to 90 cm	Organic – medium Exch. cations – medium
IIb	Sands to clays	Some gravel	Good to 60 cm	Exch. cations – medium
III	Sands to clays	Gravelly	Poor	Organic – medium Exch. cations – poor
IV	Leached sand or very heavy clay	Very gravelly	Deep peat or very bad	Poor

consulted. It will be seen that while texture and chemical status play some part in this classification it is mainly the quantity of gravel or the hydromorphic condition of the soil which will determine its agronomic class. The latter classes are used in combination with water-deficit determinations to estimate adult yields as shown in Table 3.4. The general experience of applying this system has been that, where the soil classes are I or II and the water deficit neither absent nor very high, then the yield estimates prove reasonably accurate. Where, however, there is no water deficit, higher yields than shown in Table 3.4 are often obtained; and, more seriously, where high water deficits are forecast and soils poor or variable, much lower mean annual yields than predicted have sometimes been obtained, particularly in Africa. One reason for this has been the occurrence of exceptionally dry years which have been followed by years of very low yield, thus reducing the mean yields well below expectation. The presence of patches of gravelly soil has often accentuated this yield reduction. The system therefore needs to be applied with circumspection wherever unfavourable soil or climatic factors may be irregularly encountered.

Soils of the oil palm regions

Asia

Malaysia

The latosols of the Malay peninsula on which oil palms are grown are sedentary soils derived from igneous and sedimentary rocks;

their manurial needs seem to depend largely on their derivation (Panton, 1963). The majority are oxisols or ultisols under the USDA 7th Approximation (Table 3.7, A, B and C)

Soils derived from volcanic types of basic and intermediate igneous rocks of andesitic or basaltic composition are comparatively rare. These are red-brown or red in colour with deep, uniform profiles and are of two distinguishable series, the *Kuantan* series being a reddish-brown sandy clay and the *Segamat* series being a red clay (Leamy and Panton, 1966). No large areas of these soils have yet been used for the oil palm, but they are likely to be suitable in most respects. In spite of their high clay content they have an excellent crumb structure, and are very free draining.

Soils derived from acid igneous rocks, granite or granodiorite, vary with the structure of the parent rock and with topography. These are often considered the best of the commoner inland soils, and, although strongly leached and having a low quantity of exchangeable cations, are deep and of good structure. They are suitable for oil palms because of their fair water-holding capacity, depth and ready response to fertilizers, though they may dry out rather too readily in very dry weather. For oil palms there is a primary deficiency of potassium on these soils, which can easily be rectified. Responses to other fertilizers may then follow and very satisfactory yields be obtained. Most common are the *Rengam* and *Jerangau* series which are reddish-yellow sandy clays; the less common *Tampin* series is a coarser-grained yellow sandy soil which drains rather too easily.

The remaining yellow latosols are mostly derived from sedimentary and metamorphic rock and cover the largest area under cultivation. The parent material may be sandstone, shales, phyllites, schists and quartzite, the latter name having been wrongly applied in the past to cover all the soils derived from sedimentary rocks. As a result of the great variety of parent material these soils vary from free-draining sandy loams (e.g. *Kedah* series) to very heavy silty clays (*Batu Anam* series) in which drainage is more or less impeded and the periodic presence of water-tables quite high in the profile is shown by characteristic mottling. Laterite bands are not uncommon. The value of these soils for oil palm cultivation is equally variable and deficiencies of nitrogen, phosphorus and other nutrients are common. The *Serdang* series and the shale-derived *Munchong* series are the most suitable for oil palms (Table 3.7, C).

The laterite soils of Malaysia, usually allotted to the *Malacca* series, need no special description. Their use for oil palm cultivation depends on the quantity of concretionary material in the soil. The largest areas of these soils lie in the north in Kedah, and in the south-west in Malacca, Negri Sembilan and North Johore.

Large areas of oil palms lie on the coastal marine alluvial clays,

Table 3.7   *Some Malaysian profiles. Soil series of diverse origin*

A. *Basic Igneous Rock*
1. Segamat (*Basalt derived*)

Depth (inches) (cm)	2–9 5–23	9–16 23–40	16–26 40–63	26–37 65–93	37–48 93–122
Clay (%)	90	93	95	95	95
Silt (%)	7	4	2	2	2
Sand (%)	3	3	3	3	3
C (%)	1.76	0.88	0.66	—	—
N (%)	0.22	0.10	0.09	—	—
pH	5.1	4.9	5.0	5.0	5.0
Exch. K (meq per 100 g)	0.17	0.08	0.05		
Exch. Ca (meq per 100 g)	0.25	0.08	0.08	<0.05	<0.05
Exch. Mg (meq per 100 g)	0.93	0.59	0.50	0.42	0.40
Cation exch. capacity (CEC), (meq per 100 g)	14.9	11.1	10.6	8.6	6.8
P, easily sol. in NaOH, (ppm)	89	82	78	55	35

2. Kuantan (*Basalt derived*)

Depth (inches) (cm)	0–3 0–8	3–12 8–30	12–24 30–61	24–36 61–91	36–48 91–122
Clay (%)	63	70	71	69	67
Silt (%)	10	9	10	10	10
Sand (%)	27	21	19	21	23
C (%)	3.14	1.21	0.77	0.77	0.49
N (%)	0.15	0.04	0.04	0.03	0.03
pH	4.2	4.1	4.7	4.7	4.6
Exch. K (meq per 100 g)	0.19	0.07	0.06	0.06	0.06
Exch. Ca (meq per 100 g)	0.42	0.03	<0.05	<0.05	<0.05
Exch. Mg (meq per 100 g)	0.74	0.20	0.16	<0.05	<0.05
CEC (meq per 100 g)	20.2	11.5	10.1	9.2	8.2
P, easily sol. in NaOH, (ppm)	177	203	242	223	287
P, conc. HCl sol. (ppm)	885	750	825	845	1,260

B. *Acid Igneous Rock*
3. Rengam (*Granite derived*)

Depth (inches) (cm)	0–3 0–8	3–12 8–30	12–24 30–61	24–36 61–91	36–48 91–122
Clay (%)	43	46	58	65	66
Silt (%)	6	6	2	4	4
Sand (%)	51	48	40	31	30
C (%)	1.49	0.71	0.49	0.40	0.46
N (%)	0.16	0.11	0.07	0.06	0.06
pH	4.6	4.2	4.2	4.5	4.5
Exch. K (meq per 100 g)	0.41	0.17	0.12	0.09	0.09
Exch. Ca (meq per 100 g)	0.08	<0.05			
Exch. Mg (meq per 100 g)	0.33	<0.05			
CEC (meq per 100 g)	7.3	5.0	5.3	6.0	6.3
P, easily sol. in NaOH, (ppm)	39	22	19	19	18
P, conc. HCl sol. (ppm)	161	113	121	114	124

Table 3.7   (Cont'd)

4. Jerangau (*Granodiorite*)

Depth (inches) (cm)	0–3 0–8	3–12 8–30	12–24 30–61	24–36 61–91	36–48 91–122
Clay (%)	36	50	57	59	57
Silt (%)	4	3	4	2	4
Sand (%)	60	47	41	39	39
C (%)	4.23	1.33	0.83	0.64	0.52
N (%)	0.21	0.09	0.06	0.05	0.05
pH	4.0	4.2	4.2	4.4	4.5
Exch. K (meq per 100 g)	0.20	0.08	0.07	0.06	0.06
Exch. Ca (meq per 100 g)	<0.05				
Exch. Mg (meq per 100 g)	0.33	<0.05			
CEC (meq per 100 g)	13.7	8.7	7.5	5.8	5.6
P, easily sol. in NaOH, (ppm)	72	51	48	60	29
P, conc. HCl sol. (ppm)	211	221	224	239	232

C. Sedimentary Rock

5. Serdang (*Sandstone*)

Depth (inches) (cm)	0–3 0–8	3–12 8–30	12–24 30–61	24–36 61–91	36–48 91–122
Clay (%)	22	34	36	42	42
Silt (%)	2	2	2	2	2
Sand (%)	76	64	62	56	55
C (%)	1.27	0.52	0.40	0.27	0.30
N (%)	0.10	0.06	0.04	0.03	0.03
pH	4.7	4.5	4.5	4.6	4.7
Exch. K (meq per 100 g)	0.14	0.10	0.08	0.08	0.10
Exch. Ca (meq per 100 g)	0.08	0.05	<0.05		
Exch. Mg (meq per 100 g)	0.42	0.42	<0.05		
CEC (meq per 100 g)	7.3	6.8	6.3	6.8	6.7
P, easily sol. in NaOH (ppm)	37	38	39	39	54
P, conc. HCl sol. (ppm)	62	75	88	97	105

6. Munchong (*Shale*)

Depth (inches) (cm)	0–3 0–8	3–12 8–30	12–24 30–61	24–36 61–91	36–48 91–122
Clay (%)	63	69	71	70	73
Silt(%)	8	4	6	6	6
Sand (%)	29	27	23	24	21
C (%)	2.75	0.41	0.17	0.07	0.04
N (%)	0.26	0.11	0.09	0.07	0.04
pH	4.3	4.3	4.7	5.1	5.5
Exch. K (meq per 100 g)	0.30	0.12	0.09	0.09	0.09
Exch. Ca (meq per 100 g)	0.04	<0.05			
Exch. Mg (meq per 100 g)	0.42	0.30	0.33	0.33	0.33
CEC (meq per 100 g)	11.0	10.0	9.8	8.0	6.6
P, easily sol. in NaOH, (ppm)	32	18	15	14	11
P, conc. HCl sol. (ppm)	92	92	76	81	—

Table 3.7 (Cont'd)

D. *Alluvium*

7. Briah (*River flood plain alluvium*)

Depth (inches) (cm)	0–4 0–10	4–14 10–36	14–43 36–109
Clay (%)	62	65	62
Silt (%)	35	33	31
Sand (%)	3	3	6
C (%)	3.34	0.32	0.23
N (%)	0.28	0.09	0.06
pH	4.5	4.5	4.4
Exch. K (meq per 100 g)	0.52	0.14	0.23
Exch. Ca (meq per 100 g)	1.43	0.50	1.18
Exch. Mg (meq per 100 g)	2.20	1.94	5.56
CEC (meq per 100 g)	25.8	21.0	20.0
P, easily sol. in NaOH, (ppm)	79	73	77
P, conc. HCl sol. (ppm)	225	115	100

8. Selangor (Marine clay)

Depth (inches) (cm)	0–6 0–15	6–12 15–30	12–24 30–61	24–41 61–104	41–53 104–135
Clay (%)	80	79	81	70	68
Silt (%)	18	17	17	18	20
Sand (%)	2	4	2	12	12
C (%)	1.31	1.04	0.82	1.01	1.44
N (%)	0.20	0.18	0.11	0.11	0.09
pH	4.7	4.3	4.3	5.4	7.7
Exch. K (meq per 100 g)	1.57	0.93	0.78	0.80	2.34
Exch. Ca (meq per 100 g)	4.8	3.9	5.1	8.1	10.7
Exch. Mg (meq per 100 g)	14.2	10.0	10.7	13.8	17.2
CEC (meq per 100 g)	32.5	32.5	30.0	30.9	52.0
P, easily sol. in NaOH, (ppm)	65	84	125	128	42
P, conc. HCl sol. (ppm)	226	207	244	322	335

described by Panton (1963) as 'low humic gley' soils owing to their gley horizons developed close to the surface (Table 3.7, D). The clay fraction is predominantly montmorillonitic. Typically these soils have a friable brown organic clay topsoil overlying a brownish-grey subsoil with, below, a horizon of grey silty clay with yellow and red mottling. Below this is a permanently wet blue-grey horizon with a pH as low as 3. Initially, drainage is the most important factor in the usage of these soils for oil palms and in fact, it alters these soils considerably. Not only does any superficial peat layer disappear, but the gleyed and waterlogged horizons are altered. It has been recorded that if, on inspection, the gleyed horizon remains closer to the surface than 90 cm (3 ft), drainage needs to be improved. Although there is a wide variation in fertility there are large areas of coastal clay which possess high reserves of nutrients, and, in spite of high production, they do not respond to fertilizers so readily as inland soils. Areas are also found which provide excellent topsoil for prenursery beds or as a filling for polythene bags.

The broad areas of river alluvium are not topographically separated from the coastal clays, but the soils are kaolinitic sandy clays whose properties are in most respects nearer to those of the sedentary rather than to the marine clay soils.

Adjacent to the coastal clays, areas of 'acid sulphate' soils formed under brackish water conditions are found. These soils are formed by the oxidation of pyrite, FeS_2, a stable constituent of anaerobic marine sediments. Once air penetrates, the pyrite oxidizes to form sulphuric acid and the pH falls from just below 7 to values of 3 or less. There are said to be over 100,000 hectares of acid sulphate soils in the coastal lowlands of peninsular Malaysia and larger areas in other South-east Asian territories, including Thailand and South Kalimantan. Usage has been limited owing to lack of knowledge of how to treat them, and previous recommendations to drain and apply large dressings of lime led to a rapid deterioration of both coconuts and oil palms (Bloomfield and Powlson, 1977). It has now been shown that only by waterlogging the acid sulphate horizon can oil palms be restored to reasonable health (Poon, 1977; Poon and Bloomfield, 1977). With the raised water-table, the roots, instead of concentrating near the bole, grow towards the water-table and, when this falls in dry weather, they are able to penetrate the acid horizon to reach the water. Where the soil has been previously drained, the roots are unable to do this. The effect of waterlogging on the acid sulphate horizon is quite rapid and, after 2–3 years, improved yields, approaching those on the best marine clays, are obtained.

Accounts of Malaysian soil series and a good appreciation of Malaysian soils for oil palm planting have been provided by Leamy and Panton (1966) and Ng Siew Kee (1965).

In *Sabah* oil palm development has taken place on riverine and coastal alluvium, on reddish-yellow latosols derived from sedimentary sandstones and shales and, to the north-east, on brown, reddish-brown and red latosols derived from a variety of rocks formed from lavas of successive volcanic eruptions. Yellow-brown soils derived from volcanic ashes are also found in this area, the soils of which have been surveyed and described by Paton (1963). Soils derived from basalt show some of the characteristics of the allied Malaysian soils; they are friable, uniform, reddish-brown clays which drain freely. Some of these soils have low contents of exchangeable cations but can be very productive; the high clay content, good structure and possibilities of replenishment from parent sources appear to compensate for their relative paucity of nutrients (Table 3.8). Marked year-to-year yield variations have been experienced and are thought to follow water-balance variations; Paton (1963) has described these soils as being extremely porous and almost 'fluffy' and drought effects might therefore be expected early in any period of low rainfall, especially where the soil is shallow and overlies boulders.

Table 3.8 *Profile analysis: Brown latosol developed over Quaternary lava* Sabah, Malaysia. Table subfamily. Forest, rainfall >2,000 mm

Depth (inches) (cm)	0–4 0–10	4–15 10–38	15–33 38–84	33–72 84–183
Clay (%)	36	76	83	78
Silt (%)	22	21	14	20
Sand (%)	25	0	0	0
pH	5.2	5.0	5.0	5.2
Exch. K (meq per 100 g)	—	0.12	0.20	0.46
Exch. Na (meq per 100 g)	—	0.09	0.08	0.08
Exch. Ca (meq per 100 g)	—	0.57	0.19	0.25
Exch. Mg (meq per 100 g)	—	0.51	0.09	0.34
Total exch. cations (meq per 100 g)	—	1.29	0.56	1.13
Exch. cap. (CEC) (meq per 100 g)	—	11.7	10.8	10.8

Source: Paton (1963).

Indonesia: Sumatra

No recent systematic account of the soils of Sumatra has been published. These soils are (a) 'liparitic' latosols (also described as red-yellow podzolic soils) derived from recent rhyolitic tuffs, (b) latosols derived from older volcanic deposits (basaltic, andesitic, etc.), (c) latosols derived from Tertiary sediments of both marine and riverine origin, (d) black soils developed over Tertiary coral and shells, (e) alluvial soils, sometimes covered with peat, of the coastal plain and river estuaries (Venema, 1952; Dell and Arens, 1957; Junus Dai and Soepraptohardjo, 1973).

In the east coast oil palm region soils derived from rhyolitic volcanic acid tuffs have a relatively thick humic layer, are red to yellow in colour, and loose and crumbly in structure. They are ultisols and in Indonesia are commonly known as red-yellow podzolic soils (Mohr *et al.*, 1972), and they are the most important and widespread soils supporting oil palms in the region. Their mineral content was high and the supply of potassium good, but they have been depleted of nutrients by long cultivation. They are characterized by often containing large quantities of minute quartz grains. On the less well-drained sites the soils tend to be more yellow.

South of the volcanic areas, and in Atjeh, there are large areas of latosols developed over Tertiary sediments of sandstone and shale. These are sandy clays or sandy clay loams, yellow in colour and commonly referred to in Indonesia as yellow podzolic soils; they are poor in nutrients, particularly phosphorus (Venema, 1952), and oil palms frequently show signs of magnesium deficiency. They are, however, suitable for oil palms, when well fertilized, except where impervious clay layers are present or where there is an excessive quantity of laterite gravel. These soils sometimes grade into yellow-grey soils of very varying texture and with some plinthite. Mixed soils of volcanic and sedimentary origin occur between the separate areas of the two derivations. Soils developed over older volcanic rocks are also encountered; they are more variable and are usually well supplied with nutrients.

In the lower-lying areas of the Sumatran east coast plain the successful growing of the oil palm depends, as on the Malaysian coastal clays, on good drainage. A wide range of soils are encountered, including humic gley and grey hydromorphic soils, sandy regosols and peat. Some areas of 'acid sulphate' soils, similar to those on the west coast of Malaysia, also occur.

The soils developed over coral and black earth are mixed with volcanic ash or other volcanic material. Suitability depends largely on the depth of the soil.

Africa

The general statement has been made that freely drained tropical soils have a uniformity of profile, are deep and of a clayey texture (Watson, 1962). This is because in the process of soil formation from rock in a climate suited to the oil palm the silica and cations tend to be leached out leaving high proportions of kaolinitic clay and hydroxides of iron and aluminium. Weathering is so active that most mineral particles either remain coarse or form clay; low silt to clay ratios are characteristic.

As will be noted in Table 3.7, most soils used for oil palms in Asia are clays. This is not so in the principal areas in Africa though the same soil-forming processes have been at work. The basic reasons for the formation of large areas of sandy clays or clayey sands, often with less than 20 per cent clay in the upper layers of the soil, are not entirely clear but are related to parent material. In Zaire for instance large expanses of country both in the north and south have parent material consisting of coverings of wind-borne sand with comparatively low proportions of clay (*nappes de recouvrement éolien*) (Sys, 1961). In West Africa, parent material of coarse unconsolidated sandstone is common and the amount of clay in the soil may sometimes be correlated with the extent to which the sandstones are interbedded with clay (Vine, 1956).

There is one other important contrast between the oil palm regions of Asia and Africa. While in the former regions the climate tends to be uniform and changes of parent material are frequent and abrupt, in Africa very extensive areas are covered with the same parent material and there are substantial climatic differences within these areas. Thus, while differences due to parent materials are evident over the continent as a whole, differences due to climatic variations within the oil palm regions assume as great or greater importance. Furthermore, much of Africa has been land for a very long time and soil profiles may result from several cycles of weathering and erosion of parent materials overlying one another (Paton, 1961).

Zaire

The oil palm is cultivated mainly in the north of the country between 1° S and 3° N in the provinces of Orientale and Equateur, and in the south in Kinshasa and Kasai provinces, at latitudes 4–7° S. The northern region has an even climate, but in the south there is a dry season lasting 3–4 months. The soils of these areas have been described by Kellogg and Davol (1949) and more recently by Sys (1961).

The important soils of the *northern region* are latosols, designated by Sys as hygro-kaolinitic ferralsols, which are free draining, uniform and low in mineral reserves.

The most widely used of these latosols is of the *Yangambi* type (Table 3.9) which is derived from a 100–125 cm thick wind-borne covering which has a higher clay fraction than the other parent-material sands of Zaire. The soil is developed on gently undulating low plateau country, and, as is common in such terrain, the soils have a higher clay content towards the top of the slopes, becoming more and more sandy as the valleys are approached. Clay tends to be washed down the profile so that the lower horizons are appreciably less sandy. The higher parts of the plateau are considered very suitable for the oil palm. There is some colour variation from reddish-yellow to yellow, but typically a Yangambi forest soil is distinctly brown in the top 30 cm shading downwards to brownish-yellow at greater depths.

Latosols of the *Low Plateau* are found in the south-west of the region interspersed with swamp country along the Congo tributaries. These soils resemble the yellower of the Yangambi type soils in both their physical and chemical properties; the rainfall is somewhat higher in the region of these soils than in that of the Yangambi soils.

Table 3.9 *Profile analysis: Yangambi latosol (ferralsol)*
Zaire Forest. Annual rainfall 72 inches (1,822 mm)

Horizon and depth (cm)	A1–A3 20	B1 45	B2 65	C 120
Clay (0–0.002 mm) (%)	26.3	30.0	38.7	35.6
Silt (0.002–0.05) (%)	2.1	2.0	2.1	1.7
Sand (0.05–2.0) (%)	71.6	68.0	59.2	62.7
C (%)	1.2	0.5	0.4	0.3
N (%)	0.10	0.6	0.04	0.03
pH	4.6	4.5	4.4	4.5
Exch. Ca (meq per 100 g)	1.2	0.6	0.4	0.4
Exch. capacity (meq per 100 g)	3.5	4.5	4.6	4.3

Note: No data for exchangeable K or Mg are given for this profile. In a similar profile analysis by Kellogg and Davol exch. K was 0.2–0.3 and exch. Mg 0.1 meq per 100 g at all positions in the profile down to 72 inches (183 cm).

Source: Sys (1960).

North of the Yangambi soils there is some oil palm cultivation on latosols derived from a diversity of rocks both igneous (granite) and sedimentary; colour, texture and the level of exchangeable cations depend mainly on the parent material and vary considerably. Laterite layers are common in some of these soils.

Recent alluvial soils are found in this region near to the Congo and its tributaries and they cover a very wide area. The use of these soils for the oil palm depends on whether they can be drained. The clay fraction varies considerably and may be very high (70–90 per cent). Gley horizons are common. The cation exchange capacity is relatively high.

Also to be found in this northern region are some areas of soils derived from Salonga sand and Karroo rocks and which are described below.

The oil palm districts of *southern Zaire* between the Kwango and Kasai rivers are dominated by very poor sandy latosols, designated hygro- and hygro-xerokaolinitic arenoferrals by Sys, derived from the blown Kalahari and Salonga sands. The *Kalahari sand* ridges found throughout the main oil palm area are grass-covered and unsuitable for the oil palm mainly on account of their very low water-holding capacity; the clay fraction is about 6–8 per cent. The *Salonga sands* lie within this area and to the north; though not quite so lacking in clay they are also unsuitable in this seasonal climate for oil palms.

Lying between the rounded Kalahari sand ridges and interspersed with the Salonga soils are the *Karroo latosols* (Sys's ferrisols). These soils, derived from sedimentary rock, are of a red colour and intrinsically fertile, but because of the topography and mixing with the sandy soils, satisfactory areas are limited. While very large areas of productive palm groves have developed in this region, plantations have been badly affected by disease and latterly very little planting has been done. When this soil appears in the northern oil palm region better use can be made of it.

West Africa

There being considerable climatic variations in the oil palm belt along the coast of West Africa, soils in this region are found to vary both with parent material and with climate. The majority of soils contain a high proportion of sand, and their free drainage and drying out in the dry season are important factors in oil palm cultivation in the region; avoidance of soils with less than 15 per cent clay below 40 cm has been recommended (Olivin and Ochs, 1978). Lateritic soils are found either very locally or in certain large expanses, e.g. in Sierra Leone. Almost all soils, except the best in Ghana, i.e. those usually employed for cocoa, are deficient in exchangeable cations, most conspicuously in potassium and magnesium. The amount of information available on the soils of the territories of West Africa varies considerably and published analytical data relate mainly to Nigeria and Ghana.

Nigeria and West Cameroon

The oil palm areas of *West Cameroon* contain soils derived from ancient basaltic flows. Rainfall being very heavy, some of these soils are not as rich as might otherwise be expected, and their clay content is surprisingly low. Marine alluvial soils exist seaward of the Cameroon mountain in areas of moderate and very heavy (8,000 mm) rainfall. These soils (Brzesowsky, 1962) are not as heavy as the Malaysian marine clays and they appear to be much affected by small differences of topography. While some are relatively free-draining, others are waterlogged gley soils requiring deep drains to bring them into a condition suitable for the oil palm.

On the west side of *West Cameroon* and across the border in *Nigeria*, estates have been established on latosols over basement complex rocks which consist mainly of granite and mica quartzose schists, but which have also been influenced by gravelly drift. The profiles are immature and parent material in various stages of breakdown is much in evidence. The soils typically contain quantities of transported gravel and may in places be shallow. Some of the gravel is concretionary ironstone, but does not appear to be developing *in situ*. Hill creep is in evidence. In parts of some profiles the stones and gravel may exceed 50 per cent. Of the remaining soil, the clay fraction is usually higher than in other West African soils (Tinker, 1962a).

When these soils are of a fair depth they are suitable for the oil palm, but they are very low in calcium, and are deficient in magnesium and phosphorus; in contrast to other West African soils, potassium is normally in good supply in the higher parts of the profile. Unfortunately the areas covered by these soils tend to be rather hilly (Table 3.10).

Table 3.10   *Profile analysis: Latosol derived from basement complex granite, hilltop*

Nigeria. Recently felled forest. Annual rainfall 115 inches (2,920 mm)

| Depth (inches) | 0–2 | 2–12 | 12–23 | 23–39 | 39–60 | 60–80 |
(cm)	0–5	5–30	30–58	58–99	99–152	152–203
Stones and gravel (%)	17	23	50	44	28	21
Clay (%)	7	15	27	23	43	42
Silt (%)	3	5	3	7	6	7
Fine and coarse sand (%)	90	80	70	70	51	51
C (%)	1.7	0.9	0.7	0.4	0.4	0.4
N (%)	0.08	0.05	0.04	0.03	0.03	0.03
pH	3.6	5.2	5.4	5.5	5.3	5.5
Exch. K (meq per 100 g)	0.20	0.10	0.18	0.06	0.10	0.09
Exch. Na (meq per 100 g)	0.40	0.30	0.35	0.28	0.52	0.45
Exch. Ca (meq per 100 g)	0.96	0.20	0.16	0.06	0.24	0.22
Exch. Mg (meq per 100 g)	0.48	0.16	0.24	0.12	0.24	0.18
Total exch. cations (meq per 100 g)	2.04	0.76	0.93	0.52	0.10	0.94
Exch. capacity (meq per 100 g)	4.9	4.6	5.5	4.3	7.0	6.9

Source: Tinker (1962a).

The great areas of palm groves and of oil palm planting in Nigeria lie on Cretaceous and Eocene unconsolidated false-bedded coarse sandstones interbedded in varying degree with layers of clay. This great region, surrounding the Niger delta and running some 500 km from west of Benin City to Calabar, still provides nearly half of the palm kernels and a sixth of the palm kernel oil of commerce though a much reduced proportion of the palm oil; it is as well, therefore, that the terms often used loosely in describing it should be correctly defined.

Benin sands: This is a geological term that has been used for the parent material–false-bedded sandstones with subordinate beds of clay, of Eocene or Eocene and Cretaceous age.

Acid sands: The latosols developed over the Benin sands. These have been divided into two main groupings (Vine, 1956), the *Benin fasc* and the *Calabar fasc* which develop respectively in areas of high rainfall (*c.* 1,750 mm) with a severe dry season, and in areas of higher rainfall (2,000 mm) with a less severe dry season. These soils vary from slightly clayey sands to sandy clays, but in the top 30 cm the clay content rarely rises to 20 per cent and in the next 60 cm is rarely above 30 per cent. Attempts have been made further to classify these latosols into soil series by their clay content. In other respects the two groupings have been described as follows:

> The most obvious distinguishing characteristic is colour, Benin fasc soils being predominantly reddish, Calabar fasc predominantly yellow to yellowish-brown. Both fascs have a poorly differentiated profile, with reddish or yellowish-brown topsoil layers which shade off more or less sharply into reddish or yellowish subsoil respectively. The dominant colour becomes clearer and brighter down the profile. From about 6 ft (1.80 m) downwards the subsoil is of practically uniform appearance to a considerable depth, usually at least 20 ft (6 cm) (Tinker and Ziboh, 1959).

The differences between these two fascs have been compared with the differences between Charter's (1956) ochrosols and oxisols and in general the comparison holds good, particularly in respect of the ratio of calcium plus magnesium to potassium and in percentage saturation.

Thus the Benin fasc soils (Table 3.11) tend to be deficient primarily in potassium and the Calabar soils (Table 3.12) in potassium and magnesium. The satisfying of these requirements in the

Table 3.11 *Profile analysis: Latosol, Acid Sands, Benin fasc*
Nigeria. Oil palms. Annual rainfall 72 inches (1,857 mm)

Depth (inches) (cm)	0–3 0–8	3–6 8–15	6–9 15–23	9–12 23–30	12–18 30–46	18–24 46–61	24–36 61–91	36–48 91–122
Clay (%)	15	13	16	17	10	15	18	23
Silt (%)	2	1	1	1	1	2	1	1
Sand (%)	83	86	83	82	89	83	81	76
C (%)	1.3	0.7	1.0	0.3	0.4	0.3	0.3	0.2
N (%)	0.10	0.06	0.06	0.06	0.04	0.04	0.04	0.04
pH	5.7	6.5	6.4	6.0	6.4	6.1	6.1	5.7
Exch. K (meq per 100 g)	0.19	0.07	0.08	0.05	0.03	0.04	0.03	0.02
Exch. Na (meq per 100 g)	0.04	0.01	0.04	0.02	0.01	0.02	0.01	0.02
Exch. Ca (meq per 100 g)	1.56	0.72	0.66	0.41	0.16	0.16	0.16	0.16
Exch. Mg (meq per 100 g)	0.34	0.34	0.36	0.28	0.04	0.04	0.12	0.08
Total exch. cations (meq per 100 g)	2.13	1.14	1.14	0.72	0.24	0.26	0.32	0.28
Exch. capacity (meq per 100 g)	5.3	3.8	4.4	4.5	3.3	3.6	3.4	2.9

Source: Tinker and Ziboh (1959).

Table 3.12   *Profile analysis. Latosol, Acid Sands, Calabar fasc*
Nigeria. Oil palms. Annual rainfall 97 inches (2,452 mm)

Depth (inches) (cm)	0–3 0–8	3–6 8–15	6–9 15–23	9–12 23–30	12–18 30–46	18–24 46–61	24–36 61–91	36–48 91–122
Clay (%)	7	8	7	7	10	18	24	20
Silt (%)	2	3	4	1	1	1	2	1
Sand (%)	91	89	89	92	89	81	74	79
C (%)	1.5	0.9	0.6	0.5	0.4	0.4	0.3	0.3
N (%)	0.09	0.06	0.05	0.04	0.03	0.04	—	—
pH	4.1	4.3	4.4	4.1	4.6	4.5	4.3	4.5
Exch. K (meq per 100 g)	0.08	0.04	0.03	0.03	0.02	0.02	0.02	0.02
Exch. Na (meq per 100 g)	0.04	0.04	0.02	0.03	0.02	0.01	0.02	0.04
Exch. Ca (meq per 100 g)	0.13	0.06	0.03	0.19	0.06	0.06	0.06	0.13
Exch. Mg (meq per 100 g)	0.12	0.07	0.05	0.07	0.00	0.00	0.02	0.02
Total exch. cations (meq per 100 g)	0.37	0.21	0.13	0.32	0.10	0.09	0.12	0.21
Exch. capacity (meq per 100 g)	5.8	5.0	3.9	3.5	3.4	4.8	4.9	4.0

Source: Tinker and Ziboh (1959).

case of the oil palm leads to a demand for other nutrients (see Ch. 11). In other respects these soils have both advantages and disadvantages in oil palm cultivation. Firstly, they give a deep unrestricted rooting medium and thus allow the oil palm's anchoring roots to penetrate into the more clayey portion of the profiles; secondly, applied nutrients become easily available to the palm. On the other hand nutrients are very easily leached, particularly in the Calabar fasc, and it is difficult to rebuild the organic matter content of the topsoil if this has once been lost. Much of the region has been subject to cropping which rapidly reduced both the cation content and the organic content of the soil.

A further defect which is important in the circumstances of West Africa is the rapidity with which the upper, most sandy layer of the soil can dry out in the dry season. As such a large part of the palm's root system is in the superficial layers, water and nutrient uptake becomes severely restricted after some weeks of unbroken dry weather. These edaphic factors have had considerable influence on agronomic practices in the region of the Acid Sands soils.

The areas of these soils have been discussed in considerable detail by Vine (1956) and by Tinker and Ziboh (1959). In some areas quite a sharp distinction can be found between the Benin and Calabar fasc soils but elsewhere 'intergrades' occur. As with the Zaire soils, which are somewhat heavier than the Acid Sands, soils of the plateaux tend to have a higher clay content than soils on the long slopes towards rivers. It is a common experience that oil palm yields are higher on the plateaux.

Within the region of the Acid Sands soils, lateritic soils occasionally occur and one area is known where a layer of compacted ironstone is so thick that it acts as a barrier to root penetration. Such areas must of course be avoided. In general, the best areas of Acid Sands soils will be those which have not been subjected to severe cropping with annuals; are flat and therefore do not suffer the degradation of hill slope soils; are free from laterite, but are not subject to the very intense dry season characteristic of the north of the region. With increasing populations such ideal conditions are now difficult to find.

To the west of the great areas of Acid Sands soils lie areas of basement complex soils with a rainfall which is marginal for the oil palm. Much of the area is suitable for cocoa and attempts to establish oil palm plantations have met with many difficulties.

Benin (Dahomey)

As one moves through the coastal regions from the Niger delta towards eastern Ghana, the rainfall decreases and the climate soon becomes marginal for the oil palm. Benin (previously Dahomey) is peculiar in that, in spite of unfavourable climatic conditions, it is still an important producer of oil palm products. Yields are naturally low and, in general, the soils which exist do not alleviate the adverse conditions. The most frequently encountered soil, derived from sedimentary deposits, is the *terre de barre*, characterized by 2.5–4.0 cm of grey-brown sandy topsoil overlying a red deep clay subsoil with clay percentages varying from 30 to 65 per cent (Furon, 1950; Surre and Ziller, 1963). This soil bakes very hard in the long dry season and the effective rooting volume is therefore low.

In some areas, on the edge of plateau country, underground water flows in from the plateau above an impermeable horizon. This natural irrigation helps to prolong the normal growth of the palms well into the dry season and results in much higher yields being obtained. Other soils in Benin are lateritic with compacted ironstone concretionary layers about 60 cm deep.

Exchangeable cations in the topsoil in Benin soils are somewhat higher than in the Acid Sands of Nigeria, but exchangeable potassium is low. Average figures have been given as: K 0.1, Mg 0.5–1.0, Ca 4.0–6.0, total 5–7 meq per 100 g.

Ghana

Van der Vossen (1969) has divided the rain forest zone of Ghana into unfavourable or marginal areas, suitable areas and favourable areas, with annual water deficits of above 400 mm, 250–400 mm and below 250 mm respectively, Over the greater part of the two latter areas the soils are derived from Pre-Cambrian rocks, mainly phyllites and granites, but in the extreme south-west there is an area of Tertiary sandstones contiguous with that of the Ivory Coast. Charter's ochrosols are developed in the areas of lower rainfall (900–1,800 mm) and form the main cocoa soils of the country. Though some ochrosols are to be found in Van der Vossen's 'favourable' regions most of their soils are the more strongly leached oxisols with pH below 5 or ochrosol–oxisol intergrades (Brammer, 1962). Thus the areas most likely to be used for oil palms lie outside the regions most favoured for cocoa. Satisfactory and sustained yields have been obtained on ochrosol–oxisol intergrades over phyllites but on oxisols derived from Tertiary sands the maintenance of satisfactory yields has presented problems (Van der Vossen, 1970). There is a marked phosphorus deficiency in these soils and in some areas consolidated concretionary ironstone layers are found at varying depths.

Ivory Coast

Although nine-tenths of the Ivory Coast is covered by soils derived from Pre-Cambrian rocks, the bulk of the oil palm development is on soils derived from Tertiary sandstones in the south-east of the country. However, in other parts of the coastal region soils derived from the older· rocks are found in a climate suited to the oil palm. Very few of the soils in the high rainfall area are red; the majority, whether overlying the sandstone or derived from older rocks, are typically yellow, or reddish-yellow latosols (Leneuf and Rion, 1963). The sandstone soils, which are among the most productive in West Africa, show a remarkable uniformity. The topsoil is a brown clayey sand and the largely featureless profile becomes more clayey with depth to 1.5 m, below which the clay content usually becomes more or less constant at around 20 per cent. Exchangeable cation content is low, particularly that of potassium. Table 3.13 shows a typical profile analysis.

   Soils over granites and schists, often in regions of secondary forest and derived savannah, tend to contain quantities of ironstone concretions and quartz gravel.

Sierra Leone

For oil palm cultivation Sierra Leone is handicapped by having too seasonal a climate. A severe dry season lasting for 3–4 months is followed by many months of excessive rainfall. The dry season is even more pronounced in the north and north-east, and so only the south of the country can be considered for oil palms. Here the soils

Table 3.13 *Profile analysis: Yellow latosol over Tertiary sandstone*
Ivory Coast. Secondary forest

Depth (cm)	0–3	3–10	10–20	30–60	100–130	150–160	200–230
Clay (%)	10.5	11.5	13.2	14.7	19.0	20.2	19.0
Silt (%)	2.7	2.7	3.5	2.2	2.2	1.7	2.0
Sand (%)	83.0	83.8	81.9	81.5	75.6	74.5	75.8
pH	5.3	4.6	4.5	5.1	4.9	4.9	5.1
C (%)	2.8	1.6	1.1	—	—	—	—
N (%)	0.6	0.1	0.1	—	—	—	—
Exch. K (meq per 100 g)	0.07	0.04	0.03	0.02	0.02	0.02	0.03
Exch. Na (meq per 100 g)	0.01	0.01	0.01	0.04	0.01	0.01	0.02
Exch. Ca (meq per 100 g)	1.07	0.54	0.54	0.24	0.53	0.60	0.40
Exch Mg (meq per 100 g)	0.60	0.30	0.30	0.18	0.22	0.12	0.12
Total (meq per 100 g)	2.59	1.31	1.30	0.72	1.13	1.07	0.83

Source: Leneuf and Rion (1963).

range from latosols derived from unconsolidated recent sandy sediments (*Bullom sands*) on the coast, through a yellow or brown sandy clay loam heavily filled with concretionary gravel overlying crystalline schists and gneisses of the *Kasila* series, to a wide variety of soils, some lateritic, some not, derived from basement complex granite and acid gneiss and lying well inland and near the Liberian border (Tinker, 1962b; Odell *et al.*, 1974).

Sierra Leone has had a sad history of oil palm planting, some very unsuitable areas in respect of both soils and climate having been chosen and later abandoned. The main problems have been to avoid lateritic areas and regions of high water deficit. The former problem is made more difficult by the tendency for the undesirable *Kasila* series soils to outcrop in areas of relatively good soil on the *Bullom sands*; and climatic appraisal has been complicated by a tendency, during the last two decades, for the dry seasons to become more intense (Hartley, 1979). Soils overlying the basement complex granite and gneiss vary greatly with topography, the hill and slope soils (mainly *Vashun* and *Waima* series) being either shallow and gravelly or deeper and very gravelly. By contrast, the *Baoma* series is almost gravel free and, together with the colluvial footslope soils of the *Pendembu* series and the river alluvial *Moa* series, comprise the best soils for oil palms in this region (Sivarajasingam, 1968; Stark, 1968).

Sierra Leone has provided several warnings of the need to avoid gravelly soils, especially in regions with pronounced dry seasons. At

Njala, a fertilizer experiment on good river alluvium gave more than double the bunch yield of a similar experiment on the contiguous *Njala* series lateritic soil (Walker and Melsted, 1971). On a plantation supposedly lying on soil derived from *Bullom sands*, adult yields were half that originally projected owing mainly to outcrops of *Kasila* series derived soils and to unexpectedly high water deficits. Across the border in *Liberia* there have been similar experiences.

America

D'Hoore (1956) has drawn attention to the pedological differences between tropical Africa and America and his comments have considerable relevance to oil palm cultivation. Broadly, tropical Africa contains wide depressions filled with sediments of continental origin and fringed with crystalline basement complex. In America, the important 'depressed regions' are ancient marine basins filled in by sediments of which only the upper layers are of comparatively recent continental origin. Examples of interest for oil palm planting are the Urabá Gulf region and the Magdalena valley of Colombia, the Maracaibo region, and the upper parts of the Orinoco and Amazon basins. Continental basins comparable with those of Africa are, however, to be found in the regions of the Brazilian and Guiana Shields.

Perhaps the outstanding feature of the lowland soils of tropical America is the constant provision from the Andes chain and the ranges of Central America of nutrient elements being fed to the depressions by erosion. Thus the majority of areas where oil palm planting is being expanded are either flat and covered with alluvial or colluvial soils or are provided with young soils of recent volcanic origin.

The several areas of oil palm development on the narrow plain to the west of the Andes are subjected to very varying rainfalls. In *Ecuador* the soils have developed over successive deposits of volcanic ash under conditions conducive to erosion and leaching. The young soils (regolatosols) so developed are highly porous. Organic layers are found buried at depth by later deposits, and it is characteristic that horizon differentiation of the profile has seldom proceeded very far before the soil is buried under a fresh accumulation of detritus. Some soils have an exceptionally high organic content. Nutrient status varies considerably, but is high in most localities though the magnesium supply is often inadequate.

The soils of the heavy rainfall Pacific areas of *Colombia* are quite different. River alluvium of fair fertility is encountered in valleys, but elsewhere the soils are leached latosols derived from Tertiary sedimentary rocks and their nutrient status may be expected to be only fair. They are usually reddish-yellow sandy clays or silty clays and drainage is often a problem. Soils derived from similar sedimentary deposits are also found in parts of the Magdalena and Zulia valleys, in the Urabá Gulf area and on the eastern side of the Andes; these are less fertile than the broad expanses of alluvial soils of these areas which are well suited to oil palm cultivation. Colombia contains wide flat areas of suitable soils, mostly alluvial, interspersed with areas, sometimes hummocky or peaty, which are less suitable. With the high rainfall and the erosion from the mountains which is still occurring on a large scale, drainage of the alluvial soils is likely to remain a problem. Oil palms also grow well in northern Colombia on the alluvial soils when irrigation water is available in the dry season.

Good soils on old alluvial terraces are found in *Peru* in the Huallaga valley (Lauzeral, 1980) of the upper Amazon basin. These soils, being near the Andes, are quite different from the soils of the middle and lower Amazon basin in Brazil.

The landscape of tropical *Brazil* is dominated firstly by ancient erosion surfaces bounded by scarps and giving place to younger erosion surfaces at a lower level, and secondly by the Amazon basin. The soils of the areas where oil palms are found in groves or are being planted have not yet been investigated in any detail, though Ooi *et al.* (1981) have recently provided an outline account. These authors consider that, while the Amazon river alluvial soils (*varzea*) offer best prospects as regards nutrient supply, their usage for oil palms is limited owing to the danger of flooding. Oil palm development is therefore likely to be concentrated on the latosols of Amazonas, Pará, Amapá and southern coastal Bahia. These soils as a whole are characterized by acidity, low cation exchange capacities and high phosphate fixation, but they vary widely in texture and water-holding capacity. In Bahia, suitable areas in the coastal strip are limited by topography and irregularities in the rainfall pattern, while at the north end some very difficult heavy red clays (*massape*) are encountered. In Pará, near Belém, yellow sandy loams, with the clay fraction increasing down the profile, predominate, and the main deficiency is phosphorus. The area of greatest potential in the long term is considered to be in Amazonas on the upper Amazon where there is an even rainfall, level land and suitable, though not yet adequately surveyed, latosols.

In *Central America* the oil palm is being grown on coastal alluvium on the Pacific side in Panama, Costa Rica and southern Mexico, and on the Caribbean side in Panama and Honduras. River alluvial soils of the Aguan valley are also being used in Honduras, and there is some planting on volcanic soils in Costa Rica.

Environmental aspects of oil palm development

Prospection and surveys

With the considerable expansion of oil palm planting systematic methods of prospecting for suitable territory and for the siting of plantations have been developed. Ochs and Olivin (1969) have described the methods and organization of semi-detailed surveys for determining the feasibility of an area for oil palm planting and detailed surveys for determining land usage and layout. In most cases such surveys have been preceded by a reconnaissance prospection. These authors point out that their methods differ considerably from soil surveys in the strict sense since they are directed to the eventual provision of detailed plans of the allotment of land for fields, roads, mills, etc. as well as the delimitation of areas unsuitable for planting.

In the reconnaissance prospection, existing soil, meteorological and topographical documents are studied, available paths or paths to be cut for the prospection are chosen, and reconnaisance maps are constructed. Estimates are made of a 'utilization coefficient' based on the proportion of Class I and II soils (see p. 106) and a 'coefficient of subdivision by classes' which is a measure of the homogeneity of the area. The utilization coefficient is only a comparative measure at this stage and does not represent the proportion of the land which, after detailed survey, will actually be used for planting.

Broadly similar methods have been advocated (Leamy and Panton, 1966) and employed in Malaysia, e.g. for the large Jenka Triangle Development scheme (Anon., 1967), though the recognition of soil series takes the place of the agronomic classification of Olivin (1968), and plantation siting is determined by reference to topography and an experienced assessment of the value of the series. Layout planning is described in Chapter 8 (p. 359).

Ecology

In recent years there has been much ill-informed comment on the dangers of opening large tracts of tropical forest for plantation monoculture. For oil palms, Olivin (1980) has reviewed the possible effects on climate and soil, the significance of changes in flora and fauna, and the dangers accompanying some agricultural practices. He claims, on the evidence of two instances (in Colombia and the Ivory Coast), that large clearings have initially reduced annual rainfall which, however, returned to its usual magnitude after the palms and covers were well established. It is difficult, however, to assert on this slender evidence that this will be a normal phenomenon. As for soils, Olivin claims that, with good cultural practices, no additional erosion should occur, and that soil stability is reached after a decade, with chemical status 60–90 per cent of that under the original forest. Where the vegetation was derived savannah rather than forest (e.g. near Dabou in the Ivory Coast, and Amapá, Brazil), soil improvement follows oil palm establishment. There are, of course, profound changes in fauna and flora, but a new biological balance, favourable to the crop and not unfavourable on any general grounds, can be established. Pests and diseases, whose spread is not initially checked or balanced by natural enemies, and which are encouraged by the contiguity of large areas, remain the main dangers. Pollution from insecticides, fungicides or herbicides, excessive use of fertilizers, and the discharge of mill effluents (see p. 721), are also serious dangers to the local environment but not ones incapable of proper control.

References

Anon. (1967) *The Jéngka Triangle Report.* Tippetts–Abbott–McCarthy–Stratton and Hunting Technical Services Ltd.

Aubert, G. (1964) The classification of soils as used by French pedologists in tropical or arid areas. *Sols afr. (African Soils)*, **9**(1), 107.

Bloomfield, C. and **Powlson, D. C.** (1977) The improvement of acid sulphate soils for crops other than padi. *Malaysian Agric. J.*, **51**(1), 62.

Brammer, H. (1962) Soils. In *Agriculture and land use in Ghana*, p. 88, Oxford University Press.

Broekmans, A. F. M. (1957) Growth, flowering and yield of the oil palm in Nigeria. *J. W. Afr. Inst. Oil Palm Res.*, **2**, 187.

Brzesowsky, W. J. (1962) Podsolic and hydromorphic soils on a coastal plain in the Cameroon Republic. *Neth. J. agric. Sci.*, **10**, 145.

Bunting, B., Georgi, C. D. V. and **Milsum, J. N.** (1934) *The oil palm in Malaya.* Mal. Planting Manual No. 1, Kuala Lumpur.

Chapas, L. C. and **Rees, A. R.** (1964) Evaporation and evapotranspiration in Southern Nigeria. *Quart. J. roy. met. Soc.*, **90**, 313.

Charter, C. F. (1956) The nutrient status of Gold Coast forest soils, *Report of cocoa conference*, London, Sept. 1955, 40.

Charter, C. F. (1957) The aims and objects of tropical soil surveys. *Soils Fertil.*, **20**.

Coulter, J. K. (1964) Soil surveys and their application in tropical agriculture. *Trop. Agric. Trin.*, **41**, 185.

Dell, W. and **Arens, P. L.** (1957) Inefficacité du phosphate naturel pour le palmier à huile sur certains sols de Sumatra. *Oléagineux*, **12**, 675.

D'Hoore, J. (1956) Pedological comparisons between tropical South America and Tropical Africa. *Sols afr. (African Soils)*, **4**, 4.

D'Hoore, J. L. (1965) Soil classification in Africa. Symposium on Soil Resources of Tropical Africa, London. African Studies Ass. of the UK. Mimeograph.

Furon, R. (1950) *Géologie de l'Afrique.* Payot, Paris.

Hartley, C. W. S. (1973) The expansion of oil palm planting. In *Advances in oil palm cultivation*, eds R. L. Wastie and D. A. Earp, Incorp. Soc. of Planters, Kuala Lumpur.

Hartley, C. W. S. (1979) The oil palm in Sierra Leone. *Oil Palm News*, **23**, 2.

Henry, P. (1957) Quoted by Micheau, P. *Oléagineux*, **16**, 523.

IRHO (1969) Recherches sur l'économie de l'eau à l'IRHO. L'eau et la production du palmier à huile. *Oléagineux*, **24**, 389.

Junus Dai and **Soepraptohardjo, M.** (1973) Soil survey for agricultural development in particular reference to the rationalized land use program in North Sumatra. Soil Research Institute, Bogor. Mimeograph.

Kellogg, C. E. and **Davol, F. D.** (1949) An exploratory study of soil groups in the Belgian Congo. *Publs INEAC*, Série Sci, No. 46.

Lauzeral, A. (1980) Les sols d'Amérique Latine et la culture du palmier à huile. *Oléagineux*, **35**, 11, 477.

Leamy, M. L. and **Panton, W. P.** (1966) *Soil survey manual for Malayan conditions*. Bull. 119. Div. of Agriculture. Min of Agric. & Coop., Malaysia.

Lenuef, N. and **Rion, G.** (1963) Red and yellow soils of the Ivory Coast. *Sols afr. (African Soils)*, **8**, 451. (CCTA Publication.)

Mohd Tayeb Hj Dolmat, Hj Abdul Halim Hassan and **Zin Z. Zakaria** (1982) Development of peat soil for oil palm planting in Malaysia – Johor Barat agricultural project as a case study. *PORIM Bull.*, **5**, 1.

Mohr, E. C. J., van Baren, F. A. and **van Schuylenborgh, J.** (1972) *Tropical soils; a comprehensive study of their genesis*, 3rd edn Mouton, The Hague, 481 pp.

Ng Siew Kee (1965) The potassium status of some Malayan soils. *Malaysian agric. J.*, **45**, 143.

Nienwolt, S. (1965) Evaporation and water balances in Malaya. *J. trop. Geogr.*, **19**, 34.

Ochs, R. (1963) Recherches de pédologie et de physiologie pour l'étude du problème de l'eau dans la culture du palmier à huile. *Oléagineux*, **18**, 231.

Ochs, R. and **Olivin. J.** (1969) Etude pour la localisation d'un bloc industriel de palmiers à huile. II. Les techniques de prospection. *Oléagineux*, **24**, 125.

Odell, R. T. *et al.* (1974) Characteristics, classification and adaptation of soils in selected areas of Sierra Leone. *Univ. Illinois, Bull.*, 748; *Univ. Njala, Bull.*, 4.

Olivin, J. (1968) Étude pour la localisation d'un bloc industriel de palmiers à huile. I. Les critères de jugement. *Oléagineux*, **23**, 499.

Olivin, J. (1980) Relation entre l'écologie et l'agriculture de plantation. *Oléagineux*, **35**, 2, 65.

Olivin, J. and **Ochs, R.** (1978) Propriétés hydriques des sols et alimentation en eau des oléagineux perennes en Afriques. *Oléagineux*, **33**, 1.

Ollagnier, M. *et al.* (1970) The manuring of oil palms in the world. *Fertilité*, **36**, 3.

Ooi, S. C. *et al.* (1981) The oil palm industry in Brazil – current status and future potential. In *Oil palm in agriculture in the eighties* (Malaysian Oil Palm Conference).

Owen, G. (1951) A provisional classification of Malayan soils. *J. Soil Sci.*, **2**, 20, and *J. Rubb. Res. Inst. Malaya*, **13**, com. 274.

Panton, W. P. (1963) 1962 soil map of Malaya. Dept of Agriculture, Federation of Malaysia. Mimeograph.

Panton, T. R. (1961) Soil genesis and classification in Central Africa. *Soils Fertil.*, **24**, 249.

Paton, T. R. (1963) *A reconnaissance soil survey of soils of the Semporna Peninsula, N. Borneo*. HMSO.

Poon Yew Chin (1977) The management of acid sulphate soils and its effect on the growth of the oil palm (*Elaeis guineensis* Jacq.). *Malaysian Agric. J.*, **51**(2), 124.

Poon, Y. C. and **Bloomfield, C.** (1977) The amelioration of acid sulphate soil with respect to oil palm. *Trop. Agric. Trin.*, **54**, 289.

Rasmussen, A. N. *et al.* (1981) Establishment of oil palms on deep peat from jungle. In *Oil palm in agriculture in the eighties* (Malaysian Oil Palm Conference).

Sivarajasingam, S. (1968) Soil survey of the Eastern Province, Sierra Leone. *FAO Rpt No. TA 2584*, Rome.

Sparnaaij, L. D. (1960) The analysis of bunch production in the oil palm. *J. W. Afr. Inst. Oil Palm Res.*, **3**, 109.

Sparnaaij, L. D. *et al.* (1963) Annual yield variation in the oil palm. *J. W. Afr. Inst. Oil Palm Res.*, **4**, 111.

Stark, J. (1968) Soil and land-use survey of part of the Eastern Province, Sierra Leone. *FAO Rpt No. TA 2574*, Rome.

Surre, Ch. (1968) Les besoins en eau du palmier à huile. Calcul du bilan de l'eau et ses application pratiques. *Oléagineux*, **23**, 165.

Surre, Ch. and **Ziller, R.** (1963) *Le palmier à huile.* Maisonneuve et Larose, Paris.

Sys, C. (1960) Carte des sols du Congo Belge et du Ruanda-Urundi. Congo Belge et Ruanda-Urundi. A. Sols. *Publs INEAC*, Brussels.

Sys, C. (1961) La cartographie des sols au Congo. Ses principes et ses méthodes. *Publs INEAC*, Série Tech., No. 66.

Thornthwaite, C. W. (1948) An approach towards rational classification of climate. *Geogr. Rev.*, **38**, 55.

Thornthwaite, C. W. and **Mather, J. R.** (1955) The water balance. *Publs Clim. Drexel Inst. Technol.*, **8**(1), 104 pp.

Thornthwaite, C. W. and **Mather, J. R.** (1957) Instructions and tables for computing potential evapotranspiration and the water balance. *Publs Clim. Drexel Inst. Technol.*, **10**(3), 185–311.

Tinker, P. B. H. (1962a) Some basement complex soils of Calabar Province, Eastern Nigeria. *J. W. Afr. Inst. Oil Palm Res.*, **3**, 308.

Tinker, P. B. H. (1962b) Tour of Sierra Leone by WAIFOR soil chemist. Mimeographed paper, WAIFOR.

Tinker, P. B. H. and **Ziboh, C. O.** (1959) A study of some typical soils supporting oil palms in Southern Nigeria. *J. W. Afr. Inst. Oil Palm Res.*, **3**, 16.

Vanderveyen, R. (1952) *Notions de culture d'Elaeis au Congo Belge.* Brussels.

Van der Vossen, H. A. M. (1969) Areas climatically suitable for optimal oil production in the forest zone of Ghana. *Ghana J. agric. Sci.*, **2**, 113.

Van der Vossen, H. A. M. (1970) Nutrient status and fertility responses of oil palms on different soils in the forest zone of Ghana. *Ghana J. agric. Sci.*, **3**, 109.

Venema, K. C. W. (1952) Oil palm production and oil palm soils in Indonesia. Part 11. The Atjeh soils, *Potash Rev.*, subject 27, Jan. 1952.

Vine, H. (1956) Studies of soil profiles at the WAIFOR Main Station and at some other sites of oil palm experiments. *J. W. Afr. Inst. Oil Palm Res.*, **1**(4), 8.

Walker, W. M. and **Melsted, S. W.** (1971) Effect of N, P, Mg, K and soil upon oil palm yields in Sierra Leone. *Trop. Agric. Trin.*, **48**, 237.

Watson, J. P. (1962) Leached pallid soils of the African plateau. *Soils Fertil.*, **25**, 1.

Chapter 4

Factors affecting growth, flowering and yield

As the oil palm normally develops its leaves at a regular rate and produces only one inflorescence, which may be male, female or hermaphrodite, in each leaf axil, a study of the variations in leaf and flower development, and the distinguishing of their genetic and environmental causes, is clearly of prime importance in helping to maximize yield. Much of the early physiological work was undertaken in Africa (Prévot, 1963) and was directed to the solution of specific problems. In the last 30 years, however, much more attention has been given to physiological processes both in Africa and Asia and, in particular, the use of growth analysis has given a greater understanding of the palm's potential. It is proposed in this chapter to describe and discuss, in the light of these studies, the factors affecting the palm's growth and development. Germination is discussed in Chapter 6.

The rate of growth of the oil palm

The yield of crops depends largely on the rate of development and the maintenance of the leaf surface (Milthorpe, 1956). This is very clearly shown in the oil palm where production will depend on the number and size of leaves produced and the proportion of them providing female inflorescences; and, as will be seen later, the maintenance of the highest possible leaf surface is an important factor in determining the sex ratio, which is the proportion of female to total inflorescences.

Physiologists have considered the rate of expansion of the total leaf surface from germination in two phases: the time from emergence to the unfolding of the first leaf, and the rate of growth of the leaf surface relative to the area already present. In the oil palm the first phase may occupy 4–6 weeks and measurement of growth in the second phase is difficult until at least the two-leaf stage has been reached some 2–3 months after germination.

The relative growth rate of leaves depends both on the rate of differentiation of new primordia, which in the oil palm are contained in the single apex, and on the rate of expansion of these primordia. These rates may be affected by many external influences, but there is evidence that the final leaf area is determined very largely by occurrences at the primordial stage. It has been concluded therefore that adverse conditions, particularly at germination or at the time of early growth, will be reflected in subsequent poor development. On physiological grounds, therefore, optimum conditions during germination and early seedling growth are of great significance.

For crop studies, one is not concerned only with the growth of leaf area, for it is assimilation and the weight of produce which are of ultimate importance. The net assimilation rate over a given period is the rate of increase of dry weight per unit leaf area and it is usually expressed in grams per square decimetre per week (Briggs *et al.*, 1920). Though net assimilation rate was at first thought to be relatively constant (Health and Gregory, 1938) later studies have shown that both this factor and the development of the leaf surface may vary considerably. (Watson, 1947, 1956). Such studies were started with the oil palm in West Africa in the 1950s but use has now been made of growth analysis methods in Malaysia, and these are helping in the understanding of the palm's development and reaction to environmental changes.

The growth of young seedlings

The young seedling is at first dependent on the kernel, and removal of this organ before 5 weeks after germination causes a severe reduction in relative leaf area growth rate* which lasts for about 3

---

* The mean values of the growth analysis functions employed are calculated as follows:

1 *Relative growth rate*: rate of increase in dry weight per unit of time.

$$\frac{\log_e W_2 - \log_e W_1}{t_2 - t_1}$$ per cent per day. Symbol – R_W.

2. *Net assimilation rate*: rate of increase of dry weight per unit leaf area.

$$\frac{(W_2 - W_1)(\log_e A_2 - \log_e A_1)}{(t_2 - t_1)(A_2 - A_1)}$$ .g per dm^2 per week. Symbol E_A.

3. *Leaf area ratio*: leaf area per unit weight.

$$\frac{(A_2 - A_1)(\log_e W_2 - \log_e W_1)}{(\log_e A_2 - \log_e A_1)(W_2 - W_1)}$$. cm^2 per g. Symbol F.

4 *Relative leaf area growth rate*: rate of increase in leaf area per unit of time.

$$\frac{(\log_e A_2 - \log_e A_1)}{t_2 - t_1}$$ per cent per day. Symbol – R_A.

Continued overleaf

weeks. Removal of the kernel at 7 weeks has only a small effect and it is believed that kernel reserves are completely exhausted by about 14 weeks after germination (OPGL, 1968–72). Rees (1963a) determined in Nigeria the net assimilation rates, relative growth rates and the leaf area ratios for prenursery seedlings grown in full daylight and under three levels of shade. It should be observed that the leaf area ratio (F) is the relative growth rate (R_W) divided by the net assimilation rate (E_A). The results showed that with plants not already adapted to shade, i.e. not maintained under the various levels of shade from germination, there was a reduction of net assimilation rate and relative growth rate and an increase in leaf area ratio with increasing shade. When adapted to shade these relationships were no longer apparent although in the highest degree of shade (11 per cent of full daylight) the net assimilation rate was always reduced. The plants were transplanted at the two- to three-leaf stage and this had an effect on the net assimilation rate, reducing the latter to a very low figure. This result, repeated with nursery seedlings (see later), suggests an effect of death of part of the root system and shows why the correct handling of young seedlings is of prime importance for the continuance of normal assimilation and growth.

This study also suggested that, even at an early stage, the small seedling palm will develop more efficiently in full daylight (for net assimilation rate is a measure of photosynthetic efficiency), but that it can adapt itself to partial shade. This fact may be important for the survival of the palm in palm groves and farming associations, since temporary conditions of partial shade may give the seedling the opportunity to grow and establish itself above the surrounding vegetation. Under conditions of deep shade, however, the seedling will be unable to compete.

Nursery seedling growth

The first estimations of net assimilation rates of nursery seedlings were made in the course of a field nursery growth study by Wormer (1958) using Deli crosses planted in the relatively dry conditions of Benin (Dahomey). The values obtained ranged from 0.04 to 0.30 g per dm^2 per week. Plants about 21 months of age from germination

---

5. *Leaf area index*: area of leaf laminae per unit area of land.

$$\frac{\text{Leaf area}}{\text{Ground area}} \text{ (ratio). Symbol} - L \text{ or } LAI.$$

6. *Crop growth rate*: rate of increase of dry matter per unit area of land $= EL$. Symbol $- C$ or CGR.

W_1 and W_2 and A_1 and A_2 are total dry weights and leaf areas, respectively, at times t_1 and t_2.

had developed an average of 68 leaves of which 18 had died, 15 were fully grown, 8 were in course of rapid growth and the remaining 27 were less than 2 cm long. The first visible inflorescence bud was to be found in the axil of leaves ranging from the twenty-fifth to the forty-first in order of growth. There was a suggestion that the relative growth rate was, for a period of 10 months, correlated with light intensity. A leaf area index (leaf area divided by ground area) of 4.45 was reached after 13 months (56 weeks) in the nursery (17 months from germination) at a spacing of 80 cm triangular. Higher values, up to 7.5, were recorded for older plants. It should be observed that these estimations were carried out with material of Far Eastern origin in a watered nursery in a very dry and seasonal part of the West African oil palm belt (average annual rainfall 1,231 mm).

The growth of nursery seedlings of African origin was studied in two stages by growth analysis methods in a field nursery in Nigeria where the annual rainfall is over 1,900 mm and falls almost entirely in one prolonged wet season of 7–8 months' duration. The values are not easily comparable with those of Benin since, apart from the climatic and genetic differences already referred to, the Benin results were based on the fresh weight of the aerial parts of the plants only. Whole plant dry weight increased from under 1 g to nearly 42 g, and leaf area from 82 to 2,035 cm^2 in 28 weeks. Leaf area index had reached 0.55 by the end of the wet season (31 weeks) with a spacing of 61 cm (2 ft) square. Net assimilation rate rose from 0.16 to 0.31 g per dm^2 per week. These results again suggest that the imbalance of the shoot-to-root ratio following transplanting leads to a lowering of the net assimilation rate. The least possible disruption of plant and soil during transplanting will shorten the period of low net assimilation rate and low leaf area ratio. Values of the former function (E) and of relative growth rate are again low, but there is a time trend of increasing E associated with increasing solar radiation towards the end of the wet season. Values then reach about 0.3 g per dm^2 per week (Rees and Chapas, 1963a).

With watered older nursery plants in the dry season net assimilation rates are very similar, indicating that at this stage E is largely unaffected by age or season except for the association with solar radiation already mentioned (Rees, 1963b). Values of growth functions of a nursery planted in early May and sampled between October and the following March are given in Table 4.1. The plants were spaced at 61 cm (2 ft) square and it will be seen that a leaf area index of 4.1. was reached in 10 months.

The importance of correct spacing is well illustrated by the relationship between crop growth rate and leaf area index (Rees, 1963c). With closely spaced plants the net assimilation rate soon starts to fall and reaches zero at a leaf area index of about 5.4. With

Table 4.1　*Growth from October to March in a field nursery planted in May*

Dry weight, leaf area and leaf-area ratio (F) for the sampled plants; mean leaf-area index (L) at each sampling; net assimilation rate (E_A), relative growth rate (R_W) and crop growth rate (C) for each period

Periods of 4 weeks	Sample	Mean dry weight per plant (g)	Mean leaf area per plant (dm²)	F (cm² per g)	L	E_A (g per dm² per week)	R_W (% per day)	C (g per dm² per week)
	1	55.2	24.0	43.4				
					0.6			
1	1	(53.7)	23.0			0.271	1.70	0.22
	2	8.49	37.4	44.1				
					1.0			
2	2	(89.7)	40.7			0.238	1.51	0.29
	3	135.3	60.2	44.5				
					1.4			
3	3	(111.8)	45.8			*	*	*
	4	139.6	69.2	49.6				
					1.8			
4	4	(132.0)	63.8			0.289	1.74	0.55
	5	211.3	76.5	36.2				
					2.0			
5	5	(188.4)	69.7			0.220	1.15	0.55
	6	258.0	92.2	35.7				
					3.0			
6	6	(369.8)	131.2			0.175	0.84	0.62
	7	456.0	152.9	33.5	4.1			

* Anomalous results for this sampling not included.

Note: Dry weight figures in brackets relate to plants whose leaf area was measured at one sampling time, but which were harvested at the next sampling time. The weights in these cases were estimated from the leaf area.

Source: Rees (1963b).

crop growth rate, which is the product of the net assimilation rate and the leaf area index and thus gives the rate of increase of dry matter per unit area of land, the relationship is curvilinear, so that an optimum leaf area index for crop growth appears. These relationships are shown in Fig. 4.1. Beyond the optimum a point will be reached where the net rate of growth of leaves is zero, new leaf production being balanced by the death of older leaves and the rate of production of non-photosynthetic parts falling to zero.

The practical factors determining the best time for transplanting to the field will be considered in Chapter 9, but it is clearly advisable

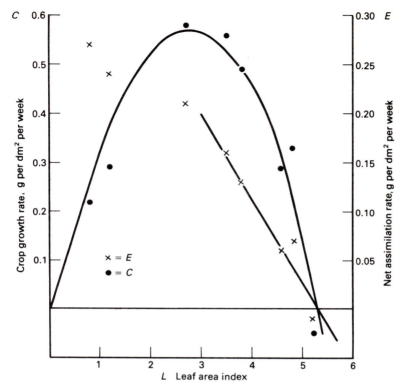

Fig. 4.1 Relationship between leaf area index (*L*) and crop growth rate (*C*) and net assimilation rate (*E*) in a field nursery spacing experiment (Rees, 1963c).

that the plants should be growing at their maximum rate. *L* for optimum growth per plant in the Nigerian trial was under 3.0. If spacing is too close, maximum growth per palm will occur when the palms are either too small for transplanting or, as would be the case in West Africa, when the season is unsuitable for transplanting. It was calculated that, in West African field nurseries, to attain optimum plant growth at a leaf area index of about 2.0 in mid-April, the maximum spacing should be about 71 cm (2.8 ft) square. Field experiments confirmed this.

It is important to note a distinction between crop growth rate as measured here and as measured in adult palm studies to be discussed later. In seedling investigations *C* has been measured as the change in the biomass (the standing crop) during a period of time, in the same manner as is usual in studies with annual crops. This is not the same as the total organic matter incorporated in the period since some of the leaves produced will have died off and not been measured. With adult palms, however, the crop growth rate is measured as the total dry matter production over a period of time. It is proposed here to distinguish this from C by using the symbol CGR.

From the growth data obtained in the Nigerian field nursery and figures of the total hemispherical radiation, an estimate of carbon fixation was made (Rees, 1963b). It was found that, compared with temperate climate plants, the oil palm is comparatively inefficient, utilizing only 0.8 per cent of total radiation and 1.8 per cent of light energy. In terms of carbon fixation (assuming carbon to be 40 per cent of dry weight) the annual value is 3.3 g carbon per m^2 per day.

Some studies of seedling growth in 'polybag' nurseries have been carried out in Malaysia (OPGL, 1968–72). Net assimilation rates in the first year of growth were about the same as in Nigeria (0.23–0.25 g per dm^2 per week) and crop growth rates (C – biomass increase) have been similar. However, with older nursery polybag seedlings, at 12–18 months, the leaf area index for maximum C was found to be around 8, which is very considerably higher than shown in Fig. 4.1. The explanation may lie in the difference of age of the nursery palms since those in the Nigerian experiments were 5–10 months old with leaves at first at the bifurcate or partially fused stages whereas by 12 months leaves have become fully pinnate, there is greater light penetration and the canopy is gradually approaching that of the adult palm. One might expect therefore that L for maximum growth would also be nearer that of the adult palm. The trend of C with age remains to be studied in detail, however.

Before the introducton of 'polybag' nurseries a good deal of work was done in West Africa on water requirements and consumptive use in field nurseries (Wormer and Ochs, 1957; Rees and Chapas, 1963b). Normally, there was a sufficiency of rain falling on a nursery during the first 6 months, but thereafter, with the onset of the dry season, the decreasing precipitation needed to be augmented by irrigation. In the early part of the nursery period most of the water loss was, of course, from the soil, but later, when the young palms covered the soil and where mulch was applied, the loss was mainly from the plants. In Nigeria, data from an evapotranspirimeter showed that the annual consumption of water in an irrigated nursery was around 1,300 mm, and that the evapotranspiration rose to a peak of 152 mm per month in the late part of the dry season (Rees and Chapas, 1963b). Early in the dry season stomatal closure was a significant factor in restricting transpiration, and Ochs (1963a) suggested the use of stomatal opening measurements as a means of determining water need.

Work on similar lines to the above has not been done with polybag seedlings, but Quencez (1974, 1975) has estimated the water requirement per bagged plant at 4 mm per day for the first 2 months, rising to 10 mm after 6 months in the nursery. With poly-bags, frequency of application of water becomes as important as quantity, since the volume of available soil is so limited. Chin and Nair (1980) showed that week-old germinated seed in prenursery polybags receiving 500 cc water at various time intervals did not survive if the interval was more than 2 days. With month-old seed-lings there was a linear relationship between watering time interval and dry weights of both shoot and roots; but, as the effect on roots was the more pronounced, the root/shoot ratio decreased as the time interval increased. Application of the growth regulator Fusi-coccin on the leaves slightly reduced the water-stress effect of the longer time intervals.

Growth of the transplanted palm

Under drought the palm does not wilt. The leaf has a high pro-portion of lignified tissue and the cells of the epidermis have a thick cuticle and overlie a hyperdermis, which is more highly developed on the upper or adaxial surface. The stomata are on the lower or abaxial surface at a mean density of 146 per mm^2 (WAIFOR, 1956). They are semi-xeromorphic, that is to say they have a structure which is adapted for the prevention of desiccation over long periods. The guard cells of the stomata are thick-walled with external thick-ened ridges which lie pressed together for their whole length when the stomata close; at the same time subsidiary cells meet between the guard cells and the stomatal cavity.

Unwatered palms in drought conditions in a nursery show pronounced midday stomatal closure which prevents death. On transplanting to the field, however, root damage may be such that even the semi-xeromorphic structure of the leaves is insufficient to prevent drying out and subsequent death of the plant. Young plants transferred to the field in very dry weather often show a character-istic blackening of the youngest fully opened leaf and it may be presumed that this is because this leaf has not yet fully developed its protective structure. It has been noted that, in young palms, stomata on old leaves open more widely than those on younger leaves.

The main objective in transplanting, therefore, must be to transfer an intact, undisturbed and developing root system so that the transpiration stream may be returned to normality as soon as possible. The means of achieving this will be described in Chapter 9.

A well-transplanted palm will start to develop new leaves very soon after transplanting. With very large seedlings, these may initially be smaller than those last developed in the nursery; this is due to a temporary imbalance between the leaf and the root systems, but does not in any way invalidate the planting of large seedlings, as will be shown in Chapter 9.

As the root system of a recently transplanted palm is necessarily small, periods without rain delay its development. In soils retentive of moisture and where competition from surrounding covers is kept at a minimum, rainless periods of a few weeks may have little adverse effect, but on sandy or stony soils even in the non-seasonal climate of the Far East, growth may be checked and premature withering of the older leaves may be caused by quite short periods without rain. Of more serious consequence are the first few dry seasons in the seasonal climate of West Africa. Here again the defence of stomatal closure comes into play and some studies have been made of the process and causes of stomatal movement in young palms.

Normally, a sharp, light-induced opening of the stomata at dawn is followed by maximum opening from 09.00 hours to a little after 17.00 hours. The stomata then close equally rapidly at nightfall. Opening is most easily estimated by a modification of the Molisch infiltration method using mixtures of isopropanol and water as follows:

Solution number	1	2	3	4	5	6	7	8	9	10	11
% isopropanol	0	10	20	30	40	50	60	70	80	90	100
% water	100	90	80	70	60	50	40	30	20	10	0

When closed, solution 11 fails to penetrate the stomata and the degree of closure is said to be 12. When fully open to 4 μm diameter, solution 4 penetrates, penetration being detected with the naked eye by the appearance of scattered dark spots of intercellular flooding. The degree of opening may be recorded on the 4–12 scale or this may be transformed to percentages.

Towards the end of a dry season, the typical curve of opening (see Fig. 4.2) is altered by pronounced though partial closure during the midday period. This is followed by a short period of reopening before final closure. During the transition from dry to wet season, slight midday closure occurs but this is not evident at the transition from wet to dry season as the effect of the drought does not show itself until well into the dry season. Rees (1957) showed that late in the dry season in Nigeria opening will drop to 25 per cent of the maximum possible and rise again to around 45 per cent when wet weather sets in.

Studies of stomatal movement and of aperture have been made in relation to evaporation, transpiration, soil humidity, light intensity and temperature. During the opening period early in the day and the closure period in the evening, aperture is positively correlated with log light intensity. The extent of midday closure is correlated with total daily evaporation and this relationship continues after rain even though the level of closure is reduced. The

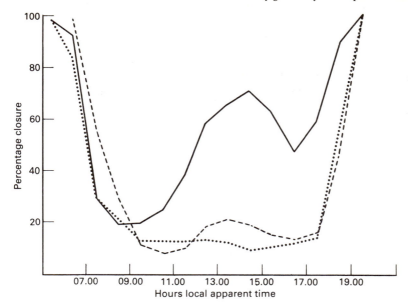

Fig. 4.2 Typical daily curves of stomatal movement (Rees, 1961): (———) dry season; (. . .) wet season; (---) intermediate.

relationship of aperture with soil humidity and transpiration follows the expected course: soil desiccation is accompanied by reduced transpiration and reduced aperture. In pot studies in Benin (Dahomey) transpiration has been shown to fall to a very small quantity as the permanent wilting point is reached, but to rise to its maximum, depending on the hour of the day, at a point somewhat nearer to the wilting point than to field capacity (Wormer and Ochs, 1959).

A relationship between midday stomatal closure and temperature has been found in 4-year-old palms in the West African dry season (Rees, 1958, 1961). Watered and unwatered palms were used, and the aperture-temperature curves were not the same in each case, showing that two factors, temperature and water strain, influence midday closure (Fig. 4.3). Closure is marked between 30 and 35 °C shade temperature under the palms. It appears that when water is not limited there is some closure at high temperatures, but when soil moisture falls below a critical level there is a more sudden increase of closure which disappears following watering. Two special circumstances influence these relationships. Firstly, watering at the end of the dry season does not reduce the extent of the closure from the 'dry' to the 'wet' values; this is believed, from observations, to

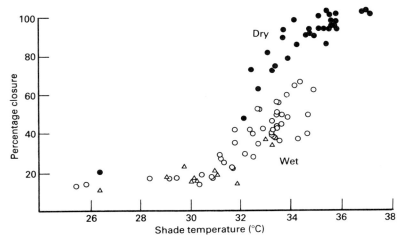

Fig. 4.3 Midday stomatal closure in watered and unwatered palms, and air temperature under the palms: (●) a palm under dry conditions; (○) a palm under wet (watered) conditions during the same dry-season period; (△) the first palm, previously under dry conditions, under wet conditions later in the season (Rees, 1961).

be due to the death of much of the root system in the upper layer of the soil during a long dry season. Secondly, low temperatures with very low humidity, as experienced in the 'Harmattan' period in West Africa, cause closure.

It is clear from the above that transpiration in the young palm can be much reduced in areas subject to a dry season. This inevitably delays growth and development, and it is perhaps surprising that there is such comparatively little difference between seasonal and non-seasonal regions of similar total rainfall in the development of the young palm and in the length of time it takes to come into bearing. Planting early in the season by sound techniques helps a plant to establish a deep root system by the time the dry season arrives, and this will clearly reduce the period during which the plant will suffer midday water stress. It is probably this fact more than any other that has enabled plantations in West Africa to be brought into bearing some $1\frac{1}{2}$ years earlier than previously, though a largely unconscious selection of planting material of precocity must not be overlooked.

While much can be done to combat the effects of successive dry seasons, a more uniform high rainfall will always be more conducive to a sustained high growth rate, and soils retentive of the rainfall they receive are to be recommended. Ochs (1963b) has suggested that such soils can be detected by tests of stomatal closure and has shown that stomatal movement is affected by the type of cultivation employed. At Pobé a relationship was established between production from a number of soil types and a 'drought index' based on the number of days in the dry season on which the stomata reached a certain degree of closure. It was also shown that a *Pueraria* cover caused earlier and more complete stomatal closure than was found in plots cultivated with maize or left bare of cover during the dry season. In a root-cutting experiment in the Ivory Coast it was shown that stomatal closure is related to the quantity of lateral primary roots severed (Ruer, 1969). This emphasizes the importance of water supplies to the oil palm and suggests that the leaf area of a cover should be reduced before the dry season and that inter-row cultivation must not be allowed to damage the roots.

It has been shown that midday stomatal closure can occur in the more even climate of Malaysia and, as in Africa, the degree of closure depends both on water stress and temperature. Corley (1973a), who measured stomatal resistance to diffusion of water vapour, estimates that the light-saturated rate of photosynthesis is reduced by over 75 per cent when stomatal resistance exceeds 20 sec per cm, and he considers that yields might be reduced by 10 per cent if stomatal closure occurred on 40 days in the year. He also showed that irrigation significantly reduced stomatal resistance.

The rapid growth that can be attained by the transplanted palm under the best conditions of soil and climate is exemplified by data from the coastal alluvium of Malaysia (Corley *et al.*, 1971a). Table 4.2 shows the total dry weights, dry matter production and growth parameters of transplanted palms in the period between transplanting and coming into bearing. Crop growth rate (*CGR*) here is the total organic matter incorporated and not, as with nursery seedlings, a biomass change.

It will be noted that these palms already had a small production of bunches 30 months after transplanting to the field. Although the leaves did not yet overlap at 30 months leaf area index was already 3.5.

The growth of the adult palm

A study of the growth of the adult palm is of particular importance in view of the large differences in bunch yields found between regions where the palm is planted as an economic crop. Superficial observation does not suggest that there is any obvious association, from region to region, of leaf production with yield. Well-grown fields of palms in Africa yielding less than half the bunch weight of fields in the Far East often give the appearance of having a similar or even denser canopy at the same spacing. Studies of growth and dry matter production in the adult palm were first carried out in

Table 4.2   *Dry weights and growth of transplanted palms on coastal alluvium in Malaysia*

Months from transplanting	Leaves	Trunk	Roots	Bud	Male flowers	Bunches	Total
	Mean total dry weights: kg per palm						
18	54.0	4.7	8.8	2.8	—	—	70.3
30	132.7	13.8	15.9	5.9	—	—	168.3
	Dry matter incorporated: kg per palm per annum						
18	40.4	4.0	8.0	2.5	—	—	54.9
30	96.1	9.1	7.1	3.1	2.3	43.4	161.1

	Leaf area (m^2)	E_A (g per dm^2 per wk)	L	CGR (tonnes per hectare per annum)
	Growth parameters (density 148 per hectare)			
18	77.4	0.137	1.14	8.10
30	235.0	0.132	3.47	23.8

Source: Corley *et al.* (1971a).

Nigeria and later work in Malaysia shed light on the differences of production between the two regions.

For the purpose of growth studies the palm may conveniently be divided into:
1. Leaflets;
2. Leaf rachises including petioles;
3. Unopened spear leaves;
4. Growing point or 'cabbage';
5. Trunk;
6. Roots;
7. Inflorescences and fruit bunches.

Accurate measurements of these organs have been made for palms of different ages in both West Africa and Malaysia and the rate of dry matter production, crop growth rate and other growth parameters have been estimated from the data (Rees and Tinker, 1963; Corley *et al.*, 1971a). Some of the results obtained are compared in Tables 4.3 and 4.4. For obvious reasons destructive methods cannot be widely employed, but Corley *et al.* evolved non-destructive methods of estimating annual dry matter production which can be used in comparisons of growth between progenies, environments or agronomic treatment, and which make possible a considerable extension of growth analysis methods in oil palm investigations. These methods cover only leaves, trunk and bunches, but these organs were found to constitute over 96 per cent of total annual dry matter production. High correlations found between

Table 4.3 *Dry weight of vegetative parts and leaf area of producing palms of different ages in Nigeria and Malaysia (kg per palm)*

Age	Country	Leaves	Trunk	Spear and cabbage	Roots	Total	Leaf area per palm (m^2)
7	Nigeria	67.4	82.3	4.5	—	(154.2)	202.8
6.5	Malaysia	118.4	78.9	8.8	40.6	246.7	238.9
10	Nigeria	90.7	171.0	5.9	—	(267.6)	254.6
10.5	Malaysia	161.1	145.3	10.8	49.0	366.2	330.3
14	Nigeria	86.0	238.7	4.5	—	(329.2)	252.0
14.5	Malaysia	168.2	232.6	11.3	68.9	481.0	309.5
17	Nigeria	95.0	280.0	10.7	128.0	402.4	297.0
17.5	Malaysia	140.2	290.7	11.9	61.6	504.4	347.9
20	Nigeria	150.0	439.0	9.0	—	—	426.8
22	Nigeria	111.4	389.0	10.8	—	—	448.0
27.5	Malaysia	115.4	300.5	8.8	130.8	555.5	251.9

Sources: Rees and Tinker (1963); Corley *et al.* (1971b).

Table 4.4 *Annual dry matter production, net assimilation rate (E), leaf area index (L) and crop growth rate (CGR) of palms of different ages in Nigeria and Malaysia*

Age	Country	Annual DM production, per palm							E	L	CGR
		Leaves (kg)	Trunk (kg)	Roots (kg)	Bud (kg)	Male flowers (kg)	Bunches (kg)	Total (kg)	(g per dm^2 per week)		(t per ha per annum)
7	Nigeria	55	—	—	—	—	25	—	—	3.0	—
6.5	Malaysia	109	18	2	1	6	100	236	0.189	2.9	28.6
10	Nigeria	62	31	—	—	—	35	128	0.097	3.7	18.8
10.5	Malaysia	136	21	2	1	6	108	274	0.156	4.0	33.4
14	Nigeria	54	19	—	—	—	29	102	0.077	3.8	15.0
14.5	Malaysia	116	22	5	0	6	90	239	0.149	3.8	29.2
17	Nigeria	66	22	—	—	—	48	136	0.088	4.4	20.1
17.5	Malaysia	119	19	5	0	6	108	257	0.142	4.2	31.3
20	Nigeria	86	18	—	—	—	30	134	0.068	5.6	19.4
22	Nigeria	84	17	—	—	—	28	129	0.077	4.8	19.0
27.5	Malaysia	92	1	5	0	6	48	152	114	3.1	18.2

Sources: Rees and Tinker (1963); Corley *et al.* (1971a).

certain measurements and the actual dry matter content of the various organs led to the following formulae being adopted:

Leaf area, L = b (n × lw), where *n* = number of leaflets, *lw* = mean of length × mid-width for a sample of the largest leaflets, and *b*, the correction factor = 0.55. This factor was found to range from 0.51 to 0.57 with age groups from 1–2 to 8–11 years (Hardon *et al.*, 1969), but for most comparisons the figure of 0.55 is used.

Leaf dry weight, kg, W = 0.1023*P* + 0.2062, where *P* = petiole width × depth in cm². The width and depth of the petiole are measured at the junction of the rachis and petiole, i.e. the point of insertion of the lowest leaflet.

Trunk dry weight, Tr = *VS*, where *V* is the volume of the trunk and *S* is the density of dry trunk in g per litre. The annual growth increment of a trunk is represented by a cylinder the volume of which is estimated from its diameter, after removal of the leaf bases, and its height. Thus $V = \pi (d/2)^2 h$, while the density $S = 7.62T + 83$, where *T* is the age of the palm in years from transplanting. While this method is accurate for measuring dry weight increases it is not directly applicable for measuring the total dry weight of a palm since the base is conical and the density of dry trunk has been found to increase from apex to base.

Bunch dry weight, D = 0.5275*F*, where *F* is fresh bunch weight in kg. There may be some difficulty in applying this formula owing to differences of fruit-to-bunch ratio and the effect of the latter on percentage dry matter. Where large differences in fruit-to-bunch ratios exist the formula $D/F = 0.3702X + 0.2865$, where *X* is the fruit-to-bunch ratio, may be used.

For use in treatment or progeny comparisons there is no reason to believe that the above formulae will not have general application. Nevertheless for greater accuracy the correction factors should be checked and adjusted before adopting them in other parts of the world.

From the gross weight of plant parts, either directly measured or estimated through the above formulae and from records of bunch and leaf production and trunk height increment, it is possible to estimate the annual production of leaf, trunk and bunch dry matter. Measurement of root production presents some difficulty. Increases in root dry weight have been estimated in Malaysia from estimates of the total weight of roots of palms of different ages, but this takes no account of the production of roots which have 'replaced' old roots which have died. Male inflorescence production can be estimated from inflorescence counts and sample dry weights. There will also be a small increase in spear and cabbage ('bud') weight until the palm reaches 15–20 years of age. The sum of all these yield components in terms of dry matter can then be related to leaf area to give the net assimilation rate, *E*, or to the land area to give the crop growth rate, *CGR*. Table 4.4 shows the annual dry matter production of the Nigerian and Malaysian palms analysed in Table 4.3.

Figures from Malaysia and Nigeria have been presented together in Tables 4.3 and 4.4 to show the differences of dry matter content and incorporation and of growth parameter values that may be found between palms on areas with marked deficits and those on first-class soils with little or no water deficit. The figures must, however, be viewed with caution and some further differences between the two areas chosen for sampling should be emphasized. Firstly, in the Nigerian area the density was 147.8 per hectare (29 ft triangular) while in the Malaysian area it was 121.3 per hectare (32 ft triangular); secondly, the area containing the older Nigerian palms was shown to be potassium deficient.

From Table 4.3 it will be noted that trunk dry weights were similar throughout. The dry weight of leaves was markedly higher in Malaysia, but two factors must be taken into account. Though certain other work does not suggest this, it is possible that wider spacing may have favoured greater leaf dry matter production. Secondly, differences of pruning policy may have affected the comparison. The large difference of root dry weight of 17-year-old palms may partly be attributable to differences of extraction techniques used or to soil differences; with the closer spacing and lower overall productivity the Nigerian root system might have been expected to be of lower total weight per palm.

Turning to dry matter production per annum (Table 4.4), account must again be taken of the differences of density. Incorporation of dry matter in the trunk was higher with the Nigerian palms both on a per palm and a per hectare basis. Leaf production was considerably lower on a per palm basis but differences were much less on a per hectare basis particularly at the higher ages. Bunch dry matter production was two to four times higher in the Malaysian palms on a per palm basis and up to three times higher on a per hectare basis.

With the higher density, leaf area indices (L) tended to be higher with the Nigerian palms, but the lower dry matter production, particularly in bunches, gives rise to considerably lower net assimilation rates (E) and crop growth rates (CGR). Among the Nigerian palms the highest L value recorded was 6.2 in a 20-year-old palm.

In Malaysia, correlations were found between L and both CGR and bunch yield in a study with twelve progenies (Hardon *et al.*, 1969), and it was suggested that one method of increasing yields would be to promote a higher L by higher-density planting (Corley *et al.*, 1971a). Experience with early experiments shows however that there are obvious limitations to this method. Rees (1963c) pointed out that bunch or oil yield could be increased either by increasing CGR without change in the proportions of bunch in the total crop increase, or by increasing the proportion of bunch without change in *CGR*, and in recent growth analysis in Malaysia interest has centred largely on the apportionment of total dry matter production between bunches and the vegetative organs (leaves and trunk). Comparisons between Malaysian and Nigerian data have shown that the large differences in *CGR* are mainly attributable to incorporation of larger quantities of dry matter in bunches and that the vegetative dry matter (*VDM*) production is broadly similar (Ng Siew Kee *et al.*, 1968; Corley *et al.*, 1971a). In Table 4.5 the *VDM*, *CGR*, *E*, *L* and bunch index (*BI*) of various groups of palms are compared. The table includes data from a density experiment in Malaysia which has helped to throw further light on the growth pattern of the oil palm (Corley, 1973b).

Under the good growing conditions of Malaysia *CGR* increases rapidly to about 30 tonnes per hectare per annum at around 7 years at usual plantation spacings, but at high density a *CGR* of up to 40 tonnes per hectare per annum is attained. However, it has been shown that this increase of *CGR* with density is achieved through the constancy of *VDM* per palm, and the higher *VDM* per hectare more than compensates for decreasing bunch dry matter (*Y*) per palm until a ceiling is reached at densities of over 300 palms per hectare with *L* reaching 10–12 (Fig. 4.4). While *CGR* increases curvilinearly with density, *VDM* per hectare increases linearly over the same range.

Two other important facts about *VDM* production have emerged. Firstly, as will be seen from Table 4.5, there is little evidence of lower *VDM* in areas of lower *Y*. Secondly, pruning experiments in Malaysia have shown that *VDM* is not reduced even by severe pruning (OPGL, 1968–72). It has therefore been postulated that the palm first satisfies its requirements for vegetative growth and only when conditions make additional assimilation possible can 'excess' dry matter be diverted to bunches. Adverse conditions for bunch production, whether these be water deficits as in Nigeria or high densities in which the palms are in severe competition with each other, result in a lower *E*, and this is reflected in a lower *BI*, which is the ratio of *Y* to *Y* + *VDM*. Thus in any particular situation the optimum density for *CGR* will be higher than for *Y* (Fig. 4.4).

When considering *CGR* and photosynthetic efficiency the high energy content of oil synthesized in the bunches must be taken into account. While the highest density will produce the highest *CGR*, this density will not provide the highest photosynthetic efficiency since the dry matter produced contains a relatively low proportion of oil-bearing bunches. Thus for 7-year-old palms in Malaysia when *L* at optimum *CGR* was 9.6, *L* at optimum photosynthetic efficiency was only 8.3 (Corley, 1973b).

At high densities leaf area per palm is reduced. This is due to a reduction in the number of leaves per palm, probably because

Table 4.5   *Bunch and vegetative dry matter (Y and VDM) production, bunch index (BI) and growth parameters of groups of palms in Nigeria and Malaysia*

Country	Age	Density (palms per hectare)	Y (kg per palm per annum)	VDM*		CGR (tonnes per hectare per annum)	BI	L	E (g per dm² per week)
				(kg per palm per annum)	(tonnes per hectare per annum)				
Nigeria	10–17	147.8	37	85	12.6	18.0	0.30	3.9	0.087
Nigeria	20–22	147.8	29	103	15.2	19.5	0.22	5.2	0.073
Malaysia (coastal)	6.5–17.5	121.3	103	132	16.0	28.5	0.44	3.6	0.160
Malaysia (coastal)	27.5	121.3	47	93	11.2	17.0	0.34	3.1	0.114
Malaysian progenies on inland soil									
Low *BI*	10	138.1	44	124	17.2	23.2	0.26	4.1	0.112
High *BI*	10	138.1	122	101	13.9	30.8	0.55	5.1	0.117
Mean of 10	10	138.1	93	107	14.8	27.6	0.46	4.7	0.115
Malaysian density experiment on inland soil									
Low density	7	112	105	88.4	9.9	21.7	0.54	3.9	0.107
Normal density	7	145	103	93.1	13.5	28.4	0.53	5.1	0.106
High density	7	184	82	93.5	17.2	32.3	0.47	6.5	0.096
High density	7	227	59	89.0	20.2	33.6	0.40	7.2	0.090

* Leaf and stem DM only.

Sources: Rees and Tinker (1963); Corley *et al.* (1971a); Corley *et al.* (1971b); Corley (1973b).

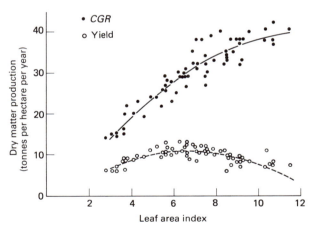

Fig. 4.4   Crop growth rate, yield and leaf area index of 6-year-old palms in Malaysia, the points being plot means and lines fitted to quadratic curves (Corley, 1973b).

leaves at the base of the canopy die when shaded below their photosynthetic compensation point. Leaf area per leaf and leaf area ratio (*F*), however, are relatively constant with density though varying with age. A relationship has been found between optimal density for current yield and leaf area per leaf and this is discussed in Chapter 9 (p. 401) when the practical aspects of determining density are considered. Corley (1973b) has presented diagrammatically the relationship between L and CGR, Y and VDM per hectare and concludes that Y could be increased by increasing E, through changes in photosynthetic rate or canopy structure, or by changing the VDM/L linear relationship as shown in Fig. 4.5. It will be seen that if the VDM/L relationship is depressed from line 1 to line 3, both Y and the L for optimum Y are increased. VDM per unit L is related to leaf area ratio (F – taken as new leaf area produced (m^2) per kg VDM), so if VDM is reduced both F and BI will increase with the increase in yield which is shown in Fig. 4.5. Corley concludes that VDM, F and E are therefore the important parameters through which yield is determined. The exploitation of these findings in the improving of yields by selection and breeding is discussed in Chapter 5 (p. 275), but it will be plain from the above that BI will be an important parameter for bunch yield selection. Breure (1986) considers that reducing the period of rapid VDM increase (i.e. selecting palms with rapid leaf expansion to their maximum) may be equally important. In a density experiment in Papua New Guinea in which he estimated CGR, VDM and Y, he also calculated, as canopy efficiency (e), the assimilates incorporated above ground per unit of intercepted radiation. From these data he concluded that the form of the canopy, which is much affected by

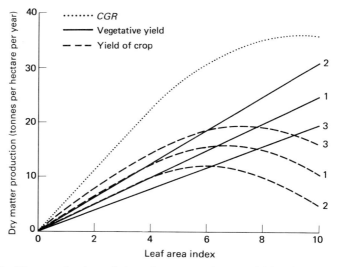

Fig. 4.5 Diagram showing relation of crop growth rate, yield and vegetative dry matter production to leaf area index, and the effect of changing VDM with constant CGR (7-year-old palms – Corley, 1973b).

density, would influence light interception and that the latter's increase or decrease would have different effects on gross photo-synthetic production in different density ranges (see pp. 250 and 280).

Growth analysis is now being employed in the interpretation of fertilizer responses. Results from Malaysia suggest that, in general, fertilizers increase *CGR* through increases in *L* (Corley and Mok, 1972). This results in an increase of both *VDM* and *Y* with no alteration in *BI*. However, nitrogen also increases net assimilation rates and the number of new leaves per palm but may reduce *F*. The lack of fertilizer effect on *BI* is of particular interest in view of the fact that this factor is strongly influenced both genetically and by water deficit.

The responses of *CGR* to fertilizers without change of *BI* has stimulated interest in the relationship of *VDM* to *Y* (perhaps as an alternative to leaf analysis). It was noticed that *VDM* reaches around 130 kg per palm per annum in the highest yielding plots of fertilizer experiments and on the best coastal soils in Malaysia (see Table 4.5). It was also shown that in fertilizer experiments correlations between *VDM* and *Y* can be as high or higher than correlations between leaf nutrient levels and yield (Lo *et al.*, 1972). However, the relation between *VDM* and *Y* at the higher levels of *Y* is not convincing and much more work will be required before use can confidently be made of this relationship. Growth analysis in its practical relation to nutrition and density will be discussed further in Chapter 11 (p. 534).

Further work on the growth of the oil palm is likely to involve interactions between genotypes, densities and fertilizers and different climatic conditions.

Perhaps the most remarkable feature of the growth of the oil palm is its leaf area duration (*D*). This is the integral of the leaf area index over the growth period; it is used as a measure of photosynthetic potential, and is measured in units of time. It thus takes into account both the magnitude of the leaf area and its persistence in time. Watson (1947) describes it as a measure of the ability of the plant to produce and maintain leaf area and hence of its whole opportunity for assimilation, and he points out that in conditions of constant net assimilation rate dry matter accumulation would be proportional to leaf area duration.

With perennials with a fairly constant leaf area index, as in the oil palm, *D* is calculated by multiplying *L* by the 52 weeks of the year. Thus when the *L* of normally spaced plantation palms is about 5.0, *D* will be 260. This is much higher than any value obtained from other crops, though values for other evergreen perennials come closest to it.

Mention has been made of photosynthetic efficiency in a Malaysian density experiment. The data were based on estimates, from hours of bright sunshine, of total radiation of 376, 383 and 377 cal per cm^2 per day in the 3 years studied (Corley, 1973b). On this basis the highest photosynthetic efficiency achieved by any treatment was 3.2 per cent of photosynthetically active radiation (taken as 40 per cent of total). However, the efficiency at optimum oil yield per hectare was around 3 per cent for 7-year-old palms, still considerably higher than an estimate of 1.2 per cent for adult palms in Nigeria (Rees, 1963b). Comparisons between West Africa and Malaysia remain somewhat unsatisfactory, however, owing to the variation of radiation estimates. An estimate of 470 cal per cm^2 per day has been used in Malaysia for coastal areas (Corley *et al.*, 1971a), giving an efficiency of only 2.2 per cent, while in Nigeria the mean of 6 years of Kipp Solarimeter measurements was 360 cal per cm^2 per day with very considerable variations between months (see p. 109). It may reasonably be concluded, however, that where growing conditions are good the efficiency of conversion of light energy absorbed is as high as in most agricultural crops.

Leaf production in the adult palm in relation to climate

Adult palms in the seasonal climates of West Africa produce from 18 to 27 leaves per annum. Early records from Malaysia gave 20 as a normal leaf production for 10-year-old palms (Bunting *et al.*, 1934). Annual production was recorded as 16–20 for Cameroon, 20–24 for Sumatra (Fickendey and Blommendaal, 1929) and 18–26 for Zaire (Beirnaert, 1935).

In West Africa and in southern Zaire the dry season is sufficiently intense to restrict the opening of leaves. Whereas the normal rate of opening is two per month, opening may be halted for 2–3 months and six or more leaves may elongate in the crown but fail to open. This is a common sight in West Africa at the end of the dry season, but is only rarely seen in the Far East or America. While there is no delay in the laying down of leaf primordia to correspond with the dry-season delay in leaf opening, there is some evidence that, within West Africa as a region, annual leaf production is low in areas with an annual rainfall as low as 1,250 mm. In Nigeria, palms of similar genetic origin gave the following averages of leaf production for the thirteenth to nineteenth year after planting:

	Place		
	Umuahia	Ogba	Ibadan
Average annual rainfall (mm)	2,108	2,032	1.219
Average annual leaf production	23.1	23.1	20.5

Flowering and yield in the oil palm

Flowering

Although many early observations were made concerning the seasons of highest or lowest production, the first study of the flowering and production of the oil palm in relation to climatic factors and leaf production was that of Mason and Lewin (1925) in Nigeria. Unfortunately their study covered periods of only $1\frac{1}{2}$ and 3 years at a centre (Moor Plantation, Ibadan) of low rainfall. However, these workers were able to recognize the importance of sex ratio (ratio of female to total inflorescences, though Mason and Lewin used male to total inflorescences in their paper). They also noted that inflorescence abortion took place between the central spear stage and anthesis and was most pronounced in the dry season. A number of much more comprehensive studies have now been carried out, first in Africa, later in Malaysia. There is a great need for flowering studies in tropical America since the palm is grown there under greater extremes of rainfall, temperature and daylength than elsewhere.

Leaf development

The progress of leaf and inflorescence development is shown by dissection of the adult palm. Forty to fifty leaves are found between the stage of initiation in the bud and the central spear stage (the stage at which the leaflets are starting to open at the top of the spear leaf in the centre of the crown). Palms in Nigeria with forty-five to fifty leaves before the spear had an annual production rate of twenty-two to twenty-four leaves; 2 years therefore elapsed between a leaf's initiation and its becoming a spear leaf (Broekmans, 1957). As the palm may carry up to forty fully grown leaves, $3\frac{1}{2}$–4 years may elapse between initiation and death. In non-seasonal climates leaves will open regularly at the rate of about two a month. There is evidence that development is more rapid with younger bearing palms in favourable climates. Four-year-old palms in Malaysia had only thirty-six to forty-four leaves younger than the spear (OPGL, 1968–72).

In seasonal climates, as mentioned earlier, opening is delayed by the dry season and leaf production is then severely reduced. What occurs within the crown can be seen from Fig. 4.6 (after Broekmans). Normally, very little growth is made by the leaf until it reaches the stage of being sixth or seventh in order from the spear leaf when it may be said to be still in phase 1 of its development. Phase 2 is a stage of very rapid development from a length of a few centimetres to a fully grown spear leaf of, perhaps, 7 m in length. There will be six or seven leaves in this stage of active growth with

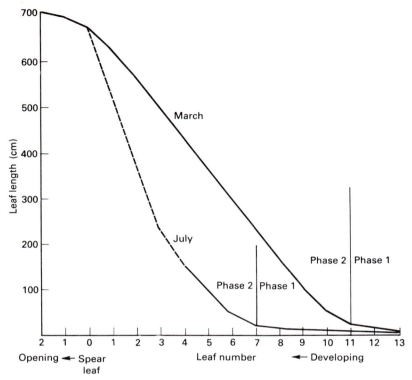

Fig. 4.6   Curves of leaf development in the crown in the dry (March) and wet (July) seasons in Nigeria (Broekmans, 1957).

a spear leaf opening every fortnight. This is illustrated by the July curve. After several months of drought the central spears will cease to open normally and leaves which would have passed over into the fully opened state if conditions had been favourable will remain in phase 2. Those leaves in phase 1 are small and drought conditions do not therefore affect their development. They pass over from phase 1 to phase 2 at the usual rate, but their growth rate becomes progressively retarded. Though the ones next to the spear may elongate to nearly its length, they, like the spear, remain 'waiting' and unopened until rain falls, when several leaves will open in quick succession. As will be seen from the March curve in Fig. 4.6, up to ten leaves may accumulate in phase 2; while the usual number per month have entered phase 2, none have left it, to become opened leaves, for several months

Leaf production determines potential bunch production and factors affecting leaf production will affect actual bunch production.

Dissection of adult palms has shown that an inflorescence primordium is produced in the axil of every leaf, and that normal though slow development takes place at least up to a time between the spear-leaf stage and the stage of anthesis.

With regard to seedlings, Beirnaert (1935) claimed that leaves produced in the first 2 years were an exception to the rule that all leaves have an inflorescence bud. Dissection has shown, however, that inflorescence buds are present in every leaf axil (OPGL, 1968–72). Though the first primordia must normally abort, the production of normal inflorescences after a palm has been only 1 year in the field indicates that inflorescence buds are being extensively produced in the axils of leaves in the nursery and that the whole process of development is much faster than with the adult palm.

Leaf production in a young palm increases rapidly to a maximum before falling off to the levels already mentioned (Sparnaaij, 1960). Examples of leaf production in *dura* fields planted in Nigeria in 1940 and 1941 are given below:

Age in years	2	3	4	5	6	7	8	9	10	11	12	13
1940 planting	18	27	27	29	29	28	25	23	23	24	21	—
1941 planting	—	—	27	30	31	28	25	24	25	23	23	23

In Malaysia, leaf production of Deli palms increases to thirty-two to thirty-three per annum at 3–4 years, levelling off at between twenty-two and twenty-six per annum at 7–8 years of age in the field. In Nigeria, *tenera* palms within a progeny open up to 3 per cent more leaves per annum than the *dura* palms while *pisifera* palms show a similar increase over *tenera*. In Malaysia the tendency of *tenera* palms to open more leaves has also been noted.

Inflorescence initiation

The inflorescence bud is microscopically visible in the axil of the fourth leaf from the apex (Henry, 1955). It is therefore initiated less than 2 months later than the leaf and its development is subject to all the same influences. The inflorescence bud will reach the central spear stage after 18–24 months and a further 9–10 months will elapse before the flowers open (anthesis). Initiation to anthesis therefore takes about 27–35 months and to fruit ripening 33–40 months. Although the rate of inflorescence initiation is the same as rate of leaf initiation, over short periods the number of inflorescences reaching anthesis may, owing to varying rates of development, exceed the number of new leaves opening (Corley, 1977).

Sex differentiation, sex ratios and inflorescence abortion

Sex differentiation is the most important process in the development of the oil palm and it is interesting to note that botanists were already speculating on the secrets of this process many years ago. O. F. Cook (1940) wrote:

> In the life of a dioecious plant a single determination of sex may be supposed to take place, and that at the beginning of embryonic development, as with the higher groups of animals, while in a monoecious plant like *Elaeis* or *Alfonsia* [*E. oleifera*] a series of sex-determinations is called for, in advance of the development of the successive inflorescences. . . . The production of two kinds of inflorescences in the monoecious palms might also be considered as an alteration of sex and from that point of view may be worthy of special study. . . . It seems not impossible that such alterations might be influenced by external conditions or by fertilizers applied to the soil. . . .
>
> As in other groups of plants, the changes from bisexual to unisexual flowers presumably are accomplished by suppression or reduction, the pistils being suppressed in forming male flowers, and the stamens suppressed in forming female flowers. . .
>
> Sexual specialization of the palms is accompanied by simplification and reduction of inflorescences forming greater protection of the reproductive system. . . . The palms that have the inflorescence reduced to a cluster of short simple branches, as in *Elaeis* and *Alfonsia* [*E. oleifera*], would represent rather advanced stages of specialization.

High yield demands the differentiation of a high proportion of female inflorescences. If the stage at which differentiation takes place is known, the factors influencing it can the more easily be sought. The time of sex differentiation is, however, only gauged by inference. Eye examination of buds dissected out shows only that sex is determined at least 12–18 months before anthesis; determining a time of sex differentiation earlier than this is more difficult, and it may of course vary. Broekmans (1957) argued that

> as inflorescences reach the anthesis stage in approximately the same order in which they were initiated, they have also passed through the stage of sex-differentiation in the same sequence. Therefore the seasonal changes in the sex-ratio of inflorescences at anthesis reflect the conditions which have prevailed during the stage of sex-differentiation.

Now it so happens that in the seasonal climate of Nigeria changes in sex ratio are well marked (see Fig. 4.7), a peak of sex ratio occurring during the dry and relatively sunny months of January to March and a fall occurring during the second half of the year when there is constant rain and overcast conditions. Thus high sex ratio and conditions of high light intensity correspond.

Several attempts have been made to interpret these changes. Beirnaert (1935) suggested that the ratio of photosynthesis to mineral uptake was a determinant, high levels of photosynthesis being expected during periods with much sunshine which, in Africa, would be during the dry season and at the beginning and end of the wet season. This hypothesis was based largely on the observation that shading or pruning reduced sex ratio.

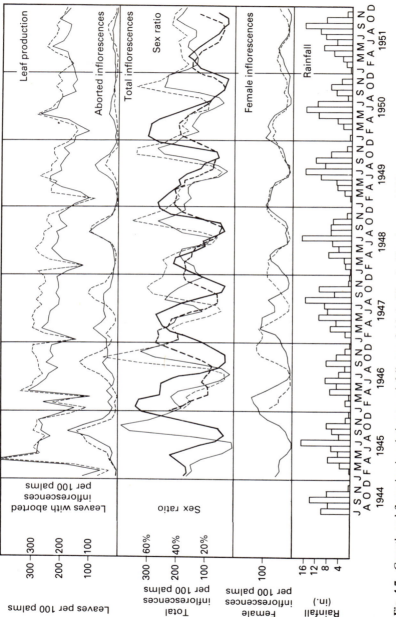

Fig. 4.7 Growth and flowering in relation to rainfall: two fields in Nigeria (Broekmans, 1957).

Broekmans (1957) developed this hypothesis a stage further by pointing out that, as his dissections indicated that sex differentiation must occur some time between initiation (at around 33 months prior to anthesis) and 18 months before anthesis, the most likely time would be 24 months, since only under these circumstances would the supposed effect of high sunshine in inducing a high sex ratio be seen in flowering records 2 years later; and, as mentioned above, he demonstrated (Fig. 4.7) that periods of high sex ratio at anthesis did in fact consistently fall in the dry season when sunshine was at its maximum. Broekmans further considered that the severity of the dry season would be in opposition to the beneficial influence of high sunshine and that therefore if rainfall in the dry season was very low this would be reflected in low maxima of sex ratio 2 years later. He was able (Table 4.6) to demonstrate in two fields in Nigeria that dry-season rainfall was positively correlated with both maximum and average sex ratio in the sex ratio cycle 2 years later. A correlation was also found with number of female inflorescences. The sex ratio cycle used in these determinations was from one minimum to the next (see Fig. 4.7) and the period varied from 10 to 14 months. In this Nigerian work it was assumed that abortion of inflorescences would not affect the apparent sex ratio as determined at anthesis; it will be shown below, however, that this assumption was not justified.

Corley (1977) gauged the timing of sex differentiation in Malaysia

Table 4.6 *The relationship between dry-season rainfall (November to April) and maximum and average sex ratios and number of female inflorescences per sex ratio cycle 2 years later*

Rainfall, Nov.–April			Sex ratio cycle	Experiment 2–1			Experiment 6–2		
				Sex ratio		♀ Inflorescences	Sex ratio		♀ Inflorescen
Period	(in)	(mm)	Period	Max. (%)	Av. (%)	No.	Max. (%)	Av. (%)	No.
1941–2	15.4	391	1943–4	75.2	60.6				
1942–3	12.1	307	1944–5	43.3	22.5	3.8			
1943–4	15.3	389	1945–6	65.0	38.1	8.4			
1944–5	12.0	305	1946–7	55.6	28.0	7.6			
1945–6	12.0	305	1947–8	42.4	25.8	4.1	37.3	27.1	6.5
1946–7	16.6	422	1948–9	53.8	32.8	6.4	44.4	28.1	7.2
1947–8	14.0	356	1949–50	60.7	29.4	6.1	39.6	23.7	5.5
1948–9	9.1	231	1950–1	33.7	18.2	3.0	31.3	18.6	4.0
1949–50	9.3	236	1951–2	33.3	17.9	3.2	29.7	20.9	4.2
1950–1	14.4	366	1952–3				48.0	27.9	5.9
1951–2	13.5	343	1953–4				32.4	21.6	4.7
1952–3	16.2	411	1954–5				36.6	24.1	4.8
Corr. coefficient, rainfall: sex ratio or ♀ inflorescences				0.83 $P =$ 0.01	0.74 $P =$ 0.05	0.75 $P =$ 0.05	0.73 $P =$ 0.05	0.70 $P =$ 0.1	0.63 $P =$ 0.1

Source: Broekmans (1957).

by dissection and from experiments with pruning and thinning treatments, and he concluded that it occurred between 16 and 24 months before anthesis and that the shorter intervals, i.e. less than 24 months, were associated with high rates of leaf production, particularly in young palms.

Investigations of floral abortion have been carried out in Zaire, Nigeria and Malaysia. In the axil of leaves which have not produced an inflorescence at anthesis the remains of an inflorescence can be found. The stage of abortion is nearly constant; in an investigation in Nigeria 80 per cent of abortive inflorescences in African palms measured 6.0–12.9 cm, in Deli palms 5.0–11.9 cm. Deli palms in West Africa show a much higher abortion rate than African palms, but there are large progeny differences. Figure 4.8 shows the progression in length of inflorescences subtended by the leaves of two Deli palms. The aborted inflorescences, it will be seen, are all of similar size. It is clear that the time of abortion coincides with the time at which the inflorescence is just starting to develop at a rapid rate. This is the critical stage in inflorescence development, and its determination in time makes it possible to seek the causes of abortion. The curve in Fig. 4.8 shows that the critical stage is entered at the eighth or ninth leaf from the central spear, anthesis being reached at about the seventeenth to twentieth leaf. As the central spear stage to anthesis occupies 8–10 months, and leaves are being produced at the rate of about two a month, the critical stage is $4\frac{1}{2}$–$5\frac{1}{2}$ months before anthesis. This period has been confirmed in Malaysia (Corley *et al.*, 1971c).

In Nigeria, Broekmans (1957) found that abortion was highest in very young palms (about 25 per cent), falling to rates between 5 and 11 per cent with age, and that abortion was at a peak with leaves which had been at the spear stage around September (Fig. 4.7). Clearly, therefore, the conditions causing the maximum abortion were those associated with the dry season, and limited water supply is likely to be the main cause. In the dry season there is also a delay in inflorescence production corresponding to the delay in phase 2 of leaf production (see Fig. 4.6). This delay, together with the relatively high abortion rate, leads to a low production of inflorescences at anthesis 8–10 months after the spear stage (September) and $4\frac{1}{2}$–$5\frac{1}{2}$ months after the dry season (January to February), i.e. in June and July. This is followed by a maximum of flowering in October and November, corresponding to the maximum of leaf production which accompanies the beginning of the rains.

Typical abortion rates in West Africa are illustrated by the following data:

Age (years)	4	5	6	7	8	9	10	11	12
Progeny A, abortion %	24	24	19	18	11	17	11	5	7
Progeny B, abortion %	38	35	28	21	15	20	17	6	11

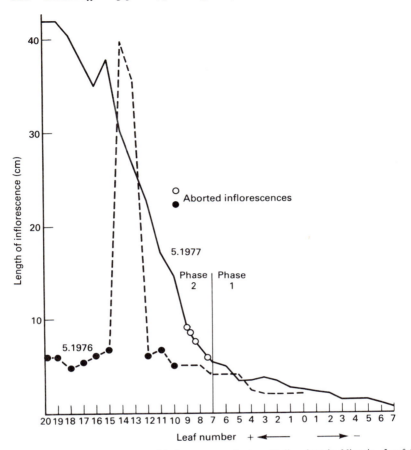

Fig. 4.8 The development of inflorescences in two Deli palms in Nigeria. Leaf 0 is the central spear (Broekmans, 1957).

In Malaysia abortion rates are often low but vary considerably both between areas and individual palms. Corley *et al.* (1971c) quote four palms where abortion over a 2-year period averaged 12 per cent, but in one palm, which produced forty-six female inflorescences, was only 4 per cent. In experiments on coastal clays a mean annual abortion rate of 10 per cent was found but the range was from 2 to 28 per cent. In some Deli *dura* the rate was as high as 50 per cent (Gray, 1969).

Broekmans considered the possibility that the sex ratio of aborted inflorescences might differ from the ratio in which inflorescences are differentiated. However, he assumed that, since aborted inflorescences usually occur in an unbroken series and the sex ratio at time of highest abortion appeared to be about average, the sex ratio of aborted inflorescences would also be about average. In Malaysia it has now been shown that there may be a preferential abortion of female inflorescences. Palms which were pruned to sixteen leaves showed a marked fall in female inflorescence production and apparent sex ratio compared with palms on which forty leaves were maintained (Table 4.7). Since sex is determined much earlier than 4–6 months before anthesis but abortion occurs around that time it must be concluded that female inflorescences were preferentially aborted (Corley *et al.*, 1971c).

Table 4.7 *Effect of pruning to sixteen leaves on female inflorescence production and sex ratio (Corley, 1971c)*

Leaves	Period after pruning (months)	No. inflorescences per palm		Sex ratio (%)
		Male	*Female*	
16	1–3	2.39	3.21	57.3
40	1–3	1.95	3.30	62.9
16	4–6	1.01	1.60	61.3
40	4–6	0.99	3.35	77.2

In an area planted at double normal density in Nigeria the degree of shading was altered for certain palms by heavy pruning of 0, 1, 3 or 4 adjacent palms, and the effect of this treatment on apparent sex ratio and abortion was observed. Pruning was done every year either in December, April or August. Reducing the shade by pruning surrounding palms reduced abortion and increased apparent sex ratio (Sparnaaij, 1960). However, Corley *et al.* (1971c) showed that, for palms with surrounding palms pruned in December (i.e. early in the dry season), if it is assumed that all the aborted inflorescences were female, reducing shade had no effect on sex ratio. If this assumption is applied to palms with surrounding palms pruned in April or August it is found that there is actually a rise in sex ratio with increasing shade. The data, set out in Table 4.8, must be viewed with some reserve since there were only twenty-four palms under observation per treatment, but the results, taken together with the effects found in pruning experiments, strongly suggest that both environmental and cultivation changes will have their primary effect on abortion and that effects on sex differentiation are complex and difficult to trace.

The main causes of high abortion rate have been shown to be those factors which reduce dry matter production in the palm, viz. over-pruning or defoliation by pests, mutual shading through high-density planting or moisture stress (Corley, 1973c). There is no direct evidence that abortion rate is directly controlled by carbo-

Table 4.8   *Effect of degree of surrounding shade on inflorescence abortion, apparent sex ratio and sex ratio adjusted on assumption that all aborted inflorescences were female (percentage)*

Month of pruning	No. of surrounding palms pruned	2nd year after initial pruning			3rd year after initial pruning		
		Abortion	Sex ratio		Abortion	Sex ratio	
			Apparent	Adjusted*		Apparent	Adjusted*
Dec.	4	12.7	25.5	35.0	12.1	26.3	35.2
	3	12.5	18.5	28.7	23.4	14.1	34.2
	1	18.3	19.4	34.1	21.2	15.3	33.2
	0	17.8	14.6	31.3	25.7	13.7	35.9
April	4	21.6	23.1	39.7	27.1	28.5	47.9
	3	31.4	22.5	46.8	27.0	24.1	44.6
	1	34.0	22.7	49.0	37.0	19.5	49.3
	0	40.5	22.4	53.8	42.3	21.1	54.8
August	4	6.8	14.6	20.4	11.6	20.0	29.3
	3	11.0	12.3	21.9	19.6	14.6	31.3
	1	17.8	15.8	30.8	25.6	16.9	38.2
	0	35.8	19.1	48.1	34.1	15.0	44.0

* Assuming all aborted inflorescences were female.

Source: From data of Sparnaaij (1960).

hydrate supply though it seems reasonable to suppose that Beirnaert's suggestion, already mentioned (p. 162), is more applicable to abortion rates than to true sex ratio. Auxin applications have been followed by decreased abortion but this effect may be secondary since the initial response has been reduced yield through the production of small parthenocarpic bunches (OPGL, 1968–72).

It will be clear from the above discussion that all previous data and deductions on sex ratio must be viewed with the probability in mind that aborted inflorescences were either all female or predominantly female. Nevertheless it has been possible to draw some conclusions about the time of sex differentiation and the factors influencing it from the data available in Nigeria and Malaysia. If Broekmans's data in Fig. 4.7 are examined it will be seen that the peak of inflorescence production in Nigeria falls in October/November at a time when female inflorescence production has not yet risen and when abortion (as indicated at time of anthesis) is low. Thus this peak is largely of male inflorescences and it is this fact which results in low sex ratios at that time. Corley (1973c) has suggested that this strongly male flowering owes its origin to the dry season 20 months earlier and states that in Malaysia abnormal peaks of male inflorescence production have been encountered 19–21 months after severe drought. Support has been given to this thesis by two experiments in Malaysia in which high-density plantings were thinned. It was found that, even when it was assumed that all aborted inflorescences were female, there was a significant effect on sex ratio 17 and 20 months respectively after thinning. Dissections within the areas of these experiments showed that 17–20 months before anthesis was also the time of initiation of the first bract subtending the spikelets. From this finding and other data it has been possible to draw up a schedule (Table 4.9) of inflorescence development for Malaysia.

Table 4.9   *Stages and time-scale of inflorescence development*

Stages of development	Range of months before anthesis
0 Inflorescence initiation	27–35
1 Outer spathe initiation	20–28
2 Inner spathe initiation	18–26
3 Initiation of first bracts subtending spikelets; sex differentiation	17–25
4 Initiation of fourth bract	15–20
5 Spikelet initiation	11–15
6 Spikelet differentiation distinct	8–11*
7 Abortion	3–6
8 Anthesis	0

* Broekmans claimed to determine differentiation by eye up to 18 months and Beirnaert, by microscopical observation, up to 20 months before anthesis. Van Heel *et al.* (1986), however, found morphological differentiation only detectable at leaf −4 to −2, i.e. about 10 months before anthesis.

Source: Adapted from data of Corley (OPGL, 1968–72).

The electron microscopy work of Van Heel *et al.* (1986) confirmed Corley's finding that the intervals between stages vary considerably with environment and age. These authors found that morphological definition of the inflorescence's sex was only possible when the first bracts are initiated on the spikelet primordia, but they considered that the morphogenetic impulse causing sex differentiation acts before this, though not before the first appearance of the axillary meristems giving rise to the spikelets. This suggests therefore, though it does not confirm, that sex differentiation takes place at the time of formation of spikelets in the subapical bracts. Van Heel *et al.*'s data indicated that first bract initiation ranges from the twenty-ninth to the eighteenth leaf before the spear (−29 to −18), which approximately corresponds to 17–23 months before anthesis, while spikelet initiation ranges from leaf −20 to leaf −2, corresponding to 10–19 months before anthesis.

Little is known about development of the inflorescence in very young palms, most of the work having been done with palms nearing or having reached their adult yield. If leaf production rises to thirty-six per annum in young palms sex differentiation may be expected to take place only some 11–19 months before anthesis, or even sooner if there are fewer leaves between the spear leaf and the leaf with an inflorescence at anthesis than is usual with the adult palm. Broekmans (1957) recognized that in young palms the period from sex differentiation to anthesis might be shorter than in adult palms and he drew attention to the fact that the peak in sex ratio may coincide with the peak in total inflorescences (due to minimal abortion) in October to November, and that this in turn gives rise to the large peaks of bunch production found in very young fields in Nigeria early in the year. This is accentuated by the fact that the period from anthesis to ripening tends to be shortened in the dry season.

Recent dissections in Malaysia of palms 15 months old in the field have shown that the bract initiation stage was 9–12 months before anthesis; the results of castration (ablation) experiments also suggest that for these very young palms the time between sex differentiation and anthesis may be as short as 9 months (OPGL, 1968–72).

The development of leaves and inflorescences in the adult palm producing two leaves per month and a young palm producing three leaves per month is illustrated diagrammatically in Fig. 4.9. The

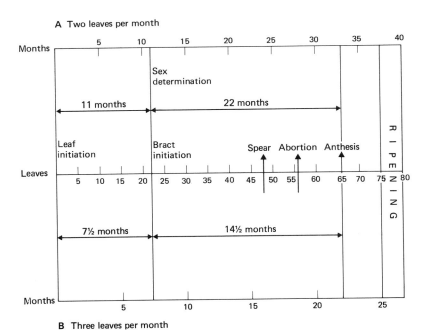

Fig. 4.9 Development of leaves and inflorescences on the assumption that: **A**, the adult palm is producing two leaves per month and sex differentiation is 22 months before anthesis; **B**, the young palm is producing three leaves per month.

approximate nature of this diagram will be clear from what has already been said.

Curiously, the peak of yield and hence of sex ratio seems to be earlier in tall grove palms in West Africa than in adult planted palms. The explanation for this is almost certainly that these tall old palms have a slower rate of leaf production than adult plantation palms. Zeven (1965) showed that in typical areas of dense and degraded groves the leaf production of palms receiving direct sunlight was only around $1\frac{1}{2}$ per month. The period from differentiation to anthesis and to bunch production would therefore be prolonged by 9–10 months and this prolongation would bring bunch production into the early months of the *following* year.

Flowering cycles

Leaf and flowering data so far considered in relation to age or season are based on populations of palms; but any influence of climatic factors impinges on individual palms which are subject to successive periods, often termed cycles, of male and female flower production. These periods alternate, sometimes with hermaphrodite inflorescences appearing between a run of male and a run of female inflorescences. The lengths of the male or female periods vary widely, and in the higher-yielding regions and progenies the female periods tend to be long drawn out. From the examination of certain data from Zaire and Nigeria, Haines and Benzian (1956) concluded that periods of 4 or 5 months predominate. This is equivalent to runs of eight to ten inflorescences. Other workers have found that female phases follow a Poisson distribution with short phases predominating and single-inflorescence phases common, and they conclude that inflorescence sex is semi-randomly determined (Hemptinne and Ferwerda, 1961; Corley, 1977).

The flowering characteristics of the individual will be determined by its genetic constitution, its position in the field, particularly as regards the growth of its neighbours, and by other factors. Haines and Benzian detected cycles of high yield in individual palms and populations at Ndian Estate in West Cameroon (high rainfall, but distinct dry season); for the individual palms 3-year cycles were detected and it was postulated that this phenomenon could be accounted for by the 'fundamental' 9-month male–female cycle producing a 'beat' of 3-year periodicity when the seasonal cycle of 12 months was superimposed. For the whole population, however, a 5-year cycle was discerned. In this case it was supposed that a high nutritional demand during the ripening of bunches in a female period might lead to the differentiation of large numbers of male inflorescences which in their turn would reach anthesis in 24 months leading to low demand during the following months, i.e. 'a reversed yield movement would occur 30 months later'. This is its turn would lead to female differentiation and the high yield to high yield period would then be 30 + 30 months, or 5 years.

Corley (1977) found 40 per cent of palms showed an approximate 3-year yield cycle but this could not be generally attributed to a cycle of sex ratio. He concluded that the only sex ratio cycle undoubtedly occurring was an annual one caused by exogenous factors and that evidence for endogenous cycles was lacking.

There is also, however, the effect of possible abortion cycles on cycles of flowering, and it has been postulated that large numbers of developing bunches from 1 to 3 months after anthesis can induce high rates of abortion which would be reflected in a low number of bunches at this same stage of development some 7–8 months later, thus giving rise to a 15-month abortion cycle. Some evidence of this has been found in a pruning trial in Malaysia but the unpruned control palms followed the normal 12-month abortion cycle (Corley, 1977).

Nutritional and hormonal aspects of flowering

It has already been mentioned that Beirnaert (1935) suggested that an increased female inflorescence production or high sex ratio was determined by high photosynthetic activity or high carbon/nitrogen ratios. Sparnaaij (1960) supposed, from this, that nitrogen applications would reduce sex ratio. However, he demonstrated only that two progenies in Nigeria reacted differently to different levels of nitrogen, and in Malaysia it has often been shown that nitrogen applications increase bunch production.

Very little work has been done on the hormonal control of leaf and flower production. In Malaysia growth regulators (2,4,5-TP, GA_3 and NAA) were found to reduce yield through increasing parthenocarpy. This was followed by decreased abortion and, after 3–4 years, increased sex ratio not accounted for solely by the decreased abortion (OPGL, 1968–72). Various interpretations can be put on these effects.

Applications to nursery seedlings which were continued on the plants after transplanting to the field showed that while Ethrel delayed flowering, GA_3 and CCC caused earlier flowering. CCC increased the number of male inflorescences appearing before the first female; this would be expected with earlier flowering because young palms normally have an initial run of males. GA_3 also produced earlier female flowering.

It has often been pointed out that the oil palm has many means – abortion, sex change, decreased or increased bunch size, leaf production changes, stem girthing – whereby adjustments are made to changes in photosynthetic supply. The successive production of large numbers of bunches, i.e. a long female cycle, has effects both on the abortion rate (as mentioned above) and perhaps also on the sex ratio of succeeding inflorescences, and this is an important factor responsible for the wide palm-to-palm variation in length and timing of male and female cycles. It might thus be expected that the palm would compensate for any growth regulator treatment effects by various means, but particularly by abortion and sex ratio changes, unless the rate of photosynthesis is partly dependent on the demand for dry matter by developing bunches. Corley (1973c) has pointed out that if photosynthesis is increased by demand then increasing the latter through treatments to reduce abortion and increase sex ratio and bunch size will itself be effective. If not, however, present understanding of the physiology of the oil palm suggests that, apart from the selection of genetically productive palms, the supply and efficient use of dry matter can best be assured by maintaining a suitable leaf area index through planting at optimal density, controlling leaf pests and diseases, manuring sufficiently, maintaining soil moisture supplies and avoiding over-pruning.

Bunch production, pollination and bunch failure

There are a number of factors, physiological and pathological, which affect the proportion of inflorescences reaching anthesis which will set and ripen their fruit. The rotting of bunches between anthesis and ripening is termed *bunch failure*. This term covers failure from whatever cause and should not be confined, as suggested by some authors, to failure of 'adequately pollinated bunches' since paucity of pollen is a major cause of bunch failure.

Bunch failure is more common in young than in mature palms, but with the majority of *pisifera* palms it tends to be extensive throughout the palm's life. It is associated with high sex ratios whether in the *pisifera* or in other fruit forms and in the past has been much more frequent in Asia than in Africa.

Pollination has been described in Chapter 2. During the long period when it was thought that the oil palm was mainly wind-pollinated, much effort was made in Asia to combat bunch failure by elaborate systems of assisted pollination (Hardon and Turner, 1967; Gray, 1966), and there was much theorizing on when or whether failure was caused by lack of pollen or by such other factors as 'over-bearing' or the Fruit-rot fungus *Marasmius palmivorus* (q.v.). Now that insect pollinators have been introduced from Africa into Asia and America it will take time to observe and assess any residual bunch failure which occurs. Such failures may, of course, be an early or late symptom of disease or pest attack but, in the absence of diseases and pests, there will still be cases of rotting of bunches at any time between pollination and ripening, though most commonly at 2–4 months after anthesis. It seems probable that these failures originate in physiological stress, since they are often associated with drought or very heavy bearing at an early stage in the plant's bearing life (Turner, 1981), or they can be induced by severe pruning or insect defoliation (Corley, 1973c, 1977). Partial bunch failure, termed Bunch End Rot, may also have the same causes (Turner and Bull, 1967).

In the areas of high sex ratio and serious bunch failure attention was naturally concentrated on pollination. Bunch failure also occurs, however, in areas in Africa where an abundant natural supply of pollen and pollinators exists. Cases of serious bunch failure have occurred, e.g. in Ghana, in young areas surrounded by forest and in these areas scarcity of pollen or pollinators are likely to be contributory factors, but in the majority of situations such vast numbers of pollen-producing palms and insect carriers exist that pollen supply cannot be held responsible. In these situations bunch failure rarely exceeds 10–15 per cent with adult palms though it is usually higher in young palms. The percentage survival of female and hermaphrodite inflorescences in two progenies in a field in Nigeria was as follows:

Year of harvesting	1	2	3	4	5	6
Progeny A, survival %	54	66	78	73	81	85
Progeny B, survival %	45	64	90	87	88	93

Though variations such as those shown above do occur, significant differences in bunch failure between progenies have not been found (Sparnaaij, 1960).

The effect of pollination failure and subsequent Bunch Rot on the behaviour of the palm may be seen in the behaviour of the infertile *pisifera* and of palms which have been subjected to ablation of inflorescences. In the former case the palm tends firstly to divert its assimilates to the production of more and larger leaves and a more massive trunk and secondly to continue to produce predominantly female inflorescences. Similarly, the ablation of female inflorescences initially increases vegetative growth and the number of female inflorescences produced.

Pollination

Imperfect pollination was noted in Asia from the start of the plantation industry, and assisted pollination was practised on young palms in Sumatra before 1920 (Hardon, 1973). As the palms grew taller and the proportion of male inflorescences increased, the practice was abandoned and did not come to the fore again until the spate of new *tenera* plantings was under way in the early 1960s. It is now clear that *Thrips hawaiiensis* provided adequate, though not optimal, pollination in established plantations in Indonesia and Malaysia, though male inflorescence production was, age for age, much lower than in Africa where weevils of *Elaeidobius* species were in abundance as pollinators (Syed, 1979). Although the role of insects was not realized until 1979, investigations from 1960 on factors influencing pollination and on the practice of assisted pollination provided some useful information on the effects of varying the pollen supply (Gray, 1969; Lawton, 1981).

The impetus to solve the problems of pollination came first from the poor fruit set and bunch rotting in the young areas of Malaysia, but later a solution became more urgent in territories such as Sabah and Papua New Guinea where it was reported that 'where no pollination is done, no fruit is produced' (Hardon, 1973, discussion of paper). It was later seen (Syed, 1979) that although *Thrips hawaiiensis* was present in Sabah it did not visit the oil palm and so it was concluded that the species found in peninsular Malaysia and Indonesia was an oil palm strain; pollination in Sabah seemed to depend on the erratic visitations of a moth (*Pyroderces* sp.) and on dispersal by wind.

Malaysian experiments were incomplete because they did not study fruit set and kernel production (Hardon, 1973), but they did show that supplementing the supply of pollen by artificial means initially increased bunch yield by reducing the amount of bunch failure and increasing bunch weight. Subsequently, however, more male and less female inflorescences are produced and the total bunch yield over a period of 6–8 years may not be increased by assisted pollination (Gray, 1966). With hindsight it can be said that, except in areas where an almost total absence of pollinating agencies gives rise to fields of rotting bunches (a situation which perpetuates the palm's female cycle and so reduces the pollen supply still further), it was seldom if ever possible to assess the value of assisted pollination. Some plantation companies 'never considered pollination necessary', others took the view that, though necessary, it was difficult to decide 'when one can afford to stop doing it', and it was admitted that experiments gave little guidance on this point (Hardon, 1973), discussion).

Papua New Guinea, before the introduction of *Elaeidobius kamerunicus*, proved to be a country were young palms were unusually subject to Bunch Rot. Here it was shown that bunch failure was related to both rainfall and the number of male inflorescences. However, there were also palms which, by virtue of their non-expanding stigmas and tightly packed spikelets, produced a more than average number of rotting bunches (Lawton, 1981). Clearly, this undesirable anatomical feature will still need to be eliminated.

The short history of pollination since the introduction of *E. kamerunicus* into Malaysia in July, 1980, has been an interesting and lively one (Syed *et al.*, 1982). After specificity tests, the weevil was released on two plantations in February, 1981, and both population increase and spread were very rapid. Later there were releases in Indonesia, Sabah, Papua New Guinea and Colombia. The population of *Thrips hawaiiensis* in Malaysia decreased and then stabilized at a low level. After a year the population of *E. kamerunicus* per male spikelet was of the same order as that of the total of all species of *Elaeidobius* in Cameroon.

The effects of the introduction of *E. kamerunicus* were similar to, but much greater than, the effects of assisted pollination. The initial effects are illustrated in Table 4.10; the most immediate of these was an increase in fruit set, resulting in higher bunch yields, mean bunch weight and fruit-to-bunch ratios. The latter in turn gave rise to smaller fruit but a higher percentage of mesocarp to bunch and, through a reduction in the number to parthenocarpic fruit, a much higher percentage of kernel to bunch, indicating that the low kernel extraction rates which had always been a feature of Malaysian mills had been due to poor fruit set. In some cases both oil to mesocarp and oil to bunch decreased, though oil plus kernel to bunch increased.

Table 4.10   *Initial effects of the release of* Elaeidobius kamerunicus *in Malaysia on bun* composition

Plantings	1973–6		1959–66	
Period	Pre-release Apr.–June 1981	Post-release Dec. 1981–Mar. 1982	Pre-release Mar.–June 1981	Post-release Dec. 1981–Mar. 1982
Mean bunch weight (kg)	10.7	13.6	23.5	26.9
Fruit set (%)	47.8	76.0	53.4	71.2
Mean fruit weight (g)	—	—	11.2	7.7
Mesocarp to fruit (%)	76.5	74.8	74.8	70.6
Shell to fruit (%)	15.7	15.0	16.1	17.8
Kernel to fruit (%)	7.8	10.2	9.1	11.5
Oil to mesocarp (%)	49.1	48.7	48.7	47.4
Fruit to bunch (%)	60.4	68.3	60.4	64.4
Mesocarp to bunch (%)	46.2	51.1	45.2	45.5
Kernel to bunch (%)	4.7	7.0	5.5	7.4
Oil to bunch (%)	22.7	24.9	22.0	21.5

Source: Syed *et al.* (1982).

When a palm is forced to set more fruit than previously, and so divert more of its assimilates to bunch production, it will be drawing on its reserves unless its photosynthetic efficiency can be increased, which is unlikely in the short term (Foster *et al.*, 1984). Under these circumstances a reaction to the increased bunch production through greater abortion and then a lower sex ratio could be expected, and this is what actually occurred in Malaysia where yields rose sharply in 1982, fell heavily in 1983 and reverted to normal in 1984.

An indication of the more permanent effects of weevil introduction has been presented by Donough and Law (1987). They compared yields and bunch analyses for periods of 4 years before and after *E. kamerunicus* release, using fields of adult palms, i.e. palms which were in their 'plateau yield' period (eighth to eighteenth year) during both 4-year spans. The years of recording were 1977–80 and 1982–5, so the latter period included the 2 years of wide yield fluctuations. The fields were in Johore state of peninsular Malaysia, where in the first period *Thrips hawaiiensis* was the pollinator, and in Sabah, where there was no major pollinating insect and assisted pollination was being practised. Table 4.11 shows that the initial effects already noted have been largely maintained but were more pronounced in Sabah than in Johore since the *Thrips* pollination in the latter territory was clearly more efficient than the assisted pollination in the former. The main permanent features of

Table 4.11   *The effect of the release of* E. kamerunicus *in Johore, Malaysia, and Sabah: yield and bunch analysis of adult palms over 4-year periods*

Plantings of 1966–8	Bunch production	No. of bunches	Mean bunch weight	Fruit set*	Fruit to bunch*	Kernel to bunch*	Palm oil to bunch*	Estimated yield per hectare[†]	
								Palm oil	Kernels
	(tonnes per hectare per annum)	(per palm per annum)	(kg)	(%)	(%)	(%)	(%)	(tonnes)	
Johore (Peninsular Malaysia)[‡]									
Pre-weevil period 1977–80	23.3	7.8	21.6	52.5	63.6	6.4	22.6	4.8	1.4
Post-weevil period 1982–5	23.9	6.2	27.8	74.4	65.1	7.4	22.4	4.9	1.7
Percentage change	+ 2.6	−20.3	+28.3	+41.7	+ 2.2	+15.2	− 0.9	+ 2.1	+21.3
Sabah[§]									
Pre-weevil period 1977–80	16.9	9.1	13.4	54.8	55.9	4.3	21.6	3.4	0.7
Post-weevil period 1982–5	20.8	8.6	17.5	56.5	56.6	4.9	22.7	4.4	1.0
Percentage change	+23.1	− 5.8	+30.6	+ 3.1	+ 1.3	+14.1	+ 5.0	+29.4	+42.9

* Analysis periods: pre-weevil period, Jan.–June 1981; post-weevil period, Oct. 1981 to Dec. 1984.
[†] Calculated yields corrected for mill extraction efficiency.
[‡] No assisted pollination in pre-weevil period.
[§] Assisted during the pre-weevil period.

Source: Donough and Law (1987).

the introduction of *E. kamerunicus* may thus be summarized as the removal of a need for assisted pollination, the production of a smaller number of larger bunches, and a substantially higher outturn of kernels per bunch and per hectare. This increase in kernels leads to a reduction of mesocarp and mesocarp oil to fruit though, owing to the increased fruit set, the mesocarp percentage to bunch may not be reduced. Wood (1985) has quoted cases of palms of different ages which showed weight and bunch composition changes following *E. kamerunicus* introduction (Table 4.12).

Table 4.12 *Changes of bunch weight and composition following the introduction of E. kamerunicus into Malaysia (Wood, 1985)*

	Percentage change of weight or composition after weevil introduction		
	Palm age		
	9 years	*10 years*	*14–24 years*
Mean bunch weight	+10	+30	+14
Fruit to bunch	+ 7	+12	+ 6
Mesocarp to fruit	− 4	− 3	− 6
Mesocarp to bunch	+ 3	+ 9	—
Oil to mesocarp	− 4	− 2	− 1
Oil to fruit	− 8	− 5	− 7
Kernel to bunch	+35	+36	+29

A lack of interest in the kernel has always been a feature of the Malaysian industry, and a larger proportionate production may force producers to give it more attention, to their ultimate advantage.

Fruit form and bunch production

There are differences in bunch production between the *dura, tenera* and *pisifera* forms which owe their origin to differences in sex ratio. A typical example from Nigeria is as follows:

	Dura	*Tenera*	*Pisifera*
Sex ratio over a 7-year period	27.9	31.1	45.7

As a result of these sex ratio differences *dura* palms produce fewer but heavier bunches than *tenera* palms. The production of shell-less *pisifera* fruit depends on the ability of the bunches to survive to maturity. The so-called infertile *pisifera* produces rotten bunches in which only a very few fruit set, and it has a very high sex ratio.

Under West African conditions the Deli palm produces a small number of very large bunches, having a very low sex ratio. Progeny differences are, of course, large. Most progenies show a tendency to reduced sex ratio on moving into dried districts though some, by maintaining a high sex ratio, seem to be adapted to drier conditions.

Yield variation within and between years: yield prediction

Much has already been said about annual variation in sex ratio and abortion, and these factors are largely responsible for annual bunch yield variations. In seasonal climates the annual yield usually has only one peak, the time of the peak depending, as has already been shown, on the age and rate of leaf production of the palms and, in the adult plantation palm, on climatic conditions about 28 months before fruit ripening. In non-seasonal climates, there are occasionally two peaks of production in the year, though one tends to be much the more prominent; there is considerable variation in the magnitude of the peaks. In Fig. 4.10 the monthly production of adult fields in different territories is illustrated.

Yield prediction can only be attempted if a good deal is known of the factors controlling yield. The general relationship between climate, and in particular the water deficit, and yield levels has been discussed in Chapter 3 (p. 102). Annual variations have tended to be greater in Africa than in the Far East and climatic variations more clear-cut, and for these reasons a greater scientific interest was at first taken in annual yield variation in Africa. Broekmans's (1957) work on the relation of the dry season to sex ratio and abortion has already been discussed; his work was preceded by a short rainfall/yield study at La Mé by Devuyst (1948) and followed by a number of other studies in Zaire and West Africa (Hemptinne and Ferwerda, 1961; Bredas and Scuvie, 1960; Michaux, 1961; Ruer, 1966).

Sparnaaij *et al.* (1963) were able to reconcile the deductions drawn by these authors and to provide parameters which were much more closely correlated with future yield; they postulated that the carbohydrate status of the plant largely depends on the sunshine which has been available, provided that the plant's metabolism has not been restricted by some other factor. They therefore directed their attention to obtaining reliable estimates of 'effective sunshine' which they defined as total sunshine during the period of water sufficiency plus a fraction of the sunshine in the drought periods. It will be realized that such measures of effective sunshine must, in seasonal climates, be directly related to dry-season rainfall; but they proved much more sensitive as prediction factors than did the rainfall data of previous authors.

It has been seen that, in adult plantations, the stage of anthesis is some 17–25 months after sex differentiation. Ripening takes place

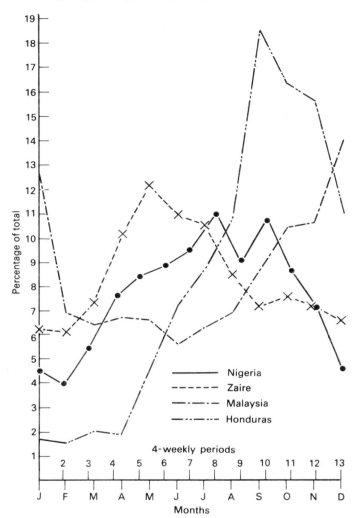

Fig. 4.10 Monthly or 4-weekly distribution of bunch yield in adult fields in Nigeria, Zaire, Malaysia and Honduras. Means of 5 years, except for Honduras where the figures are means of 12 years and relate to 4- or 5-week periods.

$4\frac{1}{2}$–6 months after anthesis, i.e. some 21–31 months after sex differentiation. Broekmans's relationship of rainfall and bunch number was for the period November to April (roughly covering the dry season) and the yield during the second calendar year following, for example:

November–April rainfall	Number of bunches in calendar year
1960–1	1963

The exact weather period taken for corrrelating with yield is not of great significance, but as the period of high sex ratio lies in the dry season in the seasonal climate of West Africa, the significant period must 'straddle' the dry season 2 years earlier. Sparnaaij *et al.* (1963) chose the period 1 September to 31 August to compare with the calendar year 28 months later. In the first place a correlation was found between annual sunshine data and yields 28 months later, though the correlation was not high. A more definite relationship was found when the sunshine figures for the drought periods were excluded, and it was apparent that the useful sunshine contribution of the drought periods could only result in minor differences. It was realized that the physiological effect of drought on the carbohydrate status of mature oil palms not having been investigated, any estimate of the contribution of drought-period sunshine must be tentative. Nevertheless the methods developed were highly successful. After a consideration of potential evapotranspiration (*Etp*) calculated by the Thornthwaite formula, and estimates by Penman's method and lysimeter determinations, a figure for *Etp* of 50 inches per annum or 1 inch (25 mm) per week was used in drawing up an estimate of the length of the drought period. Two estimates of effective sunshine (*ES*) in drought periods were then made as follows:

ES (*a*): Dividing the total duration of sunshine in hours during a drought period by the average water-deficit factor, assuming that the deficit increases by unity each week. A cumulative deficit factor for 5 weeks is:

$$\frac{1 + 2 + 3 + 4 + 5}{5},$$

and this is divided into the sunshine in hours.

ES (*b*): Estimation of effective sunshine by the same method but on a weekly basis, viz:

$$\frac{S_1}{1} + \frac{S_2}{2} + \frac{S_3}{3} + \frac{S_4}{4}, \text{ etc.}$$

where S_1, S_2, etc. denote the sunshine in successive weeks of drought.

In drawing up the total effective sunshine it was assumed that in the drought period:

1. Maximum water reserve in the soil was 4 inches (100 mm), this being derived from an estimate of the root constant.
2. Depletion of soil water was 1 inch (25 mm) per week irrespective of reserves.
3. Drought started when reserves fell below 1 inch (25 mm).
4. A weekly rainfall of below $\frac{1}{2}$ inch (13 mm) had no effect on drought.
5. A weekly rainfall of $\frac{1}{2}$–$1\frac{1}{2}$ inches (13–38 mm) did not break the drought but did not increase the deficit factor.
6. A weekly rainfall of over $1\frac{1}{2}$ inches (38 mm) broke the drought.

A simplification of the effective sunshine calculations and one which is probably only suited to Nigeria where it may be assumed that the November to April weekly dry-season sunshine is constant, was also successfully employed. This is termed 'active weeks' and is calculated by deducting the dry weeks from the 26 weeks (weekly units) of the November to April season and then adding for each drought period (as defined at 3 above) $1\frac{1}{2}$, 2, $2\frac{1}{2}$ or 3 units according to whether the length of the drought period was 2, 3–5, 6–8 or 9+ weeks. Highly significant correlations between all these factors and annual yield 28 months later were found at Benin in Nigeria for a 12-year period and the highest correlations were found with active weeks.

Short-term climatic influences were also included, i.e. the influence on abortion of the dry season immediately preceding the harvest which is gathered at the end of one calendar year and the beginning of the next. This is the time, in West Africa, of minimum yield and Broekmans (1957) found a significant correlation between dry-season rainfall and bunch yield in the time of minimum harvest about 1 year later. As, however, this is the time of minimum yield, it is unlikely to have a large effect on the total calender-year yield and the adding to the effective sunshine data of rainfall data for the yield year and the preceding year (both of which may have an effect on abortion in the calendar year of yield) did not appreciably improve the correlations. The correlations obtained by the various methods at Benin are given in Table 4.13 (with significance denoted by asterisks).

The relationships are also shown in graphical form in Fig. 4.11.

As mentioned in Chapter 3 (p. 106), the time interval of 28 months used by Sparnaaij *et al.* (1963) has been employed in determining general relationships between water deficits and yield while the parameter effective sunshine itself has been found useful in investigations of the relation between environmental and nutritional factors (Ruer, 1966) (p. 511). Both effective sunshine and active weeks are in fact good drought indicators. A negative correlation has been found between the annual yield data employed in Fig. 4.11 and the water deficit during the dry season 2 years earlier, but this

Table 4.13   *Correlations between bunch yield and some climatic parameters in Nigeria*

Correlations with yield 28 months later and sex ratio 24 months later	Effective sunshine hours (a)	Effective sunshine hours (b)	Effective sunshine (b) + 2 years' dry-season rain	Active weeks
Number of bunches per palm	0.70*	0.70*	0.71**	0.79**
Weight of bunches per palm	0.84***	0.84***	0.88***	0.95***
Sex ratio	0.77**	0.78**	0.79**	0.79**

Source: Sparnaaij *et al.* (1963).

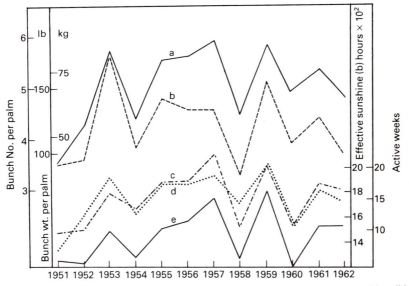

Fig. 4.11   Relation between number and weight of bunches, effective sunshine (b) and active weeks during the period 1951–62: **a**, mean weight of bunches per palm; **b**, mean number of bunches per palm; **c**, effective sunshine (b) plus 2 years' dry-season rain; **d**, number of active weeks; **e**, effective sunshine (b) (Nigeria – Sparnaaij *et al.*, 1963).

correlation ($r = -0.74$), though significant, is less close. Figure 4.12 shows the relation between yield per palm, water deficit and effective sunshine (*a*) and the water deficit/yield relationship may be compared with that given in Fig. 3.2. Another method of forecasting

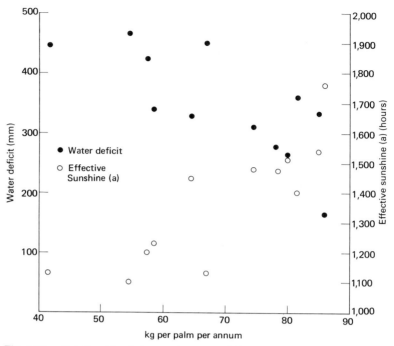

Fig. 4.12   Relationships between the water deficit during the dry season and the effective sunshine measured from 1 September to 31 August and bunch yield during the calendar years 28 months later (Nigeria – yield data from Sparnaaij *et al.*, 1963; water deficit computation by IRHO methods).

yield from climatic factors has recently been described from the Ivory Coast (Dufour *et al.*, 1987). Here again, water deficit was found to be the most significant factor. In taking the deficits found from the thirtieth month preceding the start of a year's harvesting up to the sixth month after its start, high correlations between annual yield and total water deficit were found.

In Asia, studies of the effect of climatic factors on yield have been much more recent. The approach has necessarily been different; the complete absence of a calculated water deficit being usual, interest has largely centred on the effect of such drought periods as may from time to time occur. However, the frequency of these periods varies widely both with geographical situation and time. For instance, seven estates in Sumatra between latitudes 2°30' and 3°35' N had measurable water deficits in a mean of 5.2 out of the 8 years 1968–75, whereas three estates south of this region showed deficits in the same period in a mean of only 2 years. In the same region there is some evidence also that periods of more severe drought have been more frequent in recent years than towards the beginning of the century. The Asian data so far examined suggest that a deficit has a greater effect, through increases in the abortion rate, in the year after the deficit than it has, through alterations to the sex ratio, in the following year. Nevertheless the general pattern of low yeilds 8–12 and 21–26 months after a drought period, with an increase in male flowering in the latter period, are found (Turner, 1977). There is also a suggestion that sudden drought in a normally water-sufficient area may cause abortion of inflorescence buds at the time of sex differentiation, i.e. 17–26 months before anthesis according to palm age. The same study also showed the importance of soil type and drainage intensity in periods of sudden drought in Asia. On the coastal clays in Malaysia plantations which are able to maintain high water-tables can be largely unaffected by periods of drought even where calculated water deficits are as high as 300–600 mm.

Two detailed studies aimed at determining climatic effects and hence predicting yields have recently been reported from Asia, one based on mathematical models and the other employing an exploratory identification analysis. Both studies used the parameter *LAG* for the time-lag in months from a climatic factor determination to time of harvest, and the various *LAG* periods used could also of course be identified as the number of months before harvest when initiation, sex determination, spear opening, etc. took place. The climatic factors used in the first studies (Robertson and Foong, 1977; Foong, 1981) included soil water deficit, solar radiation, maximum and minimum temperature, and, not unexpectedly, these factors were found to be important at *LAG* 34–39, 25–27, 16–21 and 7–9, corresponding to initiation, sex differentiation, spear elongation and abortion. An interesting feature of this study was the introduction of 'interbunch competition', i.e. the nutrient competition effect of developing bunches (Corley, 1977) on subsequent yield at various *LAG*s. Results showed that the development of large quantities of bunches may lead to increased abortion.

The studies of Ong (1982) in southern Malaysia were confined to rainfall, dry spells (number of days without rain), temperature and sunshine, and high correlations were found for many *LAG* periods between yield and both rainfall and dry spells. Some of these were considered to be spurious until temperature and hours of sunshine were also taken into account, and later studies indicated that a large diurnal temperature range favoured floral abortion. This important finding may partly explain the fact that the highest yields are obtained in areas of high even temperature (see p. 110). At the same time single and partial correlations of yield with rainfall, temperature and sunshine indicated that both rainfall and warm sunny weather favoured female sex differentiation.

References

Beirnaert, A. (1935) Introduction à la biologie florale du palmier à huile. *Publs INEAC*, Sér, Sci., No. 5.

Bredas, J. and **Scuvie, L.** (1960) Aperçu des influences climatiques sur les cycles de production du palmier à huile. *Oléagineux*, **15**, 211.

Breure, C. J. (1986) The effect of different planting densities on the yield trend of oil palm. *Exptl Agric* , (in the press).

Briggs, G. E., Kidd, F. and **West, C.** (1920) A quantitative analysis of plant growth. *Ann. appl. Biol.*, **7**, 103.

Broekmans, A. F. M. (1957) Growth, flowering and yield of the oil palm in Nigeria. *J. W. Afr. Inst. Oil Palm Res.*, **2**, 187.

Bunting, B., Georgi, C. D. V. and **Milsum, J. N.** (1934) *The oil palm in Malaya.* Kuala Lumpur.

Chin, H. F. and **Nair, K. C. Sekaran** (1980) Effect of moisture stress on growth of oil palm (*Elaeis guineensis*) seedlings. *Planter, Kuala Lumpur*, **56**, 3.

Cook, O. F. (1940) Oil palms of Florida, Haiti and Panama. *Natn. hort. Mag.*, **19**, 10.

Corley, R. H. V. (1973a) Midday closure of stomata in the oil palm in Malaysia. *MARDI Res. Bull*, **1** 2) 1–4.

Corley, R. H. V. (1973b) Effects of plant density on growth and yield of the oil palm. *Expl. Agric.*, **9**, 169.

Corley, R. H. V. (1973c) Oil palm physiology – a review. In *Advances in oil palm Cultivation*, p. 37, Incorp. Soc. of Planters, Kuala Lumpur.

Corley, R. H. V. (1977) Oil palm yield components and yield cycles. In *International development in oil palm*, ISP, Kuala Lumpur.

Corley, R. H.V., Gray. B. S. and **Ng Siew Kee** (1971a) Productivity of the oil palm (*Elaeis guineensis* Jacq.) in Malaysia. *Expl. Agric.*, **7**, 129.

Corley, R. H.V., Hardon, J. J. and **Tang, Y.** (1971b) Analysis of growth of the oil palm (*Elaeis guineensis* Jacq.) I. Estimation of growth parameters and application in breeding. *Euphytica*, **20**, 307.

Corley, R. H. V., Williams, C. N. and **Rajaratnam, J. A.** (1971c) Developmental and physiological aspects of sex-expression and yield in the oil palm. University of Malaysia Seminar, 1971, Mimeograph.

Corley, R. H. V. and **Mok, C. K.** (1972) Effects of nitrogen, phosporus, potash and magnesium on growth of the oil palm. *Expl. Agric.*, **8**, 347.

Devuyst, A. (1948) Influence des pluies sur les rendements du palmier à huile enregistrés à la station de La Mé de 1938 à 1946. *Oléagineux*, **3**, 137.

Donough, C. R. and **Law, I. H.** (1987) The effect of weevil pollination on yield and profitability at Pamol Plantations. Int. Oil Palm Conference, Kuala Lumpur, 1987.

Dufour, O., Frere, J. L., Caliman, J. P. and **Hornus, P.** (1987) Description of a simplified method of production forecasting in oil palm plantations based on climatology. *Int. Oil Palm Conf.*, Kuala Lumpur, June 1987.

Fickendey, E. and **Blommendaal, H. N.** (1929) *Ölpalme.* Deutscher Auslandverlag, Hamburg and Leipzig.

Foong, S. F. (1981) An improved weather model for estimating oil palm fruit yield. In *The oil palm in agriculture in the eighties*, ISP, Kuala Lumpur.

Foster, H. L. *et al.* (1984) The effect of the introduction of the weevil (*Elaeidobius kamerunicus*) on the yield performance, nutrition and physiology of the oil palm in peninsular Malaysia. Symposium on Impact of the Pollination Weevil on the Malaysian Oil Palm Industry, PORIM and MOPGC, Kuala Lumpur. Typescript.

Gray, B. S. (1966) The necessity for assisted pollination in areas of low male inflorescence production and its effect on the components of yield of the oil palm. *Planter, Kuala Lumpur*, **42**, 16.

Gray, B. S. (1969) A study of the influence of genetic, agronomic and environmental factors on the growth, flowering and bunch production of the oil palm on the west coast of Malaysia. Thesis, University of Aberdeen.

Haines, W. B. and **Benzian, B.** (1956) Some manuring experiments on oil palms in Africa. *Emp. J. exp. Agric.*, **24**, 137.

Hardon, J. J. (1973) Assisted pollination in the oil palm – a review. In *Advances in oil palm cultivation*, ISP, Kuala Lumpur.

Hardon, J. J. and **Turner, P. D.** (1967) Observations on natural pollination in commercial plantings of oil palm (*Elaeis guineensis*). *Expt. Agric.*, **3**, 105.

Hardon, J. J., Williams, C. N. and **Watson, I.** (1969) Leaf area and yield in the oil palm in Malaysia. *Expl. Agric.*, **5**, 25.

Heath, O. V. S. and **Gregory, F. G.** (1938) The constancy of the mean net assimilation rate and its ecological importance. *Ann. Bot.*, NS, **2**, 811.

Hemptinne, J. and **Ferwerda, J. D.** (1961) Influence des précipitations sur les productions du palmier à huile. *Oléagineux*, **16**, 431.

Henry, P. (1955) Morphologie de la feuille d'*Elaeis* au cours de sa croissance. *Revue gén. Bot.*, **62**, 319.

Lawton, D. M. (1981) Pollination and fruit set in oil palm. In *The oil palm in agriculture in the eighties*, ISP, Kuala Lumpur.

Lo, K. K., Chan, K. W., Goh, K. H. and **Hardon, J. J.** (1972) Oil palm – the effect of manuring on yield, vegetative growth and leaf nutrient levels. In *Advances in oil palm cultivation*, p. 324, Incorp. Soc. of Planters, Kuala Lumpur.

Mason, T. G. and **Lewin, C. J.** (1925) Growth and correlation in the oil palm (*Elaeis guineensis*). *Ann. appl. Biol.*, **12**, 410.

Michaux, P. (1961) Les composants climatiques du cycle annuel de productivité du palmier à huile. *Oléagineux*, **16**, 523.

Milthorpe, F. L. (1956) The relative importance of the different stages of leaf growth in determining the resultant area. In *The growth of leaves*, pp. 141–8, Butterworth, London.

Ng Siew Kee, Thambo, S. and **De Souza, P.** (1968) Nutrient content of oil palms in Malaya. II. Nutrients in vegetative tissues. *Malay. agric. J.*, **46**, 332.

Ochs, R. (1963a) Utilisation du test stomatique pour le contrôle de l'arrosage du palmier à huile en pépinière. *Oléagineux*, **18**, 387.

Ochs, R. (1963b) Recherches de pédologie et de physiologie pour l'étude du problème de l'eau dans la culture du palmier à huile. *Oléagineux*, **18**, 231.

Ong, H. T. (1982) System approach to the climatology of oil palm. 1. Identification of rainfall and dry spell aspects. 2. Identification of temperature and sunshine aspects. *Oléagineux*, **37**, 93 and 443.

OPGL (Oil Palm Genetics Laboratory) (1968–72) *Progress reports, 1968–72*, Layang-Layang, Malaysia.

Prévot, P. (1963) Données récentes sur la physiologie du palmier à huile. *Oléagineux*, **18**, 79.

Quencez, P. (1974, 1975) Arrosage par aspersion des palmiers à huiles en sacs de plastique. *Oléagineux*, **29**, 405; **30**, 355, 409.

Rees, A. R. (1957) *WAIFOR Fifth Annual Report, 1956–7*, p. 114.

Rees, A. R. (1958) Field observations of midday closure of stomata in the oil palm, *Elaeis guineensis* Jacq. *Nature, Lond*, **182**, 735.

Rees, A. R. (1961) Midday closure of stomata in the oil palm, *Elaeis guineensis* Jacq. *J. exp. Bot.*, **12**, 129.

Rees, A. R. (1963a) An analysis of growth of oil palm seedlings in full daylight and in shade. *Ann. Bot.*, NS, **27**, 325.

Rees, A. R. (1963b) An analysis of growth of oil palms under nursery conditions. II. The effect of spacing and season on growth. *Ann. Bot.*, NS, **27**, 615.

Rees, A. R. (1963c) Relationship between crop growth rate and leaf area index in the oil palm. *Nature, Lond.*, **197**, 63.

Rees, A. R. and **Chapas, L. C.** (1963a) An analysis of growth of oil palms under nursery conditions, I. Establishment and growth in the wet season. *Ann. Bot.*, NS, **27**, 607.

Rees, A. R. and **Chapas, L. C.** (1963b) Water availability and consumptive use in oil palm nurseries. *J. W. Afr. Inst. Oil Palm Res.*, **4**, 52.

Rees, A. R. and **Tinker, P. B. H.** (1963) Dry matter production and nutrient content of plantation oil palms in Nigeria. I. Growth and dry matter production. *Pl. Soil*, **19**, 19.

Robertson, G. W. and **Foong, S. F.** (1977) Weather-based yield forecasts for oil palm fresh fruit bunches. In *International development in oil palm*, ISP, Kuala Lumpur.

Ruer, P. (1966) Relations entre facteurs climatiques et nutrition minérale chez le palmier à huile. *Oléagineux*, **21**, 143.

Ruer, P. (1969) Système racinaire du palmier à huile et alimentation hydrique. *Oléagineux*, **24**, 327.

Sparnaaij, L. D. (1960) The analysis of bunch production in the oil palm. *J. W. Afr. Inst. Oil Palm Res.*, **3**, 109.

Sparnaaij, L. D., Rees, A. R. and **Chapas, L. C.** (1963) Annual yield variation in the oil palm. *J. W. Afr. Inst. Oil Palm Res.*, **4**, 111.

Syed, R. A. (1979) Pollinating insects of oil palm. Commonwealth Institute of Biological Control. Typescript report.

Syed, R. A. *et al.* (1982) Insect pollination of oil palm: introduction, establishment and pollinating efficiency of *Elaeidobius kamerunicus* in Malaysia. *Planter, Kuala Lumpur*, **58**, 547.

Turner, P. D. (1977) The effects of drought on oil palm yields in South-east Asia and the South Pacific region. In *International development in oil palm*, ISP, Kuala Lumpur.

Turner, P. D. (1981) *Oil palm diseases and disorders.* Oxford Univ. Press, pp. 197–201.

Turner, P. D. and **Bull R. A.** (1967) *Diseases and disorders of the oil palm.* Incorp. Soc. of Planters, Kuala Lumpur.

Van Heel, W. A., Breure, C. J. and **Menendez, T.** (1987) The early development of inflorescences and flowers of the oil palm (*Elaeis guineensis* Jacq.) seen through the scanning electron microscope. *Blumea*, **32**, 67.

WAIFOR (1956) *Fourth Annual Report, 1955–6*, p. 102.

Watson, D. J. (1947) Comparative physiological studies on the growth of field crops I. *Ann. Bot.*, NS, **11**, 41.

Watson, D. J. (1956) Leaf growth in relation to crop yields. In *The growth of leaves*, pp. 178–90, Butterworth, London.

Wormer, Th. (1958) Croissance et développement du palmier à huile. *Oléagineux*, **13**, 385.

Wormer, Th. M. and **Ochs, R.** (1957) Humidité du sol et comportement du palmier à huile. *Oléagineux*, **12**, 81.

Wormer, Th. M. and **Ochs, R.** 1959) Humidité du sol, ouverture des stomates et transpiration du palmier à huile et de l'arachide. *Oléagineux*, **14**, 571.

Wood, B. J. (1985) Some consequences of weevil pollination of the oil palm in South East Asia. *Planter, Kuala Lumpur*, **61**, 423.

Zeven, A. C. (1965) Oil palm groves in Southern Nigeria: Part 1. Types of grove in existence. *J. Nigerian Inst. Oil Palm Res.*, **4**, 226.

Chapter 5

Oil palm selection and breeding

Oil palm breeding has as its aim the production of plants giving the maximum quantity of palm oil and kernels per hectare. The oil palm breeder must therefore concern himself with all the circumstances of bunch production, with the quantity of oil and kernels per bunch, and with resistance to disease. The monoecious character of the palm makes cross-pollination general and, as with other cross-fertilized plants, the value of parents will therefore ultimately be determined by the performance of their progeny in crosses. Two products, oil and kernels, must be taken into account, and in countries where palm kernel oil is extracted the separate values of three products may have to be considered. Early breeding was directed towards palm oil production since in high-quality *tenera* fruit the quantity of oil is four to eight times that of the kernels; this was, however, insufficient reason for breeding entirely for oil, as will be shown later.

During the last decade propagation of the palm by tissue culture was achieved in two laboratories and very recently several other laboratories have also started to produce clonal plantlets, or ramets. Clonal trials have now been laid down in at least eight countries and two laboratories are now geared to 'factory' production of ramets. The rate at which these developments will proceed, how far and how soon ramet production will supersede seed production, and what effects cloning will have on breeding programmes are all matters which, though widely debated, cannot yet be exactly predicted (see pp. 281 and 286).

The early history of selection

The establishment of plantations on a large scale was an achievement of the 1920s, both in Zaire and in the Far East, and workers in these territories were not slow to start investigations on the improvement of the crop by selection and breeding. In Zaire, selection was in the hands of the Institut National pour l'Etude

Agronomique du Congo Belge (INEAC), while in the Far East the work was undertaken by the large plantation companies of Indonesia and Malaysia and by the Algemene Vereniging van Rubberplanters ter Oostkust van Sumatra (AVROS) and the Department of Agriculture, Malaya.

Owing to the great differences in the material being used, the approach to improvement in the two regions was also different. In Africa the poor quality of the *dura* fruit was apparent, and although there was some hesitance in breeding exclusively for *tenera* palms, the presence of fine specimens of the latter led to an early concentration on the discovery and reproduction of high-quality *tenera* material. This policy reaped its reward with remarkable rapidity.

In the Far East, the relatively high quality of the illegitimate *dura* progeny emanating from the Deli ornamental avenues tended largely to confine the very early work to mass selection* for the provision of seed for further planting. In 1922 AVROS affirmed that the Deli type must remain the standard oil palm for Sumatra until breeding had done its work with the offspring of newly imported varieties. Under these circumstances the general procedure was for mass selection to be carried out for extending plantings, and for separate breeding programmes to be undertaken both with outstanding Deli individuals and with imported material.

Indonesia (Island of Sumatra)

Little information exists regarding the yield and quality of the original unselected plantation population of Deli palms and the few publications from Indonesia were concerned largely with crosses between Deli and imported material. Schmöle (1930) accepted Blommendaal's figures of 62–63 per cent mesocarp, 30 per cent shell and 7–8 per cent kernel to fruit as being 'average components' in 1929, but committed himself to the somewhat contradictory statement that the Deli was of more or less constant standard with a percentage shell varying from 20 to 40. He also suggested 100 kg per palm as an average annual yield of the first generation on good soil. It may be remarked that 40 per cent shell is quite a usual figure for African *dura* fruit, and the composition of the early Deli material may not have been so 'standard' as is often claimed.

The mystery of the origin of the four Buitenzorg (Bogor) palms has been referred to in Chapter 1; but the descendance and selection of their progeny in the 70 years between their coming into

---

* The term is used according to Hayes *et al.*'s (1955) definition: selection of some desired character or characters, where progeny of the plants selected are grown in bulk. With the majority of mass selection in the oil palm, even in the early days, controlled pollination was used.

bearing and the establishment of planations are also somewhat mysterious. It has been stated that the similarity of the four palms is 'generally acknowledged', but only two palms remain and there are no published figures of their fruit and bunch analysis. Recent observation suggests that the fruit composition was not outstanding. Illegitimate progeny of these old palms, planted in West Africa in 1958, indeed showed a remarkable similarity in appearance and in the form and set of the bunches, but the fruit analyses were not similar to the modern Deli, being more akin to some of the *dura* found in the Africa groves. The progeny of later generations of Delis planted in West Africa have always shown typical Deli composition.

The standards of selection adopted when planting the first estates from illegitimate seed taken from the ornamental avenues which had been established in the last two decades of the nineteenth century are not known. Stoffels (1934) stated that the best mother palms used for the original Pulu Radja estate and Tamiang plantings were those in the Saint-Cyr and Tandjong Morawa avenues, but that, owing to an insufficiency of planting material, seed had also to be taken from poorer specimens in the gardens of Medan.

In spite of the good quality of the Deli material there was interest in importations from Africa right from the start. There are records of importations to Bogor in 1914–15 and to several estate groups and AVROS a few years later. Some of this material proved an embarrassment as it was of inferior quality; and much of it, being poor African *dura*, was eliminated at an early stage not only from Government selection stations, but also from the estates of plantation companies. As mixed plantations existed in the early days it seems most likely that some subsequent plantations were 'contaminated' with characters inherited from this material.

One of the first attempts to select within Deli material through the evaluation of proper records was made at Marihat Baris estate where 2,000 palms, planted in 1915, were recorded from the time they came into bearing. In late 1922 self-pollination of fifteen palms was carried out in cooperation with APA (AVROS) and progenies were planted in the selection areas at Sungei Pantjur and later at Polonia during the period 1924–31. The factors sought were high bunch yield, high bunch number, mesocarp thickness of 4 mm or more, and freedom from Crown disease. The selected palms were selfed and crossed for seed production as well as for the laying down of the F_1 generations at the selection stations. It has been reported that, in the event, this selection and breeding gave rise to high-yielding palms with comparatively few, though heavy, bunches. The selections were made at a very early age and the progeny were characterized by early bearing, suggesting that precocity was a factor unconsciously selected and inherited. It is possible that the exceptionally early bearing of some later fields in Sumatra and Malaysia may be partly due to this factor and not entirely to improved agronomy as is usually assumed.

Selection undertaken by estate groups led to progenies with larger numbers of relatively small bunches, but the total oil yield per palm proved to be of about the same order. Examples of this work may be given from two estates. Fifty parents were selected from 24,500 palms which had been individually yield recorded. They were originally selected for high yield, bunch weight, fruit-to-bunch ratio and mesocarp percentage. Oil determinations were introduced later. These good characters were in large measure reproduced in their progenies (F_1) from which up to 4 per cent of the palms were selected for legitimate seed production for further estate plantings. The averaged data are shown in Table 5.1. It will be seen that both in yield and fruit characters the standard of the selected palms, which were the élite of their generation, was reproduced or even bettered in the progeny; in the case of estate A the F_1 generation was a replanting whereas the F_0 was on virgin soil.

These figures illustrate the standard of Deli material being obtained by selection in Sumatra in the 1930s, and they show how the Deli palm obtained its reputation both as a high yielder and as a palm of good and uniform composition.

Another plantation group, SOCFIN (Société Financière de Caoutchouces), followed initial Deli selection with two generations of selfing or with one generation of selfing followed by an F_2 of crosses. Few details are available for this selection and breeding work at Mopoli and Bangun Bandar, but in the F_1 selfs progeny yields 50–60 per cent above those of the initial populations, and in the F_2 selfs and crosses 20, 12 and 40 per cent above the F_1 populations, were claimed (Carrière de Belgarric, 1951). These figures are impressive, particularly those of the F_1 selfs, but it is of course impossible to tell what proportion of the yield increases must be attributed to improved planting methods or the use of more fertile areas. The yields and bunch analysis of the selected F_1 palms at the two centres were strikingly similar. Bunch yield was about 200 kg per palm per annum, fruit-to-bunch ratios were 60–61 per cent and mesocarp to fruit 68–69 per cent. Further and more detailed infor-

Table 5.1 *Selection of Deli* dura *in Sumatra (means per palm per year)*

Estate	Generation	Years from planting	Weight of bunches (kg)	No. of bunches	Wt. per bunch (kg)	Fruit to bunch (%)	Mesocarp to fruit (%)	Oil to mesocarp (%)	Oil to bunch (%)
A	F_0 selections	6–14	224	10.5	21.3	66.1	67.2	48.8	21.6
A	F_1 progenies	11–17	213	9.2	23.2	67.4	65.7	49.8	22.0
B	F_1 progenies	11–17	236	10.0	23.6	64.9	67.5	51.1	22.3

mation on the origin of blocks of Indonesian Deli palms, and on the predigrees of individual Delis, have been given by Lubis (1984) and Rosenquist (1985).

Serious breeding work with material imported from Africa was started by APA soon after the First World War, imports being received from many parts of the continent; trials of Deli crosses with Deli × import *dura* and *tenera* and with crosses within African material were put in hand at once and the majority of progenies were planted at Sungei Pantjur and Polonia (Pronk, 1955). The high sex ratio and low bunch weight of the imported material were soon apparent.

The pedigree of the important Sumatran imported material is confusing, so it is set out schematically in Fig. 5.1.

An existing record shows that the well-known SP540 *tenera* at Sungei Pantjur was part of a consignment of seed sent by the Director of the Eala Botanic Gardens in Zaire. The seed was recorded as var. 'Djongo', indicating that it came from the famous Djongo *tenera* palm (see p. 202) which was to give rise to a high proportion of the best *tenera* lines in its own country. It is of interest to note here that seed from Eala was imported into Colombia some 15 years later and provided one parent of unquestionably 'Yangambi-type' *tenera* to be found at Calima. Thus it is almost certain that a great quantity of good *tenera* in all three continents is descended from the Djongo palm.

Further crossings and plantings at the F_1 and F_2 stations of Sungei Pantjur, Polonia, Karang Inoué and Aek Pantjur were made during the middle and late 1930s at a time when the inheritance of fruit-form characters was still not understood. At Polonia the planting '820' was of selfs of SP540, many of which gave a fruit analysis superior to that of the parent palm. At Sungei Pantjur the selection areas were cut out, leaving only SP540 itself and a high-yielding and high-quality Marihat Deli, SP2041; these were extensively used as parents. At Karang Inoué planting '827' was a crossing of SP540 with a Deli, Pol. 3003, of Marihat origin selected by Schmöle. The analyses of the resulting *dura* and *tenera* are of interest as showing the high quality of *tenera* which could then be produced, and the 'down-grading' effect on the *dura* half of the progeny of the Deli × African cross:

	Mesocarp to fruit (%)	Shell to fruit (%)	Kernel to fruit (%)
Dura	53.8	35.5	10.7
Tenera	83.5	11.1	5.4
Schmöle's Deli 'average components'	62.5	30.0	7.5

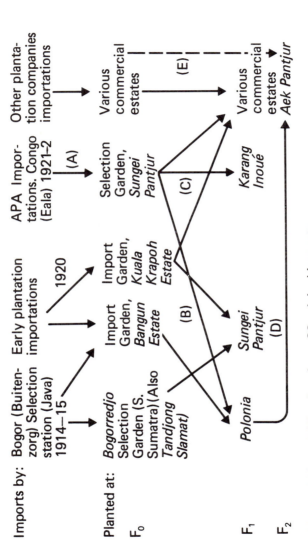

Fig. 5.1 The path of 'Import' *tenera* in Sumatra.

At Aek Pantjur there were plantings in which progenies from crossings in the F_1 of SP540 were used, and comparisons made between Deli, Import *dura* and Import *tenera*, and Deli × African crosses. In Sumatran parlance the terms Import *dura* or Import *tenera* indicated descendance from importations after 1900 with no Deli crossing. Deli Import *dura* and Deli *tenera* indicated *dura* or *tenera* palms resulting from a Deli × Import cross. In this case (Table 5.2) the fruit of the Import *dura* resulting from SP540 F_1 selfings and crosses were, curiously, of higher quality, as judged by mesocarp percentage, than the fruit of either Deli or Deli Import *dura*, indicating either that SP540 was quite exceptional in the quality of its African *dura* parentage or that there was some out-pollination with Deli in the SP540 F_1 generation (Pronk, 1953, 1955). In these trials, and in a similar one on an estate, the Deli palms gave the highest weight per bunch, with the Deli Import *dura* and the Deli *tenera* intermediate between Deli and Imports. Imports always gave a higher number of bunches and, at Aek Pantjur, the crosses gave the highest overall bunch yield. *Pisifera* derived from SP540 were always infertile, but some *pisifera* from Bangun, including Pol. 3184 which is to be found in several Suma-tran and Malaysian pedigrees, were largely fertile.

Table 5.2   *Fruit and bunch analysis of Deli, import* dura *and* tenera, *and of crosses planted at Aek Pantjur 1939–40. Analyses 1950–2*

Category of palm	Fruit to bunch ratio* (%)	Weight per normal fruit (g)	Mesocarp to fruit (%)	Nut to fruit (%)	Number of palms
Deli *dura*	58.8	15.4	56.8	43.2	102
Deli Import *dura*[†]	57.4	20.1	59.3	40.7	109
Import *dura*[‡]	55.3	20.9	64.2	35.8	49
Deli *tenera*[†]	53.2	15.6	80.7	19.3	101
Import *tenera*[‡]	56.2	18.5	87.9	12.1	86

* Normal + parthenocarpic fruit.
[†] derived from Deli × *tenera* crosses.
[‡] derived from *tenera* selfs and crosses, F_2 from SP540.

Source: Pronk (1953).

Some *tenera* programmes were carried out by estate groups. In one case the material was derived from illegitimate Deli *tenera*. Further crossing with Deli resulted in improvement in the mesocarp percentage of the *tenera*. In other cases the *tenera* were derived from the Kuala Krapoh Import Gardens. The figures in Table 5.3 show the variable standards of yield and analysis obtained.

Table 5.3   *Some yields and bunch analyses of Deli* tenera *in early Sumatran estate plantings (means per palm per annum and mean percentages)*

	Years from planting	Weight of bunches (kg)	No. of bunches	Weight per bunch (kg)	Fruit to bunch (%)	Mesocarp to fruit (%)	Oil to mesocarp (%)	Oil to bunch (%)	Oil per palm (kg
Estate B									
F₀ Illegitimate Deli *tenera*	8–11	224	12.9	17.4	59.5	83.5	51.4	25.6	57
F₁ *tenera* from D × T cross	9–14	231	12.8	18.0	59.4	86.4	50.1	25.8	60
Padang Pulau Best 5 palms in F₀	11–14	128	—	—	55.6	81.1	47.5	21.4	27
Bangun Bandar Best 10 palms in F₁	6–9	196	—	—	62.6	78.9	56.2	27.8	54

Malaysia

The early estates in Malaysia obtained open-pollinated seed from the ornamental avenues (Plate 5.1) and other plantings in Sumatra and so the Deli palm became the established plantation oil palm. The history of the first plantings is a little confused but the first estate, Tennamaram, planted in 1917, seems to have obtained its seedlings from Sumatra via a nursery at Rantau Panjang estate, while the second, Elmina, obtained seed from an avenue planted at Rantau Panjang in 1911–12 (Jagoe, 1952a).

In 1922 an avenue of eighty-six palms was established at the Central Experiment Station at Serdang. The seed came from palms in Kuala Lumpur of unknown origin, but of Deli type. These palms showed considerable variation in yield and quality, but the ten best palms were widely used for seed production and their descendants have found their way into later commercial breeding programmes in Malaysia and into programmes in Africa. It is of interest, therefore, to note the data on which the selection of these ten palms was based (Table 5.4, Jack and Jagoe, 1932). The bunch yields are of little value for comparison with the yields of palms in fields, since the former were from palms in single, widely spaced rows and had received assisted pollination for several years. Fruit analysis data are of interest, however, since they approximate to Schmöle's 'average components' and show a higher kernel percentage than has in later years been extracted from commercial Deli fruit (Plate 5.2).

Though many estates made use of the Serdang material, others made special importations of Sumatran seed. An important group of estates in Lower Perak, for instance, was planted with the Marihat material.

Pl. 5.1   A typical avenue of oil palms from which so much of the industry of the Far East developed. This avenue is in Perak, Malaysia. (T. Menendez)

Table 5.4   *The ten selected Serdang Avenue Deli* dura *palms*

Palm No.	Bunch yield per annum (kg)	Fruit to bunch ratio (%)	Mesocarp to fruit (%)	Kernel to fruit (%)
27	234	67.2	60	7.5
37	190	67.6	58	6.9
3	193	65.7	64	6.9
7	192	65.1	62	8.0
5	187	66.0	63	6.7
9	188	62.8	60	8.0
23	187	61.2	56	9.3
57	161	66.3	64	6.7
65	151	68.1	61	7.6
67	151	59.3	67	6.1
Mean	184	64.9	61	7.4

Source: $2\frac{1}{2}$ years' data (Jack and Jagoe, 1932).

Pl. 5.2 The Serdang avenue palms in 1962. (T. Menendez)

The oil palm industry made a slower start in Malaysia than in Sumatra and yield studies and selection were not under way until the 1930s; serious trial of the value of importations did not start until after the Second World War. Jack and Jagoe (1932) studied a block of 589 10-year-old palms at Elmina estate and found that in 1930 bunch yields varied from 0 to 309 kg per palm with a mean of 117 kg. The standard deviation was 121 and the coefficient of variation was 47 per cent. There seemed, therefore, to be a considerable field for selection. Even the twenty best palms which formed the preliminary selection still showed much variation in yield and fruit composition: bunch yield per annum varied from 170 to 294 kg, mesocarp to fruit from 55 to 76 per cent, shell to fruit from 20.1 to 35.8 per cent and kernel to fruit from 2.9 to 10.6 per cent. No attention appears to have been given to fruit-to-bunch ratios, a figure of 60 per cent being taken as normal.

One feature of the Elmina data which was not noted at the time was the negative and highly significant correlation (−0.72) between mesocarp and kernel percentage. Whether this correlation existed in the population as a whole is not recorded but it would appear that when shell percentage to fruit was in the 20–25 per cent range mesocarp percentages were high and kernels very small, whereas with large or even medium-sized kernels, shell percentage was usually over 30 per cent. This suggests a fair uniformity of shell thickness in selected Delis.

It was also noted at Elmina that high-yielding palms might sometimes be in pairs or small groups and that therefore environmental factors as well as heredity were concerned in their performance. This 'position effect' was later to be taken into account in making the final Elmina selections and in several other selection programmes.

Jagoe (1952b) described how in the course of his selection work at Elmina two of the palms under particular observation were seen to have an unusually large girth and a slow height increase in spite of a high rate of leaf production. One of these palms, the well-known Dumpy E206, was included among the final selections. In yield and some other characters the palm was not quite up to the selection standards set for other palms, although the bunches were large and the fruit-to-bunch percentage above average. Its position in the field was unfavourable and its yield was not therefore thought to have reached its full potential. Selfed and crossed progeny of E206 were planted at Serdang just before the Second World War; this was fortunate as the original mother palms on Elmina estate were all cut out during the Japanese occupation of Malaysia (Plate 5.3).

The selfed progeny of Dumpy E206 was quite uniform and the larger girth and small annual height increment were transmitted to a remarkable degree as shown in Table 5.5.

The bunch yield of E206 selfed progeny proved to be high where not overshaded by other palms, but the fruit-to-bunch ratio varied considerably and there were abnormalities of fruit set. The intermediate girth and height data of the Dumpy crosses suggested that dumpiness was an expression of quantitative genetic factors and that E206 and its selfed progeny exhibited a high degree of homozygosity in this respect though not in fruit characters. Some of the latter were of a very high order in the E206 × tall crosses, so a programme of 'back-crosses' between these and their half-sibs, the E206 selfed progeny, was started. True back-crosses could not be made owing to the destruction of the parent palm. The so-called 'back-crosses', together with F_2 Dumpy selfs, were distributed to a number of estates and considerable populations of these palms were established.

Pl. 5.3   E206 Dumpy F₁ self at Serdang. 26 years old. Note the characteristic 'blunt' leaf tip of this progeny.

Selection within Deli populations was only undertaken before the Second World War by two plantation groups. In the larger, selection was undertaken firstly among palms owing their origin to Serdang Avenue palms. It has been recorded that palms Nos 5, 7 and 23 were of most value and their progenies were most extensively used (Chemara, 1960; Breure *et al.*, 1982; Rosenquist, 1985).

Table 5.5   *Height and girth data of Dumpy E206 and its progeny*

	Girth at 122 cm (4 ft) (cm)	Height to base of crown (cm)	Age (Years)
Original E206	287	335	15
'Normal' palms at Elmina	226	518	15
F₁ Selfed E206	272	292	12
Cross E206 × E268	249	406	12
Cross E206 × E152	221	411	12
Selfed E152	206	417	13
Selfed E268	249	475	13

Table 5.4 shows that the fruit quality of these palms was in no way outstanding, shell percentage being 30 per cent or more in each case. Selections were also made among large numbers of palms grown from Sumatran open-pollinated Deli seed. In the period 1935–40, 338 progenies were laid down by inter-crossing 175 palms of Serdang legitimate and Sumatran illegitimate origin and planting in 'genetic blocks', the term being used to signify blocks of palms of known parentage. Individual yield recording of the parents was undertaken, but fruit and bunch analysis was lacking or incomplete.

Selection undertaken by another plantation group has not been described in such detail and the basic material was of different origin from that described above, being legitimate and illegitimate Deli material selected for production at a single source in Sumatra. This F₁ material was planted between 1928 and 1940 by progenies of 100–150 palms each. During the latter part of this period crosses were made between the best palms to form a further generation and certain of these progenies were kept after the Second World War to form the basis of future Deli breeding programmes; the latter included more than a hundred hectares of inbred material and crosses. The commercial Deli seed issued was then obtained from palms whose progenies had done well in selfings and crosses (SOCFIN, 1963). In general the standards adopted led to the breeding of progenies with comparatively few, but large, bunches. Differences in this respect are well illustrated by a trial of two Deli progenies from the Serdang Elmina collection with two progenies each from the two plantation groups. The data given are for 1962 when the palms were 10 years old:

Jerangau Station	Number of bunches	Weight per bunch (kg)	Weight of bunches (kg)
Two Serdang progenies	8.9	22.2	197
Two plantation 'A' progenies	12.5	17.8	210
Two plantation 'B' progenies	9.1	18.8	171

One of the two Serdang progenies (14/2 × 10/8) was outstanding in its consistent production of large bunches; in the whole bearing period from 1956 to 1962 the bunch weight was 30 per cent higher than the mean bunch weight of the plantation progenies.

It is disappointing that it is not possible to measure the degree of improvement in yield and quality of the Deli palm in the inter-war period. More exact analytical work was carried out in Sumatra than in Malaysia and larger numbers of palms were recorded. Claims were made that bunch production in the F_1 was 50 per cent above the production of the fields from which the parents were selected, but how far improved cultural practice was responsible for enhanced yield it is impossible to say. There are no reports, as in Africa, of trials of selected versus unselected material. The general standard of later production does not suggest that the recording of tens of thousands of palms was likely to produce much better progenies than the recording of several hundred, and in view of the restricted origin of the Deli this is perhaps not surprising.

Zaire

The history of selection and breeding in Zaire is the history of the emergence of the *dura* × *pisifera* cross. This discovery closed one epoch of selection and heralded another. The early epoch in Zaire was, however, quite different from that of the Far East, and it was the realization of the value of the *tenera* by Ringoet and the meticulous and inspired work of Beirnaert, carried through under the difficult conditions of Central Africa, that made it possible for the *tenera* era to develop within 30 years of the establishment of the first plantations.

Beirnaert (1933a,b,c), whose death in 1941 deprived the oil palm world of its leading figure, described the early Zaire selection work in both its practical and theoretical aspects in considerable detail.

Ten open-pollinated *tenera* bunches were used for the establishment of the *Palmeraie de la Rive* at Yangambi in 1922. One was from a palm – the famous Djongo (= the best) – at Eala which had 55 per cent mesocarp and 30 per cent oil to bunch, and the other nine were from groves at Yawenda. The planting was at close spacing and *dura* and poor types of *tenera* were gradually eliminated. Further fields were planted in 1924 and 1927 with open-pollinated *tenera* seed from a plantation at Gazi where 16,000 palms had been under observation. Fields planted in 1929 and 1930 were from seed of *tenera* palms taken in the *Rive* after the elimination of *dura* palms; this was a second generation of illegitimate *tenera* material.

After the thinning out, the remaining palms in these fields were subjected to a system of 'control', i.e. they were yield-recorded, and those proving to be high yielders were evaluated by an elaborate process of bunch analysis, the sum of the exterior, interior, parthenocarpic and abortive fruits being taken as the measure of the numbers of flowers produced by the inflorescence. The examination of all these data resulted in the choice of possible mother palms for breeding; these were known as *candidats arbres-mères*, or CAM. The work was carried on in stages with the elimination of many palms at each stage. Finally, after 3 years of observation when the palms were mature, the following minimum standards were laid down and resulted in the choice of actual *arbres-mères* (AM):

1. Average bunch production   140 kg per annum
2. Mesocarp percentage to fruit   75 per cent
3. Average weight of exterior fruit   12 g
4. Weight of kernel   1 kg
5. Yield of palm oil   33 kg per annum
6. Normal growth

These standards were not applied rigidly and exceptional performance in one respect allowed a palm to be selected even though it was below standard in another. This method of selection was based on certain theoretical and practical concepts (Beirnaert, 1933a). Beirnaert drew attention to the fact that, at that time, the mode of inheritance of characters combining to give a high oil yield was unknown. The parents would, however, undoubtedly be polyhybrids and selfing palms or crossing two palms with similar characters would lead to the production of individuals superior, inferior or equal to the parents. If, it was argued, parents showing high values in some respects and low values in others were crossed with parents showing the inverse of this, then a great many favourable genes and some homozygosity could be expected in the progeny. The crossing of palms with complementary characters was therefore advocated and selfing was only to be permitted with palms showing a good performance in all respects. Selfing was recommended for the F_2 generation using palms with the expected high performance in each desirable character.

The standards laid down are of much interest. The low mesocarp percentage standard reflects the initial difficulties in finding high-quality *tenera*; later breeding raised this figure considerably. Bunch production was set at an exactingly high figure for conditions in Zaire and among the original selections it was not often achieved. Large fruit and a kernel of at least 1 g in weight were thought to be advantageous. It was postulated that 'the best type is that which gives the largest absolute quantity of oil; in other words, the fruit which has the greatest absolute quantity of mesocarp and kernel'. This can be disputed. Oil yield is the product of bunch weight and oil per bunch. The absolute quantity of mesocarp in the individual fruit (as distinct from its percentage) is, theoretically, of no consequence. Ten bunches weighing 10 kg each and giving 22 per cent oil to bunch can provide 22 kg of oil whether the individual fruit weigh 8 or 16 g, and there appears, neither on theoretical nor experimental grounds, to be any virtue in large fruit *per se*.

However, Beirnaert also considered that a large number of flowers per bunch was an important factor and, naturally, if a narrow standard for this factor was to be laid down then the size of fruit would be important. In fact, however, the number of flowers per inflorescene in the CAMs varied widely and the size of fruit and the number of flowers tended to be inversely related. There was also a considerable variation in the number of flowers setting fruit. It was admitted that larger fruit size may sometimes be indicative of unfavourable characters, e.g. small numbers of bunches or poor fruit set, but it was claimed that larger fruit would have advantages in milling.

All these considerations led to the selection of a distinctive type of fruit, the *Yangambi-type*, which was large, ovoid, had a large thin-shelled nut placed a little above centre and a wide basal portion of mesocarp (Fig. 5.2) These characters of the early Yangambi-type *tenera* fruit have made them relatively easy to identify and descendants of the first selections can be found in many parts of the world, e.g. West Africa, Malaysia, Colombia. Very small numbers of this type of fruit existed in the original material. For instance, of 200 palms with 70–80 per cent mesocarp the distribution of mesocarp weight per exterior fruit was as follows:

Weight of mesocarp in exterior fruit (g)	4–8	8–12	12–16	16–18
Frequency (%)	36	51	11	2

The frequency distribution of mesocarp percentage of a group of 167 palms in the *Rive* illustrates the quality standards of *tenera* palms existing at that time and the potentiality of the best of them:

Percentage mesocarp to fruit	65–70	70–75	75–80	80–85	85–90	90–95
Frequency (%)	6	13	29	33	17	2

Table 5.6 gives data of certain of the Yangambi mother palms recorded between 1931 and 1933 and which were stated in 1941 to be among the best represented in the F_1 fields of the INEAC which were planted from 1934 onwards. Of the selected palms in the *Palmeraie de la Rive* (R) over 70 per cent were descended from the Djongo of Eala.

It is of interest to compare these selections of illegitimate material coming from original open-pollinated selections with the mean of the F_0 generation of Deli palms in Table 5.1 and with individual palm data in Table 5.4. All three lots of palms represent a starting-point in their respective countries for selection and for crossing and

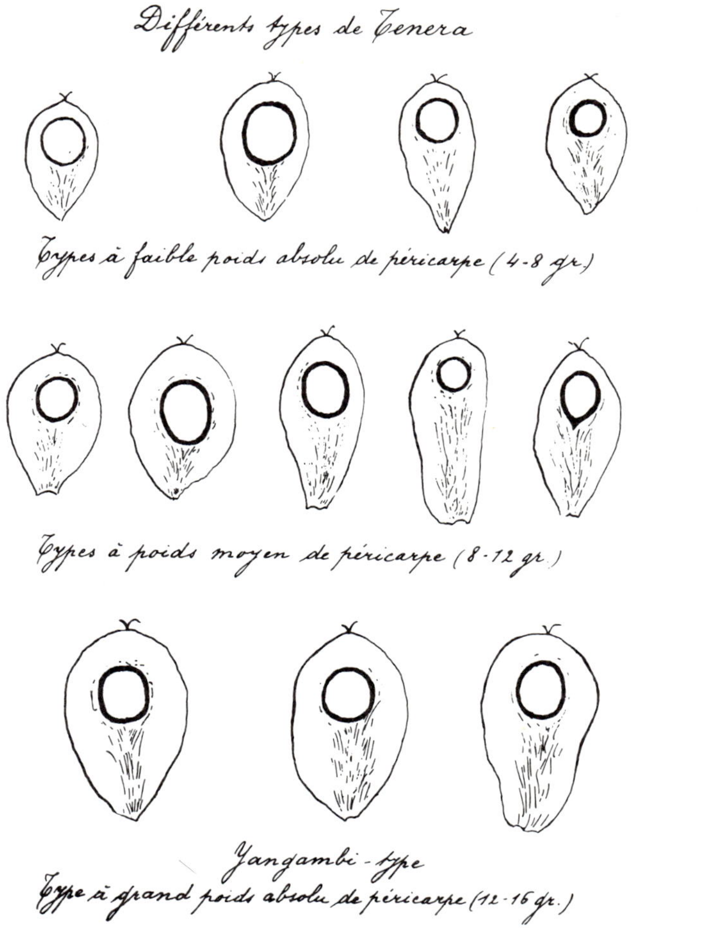

Différents types de Tenera

Types à faible poids absolu de péricarpe (4-8 gr)

Types à poids moyen de péricarpe (8-12 gr)

Yangambi-type
Type à grand poids absolu de péricarpe (12-16 gr)

Fig. 5.2 Beirnaert's drawings of different types of Congo (Zaire) *tenera* (1933a).

selfing to breed an F_1 generation. The Deli *dura* fruit characters were much more uniform than the Zaire *tenera* characters. Yield, mainly affected by environment, was higher in the Deli as was fruit-to-bunch ratio and weight per bunch, but mesocarp percentage was naturally much lower and little notice was taken of the kernel. These two groups of palms were soon to converge.

Selection and breeding came to an end at Yangambi in 1959 but fortunately several of the best lines had been established at Binga where this population (which Rosenquist (1985) terms a breeding population of restricted origin or BPRO) has been bred pure to the F_3 generation. Eight of the parents listed in Table 5.6 are now

Table 5.6   *Some data of the best represented* tenera *parents in the* F_1 *T × T generation at Yangambi (Beirnaert, 1933a)*

Palm No.	Weight of bunches per annum 1931–3* (kg)	Weight per fruit (exterior) (g)	Weight per kernel (g)	Number of flowers per bunch	Mesocarp to fruit (%)	Kernel to fruit (%)	Fruit to bunch (%)	Oil per palm (kg)
5 R	120	10	0.5	—	89	5.5	55	28
7 R	161	18	1.4	700	85	7.5	67	42
16 R	136	10	1.2	1,180	79	11.0	—	33
19 R	136	20	1.4	540	87	7.0	—	34
25 R	136	9	0.6	900	87	6.4	66	31
37 R	117	11	0.7	670	88	7.0	62	27
68 R	152	19	1.1	980	82	6.2	65	39
79 R	122	7	0.9		75	13.7	54	25
85 R	114	12	1.1	1,180	79	8.8	63	29
121 R	137	16	1.6		77	10.2	60	34
130 R	149	16	1.4	1,100	81	11.2	66	34
176 R	?	13	1.1		78	12.0	62	24
229 R	153	16	1.3	560	83	8.8	66	42
255 R	101	11	0.9		88	7.2	56	25
261 R	133	8	0.8		76	11.0	51	25
267 R	117	9	0.7	2,400	73	8.7	57	23
397 R	164	11	1.2		81	9.5	70	45
302/1	158	10	1.1	1,000	80	10.5	62	38
53/3	218	10	1.5	2,300	75	15.0	63	47
244/4	163	10	1.2	900	72	12.7	60	33
171/6	218	15	1.0	1,600	85	6.5	44	37
Mean	145	12.4	1.1		81	9.4	60	33

* Partly estimated from 3 years' data.

represented at Binga. Yangambi material has also reached the Ivory Coast, Nigeria, Cameroon, Malaysia, Indonesia and Colombia.

West Africa

In the *Ivory Coast* a start on similar lines to that of Zaire was made: the progeny of sixty open-pollinated *tenera* palms were planted at La Mé between 1924 and 1930. The grove mother palms had very poor composition, but the best of the progeny were selfed between 1934 and 1938 to give an F_1 generation which proved to be of low production and was not further used, though some of the parents were later to be employed in postwar breeding schemes of the IRHO.

In *Nigeria*, the early work was concentrated on a population of some 800 palms of different forms and types planted at Calabar in

1912–16 with later supplying. *Dura, tenera, virescens* and mantled fruits were used to provide seed and, except for the *tenera* planting, only one parent of each form or type was used (Smith, 1929). Palms of all forms and types appeared in the progeny and green-fruited *dura* and *tenera* were common. Table 5.7 shows the number of palms recorded and selected and gives data on the yield and quality of the latter.

Table 5.7 *The Calabar parents. 1922–8 data*

Form and type	Number of palms recorded	Number selected	Bunch weight of selected palms		Fruit to bunch	Mesocarp to fruit	Shell to fruit	Kernel to fruit
			per palm per annum (kg)	per bunch (kg)	(%)	(%)	(%)	(%)
Dura,								
nigrescens	376 } 382	9	96	11.8	65	49	32	12
virescens	6							
Tenera,								
nigrescens	36 } 43	10	67	9.2	56	64	18	10
virescens	7							
Mantled	24	3	74	5.6	50	—	—	—
(*Poissoni*)								
Total or mean	449	22	80	—	—	—	—	—

Note: The fruit analysis relates only to eight *dura* and eight *tenera*.

The much higher bunch yield of the *dura* palms led to the assumption that *dura* would in general yield more highly than *tenera* palms; however, this result appears to have been quite fortuitous and no doubt due to the qualities of the single original grove *dura*. Certainly the original selection from the groves was fortunate in its *dura* and unfortunate in its *tenera*; the latter gave rise to comparatively low-yielding palms with poor fruit-to-bunch ratios and low mesocarp percentages, but large kernels. A series of selfs and crosses of these palms was planted on four stations in Nigeria in 1930 and sterile *pisifera* palms soon appeared in the *tenera* × *tenera* material. The *dura* continued its superiority as a yielder into this F_1 generation and this result biased seed distribution in favour of the *dura* for quite a long period. Although there was later shown to be little justification for this bias, it did prevent a period of large-scale issue of *tenera* × *tenera* seed leading to 25 per cent steriles, as had been the case in Zaire.

Another prewar selection of *dura* and *tenera* parents was carried out among palms of grove origin at Aba. Controlled pollination was not fully successful and populations of several thousand largely out-pollinated palms were established at the Oil Palm Research Station near Benin in 1939–41. This formed a useful source of further breeding material and, with the large numbers of *tenera* produced, gave no support to the theory that the latter were inherently lower yielding than *dura* palms. The *tenera* progeny were derived from four Aba *tenera* parents. The *dura* material proved to be of little interest.

Large quantities of seed issued within West Africa owe their origin to these two programmes and to later generations from them. Mention should be made of the outstanding Calabar *dura*, CA256 (or 551.256), which produced good selfed progeny and entered into many crosses; its bunch·analysis, exceptional for African *dura*, was as follows:

Fruit to bunch	Mesocarp to fruit	Shell to fruit	Kernel to fruit	'Loss'
63.5%	52.2%	31.2%	10.3%	1.9%

Early analysis in Nigeria was done by a boiling and pounding method which led to a 'loss' figure of 2–4 per cent appearing in the analysis; mesocarp percentages were thus lower by a few per cent than figures obtained by present methods. The frequency of green-fruited specimens in Nigerian seed issues is largely due to the inheritance of this character from two Calabar *dura*, CA(551).341 and 375, of good composition, and from some green *tenera* crosses and selfs.

The emergence of the *dura* × *pisifera* cross

The hybrid nature of the *tenera* was discovered by Beirnaert when he examined the fruit form of the progeny of his selfed and crossed Arbres-mères. The latter had been selfed, and also large numbers of complementary crosses, as already explained, had been made. The majority of the parents used are shown by their numbers in Table 5.6. For several years it had been apparent that a substantial proportion of the progeny of seed issued to estates was sterile. A count of 29,454 palms in fifteen blocks at Yangambi showed that 24.34 per cent were *pisifera* and this, overall, did not differ significantly from the 'expected' 25 per cent. In the examination of the descendants of twenty-five mother palms only seven had percentages of *pisifera* differing significantly from 25 per cent. The range was from 19.74 to 33.20 per cent. Examination of the productive palms showed that the *tenera* always constituted 50 per cent of all the palms, while the percentage of *dura*, though usually 25 per cent, varied in the same manner as the *pisifera*. This variation is discussed on p. 236.

Examination of the progeny of *dura* × *tenera* crosses showed that there were no *pisifera* and for the majority of these crosses the segregation was *dura* 50 per cent, *tenera* 50 per cent. It was also noted that *tenera* seed taken from grove palms surrounded, as to about 90 per cent, by *dura* palms gave progeny in the proportion:

357 *tenera* (50.50%): 336 *dura* (47.52%): 14 *pisifera* (1.98%),

the latter being accounted for by assuming that a small quantity of *tenera* pollen might naturally fertilize female *tenera* inflorescences.

  Dura × *pisifera* adult fields did not exist at that time, but the hybrid nature of the *tenera* became clear from the above results. By 1938 it was known that as much as 25 per cent steriles could be expected in *tenera*-derived material and it was becoming plain, after the discarding of some far-fetched theories, that there was a 'correlation between the shell thickness of the parent and the percentage of *pisifera* palms in the progeny' (Toovey, 1938). The true position was realized in 1939, but a full explanation of the presence of steriles was not published until the issue of M. Beirnaert's Le problème de la stérilité chez le palmier à huile' in the first wartime Leopoldville edition (1940) of the *Bulletin agricole du Congo Belge*. In this little-read paper M. Beirnaert (1940) not only showed clearly the inheritance of the shell-thickness character, but he also brought forward evidence against the theory, current in French West Africa at that time, that the *tenera* was a degenerating form of the oil palm. At the same time he stated without hesitation the steps to be taken to prevent further sterility, namely that *dura* × *tenera* and *tenera* × *dura* seed should replace *tenera* × *tenera* seed for a short period, after which full *tenera* production should be assured by the issue of *dura* × *pisifera* seed. The full information obtained from the Yangambi F_1 plantings was published in 1941, after M. Beirnaert's death (Beirnaert and Vanderweyen, 1941).

  Confirmation of the 'Congo theory', as it was at first called, was not long in coming from other territories, though publication was tardy. The Calabar *tenera* selfs and crosses in Nigeria in due course showed their general agreement (Hartley, 1957), and *dura* × *pisifera* crosses including Delis as parents were seen to provide *tenera* progeny (Vanderweyen, 1953; Toovey and Purvis, 1956). In Sumatra many *dura* × *tenera* and several *dura* × *pisifera* crosses had been carried out before the Second World War and instances of 50 per cent *dura* and 50 per cent *tenera* in the progeny of the former were already known. Confirmation of *tenera* production from *dura* × *pisifera* crosses had, however, to await the postwar period. Errors of pollination technique were not uncommon in West Africa, and were occasional in Sumatra. The success of the huge programme in Zaire was undoubtedly due to the very high standard of skill and discipline of the special workers, recruited from outside Zaire, who undertook the pollinations, germination and planting.

The techniques of selection and breeding

To select individual palms for breeding it is necessary to measure their bunch yield and to analyse the bunches for their oil and kernel content. To breed from selected palms controlled pollination must be undertaken.

The recording of bunch yield presents no real problem provided supervision is sufficient to ensure the observance of a few simple rules. Harvesting must be regular, a fixed standard of ripeness must be employed, and bunches must be weighed at the foot of the palm and marked or labelled there with the palm number before they are removed for analysis and processing. Their loose fruit must be carried with them. The harvester uses a field notebook with duplicate sheets to record the field number, the date, the palm number, the number of bunches (occasionally more than one bunch is ripe on a single palm on the same day), and the weight of the bunch or bunches. The information is then transferred to individual palm record books or a computer.

Bunch analysis

A special building is required as an analysis or field laboratory where bunch composition is determined; the one illustrated in Fig. 5.3 was designed for the purpose. The procedures used have not differed widely one from another; the ones described here are an adaptation of those used at Yangambi in Zaire and have the advantage of being suited to the handling of up to 3,000 bunches per month at a labour expenditure of less than 0.2 man-days per bunch without mesocarp oil analysis, the procedures for which will be mentioned later (Blaak *et al.*, 1963; Sparnaaij, 1958).

Analysis must be carried out so that the minimum moisture loss occurs during the process. Bunches must be form-determined (*dura, tenera* or *pisifera*) on their arrival with their loose fruit at the laboratory, and they must be dealt with on the same day. After weighing the bunch on a Berkel scale the spikelets are removed and weighed and a random 5 kg sample is then taken. The weighed stalk and remainder of the spikelets are discarded; these weights are all recorded on a card which then accompanies the 5 kg sample which should be immediately processed.

On the spikelets there are to be found (i) fully fertile fruit, (ii) parthenocarpic fruit in which an oil-bearing mesocarp has developed although there has been no fertilization and hence no embryo,

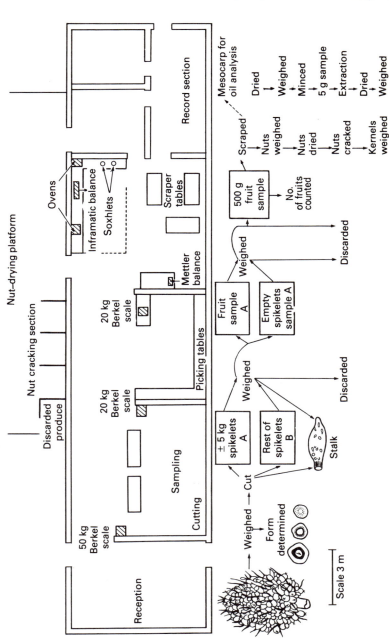

Fig. 5.3 A plan of a field laboratory for fruit and bunch analysis, with diagrammatic representations of the processes.

kernel or shell development, and (iii) infertile fruit which are small, colourless and non-oil bearing. The latter, which form a very small proportion of the total weight, are included in the weight of the empty spikelets. (i) and (ii) are sometimes estimated separately, and for exact estimations of mesocarp, mesocarp oil and kernels to bunch, separate estimations are necessary.

From the spikelet sample the ratio of fruit to empty spikelets and hence the *fruit-to-bunch ratio* can be calculated, i.e.

$$\text{Fruit-to-bunch ratio} = \frac{(\text{Normal} + \text{parthenocarpic fruit}) \times 100}{\text{Total bunch weight}}$$

For fruit analysis it has been shown that a 500 g sample is sufficient for bunches weighing less than 25 kg and even for bunches above that weight this size of sample has proved satisfactory. Normal fruits are taken by a sampling procedure and put on a scale until a 500 g sample has been obtained. The number of fruits is recorded and so the *weight per fruit* can be estimated. The fruit sample should be weighed soon after the bunch has arrived but a delay after weighing the 500 g sample is not important. After the mesocarp has been cut and scraped from the fruit by a sharp knife the nuts are weighed following 1 day of surface drying, and the *mesocarp percentage to fruit* is then calculated by difference.

The nuts are subsequently air dried for 3 or 4 days to facilitate cracking. After cracking, the kernels are weighed and the shell weight determined by difference. *Kernel-to-fruit* and *shell-to-fruit* percentages may thus be obtained. This is not an entirely accurate method as it has been shown that the kernels also lose weight during the drying process, and that the percentage of kernels to fruit may, therefore, be underestimated. In a study in Malaysia (Mollegaard, 1970) the following initial distributions of moisture and dry matter in fresh nuts were found:

	Per cent of fresh nuts				Moisture	
	Moisture in kernels	Dry kernels	Moisture in shell	Dry shell	Per cent of fresh kernels	Per cent of fresh shell
Deli *dura*	6.7	19.0	11.8	62.5	26.1	15.9
Tenera	14.8	37.1	9.1	39.0	28.5	18.9

Oven drying at 80 °C caused a more rapid moisture loss from shell than from kernels in both fruit forms, but particularly in the case of *dura*. After 4 hours' drying, moisture in the kernels was reduced to 5 per cent with *dura* and 9 per cent with *tenera* fruit, and it was calculated that if the kernel weights after this amount of drying were to be accepted in the fruit analysis then the kernel percentage to fruit would be underestimated by about 1 in the case of *dura* and 1.5 in the case of *tenera*. This work suggests that, with the usual 3 to 4 days air drying, underestimation of kernel percentages (and overestimation of shell) will not be so large; but it should nevertheless be recognized that there will be some underestimation of kernel content in all methods relying on partial drying before cracking and on calculating shell content by difference.

These methods of bunch analysis, which have been in general use for nearly 30 years, were recently the subject of a number of critical studies, and for a detailed appraisal the paper of Rao *et al.* (1983) should be consulted. These authors found the established system generally satisfactory but stressed the importance of giving special attention to the following points: (1) the maintenance of a uniform ripeness standard; (2) the close cutting of bunch stalk to the point of lowest spikelet insertion; (3) the weighing of prepared bunches immediately after harvest; (4) the use of a special random spikelet sampling procedure (Law, 1984); (5) the reduction of the fruit sample to 250 g or twenty-five to thirty fruit, its early weighing, and storage in thick-gauge polythene bags if mesocarp scraping is to be delayed.

Regarding the problem of obtaining an accurate kernel figure, Rao *et al.* (1983) recommend complete nut drying at 105 ° overnight and the recomputation of the calculated percentage kernel to fruit to commercial moisture levels by adding 7 per cent. This avoids the problem of different rates of moisture loss of *dura* and *tenera* nuts and of shell and kernel, but leaves an additional moisture loss figure which is not attributable only to shell, e.g.:

Tenera	*Fruit*	*Fresh nuts*	*Fresh mesocarp (by diff.)*	*Dry nuts*	*Dry kernels*	*Dry shell*	*Kernels with 7% moisture*	*Additional nut moisture loss*
Weight (g)	250	50	200	38	18.5	19.5	19.89	10.61
Per cent of fresh fruit	100	20	80	15.2	7.4	7.8	7.96	4.24

Oil analysis of the mesocarp may be undertaken by:
1. Indirect estimation through a knowledge of the water content.
2. Direct extraction.
3. A specific gravity determination.

Indirect estimation. An indirect method of estimating mesocarp oil content has been based on the assumption that a linear relationship exists between moisture content and oil content with a fixed dry matter percentage. Various formulae have been put forward, the first, Y (oil/fresh mesocarp) $= 87.38 - 1.08X$ (water/fresh mesocarp) being that of Vanderweyen *et al.* (1947) in Zaire. Later investigators in West Africa considered the method to be admittedly rough and obtained equations so near to

$$Y = 84 - X$$

that this approximation, which assumes a constant fibre percentage of 16, was considered satisfactory (Dessasis, 1955; Chapas *et al.*, 1957). No significant difference was found between the mean fibre content of *dura* and *tenera* pulp.

For this method the cut mesocarp obtained by the previous sampling methods may be reduced by quartering to duplicate samples of 100–150 g. These are accurately weighed and dried in an oven at 105 °C to constant weight. This takes 18–24 hours. Studies in West Africa (Blaak *et al.*, 1963) showed that there can be variations from progeny to progeny of 2–3 per cent in the mean percentage of fibre to mesocarp, and that seasonal within-progeny variations can be even greater. The indirect method is therefore unlikely to be sufficiently accurate for regular use in selection and breeding.

Direct extraction. Direct extraction by Soxhlet extractors is now widely practised both for bunch analysis and mill control. Large Soxhlets are used and adequate techniques and sampling procedures have been worked out. In using large Soxhlet extractors (600 ml capacity), the fresh mesocarp content of the 500 g fruit sample is calculated by subtracting the nut weight from the fruit weight. This avoids error due to moisture loss during depulping. Oven-drying of the mesocarp is then undertaken at 105–110 °C for 24 hours and duplicate 5 g samples are prepared on an inframatic moisture balance in filter paper containers plugged with cotton wool. The infra-red lamp fitted above the balance, which is designed to measure percentage loss, prevents moisture uptake during weighing. The duplicates may then be put with other samples into separate Soxhlets for extraction. Twenty-four samples can be housed in each Soxhlet and extraction with isohexane proceeds for 18 hours. After air drying until all isohexane has evaporated the residual fibre is oven-dried for 24 hours, weighed and the *mesocarp oil content* is calculated from the loss in weight. Duplicates differing by more than 3.1 per cent are discarded.

Using two ovens, one Paladin mincer, one inframatic balance, two 600 ml Soxhlets with electric heating mantles, forty-eight 1-litre aluminium containers with tight-fitting lids for dry samples and one distillation apparatus for isohexane recovery, over 6,000 duplicate samples, representing bunches, can be oil-determined per annum (Blaak *et al.*, 1963).

A cold solvent extraction method has been developed and it is claimed that where large numbers of analyses are required this method is much less expensive (Blaak, 1970). Large quantities of petroleum ether 60–80 ° are used and the solvent is recovered by distillation. The method relies on eight changes of solvent over a period of 4 days, but to reduce distillation the solvent is transferred from one partially extracted lot of samples to another on a shelf system.

A recent critical study of oil extraction methods showed that the most complete extraction is obtained when the dried mesocarp, after grinding, is passed through a 3 mm mesh sieve (Rao *et al.*, 1983), and this is now generally recommended.

Oil determination by measurement of specific gravity. A method of measuring mesocarp oil content by specific gravity determinations is in use in some laboratories. This *oléomètre* method is not claimed to be as accurate as Soxhlet extraction but it has the advantage that each of a large number of determinations can be completed in a relatively short time. If a given weight of solvent of known specific gravity is added to a given weight of dried mesocarp and the mixture is ground in a high-speed grinder, the oil and solvent may be filtered out. The specific gravity of the filtered mixture will be found to depend on the amount of oil in the solvent and hence in the original sample. The solvent used must have high extraction properties, a high evaporation point and a specific gravity as different from that of the oil as possible.

Results from the *oléomètre* tend to be slightly lower than from Soxhlet determinations and, by occasional Soxhlet checks, correction can be made. For comparisons, however, the *oléomètre* method is quite satisfactory (Servant and Henry, 1963).

Records. As previously mentioned, all weighings should be recorded on a card which moves with the bunch and samples, and on which all the weighings are first recorded and percentage determinations subsequently made. After fruit scraping, the mesocarp and nut samples go different ways; a separate card must therefore be used for oil analysis, or a perforated card can be employed, the bottom portion being torn off to go forward with the wet pulp. This method may assist in reducing errors of identification. The information from the bunch analysis card or cards is later transferred to the individual palm record card, an example of which is shown in Fig. 5.4.

Controlled pollination

The production of plants of known parentage is a long procedure in the oil palm. A very large female inflorescence has to be isolated, pollen has to be collected from a large male inflorescence without

PALM NO. 364						PARENTAGE Calabar *tenera, virescens*					
EXPT. 32-2						TYPE & FORM *Tenera*					
PLANTED 194						POSITION					

FRUIT AND BUNCH ANALYSIS

Date harv	No. of anal.	Total bunch wt. kg	Wt. of fruit kg	% F/B	500 g Sample No. of fruit	% Mes.	% Shell	% Kern	% Oil/m indirect	% Oil/m soxhlet
1955–6*	13	291.72	175.96	60.3		82.8	9.0	5.1		
22.5.59		45.000	27.9000	62.0	67	90.2	6.4	3.4		51.0
8.9.61		13.330	7.603	57.0	79	91.6	5.6	2.8	53.4	50.1
18.9.61		17.150	10.588	61.7	59	89.0	6.8	4.2	51.2	52.7
29.8.62		21.600	15.298	70.8	66	89.8	6.2	4.0		
5.10.62		16.625	9.965	59.9	81	87.4	7.4	5.2	50.1	52.7
23.4.63		9.800	4.786	48.8	63	89.4	6.6	4.0	51.6	54.4
Mean				60.1	69	89.6	6.5	3.9	51.6	52.2

*Old method of analysis (boiling and pounding)

BUNCH PRODUCTION

Year	No. bunches	Wt. bunches kg	Year	No. bunches	Wt. bunches kg	Year	No. bunches	Wt. bunches
1946	5	15.9	1953	9	172.3			
1947	15	79.8	1954	5	84.4			
1948	7	58.0	1955	10	218.1			
1949	10	122.0	1956	5	106.6			
1950	10	132.0	1957	5	96.1			
1951	18	59.0	1958	5	96.1			
1952	7	59.9						

REMARKS: Selected for the Main Breeding Programme in 1958

Fig. 5.4 Individual palm record card.

contamination, and the number of stages of the work from isolation and collection to the final transference of the plant from the nursery to the field is so great that very special measures have to be taken to prevent errors both of technique and identification.

When crossing *dura* with *tenera* or *pisifera*, or crossing *tenera* with *tenera* or *pisifera*, the proportions in which the fruit forms segregate in the progeny are known and therefore provide some check on the work done. It is not, however, possible to detect errors of pollination in *dura* × *dura* crosses, particularly in the Deli palm (except when other forms appear) unless some other distinctive heritable factor, e.g. dumpiness, is involved. In the early days of oil palm breeding many bunches said to be control-pollinated were in fact out-pollinated due to the crude methods used or lack of constant supervision. The discovery of the facts of fruit form inheritance provided a further stimulus for the devising of sound methods of controlled pollination and, although the essential requirements have remained unaltered, many refinements of technique have been introduced. It is not proposed here to describe in detail all the methods used, but only to state the essential requirements and to describe up-to-date methods.

The collection and storage of pollen

Male inflorescences must be bagged 7 days before the flowers open. Pollen-providing palms must therefore be visited frequently and the progress of male inflorescence production noted. At anthesis the inflorescence will be in the axil of the seventeenth to twentieth leaf from the central spear. The spathes of the inflorescence are cut away and the spikelets are thoroughly sprayed with a 40 per cent formaldehyde solution diluted by reducing to 1 part in 10 parts of water. This solution kills any foreign pollen or insects adhering to the spikelets. The inside of the bag to be used is similarly treated, the mouth being held downwards during the operation. Before carrying out any work connected with pollen the hands should also be sterilized with methylated spirits.

Bags have been made of many kinds of material, strong brown paper or heavy canvas being common. The former is subject to damage in the field while the latter requires to be impregnated with some substance to ensure that pollen does not pass through the cloth. In Zaire, where results showed that the standard of isolation was very high, khaki drill was dipped in rubber latex and vulcanized by dusting with sulphur. One corner was open along the seams but was tied up tightly with cord when not opened for observation or pollination. In Nigeria canvas bags 24 inches (61 cm) long by 18 inches (46 cm) wide and soaked in old engine oil to make them impermeable were used, but they were superseded by Terylene bags. Both types of bag have celluloid windows for observation, one measuring 4 inches (10 cm) square and set in the middle of the front and one 2 inches by 1 inch (5 × 2.5 cm) at the top of the rear of the bag as placed on the bunch; but bags used for male inflorescences may measure 24 inches by 16 inches (61 × 41 cm) and be provided with a window in the front only.

In the bagging operation a collar of cotton lint, well sprayed with formalin and dusted with insecticide, is placed around the peduncle, and the mouth of the bag is tied over this collar with strong twine. The bag may need supporting by tying its upper corners to nearby leaf petioles.

The pollen has to be removed under sterile conditions and dried before storing. By using a bag of Bondina Terylene material the male inflorescence may be dried and shaken out while still in the bag, thus obviating the need for removing it from the bag and shaking out the pollen in boxes or chambers where, in spite of formalin spraying, contamination may take place. After withdrawing the empty inflorescence, the pollen is dried at 35–40 °C for 24 hours in the bag. The pollen may then be sieved through a special sieving chamber and passed straight into a storage tube without any exposure to the outside atmosphere.

In a breeding programme, pollen may need to be stored for several months. Testing pollen for viability is usually done by germination of the grains on a maltose agar medium, but it has been shown that actual germination on stigmas is higher than on the artificial medium, and that good fruit set may be obtained when viability, as shown by laboratory tests, is as low as 10 per cent (Broekmans, 1957b). Pollen may be stored satisfactorily to this level of viability over calcium chloride at tropical room temperatures for 6–8 weeks. Prolongation of this period may be obtained by refrigeration to ± 5 °C or by storage in a partial vacuum (Devreux and Malingraux, 1960; Henry, 1959). Bunches giving low germination may contain seed with atrophied embryos, and a relationship has been found between the number of the latter and the length of time pollen is stored. This relationship is particularly clear with pollen stored for 1, 2, 3 and 4 months by ordinary methods. Where a partial vacuum is used, the effect of time of storage on germination is less pronounced, and when a complete vacuum is used the deleterious effect of storage is eliminated. The cause of this effect is unknown, but there seems no doubt that the use of differently stored pollen, or even of genetically different pollen similarly stored, may have differential effects not only on germination but also on early seedling growth. Such effects have been particularly noted with seed produced by fertilizations with stored pollen sent from one part of the world to another.

Standard methods and apparatus have now been adopted at seed-producing centres for vacuum drying pollen after initial drying over calcium chloride or in an incubator (Hardon and Davies, 1969; Bénard and Noiret, 1970). The pollen is sealed in ampoules and stored in deep-freezers at −18 to −20 °C. Rehydration for use takes about 24 hours; a sharp fall in viability is usual after 10 days, as with fresh pollen.

Preparation of the female inflorescence

A female inflorescence should be bagged at least 1 week before the first flowers are expected to open. Gauging this time is a matter of experience but entails regular inspections of each bunch. If there is any premature opening of the flowers, the bunch must be discarded. A flower is receptive when the lobes of the stigma are well separated but are still pink. On changing to red the stigma is believed to be no longer receptive, but to be safe the bag should be left on the bunch for 3–4 weeks after pollination; the stigma will then be black. Observations have shown that the receptive period for the flowers on an inflorescence normally lasts for 36–48 hours. Flowering starts at the base of the inflorescence and continues to the top. Occasionally receptive flowers may be found on an inflorescence for a week. Usually, therefore, if pollination is carried out on 3 successive days, maximum fertilization will result.

Spraying the inflorescence and bagging is carried out in the same manner as for the male inflorescence, i.e. a cotton-wool collar well sprayed with formalin and dusted with insecticide is placed round the stalk and the bag tied over it with strong twine and supported above by tying to adjoining petioles. Labelling is done at bagging, and the pollination particulars should be added to the label at the time of pollination. Canvas bags soaked in engine oil and fitted with celluloid windows as already described are suitable, but the lighter Bondina Terylene bags have also been used in West Africa. For female inflorescences they measure 48–51 cm by 73–76 cm and are provided with windows at the front and at the top of the rear side of the bag. Tests have shown that isolation with these bags is complete even though they are air porous. These tests take the form either of isolation without pollination or isolation followed by the inducing of parthenocarpic development of fruit by application of 2,4,5-T at 100 ppm. The latter method is to be recommended as the inducing of parthenocarpy ensures the development of any flower which may have been fertilized but which might fail to develop on a mainly non-developing bunch. Bunches treated with 2,4,5-T have to be examined fruit by fruit to detect any developing kernel and embryo.

Some doubt has been thrown on the efficacy of Bondina Terylene under Malaysian conditions. It has been possible to blow pollen through the bags under pressure and small quantities of pollen have been detected inside the bags after exposure to heavy rain.

Although double bagging may be resorted to, it is now more usual to treat the bags with a dilute (1 : 20) spray of acid-stable emulsified wax (Menendez, 1969).

Pollination

In bags with celluloid windows there are small round holes covered over with adhesive tape. When pollination is to be carried out the pollen may be placed in a test-tube, carrying a cork fitted with two L-shaped pieces of glass tubing stopped with cotton wool; or a bulb puffer with a projecting nozzle may be used. The puffer should be of single-spout type which does not inhale plantation air through the bulb and is only allowed to inhale through the spout when in the bag. Before pollination the pollinator sprays the bag with formalin and sterilizes his hands with methylated spirits. Each hole is then opened in turn and the pollen blown (in the case of the test-tube by blowing into one L-shaped piece while inserting the other in the hole) into the bag while the nozzle or glass tube is so manipulated as to ensure that pollen reaches every part of the inflorescence. The adhesive tape is replaced as the nozzle or tube is withdrawn. The bag is then shaken to help the pollen settle on all the flowers. Pollination is normally carried out in the morning (Plate 5.4).

Palms being used as seed-producers need a good deal of attention in order to facilitate pollination. Old leaf bases and debris have to be removed and the fibres and spines on the leaves subtending or near the bunch need to be cut away. Spines should be removed up to the lowest leaflet only and as little damage as possible must be done to the leaves since any pruning would be likely to reduce further female inflorescence production. However, light cuts may be made on each side and on the under surface of the leaf petiole of the subtending leaf so that it may be gently forced downwards to make access to the female inflorescence easier. If carefully done this does not seem to harm the leaf as a photosynthetic organ and premature withering will not take place.

Variation in the oil palm

In the 1920s it was considered that, in selection in the oil palm, it was necessary to differentiate between the Deli palm of the Far East and the palms of Africa on account of their difference of variation. Not only was the fruit form of the Deli constant, but variation of shell percentage did not seem to be very great. As late as 1930 it was stated that it was not known whether such variations as existed within the Deli were the result of genetic differences or of variations in environmental conditions (Schmöle, 1930). Nowadays the term Deli must be taken simply to indicate that the material concerned

Pl. 5.4 Controlled pollination: **A**, pollinator with bag for isolating the inflorescence; **B**, spraying formaldehyde to kill extraneous pollen before pollination takes place.

emanates from the *dura* populations of the early plantations of Sumatra and Malaysia.

As to African material, both in Africa and within the small quantities of material exported from Africa to other countries, it was obvious that very great variation existed, particularly in fruit and bunch characters, and Belgian and other workers set about the examination of these variations with great thoroughness (Beirnaert, 1933b; Smith, 1930, 1933).

This section describes variation as it exists within fields and progenies. The genetic element in this variation and its employment in selection and breeding are discussed later.

Bunch production

In Sumatra it was found that in reasonably uniform plantations the bunch yield of the highest yielding palms was twice or three times the average yield of all the palms. Very similar variation was found in Zaire where in a field of 2,000 illegitimate palms 0.8 per cent of these gave a yield a little less than three times the average of the field (Beirnaert, 1933b). At the other end of the frequency curve as much as 15 per cent of the palms may be giving less than 1 per cent of the yield (Yeow *et al.*, 1981) and of these some will not be bearing at all. The plant breeder's aim is to reduce this palm-to-palm variation in production and to eliminate those factors causing low yield. The variation is not, of course, entirely due to genetic factors. Most fields contain several non-producers; these are often palms which have been damaged or supplied in the early years and the shade thrown by surrounding palms and competition for nutrients has retarded their growth and inhibited flowering. Even when these and other environmental factors which lessen uniformity are taken into account, however, there is still a very large variation in palm yields.

A few examples will now be given of the degree of variability to be found in fields of palms in Africa and Malaysia which may be regarded as 'foundation stock' or F_1 from illegitimate material. These examples and the differences to be seen between them are purely illustrative; they suggest trends rather than establish firm disparities.

In Africa, coefficients of variation per palm for annual bunch yields have been computed for several fields. In Nigeria these coefficients varied from 40 to 70 per cent whether the material was illegitimate and unselected or from selfings of individual *dura* (Webster, 1939; Chapas, 1961). Coefficients for combined years' yields tend to be lower, and in the Ivory Coast Gascon *et al.* (1966) found that for 4-year totals they lay between 21 and 26 per cent for groups of more than 200 palms of different origins. Single-year yields are often positively skew, this being most marked in the early years; when totalled over 4 years the distribution becomes normal. Coefficients also tend to decrease with age.

Figure 5.5 shows the frequency distribution of mean annual yields per palm in 1961–3 of a field planted in 1952 with a large number of *dura* × *pisifera* crosses, together with the frequency distribution of two separate crosses within the field. The coefficients of variation of the latter are 31.5 per cent (progeny 24) and 46.5 per cent (progeny 40). In this case the per palm yield variation in progeny 24 is clearly less than for the whole field while the variation in progeny 40, though higher yielding than progeny 24, appears to be considerably wider. The variability of number of bunches per palm was a little higher than that of bunch yield, being 36.0 per cent for progeny 24 and 48.1 per cent for progeny 40.

In examining variability data of weight per bunch the relationship of bunch number to bunch weight must be borne in mind. Total bunch yield is the product of number of bunches and weight per bunch, the former depending on leaf production, sex ratio and

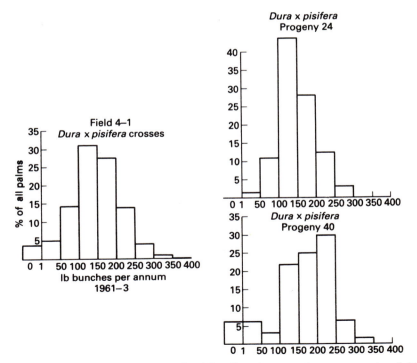

Fig. 5.5 Bunch yield: frequency of weight of bunches per palm per annum in a field in Nigeria and of two progenies within it.

abortion rates. Bunch number and weight per bunch are negatively correlated; when a palm sets a small number of bunches these tend to develop to large size. Estimates of mean bunch weight per palm are thus not based on the same number of observations. But bunch weight does nevertheless seem to be strongly affected by heredity. In the case of the two Nigerian progenies already referred to the coefficients of variation per palm in weight per bunch over the period 1961–3 were 28.7 and 26.9 per cent against a coefficient of 42.6 per cent for a similar-sized random sample from the field as a whole. The histograms in Fig. 5.6 show the altering variation in 1962 and 1963; it will be seen that the variation is less in both years for these progenies than for the field as a whole. Thus in the case of these progenies the comparatively low variation in weight per bunch seems to be counteracting the high variation in bunch number to give the variations in bunch yield per palm seen in Fig. 5.5. In the Ivory Coast, with 4-year yields, coefficients of variability also tended to be higher for number of bunches than for bunch yield, but coefficients for weight per bunch were also higher (Gascon *et al.*, 1966).

Some data are available from Malaysia both for old fields of 'foundation stock' of Deli material and for recent plantings of *dura*

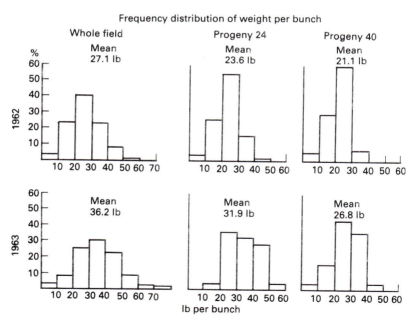

Fig. 5.6 Frequency distribution of weight per bunch of palms in a field in Nigeria and of two progenies in the field.

× *pisifera* material. A Deli field of thirty-six F₁ progenies was derived from thirty-two parents; many progenies thus had at least one parent in common with other progenies. Moreover thirteen parents were themselves derived from ten Serdang avenue palms. Variation in the individual palm yields for single years was not as noticeably skew as in West Africa, and the variation in mean annual yields per palm over a 3-year period was rather less than in any of the West African fields cited. Figure 5.7 shows the bunch yield frequency curve for 3 years and for a single year; the lower variability in the 3-year means is apparent. Bunch number variability was similar.

Variability of bunch yield within the Deli progenies was lower than in the Nigerian *dura* × *pisifera* progenies cited, even when there had been inbreeding. In one F₁ progeny the coefficient of variation for 3 years' per palm yields was only 21.5 per cent while in an F₂ progeny derived by selfing Serdang palm S5 and crossing sibs in the F₁ generation the coefficient of variation was still only 26.0 per cent. The coefficient of variation in a 5 per cent sample of the field as a whole was 30.2 per cent. Though considerable yield variability still exists in this Deli material there is certainly a suggestion of less variability than in African material.

Variability in weight per bunch also appears to be less in the Deli palm in Malaysia than in African palms in West Africa. The two progenies cited above had mean bunch weights of 16.9 and 17.2 kg

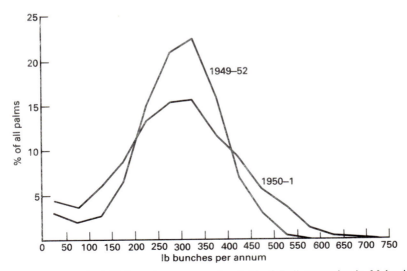

Fig. 5.7 Bunch yield frequency curves of a field of Deli progenies in Malaysia: single year (1950–1), and 3-year mean (1949–52).

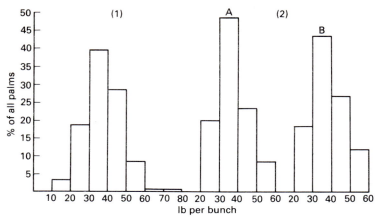

Fig. 5.8 Frequency distribution of weight per bunch: **1**, 7 per cent sample of palms in a field of Deli progenies in Malaysia; **2**, two progenies – A and B – in the field. Three-year means per palm

respectively over a 3-year period and gave coefficients of variation of 23.0 and 24.3 per cent. The field as a whole showed a sample coefficient of variation of 26.3 per cent (Fig. 5.8).

Yeow *et al.* (1981) studied several fields of *dura* × *pisifera* material on coastal clay soils in Malaysia over periods of 2–9 years. Coefficients of variation for individual palm bunch yields were lower than in the Deli field described above, even when only 2 years' (as against 3 years') data were used. In one density trial the coefficient for palms at normal density of 145 per hectare was 23.6 per cent for the 2-year period, but rose as density increased, reaching 36.5 per cent at 371 palms per hectare.

The yield frequency distribution of another field is shown in Fig. 5.9. For this field coefficients of variation were computed, for a 7-year period, for low-yielders giving less than 100 kg, for medium-yielders giving 100–200 kg, and for high-yielders giving over 200 kg per palm per annum. The latter had a coefficient of only 6.3 per cent, while the low-yielders had a coefficient of 49.5 and the medium-yielders 14.5 per cent. The high-yielders maintained their superiority over other groups throughout the 7-year period, and the low-yielders showed no signs of improvement.

Variation in production with age and fruit form

Bunch number decreases rapidly as the palm ages, both on account of a lower leaf production and a reduction of the sex ratio, while weight per bunch increases even more rapidly until the product of the two factors levels out at 'maturity'. This stage was reached in the earlier fields of palms when they had been bearing between 6

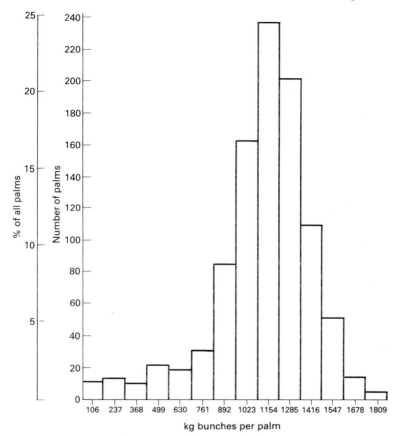

Fig. 5.9 Bunch yield frequency distribution for 8 years' yield for *dura* × *pisifera* material in Malaysia (Yeow *et al.*, 1981)

and 8 years, but with better planting methods and more precocious planting material a mature total bunch yield may now be reached much earlier.

Yield progression with age has been different in areas with little or no water deficit and those with strong dry seasons. In the latter, modern material tends to progress steadily to a mature yield at the eighth or ninth year after planting; if conditions are favourable optimum yields may be obtained earlier, but any unusually severe dry season may delay full bearing. In areas of no water deficit modern *tenera* material may advance very rapidly. to a mature yield and bunch production of around 20 tonnes per hectare in the second bearing year has not been uncommon in recent times.

While precocity of bunch yield has undoubtedly been increased by improved cultural practices, a strong genetic component has been established by Blaak (1972).

Bunch number depends largely on sex ratio, a character which varies widely between individual palms, between progenies and between palms of different fruit form. In Africa *pisifera* palms have a much higher sex ratio than *tenera* palms while *tenera* palms have a higher sex ratio than *dura* palms. Two cases in Nigeria may be quoted (Broekmans, 1957c):

Field	Planted	Sex ratio in			Period of observation	Total number of palms
		dura (%)	*tenera* (%)	*pisifera* (%)		
1–1	1939	27.9	31.1	45.7	1943–9	814
6–2	1941	29.4	31.4		1945–55	1,636

Although there are wide variations, palms of African origin when planted in Africa or in the Far East usually give a higher sex ratio and number of bunches than Deli palms. Examples from fields in Nigeria (Broekmans, 1957c) and Sumatra (Pronk, 1955) are given in Table 5.8.

Table 5.8 *Sex ratio or number of bunches per palm of Deli and African palms in Nigeria and Sumatra*

		Sex ratio		
Period	Planted	Deli palms	African palms	
Africa (Nigeria)				
1946–9	1941	29.1	38.0	
1950–4	1941	20.4	33.5	

		Number of bunches per palm		
Period	Planted	Deli palms	African dura	African tenera
Far East (Sumatra)				
1949–54	1939–40	32.8	56.2	56.6

Individual palm bunch weights will vary from a few kilograms when first in bearing to 10 to over 100 kg in weight when mature, depending on locality, fertility and inherited characters. A typical well-grown mature field in Africa may have bunches averaging 20 kg. In a typical productive mature field in Sumatra or Malaysia, Deli *dura* × *pisifera* bunches may average 20–30 kg with a range from 6 to 80 kg per bunch. In very productive soils such as the coastal clays of Malaysia, average bunch weights as high as 30–40 kg may be attained. The variation will be from palm to palm, progeny to progeny and between fruit forms. Deli bunches are heavier than African bunches both in the Far East and in Africa; for example, in a Deli and African progeny trial in Nigeria and in a trial in Sumatra (Pronk, 1955) the bunch weights at 10–12 years old were as follows:

Year of record	Planted	No. of palms	Deli (kg)	African (kg)	
Nigeria 1952–3	1941	1,182, 498	14.8	12.5	
			Deli (kg)	Dura (kg)	Tenera (kg)
Sumatra 1949–54	1939–40	487, 56, 104	18.6	12.0	10.4

It is clear that it is characteristic of the Deli palm to produce fewer but heavier bunches than the general run of African palms from whatever region they may come. Exhaustive studies in Zaire and elsewhere confirm that *dura* palms give a lower mean number of bunches but a higher mean weight per bunch than *tenera* palms.

Bunches vary considerably in the length of the spines at the end of the spikelets, African palms tending to have long spines and Deli palms short spikes. Many African palms can be found with short spikes however, for example, the Yangambi-type *tenera*.

Fruit production per bunch

The percentage of fruit per bunch is an important yield factor. The usual range is between 60 and 65 per cent but fruit-to-bunch (F/B) ratios below 60 per cent are not uncommon while progenies with percentages of over 70 per cent are sometimes found. With some palms the fruit-to-bunch ratio is a consistent character. With other palms this is far from the case. Young palms often show very low F/B ratios sometimes amounting to no more than 40 per cent. Fruit-to-bunch ratios of *pisifera* palms depend on their fertility. Complete Bunch Rot, or a bunch with only a few fruit, is the rule in the so-called infertile *pisifera*. In fertile *pisifera*, i.e. those setting sufficient fruit to give a recognizable non-rotten bunch, ratios may be as low as 25 per cent and only very rarely reach 60 per cent. *Tenera* palms have, on average, a lower F/B ratio than *dura* palms. Shell weight is partly responsible for this form difference.

The percentage fruit to bunch depends partly on the number of spikelets which, as already mentioned (p. 74), varies widely. There is no difference between the mean number of spikelets of *dura* and *tenera* bunches, but the mean number of spikelets on *pisifera* bunches is significantly higher than on bunches of the other fruit forms. The number of flowers per spikelet differs to a very small degree between fruit forms, the ascending order of number being *dura, tenera, pisifera*. Fruit set, however, tends to be higher in the *dura* than in the *tenera* (Beirnaert and Vanderweyen, 1941).

Fruit type and form variations

The main fruit types (external appearance) and fruit forms (internal structure) have been described in Chapter 2. For the inheritance studies shortly to be described, the form and type classification is not entirely satisfactory since colour, though it may be noted externally, is an internal factor of chemical composition. The classification used by Belgian workers in Zaire has much to commend it and is set out below with an indication of when a 'type' or 'form' is concerned:

1. Presence or absence of supplementary carpels
   (*Poissoni* or 'mantled')   2 types
2. Presence or absence of shell and/or of fibre
   ring (*dura, tenera pisifera*)   3 forms
3. Presence or absence of anthocyanin in the
   exocarp (*nigrescens, virescens*)   2 types
4. Presence or absence of carotenoids in the
   mesocarp at maturity (*albescens*)   2 types

These characters being independent, there can be eight types or type combinations of fruit, and twenty-four type-form combinations. The variety of specific fruit characters is therefore large before one begins to consider the variation within each character. Fruit form and type differences do not give rise to 'varieties' in the botanical or systematic sense; the differences are ones of horticultural form.

The *Poissoni* or mantled character has already been described (p. 80). It may be added here that from six to ten supplementary carpels are formed; often only shell is found within the flesh but sometimes endosperm is also present. There is a well-known illustration of Janssens (1927) in which a complete shell and kernel, perhaps with embryo, is shown in one of the carpels and described as an 'exterior fruit' (see Fig. 2.11). From the point of view of the plant breeder the *Poissoni* must now be regarded as an aberrant type which he would wish to keep out of his collection.

Something has already been said about the variations within the *nigrescens* and *virescens* types (p. 80–81) and their colour changes on ripening. Ripe fruit of the Deli palm normally appear to be of a paler orange colour than that of most African fruit, and oil from the Deli is easier to bleach. Carotene contents of plantation palms (estimated as α-carotene) have been found to vary in Nigeria from

200 to 1,100 ppm of extracted oil. Oil from grove palms has varied from 600 to 2,875 ppm carotene. The reasons for these differences of carotene content are still obscure, but may be connected with degree of ripeness and over-ripeness, and with the incidence of light on the bunches. There is evidence of differences between progenies in carotene content (Nwanze, 1961) and between oil from African palms and oil from Deli palms. The oil from palms considered to be *albescens* usually contains less than 60 ppm carotene but as much as 90 ppm is sometimes found.

Variation within fruit forms

In Chapter 2 the *dura* and *tenera* forms were described as follows: *dura* – shell usually 2–8 mm thick though occasionally less, low to medium mesocarp content (35–55 per cent but sometimes, as in the Deli palm, up to 65 per cent), *no fibre ring; tenera* – shell 0.5–4 mm thick, medium to high mesocarp content (60–96 per cent but occasionally as low as 55 per cent), *fibre ring.* The fibre ring only gradually came to be recognized as the ultimate criterion. Previously shell thickness was taken as the main classifying measurement, Beirnaert and Vanderweyen defining the *tenera* as fruit with a shell thickness of 0.5–2 mm. The remarkably consistent segregations in these authors' classic work were obtained through using *tenera* of low shell thickness (i.e. below 2 mm), thus assuring that only genuine *tenera* were being used and that the progeny would not be hard to classify. Later, it was recognized in Zaire that *dura* of shell thickness less than 3 mm, and *tenera* of shell thickness up to 3 mm, could be found (Nigerian Conf., 1949). Studies in West Africa have confirmed that the overlap of the range of thickness of shell of the *dura* and the *tenera* may be as much as 2 mm. Shell measurements are made, preferably with sliding calipers, at two opposite points of the shell where it has its maximum diameter.

In grove populations a slightly skew distribution of shell thickness is found. Very thick-shelled fruit is more common than the very thin-shelled fruit. This may be seen in the histogram in Fig. 5.10(a). The distribution of shell thickness among Deli palms in Sumatra is remarkably similar, only the small percentage with less than 2.5 mm thickness and the very thick shells being almost absent. The histogram (b) shows that Deli *dura* with less than 2.5 mm shell thickness exist, however. It is the thickness of mesocarp, not differences of shell thickness, that normally distinguishes the Deli *dura* from the African *dura*, though African *dura* with the composition of good Deli *dura* are to be found.

Figure 5.10(c) shows the wide distribution of shell thickness to be obtained from the illegitimate progeny of grove *tenera*. A selection of such material closely conforms in practice to random *tenera*

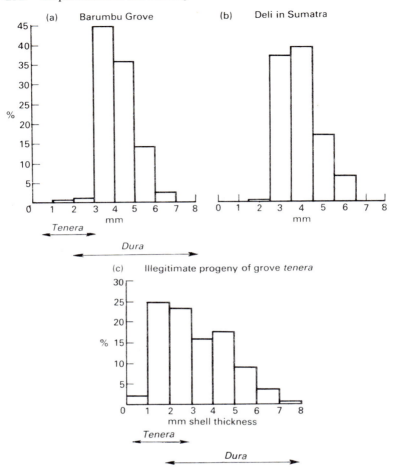

Fig. 5.10 Distribution of shell thickness in: (**a**) a Congo grove; (**b**) an early field of Deli palms in Sumatra; (**c**) illegitimate progeny of Congo *tenera* palms. (Drawn from the data of Beirnaert and Vanderweyen, 1941.)

× *dura* crosses since the majority of the pollen in the groves is of *dura* origin. *Tenera* and *dura* peaks can be distinguished, but owing to the wide variation in shell thickness of the parents there is considerable overlapping.

By weight, shell in the fruit is found to vary in Africa as follows (Sparnaaij *et al.*, 1963): *dura* 25–55 per cent shell; *tenera* 1–32 per cent shell. Mesocarp variation has been given at the beginning of this section. Contrary to general belief the mean kernel content of *tenera* palms is the same as that of *dura* palms. Indeed, with nearly

7,000 *tenera* and over 3,000 *dura* Beirnaert and Vanderweyen (1941) found the average kernel percentage to fruit actually higher in the *tenera*, though this was due entirely to the higher kernel content of *tenera* interior fruit. The usually lower kernel content of *tenera* fruit now encountered is due to selection and breeding for high mesocarp content; this has reduced both the shell and kernel contents. *Tenera* with large kernels still exist however. Sparnaaij *et al.* (1963) give the range of kernel content for *good* African fruit as: *dura* 7–20 per cent; *tenera* 3–15 per cent. With Deli *dura* the range is 4–11 per cent though percentages below 7 are uncommon.

Variation in weight and shape of fruit has already been described (p. 180 and Plate 5.5). Beirnaert's preference for large fruit, which led to the production of the Yangambi-type *tenera*, was discussed on p. 203. Variation in the size and conformation of *pisifera* fruit is similar to that of *tenera* fruit, but in the palm as a producer the chief variable is fruit set.

No very detailed studies have been made of the variation of the oil content of the mesocarp, but Bénard (1965) has shown that in Deli palms the percentage oil to pulp, though often well below 50 per cent, can reach 56.6 per cent, and that there can be marked differences between groups of palms of distinct origins.

Pl. 5.5 The wide variety of shape and size of African fruit.

Variation in sterility of fruit

Sterility occurs predominantly in the *pisifera*, the majority of palms of this fruit form bearing few fertile fruit and showing a strong vegetative development. *Pisifera* have been classified as: (i) fertile – these palms are relatively few and their value in breeding is discussed on p. 277; (ii) showing partial sterility – small numbers of fertile fruit per bunch, vegetative development less than in (iii); and (iii) sterile – giving, occasionally, a few fruit, but the bunches rotten and vegetative development strong. Intermediates between these categories exist and palms tend to become less infertile as they age. In sterile fruit there is no development of the ovule, or ovular development is retarded. Abnormalities of the tissues surrounding the ovule also occur (Henry and Gascon, 1950). Sterility sometimes occurs in the *tenera* and has been reported in the *dura*.

Variation in vegetative characters in relation to breeding

Marked differences of vegetative development and leaf habit have always been apparent in the oil palm but until comparatively recently few systematic studies have been made. The work of de Berchoux and Gascon (1965) may be cited as an example of the wide differences existing between populations. They studied a Deli progeny from Johore Labis estate, Malaysia (JL1133 × JL1113), a local Ivory Coast La Mé *tenera* × *tenera* cross (L2T × L11T), and a cross derived from Yangambi (Zaire) material (2469B × 1473B), all growing at La Mé. It is presumed that these progenies were chosen as being typical of the material they represented; forty-two to fifty palms per progeny were studied. While the expected difference between Deli and African material emerged, i.e. the Deli had fewer but longer, wider and heavier leaves, more leaflets and wider petioles, the differences between the Ivory Coast and Yangambi progenies are of particular interest. About the same number of leaves were produced by both progenies, but the Yangambi progeny's leaves were about 40 per cent heavier and the weight of the leaflets and their number approached those of the Deli (though the rachis was lighter). Leaf length was, however, no greater in the Yangambi than in the Ivory Coast progeny and the advantage of the former was found in its wide leaves and larger number of leaflets. Some of the data are shown in Table 5.9. The results suggest that the early selection in Zaire provided palms of vegetative development somewhat different from that of the Deli palm but of equal photosynthetic potential; but in view of the limited nature of this study general deductions cannot be drawn.

Through the growth analysis techniques developed by Corley in Malaysia it has been shown that important differences exist between individual palms and progenies in leaf area and vegetative dry matter production and in the growth parameters leaf area index and net assimilation rate. The physiological aspect of this work has been discussed in Chapter 4 and its value in breeding will be considered later in this chapter (see p. 249).

Table 5.9 *Extremes of vegetative development at La Mé, Ivory Coast. Some data of Deli, Ivory Coast and Zaire progeny*

Derivation and progeny	Deli dura	African tenera crosses	
	SOC1386 (JL1133 × JL1113)	La Mé LM6 (L2T × L11T)	Yangambi YA3 (2469B × 1473B)
Annual leaf production, 1960–3	19.3	24.2	24.4
Number of leaves on the palms, per palm, mean of 4 years	34.6	40.6	41.6
Weight of 17th leaf, kg, 4-year average	8.1	4.8	6.6
Weight of leaflets, 17th leaf, 1963, kg	3.1	2.3	3.0
Length of 17th leaf, mean of 2 years, m	6.60	5.63	5.64
Width of 17th leaf at mid point, m	2.15	1.78	2.07
Number of leaflets on 17th leaf, 1963	350	309	344
Width of petiole, 17th leaf, 1963, cm	9.0	6.7	8.2

Source: De Berchoux and Gascon (1965).

The inheritance of bunch and vegetative characters

To effect an improvement of the yield of products of the oil palm it is necessary to have a knowledge not only of the variability of production factors, but also of the relative extent to which these factors are inherited. Variance due to genetic factors may be exhibited through:

1. The simple Mendelian inheritance of specific qualitative characters;
2. Additive genetic variance;
3. Non-additive genetic variance. ·

Inheritance of specific fruit characters

Presence and absence of shell

An account has already been given (p. 202) of the emergence of the *dura* × *pisifera* cross in the study of the inheritance of fruit forms in *tenera* × *tenera* selfs and crosses. This cross may be illustrated as follows:

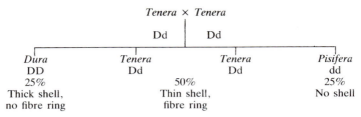

The *tenera* is thus a hybrid (Dd) between the thick-shelled *dura* and the shell-less *pisifera* forms and, apart from its generally thinner shell, it exhibits itself by the ring of fibres embedded in the mesocarp and running, at varying distances from the shell, longitudinally from the top of the fruit to its base.

It follows, therefore, that other fruit form crossings or selfing will result in the following segregations:

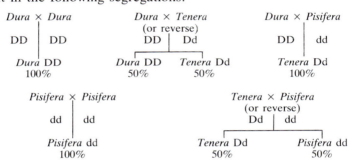

In the original Zaire *tenera* × *tenera* selfings and crossings certain divergencies from the 1 : 2 : 1 D : T : P ratio were found (Beirnaert and Vanderweyen, 1941). While *tenera* progeny remained nearly constant at 50 per cent of the total, a few *tenera* mother palms gave as much as 35 per cent or as little as 15 per cent *dura* with a correspondingly low or high proportion of *pisifera*. Such cases can emerge through contamination of *tenera* pollen with small quantities of either *dura* or *pisifera* pollen, e.g.

	D	T	P
× *Tenera* =	25	50	25
Tenera (Mother palm)			
× *Dura* =	30	30	
Total (%)	34.4	50.0	15.6

But detailed examination showed that the cases in Zaire not only occurred in certain *tenera* × *tenera* lines, but when *tenera* which produced about 15 per cent *pisifera* in selfings (Type II) were crossed with *tenera* giving the normal 1 : 2 : 1 segregation (Type I) the segregation was intermediate, i.e. 20 per cent *pisifera* and 30 per cent *dura* were produced. Similarly *tenera* producing 15 per cent *dura* and 35 per cent *pisifera* in selfings (Type III) when crossed with *tenera* giving normal segregation (Type I) gave 20 per cent *dura* and 30 per cent *pisifera*. Finally *tenera* giving 15 per cent *pisifera* (Type II) when crossed with *tenera* giving 35 per cent *pisifera* (Type III) gave the normal 1 : 2 : 1 segregation. A few examples from the work of Beinaert and Vanderweyen are given in Table 5.10.

Table 5.10   *Segregation from different types of* tenera *in Zaire (Beirnaert and Vanderweyen, 1941)*

Tenera × Tenera	Segregation of fruit forms (%)			Number of palms
	Dura	Tenera	Pisifera	
Type II selfs and crosses	32.6	51.3	16.1	353
Type III selfs and crosses	14.8	49.2	36.0	390
Type II × normal (Type I)	28.8	50.9	20.3	645
Type III × normal (Type I)	19.5	51.4	29.1	492
Type II × Type III	74.2		25.8	275

These variations from the normal have not as yet been recorded on a large scale elsewhere and no fully satisfactory explanation has been provided, although de Poerck (1942) suggested a hypothesis based on the possible presence of a series of alleles modifying the proportions of the three fruit forms in *tenera* × *tenera* progeny.

Presence or absence of pigmentation of the exocarp

Inheritance of the presence of anthocyanins in the exocarp has been studied in Zaire and Nigeria. *Nigrescens* or ordinary fruit are black capped, but vary considerably in appearance (p. 81). In green fruit (*virescens*) absence of anthocyanin does not appear to be absolute; there is evidence of traces of an anthocyanin which may be distinct from the one normally encountered in the ordinary fruit.

There is evidence that the green-fruited character is monofactorial and dominant, but the number of green fruited palms found in natural populations is so small that it might normally be expected to be found in the heterozygous condition. In Zaire a green-fruited palm, assumed to be heterozygous, gave the following progeny (Beirnaert and Vanderweyen, 1941):

(1)
Virescens × *Virescens*

Bb | Bb

BB   Bb   Bb   bb
75% *virescens*   25% *nigrescens*

(2)
Virescens × *Nigrescens*

Bb | bb

Bb   bb
50% *virescens*   50% *nigrescens*

In Nigeria an open-pollinated green-fruited bunch gave rise to 46

per cent *virescens* ahd 54 per cent *nigrescens*, and nine *virescens* ×
nigrescens crosses gave 54 per cent *virescens* and 46 per cent *nigrescens*, both results appearing to confirm that these *virescens* were
heterozygous and that the factor is dominant. Five other crosses
gave anomalous results, possibly due to out-pollination.

Presence or absence of carotenoids

The *albescens*, or 'white-fruited', type can be 'ordinary' or green-
fruited as to its exocarp character. It was early noted in Zaire that
illegitimate *albescens* gave either no *albescens* or 1–5 per cent *albescens* in the progeny. It was therefore supposed that absence of
carotenoids was a recessive character (Vanderweyen and Roels,
1949). As already noted, there is not a complete absence of carotene in the *albescens* palm.

Presence or absence of supplementary carpels

This character, *Poissoni*, 'mantled' or *diwakkawakka*, appears to be
inherited monofactorially and to be dominant. Certain mantled
palms in Zaire gave 100 per cent mantled progeny when selfed;
others gave 75 per cent. Crosses between mantled and ordinary
palms gave 50 per cent mantled and 50 per cent ordinary (Nigerian
Conference, 1949). The inheritance was therefore assumed to be as
follows:

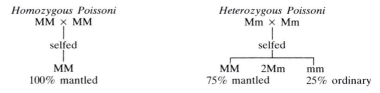

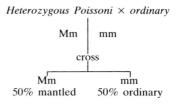

In Nigeria mantled *tenera* palms showing a high degree of out-
pollination also showed approximately 50 per cent of each type in
their progeny. Zeven (1973) quotes other illegitimate progeny in
support of dominance. In Sumatra, although most results supported
dominance, two *Poissoni* selfs gave only 50 per cent *Poissoni*
progeny instead of the expected 100 or 75 per cent; other anomalous
results were the appearance of 25 per cent *Poissoni* palms in a
normal *tenera* self, and the sudden change of an ordinary *tenera* to
the production of mantle-fruited bunches (Janssen, 1959; Pronk,
1955).

The *Poissoni* character was regarded with favour in early breeding work in Sumatra. There was evidence that *Poissoni* Deli *tenera* and even *Poissoni* Deli *dura* would produce a higher proportion of mesocarp than the ordinary Deli. Extensive crossings with *Poissoni tenera* and *dura* were therefore made. Later, when the inheritance of the *tenera* characters was understood, *Poissoni* palms dropped out of use, and the recent appearance of mantled fruit in some tissue culture clones caused some alarm (Corley *et al.*, 1986).

The inheritance of specific vegetative characters

The only specific vegetative characters which have been given attention are the *idolatrica* leaf form and trunk 'dumpiness'. With regard to the former, evidence is conflicting and it has been thought to be both dominant and recessive. The former theory is suggested by the fact that an open-pollinated bunch from an ordinary palm in Zaire which had some *idolatrica* neighbours gave 16.7 per cent *idolatrica* progeny. Of thirty-four progeny of a selfed *idolatrica* in Zaire, twenty had been identified as *idolatrica* when they were quite young seedlings (Beirnaert and Vanderweyen, 1941). Intermediate forms were found in Nigeria, but controlled pollinations also suggested dominance. The breeding true of a selfed *idolatrica* palm in Sumatra has, however, suggested that the character might be recessive (Fickendey, 1944).

Of much greater importance is the inheritance of dumpiness. This has already been described in the case of the Malaysian dumpy E206; the growth of the F_1 selfs suggested a high degree of homozygosity for this character in the original parent. This supposition was supported by results from the F_2 generation in which F_1 sibs from the original E206 selfing gave progeny with uniformly small height increment and large girth. Thus the F_2 on one estate had an annual height increment of 16.3 cm with girth of 358 cm at 7 years of age, while non-dumpy palms under equally fertile conditions had shown annual height increments of 22.9–35.6 cm at 5–8 years with girths of 256–267 cm (Haddon and Tong, 1959). The dumpy character was not strongly transmitted in the F_1 dumpy × normal cross but in the so-called back-cross (F_1 dumpy × normal) × F_1 dumpy self, the characters were inherited as shown below:

Estate	Progeny	Age	Annual height increment (cm)	Girth (cm)
H	Dumpy 'back-cross'	8	30	302
	Normal Deli	8	36.6	257
Jerangau	Dumpy 'back-cross'	5	25.4	267*
Exp. St.	Normal Deli	6	36.8	267

* At 90 cm; otherwise at 120 cm.

Studies in the Ivory Coast with groups of progenies of Malaysian, Zaire and Ivory Coast origin showed that measures of leaf production, internode length and stem growth tend, in crosses between these groups, to be intermediate to the same measures in crosses within the groups (Noiret and Gascon, 1967). It was also shown that variation in height within a progeny is itself very variable, so uniformity in height is also an important selection criterion (Jacquemard, 1979).

With the movement of the oil palm into higher latitudes the resistance of the stem and stem base to strong winds has assumed some importance. In general the palm stands up well to the winds preceding tropical storms in West and Central Africa and South-east Asia. Oil palms of all ages survived the severe force of hurricane Fifi in Honduras in 1974, when adjoining banana plantations were completely flattened. In Colombia a sudden tornado in the Magdalena valley in 1968 demonstrated that there were distinct progeny differences in resistance to high winds (Taillez and Valverde, 1971). In general Deli *dura* × La Mé *pisifera* progenies were less disturbed than Deli *dura* × Yangambi *pisifera* progenies. Most affected were palms of $2\frac{1}{2}$–$4\frac{1}{2}$ years of age in the field, a period when there is likely to be the greatest imbalance between the trunk and canopy and the roots. At all ages there was a significant correlation between height of trunk and the amount of disturbance (measured by the angle of inclination). This provides an additional reason for breeding palms which grow slowly in height.

Heritability and genotypic values

Some of the specific characters of fruit form and type whose inheritance has been discussed above are governed by single gene pairs and they are referred to as qualitative characters. It has been seen in the previous section, however, that characters such as bunch yield, bunch number, weight per bunch, F/B ratio and many of the factors of fruit composition show continuous variation. These characters, which are known as quantitative or metrical characters, do not segregate in discrete classes and they are usually governed by a large number of genes.

The variability observed in quantitative characters is due partly to differences in genetic factors and partly to differences of environment. The concept of heritability was developed to enable the relative importance of genetic and environmental factors to be determined and to show whether selection progress is relatively easy or difficult to obtain. Heritability is the genetic variance in relation to the observed, or phenotypic, variance. The genetic variance has an additive portion, which is caused by genes acting independently in affecting the character concerned, and a non-additive portion which is concerned with general and specific combining ability.

The heritability of any character may be measured by the regression of the progeny values on the values of the parents. Where the relationship is between progenies and, in each case, two known parents, the mid-parent values are used. When data of single parents only are available, heritability is taken as $2b$ where b is the regression coefficient (Falconer, 1960). It should be realized that heritability figures are likely to differ from environment to environment, from population to population and from generation to generation.

The first studies of heritability were those of Blaak (1965) and Menendez and Blaak (1964) working in Nigeria on material established by Toovey and others, but more recently several other studies have been made, both in Africa and Malaysia, and some of these have embraced vegetative characters as well as those of fruit and bunch. Before discussing these studies an example from Nigeria of heritability in the oil palm may first be given.

Twenty-five parent *tenera* palms were pollinated with a mixture of *tenera* pollen from those of their number which had the highest fruit quality. The regression of the mean mesocarp-to-fruit percentage of their *tenera* progeny (y) on the mesocarp percentage of the parents themselves (x) was calculated from the data to be:

$$b_{yx} = \frac{82.17}{263.64} = 0.3116 \pm 0.080$$

Heritability $2b_{yx} = 0.623$, or 62.3%

This regression is highly significant and it shows that, for this population under the existing environment, a difference of 1.0 per cent in the mesocarp content of the female parents may be expected to give a difference of 0.312 between the mean mesocarp content of the progenies. This is illustrated by the regression line in Fig. 5.11. Although the male parents were not known and therefore the

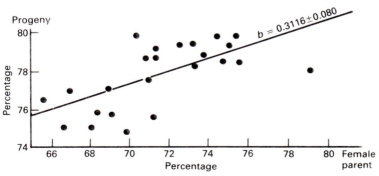

Fig. 5.11 Regression of the mean mesocarp to fruit percentage of progeny on the mesocarp to fruit percentage of female parents: Ufuma, Nigeria.

heritability could only be calculated from the single parents, the use of mixed *tenera* pollen from high-quality parents appeared to have the effect of raising the mean level of mesocarp content from 71.4 per cent in the parents to 77.7 per cent in the progeny.

Bunch composition heritability

The work on heritability in Nigeria (Menendez and Blaak, 1964) showed that most fruit and bunch characters are strongly inherited and that therefore selection for those characters leading to high oil to bunch is justified. Supporting data were provided from the Ivory Coast in crosses between Deli palms and palms of Zaire and Ivory Coast origin (Noiret *et al.*, 1966).

Fruit-to-bunch ratio

This is an extremely important character and very variable. It has assumed an added importance with the advent of the *dura* × *pisifera* cross since *tenera* bunches have on average a lower fruit-to-bunch ratio than *dura* bunches; this is largely due to the increase in mesocarp to bunch failing to compensate in weight for the lower shell to bunch. However, if full advantage is to be taken of the fruit composition of the *tenera*, then the F/B ratios must be maintained and a high account must be taken of this in the *dura* parent. Unfortunately this character is the least strongly inherited and environmental factors play a large part in fruit set. Within-field heritabilities varying from 0.19 to 0.34 were recorded in Nigeria.

With later material in Nigeria Van der Vossen (1974) found significant fruit-to-bunch heritabilities for *tenera* of 0.55 with *tenera* × *tenera* progenies and of 0.18 within *dura* × *tenera* progenies. In the Ivory Coast overall heritabilities of fruit to bunch were low and parent/progeny correlations insignificant. However, within certain groups of parents, e.g. in the La Mé *tenera* × Deli lines, heritability was found between *tenera* parents and their *tenera* offspring, and Van der Vossen has concluded that the generally lower heritabilities found by Meunier *et al.* (1970) are explained by the lack of distinction between fruit forms. These authors consider, however, that the discovery of heritability in certain groups of progenies where a character is only weakly heritable may indicate non-additive variance or specific combining ability. This view agrees with Ooi's (1975) finding with Deli material in Malaysia. Other data gave heritabilities of 0.36 and 0.55 (OPGL, 1967–71).

Gascon and de Berchoux (1963) found a fairly high and significant correlation at Le Mé in the Ivory Coast between the fruit-to-bunch ratios of *dura* and *tenera* in the same progeny. Twenty-five progenies were Deli *dura* × *tenera* crosses, while of the remainder thirty-six were of West African and twenty- five of Zaire origin. The regression lines showed that although *dura* palms had a higher F/B

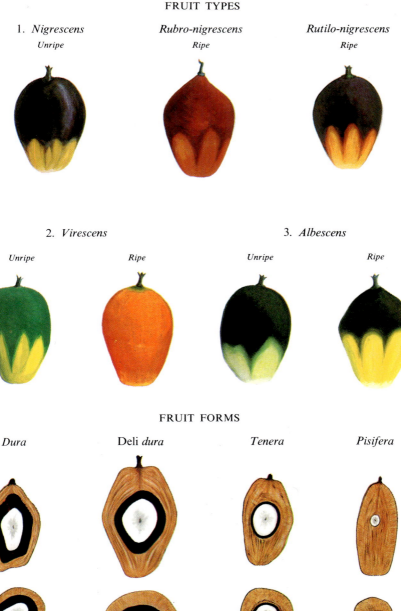

FRUIT TYPES

1. *Nigrescens*
Unripe

Rubro-nigrescens
Ripe

Rutilo-nigrescens
Ripe

2. *Virescens*

Unripe

Ripe

3. *Albescens*

Unripe

Ripe

FRUIT FORMS

Dura

Deli *dura*

Tenera

Pisifera

Pl. I   The fruit types (external appearance) and forms (internal structure) of the oil palm.

Pl. II *Elaeis guineensis:* palm with a heavy crop at only 4½ years old in Guadalcanal, Solomon Islands.

Pl. III *Elaeis oleifera:* palm with many ripe and unripe bunches in the Sinú valley, Colombia.

Pl. IV The fruit of three *tenera.* **A.** Long-shaped Nigerian *tenera* 24.2864. **B.** A large-kernel *tenera,* 1.2229. **C.** *Tenera* from the cross *Elaeis oleifera* x *E. guineensis, pisifera* (note no fibre ring).

B

C

Pl. V   Propagation of the oil palm by tissue
culture from root tissue:
A. Primary callus;
B. Embryoids;
C. Embryoids with shoot.

D

E

Pl. V (cont'd)   **D.** Plants ready for despatch;
E. Young tissue-cultured plants in a prenursery.

Pl. VI   Seedling showing acute
N-deficiency symptoms on waterlogged soil in Brazil.

A

II Magnesium deficiency symptoms: (J.M.A. Sly):
ursery seedling in Nigeria (above).
oung field plants in Colombia (right).
range frond of the adult palm leaf, Sierra Leone.

B

C

Pl. VIII Symptoms which have been attributed to potassium deficiency (J.M.A. Sly):
A. Confulent Orange Spotting

A

B   C

Leaves and a leaflet from palms showing Mid-Crown Yellowing. Some of the lesions in C are similar to the Mbawsi symptom.

D

Pl. IX   Two types of Anthracnose of prenursery seedlings:
A. *Botryodiplodia palmarum*.

B. *Melanconium* sp.

Pl. X   A third type of Anthracnose of prenursery seedlings:
C. *Glomerella cingulata.*

Pl. XI *Cercospora elaeidis* infection of the terminal leaflet of a heavily infected 6-month-old seedling.

Pl. XII   Advanced stage of Blast disease (J.M.A. Sly).

Pl. XIII   Patch Yellows (J.M.A. Sly).

Pl. XIV   A leaf
(A) and individual leaflets
(B) infected with *Pestalotiopsis* sp. in Colombia.

B

Pl. XV   The 'Peat Yellows' condition on palms growing on peat soil in Malaysia.

Pl. XVI Nettle caterpillars in Malaysia:
A. *Setora nitens*.

B. *Darna trima.*

Pl. XVII Nettle caterpillars in Nigeria: *Parasa*
sp.

ratio than the *tenera* in the same progeny, the difference between the two values diminished as the F/B ratio rose, e.g. a *dura* F/B ratio of 50 per cent corresponded to a *tenera* F/B ratio of 43 per cent, but at the higher end of the regression line the correspondence was *dura* 75 per cent: *tenera* 71 per cent. This finding increases further the importance of high F/B ratios in *dura* parents.

Mean fruit weight

Meunier *et al.* (1970) showed that this is perhaps the most heritable of all the bunch composition characters, with a mean heritability of 1.01 on the mid-parent values.

Mesocarp to fruit

In Nigeria, heritabilities from 0.62 to 0.96 were recorded (Menendez and Blaak, 1964; Van der Vossen, 1974); in the Ivory Coast the range was 0.52–0.56 (Meunier *et al.*, 1970), and in Malaysia 0.59 was recorded for Deli material (OPGL, 1967–71). Significant correlations were found, as with fruit-to-bunch ratios, between *dura* and *tenera* mesocarp percentages in the Ivory Coast progenies (Gascon and de Berchoux, 1963). As these included both Deli × African and pure Africa crosses, the correlations might have been even higher if the two groups had been considered separately. In the Deli × African *tenera* cross the mesocarp of the *dura* progeny tends to be intermediate between the high Deli range and low African range; but large Deli fruit with thick shells and mesocarp over 60 per cent do not provide *tenera* progeny with a higher mesocarp percentage than that of the *tenera* progeny of many Africa *dura* with mesocarp below 50 per cent. A standard African *dura* × *tenera* cross (6.594 × 5.1450) used in seven fields in Nigeria gave a mean mesocarp percentage of 45.7 (range 43.1–47.3) for *dura* progeny and 81.7 per cent (range 80.3–82.9) for *tenera* progeny, while in a comparison between two Nigerian *dura* × *tenera* crosses and two Ivory Coast *tenera* × Deli crosses the mesocarp percentages were:

	Dura	Tenera
Nigerian *dura* × *tenera*	46.7	87.2
Tenera × Deli *dura*	47.4	75.1

It seems likely that the inheritance of mesocarp to fruit must be considered in conjunction with nut size and shell thickness. More will be said on this subject when additive genotypic or breeding values are considered (see p. 248).

Shell to fruit

The heritability of shell percentage from *dura* mother palms to *dura* progeny was found to be very high in Nigeria: 0.83–1.32. From *tenera* to *tenera*, heritability was still 0.62 in the earlier and 1.06 in the later Nigerian determinations. Heritability from *dura* to *tenera* in *dura* × *pisifera* crosses was lower, at 0.53, but still highly significant. Shell percentage heritability is clearly of the greatest importance as within the single fruit it is high oil plus kernel that is required. Shell thickness has also been shown to have a high heritability (Menendez, 1965). This and inheritance from *dura* to *tenera* forms are considered further on p. 248.

Kernel to fruit

Heritability in Nigerian material was found to be relatively high, varying from 0.41 to 0.74. In the Ivory Coast values of 0.29–0.65 were recorded (Meunier *et al*, 1970) and the correlation of within-progeny *dura* and *tenera* kernel to fruit percentage was high ($r = 0.81$) and significant (Gascon and de Berchoux, 1963). Kernel percentages were usually lower in the *tenera*, but at the higher end of the regression line this difference tended to disappear.

There seems no reason to believe that kernels cannot be as large, or form as high a percentage of the fruit, in *tenera* as in *dura*. The early Zaire selections had large kernels, but since those days there has been a tendency to increase the mesocarp at the expense of both kernel and shell. This is discussed further on p. 260–1.

Oil to mesocarp

In the early days of oil palm selection and breeding too little notice was taken of this factor in the mistaken belief that it approximated to 50 per cent and was not very variable. More efficient techniques for estimating oil-to-mesocarp percentages have shown that this is not the case, however, and it must be regarded as a potentially important selection factor. There are of course practical difficulties in its assessment:

1. The bunch must be cut fully ripe and only ripe fruit must be used in the estimation.
2. The oil content of the fruit of young palms is always low; the mature level is often not reached until the fourth or fifth year of bearing.
3. There are seasonal variations in oil percentages, fruit in the peak production months tending to have a higher oil content than fruit harvested at other times.

When these variables have been accounted for, oil percentage of mesocarp may still vary between palms from 40 to 62 per cent. In West Cameroon an estimate of 0.56 heritability has been made for D × T material and in Malaysia 0.48 has been recorded (Ooi and Abdul Wahab, 1977), but in both the Nigerian material (Van der Vossen, 1974) and in the Ivory Coast (Noiret *et al.*, 1966) oil-to-mesocarp heritability did not reach a significant level though in the latter case a few individual parents seemed to transmit relatively high oil to mesocarp to their progenies. Van der Vossen (1974) has studied the differences of oil and dry-fibre content of *dura* and *tenera* full sibs. He found that, with inclusion of the fibre of the fibre mantles, the dry-fibre content of the *tenera* was higher than that of *dura* fruit (Table 5.11). Nevertheless there was not a corresponding difference in the oil content of the whole mesocarp because the oil content of mesocarp outside the fibre mantle was found to be significantly higher in the *tenera* than in the *dura*. Highly significant correlations were found between the *dura* and *tenera* full sibs in both oil and fibre contents, and estimations of the proportion of genetic variance to total variance (wide sense heritability) suggested that both oil and dry fibre to mesocarp are heritable though dry-fibre heritability is likely to be higher.

Table 5.11   *Oil and dry fibre content of fresh mesocarp in full-sib progenies* * *(per cent)*

	Tenera				Dura	
	$Oil^{\dagger}$	$Oil^{\ddagger}$	$Fibre^{\dagger}$	$Fibre^{\ddagger}$	Oil	Fibre
Mean	52.6	48.7	15.0	16.8	49.8	15.3
Range	47.3–56.7	42.0–52.3	14.1–17.2	15.3–19.0	44.2–55.6	14.4–16.6
Coeff. of variation	4	5	5	5	5	4

* The data are for twenty-nine progenies. Computations for twenty-one progenies showed significant within-progeny correlations between *dura* and *tenera* for both oil and fibre, and significant differences between the fruit forms in fibre content of the whole mesocarp, in oil of the mesocarp outside the fibre mantle (in favour of the *tenera*), and oil in the whole mesocarp (in favour of the *dura*).
† Oil or fibre outside the fibre mantle.
‡ Oil or fibre in whole mesocarp.

Source: Van der Vossen (1974).

This work suggests that progress in oil-to-mesocarp selection is only likely to be made when determination methods are improved, and Van der Vossen recommended that dry-fibre contents should be determined by a limited number of direct analyses. As this factor has been found to be relatively constant, within palms, irrespective of age and stage of ripeness, the data obtained can then be used in conjunction with the indirect method (i.e. water content determinations) applied to mature palms and fully ripe fruit.

Bunch yield inheritance

Until recent studies were undertaken the inheritance of bunch yield was only inferred. Firstly, increases in yield from one generation to the next following the selection of high-yielding parents were seen in many parts of the world to be marked but there was no means of telling how far this was due to better agronomic practices and how far to heredity. Secondly, comparisons in replicated trials showed that mixed progeny from selected parents gave a higher bunch yield than the mixed progeny of unselected material (Hartley, 1957; Broekmens, 1957a). In a series of trials of this nature in Nigeria significant increases in yields over 4-year periods of between 12 and 50 per cent were obtained, but there was a tendency for these differences to be less marked after about 8 years of bearing. This suggested that selection may have been partly, and unconsciously, for precocity. In five trials planted over successive years a mean annual yield increase of selected over unselected material of 22 per cent was obtained during the period from bearing to the fifth to eighth bearing year. That precocity is heritable was clearly shown by Blaak (1972).

Heritability of bunch yield was studied in Nigeria by means of a number of parent–progeny comparisons (Blaak 1965). In all these comparisons the yield of only one parent was available; one study concerned *dura* × *pisifera* crosses, while the other two concerned *dura* parents from two separate groves pollinated with a mixture of *dura* pollen. The heritabilities of bunch yield were:

Field 4–1	*dura* × *pisifera* crosses.	
	dura parents – *tenera* progeny	0.29
Field 15–1	Ufuma *dura* parents – *dura* progeny	0.36
Field 15–1	Aba *dura* parents – *dura* progeny	0.23

In more recent studies of Ivory Coast and Nigerian material the heritability of bunch number (0.50–0.51) was found to be higher than for bunch weight (0.13–0.21), while values for total bunch yield were extremely low and insignificant (Van der Vossen, 1974; Meunier *et al.*, 1970). Within the Deli population in Malaysia low heritability for bunch yield was also suggested by the absence of additive genetic variation (Thomas *et al.*, 1969), and the few estimates that have been made confirm this. Ooi *et al.* (1973), using data from four Deli *dura* breeding experiments, found that while there were significant differences between females within males in progeny bunch yield and its components, the genetic variability present was largely non-additive. These findings imply severe limitation to increases in bunch yield potential and this is discussed further in the next section of this chapter.

Although mean bunch weight is only weakly heritable, in the Ivory Coast it was shown that individual parents could transmit a high mean bunch weight to their progeny in a range of crosses (Noiret *et al.*, 1966). In general, bunch number and mean bunch weight are negatively correlated, and this fact accounts for the low heritability of bunch yield.

Genotypic or breeding values

To make full use of the ability of the plant to transmit characters from one generation to the next it is necessary to disentangle the genetic component of variance from the environmental component. It has been shown that the genetic component in the oil palm is largely though not entirely additive (Gascon *et al.* 1966; Sparmaaij, 1969), and that therefore the mean of a full-sib progeny for any character will give a reliable estimate of the genotypic value of the mid-parent, i.e. the mean of the values of the two parents. If crosses are made between three palms and if the mean values for any character for each of the three crosses is known then the genotypic value for each parent can be calculated. The example given by Van der Vossen (1974) may be used, viz:

Kernel to fruit.   Progeny $4.3488 \times G145 = 9.4$ per cent
$2,381D \times G145 = 12.4$ per cent
$4.3488 \times 2,381D = 10.0$ per cent

Designating the parents T_1, T_2 and D_3,

$$\frac{A\ (T_1) + A\ (D_3)}{2} = 9.4,$$

$$\frac{A\ (T_2) + A\ (D_3)}{2} = 12.4$$

and

$$\frac{A\ (T_1) + A\ (T_2)}{2} = 10.0,$$

where $A =$ the genotypic value. Thus $A\ (T_1) = 7.0$, $A\ (T_2) = 13.0$ and $A\ (D_3) = 11.8$, and the genotypic value for kernel to fruit for any other palm can be determined if the mean value in its cross with any of the above three palms is known, e.g. $T_1 \times T_4$: progeny $4.3488 \times 14.892 = 8.0$, so $A(T_4)$, i.e. the genotypic value of palm $14.892 = 9.0$.

Using Nigerian data Van der Vossen (1974) found that there was a strong or medium correlation between parental phenotypic and genotypic values, particularly *tenera* values, for bunch quality factors and for number of bunches. Correlation coefficients for mean bunch weight were lower and for bunch yield were barely or not significant. These findings conform in general with those of the heritability studies and were confirmed in a more detailed assessment of material from the same programme; genotypic values for quality factors calculated from different sets of values showed close agreement with each other.

Genetic correlations between bunch yield and quality components

It has been generally held that all components are inherited independently. However, in the case of bunch quality components it is obvious that a change in the percentage of one directly affects the percentages of the others. It is not surprising therefore that negative correlations are found between the additive genotypic values of mesocarp and shell. However, the correlation between kernel and shell is positive. Of greater interest is the negative genotypic correlation found between number of bunches and mean bunch weight. The significance of this finding for breeding policy is discussed later in this chapter.

Inheritance from the *dura* to the *tenera*

The correlations found in the Ivory Coast between bunch quality values in the *dura* and *tenera* have been mentioned in the previous pages (Gascon and de Berchoux, 1963). With more recent material, however, the correlations have been somewhat lower. Van der Vossen (1974), using Nigerian data, estimated the *tenera* additive genotypic values of *dura* parents and found that, except for kernel percentage, there were no significant correlations between the phenotypic value of a *dura* parent and its *tenera* genotypic value, and heritability estimates from *dura* to *tenera* based on the regression of genotypic on phenotypic values were equally insignificant. In contrast, the correlations and heritabilities from *dura* to *dura* and *tenera* to *tenera* were, as already described, marked and significant (Table 5.12).

With mesocarp percentage these findings confirmed the anomalies found between Deli *dura* and African *dura* in their *tenera* progeny of which mention was made on p. 243. It will be seen in Table 5.12 that although the *dura* mesocarp values for the Deli are higher than for African *dura*, the genotypic (A) *tenera* values are similar for both groups. Thus the superior fruit composition of the Deli palm is of no advantage over African *dura* in breeding *tenera* material.

The explanation of this phenomenon has been provided by Van der Vossen (1974) working on an assumption of Sparnaaij (1969) that both the shell and the fibre mantle of the *tenera* must be taken into account. Volumetric determinations showed that if only that part of the mesocarp lying outside the fibre was accounted in the *tenera*, then the correlation for mesocarp volume between *dura* and *tenera* full sibs was high (0.82***). Moreover the correlation was even higher (0.94***) between shell volume in the *dura* and shell + fibre mantle volume in the *tenera*, although the correlation between shell only in the two fruit forms was insignificant. Thus although the amount of lignification (*dura*, 100 per cent; *tenera*, partial, and *pisifera*, none) is inherited monofactorially, the shell thickness or percentage is determined by polygenes and is exhibited in the *dura* by the shell itself and in the *tenera* by the shell plus fibre mantle.

Table 5.12 *Correlations between phenotypic and genotypic values for yield and bunch quality components*

Character	Fruit form comparison	Correlation coefficient	Mean values			
			tenera	dura	Deli dura	d + t
Number of bunches	t + d	0.632***				P 9.8
						A 9.2
Mean bunch weight (kg)	t + d	0.441***				P 6.6
						A 5.1
Bunch yield (kg)	t + d	0.297*				P 59.2
						A 46.0
Single fruit weight (g)	P_t–A_t	0.567**	P 8.0			
			A 6.5			
Fruit to bunch (%)	P_t–A_t	0.577***	P 63.8	P 66.7	P 68.2	
	P_d–A_d	0.179	A_t 65.6	A_t 66.6	A_t 72.0	
	P_d–A_t	0.045			A_d 67.7	A_d 72.5
Mesocarp to fruit (%)	P_t–A_t	0.912***	P 83.1	P 55.9	P 59.7	
	P_d–A_d	0.872***	A_t 79.0	A_t 77.2	A_t 77.8	
	P_d–A_t	0.249			A_d 50.7	A_d 55.9
Shell to fruit (%)	P_t–A_t	0.924***	P 9.0	P 33.0	P 30.9	
	P_d–A_d	0.744***	A_t 12.5	A_t 12.9	A_t 13.8	
	P_d–A_t	0.176			A_d 37.5	A_d 34.7
Kernel to fruit (%)	P_t–A_t	0.718***	P 7.9	P 11.1	P 9.4	
	P_d–A_d	0.584***	A_t 8.5	A_t 9.9	A_t 8.4	
	P_d–A_t	0.541***			A_d 11.8	A_d 9.5

P = Phenotypic value, A = genotypic value. d and t indicate whether the values are *dura* or *tenera*. * P <0.05, ** P = <0.01, *** P = <0.001.

Source: Van der Vossen (1974).

This work also showed that, volumetrically, the ratio of shell to shell + fibre mantle in the *tenera* varied from 0.29 to 0.50, the latter figure being usual where the kernel is relatively large and there is the same kernel-to-fruit ratio in both *dura* and *tenera* full sibs. When the kernel-to-fruit ratio is much lower in the *tenera* sibs, as appears to be the case where the kernel percentage of one parent differs markedly from that of the other parent, then the ratio shell/shell + fibre mantle is much lower.

Van der Vossen concludes that the *tenera* genetic value of a *dura* for shell to fruit (and mesocarp to fruit) can only be obtained from the mean of its *tenera* full sibs or through *dura* × *tenera* crosses.

The inheritance of quantitative vegetative and growth characters

In discussing crop growth rates and net assimilation rates in Chapter 4 (p. 156) it was suggested that increased yields must be sought in increases of net assimilation rates and the diversion of increases in dry matter to bunch production, and that these increases would be likely to be attained by selection and breeding. In three progeny trials in Malaysia which involved the crossing of each of a number of *tenera* palms with a number of Deli *dura*, measurements were made of bunch yield (Y), vegetative dry matter production (VDM), bunch index ($BI = Y/Y + VDM$), net assimilation rate, leaf area index (L), leaf area and leaf area ratio (F) (Hardon *et al.*, 1972). Estimates of genetic variability were computed for each parameter. The results suggested that heritability may be high for VDM, BI and L. Heritability of the other factors, though not negligible, was less consistent. Breure and Corley (1983) reviewed these and further data from Malaysia (Ooi, 1978; Tan, 1978) and Papua New Guinea on heritability of growth parameters calculated by various means. They concluded that, for vegetative measurements, heritability can be high, but that for growth parameters derived from bunch yield, i.e. total dry matter per palm and *BI*, it is lower. However, they found that palms selected for high *BI* in their early years maintained both a high *BI* and a high bunch yield in later life, and they concluded that, for higher-density planting, selection for high *BI* would be more effective than selection for yield.

The wide differences between Deli palms and certain populations of African palms were described on p. 228. De Berchoux and Gascon (1965) found that crosses between palms of these two populations gave progeny which were intermediate in respect of number of leaves per palm, weight of leaf, weight and number of leaflets, length of leaf, breadth of leaf and breadth of petiole.

Methods of selection and breeding

Prospection and conservation

In spite of the vast populations existing in the palm belt across West and Central Africa, the material used for the great expansion of recent decades had a narrow genetic base. The need for new material of sufficient genetic variability to improve and safeguard the crop has been widely acknowledged for a long time, but operations in Africa were, with one exception, hampered by the modest facilities available and the view that more immediately promising results would be obtained by the use, in crosses, of the restricted lines already available; and in the Far East the industry was satisfied, for more than a third of a century, with the Deli palms in its possession.

The earliest important prospection of oil palms to obtain high-quality *tenera* progeny was that carried out in Zaire in the 1920s and already described. Postwar prospection in Zaire included later generations of the same material, palms of local origin on estates and palms in grove areas. Provisional CAMs, which were largely *tenera* on estates and *dura* in the groves, were selected after fruit and bunch analysis and yield recording (Vanderweyen, 1952). In the area of northern Zaire only one *tenera* palm would be selected as a CAM out of up to 35,000 palms examined (Pichel, 1957). The rigour of this selection may be gauged from the fact that while 'average' Yangambi *tenera* had mesocarp-to-fruit and oil-to-bunch contents of 70 and 22 per cent respectively, selected mother palms at two centres had contents of 92.5 per cent mesocarp to fruit and over 32 per cent oil to bunch. In southern Zaire (Kwango) about 420 hectares were searched, but only seventeen provisional selections were made. Of these, the *tenera* selected showed a fruit and bunch analysis very similar to those obtained in the north (Desneux, 1957).

In Nigeria only very small areas were prospected. These consisted of the palms at Calabar derived from fruit of a range of forms and types, an old grove area at Aba covering 11 hectares and another grove area of 49 hectares at Ufuma. Both the Calabar and Aba plots were in areas of poor composition and very rigorous standards could not, because of the small number, be applied in prospection. The Ufuma grove was, however, chosen because of the very high proportion of *tenera* palms (43 per cent). This grove underwent various thinning processes, but finally about 2 per cent of the original stand was selected. Fruit analysis was not impressive; shell percentages varied from 11 to 19 and mesocarp was never over 79 per cent, estimated by the boiling and pounding method (equivalent to 85 per cent by modern analysis methods). The selected palms were pollinated by mixed pollen from the best of their number, and the resulting progenies gave a mean mesocarp-to-fruit percentage 6.3 above that of the parents. This small prospection was clearly too restricted to capture a wide variation of material.

In the early 1960s a preliminary prospection of *dura* and *tenera* fruit in many grove areas in eastern Nigeria was undertaken. Table 5.13 shows the mean *tenera* values in eight of the areas chosen. The data showed that there was still to be found in eastern Nigeria *tenera* material of very high quality and, no doubt, of potentially high yield. The variations between areas, and even between villages, were interesting. Some areas did not appear to contain appreciable numbers of high-mesocarp fruit, but the kernel percentage was high; in others, e.g. Aba, many samples had high kernel and rather high shell percentages, but everywhere there was a wide range of shell thickness. Fruit size varied greatly and although no very large fruit emerged in this survey, fruit well up to the size of Deli fruit were frequently encountered. Some of the *dura* fruit seen were very similar to Deli fruit and there was a great variation in spine length in the spikelets.

Table 5.13 *Analyses of* tenera *fruit samples obtained during a prospection in eastern Nigeria*

Area	Mean wt per fruit (g)	Mean mesocarp (%)	Mean kernel (%)	Mean shell (%)	Range of shell (%)
Awka	9.5	78.8	10.7	10.6	4–20
Ufuma	9.4	81.1	11.7	6.9	2–18
Nnewi	9.6	80.0	11.3	8.6	2–25
Owerri	11.3	83.4	9.8	6.7	1–24
Okigwi	10.2	82.5	10.2	7.3	2–18
Aba (south)	11.3	71.7	17.3	11.0	4–25
Umuahia	9.9	84.0	7.8	8.2	
Ikot Ekpene	8.4	81.5	10.4	8.2	

In the Ivory Coast, production material was developed from limited local prospections in that country and Benin, and from progenies of Zairean and Malaysian origin. Material designated as 'La Mé origin' was developed from twenty-nine palms of a grove at Bingerville while 'Pobé origin' stemmed from thirty-eight *tenera* selected in groves near Porto-Novo (Meunier, 1969).

A systematic prospection was later carried out by Meunier (1969) in groves in the Ivory Coast. These groves are in general less dense and more scattered than those in Nigeria. About 100 palms were taken at random in each of eleven areas. The proportion of *tenera* varied from 0 to 41 per cent and only seven *pisifera* were found. *Virescens* palms were rare, only one *albescens* was found and no mantled palms were encountered. Areas with the highest proportion of *tenera* tended to have the best mean fruit composition. Where *dura* of good composition were found, the *tenera* were also of good composition. There were wide differences between groves in bunch and fruit weight as well as composition and populations could be grouped according to their fruit and bunch characteristics although members of a group were not necessarily adjacent to each other. Some data from this prospection are given in Table 5.14. This is not directly comparable with the Nigerian data in Table 5.13 since in the latter prospection the palms were not taken at random.

It was concluded from the Ivory Coast prospection that the Yocoboué–Sassandra group was close to the imported Zaire material ('Sibiti') as far as mesocarp to fruit and bunch weight were concerned, whereas other groups were similar to the original F_0 La Mé material from Bingerville or to the Pobé material from Benin. Sufficient variability was shown to exist for improving the *tenera*/*pisifera* stocks at present being used in breeding and selection.

In the early 1970s a much greater interest began to be taken in

Table 5.14 *Some data from a random selection of grove palms in the Ivory Coast*
Mean bunch weight and composition of palms from groves grouped according to their fruit and bunch characteristics

Grove group	Fruit form	Mean bunch wt (kg)	Fruit to bunch (%)	Mesocarp to fruit (%)	Shell to fruit (%)	Kernel to fruit (%)	Mean fruit wt (g)
Yocoboué–Sassandra	*Dura*	14.7	58.4	47.7	39.5	12.8	7.0
Abobo–Dabou–Bingerville	*Dura*	11.0	55.3	43.3	42.9	13.8	7.5
Tabou–Danané	*Dura*	10.7	53.3	45.8	41.4	12.8	8.7
Soubré–Man	*Dura*	10.0	55.1	36.2	46.8	17.0	6.0
F$_0$ of La Mé*	*Dura*	10.1	60.9	40.2	45.2	14.6	7.4
Yocoboué–Sassandra	*Tenera*	12.8	54.1	68.4	20.5	11.1	6.2
Abobo–Dabou–Bingerville	*Tenera*	9.9	52.4	64.2	22.7	13.1	6.0
Tabou–Danané	*Tenera*	9.2	47.1	62.6	24.6	12.8	6.9
Soubré–Man	*Tenera*	8.4	52.3	50.0	33.6	16.4	4.6
F$_0$ of La Mé*	*Tenera*	9.2	56.6	61.6	25.3	13.1	6.0

* Mean composition of original 'La Mé' F$_0$ open-pollinated material from Bingerville for comparison.

Source: Meunier (1969).

widening the genetic base on which the rapidly expanding industry should in the future be maintained. Concern was also felt about the loss of foundation material; not only were large areas of potential Deli mother palms being replaced by plantations of *tenera* hybrids, but the great grove areas of Africa were likely to contract. With the establishment of national oil palm research in Malaysia, specific suggestions were made for a wider collection and preservation of material (Hartley, 1971), and this led to international cooperation in prospection. Later, the International Board for Genetic Resources (IBPGR), established in 1974, accorded the oil palm second priority (first priority being given mainly to annual food crops), and the Board agreed to give assistance in the collection of material and to identify areas for future collections of both *Elaeis guineensis* and *E. oleifera* (Murthi Anishetty, 1985).

Hardon (1985) reviewed the long-term problems of oil palm conservation and considered that although insufficient is known about inter- and intra-population variation, the widespread distribution of the palm is probably of recent origin and has not yet led to any distinctive population differentiation. Within-population and within-site variation is, however, high, and so he believed that the generality of oil palm material, considered as an *in situ* gene pool, is not in any imminent danger of being eroded. However, he recognized that the future of *in situ* conservation of the oil palm cannot, owing to the economic problems of the regions involved, be predicted, and, while acknowledging the part played by recent collections (see below), he thought that the time is now ripe for a systematic examination of populations, their geography and genetic variation prior to embarking on an organized selective conservation programme.

The first collection in the new phase of prospection was undertaken in 1974 jointly by the Malaysian Agricultural and Development Institute (MARDI, to be succeeded by the Palm Oil Research Institute of Malaysia (PORIM)) and the Nigerian Institute for Oil Palm Research (NIFOR) and was followed by prospection in Cameroon and Zaire (Rajanaidu, 1984). Whereas previous prospections had as a prime object the capturing of high-quality material for introgressing into already existing breeding programmes, the purpose of these programmes was to obtain a largely random sample of grove material across the geographical range of the species. This method was followed, in preference to any form of selection, firstly because desirable genotypes cannot be predicted for the future since selection criteria are subject to change, and secondly because the performance of a grove palm does not give a fair indication of its genetic potential (Arasu and Rajanaidu, 1977; Rajanaidu *et al.*, 1979; Rajanaidu, 1985a). Such a programme also serves as a start of properly managed *ex situ* conservation and can become a nucleus for further *ex situ* conservation if genetic erosion becomes more serious (Hardon, 1985). The practical drawbacks are that a large amount of land is needed and the incorporation of the material in breeding programmes is delayed for a generation.

Table 5.15 gives details of the PORIM collections from West and Central Africa. At a later date 95 samples (58 *dura* and 37 *tenera*) were obtained from the Kigoma (Lake Tanganyika) region of Tanzania, and 17 samples (fruit form unidentified) from Madagascar where the fruits were extremely small (Rajanaidu and Rao, 1987). In Nigeria, human interference leading to a much higher proportion of *tenera* in areas of high human population was evident. The variance for all bunch and fruit characters between palms within sites was much greater than between sites or areas. Nevertheless

Table 5.15   *PORIM collections in Nigeria, Cameroon and Zaire*

Country	Sites	Palms sampled		Distribution of forms at sites*		
		Dura	Tenera	Tenera (%)	Pisifera (%)	Virescens (%)
Nigeria[†]	45[‡]	595	324	0–62	0–4	0–10
Cameroon	32	58	37	0–19	Nil	0–36
Zaire[§]	56	283	86	0–13	Nil	Nil

* Sites with more than 20 palms observed.
[†] Prospection in conjunction with NIFOR.
[‡] From 21 areas in 6 states.
[§] Prospection in conjunction with Unilever.

there were, for most bunch and fruit characters of both *dura* and *tenera*, significant differences between sites though not between areas. In Cameroon there were similar differences between sites, while in Zaire, though most of the variance was again between palms, there was an exceptionally high between-site variance for mesocarp percentage to fruit. In general, the samples from Zaire showed the highest single fruit weight, while the Nigerian samples gave the highest percentage mesocarp (Rajanaidu, 1985a).

In Nigeria, there were significant correlations between *dura* and *tenera* for all characters measured except fruit length, nut diameter and shell thickness (Rajanaidu, 1974). The lack of correlation in the case of shell thickness is of particular interest in the light of the work on mesocarp and shell by Van der Vossen (1974) already described (p. 248). The overlap of shell thickness between the *dura* and *tenera* was quite marked: some *tenera* were recorded with shell thickness of 2–5 mm and *dura* with thickness of less than 1 mm. Although the site mean for *dura* shell thickness was usually 1.9 mm and above there were a substantial number of *dura* with shells of between 1 and 2 mm thickness. An early evaluation of the Nigerian material planted in Malaysia showed that there were outstanding progenies and individual *tenera* in the population both in respect of palm oil to bunch and height increment. Bunch yields of these palms were high, while the majority were shorter than currently planted *dura* × *pisifera* palms of similar age (Rajanaidu and Rao, 1987).

Outside Africa no large prospections have been undertaken. In the Brazilian groves, however, collections and bunch analysis have shown that high-quality *tenera* and *dura* palms also exist there; these may well be adapted to the environment of coastal Bahia (Hartley, 1968). There have been no recent collections of Deli *dura*; the inbred nature of the present Deli population has been repeatedly stressed and, in the Far East, evidence of its lack of genetic variance was shown in studies in Malaysia (Thomas *et al.*, 1969; Ooi *et al.*, 1973).

Lawrence and Rajanaidu (1985) considered the strategy of sampling natural oil palm populations in the light of an early evaluation of the PORIM/NIFOR prospection material. Estimates of variance components, inter-class correlation coefficients, and maximum and minimum estimates of heritability for two vegetative characters (rachis length and total leaf area) and three yield components (bunch weight, mesocarp to fruit and oil to dry mesocarp) showed that most of the variation was between plants within progenies (i.e. plants raised from single, open-pollinated bunches), though for mesocarp to fruit and to a lesser extent for rachis length there was also considerable between-population (site) variation; and these two variables were the ones showing the highest heritabilities.

The variation between progenies (bunches) within populations was small. It was concluded that, although for the detection of variation the numbers of bunches sampled could be reduced, for a reasonable estimation of the amount of variation the twenty palms per population which was the sampling standard in the prospection should be adhered to.

Prospection for *Elaeis oleifera* material is discussed on a later page (p. 288).

Bunch production or bunch quality?

The move from *dura* to *tenera* planting gave substantial increases in oil production, but improvement within the *tenera* is likely to be of a lesser order and it has therefore been argued that more benefit is to be gained from improved bunch yield. Van der Vossen (1974) showed that if each component of oil yield were increased separately by its standard error then the returns from increases in either number or weight of bunches would be five to thirteen times greater than the returns from increases in fruit to bunch, mesocarp, mesocarp oil or kernel oil. His computations were based on calculated genotypic mean values of *tenera* selected in Nigeria and recorded over the first 3 years of bearing; moreover the material showed high heritabilities for fruit components but low heritability for bunch yield. Thus it was doubtful if representative economic conclusions could be drawn from the data, and it is probably more realistic to enquire what improvement in returns can be expected in adult fields from different increases or decreases in bunch yield, shell percentage to fruit and oil percentage to mesocarp. Progeny trials have indicated substantial bunch yield gains where individuals from separate populations have been crossed, but differences between progenies within such crosses are not of the same order. The uncertainties of bunch production breeding have been stressed by Hardon *et al.*, (1973) but further progress can be expected through widening the field of selection, maintaining genetically diverse populations and making use of the heritability of vegetative characters.

Table 5.16 shows the returns to be obtained in an area with a low level of yield such as Nigeria by improving bunch yield by 33 per cent and other factors from relatively low levels to possible high levels. Increasing yields by similar percentages in high-yield areas such as Malaysia would of course have the same proportionate effect on returns. It will be seen that increasing bunch production without improving quality is more rewarding than improving any single bunch quality factor. However, if several quality factors can be improved simultaneously then the effect of the latter will be as great or greater than that of improving bunch yield. Nevertheless, bunch quality improvement has severe limits and there are pro-

Table 5.16   *Returns to be obtained by improving bunch yield* and bunch quality*

Tons/hectare palm oil and kernels, and economic units, obtained by improving each of the factors bunch yield, F/B, shell to fruit, oil/mesocarp, while keeping the other factors constant

Bunch yield (tonnes per hectare)				Fruit to bunch (%)				Shell (%)	Mesocarp (%)	Kernel (%)				Oil to mesocarp (%)			
12	M	2.52		58	M	2.52		14.0	77.0	9	M	2.52		47	M	2.52	
	K	0.63			K	0.63					K	0.63			K	0.63	
	Eu	30.2			Eu	30.2					Eu	30.2			Eu	30.2	
13	M	2.72		61	M	2.65		11.5	80.5	8	M	2.63		50	M	2.68	
	K	0.68			K	0.66					K	0.56			K	0.63	
	Eu	32.6			Eu	31.8					Eu	30.8			Eu	31.8	
14	M	2.94		64	M	2.78		9.0	84.0	7	M	2.75		53	M	2.84	
	K	0.73			K	0.67					K	0.49			K	0.63	
	Eu	35.2			Eu	33.3					Eu	31.4			Eu	33.4	
15	M	3.15		67	M	2.91		6.5	87.5	6	M	2.86		56	M	3.00	
	K	0.78			K	0.72					K	0.42			K	0.63	
	Eu	37.7			Eu	34.9					Eu	32.0			Eu	35.0	
16	M	3.36		70	M	3.04		4.0	91.0	5	M	2.98		59	M	3.16	
	K	0.84			K	0.76					K	0.35			K	0.63	
	Eu	40.3			Eu	36.5					Eu	32.6			Eu	36.6	
Eu per cent increase lowest to highest		33				21						8				21	
Eu per cent increase by raising three or two bunch quality factors lowest to highest												61				33	

* Bunch yield levels used correspond to those in more seasonal parts of West Africa, e.g. Nigeria.
M = Mesocarp oil. K = Kernel. Eu = Economic units based on 10 Eu per tonne mesocarp oil, 8 Eu per tonne kernels.

genies in existence near those limits; but the consistent production of large quantities of only these high-quality bunches has not yet been achieved in practice by seed producers. Clearly the breeder still has to give attention, in each situation, to both bunch yield and bunch quality breeding and to try to provide sufficient high bunch quality progenies to maximize oil plus kernel production to bunch.

Oil or kernels?

The kernel has been almost entirely neglected in the breeding of the oil palm since the days of Beirnaert (1933a), who had the wisdom to record it as one of his breeding criteria. Broekmans (1957a) made a positive plea for the kernel in the introduction of his 'C' factors: C1 was the content of kernel plus mesocarp to bunch, i.e. $(1 - \text{shell}/\text{fruit}) \times \text{F/B}$. Though often appearing in tables of fruit composition, kernel production has seldom seemed to have been a positive aim. In fact, however, at prices which have been ruling for the last few decades, kernels make a contribution of great importance to the economy of a plantation though the significance of this contribution is masked by the higher price per tonne of palm oil. The ordinary producer contends that he wants the maximum oil to bunch as palm oil has a high price, let us say $500 per tonne. He does not want any of this replaced by kernels at, say, $400 per tonne. Assuming F/B, shell to fruit and oil to mesocarp remain constant, an increase of 3.25 in the kernel-to-bunch percentage (5 per cent kernel to fruit) would lead to the theoretical changes in gross return per 100 tonnes bunches shown in Table 5.17.

Table 5.17   *Theoretical returns from 100 tonnes bunches containing fruit with (A) High oil, or (B) High kernel contents*

	Tonnes per 100 tonnes bunches	
	A (High oil)	B (High kernel)
Fruit to bunch (F/B)	65	65
Mesocarp/bunch (mesocarp/fruit %)	52 (80%)	48.75 (75%)
Shell/bunch (shell/fruit %)	6.5 (10%)	6.5 (10%)
Kernel/bunch (kernel/fruit %)	6.5 (10%)	9.75 (15%)
Oil to bunch	26	24.375
Gross return:	($)	($)
Oil @ $500	13,000	12,188
Kernel @ $400	2,600	3,900
Total	15,600	16,088

In practice, kernel extraction efficiency being usually a little higher than oil extraction efficiency, the difference in gross return in favour of B will be rather greater than the $488 shown. The two analyses are quite possible, indeed usual, and at the 5 : 4 price ratio the high-kernel type B is more profitable. The oil palm breeder is then presented with a difficult problem – the parameters mesocarp to bunch, oil to bunch, mesocarp + kernel to bunch (C1), even oil plus kernel to bunch, are not fully adequate. An increase in oil percentage must mean a similar increase in the non-oily half of the mesocarp and, with shell percentage unchanged, must be accompanied by a reduction of kernel by about $2 \times$ oil increase, thus obliterating the price advantage.

A number of general conclusions may be drawn.
1. If high-mesocarp fruit is being bred, then the factor oil to meso-

carp becomes of greatest importance; an increase from 50 to 55 per cent will appreciably reduce the advantage kernel has over mesocarp. Conversely oil percentages of less than 50 per cent are disadvantageous and fruit with large kernels would in this case be preferable. This brings out the importance of oil-to-mesocarp determinations which should not be masked in oil-to-bunch data.

2. It is probably not possible to reduce shell percentage to as low a figure in high-kernel *tenera* as in high-mesocarp *tenera*; therefore the highest mesocarp plus kernel percentage will be obtained by high-mesocarp fruit. If this can be combined with oil to mesocarp well above 50 per cent this type of fruit is likely to give the highest gross return. This point has not yet been generally reached in breeding however. Two high-quality palms in West Africa may be quoted in which the shell averaged 8 per cent of fruit:

Percentage:	Mesocarp	Shell	Kernel
Palm 14.437	85.8	8.0	6.2
Palm 38/0401	79.0	8.0	13.0

With similar F/B ratios and oil/mesocarp, 38/0401 would give the higher monetary return; in fact this palm had a much higher F/B ratio and a high oil content. To evaluate these two palms on mesocarp percentage or on palm oil to bunch would be gravely to underestimate the value of 38/0401. On the other hand a palm with a shell content as low as 3–4 per cent, e.g.:

	Mesocarp	Shell	Kernel
Palm 4.493	90.6	3.8	5.6 per cent

will give a very high monetary return, and an alternative composition such as

	85.0	3.8	11.2

giving a higher gross return, would be impossible to find. Even the thinnest shells usually constitute 50 per cent of the nut weight and cases like palms 4.493 and 38/0401 where the shell is 40 per cent and 38 per cent of shell + kernel are rare. In the above hypothetical alternative to 4.493 the shell is 25 per cent of shell + kernel and this is unknown.

One must conclude therefore that a high kernel percentage gives the highest gross return unless shell percentages can be reduced to about 5 per cent of fuit or below; breeding should not be dominated by considerations of mesocarp or oil to bunch or oil yield per hectare as these measures do not indicate the full and accurate production yield of the oil palm.

Breeding systems – General considerations

The oil palm is a monoecious plant which is naturally cross-pollinated. It has alternating cycles of male and female inflorescences and it is rare for pollen to be produced on a palm when female flowers on the same palm are receptive. The most important other monoecious crop is maize, and probably more is known about the genetics of maize than of any other plant. It was early found in the breeding of maize that the most satisfactory production and quality could be attained by the concentration of desirable characters in inbred lines followed by the use of these lines to provide productive hybrids (Robinson and Treharne, 1985).

Although oil palm breeders have been aware of the similarities between their plant and maize, breeding has not, for a variety of reasons, been dominated by the work on maize. It is known that Beirnaert in Zaire had for many years been considering the difficulties of following the classical methods of maize breeding (Toovey, 1938). He considered that high productivity being the consequence of the presence through Mendelian inheritance of a large number of characters, the concentration of all or some of these characters by the classical method of inbred lines would take a very long time and, though one might readily obtain homozygosity in one factor such as high number of bunches, it would be difficult to obtain by inbreeding many or any palms containing all the productivity factors. He proposed, therefore, to omit this stage and to cross within his basic material palms which had complementary characters, e.g. palms with productivity characters of the type AA Bb cc dd would be crossed with aa bb Cc DD. This and other complementary crosses would be carried out on a large scale to produce palms of the type Aa Bb Cc Dd. Later, selfing would be carried out to fix these characters in homozygous lines (Beirnaert, 1933c). By 1938, still unaware of the nature of the *tenera*, he had at his disposal about 100,000 palms, largely *tenera*, and but for the problem of sterility and the fruit-form inheritance discovery shortly to be made, selection and breeding might have gone further forward on these lines than in fact it did.

A number of factors other than longevity has tended to divert oil palm breeding away from the maize analogy and to bring the ideas and exposition of the animal breeders, through D. S. Falconer (1960), to the fore. Oil palm breeding is breeding from a pair of individuals, each occupying about seventy square metres of space, for the production of large though manageable families of hybrid offspring. Though both parents have male and female flowers, the *dura* has taken on the role of mother palm and the *pisifera* the role of father. It is through the distinctive qualities as well as the yield performance of the hybrid that parents may be chosen for continued breeding into further generations.

With outbreeding annual crops there is no possibility of using the parents after the performance of the progeny has been noted; with animals and with the oil palm the parents can be re-employed for many years and, indeed, an oil palm can be employed a great many more times than most animals. However, it must not be thought that seed production breeding can depend solely on 'proved' parents. The disadvantages of relying entirely on *progeny testing* have been outlined by Falconer (1960) as follows:

> In practice progeny testing suffers from the serious drawback of a much lengthened generation interval because the selection of the parents cannot be carried out until the off-spring have been measured. The evaluation of selection by progeny testing is apt to be rather confusing because of the inevitable overlapping of generations, and because of a possible ambiguity about which generation is being selected, the parents or the progeny. The progeny, whose mean is used to judge the parents, are ready to be used as parents just when the parents have been tested and await selection. Thus both the selected parents and their progeny are used concurrently as parents.

This certainly occurs in oil palm breeding though there are ways of avoiding the difficulties and confusion.

Inbreeding depression and heterosis

Instances of inbreeding depression were noted at an early stage in oil palm breeding. Selfed *tenera* in southern Zaire showed degeneration in their early life, giving poor germination and irregular or deformed seedlings, while at Yangambi two selfed progeny (1393B and 2381D) gave a mean bunch yield, over 2 years, 33 per cent less than that of their cross (Sparnaaij, 1958). In Nigeria, comparisons between selfs and their crosses showed reduced cumulative yields in the selfs of 13–14 per cent with Deli *dura* and from 17 to 49 per cent with *dura* (NIFOR, 1969). In Malaysia, Hardon (1970) has estimated the inbreeding coefficient, F_x, for a range of Deli *dura* progenies on the assumption that breeding from the original Bogor palms to the establishment of basic material in Malaysia was equivalent to three half-sib matings, giving an inbreeding coefficient of 0.305. Significant negative correlations between the coefficient and bunch yield were obtained, though this was composed of a strong negative correlation for mean bunch weight and a weak positive correlation for number of bunches. However, other data provided by Hardon and Ooi (1972) suggested that yield depression would not be significant where inbreeding coefficients are low (below about 0.3), particularly in the Deli *dura*; and that these low levels could be tolerated where considerations of adaptability of the material to the environment and the characteristics of the individual parents become overriding.

Data from the Ivory Coast indicated a strong inbreeding yield depression with selfing within populations, though only inter-population crosses were available for comparison (Gascon *et al.*, 1969).

Heterosis is the opposite of inbreeding depression and has been defined as the superiority of the F_1 cross over its better parent (Jinks and Lawrence, 1983). If crosses are made at random between inbred lines developed from a certain population the mean value of any character in the progeny will be the same as in the original population (Falconer, 1960). Without selection, therefore, the crossing of lines in a large population cannot be expected to make any permanent change in the population mean. Heterosis can, it is believed, be explained in terms of quantitative inheritance. Where crossing of *selected* individuals from different populations has been successful this has been due to the selection itself, to the effects of the utilization of additive genetic variance of independent characters exhibited in general combining ability and to a lesser degree to specific combining ability.

Heterosis is caused either by overdominance or by the dispersion of dominant genes, with or without non-allelic interaction. Jinks and Lawrence (1983) consider that the latter is by far the commoner cause and that there is little genetic justification for a belief in the unconditional superiority of heterozygotes in populations of outbreeding species like the oil palm. They suggest that, although heterozygotes are on average superior to homozygotes, it should be possible to extract from the F_2 of a cross between inbred lines recombinant lines equalling or exceeding the F_1 cross in their performance. This is discussed in a later section.

In general, it has been found that inbreeding reduces yield but has a greater effect on bunch weight than on bunch number. However, there have been many instances of individual selfings showing little or no inbreeding depression, while in other cases inbreeding has tended to concentrate either good characters (e.g. dumpiness) or bad characters (e.g. Orange Spotting or Crown disease). The fact that there are large yield differences between selfed progenies suggests that in some cases deleterious genes are largely absent and that the high-yielding selfs may have good general combining ability (Rosenquist, 1984a).

French workers have extolled the value of 'inter-origin' crosses but have confined their evidence to comparisons involving Deli *dura* × African *pisifera* or *tenera* crosses (Gascon and de Berchoux, 1964). The performance of their 'inter-origin' crosses is attributed to the additive effect of a favourable combination of factors for weight per bunch and number of bunches from the parents (Gascon *et al.*, 1966). Even if only mid-values are obtained in the progeny there may be an overall gain in yield, for example:

	Weight per bunch (kg)	Number of bunches per annum	Product (kg)
Deli	15.0	6.0	90.0
African	10.0	9.0	90.0
Cross (mid-values)	12.5	7.5	93.75

With young palms giving lighter bunches the increase may be greater, as exemplified by data from La Mé shown in Table 5.18.

Table 5.18 *Yield increase obtained from inter-origin crosses in the Ivory Coast (Gascon and de Berchoux, 1964)*

(5th–8th year)	Wt per bunch (kg)	No. of bunches per annum	Wt of bunches per annum (kg)	Increase over product of mid-values (%)
Deli × Deli	14.2	6.0	84.3	
African × African (Ivory Coast)	7.3	14.4	100.7	
African × Deli	11.8	10.3	119.3	9.1
(Mid-values)	(10.8)	(10.2)	(110.2)	

Source: Gascon and de Berchoux (1964).

It will be noticed that in the case of weight per bunch the cross gives a value nearer to that of the Deli. This is a common experience and is exhibited to an even greater extent in the Deli cross with selected material of Zaire origin. Two years' data with young palms gave the following results:

Origin:	Deli	Zaire	Cross	Mid-value
Weight per bunch (kg)	14.5	7.4	12.4	11.0

The comparisons here are not, of course, between selfed and out-crossed material but are of individuals from two populations one of which is considered to be partially inbred. The low production of the Deli in Africa is due to a relatively low sex ratio not fully compensated for by its large bunches. Its introduction into Africa brought factors for low sex ratio to an area where environmental conditions already favour low sex ratios and where selection is largely geared to the discovery of high sex ratio palms. Fully satisfactory exploitation of the high bunch weight character of the Deli must therefore include a search for Delis giving a much higher than average sex ratio under African conditions.

Breeding for vegetative propagation

With the successful production of plantlets through tissue culture
(see p. 281), it is probable that a good deal of planting and
replanting will, by the end of the century, be undertaken through
vegetative propagation and the planting of clones. It is not yet possible to foretell how far yields and extraction rates will be improved
simply by the selection and propagation of already existing élite
tenera palms, but there will certainly be a need for a supply of ortets
(clone sources) of steadily improving standard; soundly based
breeding programmes will therefore continue to have an essential
part to play in oil palm improvement.

Selection and breeding in practice

Africa

Practical breeding programmes in Africa have been much influenced
by a recurrent selection programme started in Zaire and planted in
1956 and which was based on the material of the early selection
programmes described at the beginning of this chapter. The basic
material was six *tenera* selections which were selfed and crossed in
all combinations, the intention being to breed from *dura* and *pisifera* in those selfs whose parents had shown good progeny in their
crosses (Sparnaaij, 1958; Pichel, 1957). The resulting *tenera* material, besides being used for production, would also be planted in
dura × *pisifera* comparative trials. This is illustrated in Fig. 5.12A.

It can at once be seen that, because the parents for production
must be *dura* and *pisifera* to give 100 per cent *tenera*, the recurrent
selection procedure is modified to allow for: (i) the recurrence to
take place from the original comparative trial of selfs and crosses
(because the oil palm is perennial there is no need to run a generation of inbreeding and then to inter-cross before choosing the next
recurring generation); (ii) seed production for plantation use to take
place in a special area; (iii) *pisifera* parents, which cannot exhibit
their potential yield and quality characteristics themselves, to be
chosen in the selfs in the comparative trial on the basis of the
performance of their *tenera* parents as parents in the crosses; (iv)
dura parents to be selected for production breeding mainly on the
performance of their *tenera* sibs. For instance, if high yielding and
quality *tenera* are found in the progeny T4 × T6 then seed would
be produced for distribution by crossing selected *dura* within the
progeny of the T4 selfing in the special seed garden with *pisifera*
pollen from the T6 selfing in the comparative trial. Two questions
arise. Can the *dura* and the *pisifera* be judged as parents from the

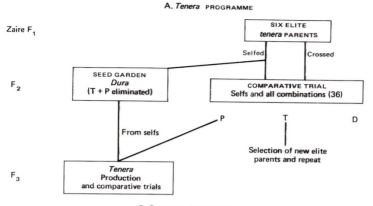

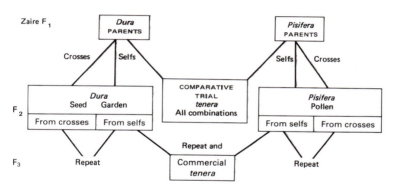

N.B. The *dura* and the *pisifera* crosses constitute later observation fields and are not part of the original programme.

Fig. 5.12 A representation of modified recurrent and reciprocal selection in the INEAC (Zaire) breeding programmes.

performance of their *tenera* parents and half-sibs? Is inbreeding necessary or desirable?

Another part of the INEAC programme involved 'purification' of the two lines by inbreeding to give homozygous *dura* and *pisifera* lines for crossing. This scheme has a number of advantages and drawbacks: (i) the comparative test was D × P and selected parents provided the actual *tenera* it was hoped to reproduce for as long a period as the parents survive; seed is limited by the number of parents successfully used in the comparative trials; (ii) seed provided from *dura* selfs and *pisifera* selfs in the next generation is not identical to that from the original tested parents; (iii) the *pisifera* line is of fertile *pisifera* unless the inbreeding is done in a *tenera* line; it is known that the thinner-shelled *tenera* palms come predominantly from infertile *pisifera*, and breeding within selfed *pisifera* lines is technically difficult; (iv) the original parents were chosen for their individual qualities as *dura* and *pisifera* palms and it is not known whether they will be the best producers of *tenera* progeny. The *dura* and *pisifera* programme is illustrated in Fig. 5.12B.

This programme is in reality a form of *reciprocal* (or *reciprocal recurrent*) selection used commonly in both animal and plant breeding. Two sources, in this case *dura* and *pisifera* palms, form the starting-point and they are crossed in comparative trials. In animal selection the parents would await the results of these trials before being selected and the selected individuals would be crossed within their own line or source to produce another generation for testing. The cycle is then repeated. In annual crop breeding the selfed or 'within-source' seed can be held over and the selected portion planted in the third year after appraisal of the comparative trials; the progeny is then inter-crossed within the same source for further between-source crossings in the next generation (Hayes *et al.*, 1955). In palm breeding all the parents used for crossing may be used in the inbreeding or 'source-breeding' programmes so that the progeny of selected palms as well as the selected palms themselves may be used in large programmes of seed production.

Other oil palm breeders in Africa have adopted forms of reciprocal selection though the source material has been predominantly *tenera* on the one side and *dura* on the other, and the methods of choice have varied. As already mentioned, Gascon and de Berchoux (1964) in the Ivory Coast adopted the Deli as almost their sole *dura* source though a few Deli × African *dura* were included in their programme. The other side of the programme consisted largely of *tenera*, but included some *pisifera*. These varied considerably in quality and were obtained from four separate geographical origins. Taking *tenera–pisifera* as source A and Deli *dura* (pure or crossed) as source B, breeding and seed supply proceeded as follows. The planting consisted of 85 Deli × Deli and 79 *tenera* × *tenera* selfings and crosses with 446 Deli × *tenera* or *pisifera* crosses made to compare the ability of the parents to provide productive progeny in crosses. Following appraisal of the results of this comparative trial the best parent *tenera* and, separately, the best parent Deli *dura* are crossed among themselves, thus providing a further selection generation. In commercial seed production the *pisifera* bred from source A are crossed with *dura* bred from source B and the progeny will thus come solely from palms whose parents have shown their performance in the comparative trials. Selection is repeated in the

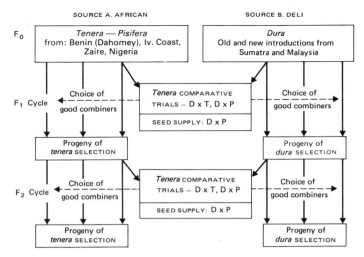

Fig. 5.13 Outline of reciprocal selection system adopted at La Mé, Ivory Coast.

second selection generation through a further comparative trial and a repetition of the selection process. The system is illustrated in outline in Fig. 5.13.

A second cycle of selection and breeding was later started which included not only selections from the first cycle but also the exploitation of introductions previously made and material of *Elaeis oleifera*. This cycle also includes crossings between the progenies of selfings made in the first cycle particularly those of Dabou Deli D10D and La Mé *tenera* L2T and top crosses of either of these palms to the progeny of the other.

Data from these programmes have been largely presented as means of hybrid populations. The Deli *dura* have been crossed with three groups of *tenera/pisifera* emanating respectively from La Mé, Sibiti/Yangambi and NIFOR, Nigeria. In general, bunch production has been higher in the La Mé and NIFOR crosses than in the Sibiti/Yangambi crosses though the quality of some of the latter has been good and fruit size is large. Shell content in the NIFOR crosses was low with percentages down to 5–8 in certain progenies. Oil to mesocarp was highest in the La Mé crosses but the NIFOR crosses gave the highest overall palm oil per hectare (Gascon *et al.*, 1969; IRHO, 1969–73).

A breeding programme adopted at NIFOR contained in its design the elements of both recurrent and reciprocal selection. It will be seen from Fig. 5.14 that *dura* seed palms can come from the

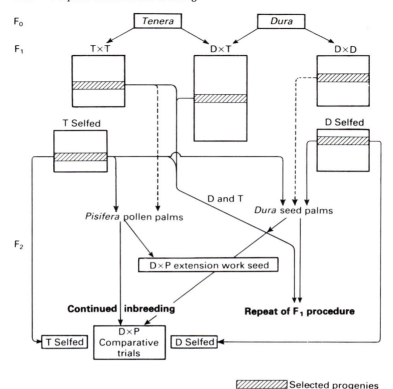

Fig. 5.14 The NIFOR breeding programme.

tenera selfs as in the Zaire programme, but they can also come from the *dura* selfs and crosses as in the Ivory Coast programme. In the former case the testing of parents will have been through the *tenera* × *tenera* trials, in the latter through *dura* × *tenera* trials; but *tenera* for production can be provided through the combined information of these tests as suggested by Sparnaaij *et al.*, (1963). For example, if *tenera* palms T1, T4 and T6 gave outstanding progeny in crosses T1 × T4 and T1 × T6 in the *tenera* × *tenera* trials, and *dura* and *tenera* palms D2, D5 and T5 gave outstanding progeny in crosses D2 × T5, D5 × T5, D2 × T6 and D5 × T6 in the *dura* × *tenera* trials, then seed for distribution can be produced by *dura* × *pisifera* crosses from the following sources:

Dura sources	*Pisifera* sources
1. F_0 palms D2 and D5	F_1 palms in T5 and T6 selfed
2. F_1 palms in T1 selfed	F_1 palms in T4 and T6 selfed

3. F₁ palms in T4 and T6 selfed   F₁ palms in T1 selfed
4. F₁ palms in D2 selfed   F₁ palms in T5 and T6 selfed
5. F₁ palms in D5 selfed   F₁ palms in T5 and T6 selfed

In examples 2 and 3, the material comes solely from *tenera* breeding, but one F_0 parent concerned (T6) has also done well in the *dura* × *tenera* trials. In examples 1, 4 and 5 the parent material has been selected from the results of both the *tenera* and *dura* breeding. It was also intended that *tenera* production palms should be provided from parents within appropriate crosses as well as within good selfs. *Dura* × *pisifera* comparative trials with parents from the *dura* and *tenera* selfings were also planned.

The NIFOR programme was designed to continue into further generations as in the INEAC and IRHO programmes, but unlike the latter the *dura* source is not confined to Deli palms. However, the latter were not crossed with African *dura* in the *dura* programme since it was thought best to determine separately the value of the best Delis available and to compare them with good African *dura*. All crosses in the trials, whether comparative or within fruit form, were either compensatory, i.e. a defect in one parent was compensated by excellence in the same character in the other parent, or 'excellence' crosses, where palms with similar outstanding characters are crossed to improve them still further. Van der Vossen (1974), who used the data to compute genetic values, pointed out that the programme is basically one of selection for good genotypes, using a special 'assortative mating' between groups of palms (Sparnaaij, 1969). More recently an appraisal of the programme has been undertaken by West (1976). Many very high quality *tenera* palms with shell percentages below 7 have emerged and the mean *tenera* analysis of five *tenera* crosses planted in 1959 and one *tenera* self may be given as examples (NIFOR, 1966).

	Fruit to bunch (%)	Mesocarp	Shell to fruit (%)	Kernel	Single fruit weight (g)
Five *tenera* crosses	69.4	81.8	9.8	8.8	7.2
2.3495 self	72.2	86.7	6.1	7.1	7.5

Material from Nigeria bred in the Ivory Coast has given *tenera* progenies containing palms with 83–90 per cent mesocarp and only 4–9 per cent shell.

The exclusive use of the Deli as mother palm in African programmes has been questioned. Van der Vossen (1974) showed that the actual bunch yields and yields predicted from genotypic values were closely similar where no inbreeding had occurred, but with the partially inbred Deli population actual yields were below those predicted. In the NIFOR material he examined, neither predictability nor high yield in the out-crossings depended on a Deli palm being one of the parents. He concluded that it was of basic importance to maintain and increase genetically diverse subpopulations and to employ only interpopulation crosses in estimating genotypic values.

Asia

The course of oil palm breeding in the Far East was severely checked by the Second World war. Breeding by AVROS was continued under the new name RISPA (Research Institute of the Sumatra Plantation Companies Association). The breeding methods adopted were akin to those of Africa in that further breeding from Deli *dura* lines was carried on simultaneously with breeding from *tenera* lines. Test crosses of *dura* × *tenera* and *dura* × *pisifera* were also carried out or planned (Pronk, 1955).

The early selfings and F_2 sib crossings of SP540 have already been described (p. 193–95). F_1 *tenera* parents Pol. 3468, 3258, 3520 and 3409 were outstanding in various characters or in their progeny, and an inbreeding programme was continued into the F_3 with these and other palms at Aek Pantjur. *Tenera* of other origins were handled in the same manner. In the course of this breeding, pure African *dura* progeny superior in some respects to selected Delis arose and programmes for the selection of these for the breeding of new generations of pure African *tenera* were put in hand. Plans were also made to breed from a few semi-fertile *pisifera* which had arisen in the later progeny of the *tenera* programmes. Although these programmes entailed the separate maintenance of Deli *dura* and African *tenera* lines, a programme of selfing Deli *tenera* was included.

In recent years Marihat Research Station has been set up to serve two groups of nationalized estates and a considerable breeding and seed production programme has been put in hand. Special efforts have been made to trace out the pedigree of the Deli and imported material bred during previous decades by commercial estate groups and RISPA. Much use is being made of *pisifera* descended from *pisifera* EX5 of Bah Jambi which had given good progeny in the past. At Aek Pantjur RISPA has planted areas of selfed dumpy E206 (from Malaysia) × SP540 or *pisifera* descended from SP540. The latter palm was still growing at Sungai Pantjur in the early 1980s. The progenies, besides showing the dumpy character, are high yielding and have large fruit. Unfortunately, however, no data have been published on the yield performance and analysis of either the parents or progenies in these Sumatran programmes, but improvement is being sought firstly through breeding programmes in which the purity of the Deli and *tenera/pisifera* lines are maintained and secondly through the introduction of considerable quantities of diverse material from overseas to combine with the local material (Lubis and Kiswito, 1975).

In Malaysia there was no institution to carry on the breeding work started by Jagoe before the Second World War and the areas available to the Department of Agriculture were not large enough for the carrying out of programmes of the size of those in Africa. Quantities of material, both in the form of pollen and seed, were, however, imported from Africa and Sumatra, and a cooperative breeding programme was started with a number of plantation companies in 1956. This has been described by Haddon and Tong (1959), and consisted of a number of small main and minor breeding lines. These included further breeding with the dumpy and other Elmina Delis, and selection within Deli *tenera* progenies for both *tenera* and *pisifera*.

Commercial breeding programmes were carried forward with two aims in view: (i) the breeding of the highest yielding and quality material for the companies themselves; and (ii) the sale of seed to an expanding industry. The early histories of two commercial programmes have already been described (p. 198f). In the first of these programmes Deli breeding was continued by back-crosses of the Serdang × Sumatran material on to outstanding parent palms of Sumatran origin (Chemara, 1960). The Deli *dura* line is shown in Fig. 5.15.

The F_2 palms were planted in 'genetic blocks' containing 338 progenies from 173 parents and totalling over 20,000 palms (Rosenquist, 1985). Bunch analysis was not possible until after the Second World War and was at first confined to fruit to bunch and mesocarp to fruit. Palms of a further generation showed high oil to mesocarp, but shell and kernel content was not given attention. These Delis were passed on to other large commercial seed producers and have been very widely used.

The *tenera* lines were from two sources: in one case *pisifera* pollen from Zaire provided an F_1 Deli *tenera* generation by crossing with the F_2 UR(A) palms. These were crossed with Serdang fertile

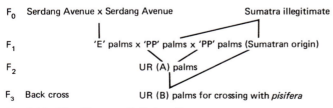

F_0 Serdang Avenue x Serdang Avenue   Sumatra illegitimate

F_1 'E' palms x 'PP' palms x 'PP' palms (Sumatran origin)

F_2 UR (A) palms

F_3 Back cross   UR (B) palms for crossing with *pisifera*

Fig. 5.15   The Chemara Deli *dura* line

pisifera to provide further *pisifera* which were used for the first *dura* × *pisifera* commercial seed. The other source was the Sumatran SP540 whose selfed progeny were crossed with other selections of Zaire origin. F_3 and F_4 generations were raised in Malaysia and Papua New Guinea and have been widely used for commercial seed.

Breeding in Malaysia was much expanded in the 1960s by a large exchange programme, organized by the Department of Agriculture in Sabah, and by the setting up of a breeding and genetics centre, the Oil Palm Genetics Laboratory (OPGL), by four of the principal plantation companies.

The Sabah programme allowed for the collection and exchange of *dura* and *tenera* crosses and selfs from many sources in Malaysia and Africa and for linked Deli *dura* × *tenera* comparative trials (Hartley, 1962; Sabah Report, 1970) as in the Africa programmes of reciprocal recurrent selection. The programme included 192 progenies but owing to difficulties of transport only 152 were provided up to the end of 1969 when the programme was virtually completed. The material was distributed to all participants and as far as Malaysia was concerned led to an increase in the availability of good Deli progenies and to a much-needed injection of *tenera/pisifera* sources. Rajanaidu *et al.*, (1985a) published some preliminary results of this programme based on 5–6 years' yields and limited bunch analysis at two centres. Nearly all the trials showed significant yield and fruit character differences between progenies. Interactions between sites and progenies were not significant. Some good progenies emerged in the *dura* × *tenera* crosses, NIFOR palm 32.005 (WT1) in particular showing good combining ability in both yield and oil to bunch when crossed with a Malaysian Deli. This cross gave the following yield and analysis of the *tenera* palms on Briah series soil at Banting:

Bunch yield per palm per annum years 4–7 (kg)	No. of bunches per palm per annum	Mean bunch weight (kg)	Fruit to bunch (%)	Mesocarp to fruit (%)	Shell to fruit (%)	Kernel to fruit (%)	Oil to mesocarp (%)
212	17.4	12.2	66.3	83.6	10.0	6.4	48.0

Tenera selfs all showed inbreeding depression in yield though the *tenera* analysis of 32.005 was good; some *dura* selfs, particularly Deli BD6b, showed no depression. This palm also did well in *dura* × *tenera* crosses. Some *tenera* × *tenera* crosses of mixed parentage gave both *dura* and *tenera* of high yield and good analysis. It was concluded that the programme contained valuable breeding material which would at the same time help to evaluate the reciprocal recurrent selection method.

The OPGL was established in Malaysia in 1965 to give a firmer genetic basis for selection and breeding and to assist and interpret the selection and seed production programmes of the contributing companies. The OPGL was taken over by MARDI in 1973 and later by PORIM. The original work was done largely in conjunction with that of the companies and, in outline, the breeding system is the one used by most oil palm organizations in Malaysia and shown in Fig. 5.16 (Rajanaidu *et al.*, 1985a). The aims were to study existing breeding populations, introduce new material, establish progeny trials, evaluate experimental designs, and study all factors related to yield (OPGL, 1966–71; Hardon and Thomas, 1968). Some of this work was mentioned in connection with prospection and heritability. With regard to selection and breeding methods, a system of family and individual mass selection in *dura* × *dura* and *tenera* × *tenera* was followed (Hardon, 1969a). *Pisifera* selection was based, in the first instance, on the performance of *dura* and *tenera* in the *tenera* × *tenera* progenies and, later, on the results of the D × P progeny tests. This entails recording another generation, and the

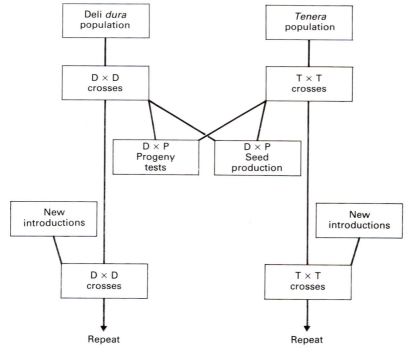

Fig. 5.16 Outline of breeding as usually practised in Malaysia (Rajanaidu *et al.*, 1985a).

results are then compared with those of the second round of selection in the *tenera* × *tenera* crosses (Hardon *et al.*, 1973).

The OPGL Deli *dura* programme consisted of a large collection from various sources in the Far East and included two replicated diallele trials designed to estimate genetic variances of yield components and environment × genotype interaction. Later, a much larger *dura* programme was undertaken to provide a nucleus of *dura* material (not only Deli) from as wide a field as possible to cross with *pisifera* from the *tenera* programme (Chan Kook Weng *et al.*, 1985). The latter programme consisted of trials of material from various sources including Zaire, NIFOR (Nigeria), Ivory Coast and Cameroon. The D × P trials were polyallele experiments designed not only to test the *pisifera* but also to estimate genetic variance in two local environments. The *dura* mother palms were chosen on their own and their progeny performance, while the *pisifera* were chosen on the performance of their *tenera* sibs. The choices were based on the usual bunch yield and analysis parameters (Plate 5.6), including oil and kernel to bunch and vegetative growth measurements.

In a second stage of this programme, started in the early 1970s, it was possible to take account of the results of the D × P progeny tests and to make use of the *pisifera* palms available from the early *tenera* × *tenera* crossings. New D × P progeny tests, covering 240 hectares, were laid down. The first round of D × P tests had shown

Pl. 5.6 *Tenera* fruit from Deli *dura* × *pisifera* crosses. The effect of different *pisifera* on shell thickness in the progeny can be clearly seen.

highly significant progeny differences but very little genotype × environment interaction (Chan Kook Weng *et al.*, 1985). It should be mentioned that this programme, though it gathered material from many parts of the tropics, was still only using material from the rather restricted lines in use around the world. Really new germplasm will come to Malaysia via the prospections already described.

An innovation in the Malaysian breeding work has been the introduction of the growth analysis parameters, particularly vegetative dry matter (*VDM*) and bunch index (*BI*), in an attempt to improve bunch yield per hectare. The heritability of these characters was referred to on p. 250. Hardon *et al.* (1973) presented some data which suggested that palms selected for high yield also have a higher *VDM* and leaf area and may therefore be more competitive and require a reduced density. Palms with a high *BI* had a leaf area equal to the mean and, of course, a low *VDM*. This suggests tolerance of competition and should allow planting at higher densities leading to a higher yield per unit area.

It appears not unlikely that much past breeding work favoured palms with a high *VDM* and leaf area and that this may account for the declining yields which have sometimes been recorded when the palms have closed in and are in competition with each other. Such competition results in a lowering of net assimilation rate and this is reflected in a lowering of the *BI* (see p. 154). Such a situation may be met by reducing density, but a method more likely to give optimum yields would be to breed palms which maintain a high *BI* at a high density.

Latin America

The great majority of the early plantations in Latin America and many of the plantings undertaken from 1959 owe their origin to a collection of seed introduced in 1926 by the United Fruit Company and planted in the Experimental Garden at Lancetilla near Tela in Honduras (Trafton, 1951; Hartley, 1965).

The collection consisted of African material already in existence at Bobos in Guatemala and seed collected by Dr Fairchild, together with Deli material collected from various sources in Java, Sumatra and Malaysia. The original collection also included out-pollinated *tenera* imported from Africa. When it was seen that the Java and Sumatra material was more productive than that from Africa or Malaysia the latter material was cut out, though not before some Malaysian illegitimate progenies had been established at San Alejo and seed from Java and Sumatra palms had been issued to Quepos in Costa Rica and Patuca in Colombia. Thus the predominantly Deli material which was passed on to other plantations in America may have carried some African *dura* characters, though the younger plantings of the United Fruit Company, and the first plantation in Ecuador, were planted with Java and Sumatra material obtained long after the African palms had been cut down. Until the late 1950s oil palm plantings in America were dominated by material from these 1926 introductions.

Other introductions into America include seed brought from Zaire by Mr Florentin Claes in 1931 and imported by Dr M. J. Rivero from Eala in 1936 and planted at Palmira, near Cali in Colombia. From these introductions Dr Patiño made *tenera* selections, and a second generation which included some legitimate progenies was established at Calima (Patiño, 1948, 1958). The *tenera* are of the Yangambi type and the whole population was used in a selection and breeding programme carried out by the Instituto Colombiano Agropecuario and including *dura, tenera* and *pisifera* from Calima, Deli *dura* originating from Honduras and *dura* × *tenera* and *dura* × *pisifera* for production all laid down at the El Mira station at Tumaco. In Venezuela the La Esperanza plantation has an interesting collection of planting material owing its origin to Yangambi, Zaire, to Sumatra and to West Africa. Exceptionally fertile and high production *pisifera* palms exist at La Esperanza through breeding from a thick-shelled *tenera*. In Ecuador a breeding programme was started with imported material and seed is being produced. Nigerian *pisifera* pollen was first employed, but locally bred *pisifera* have now been selected.

The recent spate of planting which started around 1960 was composed either of Honduras material obtained direct or second hand and crossed with *pisifera* locally established or imported as pollen, or of importations from the Ivory Coast, Cameroon, Malaysia and Surinam. In Costa Rica a breeding programme with imported material was established in the 1960s by the United Brands Company, and their experiment station became the first large commercial producer of *dura* × *pisifera* seed on the continent.

The choice of breeding systems

The role of the *pisifera*

The *pisifera* is a pollen parent and provided it yields a sufficiency of male inflorescences it can be used for the pollination of many female parents over a long period of time. The female *dura* parent is more limited in the production of progeny, producing, say, six bunches of 1,000 seeds per year in Africa and double that amount in the Far East, i.e. enough, with normal germination and nursery losses, for about 4 hectares per bunch or 25–50 hectares per year's production. By diluting the pollen with talc it is possible to obtain a very much larger number of progeny from the male parent. From a single inflorescence 50–100 g of pollen may be collected; but 0.5 g or less may be sufficient for pollinating one female inflorescence, so 100 bunches giving 100,000 seeds for the establishment of, say, 400 hectares can be produced from one male inflorescence. With the most recent techniques (Bénard and Malingraux, 1965) it is believed that as little as 0.05 g pollen may be required for a single pollination and the area covered by the progeny of one inflorescence of one pollen parent could be increased to 4,000 hectares.

Some infertile *pisifera* are shy of producing male inflorescences; there are ways of inducing these palms to produce them however, and their relative paucity will not eliminate the potential productive predominance of the male parent. For reproductive potential one male inflorescence equals at least 100 female inflorescences.

These facts tend to give greater importance to the selection of the male than of the female parent. However, the potential of the *pisifera* is harder to gauge than that of the *dura*; whereas the potential of the latter may be estimated from its own bunch production and analysis as well as those of its *tenera* sibs, the yield of the *pisifera* is more or less strongly affected by abortion, and fruit analysis gives little or no assistance. On the other hand the very fact that a *pisifera* can provide its genes (through pollen) in so much greater quantity than can a *dura* means that much more use can be made of a *pisifera* already 'proved' by its progeny than can be made of a *dura* similarly proved. Thus in breeding-cum-production programmes, *pisifera* palms which have themselves produced outstanding *tenera* progenies will still be being used when *dura* seed trees of a later generation are in production. The former may not be such potentially good parents as some of the *pisifera* of the later generation, but their worth will be more fully known.

As partially fertile *pisifera* of high bunch production exist, there has naturally been a desire to measure their production and make use of them in breeding. Several reasons for employing such fertile *pisifera* have been put forward. Firstly, with the usual almost-sterile *pisifera* it is necessary to obtain a few properly developed fruit in order to identify them with certainty as *pisifera*; infertile *tenera* and *dura* have been known and it is of course essential that these should be recognized and not confused with *pisifera*. Secondly, oil-to-mesocarp contents can be measured in fertile *pisifera* and this perhaps provides the strongest argument for their use. Against the use of fertile *pisifera* are the facts that they are limited in number, that their own bunch yield is no real indication of their power to transmit factors for bunch yield, and that they are normally related to thick-shelled *tenera* and carry heritable factors for shell thickness. In Nigeria, a negative correlation was found between the occurrence of fertile *pisifera* and the thinness of shell of their *tenera* sibs (Sparnaaij *et al.*, 1963). Two other programmes confirmed this result: in one, a set of thick-shelled *tenera* crosses, besides giving a good example of the high heritability of shell thickness, provided a large number of fertile *pisifera*; in the other, different types of *tenera* crossed with fertile *pisifera* gave the following results (Menendez, 1965):

Cross	*Fertile* pisifera *as a percentage of bearing palms*
Deli *tenera* (Deli × fertile P) × fertile *pisifera*	52
Thick-shelled *tenera* × fertile *pisifera*	36
Thin-shelled *tenera* × fertile *pisifera*	23

Doubts as to the usefulness of yield data from fertile *pisifera* are engendered by the fact that fertility varies so greatly (Henry and Gascon, 1950). *Dura* palms have on average a lower sex ratio than *tenera* palms but a higher F/B ratio. The sex ratio of *pisifera* palms is very high because they are either not setting their bunches or setting bunches with low F/B ratios. In the Zaire prospection of estates derived largely from *tenera* parents about two-thirds of nearly 900 *pisifera* identified aborted nearly all their fruit (Vanderweyen, 1952). Of 147 bunches cut from palms setting some fruit the distribution of those with over 10 per cent F/B was as follows:

Number of bunches	*Per cent fruit to bunch*
14	10–20
43	21–30
30	31–40
15	41–50
7	51+

Further analysis of the seven palms giving bunches with 51 or more per cent F/B showed that, though there were two palms with bunches of over 60 per cent F/B, regularly high ratios were not obtained from any palm. However, a remarkable *pisifera*, numbered P21, was found in southern Zaire. Bunch yield was 127 kg per year, F/B 66.7 per cent and the fruit had a large kernel constituting 10.7 per cent of the fruit; 38.7 per cent germination was obtained from the kernels (Desneux, 1958). Such *pisifera* are very rare and in general it cannot be expected that their bunch yield will reflect the yield of their *tenera* progeny.

Recently a special plea has been made for *pisifera* × *pisifera* production breeding in Malaysia (Tan Teng Lai, 1971). Two Serdang fertile *pisifera*, S112 and S29/36, were crossed. Germination was satisfactory and 377 seedlings were successfully brought to bearing. Unfortunately, although the bunch production was said to be prolific, no overall yield records were provided. Analysis of two palms showed F/B ratios of 68.7 and 71.8 per cent and kernel to fruit of 18.7 and 15.2 per cent. Oil to mesocarp was, at 36–39 per cent, lower than that of adjacent *tenera* palms of the same age. In view of the positive correlation between kernel and shell (see p. 248) it is not unexpected that fertile *pisifera*, derived as these are from thick-shelled *tenera*, will have large kernels; and milling problems might be the extraction of the mesocarp and kernel oil simultaneously and the disentanglement of the valuable palm kernel residue from the mesocarp fibre.

The use of genotypic values and combining ability

The question of whether to employ for the oil palm a system based on modifications of reciprocal recurrent selection or a system of individual and family mass selection has been a matter of some controversy. Against the use of reciprocal recurrent selection it is claimed (Hardon *et al.*, 1973) that it is mainly additive genetic variance which is to be exploited and that (a) the limited number of parents in the two base populations which can be tested may result in random loss of genetic variability, (b) the alternate cycles of progeny testing and selection may cause gene frequencies to oscillate rather than show progress, (c) the high degree of inbreeding in existing base populations, especially the Deli *dura*, does not allow a wide enough choice of parents. Great stress is put on the need to create new and genetically variable populations.

In favour of the use of a form of reciprocal recurrent selection it is argued that, though mass selection may give progress for the more heritable characters showing additive genetic variance, for the important but less heritable characters such as bunch yield it is probable that marked non-additive genetic variance is involved (Meunier and Gascon, 1972; Noiret *et al.*, 1966). Some evidence for this has been provided (Meunier *et al*, 1970) and has been referred to earlier (pp. 242 and 246). It is also argued that progenies can show combining ability which is not foreshadowed by their phenotypic values and that therefore the comparative trials have an important part to play.

What has clearly emerged from the work already cited and from assessments of the large breeding programme conducted in Nigeria (Van der Vossen, 1974; West *et al.*, 1976) is that, whereas for fruit and bunch characters additivity predominates and both phenotypic selection and the estimation of genotypic values can be usefully employed in any breeding scheme, with bunch yield and its components the results are never so clear.

Analysis of data from the Nigeria programme showed significant deviation from additivity with both bunch yield and mean bunch weight, though number of bunches was found to have a fairly high additive component. These results were held to support the claim of Meunier and Gascon (1972) that selection methods which took only additive genetic variance into account would neglect real advances to be made through utilizing specific combining ability. West (1976) found numbers of progenies which produced much higher or much lower yields than could be predicted from additive inheritance alone, and he considered that, for bunch yield and mean bunch weight, the use of the *dura* × *tenera* comparative trials was essential. However, Sparnaaij and Van der Vossen (1980) considered that greater attention should be given to both yield components, number of bunches as well as bunch weight, which, they claim, are mainly additively inherited, while yield itself will inevitably show less additivity because it is the product of two components which are negatively correlated. They therefore recommended that, in breeding for yield, special attention be given to determining genotypic or breeding values (see p. 247) of as many promising parents as possible, and that transgressive segregation be exploited by crossing parents with contrasting yield components. This has the attraction of reducing the large area that, under reciprocal recurrent selection, must be given over to *dura* × *tenera* comparisons. In some breeding programmes priority is therefore being given to the determination of breeding values of selected palms (Rosenquist, 1984b). However, it would be a mistake to neglect altogether (and particularly perhaps for bunch weight which has often been shown to be weakly inherited) the search for instances of specific combining ability. Furthermore, breeding programmes will be increasingly directed towards high bunch index (*BI*) palms for higher density planting, and insufficient work on the inheritance of this factor has been done. Mention can only be made here of the preliminary work of Breure and Corley (1983) who found from a planting density trial in Papua New Guinea that young palms selected for high early yield or for net assimilation rate (*NAR*) tended to have above average vegetative dry matter (*VDM*) requirements and height increments, whereas those selected for *BI* did not show these faults. In later studies in the same area Breure (1985, 1986) showed that the leaf-Mg level of *pisifera* parents could influence both yield and *BI* in their *tenera* progeny, and he postulated that selection for rapid ground coverage by measurement of the rate of leaf area expansion or leaf area ratio (*LAR*) would lead to improved *BI* and yield.

Sparnaaij and Van der Vossen (1980) drew attention to the misunderstanding which can arise through ascribing to superior Deli × African lines an inter-origin effect (or kind of heterosis) rather than a simple and predictable transgression through the use of palms of contrasting composition with, on the one hand, large numbers of bunches and, on the other hand (the Deli), a high mean bunch weight. It is now widely believed that much less residual variability can be found in the Deli than is available in African material (Rosenquist, 1985), and the idea either that the Deli population should be kept pure or should be used exclusively as the female parent in seed production would, it is believed, unnecessarily prolong a dependence on a limited and inbred population.

It has been proposed (Jinks and Lawrence, 1983) that oil palm breeding might revert to the use of inbred lines for the reason that it should be possible, on theoretical grounds, to provide homozygous inbred progenies with yields superior to anything now available. This idea has some attraction when considered in conjunction with cloning through tissue culture. Using about 100 seeds from the F_2 of the palms selected, the selfing would be carried through to the F_5 or F_6 by single seed descent. The scheme is complicated by the fact that it is the *tenera*, which is heterozygous in the most important respect, that is required; therefore enough seed would be needed in each generation to ensure the appearance of the required number of *tenera*.

Another suggestion is that any selection programme should include one-generation selfing of most palms employed. This would show which lines remained high yielding when inbred, and would expose characters of economic importance (Rosenquist, 1984a). Such palms might be considered homozygous for advantageous characters and contain fewer deleterious genes. There has been some evidence that high-yielding selfs also show good combining ability.

Vegetative propagation

The oil palm does not sucker like certain other Cocoids and there is no simple way of propagating it vegetatively. Work on propagation by tissue culture began in several laboratories in the 1960s (Stavitsky, 1970; Rabechault *et al.*, 1970, 1972; Smith and Jones, 1970; Smith and Thomas, 1973; Jones, 1974) and techniques were developed which make cloning possible through the production of plantlets (ramets) originally developed from an ortet (clone source). Briefly, the method is to take suitable tissue and induce the development of disorganized callus by regulators of the auxin type (Jones, 1983) and by controlled lighting and atmosphere (Noiret *et al.*, 1985). In England the explant used was at first mainly root tissue, but more recently leaf tissue has also been employed; in France young leaf tissue was the main source (Pannetier *et al.*, 1981). Cell cultures have, however, also been raised from the apical meristem, the rachis, the base of the petiole, and the inflorescence.

Tissue culture

Leaf material must be immature and close to the apex. In practice material of leaf −8, i.e. the eighth leaf younger than the spear, is usually employed, and at least 2,000 individual pieces can be excised for culturing (Noiret *et al.*, 1985). This excision causes a check to the palm's growth and there is even a risk of death. Full recovery is therefore necessary before further sampling is undertaken and a palm taking part in a breeding programme can only be used after it is no longer required for that purpose. The excising of young leaf tissue has the advantage that the tissue is completely enclosed by the leaf bases of older leaves and therefore it does not require severe disinfection. Root tissue can be easily obtained but there is a danger of taking, in error, roots from an adjoining palm. To avoid this possibility soil can be mounded up around the palm base and roots growing into it from the required palm can be taken. Roots are of course heavily contaminated and need vigorous disinfection (Wooi, 1984).

The two major laboratories in which oil palm tissue culture was first undertaken have not released details of their methods and culturing media. However, Wooi (1984) and Paranjothy (1986) have recently given limited descriptions of laboratory practice. An auxin is essential for callus initiation and 2,4-D and NAA are commonly used together with nutrient formulations such as that of Murashige and Skoog (Paranjothy and Rohani, 1982). Callus from root segments is initiated within 2 months of culturing.

Most calluses grow slowly, but a fast-growing callus can be induced by incubating under light on suitable media; this can then be subcultured with regular transfers until embryoids develop (Plate V). These white bodies will also develop on slow-growing callus cultures (Wooi, 1984) and they are then transferred to a medium which induces their proliferation and further development. Shoots appear and are removed as they grow on the proliferating material which can then be recirculated to produce more shoots (Fig. 5.17). The latter are grown in a rooting medium and, after hardening, can be planted in soil or compost. The embryoids are thought to arise when meristomatic nodules in the callus become delimited by layers of dividing cells forming an epidermal layer (Jones, 1983), and structurally they come to resemble true embryos (Ahée *et al.*, 1981). They germinate with features similar to a zygotic embryo, the shoot and root emerging from the internal meristem.

For commercial production of ramets it is important to improve the speed and frequency of embryogenesis and the production of shoots. Much progress has been made with this through experiments on media sequences, and streamlined production systems have now been set up so that ramets can be distributed world-wide and in

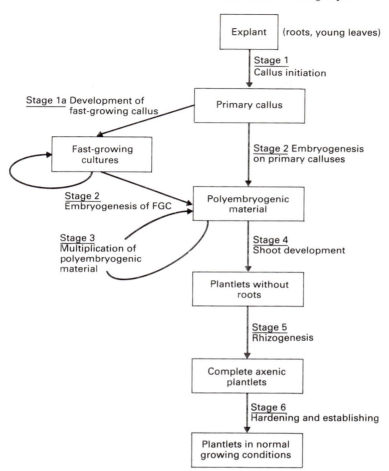

Fig. 5.17 Propagation of the oil palm by tissue culture (Jones, 1983, adapted from Loiret).

quantity for field trials and eventual commercial planting (Unifield, 1985; Pannetier *et al.*, 1981; Noiret, 1981; Noiret *et al.*, 1985).

Raising ramets in the nursery

Once a good root system has been obtained in the laboratory, a ramet can be transferred to a prenursery or to a 'ramet house' and then a prenursery. At first there were many losses due to drying out, and the percentage loss varied considerably between clones and countries of planting. It is important to maintain a high humidity

around the young leaves for at least 4 weeks and to avoid temperatures above 32 °C, but nevertheless to ensure a light transmission
of at least 17 per cent (Turner, 1984). In practice this is achieved
with mist spraying and suitable shade netting. Loss in the prenursery
and nursery is less than in the ramet house, but ramets should not
be transferred to the nursery at some prearranged time but only
when they have produced at least one new and healthy leaf. Growth
in the nursery is similar to that of seedling palms.

Testing clones in the field

The establishment of clones in the field was first achieved in
Malaysia in January 1977 (Corley *et al.*, 1979) and subsequently in
the Ivory Coast in June 1978 (Ahée *et al.*, 1981). Genetic changes
can occur during tissue culture and it is important therefore to
establish at an early date that only plants conforming to the original
material can be constantly provided. An early clone produced by
Rabéchault and Martin (1976) is stated to have given mantled fruit
(Corley *et al.*, 1986). Loiret (1981) reported that the early plantings
in the Ivory Coast were completely normal and retained their
diploid chromosome number ($2n = 32$), but he considered that the
risk of deviation remained, though this would be reduced by shortening the period on artificial media. Corley *et al.* (1979) considered
that any genetic change would be exhibited by detectable
abnormalities.

Recently, two types of abnormalities – mantled (*Poissoni*) fruit
and androgynous inflorescences – have appeared in Malaysia in
three clones in some 1981–3 plantings (Corley *et al.*, 1986). There
was also some bunch failure in one locality in these clones following
a high incidence of parthenocarpy. The unexpected appearance of
abnormal adult palms in selected clones has naturally caused some
alarm and encouraged a re-examination of tissue culture methods
so as to ensure that these abnormalities do not recur. Soh (1987)
has discussed in some detail their possible causes, but much more
experience must be gained before such discussion has more than a
theoretical basis. In the meantime Corley *et al.* (1986) suggest that
the abnormalities found on the three clones mentioned above are
unlikely to be either of environmental origin or due to somaclonal
variation, i.e. genetic changes occurring in somatic cells from, say,
leaf or root tip, or in callus cells developed from these parts; and
they consider that the problem can be overcome by adjusting the
time of culturing and the media used.

Apart from these clonal abnormalities, the phenomenon of
terminal inflorescences (a condition also found, though very rarely,
on seedlings) occurs in most clones. The percentage varies from less
than 1 to over 30, and the majority of cases are detected in culture
though development may occur as late as the prenursery or early

nursery stages. Vegetative growth ceases and the plant flowers and dies. Observation so far suggests that where flowers on the terminal inflorescence are mantled the ordinary inflorescence flowers of the same clone will also be mantled (Corley, 1987).

There must necessarily be a pause between the achievement of vegetative propagation through tissue culture and the production and marketing of clones which justify planting in preference to current *dura* × *pisifera* seed supplies, and the appearance of abnormalities has emphasized the paramount need for properly designed and sufficiently lengthy trials before any commercial planting of clones is undertaken. Some authors (Paranjothy, 1986; Noiret *et al.*, 1985) have predicted commercial production of ramets by 1990–1, but it is very unlikely that sufficient reliable data will be available to warrant the exclusive use of clonal material on any planting or replanting before the middle or end of the decade. The prospect of rapid change from seed to ramets, with consequential marked yield increases, has caught the imagination of the industry, and both tissue culture laboratories and agreements covering the exclusive use of undisclosed techniques (Noiret *et al.*, 1985) have multiplied. The appearance of abnormalities has, however, slowed the pace of events and checked the scale of ramet production and a less headlong approach may be expected. The immediate future will be occupied in considering selection criteria, searching for candidate ortets, laying down clonal trials, and continued breeding. The onset of vegetative propagation does not mean that breeding is any less needed, only that the material may be differently used.

The first clonal trials in Malaysia showed the following results in their early years (Corley *et al.* 1982; Corley, 1982; Jones, 1983):
1. A much more uniform flowering succession was followed by clonal palms than by palms of an ordinary seedling progeny, and this led to larger within-year fluctuations of yield;
2. Between-palm variance in bunch composition was nearly always much lower in clones than in seedling progenies;
3. There were remarkable and consistent differences between the clones in fruit composition and in oil composition, including carotene;
4. Some large early differences in bunch yield appeared to indicate differences of precocity.

The clones from these first trials were all from seedling ortets, so the appearance of any particularly productive clone would be entirely fortuitous. The first trials to include clones propagated from selected mature palms were planted in Malaysia and Cameroon from 1979 to 1981 and the results of 2–4 years of bearing became available in 1987 (Corley *et al.*). These authors stated that no clones had yet given sufficient data for firm conclusions to be drawn, and the importance of lengthy testing was emphasized by the finding of significant year × clone interactions. 'Year' encompasses both age and weather effects, so these interactions indicate different yield trends with age or different reactions to climatic factors or both. There was some evidence, from trials planted at different dates within a single year, that the weather effects were the more important. That differential reactions to weather will be more distinct with single clonal genotypes than with genetically heterogeneous progenies is not unexpected, and these results underline the importance of testing clones, not only for sufficient years, but also in as many environments as possible.

To obtain a reliable estimate of the enhanced bunch, palm oil or kernel yields that may be expected from a clone it should be compared with several high-yielding individual progenies and, later, with standard clones. Mixed commercial D × P seed is likely to be too variable for satisfactory comparisons. In the trials referred to above commercial comparisons were used, but in seven of the trials recently selected progenies were also included.

Some of the early results gave an indication of what may be expected from future clonal selection. One clone gave high bunch yields, compared with those of seedling controls, on inland Malaysian soils but not on coastal soils; another clone was outstanding on coastal soils though poor on inland soils. This emphasizes the importance of local testing. Another clone gave consistently very high oil to bunch which conformed to the oil-to-bunch percentages found in the ortet, confirming that environment has little or no effect on bunch quality factors.

Further programmes of clonal testing were undertaken in Asia, Africa and America during the mid and late 1980s by the two organizations first in the field with tissue culture. In one case these trials included, as controls, a standard progeny, a standard clone and commercial seed. Some semi-commercial trial plantings were also undertaken, as strips within replanted fields, to gain experience of the handling and production patterns of clonal material planted on a field scale.

Selection of ortets

Ortet selection presents problems particularly for the improvement of bunch yield since individual palm yield is often determined by microenvironment and interpalm competition effects rather than genetic factors. But for fruit composition, the components of which show high heritabilities, selection should be relatively simple. Tentative estimates have been made of possible bunch and palm oil yield increases to be obtained from early ortet selection and cloning (Hardon *et al.*, 1982; Jones, 1983), and a figure of 30 per cent has been suggested. Soh (1986), however, has criticized the concept of an immediate 30 per cent increase which, he says, has now been widely accepted by the industry. Using data from four D × P progeny trials he confirmed the low broad sense heritabilities of palm oil yields between palms within progenies and he calculated that simply choosing the highest-yielding individuals within a high-yielding progeny would be unlikely to provide increases of this order; even in cloning the highest-yielding 1 per cent of the palms only a 16 per cent increase could be expected and the risk of missing a genuine high yielder would still be appreciable. Meunier *et al.* (1987), however, consider that in a second stage of clonal selection a 30 per cent increase can be reached.

An additional means of raising yield per hectare may be the selection of ortets on combined vegetative and yield characters with the object of obtaining a lower proportionate production of *VDM* to total dry matter, i.e. a higher *BI*, and planting ramets which yield well though planted at a higher density than normal. Under Malaysian conditions, for instance, oil palms produce about 30 tonnes dry matter per hectare per annum when planted at the normal density of 143 palms per hectare, and of this 45–50 per cent is produced as bunches (Corley, 1982). The total dry matter can rise to 40 tonnes per hectare per annum at higher densities, but with mixed progenies a much lower proportion of the dry matter will then appear as bunches. If palms can be found which still produce 45–50 per cent of their dry matter as bunches when planted at, say 180 palms per hectare, the bunch yield would be substantially raised. To achieve this, high density plantings can be searched for palms which are tolerant of competition, or palms can be selected on the basis of yield per unit leaf area or *BI*.

The risks of cloning are principally of disease since some oil palm progenies have been shown to be more susceptible than others to certain diseases. Some clones have already been detected as highly susceptible to Freckle (*Cercospora*) and Crown disease (Corley, 1982). With monoclone blocks pollination difficulties might arise, particularly in the early years during periods of predominantly female flowering.

Another possible development from tissue culture is the production of clonal seed from areas planted with *dura* and *pisifera* ramets, and programmes to test this method are under way. Élite *dura* and *pisifera* which have provided superior *tenera* when crossed could be cloned and planted out in mixed stand. If these areas are large enough and isolated, open-pollinated seed could be taken, but the planting of separate *dura* and *pisifera* areas, with bagging and artificial pollination of the *dura*, would provide a more controllable and assured seed supply. Clonal seed would have the advantage of lower cost, simpler handling and, through greater genetic diversity, a more even overall yield distribution and less risk of widespread disease susceptibility. It would simply represent seed of the best progenies currently available and therefore would be expected to give a lower yield than the best clones (Unifield, 1984). As alternatives to ordinary clonal seed production, Soh (1986) has suggested the cloning of seedlings obtained from élite D × P crosses, thus sampling greater numbers of individual palms within the progeny, or cloning all except obvious poor doers within an élite progeny in the hope of capturing some clone giving enhanced yields well above that of the average of the progeny individuals.

Oil composition and the employment of *Elaeis oleifera* in breeding

Interest in breeding from *Elaeis oleifera* increased in recent years owing principally to its apparent resistance to a serious form of Bud Rot in Colombia. However, the hybrid with *E. guineensis* is also of considerable promise because of its slow growth in height, its possible special adaptation to climates with strong dry seasons, its resistance to other diseases and its production of oil with a higher proportion of unsaturated fatty acids than is found in palm oil from *E. guineensis*.

As there are no *tenera* in *E. oleifera*, breeding for production entails the crossing of selected *E. oleifera* with selected *E. guineensis* *pisifera*. There has been much haphazard planting of the hybrid in small plots in various parts of the world, mostly with *dura* or *tenera* as the male parent; but what is most needed is the establishment in all oil palm countries of base populations of *E. oleifera* for subsequent selection and crossing. *Elaeis oleifera* yields poorly under plantation conditions, but the hybrid gives bunch yields at least comparable with those of *E. guineensis*. In the cooperative breeding programme of the Department of Agriculture in Malaysia hybrids gave a mean bunch yield of 203 kg per palm per annum against 190 kg for four *E. guineensis* *tenera* progenies, while in Nigeria four hybrid progenies gave a mean of 55 kg per palm per annum over the first 5 bearing years against 56 kg for *E. guineensis* palms (Obasola, 1973).

Many localized collections of *E. oleifera* have been made (de Blank, 1952; Vallejo and Cassalett, 1974) but until recently no systematic prospection of the whole range of the palm has been attempted. A large collection was made in Costa Rica in the late 1960s and mid 1970s by the United Brands Company; 36 zones in 7 countries were covered and 326 'accessions', usually consisting of seed from a single bunch, were obtained (Escobar, 1981). From these, 4,974 palms were planted in Costa Rica while parallel collections were established in the countries of collection. Mean yields of the earlier collections from different zones varied from 76 to 156 kg per palm per annum, with coefficients of variation from 18 to 87 per cent with lots of more than seven accessions per zone. Mean bunch weights varied from 8.1 to 12.8 kg. The mean bunch composition of bunches is shown in Table 5.19. There were interesting vegetative differences: palms from the Santiago district of Panama had shorter leaves than palms of other origins and this was considered a useful feature in view of the inordinate length of the leaves of the inter-specific hybrid. Comparisons of the palms with hybrids of similar age whose female *oleifera* parent came from one zone in Panama confirmed that most of the bunch and leaf characteristics of the hybrid were inherited from the *oleifera* parent, while the leaf length, leaflet width, petiole cross-section and leaflet area of the hybrid exceed those of both parents.

Two comprehensive prospections were undertaken in the early 1980s by collectors from Malaysia, and the collections were established in that country. These prospections revealed the highly discontinuous nature of the *E. oleifera* populations and the great danger of the complete disappearance of the palm from certain areas owing to cultivation and the creation of pastures. It was often difficult to obtain a ripe bunch from a group of several palms lying at a great distance from any other groups. Rajanaidu (1985b) covered Surinam, Colombia, Nicaragua, Honduras, Costa Rica and Panama and collected one to six bunches per site where possible; random sampling was restricted since only the limited number of palms with ripe bunches could be sampled and a rule was made that the minimum distance between palms should be 30 m. Data from the progeny of these palms are not yet available but in the collection there were highly significant differences of fruit weight, mesocarp percentage and nut weight between countries and sites within countries; but the between-palm variances were small in comparison with those found for *E. guineensis* in the West and Central Africa prospection (see p. 255). There were also significant differences between sites in vegetative characters.

Only small collections of seed have reached the major oil palm countries from Brazil, but *E. oleifera* is widely distributed along the upper reaches of the Amazon and its tributaries. Owing to its dispersion over such a very large area little was known until lately of its habitats and variation (Santos *et al.*, 1985). Barcelos *et al.* (1985) surveyed 6 separate riverine regions, covering 5–13 zones per region and sampling 2–25 palms per zone. Variation appeared to be greater in Brazil than in Central America, and the population in the region running due north from Manaus to Caracarai resembled that of Guyana in the small size of its leaves and bunches. However, bunch characters elsewhere appeared generally to be superior to those of the material of other countries: fruit weight was higher and the proportion of parthenocarpic fruit much lower.

Until the collections of *E. oleifera* have been fully examined and exploited it will not be possible to see the exact potential of the

Table 5.19 Mean bunch composition of an Elaeis oleifera collection originating from 13 zones in 3 countries (Escobar, 1981)

	Fruit to bunch (%)	Fertile fruit (%)	Parthenocarpic fruit (%)	Fruit Weight (g)	Mesocarp to: Fert. fruit (%)	Mesocarp to: Parth. fruit (%)	Shell to fruit (%)	Kernel to fruit (%)	Oil to mesocarp Fertile fruit (%)	Oil to mesocarp Parth. fruit (%)	Oil to bunch (%)
	60.8	49.2	11.6	3.31	36.5	87.1	47.6	15.9	18.8	18.2	5.2
CV*%	4.7	6.8	15.9	11.7	5.5	0.8	3.9	8.5	15.4	17.8	18.3

* CV = coefficient of variation

interspecific hybrid in terms of oil and kernels and bunch quality. In Malaysia, with a Deli *dura* × *E. oleifera* cross, it was shown that parthenocarpic fruit may constitute more than half the total fruit and that the oil-to-mesocarp contents of normal and of small and large parthenocarpic fruit are not identical (Hardon and Tan, 1969; Hardon, 1969b). Mesocarp in normal *dura* fruit was 53 per cent and in large and small parthenocarpic fruit 76 and 89 per cent respectively, while kernel percentage for normal fruit was 8.8. Oil-to-mesocarp percentages are low in *E. oleifera* and in the hybrid are intermediate between those of the parent species, though there seems to be a high variability. The chromosome number of both species of *Elaeis* is 32 and there is no difficulty in crossing them. However, Hardon and Tan (1969) found that seed set, germination and seedling survival from germinated seed were all well below what is normally expected with *E. guineensis*. In examining the inheritance pattern in the hybrid it was found that there was dominance of *E. oleifera* in height increment, fruit shape and colour, parthenocarpy and arrangement of leaflets on the rachis, with over-dominance of leaf length and size of leaflets (and possibly bunch yield); intermediate values between parent species in mesocarp, shell and oil per fruit and chemical composition of the oil; and partial dominance of *E. oleifera* in spikelet morphology and size of subtending bracts. The over-dominance of leaf length and size of leaflets is of particular note, and it has been thought in the past that this would demand a reduction in planting density. However, it has been shown in many plantings that sex ratio has remained high at the normal planting density of 143 palms per hectare. The very large leaves tend to have long petioles so that the proximal leaflets are at a greater distance from the stem than in *E. guineensis*.

Meunier *et al.* (1976) crossed Deli *dura* and *tenera* with *E. oleifera* and compared the results with crosses reported from elsewhere. The crosses gave *E. guineensis* × *E. oleifera tenera* with 68–74 per cent mesocarp in normal fruit and 89–100 per cent in parthenocarpic fruit. Oil to mesocarp was raised to between 43 and 50 per cent in normal fruit and 36–43 per cent in parthenocarpic fruit. Kernels amounted to 8–11 per cent of normal fruit, but the quantity per bunch was not given; this clearly depends on the amount of parthenocarpy.

Palm oil of Colombian *E. oleifera* was found to have 78.8 per cent unsaturated fat (oleic and linoleic) and an iodine value (IV) of 85; interspecific hybrids from two separate populations of *E. guineensis* gave 62.4 and 67.0 per cent unsaturated fats and IVs of 65 and 70 respectively. Kernel oil of *E. oleifera* has a higher myristic and oleic content than that of *E. guineensis*. Meunier *et al.* (1976) found that the hybrid was closer to *E. oleifera* when the latter was the female parent but closer to *E. guineensis* when *E. oleifera* was the male parent. Ng *et al.* (1976) and Corley (1979) have compared the possibilities of improving palm oil composition by selection within *E. guineensis* and through the interspecific hybrid, and have drawn attention to the fact that *E. guineensis* itself varies widely in oil composition, with both *pisifera* and *albescens* palms giving high linoleic contents.

Studies of heritability and repeatability of oil composition have shown contrasting results, but Gascon and Wuidart (1975) found significant repeatabilities with twenty crosses in oleic and linoleic acids and believe that increases of 8–12 per cent in unsaturated acid contents by selection and breeding are feasible in *E. guineensis*.

The intermediate composition of the oil of the interspecific hybrid, indicating co-dominance, was confirmed in studies by Macfarlane *et al.* (1975) and Ong *et al.* (1981). The former study demonstrated that oil of the same composition as that of *E. guineensis* could be obtained from a palm of the hybrid × *guineensis* back-cross (GO × GG = 50 per cent GO (hybrid) + 50 per cent GG (*E. guineensis*), while the latter study showed that the observed and expected composition in F_1, F_2 and both types of back-cross gave agreement at the 5 per cent level of significance.

A programme of back-crossing the *tenera* hybrid with the *tenera* form of *E. guineensis* has been described from Nigeria (Obasola *et al.*, 1976). As might be expected, the progenies were heterogeneous though falling into three morphological categories: those appearing like *E. guineensis*; those looking like *E. oleifera*; and those combining characters of both species. The majority (85 per cent of those included in the study) were, as expected, of the first category, while those of the last category showed the most vigorous growth and had bunches more like *E. oleifera*. The fruit in general resembled that of *E. guineensis*. Not all palms could be classified but many aborting palms were thought to be *pisifera*; two fertile *pisifera* were identified. *Tenera* palms showed between 66 and 79 per cent mesocarp, with 13–26 per cent shell and 39–48 per cent oil to mesocarp, and there was much variation in fruit size and structure with differences in positioning and concentration of the fibre ring which does not appear distinctively at all in the hybrid. The mean height of the *E. guineensis*-like palms at 15 years of age was only 87 cm, a very low figure for the palms' age, but no height data of adjoining *E. guineensis* palms were recorded for comparison.

While slow growth in height, resistance to Bud Rot and the production of a high proportion of unsaturated fatty acids are clear advantages, the hybrid suffers from certain defects. It has been estimated that its low F/B ratio and low percentage oil to mesocarp give an oil-to-bunch extraction of only 17 per cent against 22–23 per cent with the best *E. guineensis tenera* commercial material (Mennier and Boutin, 1975). Many hybrids have, in Africa, been found to be heavily attacked by Freckle (*Cercospora elaeidis*) while cases of severe Orange Spotting have been attributed to the mite, *Retracus elaeis*. The hybrid is also attacked, in America, by the root caterpillar, *Sagalassa valida*, and by *Pestalotiopsis*, though the vector, *Leptopharsa gibbicarina*, does not thrive on the leaves. In the hybrid's favour, apart from resistance to Bud Rot, is the evidence of resistance to *Ganoderma* Trunk Rot (Meunier *et al.*, 1976), to a Basal Rot caused by *Ustilina* sp. and, in the case of certain crosses only, to *Fusarium* Wilt.

Breeding for disease resistance

A genetic element in susceptibility and resistance has been detected in Crown disease, Vascular Wilt, Dry Basal Rot, Lethal Bud Rot, Blast and *Ganoderma* Trunk Rot, and this is discussed in Chapter 13 under the individual diseases.

Breeding for drought resistance

Morphologically the oil palm is well adapted to resist drought (see p. 145), but the effect of drought on bunch yield is severe. Maillard *et al.* (1974) have recorded marked differences between progenies in a drought 'sensibility index'. This index is simply a numerical assessment of the drought effects recorded in a population or progeny using the formula

$$SI = \frac{10\,M + 5S_3 + 3S_2 + 2S_1}{N \text{ (Total number of palms)}}$$

where M = number of dead palms, S_1, S_2 and S_3 = palms showing three stages of drought reaction, i.e. S_1 = accumulation of unopened spears, S_2 = four to six leaves broken or collapsed, and S_3 = all lower leaves withered. These preliminary findings have encouraged both the selection of progenies showing low sensibility indices and further investigations to determine methods of measuring drought resistance through submitting seedlings to high osmotic pressures (using polyethylene-glycol) or high temperatures.

Results of breeding and selection, and choice of seed

Information on the first INEAC *dura* × *pisifera* issues in Zaire was published in 1956–7 (INEAC, 1956, 1957). This gave bunch yields per annum in the second and third years of production as varying from 9 to 13 tonnes per hectare which showed, for Africa, a startling precocity. A short account was given of the quality, in terms of oil to fruit, of seed issued at various times by the INEAC (Poels, 1959). The progeny of the seed supplied contained both *dura* and *tenera* palms until 1951 as first *tenera* × *tenera* and then *dura* × *tenera* seed was issued. A gradual improvement was claimed in the standard of the *tenera* palms being grown, and at the end of the period 1951–9 the average standard lay between two sample types which were described as having the following analysis: type 1, mesocarp to fruit 93 per cent, oil to fruit, 43.7 per cent; type 2, mesocarp to fruit 73 per cent, oil to fruit, 37.1 per cent.

Data relating to early IRHO Deli × *pisifera* material was published in the 1960s. Yields of 120–123 kg per palm per annum were cited for palms 5–10 years of age in the Ivory Coast (Bénard and Malingraux, 1965). These were calculated from experiment station progenies and partly from *tenera* in *dura* × *tenera* crosses; they are therefore not comparable with the INEAC data which relate to over 800 hectares of commercial planting. Quality data relating to some 24 hectares of Deli × *pisifera* ex-Yangambi–Sibiti progenies and 48 hectares of Deli × *pisifera* ex-La Mé showed the clear differences between the Yangambi and La Mé *tenera* ancestry. The former cross is characterized by *tenera* with large bunches and fruit and with high mesocarp-to-fruit and oil-to-mesocarp ratios; the latter by more numerous but smaller, spiny bunches still carrying the comparatively low mesocarp-to-fruit character of the La Mé selections, though having satisfactory oil-to-mesocarp and kernel-to-fruit ratios. More recently, Gascon *et al.* (1987) have presented data to show that, under the system of recurrent reciprocal selection described on p. 266, yield improvements of 18 per cent were obtained in the first or F_1 cycle, and that early yields in the second cycle suggested further improvement of the order of 10–15 per cent. Of special interest in the second cycle was the value of parents from the selfed progenies of the first cycle. These were successfully crossed with palms from other selfed progenies or with palms from a recombination of first-cycle parents.

Seed issued in Nigeria was from time to time compared with unselected material and bunch yields were shown to be some 20–25 per cent higher (Menendez, 1965).

Now (1986) that *dura* × *pisifera* seed has been issued in Africa for some 40 years and in the Far East for more than 25 years, several distinctive characters can be noted in the commercial material from different seed sources. Commercial progenies from eight sources planted in the early 1970s at ten sites in Africa and Asia came from the Ivory Coast, Zaire, Cameroon, Malaysia and Indonesia. Unfortunately Nigeria, which had some of the best quality *tenera* material at the time, was not included. Results for only three sites have been published (Rajanaidu *et al.*, 1985b), but within

Malaysia they indicated highly significant differences in yield and bunch components between seed sources, and showed no site × source interaction. The Malaysian sources were superior in yield to African sources, but the source relying on *pisifera* descended from SP540, though the highest yielding and producing a high number of bunches, also grew very rapidly in height. The two Ivory Coast sources, whose earlier issues have been mentioned above, maintained their distinctive differences, the La Mé crosses producing large numbers of small bunches and having a small height increment, the Yangambi/Sibiti crosses giving smaller numbers of larger bunches and growing faster in height. All the Malaysian sources were characterized by larger bunches than the African sources, but the latter generally had as large or a larger number of bunches; this can be partially attributed to the environment under which the parents, and particularly the male parents, have been selected. There were also significant differences in leaf length.

A rather similar trial in Costa Rica (Escobar, 1980) confirmed the high yield but very fast trunk growth of the Malaysian source with the BM119 *pisifera* (descended from SP540).

In Malaysia and Indonesia current seed supplies are exclusively from Deli mother palms crossed with *pisifera* palms of various pedigrees. Soh (1983) has classified the latter as 'Yangambi', 'AVROS', 'NIFOR', 'La Mé', 'Serdang', and 'Derived'. 'Yangambi' and 'AVROS' are essentially of the same stock by different routes, the descendance coming to a greater or lesser extent (sometimes through SP540) from the Congo Djongo palm. 'La Mé' is derived from the breeding programme of the Ivory Coast, some part of which also has, as mentioned above, Yangambi parentage. The exact origin of the 'Serdang' West African *pisifera* S27B and S29/36 is not known. 'NIFOR' *pisifera* have probably the greatest diversity of origin and some of their parents have been used in Nigeria itself and in the Ivory Coast (WA series) breeding programme. 'Derived' *pisifera* are from Deli or Dumpy Deli × Serdang or Deli × AVROS crosses.

To avoid unscrupulous selling of D × P or other seed not owing its origin to properly conducted breeding programmes, the Standards and Industrial Research Institute of Malaysia (SIRIM, 1973) has introduced an oil palm seed certification scheme (MS157: 1973) which demands, for *dura* parents, production of at least 159 kg bunches per palm per annum for 4 recorded years, with minimum mesocarp to bunch of 33 per cent, and oil to dry mesocarp 65 per cent; and for the *tenera* progeny or sibs of *pisifera* parents, the same bunch yield with mesocarp to bunch at least 40 per cent, kernel to fruit 6 per cent, shell to fruit less than 15 per cent, and oil to dry mesocarp 65 per cent. The bunch yield standards do not take account either of soil or of the fact that the yield of a mother palm, which may be inbred, often bears little relation to the yield of its D × P progeny, but apart from this the standards are unexacting.

In choosing a seed source it is important that the material be suited to the environment of the plantation. Locally bred and produced seed will clearly have an advantage here, but the soil type may be of great importance since volcanic and some other soils tend to grow palms which elongate much too rapidly. In this case a short-growing or dumpy type, even if produced in another part of the world, will be preferable to very vegetatively vigorous material whether produced locally or elsewhere. Seed suppliers have tended to give more attention to the shortness factor in recent years (Soh *et al.*, 1981), and some have produced an alternative type of seed which may be descended from the Dumpy palm E206 or other short palms. Seed purchasers should also bear in mind the following points:

1. Prospective yields in terms of palm oil per hectare are sometimes given. These are of limited value, since oil yield depends on bunch yield and the latter varies widely with environment both within and between countries.
2. The seed should, if possible, be obtained as progenies with information on parental achievement and analysis.
3. A clear definition of the terms 'proved seed' or 'proved parentage' should be obtained if these terms are used by the seller.
4. The following data relating to parents, sibs or progeny should be obtained:
   (a) Bunch yield at maturity over a given number of years with statement of location, soil, rainfall distribution, water deficit and sunshine.
   (b) Bunch analysis: F/B; mesocarp, shell and kernel to fruit; oil to mesocarp. (This enables calculation of oil plus kernel to bunch.)

References

Ahée, J. and 13 others (1981) La multiplication végétative *in vitro* du palmier à huile par embryogénèse somatique. *Oléagineux*, **36**, 113.

Arasu N. T. and **Rajanaidu, N.** (1977) Oil palm genetic resources. In *Int. development in oil palm*, Kuala Lumpur.

Barcelos, E., Santos, M. deM. and **Vasconcellos, M. E. C.** (1985) Phenotypic variation in natural populations of Caiaue (*Elaeis oleifera* HKB Cortés) in the Brazilian Amazon. Int. Workshop on Oil Palm Germplasm and Utilization, ISOPB and PORIM, Kuala Lumpur, **10**, 102.

Beirnaert, A. (1933a) La sélection du palmier à huile. *Bull. agric.* Congo belge, **24**, 359–80, 418–58.

Beirnaert, A. (1933b) Les bases de la sélection du palmier à huile. *Journée Agron. Colon.*, 124.

Beirnaert, A. (1933c) Les méthodes de la sélection du palmier à huile. *Journée Agron. Colon.*, 135.

Beirnaert, A. (1940) Le problème de la stérilité chez le palmier à huile. *Bull. agric. Congo belge* (Leopoldville edn.), **31**, 95.

Beirnaert, A. and **Vanderweyen, R.** (1941) Contribution à l'étude génétique et biométrique des variétés d'*Elaeis guineensis* Jacq. *Publs INEAC* Sér. Sci., No. 27.

Bénard, G. (1965) Caractéristiques qualitatives du régime d'*Elaeis guineensis* Jacq. Teneur en huile de la pulpe des diverses origines et des croisements interorigines. *Oléagineux*, **20**, 163.

Bénard, G. and **Malingraux, C.** (1965) La production de semences sélectionnées de palmier à huile à IRHO Principe et réalisation. *Oléagineux*, **20**, 297.

Bénard, G. and **Noiret, J. M.** (1970) Le pollen de palmier à huile. Recolte, préparation, conditionnement et utilisation pour la fécondation artificielle. *Oléagineux*, **25**, 67.

Blaak, G. (1965) Breeding and inheritance in the oil palm. Part III. Yield selection and inheritance. *J. Nig. Inst. Oil Palm Res.*, **4**, 262.

Blaak, G. (1970) L'extraction de l'huile, à froid, dans l'analyse des régimes de palmier à huile. *Oléagineux*, **25**, 165.

Blaak, G. (1972) Prediction of precocity in the oil palm (*Elaeis guineensis* Jacq.). *Euphytica*, **21**, 22.

Blaak, G. *et al.* (1963) Breeding and inheritance in the oil palm. Part II. Methods of bunch quality analysis. *J. W. Afr. Inst. Oil Palm Res.*, **4**, 146.

Breure, C. J. (1985) Relevant factors associated with crown expansion in oil palm (*Elaeis guineensis* Jacq.). *Euphytica*, **34**, 161.

Breure, C. J. (1986) Parent selection for yield and bunch index in the oil palm in West New Britain. *Euphytica*, **35**, 65.

Breure, C. J. and **Corley, R. H. V.** (1983) Selection of oil palm for high density planting. *Euphytica*, **32**, 177.

Breure, C. J., Konimor, J. and **Rosenquist, E. A.** (1982) Oil palm selection and seed production at Dami oil palm research station, Papua New Guinea, *Oil Palm News*, **26**, 17.

Broekmans, A. F. M. (1957a) The production of improved oil palm seed in Nigeria. *J. W. Afr. Inst. Oil Palm Res.*, **2**, 116.

Broekmans, A. F. M. (1957b) Studies of the factors influencing the success of controlled pollination of the oil palm. *J. W. Afr. Inst. Oil Palm Res.*, **2**, 133.

Broekmans, A. F. M. (1957c) Growth, flowering and yield of the oil palm in Nigeria. *J. W. Afr. Inst. Oil Palm Res.*, **2** 187.

Carrière de Belgarric, R. (1951) Notes sur la sélection du palmier à huile à Sumatra:. résultats obtenus par la SocFin. *Oléagineux*, **6**, 65.

Chan Kook Weng, Ong Eng Chuan, Tan Kiap Seng, Lee Chong Hee and **Law Ing Hock** (1985) The performance of Oil Palm Genetic Laboratory (OPGL) germplasm material. Int. Workshop on Oil Palm Germplasm and Utilization, PORIM, Kuala Lumpur, **10**, 162.

Chapas, L. C., Tinker, P. B. H. and **Zibah, C. O.** (1957) The determination of the oil content of oil palm fruit. *J. W. Afr. Inst. Oil Palm Res.*, **2**, 230.

Chapas, L. C. (1961) Plot size and reduction of variability in oil palm experiments. *Emp. J. exp. Agric.*, **29**, 212.

Chemara Research Station, Layang Layang (1960) Oil palm planting material, *Ulu Remis* (D × P). Mimeograph.

Corley, R. H. V. (1979) Palm oil composition and oil palm breeding. *Planter, Kuala Lumpur*, **55**, 467.

Corley, R. H. V. (1982) Clonal planting material for the oil palm industry. *Planter, Kuala Lumpur*, **58**, 515.

Corley, R. H. V. (1987) Private communication.

Corley, R. H. V., Lee, C. H., Law, I. H. and **Cundall, E.** (1987) Field testing of oil palm clones. *Int. Oil Palm Conf.*, Kuala Lumpur, 1987.

298 *Oil palm selection and breeding*

Corley, R. H. V., Lee, C. H., Law, I. H. and Wong, C. Y. (1986) Abnormal flower development in oil palm clones. *Planter, Kuala Lumpur*, **62**, 233.

Corley, R. H. V., Wong, C. Y. and Wooi, K. C. (1982) Early results from the first oil palm clone trials. In *The oil palm in agriculture in the eighties*, ISP, Kuala Lumpur.

Corley, R. H. V., Wooi, K. C. and Wong, C. Y. (1979) Progress with vegetative propagation of oil palm. *Planter, Kuala Lumpur*, **55**, 377.

De Berchoux, C. and Gascon, J. P. (1965) Caractéristiques végétatives de cinq descendances d'*Elaeis guineensis* Jacq. *Oléagineux*, **20**, 1.

De Blank, S. (1952) A reconnaissance of the American oil palm. *Trop. Agric. Trin.*, **29**, 90.

De Poerck, R. (1942) Comment expliquer les disjonctions anormales de certains *tenera*? *Bull. agric. Congo belge*, **33**, 206.

Desneux, R. (1957) Prospection des palmeraies et sélection du palmier à huile au Kwango. *Bull. Inf. INEAC*, **6**, 351.

Desneux, R. (1958) Un palmier 'Pisifera' remarquable. *Bull. Inf. INEAC*, **7**, 95.

Dessasis, A. (1955) La détermination de la teneur en huile de la pulpe de fruit d'*Elaeis guineensis*. *Oléagineux*, **10**, 823.

Devruex, M. and Malingraux, C. (1960) Pollen d'*Elaeis guineensis* Jacq. Recherches sur les méthodes de conservation. *Bull. agric. Congo belge*, **5**, 543.

Escobar, C. R. (1980) Productividad potential de diferentes cruces commercial D × P de palma Africana (*E. guineensis* Jacq.) en Coto, Costa Rica. *Turrialba*, **30**, 250.

Escobar, C. R. (1981) Preliminary results of the collecting and evaluation of the American oil palm (*E. oleifera*, HBK Cortés) in Costa Rica. In *The oil palm in agriculture in the eighties*, Kuala Lumpur.

Falconer, D. S. (1960) *Introduction to quantitative genetics*, 1st edn. Oliver and Boyd, London.

Fickendey, E. (1944) Die Züchtung der Ölpalme (*Elaeis guineensis* Jacq.). *Z. Pfl Zücht.*, **26**, 136.

Gascon, J. P. and de Berchoux, C. (1963) Quelques relations entre les *dura* et *tenera* d'une même descendance et leur application à l'amélioration des semences. *Oléagineux*, **18**, 411.

Gascon, J. P. and de Berchoux, C. (1964) Caractéristiques de la production d'*Elaeis guineensis* (Jacq.) de diverses origines et de leurs croisements. *Oléagineux*, **19**, 75.

Gascon, J. P., Le Guen, V., Nouy, B., Asmady and Kamga, F. (1987) Results of second cycle recurrent reciprocal selection trials on oil palm. Int. Oil Palm Conf., Kuala Lumpur, June 1987.

Gascon, J. P. Noiret, J. M. and Bénard, G. (1966) Contribution à l'étude de l'hérédité de la production de régimes d'*Elaeis guineensis* Jacq.: Application à la sélection du palmier à huile. *Oléagineux*, **21**, 657.

Gascon, J. P., Noiret, J. M. and Meunier, J. (1969) Effets de la consanguité chez *Elaeis guineensis* Jacq. *Oléagineux*, **24**, 603.

Gascon, J. P. and Wuidart, W. (1975) Amélioration de la production et la qualité de l'huile de *Elaeis guineensis* Jacq. *Oléagineux*, **30**, 11.

Haddon, A. V. and Tong, Y. L. (1959) Oil palm selection and breeding: a progress report. *Malay. agric. J.*, **42**, 124.

Hardon, J. J. (1969a) Developments in oil palm breeding. In *Progress in oil palm*. Incorporated Society of Planters, Kuala Lumpur, p. 13.

Hardon, J. J. (1969b) Interspecific hybrids in the genus *Elaeis*, II, *Euphytica*, **18**, 380.

Hardon, J. J. (1970) Inbreeding in populations of the oil palm (*Elaeis guineensis* Jacq.). *Oléagineux*, **28**, 449.

Hardon, J. J. (1985) Long term conservation of oil palm. Int. Workshop on Oil Palm Germplasm and Utilization, PORIM, Kuala Lumpur, **10**, 197.

Hardon, J. J., Corley, R. H. V. and Lee, C. H. (1982) Breeding and selection for vegetative propagation in the oil palm. In Improvement of vegetatively propagated plants. *Proc. 8th Long Ashton Symposium.*

Hardon, J. J., Corley, R. H. V. and **Ooi, S. C.** (1972) Analysis of growth in oil palm. II. Estimates of genetic variances of growth parameters and yield of fruit bunches. *Euphytica*, **21**, 257.

Hardon, J. J. and **Davies, M. D.** (1969) Effects of vacuum drying on the viability of oil palm pollen. *Expl. Agric.*, **5**, 59.

Hardon, J. J., Mokhtar Hashim and **Ooi, S. C.** (1973) Oil palm breeding: a review. In *Advances in oil palm cultivation*, Incorp. Soc. of Planters, Kuala Lumpur.

Hardon, J. J. and **Ooi, S. C.** (1972) To what extent should inbreeding be avoided in oil palm seed production? *Com. (Agron.)* No. 9. Kumpulan Guthrie Sdn. Bhd.

Hardon, J. J., Tan, G. Y. (1969) and **Hardon, J. J.** (1969) Inter-specific hybrids in the genus *Elaeis*, I and II. *Euphytica*, **18**, 372 and 380.

Hardon, J. J. and **Thomas, R. L.** (1968) Breeding and selection of the oil palm in Malaya. *Oléagineux*, **23**, 85.

Hartley, C. W. S. (1957) Oil palm breeding and selection in Nigeria. *J. W. Afr. Inst. Oil Palm Res.*, **2**, 108.

Hartley, C. W. S. (1962) Report on a visit to North Borneo. Mimeograph.

Hartley, C. W. S. (1965) Some notes on the oil palm in Latin America. *Oléagineux*, **20**, 359.

Hartley, C. W. S. (1968) Report on oil palm research and development in Brazil. *Comun. Tecn. CEPLAC*, **17**, 1.

Hartley, C. W. S. (1971) Proposals for oil palm research, MARDI, Malaysia. Mimeograph.

Hayes, H. K., Immer, F. R. and **Smith, D. C.** (1955) *Methods of plant breeding.* McGraw-Hill, New York and London.

Henry, P. (1959) Prolongation de la viabilité du pollen chez *Elaeis guineensis. C. r. hebd. Séanc. Acad. Sci., Paris*, **98** 722.

Henry, P. and **Gascon, J. P.** (1950) Les palmiers à huile du type *pisifera* et la stérilité. *Oléagineux*, **5**, 29.

INEAC (1956 and 1957) Rendements obtenus en plantation par l'utilisation de graines d'*Elaeis* sélectionnées à Yangambi et issues du croisements *dura* × *pisifera. Bull. Inf. INEAC*, **5**, 271 and **6**, 329.

IRHO *Rapports annuels*, 1969–73.

Jack, H. W. and **Jagoe, R. B.** (1932) Variation in fruiting ability of oil palms. *Malay. agric. J.*, **20**, 16.

Jacquemard, J. C. (1979) Contribution à l'étude de la croissance en hauteur du stipe d'*Elaeis guineensis* Jacq. Étude du croisement L2T × D10D. *Oléagineux*, **34**, 492.

Jagoe, R. B. (1952a) Deli oil palms and early introductions of *Elaeis guineensis* to Malaya. *Malay. agric. J.*, **35**, 3.

Jagoe, R. B. (1952b) The 'dumpy' oil palm. *Malay. agric. J.*, **35**, 12.

Janssen, A. W. B. (1959) Miscellaneous notes on estate agriculture in Sumatra, No. 7. Oil palm productivity and genetics. Chemara Research Station.

Janssens, P. (1927) Le palmier à huile au Congo Portuguaise et dans l'enclave de Cabinda. *Bull. agric. Congo belge*, **18**, 29.

Jinks, J. L. and **Lawrence, M. J.** (1983) The genetical basis for inbreeding depression and of heterosis: its implications for plant and animal breeding. *PORIM Occasional Paper*, 11.

Jones, L. H. (1974) Production of clonal oil palms by tissue culture. *Oil Palm News*, **17**, 1; and *Planter, Kuala Lumpur*, **50**, 374.

Jones, L. H. (1983) The oil palm and its clonal propagation by tissue culture. *Biologist*, **30**, 181.

Law, I. H. (1984) Bunch analysis techniques. Oil Palm Breeders Meeting, Unifield T. C. Ltd. Mimeograph.

Lawrence, M. J. and **Rajanaidu, N.** (1985) The genetical structure of natural populations and sampling strategy. Int. Workshop on Oil Palm Germplasm and Utilization, PORIM, Kuala Lumpur, **10**, 15.

Loiret, C. (1981) Vegetative propagation of oil palm by somatic embryogenesis. In *The oil palm in agriculture in the eighties*, Kuala Lumpur.

Lubis, A. U. (1984) Historio dan penn tentang tanaman kelapa sawit. *Pusat Penelitianat*, pp. 39.

Lubis, A. U. and **Kiswito** (1975) New perspectives in oil palm breeding in Indonesia. In *South East Asian plant genetic resources*, eds J. T. Williams *et al.*, Bogor.

Macfarlane, M., Alaka, B. and **Macfarlane, N.** (1975) Analysis of the mesocarp oils from several different oil palm hybrids. *Oil Palm News*, **20**, 1.

Maillard, G., Daniel, C. and **Ochs, R.** (1974) Analyse des effets de la sécheresse sur le palmier à huile. *Oléagineux*, **29**, 397.

Menendez, T. (1965) *NIFOR first annual report, 1964–5* pp. 59–73.

Menendez, T. M. (1969) *NIFOR third annual report, 1966–7*, p. 72.

Menendez, T. M. and **Blaak, G.** (1964) *WAIFOR twelfth annual report, 1963–64*, pp. 68–9 and unpublished data.

Mennier, J. and **Boutin, D.** (1975) L'*Elaeis melanococca* et l'hybride *E. melanococca* × *E guineensis*. Premières données. *Oléagineux*, **30**, 5.

Meunier, J. (1969) Étude des populations naturelles d'*Elaeis guineensis* en Côte d'Ivoire. *Oléagineux*, **24**, 195.

Meunier, J., Baudouin, L., Nouy, B. and **Noiret, J. M.** (1987) The expected value of oil palm clones. *Int. Oil Palm Conf.*, Kuala Lumpur, June 1987.

Meunier, J. and **Gascon, J. P.** (1972) Le schéma général d'amélioration du palmier à huile à l'IRHO *Oléagineux*, **27**, 1.

Meunier, J., Gascon, J. P. and **Noiret, J. M.** (1970) Hérédité des caractéristiques du régime d'*Elaeis guineensis* Jacq. en Côte d'Ivoire. *Oléagineux*, **25**, 377.

Meunier, J., Vallejo, G. and **Boutin, D.** (1976) L'hybride *E. melanococca* × *E. guineensis* et son amélioration. *Oléagineux*, **31**, 519.

Mollegaard, M. (1970) The effect of drying of oil palm nuts in respect of crackability of same and estimation of the percentage of shell and kernel to fruit. *Oléagineux*, **25**, 139.

Murthi Anishetty, N. (1985) Plant genetic resources – an overview of the International Board for Plant Genetic Resources IBPGR Programme. Int. Workshop on Oil Palm Germplasm and Utilization. PORIM, Kuala Lumpur, **10**, 11.

Ng, B. H., Corley, R. H. V. and **Clegg, A. J.** (1976) Variation in the fatty acid composition of palm oil. *Oléagineux*, **31**, 1.

Nigerian Conference (1949) Proceedings of the Conference on Oil Palm Research at the Oil Palm Research Station, near Benin, Nigeria, Dec. 1949. Mimeographed report.

NIFOR (1966) *Second annual report, 1965–6*, p. 80.

NIFOR (1969) *Fifth annual report, 1968–9*, pp. 46–50.

Noiret, J. M. (1981) Application de la culture *in vitro* à l'amélioration et la production de matériel clonal chez le palmier à huile. *Oléagineux*, **36**, 123.

Noiret, J. M. and **Gascon, J. P.** (1967) Contribution à l'étude de la hauteur et de la croissance du stipe d'*Elaeis guineensis* Jacq. Application à la sélection du palmier à huile. *Oléagineux*, **23**, 85.

Noiret, J. M., Gascon T. P. and **Bénard, G.** (1966) Contribution à l'étude de l'hérédité des caractéristiques de la qualité du régime et du fruit d'*Elaeis guineensis* Jacq. Application à la sélection du palmier à huile. *Oléagineux*, **21**, 343.

Noiret, J. M., Gascon, J. P. and **Pannetier, C.** (1985) La production de palmier à huile par culture *in vitro*. *Oléagineux*, **40**, 365.

Nwanze, S. C. (1961) *WAIFOR Ninth annual report, 1960–1*, p. 100.

Obasola, C. O. (1973) Breeding for short-stemmed palms in Nigeria *J. W. Afr. Inst. Oil Palm Res.*, 5(18), 43.

Obasola, C. O., Obesesan, I. O. and **Opute, F. I.** (1976) Breeding of short-stemmed oil palm in Nigeria. *Int. Agric. Oil Palm Conference*, Kuala Lumpur.

Ong, S. H, Chuah, C. C. and **Soo, H. P.** (1981) The co-dominance theory. Genetic interpretations of analyses of mesocarp oils from *E. guineensis*, *E. oleifera* and their hybrids. *J. Amer. Oil Chem. Soc.*, **58**, 1032.

Ooi, S. C. (1975) Variability in the Deli *dura* breeding population of the oil palm (*Elaeis guineensis* Jacq.). II. Within bunch components of bunch yield. *Malaysian Agric. J.*, **50**, 20.

Ooi, S. C. (1978) Variability in the Deli *dura* breeding population of the oil palm (*E. guineensis* Jacq.). IV. Growth and physiological parameters. *Malay. agric. J.*, **51**, 359.

Ooi, S. C. and **Abdul Wahab bin Ngah** (1977) Oil palm breeding – some aspects of selection. *Int. Development in Oil Palm*, Kuala Lumpur.

Ooi, S. C., Hardon, J. J. and **Phang, S.** (1973) Variability in the Deli *dura* breeding population of the oil palm (*Elaeis guineensis*, Jacq.). I. Components of bunch yield. *Malaysian Agric. J.*, **49**, 112.

OPGL (1967–71) Oil Palm Genetics Laboratory, *Annual reports, 1966 to 1970*.

Pannetier, C., Arthuis, P. and **Lievoux, D.** (1981) Néoformation de jeunes plantes d'*Elaeis guineensis* à partir de cals primaires obtenus sur fragments foliaires cultives *in vitro*. *Oléagineux*, **36**,119.

Paranjothy, K. (1986) Recent development in cell and tissue culture of oil bearing palms. *PORIM Occasional Paper*, No. 19.

Paranjothy, K. and **Rohani, D.** (1982) *In vitro* propagation of oil palm. In *Proc. 5th Int. Cong. Plant Tissue and Cell Culture*. ed. F. Fijiwara, pp. 374–88, the Jap. Assoc. for Plant Tissue Culture.

Patiño, V. M. (1984) Información preliminar sobre la palma de Aceite Africana (*Elaeis guineensis*) en Colombia. *Estación Agro-forestal del Pacifico de Calima-Buenaventura*, Serie Bot. Apl., **1** (2), Dec. 1948.

Patiño V. M. (1958) La Palma de Aceite Africana (The African oil palm). *Economia Colombiana*, Oct. 1958.

Pichel, R. (1957) L'Amélioration du palmier à huile au Congo Belge. Conf. Franco-Britannique sur le palmier à Huile. *Bull. agron. Minist. Fr. d'outre mer*, **14**, 59, and *Bull. agric. Congo belge*, **48**, 67.

Poels, G. (1959) La qualité des fruits de palme produits par les agriculteurs congolais. *Bull. Inf. INEAC*, **8**, 309.

Pronk, F. (1953) Vergelijkend tros-analytisch onderzoek van enkele typen van de oliepalm (*Elaeis guineensis* Jacq.). *Bergcultures*, **22**, 527–33 and 573–81.

Pronk, F. (1955) *De Veredeling van de oliepalm door het Algemeen Proefstation der AVROS*. Mimeograph, 29 pp.

Rabéchault, H., Ahée, J. and **Guénin, G.** (1970) Colonies cellulaires et formes embryoïdes obtenus *in vitro* à partir de cultures d'embryons de palmier à huile. *C. R. Acad. Sci., Paris*, Sér. D, **270** 3067.

Rabéchault, H. and **Martin, J. P.** (1976) Multiplication végétative du palmier à huile à l'aide de cultures de tissues foliaires. *C. R. Acad. Sci., Paris*, Sér. D, **238**, 1735.

Rabéchault, H., Martin, J. P. and **Cas, S.** (1972) Recherches sur la culture des tissus de palmier à huile. *Oléagineux*. **27**, 531.

Rajanaidu, H. (1974) Some aspects of the variation of the genus *Elaeis*. Typescript thesis, dept. of Botany, University of Birmingham.

Rajanaidu, N. (1984) PORIM's oil palm prospections in Zaire and Cameroon. *ISOPB Newsletter, Int. Soc. Oil Palm Breeders*, **1**(2), 5.

Rajanaidu, N. (1985a) The oil palm (Elaeis guineensis) collections in Africa. Int. Workshop on Oil Palm Germplasm and Utilization, PORIM, Kuala Lumpur, **10**, 59.

Rajanaidu, N. (1985b) *Elaeis oleifera* collection in central and south America. Int. Workshop on Oil Palm Germplasm and Utilization, PORIM, Kuala Lumpur, **10**, 84.

Rajanaidu, N., Arasu, N. T. and **Obasola, C. O.** (1979) Collection of oil palm (*Elaeis guineensis* Jacq.) genetic material in Nigeria. 2. Phenotypic variation of natural population. *MARDI Res, Bull.*, **7**(1), 1.

Rajanaidu, N., Ngui, M., Ong Eng Chuan and **Lee Chong Hee** (1985a) Sabah breeding programme (SBP). Int. Workshop on Oil Palm Germplasm and Utilization, PORIM, Malaysia, **10**, 175.

Rajanaidu, N. and **Rao, V.** (1987) Oil palm genetic collections: their performance and use to the industry. *Int. Oil Palm Conf.*, Kuala Lumpur, June 1987.

Rajanaidu, N., Tan Yap Pau, Ong Eng Chuan and **Lee Chong Hee** (1985b) The performance of inter-origin commercial D × P planting material. Int. Workshop on Oil Palm Germplasm and Utilization, PORIM, Kuala Lumpur, **10**, 155.

Rao, V. and 10 others (1983). A critical examination of the methods of bunch analysis in oil palm breeding. *PORIM Occasional Paper*, No. 9.

Robinson, J. B. D. and **Treharne, K. J.** (1985) Maize. *J. Inst. Biol.*, **32**(4), 199.

Rosenquist, E. A. (1984a) Notes on inbreeding in the oil palm. Oil Palm Breeders Meeting, Unifield T.C. Ltd. Mimeograph.

Rosenquist, E. A. (1984b) Crossing programme design. Oil Palm Breeders Meeting, Unifield T.C. Ltd. Mimeograph.

Rosenquist, E. A. (1985) The genetic base of oil palm breeding populations. Int. Workshop on Oil Palm Germplasm and Utilization, PORIM, Kuala Lumpur, **10**, 27.

Sabah (1970) Dept. of Agriculture, State of Sabah, *Annual report for the year 1969*. Kota Kinabalu.

Santos, M. deM., Barcelos, E. and **Nascimento, J. C.** (1985) Genetic resources of *Elaeis oleifera* (HBK Cortés) in the Brazilian Amazon. Int. Workshop on Oil Palm Germplasm and Utilization, PORIM, Kuala Lumpur, **10**, 95.

Schmöle, J. F. (1930) The selection of oil palms (*Elaeis guineensis* Jacq.). *Proc. 4th Pacif. Sci. Congress*, Java, 1929, Proc. **4**, 185.

Servant, M. and **Henry, J.** (1963) Détermination de la richesse en huile de la pulpe du fruit de palme. *Oléagineux*, **18** 339.

SIRIM (1973) Specification for oil palm seed for commercial planting. Standards Institution of Malaysia, Kuala Lumpur, MS 3.18 : 1973.

Smith, E. H. G. (1929) The oil palm (*Elaeis guineensis*) at Calabar, *8th An. Bull. Dep. Agr., Nigeria*.

Smith, E. H. G. (1930) The oil palm at Calabar. 2nd Conf. W. Afr. Agr. Officers, 1929, Paper 16, 181 (*Gold Coast Dept. Agr. Bull.*, 20).

Smith, E. H. G. (1933) Further yields from the Calabar plantation oil palms. *10th An. Bull. Dept. Agr., Nigeria* 1–18.

Smith, W. K. and **Jones, L. H.** (1970) Plant propagation through tissue culture. *Chem. and Ind.*, **44**, 1399.

Smith, W. K. and **Thomas, J. A.** (1973) The isolation and *in vitro* cultivation of cells of *Elaeis guineensis*. *Oléagineux*, **28**, 123.

Socfin Company Ltd (1963) Oil palm planting material. Mimeograph.

Soh Aik Chin (1983) *Choice of planting materials*. Incorp. Soc. of Planters, Kuala Lumpur.

Soh, A. C. (1986) Expected yield increases with selected oil palm clones from current D × P seedling material and its implications on clonal propagation, breeding and ortet selection. *Oléagineux*, **41**, 51.

Soh, A. C. (1987) Abnormal oil palm clones. Possible causes and implications: further discussions. *Planter, Kuala Lumpur*, **63**, 59.

Soh, A. C., Vanialingam, T. and **Taniputra, B.** (1981) Derivatives of the Dumpy palm: some experimental results. *Planter, Kuala Lumpur*, **57**, 227.

Sparnaaij, L. D. (1958) Oil palm breeding and selection in Belgian Congo. WAIFOR mimeographed report.

Sparnaaij, L. D. (1969) Oil Palm. In *Outlines of perennial crop breeding in the tropics*. Landbouwhogeschool, Wageningen. Misc. papers 4.

Sparnaaij, L. D. and **Van der Vossen, H. A. L.** (1980) Development in oil palm breeding: a reappraisal of present and future procedures in the light of results from the Nigerian Institute for Oil Palm Research breeding programme. *Oil Palm News*, **24**, 4.

Sparnaaij, L. D., Menendez, T. and **Blaak, G.** (1963) Breeding and inheritance in the oil palm. Part I. The design of a breeding programme. *J. W. Afr. Inst. Oil Palm Res.*, **4**, 126.

Stavitsky, G. (1970) Tissue culture of the oil palm as a tool for its vegetative propagation. *Euphytica*, **19**, 288.

Stoffels, E. (1934) Contribution à l'étude de la sélection d'*Elaeis guineensis* à Sumatra. *Revue Bot. appl. Agric. trop.*, **14**, 93.

Taillez, B. and **Valverde, G.** (1971) Sensibilité aux ouragans de différents types de croisements de palmier à huile. *Oléagineux*, **26**, 753.

Tan, G. Y. (1978) Genetic studies of some morphophysiological characters associated with yield in the oil palm (*E. guineensis* Jacq.). *Trop. Agric. Trin.*, **55**, 9.

Tang Teng Lai (1971) The possible use of *pisifera* in plantation. *Malaysian Agric. J.*, **48**, 57.

Thomas, R. L., **Watson, I.** and **Hardon, J. J.** (1969) Inheritance of some components of yield in the Deli *dura* variety of oil palm. *Euphytica*, **18**, 92.

Toovey, F. W. (1938) Report on a visit to the Belgian Congo to study certain aspects of the oil palm industry. Mimeographed report, Agric. Dept., Nigeria.

Toovey, F. W. and **Purvis, C.** (1956) Segregation from the *dura* × *pisifera* cross. *J. W. Afr. Inst. Oil Palm Res.*, **1**(4), 106.

Trafton, M. (1951) *The African oil palm in Honduras*. Bull. 2, United Fruit Co's. Trop. Res. Dept., La Lima, Honduras.

Turner, J. M. (1984) Ramet establishment and survival. Oil Palm Breeders Meeting. Unifield, T.C. Ltd. Mimeograph.

Unifield, T. C. (1984) *Proceedings of the Oil Palm Breeders Meeting*, Unifield T.C. Ltd. Bedford, 19.

Unifield (1985) *Oil palm tissue culture*. Unifield, T.C., Bedford.

Vallejo Rosero, R. and **Cassalett, C.** (1974) Perspectivas del cultivo de los hibridos interspecificos de Noli (*Elaeis oleifera* (HBK Cortez) × palma Africana de aceite (*Elaeis guineensis*, Jacq.)) en Colombia. Instituto Colombiano Agropecuario, Colombia. Mimeograph.

Van der Vossen, H. A. M. (1974) Towards more efficient selection for oil yield in the oil palm (*Elaeis guineenis* Jacq.). *Agr. Res. Rpts.* 823, Wageningen, Holland.

Vanderweyen, R. (1952) La prospection des palmeraies congolaises et ses premiers résultats. *Bull. Inf. INEAC* **1**, 357.

Vanderweyen, R. (1953) Le croisement 'Dura × Pisifera' et ses premiers resultats. *Bull. Inf. INEAC*, **2**, 123.

Vanderweyen, R. and **Roels, O.** (1949) Les variétés d'*Elaeis guineensis* Jacquin du type *albescens*. *Publs INEAC*, Série Sci., No. 42.

Vanderweyen, R., **Rossignol, J.** and **Miclotte, H.** (1947) Considérations sur les teneurs en eau et en huile de la pulpe des fruits d'Elaeis. *C. R. de la Semaine Agricole de Yangambi*. Common. No. 54, 730.

Webster, C. C. (1939) A note on a uniformity trial with oil palms. *Trop. Agric. Trin.*, **16**, 15.

West, M. J. (1976) The analysis of yield data of the NIFOR oil palm main breeding programme and the choice of new parental material. Supplementary report to the Min. of Overseas Development. Mimeograph.

West, M. J., **Ross, J. M.**, **Obasola, C. O.** and **Nekkako, H. U.** (1976) The inheritance of yield and of fruit and bunch composition characters in the oil palm – an analysis of the NIFOR main breeding programme. *Int. Dev. in Oil Palm*, Kuala Lumpur.

Wooi, K. C. (1984) Palm tissue culture. In Micropropagation of selected root crops, palms, citrus and ornamental species. *FAO Plant Production Paper*, No. 59, pp. 88–112, FAO, Rome.

Yeow, K. H., **Tam, T. K.**, **Poon, Y. C.** and **Toh, P. Y.** (1981) Recent innovation in agricultural practice: palm replacement during immaturity. In *The oil palm in agriculture in the eighties*, ISP, Kuala Lumpur.

Zeven, A. C. (1973) The 'mantled' oil palm (*Elaeis guineensis* Jacq.). *J. W. Afr. Inst. Oil Palm Res.*, **5**, 31.

Chapter 6

Germination and the preparation and storage of seed

In the early years of plantations the methods used for germinating seed often had very erratic results and germination failures were common. The industry had to await the physiological studies of Hussey (1958, 1959) and Rees (1963a), coupled with investigations of germination under natural conditions, before a full understanding was obtained of the factors involved.

Germination under natural conditions

In the semi-wild palm groves seed is distributed by a variety of means, human and animal, and fruit or seed becomes haphazardly scattered through the groves themselves and adjacent forest. Examination of seed in various sites in Nigeria from high forest to plantations, swampland, palm grove and open country (Rees, 1963a, b) showed that in the natural forest and grove sites the majority were eaten by rodents, bored by beetles (chiefly the Scolytid, *Coccotrypes congonus*, and to a much lesser extent the Bruchid, *Pachymerus cardo*), or suffered delignification of the shell to become 'white' seed (Table 6.1). While rodent and Scolytid activities are major

Table 6.1 *Condition of ungerminated seed at end of a period of 2–2½ years under different habitats*

Percentage of seed remaining in each treatment

Condition of seed	Type of habitat			
	Bare soil	Grassland	Plantation	Forest
Eaten by rodents	7.3	6.4	7.7	8.4
'White seed'	4.3	6.5	16.2	14.6
Perforated by beetles	3.0	1.3	74.7	74.9
Unperforated	85.4	85.8	1.5	2.2
Viable	3.0	10.0	3.5	1.0

Source: Rees (1963b).

factors in reducing viability under natural conditions, the germination of the remaining seed depends markedly on season and probably to a lesser extent on site and treatment of the vegetation. In general, seed remains dormant throughout the wet and dry seasons in West Africa, but germination takes place during a short period of about 6–10 weeks at the start of the rainy season; there is a lag period of around 60 days after the first heavy rainfall before germination starts.

Differences between sites, after the differential effect of Scolytid activity has been taken into account, are more difficult to detect; the differences between the curves in Fig. 6.1 are almost entirely due to perforations of the seed by *Coccotrypes*. However, it is constantly observed that large numbers of seeds germinate when forest is cut down for farming; here it is likely that the heat effect to be mentioned later is playing its part, and it is often suggested that the high-temperature requirement for rapid germination has a survival value in a region with a severe dry season. Nevertheless, 50–70 per cent of seed stored at optimum moisture content for germination, but at temperatures similar to those prevailing in forest or plantation, will germinate though germination is very slow (Rees, 1963a).

Paradoxically, in spite of the exacting conditions required for germination and the poor seed survival in natural habitats, the oil palm escapes very readily into surrounding or even remote secondary forest. Numerous instances have now been recorded of

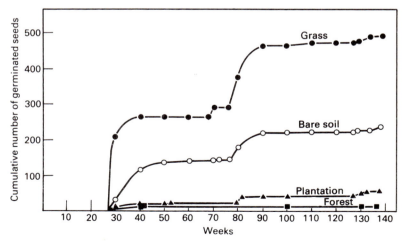

Fig. 6.1   Germination of oil palm seed planted in the open in Nigeria at the end of the wet season (November 1959) until June 1962: (●) grass planting; (○) bare soil planting; (▲) palm plantation; (■) high forest (Rees, 1963a).

fully grown palms being found in out-of-the-way areas of periodic clearing in Malaysia and Indonesia, while in Brazil escapes from the Bahia grove areas have been seen on valley sites in open dry country as far as 18° S.

Factors inducing germination

That a high temperature was needed for the satisfactory germination of oil palm seed was realized early in the history of the plantation industry (Bücher and Fickendey, 1919). Sand beds, well watered and exposed to the sun, proved sufficiently satisfactory in the Far East until valuable selected seed began to be produced. In West Africa, with its relatively sunless climate, germination was poor more often than not, and in spite of various systems of applying heat a germination of between 25 and 60 per cent was as much as could usually be expected until a full physiological study had pointed the way to more refined methods.

The seed is comprised of the shell and the kernel in which is embedded the embryo (see p. 52). The excised embryo is non-dormant and when placed on moist filter paper begins to elongate within 24 hours. Embryos still in contact with portions of the kernel will also begin to germinate provided they are free to elongate. Elongation is possible if the operculum is removed or the portion of the kernel near the cotyledon is cut away (Hussey, 1958). Under these circumstances elongation is hastened in oxygen. Fully developed seedlings have been produced from excised embryos (Bouharmont, 1959). Rabéchault (1962), using a comprehensive nutrient medium and maintaining a temperature of 30 °C, distinguished two periods of development, 'germination' and 'seedling development'. The stages of germination were distinctive and the haustorium developed into a sickle-shaped body. By the end of this period, which occupies 4–5 weeks, both a plumule and a radicle have appeared. Subsequently normal leaves and roots develop. Seedlings can be developed without the addition of β-indole acetic acid to the medium, but most satisfactory growth is obtained with concentrations of 10^{-7} to 10^{-6}. Light accelerates differentiation, but retards growth. The reaction of embryos to gibberellic acid has also been investigated (Bouvinet and Rabéchault, 1965), and has been found to be irregular. However, in a liquid medium the response is more regular and a concentration of 10^{-5} increases the speed of differentiation and the growth of the plumule and first leaves, although there appears to be no effect on root growth.

Although naked kernels cannot be conveniently germinated on a large scale, it was experiments with kernels which first opened the way to an understanding of the germination process. Hussey (1958) showed that, at high temperatures (38–40 °C), increasing the oxygen tension markedly accelerated the process. At ambient temperatures, however, increasing the oxygen tension was ineffective except when high temperature had previously been applied. The results of two of Hussey's classic experiments are given in Tables 6.2 and 6.3; These experiments not only showed the importance of oxygen supply but demonstrated that germination was inhibited until some process induced by high temperature had taken place. Subsequent physiological work showed that the optimum

Table 6.2 *Effects of oxygen on the germination of* tenera *kernels at various temperatures*

Cumulative germination percentage

Temperature		Weeks	
		4	8
25 °C	Air	0	0
	Oxygen	0	0
33 °C	Air	0	0
	Oxygen	0	0
40 °C	Air	0	5
	Oxygen	34	81
45 °C	Air	0	0
	Oxygen	0	0

Source: Hussey (1958).

Table 6.3 *Effect of high temperature and oxygen on the germination of* tenera *kernels transferred to differing treatments after a period of 4 weeks*

Cumulative germination percentage

First 4 weeks			Second 4 weeks		
Initial treatment		Germination	Transferred to		Germination
40 °C	Oxygen	32	40 °C	Oxygen	78
			28 °C	Air	67
			28 °C	Oxygen	69
40 °C	Air	0	40 °C	Air	5
			28 °C	Air	9
			28 °C	Oxygen	58
28 °C	Oxygen	0	28 °C	Oxygen	1
			40 °C	Air	3
			40 °C	Oxygen	66
28 °C	Air	0	28 °C	Air	0
			40 °C	Air	0
			40 °C	Oxygen	45

Source: Hussey (1958).

moisture and temperature conditions for germination are as follows (Rees, 1962a):

Moisture content –	*dura* seed	21–22 per cent
(dry weight basis)	*tenera* seed	28–30 per cent

These optima correspond to saturation moisture contents; Rabéchault *et al.* (1967) showed this to be 21.5 per cent overall for Deli *dura*, the approximate moisture contents of components being shell 21 per cent and kernel 23.5 per cent.

Temperature	– kernels, wet	40 °C
	– seed, dry	42 °C

The first temperature figure was deduced from the fact that germination at 36 °C was very slow, at 39.5 or 40 °C it was comparatively rapid while at a constant temperature of 42 °C no germination was observed with *tenera* kernels (Hussey, 1958). Under certain circumstances, and for short periods, seed can endure high temperatures, however. Seed exposed to a temperature of 50 °C for 48 hours shows no ill effect on subsequent germination, though this temperature applied to wet seed for more than 3 days results in loss of viability. When *dry* seed is heat-treated at 42 °C, subsequent germination at optimum moisture content is unimpaired (Rees, 1959b, 1962a).

The relationship of temperature, moisture and oxygen supply will now be considered. There was early evidence that cooling during or after heat treatment produced a flush of germination and hence that the application of heat was not necessary throughout the period of germination (Henry, 1951). This was consistent with the belief that lowering the temperature would increase the oxygen supply to the embryo tissues because of the increased solubility of oxygen at lower temperatures and the high temperature coefficient of respiration. Seeds heat-treated at 39.5 °C for 80 days at optimum moisture content show a rapid flush of germination when transferred to ambient temperatures; 50 per cent of final germination is reached after 5–6 days and germination is complete within 3 weeks. If heat is applied at moisture contents too low for germination and the seed is then simultaneously cooled and brought to the optimum moisture content, a similar flush is obtained, though this is delayed for about 3 weeks. Results of an experiment with *dura* seed demonstrate these facts (Table 6.4).

The dry heat-treated seed may be stored for up to 6 weeks before being brought to the optimum moisture content and a similar germination curve will then be obtained (Fig. 6.2). It is thus clear that the high-temperature reaction is irreversible. While it is not necessary for heat to be applied at or even near to the optimum moisture content, there is evidence that the heat treatment is not

Table 6.4 *Germination of* dura *seed given 80 days' heat treatment at 39.5 °C at a range of moisture contents*

	Moisture content	Germination after cooling at		Days to 50% of final germination at optimum moisture content after 80 days
		Original moisture content	Optimum moisture content	
	(%)	(%)	(%)	
1.	24.2	12	71	23.5
2.	21.3	95	92	6.5
3.	18.6	0	97	17.5
4.	14.2	0	97	25.0

Source: Rees (1962a).

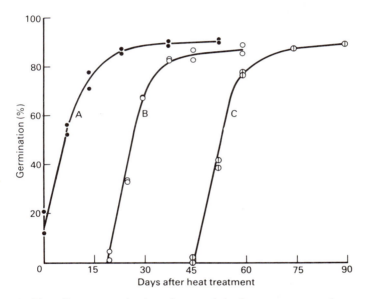

Fig. 6.2 The effect on germination of wet and dry heat treatment and storage of heat-treated seed (Rees, 1962a): **A**, seed heat-treated at optimum moisture content; **B**, heat-treated dry; **C**, heat-treated dry but stored dry for 25 days before bringing to the optimum moisture content.

satisfactorily completed if the moisture content of the seed is either very high or very low (Rees, 1962a). The figures presented in Table 6.4 show that with over-watering (treatment 1) subsequent germination is both impaired and delayed. Seed set in a germinator at 39.5 °C dry down to about 10 per cent moisture content at constant rate. If the seed is exposed to the air, this occurs in about 5 days and an equilibrium moisture content of *dura* seed of about 7.5 per cent is soon attained. When kept in polythene bags, the rate of moisture loss is of course much slower. It has been shown that seed drying down to 7.5 per cent moisture content in the germinator subsequently gives poor germination at optimum moisture content. The seed is not killed, however, and if the ungerminated nuts are given further heat treatment at a higher moisture content, normal germination will be obtained. If *dura* seed initially at 17.5–18 per cent moisture is set in polythene bags at a germinator temperature of 39.5 °C, the final moisture content will be about 8.5–9.0 per cent and the high-temperature reaction will proceed normally.

During the course of work on dry-heat treatment of seed the optimum temperature for germination was re-examined. It was found that at a temperature of 42 °C the length of the heat period could be reduced to 60 days. In practice, however, this is so near the lethal temperature that 39–40 °C is safer. Stored seed germinates more readily than fresh seed and this indicates that the high-temperature reaction is proceeding at ambient temperatures though at a much slower rate. It has been estimated by Rees (1962a) that the temperature coefficient (Q_{10}) of the high-temperature reaction lies between 3.5 and 5, and that at ambient temperatures (*c.* 27 °C) it would take about 300 days for completion. This is borne out by the results of seed storage experiments which have shown that up to 70 per cent germination may occur in seed stored at ambient temperatures and a moisture content around the optimum for germination.

The examination of the effect of heat treatment in relation to time was taken a stage further by Labro *et al.* (1964) who subjected seed stored for 6 months to high temperatures for relatively short periods of time as follows:

Temperature (°C)	Period of heat pretreatment and condition of seed
40	80 days, wet and dry
45	35, 40, 45, 50 days, wet and dry
50	6, 8, 10, 12 days, wet and dry
55	2, 4 days wet, 2, 4, 6, 8 days dry
60	3, 6, 12 hours, wet and dry

In each case the shortest pretreatment was the least deleterious. Germination percentages of between 43 and 53 per cent were obtained after 80 days following the shortest pretreatment at 45, 50 and 55 °C; 73 per cent germination was obtained after 3 hours' wet-heat pretreatment at 60 °C. A curious feature of these trials was the fact that further heat treatment of 20 days at 40 °C produced a secondary flush of germination, this being particularly marked when the first pretreatment had been carried out with *dry* seed. The germination percentage after dry-heat pretreatment at 40 °C was abnormally low (49 per cent) and the further 20 days' heat treatment raised the germination percentage to 79; this suggested that some other factor, e.g. origin of seed, storage or moisture levels, may have influenced the results of all the trials. Since the temperature effect is proceeding, though slowly, at ambient temperatures, the period required at 40 °C for this stored seed would be expected to be well below 80 days. Nevertheless, this work clearly showed that 'shock' heat treatment at temperatures lethal when applied over long periods can induce the temperature effect in oil palm seed in periods very much shorter than those needed when non-lethal temperatures are used. On the basis of these results and his own experiments, Rees (1965a) pointed out that the relation between the pretreatment temperature and log hours of high-temperature pretreatment required for satisfactory germination is linear.

It will thus be seen that oil palm seed has certain moisture, oxygen and temperature requirements for germination and that the latter requirement can be satisfied as a pretreatment. The moisture requirements (p. 308) are different for *dura* and *tenera* seed, while the oxygen requirement is best satisfied by supplying the optimum moisture for germination after the seed has been brought to ambient temperature. The three paths to germination are illustrated in Fig. 6.3. The upper path – including dry-heat treatment (A to B) – has considerable practical advantages; the middle path is most rapid; the lower path is now seldom used.

Some attempts have been made to break the dormancy of seed

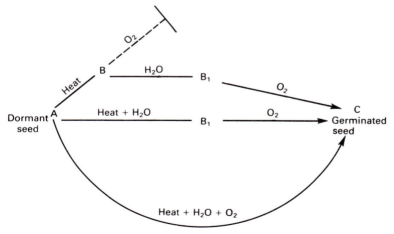

Fig. 6.3   Three paths to germination. (After Rees, 1959a).

by growth substances in place of heat treatment. These have been largely unsuccessful, though Wan and Hor (1983) produced some hastening of germination by soaking seed in a 500 μ/l solution of gibberellic acid (GA_3) prior to heat treatment.

In many cases of poor germination a proportion of the embryos are abnormal through incomplete development, malformation, necrosis or doubling or tripling of the embryo. It has been found that there is a relationship not only between the percentage of abnormality and germination, but also between abnormality and seedling development. Seed lots with more than 15 per cent abnormal embryos tend to give germination percentages of around 60 per cent while the percentage of abnormal and non-developing seedlings will be as high as 50 per cent. There is a direct relationship between the amount of abnormality and the length and method of pollen storage (Noiret and Ahizi Adiapa, 1970).

Although good overall germination will nearly always be obtained if the heat and moisture requirements are fulfilled, there are factors which may give rise to poor results. Firstly, the seed may have been subjected to bad storage conditions. Secondly, there is evidence of genetic differences in the germination of progenies, some progenies showing consistently poor germination (Henry, 1951 and 1952), though this is probably much less common than was previously supposed. On the other hand, there is also evidence of some selected material having partially lost the high-temperature requirement, and there is even one instance quoted of a batch of West African seed which gave 61 per cent germination in 8 months at a temperature of around 22 °C (Rees, 1962a).

Preparation of seed

The degree of ripeness of the fruit has no effect on germination percentage (Rees, 1962b); nevertheless it is useful to set some standard to be adhered to in practice, and it is simplest to require the bunches to be at the normal stage of ripeness for harvesting.

In the preparation of seed for germination the fruit has first to be removed from the bunch by hand picking. This is facilitated by first cutting up the bunch into its constituent spikelets and then allowing the fruit to loosen naturally for a day or two. The removal of the exocarp and mesocarp from the fruit is usually, though inaccurately, termed 'depericarping'. It can be carried out by scraping off the mesocarp with a knife, by retting in water, or by rotary screened depericarpers. The first method results in less damage to the seed, but is impracticable for large quantities.

In the retting process the fruit may be immersed in either running or standing water. Early experiments on the effects of these

methods were probably vitiated (as with other germination comparisons) by imperfect or varying germination techniques, but with the
perfecting of these techniques useful information has now been
obtained. There is no difference in final germination percentage or
in the rate of germination* following depericarping by hand or by
retting in standing or running water. The retting process is
completed earlier in running water and this method avoids the
housing of large numbers of evil-smelling retting tubs. On the other
hand a convenient constant-flowing stream or special means of
providing a flow of water may not be available for running-water
retting. Retting for 10 days is usually sufficient, but there is no effect
on subsequent germination of the length of the retting period up
to 20 days. In Africa, one species of fungus and one bacterium
predominate in the water during the retting process; towards the
end of the retting period, the fungus, which is aerobic and grows
on the surface of the water, excludes almost all other organisms.
The faster retting in running water appears to be due to these
conditions favouring the growth of the fungus against that of the
bacterium (Rees, 1962b). If retting is carried out in standing water,
the water should be changed once a day.

The retting process is completed by pounding the fruit with sand
in a wooden mortar to remove adhering mesocarp. With *dura* fruit
a clean product is obtained; with *tenera*, a quantity of the fibre
remains adhering to the smaller seed.

In centres of large-scale seed production mechanical depulpers
are used. These consist of hexagonal cages revolving at 30 revol-
utions per minute. The walls are made of expanded wire metal mesh
and water is fed to the cage to keep the fruit wet. The fibre removed
from the seed is hosed off the mesh as necessary by pressure hoses.

Depulping is followed by the drying of the seed to a moisture
content suitable for storage. Before drying, the seeds may be treated
by immersion and agitation in a solution of fungicide and bactericide
such as thiram and streptomycin for 3 minutes. There is no direct
evidence that this treatment is necessary, however, and as no
adverse effects of methods of harvesting or depulping have been
detected, it is thought that any loss of viability must be concerned
with the drying process. Drying of seed in full sunlight may be very
rapid and moisture contents below the normal air-dry percentage
may soon be reached. Although the surface temperatures reached
are unlikely to be lethal, the rate of drying may be an important
factor. Evidence for this is to be found in an experiment (Rees,
1965b) on the storage of differently dried lots of *dura, tenera* and

---

* Various terms such as the rate or speed of germination have been used in the literature. The best measure is the number of days to 50 per cent of final germination
percentage.

small *tenera* seed. Seed was dried, towards the end of a dry season, in the sun or in the shade for 1, 2 or 3 days. After 160 days' storage in 500-gauge polythene bags germination was still between 84 and 100 per cent for all lots of seed, but by the end of a year's storage viability had been almost totally lost in all sun-dried seed and in seed dried for 3 days in the shade. *Dura* seed dried for 1 or 2 days in the shade still gave 98 per cent germination; for *tenera* seed, however, more than 1 day's shade drying was deleterious. Seed should therefore be dried in the shade and care should be taken to see that even shade drying is not too prolonged. The usual practice is to spread the seed out on concrete or wooden floors or on wire racks for air-drying under shade for 2 days. With *tenera* seed a shorter period is advocated. Progenies are kept in separate trays. The seed is then ready for storage.

It may be mentioned here that no differences in viability or germination percentage have been consistently noted between seed from outer fruit and the rather smaller inner fruit, nor between seed of *dura* and *tenera* fruit form. *Tenera* seed may, however, be dried out too quickly if care is not taken and poor germination of some *tenera* samples is believed to be due to this. No seasonal effect has been generally found, nor does the age of the seed-bearing palm affect germination.

Outer seeds on a bunch tend to be larger and more are multi-kernelled. As already mentioned, there is no difference between the germination of large and small seed though very small seed often fails to germinate. A higher percentage of very small seed has no kernel or no embryo. Small numbers of 'white' seeds often appear which lack a normally lignified shell. Though these seeds germinate satisfactorily, the white-shell character, which is likely to be troublesome during shell and kernel separation in processing, may be inherited. Empty seeds and the majority of white seeds float in water and can thus be separated off and discarded.

Storage of seed

It was noticed by many early workers that viability deteriorated gradually, sometimes alarmingly, with length of storage time. The fact that stored seed germinates faster was not noticed so soon, however, probably owing to the imperfections of germination techniques. Vanderweyen (1952) stated that in Zaire there was a slight diminution of 'speed' of germination after 4 months' storage and that germination was markedly reduced after 10 months. Galt (1954), however, showed that stored seed germinates more readily than fresh seed and this has been interpreted as the effect of the high-temperature reaction proceeding at ambient temperatures.

Subsequent experiments in Zaire confirmed Galt's finding (INEAC, 1958).

A thorough investigation of storage had to await the devising of fully efficient germination techniques. Storage of seed for at least some months is often necessary because nursery seedlings of trans-plantable size may be required for planting out only at the beginning of a rainy reason. An extreme case is to be found in parts of West Africa having a single wet season where planting in the field is best carried out between March and May. In other parts of the world nurseries may be established in many months of the year, and there is more latitude in the time of planting in the field. Under these circumstances, problems of seed storage are not of such importance.

Seed has been conveniently stored on a large scale in 44-gallon (200 l) drums covered with a wooden lid. Such a drum will contain 25,000–30,000 African *dura* seed or a rather smaller quantity of Deli *dura* seed. The effect on subsequent germination of storing large quantities of seed at both ambient (*c.* 27 °C) and air-conditioned (22 °C) temperatures (i) when stored after normal air drying for $2\frac{1}{2}$ days in shade, and (ii) when stored at around the optimum moisture content (see p. 308) for germination has been investigated in an experiment (Rees, 1963c). Under all the treat-ments investigated, viability remained very high up to 12 months of storage. In the drums kept at ambient temperature with seed maintained or started at a moisture content near the optimum for germination, very considerable germination took place between 6 and 15 months of storage, amounting to between 56 and 72 per cent. Such treatment is therefore clearly impracticable.

Interest centres therefore on seed stored at ambient temperature after normal air drying to 14–17 per cent moisture or seed stored wet (21–22 per cent moisture) in an air-conditioned room at 22 °C. In the experiment referred to, where the original moisture contents were estimated visually, there was a net loss of moisture with time in the seed at the top of the drums, though this was almost negli-gible with the seed stored wet. In the middle and bottom of the drums there was a marked rise in moisture content, the levels reached being largely maintained for 12–15 months. Some results from these two treatments in the experiment are shown in Table 6.5.

The results show clearly that viability can be maintained at ambient temperatures at moisture contents too low for germination or at low temperatures under higher moisture conditions. In another storage experiment (Rees, 1965b) using these two temperatures, small lots of *dura, tenera* and small *tenera* seed were all stored at the optimum moisture contents for germination. Three storage treatments were used: (i) polythene bag under an inverted wooden box; (ii) polythene bag exposed; (iii) seed exposed to air. In all

Table 6.5 *Viability (as indicated by germination percentage) and moisture content of seed stored under two sets of conditions*

Original moisture contents*	Temperature	Position in drum	Months of storage				
			3 (%)	6 (%)	9 (%)	12 (%)	15 (%)
A. Germination percentage							
14–15%	27 °C (ambient)	Top	100	96	88	89	75
		Middle	100	95	96	95	95
		Base	100	94	95	97	96
21–22%	22 °C (air-conditioned)	Top	100	99	97	98	100
		Middle	93	94	91	92	98
		Base	86	86	90	91	96
B. Seed moisture contents							
14–15%	27 °C	Top	15.5	14.4	13.2	14.2	13.6
		Middle	21.4	18.2	18.0	18.2	19.1
		Base	20.4	19.3	17.7	18.4	18.6
21–22%	22 °C	Top	23.0	21.8	20.4	21.8	21.0
		Middle	24.5	22.1	23.0	24.3	24.6
		Base	24.6	24.4	24.4	24.8	22.4

* Visual estimation.

Source: Rees (1963c).

cases the seed dried out sufficiently to inhibit germination in storage. Between 92 and 100 per cent germination was obtained after 1 year's storage in all lots except two (one being destroyed by rats, the other unexplained). These experiments and the one referred to earlier (p. 315) clearly indicate that prestorage drying is likely to be more deleterious than any gradual drying after the start of storage; moreover storage temperature seems to be unimportant.

Where seeds are stored on a large scale without regard to careful drying, viability may be lost at over 1 per cent per week. It will be noted from Table 6.5 that this rate of loss was only approached in the top of the ambient-temperature drum when seed moisture content had fallen to below 14 per cent. Other work has shown that loss of viability during storage at about 1.2 per cent per week occurs at ambient temperatures and a moisture content at or below the air-dry level (about 14 per cent) (Rees, 1962b; WAIFOR, 1961).

In conclusion, it can be said that seed may be satisfactorily stored at ambient temperatures without appreciable loss of viability if it is carefully dried in the shade for not more than 2 days. While drum storage has proved quite satisfactory most large seed producers now prefer to store seed in polythene bags in an air-conditioned atmosphere at 22 °C. However, it must be emphasized that it has not been shown experimentally that air-conditioning has any advantage over ambient conditions. Seeds have also been satisfactorily stored on trays with air-conditioning to give a temperature of 20–22 °C and a humidity of 60 per cent.

Practical methods of germination

The requirements for a rapid and even germination of fresh seed are: (i) a temperature of 39–40 °C for 80 days at a moisture content too low for germination but not below about 14.5 per cent; (ii) transfer to ambient temperatures and the optimal moisture content for germination, i.e. 21–22 per cent for *dura* or 28–30 per cent for *tenera*. The optima for thinner-shelled *dura* may be intermediate (Odetola, 1974). These moisture contents are achieved in practice by keeping the seed as wet as possible but with no superficial moisture; alternatively samples may be taken for moisture determination by oven-drying to constant weight. Satisfactory germination is obtained with Deli seed in Malaysia after only 40 days' heat treatment, and where seed has been stored for 3 months or more shorter periods will suffice.

These requirements can best be met by keeping the seed in polythene bags and placing them in a germinator in which temperature control to within 1 °C can be assured. Electrically controlled germinators have proved most satisfactory, but germinators heated by hot water pipes can be satisfactorily operated provided some means of temperature warning is installed.

Germinators

Galt (1953) has given a comprehensive account of the germination methods and germinators in use up to 1953. Most of the germinators he describes, and the fermentation boxes and pits employed in Zaire and described by Marynen and Bredas (1955) and Desneux (1959), have fallen into disuse. For details of these historically interesting constructions their papers should be consulted. Germinators in which heat is applied within the body of the germinator may be divided into (i) fermentation pits or boxes, (ii) oven-type or flue-type germinators heated by a wood fire, (iii) boiler and hot water pipe germinators, (iv) electrically controlled germinators.

Fermentation box or pit germinators were widely used in Africa together with the use of wood charcoal as a medium in which the seeds were germinated. Mixtures of cut legumes and grasses were employed for supplying the heat of fermentation around the germinating boxes or drums. While satisfactory results could usually be obtained from the fermentation box when operated by practised hands, it was recognized (Desneux, 1959) that its greatest inconveniences were: (i) the need for constant knowledgeable supervision to maintain sufficiently high temperatures and to prevent overheating (temperatures of around 50 °C are sometimes reached); (ii) the difficulty of obtaining suitable green fermentable material. Both published results and experience have indeed shown that the germination percentages obtained were very variable; the most successful fermentation boxes were situated near to managers' houses. Desneux's pit design, which was 4 m long, 1.2 m wide and 75–90 cm deep and could accommodate 25,000–30,000 *dura* seed in five boat-shaped, charcoal-filled half-drums, was the most successful. However, even when the final percentage was satisfactory, germination proceeded slowly and often took 5–6 months to complete (Desneux, 1960).

Various crude methods of applying the heat of a wood fire to seed embedded in sand or moist wood charcoal were used in the early days of oil palm culture in West Africa. Oven-type germinators are units consisting of a number of upper and lower chambers, the latter for wood fires and the former for holding charcoal-filled seed boxes or polythene bags which are placed on slatted shelves. The chambers are separated by sheet metal covered with a cement layer. More satisfactory than the open-fire type of germinator were the various water-heated systems developed in West Africa. These germinators may be of any convenient size according to seedling requirements and are divided into the working shelter in which the boiler is situated, and the germination room. The latter is provided with large-diameter hot water pipes as in normal glasshouse practice. The boiler is of a type used for domestic hot water systems and it is fired by wood. The walls can be constructed of concrete blocks and a double ceiling must be provided. Any windows must be double-glazed. The hot water pipes run beneath the staging on which are placed the boxes containing the polythene bags of seed. Passage ways may run on both sides of central staging or down the middle of the room with staging on the sides.

Temperature control in water-heated germinators is easier than in open fire-heated germinators, but results still depend on the integrity of the operator and adequate supervision of his work. Improved control can be obtained by the use of two thermostats connected to warning lights which light up if the temperature rises above or falls below certain points. This does not entirely eliminate the 'human factor', however, and although water-heated germinators can be generally recommended, electrically controlled germinators are much more satisfactory.

A small electric germinator was first constructed at NIFOR in Nigeria in 1958 (Rees, 1959b). A description of this germinator is given below. The building is constructed of concrete blocks to the

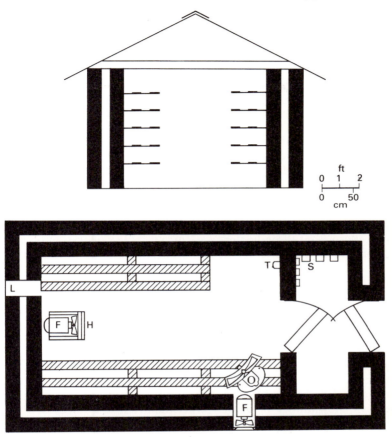

Fig. 6.4 An electrically controlled germinator (Rees, 1959b). Median section and plan: F, fan; H, heater; O, oscillating fan; T, thermostat; S, switches and relays; L, automatic louvre.

dimensions in Fig. 6.4. The walls are double with a 13-cm (5-in) air gap and the ceiling is of double asbestos sheeting. The double doors are hollow with a 13-cm (5-in) air space, and they enclose an air lock where the fuse boxes, main and subsidiary switches, relays and time switch are housed. The internal working space is 3.9 × 2.2 × 2.0 m high and there are 61-cm (2-ft) wide slatted shelves on each side of a 1.22-m (4-ft) central gangway.

The placing of the heater, the air intake louvre, the oscillating fan and the air exit fan are of great importance and an even temperature on the shelves is not obtained unless the design is followed and the apparatus operated correctly. The design originally allowed for a humidifier which was tried out because it was thought that it might be possible to germinate naked seed on the shelves without the protection of polythene sheeting. It was found, however, that seed dried out in spite of use of the humidifier and full germination of viable seed was only obtained by use of polythene bags. Moreover, the constant high humidity in the germinator led to rapid depreciation of apparatus and materials.

The germinator is heated by six 500 W strip heaters mounted in front of a 1/6 hp fan, the whole unit being placed on the floor against the back wall. Aeration is achieved periodically by a 1/6 hp exhaust fan near the top of one wall and an automatic louvre near the heater at floor level; the louvre opens when the exhaust fan is in operation. The fan and louvre are operated by a time switch. It was found that there was only a very slight drop in temperature when the extractor fan was in operation. The building is heated rapidly from cold by all six strip heaters; at 1 °C below the setting of the thermostat three heaters are cut out by the operation of one-half of a two-stage thermostat mounted on the wall. Temperature distribution is achieved by the oscillating fan on the top platform. The right positioning of this fan will remove any 'islands' of too high or too low a temperature and reduce the temperature range to 1 °C.

A small germinator of this type, holding up to 200,000 seeds in polythene bags, is inexpensive to construct. Power consumption per week is less than 300 kWh.

It should be noted that in constructing electrically controlled germinators of different sizes certain essential features must be retained, viz. double walls and ceiling, air lock with double doors and control panel, heater on floor and exhaust fan near top of shelves. The number of heater strips required, the stages of thermostatic control, and the obtaining of a uniform distribution of temperature will all require trial *before operation*, using oscillating fans in different positions until all 'pockets' have been eliminated. It is only when a uniform temperature of 39–40 °C has been achieved that germination should be started.

Germination technique

Assuming that a germinator can be consistently operated to give the correct temperature for heat treatment, then germination may be carried out by one of two techniques which may be designated (i) wet-heat treatment, and (ii) dry-heat treatment. In the former case the whole process will be shorter (95 days against 120 days), but some germination will take place during the heating period and so a less uniform set of seedlings will be produced. Watering will be unnecessary throughout the heating period with the dry-heat treatment technique so there will be considerable economies of watering time. Better results are almost invariably obtained with this technique as the need for gauging the optimum moisture content through the heat period is obviated.

Wet-heat treatment

The seed should be soaked for 5 days in buckets or pans with a daily change of water, then dried carefully in the shade for about 2 hours until adhering water evaporates, but the seed is still almost black. Polythene bags to receive this seed are most conveniently made from 69 cm (27 in) wide 500-gauge lay flat tube. This is cut into 61 cm (24 in) lengths and one cut edge is then sealed. Five hundred to 700 *dura* seeds or about 1,000 *tenera* seeds can be put in these bags, the open end of which is then secured by a stout elastic band made from cut-up cycle inner tube. The bags are then placed in boxes of convenient size (30 × 20 × 10 cm) and put in the germinator where the temperature should be 39 °C. The seed is examined twice weekly and watered with a fine spray from a small hand mist sprayer to restore the correct moisture content (Plate 6.1). This may be defined as – *as wet as possible but with no superficial moisture.*

Pl. 6.1 Small *tenera* seeds having their moisture content adjusted, on inspection, by a fine spray.

Alternatively, samples may be taken for moisture determination by oven-drying to constant weight. Moisture contents should be 21–22 per cent for *dura* and 28–30 per cent for *tenera*. Any germinated seed is removed for sowing in the prenursery or nursery.

After 80 days the boxes and bags should be taken out of the germinator and put in a cool place and the correct moisture content maintained as before. A flush of germination will start a few days after cooling and, if watering in the germinator has been done with care, germination will be complete in about 15–20 days.

Stored seed requires less heat treatment than fresh seed, so if some of the seed being germinated has been stored for a few months, more will germinate before cooling and the flush of germination of the remainder will not be so pronounced.

Dry-heat treatment

The technique of dry-heat treatment entails the heating of seed at 39 °C when it is in a condition too dry for germination, but not so dry that the heat reaction is impaired. The seed must be soaked for 5 days with a daily change of water, as for the wet-heat treatment. It should then be spread out in a single layer and dried for 24 hours in the shade. A moisture content of 17 per cent has been found satisfactory in Malaysia with Deli *dura* (Turner and Gillbanks, 1974), but a rather higher content has been suggested in Nigeria (Odetola, 1974). *Excessive drying must be avoided.* The seed is then placed in polythene bags as previously described. These must be checked from time to time to see that no seam has become unwelded nor rubber band broken.

After 80 days the seed is removed from the bags, soaked again in water for 5 days with a daily change of water, then dried slightly in the shade for about 2 hours to evaporate surface water. The amount of drying needed before returning seed to the bags has been described as 'until their colour passes from shiny to dull black' (IRHO, 1970). At this stage the seed may be retreated with TMTD and streptomycin (see p. 313). It is then returned to the bags, kept in a cool place, and examined twice a week to maintain, by means of the fine spray, the correct moisture content for germination (Plate 6.1). There will be a delay of some 10–20 days after the 80 days; germination should then be rapid and be complete in 10–15 days (Plate 6.2).

Jars of 600 ml capacity covered with 200-gauge polythene sheet may be used for germinating small lots of seed.

An additional advantage of the use of the dry-heat treatment is the avoidance of Brown Germ disease of just germinated seedlings caused by *Aspergillus fumigatus*. This fungus shows optimum growth at a temperature near that of heat treatment, but when seed is germinating at ambient temperature following dry-heat treatment,

Pl. 6.2 A flush of germination in a polythene bag, following the cessation of heat treatment.

growth is usually too slow to cause serious infection before the non-susceptible stage is reached.

Length of period of heat treatment

Though 80 days is normally required for fresh African *dura* seed, stored seed requires a shorter period and some lots of seed, whether fresh or stored, have been found to have lower heat requirements. Experience with selected Deli seed in Malaysia has shown that 40 days is normally sufficient to provide over 80 per cent germination and is now standard practice in some germinators.

To obtain the most complete and rapid germination possible with a given lot of seeds the wet-heat treatment may be combined with a sampling procedure devised by Tailliez (1970). This may be useful if seedlings are urgently required and there is reason to believe that the heat period requirement may be well below 80 days. Samples are taken from the germinator weekly and kept at ambient temperature. As soon as any such sample has attained 80 per cent germination (and this may be expected within 20 days if the heat requirement has been fulfilled) all the seed may be taken out of the germinator to ambient temperature. The method was devised for use with a wet charcoal substrate, but could also be employed with the polythene bag wet-heat method.

Germination and embryo abnormalities

Examination of the seeds of large numbers of Deli *dura* × *pisifera* crosses has shown that embryo abnormalities may consist of malformation of the whole or part of the embryo, incomplete development, double or triple embryos, a tendency to necrosis or the complete absence of an embryo. A direct relationship was found between both the percentage and speed of germination and the percentage of normal embryos (Noiret and Ahizi Adiapa, 1970). Crosses with a low percentage of normal embryos show a high variability in embryo length, suggesting that a low embryo normality and hence low germination percentages indicate that a progeny is strongly heterogeneous. While this may be so, it should also be borne in mind that both the length of time of pollen storage and the methods of pollen preparation used have an effect on the percentage of normal embryos.

Noiret and Ahizi Adiapa (1970) suggest that in seed production one cross in five should be examined for abnormalities. They test successive samples of 50 seeds and eliminate a cross unless 0, less than 4, less than 7, or less than 20 abnormal seeds are found in samples 1, 1 + 2, 1 + 2 + 3 or 1 + 2 + 3 + 4 respectively. This is equivalent to the imposition of a standard of normality of 90 per cent. If one or more crosses have to be eliminated all crosses made in the same month are checked.

Germination of *pisifera* seed

Plant breeders occasionally wish to propagate *pisifera* × *pisifera* progenies. *Pisifera* seeds are difficult to germinate because they tend to dry out rapidly and are susceptible to fungal and bacterial infection. A method developed in Malaysia has given germination percentages of fertile seeds of from 5 to 52 (Arasu, 1979). The percentage of fertile seed varied between parents and between bunches but rarely exceeded 60 per cent.

The seeds are extracted from the fruit and cleaned to remove all mesocarp, great care being taken not to nick or bruise the testa. They are washed and rinsed in 0.05 per cent sodium hypochlorite before being put into polythene bags. They are then treated in the germinator at 39–40 °C by the wet-heat method and individual seeds are removed as germinated or if showing any signs of fungal or bacterial infection. Superficial fungal growth can be removed by washing when the seeds are examined and the likelihood of penetration to the embryo is thus reduced.

In Ghana, as high a germination percentage as that already quoted was obtained simply by allowing bunches to lie under the palms during the wet season and protecting them from rodents with wire netting. In this case the mesocarp was clearly giving some protection to the kernel and the heat requirement was being fulfilled by the heat of fermentation (Wonkyi-Appiah, 1973).

Germination of *Elaeis oleifera* seed

The germination percentages obtained with *E. oleifera* seed have usually been low, and in general the methods used for *E. guineensis* have been used. It has been reported that in Colombia very satisfactory germination has been obtained by adopting the dry-heat treatment with 22 per cent moisture content for an initial 15 days, then soaking in warm water at 43 °C for 15 minutes and following this with the wet-heat treatment for 65 days (Chew Poh Soon, 1976).

Despatch of germinated or preheated seed

Seed can be sent long distances after preheating or when germinated. In Malaysia, newly germinated seed is usually placed in 200-gauge polythene bags with 200 seeds per bag (Bevan and Gray, 1966). These are packed in boxes with vermiculite at the rate of 15 bags (3,000 seeds) per box of 18 × 12 × 11 inches (46 × 30 × 28 cm). The advantage of this method is that the seed can remain in the bags for up to 10 days before planting, though the plumule and roots should not be allowed to grow so long that they become twisted.

In the Ivory Coast a box has been designed for the rapid distribution of germinated seed by motor vehicle to distant plantations where prenurseries are being established (Sérier, 1966). The box consists of a framework 50 × 30 cm in which are inserted twelve shallow trays, 3.5 cm high, each containing 250 germinated seeds embedded in sawdust. Thus a consignment of 30,000 in 10 boxes can readily be despatched in one vehicle.

Germinated seed should be despatched as soon as germination has started and if possible not later than the fourth day, when the root and plumule can just be distinguished (D of Fig. 2.1, p. 54). This will enable planting to be done up to 10 days later when the root and plumule are clearly differentiated (F–G of Fig. 2.1).

Germinated seed can be badly damaged by treatment with seed dressings containing fungicides or insecticides, and Brown Germ cannot be controlled by such methods (see p. 581). Gamma-BHC is particularly dangerous and should not be used either with seed or in prenurseries.

Seed can also be sent in polythene bags at the end of the dry-heat period and its moisture content can be adjusted to the optimum for germination on arrival at its destination.

If seed is sent from one country to another the sanitary regulations of the receiving country must be observed. Seed is normally treated with fungicides and insecticides but fumigation with methyl bromide has been found satisfactory provided the seed moisture content is not more than 10 per cent. Experiments have shown that such seed may be fumigated at 24–30 °C with 1 kg methyl bromide per 30 m^3 for up to 18 hours without affecting viability, germination or growth (Mok Chak Kim, 1970). Methyl bromide has also been employed for the fumigation of germinators between periods of usage. Mercurial dressings should not be used. Fungicide and insecticide treatment may consist of washing for 2–3 minutes in 0.2 per cent thiram with 0.1 per cent trichlorphon and including a spreader, e.g. Teepol, at 0.1 per cent. It is important to have the seed as clean as possible before treatment and it should therefore be prewashed in Teepol.

Locke and Colhoun (1973) have shown that *Fusarium oxysporum* f.sp. *elaedis*, the pathogen of Vascular Wilt disease (see p. 607), can be seed-borne. They isolated the fungus from seed despatched from West Africa to England and showed that even where routine seed disinfection had been carried out the seeds yielded large numbers of other fungi, though not *F. oxysporum*. This study showed that some current methods of disinfection are inadequate and that the most stringent procedures are required.

References

Arasu, N. T. (1970) A note on the germination of *pisifera* (shell-less) oil palm seeds. *Malay. agric. J.*, **47**, 524.

Bevan, J. W. L. and **Gray, B. S.** (1966) Germination and nursery techniques for the oil palm in Malaysia. *Planter, Kuala Lumpur*, **42**, 165.

Bouharmont, P. (1959) La culture d'embryons d'*Elaeis guineensis* Jacq. *Agricultura, Louvain*, **7**, 297.

Bouvinet, J. and **Rabéchault, H.** (1965) Recherches sur la culture 'in vitro' des embryons de palmier à huile. II. Effets de l'acide gibberillique. *Oléagineux*, **20**, 79.

Bücher, H. and **Fickendey, E.** (1919) *Die Ölpalme.* Berlin.

Chew Poh Soon (1976) Private communication.

Desneux, R. (1959) Une méthode simplifiée pour la germination des graines du palmier à huile. *Bull. Inf. INEAC*, **8**, 23.

Desneux, R. (1960) A propos d'une méthode simplifiée de germination des graines du palmier à huile. *Bull. Inf. INEAC*, **9**, 43.

Ferwerda, J. D. (1956) Germination of oil palm seeds. *Trop. Agric. Trin.*, **33**, 51.

Galt, R. (1953) Methods of germinating oil palm seed. *J. W. Afr. Inst. Oil Palm Res.*, **1** (1), 76.

Galt, R. (1954) WAIFOR report (unpublished).

Henry, P. (1951) La germination des graines d'*Elaeis*. *Rev. Int. Bot. appl. Agric. trop.*, **31**, 349.

Henry, P. (1951 and 1952) La germination des graines d'*Elaeis*. *Revue Int. Bot. appl. Agric. trop.*, **31**, 565; **32**, 66.

Hussey, G. (1958) An analysis of the factors controlling the germination of the seed of the oil palm, *Elaeis guineensis* (Jacq.). *Ann. Bot.*, NS, **22**, 260.

Hussey, G. (1959) The germination of oil palm seed: experiments with *tenera* nuts and kernels. *J. W. Afr. Inst. Oil Palm Res.*, **2**, 331.

INEAC (1958) La conservation des graines d'*Elaeis* par la division du palmier à huile. *Bull. Inf. INEAC*, **7**, 31.

IRHO (1970) Germination des graines de palmier à huile en sacs de polyéthylène. Méthode par 'chaleur sèche'. *Oléagineux*, **25**, 145.

Labro, M. F., Guénin, G. and **Rabéchault, H.** (1964) Essais de levée de dormance des graines de palmier à huile (*Elaeis guineensis* Jacq.) par des températures élevées. *Oléagineux*, **19**, 757.

Locke, T. and **Colhoun, J.** (1973) *Fusarium oxysporum* F.sp. *elaeidis* as a seed-borne pathogen. *Trans. Br. mycol. Soc.*, **60**(3), 594.

Marynen, T. and **Bredas, J.** (1955) La germination des graines d'*Elaeis*. *Bull. Inf. INEAC*, **4**, 156.

Mok Chak Kim (1970) Effects of methyl bromide on oil palm seed. *Proc. Int. Seed. Test. Ass.*, **35**, 243.

Noiret, J. M. and **Ahizi Adiapa, P.** (1970) Anomalies de l'embryon chez le palmier à huile. Application à la production de semences. *Oléagineux*, **25**, 511.

Odetola, A. (1974) A re-examination of large-scale germination of the oil palm seed at NIFOR. *J. Nig. Inst. Oil Palm Res.*, **5**, 79.

Rabéchault, H. (1962) Recherches sur la culture 'in vitro' des embryons de palmier à huile (*Elaeis guineensis* Jacq.) I. Effets de l'acide β indolylacétique. *Oléagineux*, **17**, 757.

Rabéchault, H., Guénin, G. and **Ahée, J.** (1967) Absorption de l'eau par les noix de palme (*Elaeis guineensis* Jacq. var. *Dura* Beec.) 1. Hydration des différentes parties de graines amenées à teneurs globales en eau determinées. *Cah. Orstom, sér. Biol.*, No. 4, 31.

Rees, A. R. (1959a) The germination of oil palm seed: the cooling effect. *J. W. Afr. Inst. Oil Palm Res.*, **3**, 76.

Rees, A. R. (1959b) The germination of oil palm seed: large scale germination. *J. W. Afr. Inst. Oil Palm Res.*, **3**, 83.

Rees, A. R. (1962a) High temperature pretreatment and the germination of seed of the oil palm, *Elaeis guineensis* (Jacq.). *Ann. Bot.*, NS, **26**, 569.

Rees, A. R. (1962b) Some observations on preparation and storage of oil palm seed. *J. W. Afr. Inst. Oil Palm Res.*, **3**, 329.

Rees, A. R. (1963a) Some factors affecting the germination of oil palm seeds under natural conditions. *J. W. Afr. Inst. Oil Palm Res.*, **4**, 201.

Rees, A. R. (1963b) A note on the fate of oil palm seed in a number of habitats. *J. W. Afr. Inst. Oil Palm Res.*, **4**, 208.

Rees, A. R. (1963c) A large scale test of storage methods for oil palm seed. *J. W. Afr. Inst. Oil Palm Res.*, **4**, 46.

Rees, A. R. (1965a) Private communication.

Rees, A. R. (1965b) Some factors affecting the viability of oil palm seed in storage. *J. Nigerian Inst. Oil Palm Res.*, **4**, 317.

Sérier, J. B. (1966) Transport à longues distances des graines germées de palmier à huile. *Oléagineux*, **21**, 211.

Tailliez, B. (1970) Germination accélérée des graines de palmier à huile. Technique avec substrat. *Oléagineux*, **25**, 335.

Turner, P. D. and **Gillbanks, R. A.** (1974) *Oil palm cultivation and management.* Incorp. Soc. of Planters, Kuala Lumpur.

Vanderweyen, R. (1952) *Notions de culture de l'Elaeis au Congo Belge.* Brussels.

WAIFOR (1961) *Ninth annual report 1960–1*, p. 92.

Wan, C. K. and **Hor, H. L.** (1983) A study on the effect of certain growth substances on germination of oil palm (*Elaeis guineensis* Jacq.) seeds. *Pertanika (Malaysia)*, **6**(2), 45.

Wonkyi-Appiah, J. B. (1973) Germination of *pisifera* oil palm seeds under plantation conditions. *Ghana J. agric. Sci.*, **6**, 223.

Chapter 7

The raising of nursery seedlings

Early bearing in the field is dependent on the transplanting of healthy seedlings from a nursery. A great deal of attention was therefore paid to nursery techniques, and various systems were developed in different parts of the world in response to differences of climate, soil, disease incidence and management. Damage and exposure of the root system of the palm at any stage in the nursery process is followed by a slow recovery, and so all early nursery systems called for the minimum disturbance of the roots throughout the nursery period and during transplanting to the field.

Seedlings are more difficult to raise in the seasonal climate of Africa than in the more uniform climate of the Far East and they are much more subject to nursery diseases. At first therefore more research was done on nursery methods in West Africa than elsewhere, and for a long period the standard method of raising seedlings was to plant germinated seeds in prenursery beds or pots and when the seedlings reached the four- to five-leaf stage to transfer them to a specially prepared field nursery where they grew on for about a year. In the Far East, however, satisfactory methods of raising seedlings in large polythene bags were developed in the mid-1960s and this system, with or without a prenursery, has become universal. Growing seedlings in polybag nurseries was gradually found to be more convenient and less costly than field nurseries; moreover transplanting from field nurseries requires more labour and special techniques and cannot be satisfactorily accomplished if the plants have to travel any great distance. Direct planting of germinated seed or very small seedlings in the field is possible but for three reasons is not a practical method: firstly, because the young plants cannot easily be protected in the field from insects and rodents; secondly, because such very young material will produce an uneven stand; and thirdly, because the time between the completion of field preparation and first cropping would be much longer than with the planting of large nursery seedlings.

Prenurseries

Prenurseries may be constructed of bags, baskets or beds or the prenursery stage may be omitted altogether. Few comparative trials have been carried out and healthy plants can be raised in either bags or beds, so the decision is usually dictated by conversance or managerial convenience. The prenursery methods to be adopted depend in some measure on the stage at which the planting material is distributed. Seedlings at the two-leaf stage may be sent out from the distribution centre either packed in wet soil in boxes or as bare-root seedlings packed in polythene bags. In either case the seedling will have to be planted in a small basket or polythene bag, unless the prenursery stage is omitted altogether, and this extra trans-planting causes a severe check to growth. Nevertheless, by careful handling and the use of a suitable potting mixture, an even set of seedlings can be raised from plants at the two-leaf stage; however, the direct establishment of prenurseries from germinated seeds is much to be preferred, and in this case either bags, beds or raised trays can be used.

The obtaining of a suitable potting or bedding soil may present some difficulty. A wide variety of soils and mixtures have been successfully used of which the following may be mentioned:
1. Sandy topsoil partially sterilized by heating over a fire (Nigeria).
2. Deep friable topsoil, overlying alluvial clay, undiluted or mixed with a small proportion of coarse river sand (Malaysia).
3. Stiff clay topsoil mixed with coarse river sand in proportions 3:2 (Malaysia).
4. Peat and sand mixed in equal proportions (Malaysia).
5. Sandy topsoil (Zaire).
6. Inland clay-loam topsoil mixed with river sand and well-rotted cattle manure (Malaysia).
7. Sifted forest topsoil (Ivory Coast).

In general a fertile topsoil sufficiently free-draining to prevent 'puddling' or sealing of the surface is required. In areas where soils tend to be too heavy the admixture of a proportion of coarse sand is always desirable. On the other hand many soils of good structure have been successfully used without mixing with sand and if too much sand is used the soil may break up on transplanting to the nursery.

Baskets and polythene bags

Baskets, about 25 cm deep and 10 cm broad can be used but black polythene bags of 250 gauge and of similar size (25 × 10 cm lay-flat) are now preferred. If bare-root seedlings are being planted some of the soil medium to be used is placed at the bottom of the bag and the seedling is then held in position while further soil is packed round it, layer by layer, great care being taken to see that none of the primary roots is fractured in the process. Root fracture is the main cause of blackening off of seedlings transferred to bags. Some damage to the root system is inevitable in transplanting bare-root seedlings and, for this reason only, it may be necessary to provide some light shade for about 2 weeks after planting. Where germinated seed is planted no shade is required provided the supply of water is sufficient; however, if overhead irrigation is not installed or water supplies are likely to be restricted, then a light shade of palm leaves or shade cloth erected on frames may be required for a few weeks and should be removed in stages. It should be emphasized that entirely satisfactory prenursery plants have been raised without shade and that over-shading causes etiolation and depression of growth.

Baskets rot very rapidly under constantly moist conditions and they may sometimes be in an advanced condition of decay by the time the seedlings are ready for transference to the nursery. Great care has therefore to be taken in transporting them. Black polythene bags do not have this drawback. If the bags are properly filled in the first place, the growth of the root system tends to assist the formation of a compact ball of soil within the polythene film. When the polythene is torn away just before planting, this ball retains its shape and can be inserted in the nursery planting hole without fracturing. Basket plants are planted in the nursery in their baskets.

Bags should be filled to within a centimetre from the top. The germinated seed is placed in a hole made with the finger; this hole is about 2.5 cm deep and in the centre of the bag. The seed is then covered in with soil. The radicle and plumule of the germinated seed should be clearly differentiated and care taken to see that the plumule is pointing upwards and the radicle downwards; twisted seedlings are thus avoided (Plate 7.1). A just-germinated seed, where only a 'button' is showing, may be returned to the polythene bag for a few days until it has developed further; if it is planted immediately it should be laid on its side, i.e. with its long axis horizontal. The radicle and plumule will then grow respectively downwards and upwards with the minimum of bending. One advantage of planting as soon as germination has taken place is that damage in handling is less likely to occur. A disadvantage, apart from the possibility of bent growth, is that infection by Brown Germ (*Aspergillus* sp.) may not be noticed. By the time the plumule has developed, however, any infection will be obvious and the infected seed should be discarded. Each seed should be carefully examined for abnormalities of growth and development. If such abnormalities are rigorously excluded the amount of roguing needed in the prenursery will be reduced.

Pl. 7.1   Planting a germinated seed in a prenursery.

The use of polythene bags at the prenursery stage has been found particularly useful when germination is carried out centrally and nurseries are to be established far from the germinator and near to large plantings. Bags may be arranged in plank-walled 'beds'. In the Ivory Coast beds of 20 lines of 175 bags (3,500 bags) have been found convenient. The bags are transported to the nursery sites when the seedlings have reached the five-leaf stage (Ruer, 1963).

Prenursery beds

Prenursery beds are now less commonly used. The raised beds, which are 20–23 cm deep and for convenience should not be more than 1.2 m wide, require low retaining walls of brick or wood. The ends are removable to facilitate transplanting. Such beds will be satisfactory for an estate where a finite programme of planting is going to be carried out over a period of a few years; where, however, the raising of prenursery seedlings is going to be a continuous process it is more convenient to set aside an enclosed area for growing seedlings in raised trays (Plate 7.2). A permanent system of mist irrigation can then be constructed and the seedlings can be planted and later dug up with great ease, while the marking of progenies is greatly facilitated. Experiments and experience have shown that a more uniform set of seedlings is obtained in trays than in beds. The trays should be raised 60 cm (2 ft) off the ground and be 15–20 cm deep; a convenient size is 120 × 60 cm (Anon., 1953).

The best medium for prenursery beds or trays is a sandy topsoil, but any of the mixtures described at the beginning of this section may be used. The soil should be partially sterilized before being put

Pl. 7.2   Overhead spray lines in operation over a raised-tray prenursery in Nigeria.

in the beds or trays. This is carried out quite simply by baking on a steel sheet; weed seeds and insects are killed by this treatment (Toovey, 1947). In Zaire parathion at a concentration of 8 g to 10 litres water has been used (Dupriez, 1956) and this method of insect control is also advocated in the Ivory Coast; in Malaysia, however, it is believed that the young growing point may be endangered and the application of insecticides as granules scratched into the surface soil is recommended.

Transplanting to the nursery should be at the four- or five-leaf stage and, for this, the plants must be spaced at 7.5 or 10 cm. Seedlings so spaced can be lifted and transferred to the nursery in soil blocks and their root systems are thereby protected (Plate 7.3). At

Pl. 7.3   Removing seedlings from a raised-tray prenursery in soil blocks.

7.5 cm, 128 seedlings may be raised in a tray of internal measurement 120 × 60 cm; this is equivalent to 177 seedlings per m² of bed. At 10 cm apart 81 seedlings are accommodated per m². Planting of germinated seed is carried out in the same manner as when using bags.

Maintenance of the prenursery

Shading of beds or bags is not required when overhead irrigation is supplied (see Plates 7.2 and 7.4) or when hand watering is adequate. Mulching, on the other hand, has been shown to be beneficial and mulched seedlings subsequently grow better in the nursery (Gunn *et al.*, 1961; Dupriez, 1956). The most satisfactory mulching material is finely divided bunch refuse. This may be applied soon after planting and the seedlings grow up easily through it. If bunch refuse is not available, sawdust, palm shell, groundnut husk or other fibrous material may be used.

The primary requirements for maintaining a steady growth after emergence of the first leaf are adequate watering and a balanced

Pl. 7.4 Large polythene bag prenurseries under irrigation in the Jengka Triangle area, Pahang, Malaysia.

supply of fertilizer. Mulching is a supplementary factor which conserves moisture, prevents soil compaction and, to a lesser degree, provides nutrients.

The frequency of watering depends, of course, on the amount and frequency of rain falling during the prenursery period. Adequate watering arrangements must be available, however, and provided the soil used is not too heavy and drainage is unimpeded, there is little danger of overwatering. In severe dry weather twice-daily watering will be needed at the rate of one 18 litre watering can per two trays or 2 m² of bed or, when bags are used, to 'run off'. Can-watering is inevitably uneven and in any permanent pernursery site an irrigation system should be installed. In Malaysia it is considered that irrigation equipment is justified for plantings of 400 hectares or more (Bevan and Gray, 1966). A simple system for a permanent raised tray site employed polythene tubes and spray nozzles designed to give a fine mist spray over the whole area (Sly and Sheldrick, 1961). Tubes of 1½ inches (3.8 cm) bore are cut into 9–12 m lengths and are jointed together with fittings provided by the suppliers of the tube; they are connected directly onto the water supply which should have a pressure of 20 psi (1.4 kg per cm²) or more. The tubes are supported 1 m above the seedlings; frequent supports are required (Plate 7.2). The spray line can be moved as needed or lines can be installed to cover the whole area. Pairs of two types of nozzles, inserted alternately and spaced 60 cm apart along the line, have been found to cover the area evenly. For a large temporary site ordinary rotary sprinklers as used in nurseries can also be employed in the prenursery (Plate 7.4). With hand watering, labour can be reduced by using a hose-pipe with a fine rose.

If very fertile forest topsoil is used for filling prenursery beds or bags, fertilizer application may not be needed, but such soil is rarely obtainable and, in most media used, prenursery seedlings have shown marked responses to fertilizer, particularly to nitrogen. The amounts required are very small. Experiments in Nigeria and Malaysia have shown responses to nitrogen and phosphorus, particularly as ammonium phosphate, but not to potassium. A solution of 2 oz of ammonium phosphate in 4 gallons of water (57 g in 18 litres) should be applied to four standard trays or per 4 m^2 of bed every week. If sulphate of ammonia is used 2 oz in 4 gallons (57 g in 18 litres) every 2 weeks may be applied. The beds or trays should be lightly watered afterwards to remove the fertilizer from the leaves. With bag seedlings in Malaysia urea used at $\frac{1}{2}$ oz in 1 gallon (14 g in 4.5 litres) of water per 100 seedlings per week.

Since seedlings are in the prenursery for such a short time very little weeding is usually required. Hand weeding is therefore resorted to. However, in the Ivory Coast it is claimed that application of ametryne (Gesapax) will reduce costs to 10 per cent of those of hand weeding (Taillez, 1969). This herbicide can be applied before the emergence of the plumule at a rate equivalent to 2–3 kg of the commercial product per hectare or after the emergence of the plumule but before the appearance of the first leaf. In the latter case the concentration is of importance and the herbicide is watered on at a strength of 0.1–0.3 per cent; for 10,000 seedlings an average of 18 g of the commercial product in 9 litres water is sufficient.

In Africa, Anthracnose fungi (see p. 582) may be troublesome in the prenursery and regular weekly spraying with Captan or Cuman organic fungicides starting about 6 weeks after planting has been found effective. In the Far East *Curvularia* Leaf Spot is occasionally encountered. Pests such as snails, grasshoppers and night-flying beetles are usually controlled by hand collection.

Nurseries

Early experience of transferring prenursery seedlings to a nursery of pots or baskets varied, and field nurseries were usually found to be more successful (Vanderweyen, 1950; Marynen and Poels, 1960; Gunn *et al.*, 1961). In Africa the use of baskets was found to be conducive to a high incidence of both Blast and Freckle (*Cercospora* Leaf Spot). Field nurseries thus became the rule until the mid-

1960s, when they gradually gave way to the 'polybag' nursery, and they are now rarely employed. For this reason they will be only briefly described here.

Field nurseries

Field nursery sites must be flat and adjacent to a good water supply, the drainage must not be impeded, and the soil must be friable and capable of forming a good ball of earth at transplanting time. If the area is opened from forest, felling and burning must be carried out as rapidly as possible, and concentrations of ash from the burn must be dispersed. Following this standard treatment, the area must be cultivated and manured. In the deep sandy soils of West Africa it was found that deep cultivation (25 cm) with a disc plough increased growth and that growth was further assisted by the ploughing in of organic manure, usually in the form of bunch refuse, at the rate of 125 tonnes per hectare (Gunn *et al.* 1961). On clay soils, whether inland or coastal, in the Far East, the depth of ploughing is largely determined by the depth of topsoil, it being undesirable to bring to the surface large clods of subsoil.

Experiments in Nigeria (Gunn, 1960; Gunn *et al.*, 1961) have shown that a density of more than about 17,000 plants per hectare (7,000 per acre) is disadvantageous, and square planting at 80–90 cm is advocated. Close spacing also makes disease control more difficult and this is of particular importance in Africa where Freckle is troublesome.

The layout of a field nursery will depend on whether or not irrigation equipment is to be used. With such equipment paths must be left along the lines of piping with occasional cross paths for inspection. Usually nursery sites are not re-employed, but at experiment stations semi-permanent sites have been established, parts of the site being fallowed for several years between usage. This is quite satisfactory at least for three or four rounds over a 20-year period in spite of the removal of soil and provided plenty of bunch refuse is available for mulching or ploughing in.

Mention has already been made in Chapter 4 of the water requirements of young seedlings. In West Africa the equivalent of 1 inch (25 mm) of rain per week has proved adequate for well-grown seedlings in field nurseries though newly planted seedlings require more, and in the 'Harmattan' conditions of the dry season the requirement is even higher. With hand watering, a minimum of 1 gallon (4.5 litres) per seedling per week should be given and in very dry weather this should be increased to 2 gallons (9 litres). In the Ivory Coast, irrigation in the dry season at the rate of 20 litres (about $4\frac{1}{2}$ gallons) per week per m^2, equivalent to 2 cm of rain, was recommended with a planting distance of 80 cm (Fraisse, 1962).

Later, however, recommendations based on estimations of stomatal aperture were adopted (see p. 147) (Ochs, 1963). Hand watering is costly, about 1,000 man-days per hectare per year being required under Nigerian conditions.

The mulching of field nurseries, whether with organic material or artificial mulches, improves seedling growth and reduces the need for weeding. The most effective mulch is bunch refuse, but if not available any cut vegetation may be used, large-leaved grasses being particularly suitable, or an artificial mulch of black polythene sheeting may be laid down. Mulching is most effective when applied 2–4 weeks after planting and maintained or replenished at least until the seedlings are fully covering the ground. There is some evidence that if the supply of bunch refuse is limited it is better used as a mulch than ploughed in before planting. Where soil potassium and magnesium levels are low, bunch refuse mulches may induce symptoms of magnesium deficiency (Orange Frond) and fertilizer dressings must then contain a sufficiency of magnesium.

Except where unusually fertile jungle soil is being employed for the first time, fertilizers, particularly nitrogen, are always required for maximum growth of plants in a field nursery. On the poorer sandy soils of West Africa applications of organic manure before planting, and of mulch after planting, seem to have a greater effect than fertilizers on growth. In most oil palm growing areas growth is improved by applications of N, P, K and Mg. Applications of N and K without Mg tend to induce magnesium deficiency symptoms which are common in field nurseries. Phosphorus as superphosphate has given increases of growth and enhanced the responses to nitrogen. Potassium applications improve growth as well as reducing the incidence of Freckle. Magnesium applications, although they have not been shown to improve growth, remove the symptoms of Orange Frond. A suitable fertilizer mixture contains: 1 part sulphate of ammonia, 1 part superphosphate (18 per cent P_2O_5), 1 part sulphate or muriate of potash, and 2 parts magnesium sulphate (or 1 part anhydrous magnesium sulphate).

In the Far East, fertilizer mixtures are often applied monthly, but experiments in Nigeria have shown that with field nurseries planted in April it is sufficient to apply a total of 100–120 g per plant according to the following schedule:

May	15 g per plant	Age – 1 month
July	45 g per plant	3 months
October	60 g per plant	6 months

In West Africa the dry season sets in in November and no further fertilizer dressings are usually needed. In non-seasonal climates nirogen in particular may become insufficient and two further dressings of 60 g per plant may be given at 7 months and 9 months.

Fertilizer mixtures applied to seedlings should be spread in a ring 5–8 cm away from the seedling for the first application, gradually widening to 15–20 cm from the seedling for the last application.

Polythene bag nurseries

The successful use of bags in the Far East was due to the work of Gray, who, following experience in Sabah, carried out trials with large black polythene bags filled with friable topsoil from areas of coastal alluvium. He showed that, with heavy watering, large healthy seedlings could be produced in bags weighing only about 16 kg; and he later demonstrated that it was possible to dispense with the prenursery stage by planting germinated seed straight into the large 'polybag' (Bevan and Gray, 1966).

Experiments comparing polybag and field nurseries have been carried out in both Malaysia and Nigeria. In the Malaysian experiments, there were no significant differences between palms raised in polybags and those raised in field nurseies either in growth or early bunch yield, but the survival rate, which was generally very high, was slightly higher with polybag palms. Palms raised in prenursery beds were less advanced than those raised in prenursery bags, but this was due largely to the erosive effect of heavy rain on the plank-sided beds used (Hew and Tan, 1969).

In the Nigerian experiment, polybag plants showed slower growth in height in the nursery than field nursery plants, but they had a lower Blast incidence, produced more leaves and more plants were adjudged transplantable. On planting into the field the polybag plants showed more rapid initial growth and after 18 months were significantly taller and had produced more leaves (Aya, 1969, 1971, 1974). Plants originally raised in prenursery bags grew better and had less Blast in the nursery than those grown in prenursery raised trays, but the growth difference did not persist into the field. The cultural practices used for field nurseries, i.e. irrigation, no shade and standard fertilizer treatment, were found applicable to polybag nurseries.

In the Malaysian experiments, omitting the prenursery stage and planting germinated seed direct into large polybags gave larger seedlings on transplanting to the field at 13 months from germination than were obtained from using prenurseries, but this effect did not persist for more than a year in the field.

Polybag nurseries have other advantages: their sites can be chosen for purely managerial rather than agricultural reasons, clearing and levelling only, rather than soil cultivation, is needed, weeding is much reduced and transplanting is easier. On the other hand large quantities of suitable soil often have to be transported from a distance, and more exact attention must be given to the water supply and watering. On balance, however, a bag nursery is always to be recommended except in very exceptional circumstances.

Black polythene of 500 gauge is used for the bags. Various sizes have been employed, but bags too wide in proportion to their height tend to become lopsided while those too high in proportion to their width may fall or lean over. Bags 50 cm deep by 38 cm wide, lay-flat, have been found serviceable and are in general use; they are perforated with holes spaced 7.5 × 7.5 cm on the lower half of the bag for drainage. When filled to the top they contain about 16 kg of nursery soil, the exact amount depending on the type of soil or mixture used (Plate 7.5). Shallower bags (35–40 cm deep) have been successfully used in Nigeria (Aya, 1979).

Pl. 7.5   A polybag plant about 75 cm (2.5 ft) high, 4–6 months before field planting. Bag 38 × 50 cm (15 × 20 in lay-flat).

Normally, the polybag nursery will be sited on a convenient level piece of land near to the essential water supply and, if possible, not far from the source of soil. Soil–sand mixtures should be similar to those used for prenurseries (p. 330). The bags are best filled at the source of soil and then transported to the nursery as this entails less handling of soil than filling at the nursery site. The addition of cattle manure has been shown to improve growth in the bags.

Where an irrigation system is in use the bags may be spaced at 90 cm (3 ft) triangular from the start (Plate 7.6). Closer spacing is sometimes recommended (Quencez, 1982) but will cause etiolation of larger plants. With hand watering the bags can most economically be placed close together in rows three-bags wide until the seedlings are about 6 months old from the germinated seed stage. They are then separated to 90 cm.

Polybag nurseries may be planted with either: (i) prenursery seedlings; (ii) bare-root seedlings from seed suppliers; or (iii) germinated seed.

Prenursery seedlings at the four- or five-leaf stage will be transplanted from bags or beds. In the former case the small bag is simply torn off and the seedling with its ball of soil is inserted in a hole dug in the soil in the large bag and the soil consolidated around it. With plants raised in a prenursery bed or raised tray, the method to be employed depends upon the consistency of the soil in the prenursery. If the soil holds together well, longitudinal and transverse cuts may be made in the soil between the plants 4 weeks

Pl. 7.6   A large polythene bag nursery in Malaysia.

before transplanting or at the time of transplanting. In the former case some proliferation of roots within the block of soil takes place before planting in the nursery and this may assist initial establishment and growth. At planting, the soil blocks with their seedlings are carefully lifted out of the prenursery and planted in the prepared planting holes (Plate 7.3). When the consistency of the soil prevents the lifting of an intact block, a special tool may be used. Several of these have been devised two of which are described below:

1. *Plantoir Richard*. This is a small cylinder 14 cm high and 10 cm in diameter which is pressed down over the plant into the soil of the prenursery (Klaver, 1961). The plant, which is at the four- to five-leaf stage but has been lightly leaf-pruned, must be exactly in the centre of the cylinder which has two small flaps serving as handles. After rotating the cylinder with the aid of the flaps, the cylinder and the ball of soil with the seedling may be lifted from the prenursery, the vertical roots being cut with scissors. Seedlings thus encased may be sent long distances and on arrival the cylinder is inserted in the nursery planting hole. A flat plate shaped in the form of a U and of slightly smaller diameter than the cylinder is then placed on the soil surrounding the seedling and held in place by a handle attached vertically to it. The cylinder is then withdrawn by its handles and the soil round the seedling is firmed down.

2. *Square-sided scoops*. Where a prenursery bed has a detachable plank at one end, this may be removed, and scoops which exactly fit the block of soil in which each seedling is growing may then be used to remove them. A simple three-sided scoop with a handle has been found quite satisfactory, but a more refined design includes a movable plate which lies flat against the middle side and can be pushed forward by a handle connected to it through a hole in this side. Thus, when the seedling reaches the planting hole in the scoop it is pushed out by the movable plate and drops into the hole with the soil intact.

Seedlings taken from a bed prenursery should be planted at once. If they have to be transported some distance they should be dug early in the day and use should be made of some transplanting tool or cylinder as already described. If it is impossible to carry them embedded in a block of prenursery soil then they may be very carefully lifted and dipped in a clay slurry and tied together in bundles before transport, but this method is not to be generally recommended.

In planting, holes a little larger than the block of soil to be inserted should be dug with a trowel. The soil block is then carefully inserted so that it does not shatter and the surface of the soil block is made level with the surrounding soil. Deep planting must be avoided. Seedlings not surrounded by a block of soil will require even more careful handling. The roots must be inserted in the hole dug by the trowel and the soil must be packed round them so as to ensure that they are not bent or broken.

Bare-root seedlings at the two-leaf stage used commonly to be despatched by seed suppliers but are now seldom employed. They are usually planted in a prenursery, but they may be planted direct into large polybags. Shading will be required during the first few weeks and can be gradually removed. Planting is carried out in the same manner as described above.

In the direct planting of germinated seed, methods are the same as for the prenursery, but unless irrigation water can be supplied at the heavy prenursery rate then some shade is required until the one- or two-leaf stage is reached. The success of direct planting in Malaysia has seemed to depend on this shade which can be supplied most economically by staking palm leaflets in each bag to form an umbrella over the developing seedling. The system has been reported as being more expensive in labour and water in the first 4 months than the prenursery and nursery polybag system, but experiments have shown that this disadvantage is outweighed by the reduction of overall time in the nursery by about 2 months (Bevan and Gray, 1966). It is probably inadvisable to introduce the system until a considerable skill in nursery management has been attained; nevertheless, it is now the most commonly practised nursery method in all parts of the world.

Roguing in the prenursery

The idea that a large amount of roguing in oil palm prenurseries and nurseries is required has been widespread but is not borne out by the experience of those who have maintained a high standard of cultural practice. While there may be a very small percentage of stunted or abnormal seedlings in the prenursery which can only be accounted as genetical in origin, the majority of poor doers or abnormal plants may well be the result of cultural deficiencies, usually bad planting or insufficient water. The plants which must be removed are those suffering from a twisted growth or showing symptoms of the freak conditions Leaf Roll, Leaf Crinkle or Collante which are described in Chapter 13 (p. 587). Apart from these easily distinguishable conditions (Plate 7.7), plants which should be rogued include those which show poor development, abnormally erect or dumpy habit, or have been affected either by poor planting of the germinated seed or by being left too long in the prenursery and hence become etiolated. In raised trays, a larger

Pl. 7.7   Certain abnormalities of nursery seedlings: **A**, Leaf Crinkle; **B**, Leaf Roll; **C**, Collante.

number of healthy seedlings are sometimes, owing to double seeds, obtained than the number of seeds planted.

There are fairly wide differences in the expectation of percentage transplantable prenursery seedlings. In Nigeria and Malaysia general experience has been that 95 per cent are transplantable and in the Ivory Coast the average losses from deaths and abnormalities are said to be not more than 15 per cent (Ruer, 1963). It is possible to raise both seedlings from double seeds, but it is sounder practice to discard one, retaining the other with the seed attached.

Time of planting

In seasonal climates and in areas subject to Blast disease, the time of planting a nursery is of considerable importance. Elsewhere nurseries may be laid down at any time of year though it is preferable to choose a period when rain is most likely to fall and which allows a lapse of 12–16 months before the prearranged field planting programme. Both in Zaire and in the Far East nurseries are commonly planted all the year round and nursery plants are taken for planting in the field in the wetter periods beginning in April or September.

In West Africa the field planting season runs from the very beginning of the rains, in March, to the time when the rains have fully set in in May or early June. In the Ivory Coast, where the rainfall is more even than in other parts of West Africa and completely dry months are rare, April is also preferred as the planting month even though the short dry season of August and September is more pronounced than elsewhere and October and November rainfall is high.

Apart from the consideration of field-planting time, however, early planting of nurseries is desirable in Africa for the avoidance of Blast disease. Early experiments in field nurseries in Nigeria showed that April to June was the best period for establishing a nursery from prenursery seedlings, giving the least Blast and the highest number of transplantable seedlings at field-planting time 1 year later (Gunn *et al.*, 1961). These findings were largely confirmed for polybag nurseries with the direct planting of germinated seed. Aya (1979) found that mid-April to mid-May, i.e. early in the rains, was the best time for establishing a nursery, giving a lower Blast incidence and better development up to 10 months than establishment either earlier or later.

Blast disease in Nigeria tends to attack seedlings of a susceptible, i.e. small, size and usually only between the months of October and January. *Very* small seedlings planted in October and shaded are not susceptible and for this reason a 'dry-season nursery' planted in that month was tried in Nigeria (Gunn *et al.*, 1961). This nursery can be forced forward by irrigation, mulching and more frequent fertilizer dressings and seedlings become transplantable in late May when 70–100 cm (2–3 ft) high. This is not, however, the optimum size or age for transplanting, and Aya (1979) concluded from his polybag experiments that, although dry-season nurseries are feasible, the optimum time for planting them varies with locality from September to November and the resulting seedlings will not be of transplantable size at the right time.

In South and Central America there is a great variety of seasonal climates. In northern Colombia, for instance, the dry season is at much the same time of year as in Nigeria and both transplanting to the field and nursery establishment are therefore best done in March to May. On the Pacific plain of Ecuador on the other hand, the long low-rainfall but cloudy period ends in November–December, and both field and nursery planting are best done then, before the torrential rains of January to April set in.

To summarize, in planning germination and nursery work, timing should be calculated back from the start of the planned time of field planting, as follows:

	With prenursery		Direct planting in nursery
Germination (dry-heat treatment)		3–4 months	3–4 months
Prenursery	4–5	} 14–17 months	
Nursery	10–12		12–14 months
Total		17–21 months	15–18 months

Thus if planting is planned to run from September to November, germinated seed should be ordered to arrive in batches from April to July of the previous year if a prenursery is to be used, or from July to September if there is to be no prenursery. When germination is done on the plantation seed would be set in the germinator in this case in late December or early January for the April prenursery, or in late March or early April for direct planting.

Water supply and nursery irrigation

The water requirements of polybag nurseries have been variously estimated. Account should be taken of both the utilizable water reserve of a bag (estimated to be 30–35 mm) and the increase in evapotranspiration with the growth of the seedlings (Quencez, 1982). Most authorities arrive at an overall requirement figure of between 6 and 8 mm per day (Turner and Gillbanks, 1974; Gilbert, 1979), but Quencez (1982) more properly considers the requirement to rise with age as follows: 0–2 months old, 4 mm per day; 2–4 months, 5 mm; 4–6 months, 7 mm; 6–8 months, 10 mm.

The water needs can be more accurately and economically supplied by an overhead irrigation system than by hand watering. Moreover hand watering is often difficult to supervise, and the application rate is higher than the infiltration rate, leading to over-flow and wastage. To reduce this disadvantage the total daily requirement may be applied in two halves. With hand watering in Malaysia the minimum watering is considered to be four times weekly to 'run off' using, in the absence of rain, 2–4 gallons (9–18 litres) per plant per week according to stage of growth, soil type and other factors (Bevan and Gray, 1966). With this heavy watering the soil surface needs to be disturbed from time to time to prevent compaction. Water usage is less with hand watering than with an irrigation system since in the latter case the water will be applied to the whole area and not only to the soil in the bags.

Irrigation systems for oil palm nurseries may be of the large rain-gun type, needing pressures of around 100 psi (7 kg per cm^2) and covering about 1 hectare at the rate of 25 mm per hour, or they may be of the smaller rotary type requiring lower pressures with a minimum pressure of 20 psi (1.4 kg per cm^2) and sprinkling at the rate of 2–11 mm per hour. The latter types are usually more suit-able unless very large nurseries are to be maintained over a long period of time.

The construction, layout and operation of irrigation systems have been described in great detail for Malaysia by Bevan and Gray (1969) and Gilbert (1979) and for West Africa by Coomans (1971) and Quencez (1974, 1975, 1982). The former authors base their designs on the assumed need for 0.33 inch (8.4 mm) rain or irri-gation per day after allowing for losses due to evaporation, the spacing of bags 3 ft (91 cm) triangular from the start, and the provision of water by spray lines which supply the 8.4 mm in 1½ hours.

In their design for a 25-acre (10 hectare) nursery (see Fig. 7.1) they allow for spray lines shifted seven times during the day giving 1½ hours spraying in each position. Each spray line is of 3 inches (7.6 cm) diameter served by a 6 inch (15.2 cm) main pipe from the pump house. The long beds lying between the spray-line positions are 49.4 ft (15 m) wide and divided into five sub-beds to allow for paths along which the workers shifting the lines can move. The spray lines are 705 ft (215 m) long with twenty-four sprinklers; they have a throw of about 43 ft (13 m) radius and supply 3.6 gallons (16.4 litres) per minute at 40 psi (2.8 kg per cm^2). Single-nozzle sprinklers ($\frac{5}{32}$ inch–4 mm) are suitable. A nursery of this size will provide about 130,000 seedlings which, after culling, would be sufficient for about 1,750 acres (700 hectares).

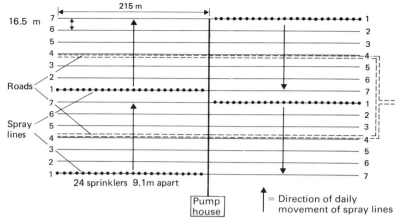

Fig. 7.1 Plan of nursery system covering 10 hectares supplying about 100,000 seedlings (Bevan and Gray, 1966).

Swire (1968) and Coomans (1971) recommended a nursery organized in two stages, first with the bags at 40 × 40 cm and later, at the end of February, at 70 × 70 cm. Under this Ivory Coast system less water was required but labour was needed for reorganizing the nursery in February, and the eventual spacing of 70 cm, in contrast with 91 cm under the Malaysian system, could lead to etiolation. Moreover the close-spaced stage included the provision of overhead palm-leaf shade as an anti-Blast measure (see later). This system has largely given place to that of Quencez (1982) in which the bags, measuring 40 × 40 cm lay-flat, are placed, without shade, at 70 cm triangular from the start. The designs recommended allow for wider-spaced sprinklers (36 m) with a further throw (26 m radius) than in the Malaysian system. To gauge the amount of irrigation water the water balance is calculated daily by the formula:

$$B(n) = B(n - 1) + R - ET,$$

where $B(n)$ = the water balance on day n, $B(n - 1)$ = balance on day $n - 1$, R = rainfall and irrigation of days n and $n - 1$, and ET is the evapotranspiration or consumption of the plants according to the age schedule given above (p. 347). The whole nursery is then watered every 2 days on a three rounds per week cycle, but with no watering on Sunday, and the duration of watering is adjusted according to the water balance calculations and the type of equipment in use. Under this arrangement 4.5 hectares of nursery are needed for a 500 hectare field planting. With the wider spacing used in the Malaysian nursery 7 hectares are needed per 500 hectares of field planting.

Post-planting cultural treatments

Shading of a nursery is never required except as an anti-Blast measure. In West Africa it has been noticed that plants shaded through the dry season increase more rapidly in height, but later, when the shade is removed and the rains have set in, they grow more slowly than unshaded plants (Gunn and Sly, 1961).

Nursery weeding is usually carried out by hand. Monthly weeding rounds are recommended in Malaysia both for the bags and the intervening ground (Bevan and Gray, 1969). Mulching is not normally practised with polybags; the surface soil is lightly scratched at the time of fertilizer application so the amount of weeding required should never be very great. In the Ivory Coast the treatment of the area, before the bags are positioned, with a pre-emergent weedkiller is advocated. Ametryne is applied at 3 kg a.i. in 300 litres water for one treated hectare. This treatment can be repeated later, provided care is taken to avoid spraying the plants (Quencez, 1982).

In polybag nurseries in the Far East monthly application of a compound fertilizer of formula N12 : P12 : K17 : Mg2 is recommended starting about 1 month after transplanting from a prenursery and increasing the rate per plant from about 5 g at 1 month to 10 g at 2–4 months, 20 g at 5–6 months, and 30 g at 7–9 months. With a single-stage polybag nursery the first 3–4 months correspond with the prenursery stage. No fertilizer is given during the first month but from the beginning of the second month to the end of the third month, i.e. from the one- to three-leaf stages, urea can be watered on weekly at the rate of about 7 g in 5 litres of water per 100 seedlings; this rate may be doubled at the four- to five- leaf stages and solid fertilizers will then be applied from about the end of the fourth month. Monthly applications are then usual, starting with about 10 g per plant and gradually increasing the rate to about 30 g at 10 months. Alternatively, solid application of a 15 : 15 : 6 : 4 mixture may start at 2 months at the rate of 1 g, increasing to 5 g at 3–4 months, when the higher K mixture given above is started (Turner and Gillbanks, 1974). At all stages care has to be taken to see that the fertilizers do not cause leaf scorch and washing off with water after application is advisable.

For West African soils Quencez (1982) has recommended, in addition to mixing the soil with some form of organic material, watering in before planting and when the bag is half full of a solution of 1 kg urea, 2 kg KCl and 1 kg kieserite in 200 litres of water. Each bag receives 1 litre of the solution, and 10 g of super-phosphate (18 per cent K_2O) is added separately. After 1 month 5 g of urea is applied monthly, increasing to 10 g at the fifth month when 10 g KCl, 10 g superphosphate and 5 g kieserite are also applied once. A further 10 g of KCl may be applied at 7 months.

Treatment of the soil with 0.1 g neutral sulphate of oxyquinoleine and 0.2 g of aldrin is also advocated in the Ivory Coast. In Nigeria Aya (1969, 1971) reported satisfactory growth by mixing the nursery soil with cattle manure and using the same fertilizer schedule as for field nurseries (p. 339).

The prevention of nursery diseases

Diseases of oil palm seedlings are described in Chapter 13, but as they have had a considerable influence on nursery practice some mention should be made of them here.

The Anthracnose fungi can be largely eliminated by wide spacing in the prenursery, careful transplanting and the use of organic fungicide sprays or by avoiding prenurseries altogether. Spraying with fungicides (see p. 583) is required weekly in the prenursery from the two-leaf stage and is continued for 6 weeks after transplanting the seedlings into the nursery. Both sides of the leaves must be covered.

Cercospora Leaf Spot (Freckle) appears in African nurseries after the danger of Anthracnose is over. Its incidence is reduced by potassium manuring. Spraying with contact or systemic fungicides (see p. 584) should be carried out fortnightly from 6 weeks after the nursery has been planted. It is particularly important that the undersides of the leaves should be sprayed as well as the upper sides.

Blast disease, described on p. 588, has been killing seedlings in African nurseries for a long time. More recently, symptoms of Blast have been seen in the Far East. The disease may appear at any time, but in West Africa it normally attacks young seedlings between October and January, i.e. from the end of the rains until well into the dry season. It has been shown that: (i) seedlings of a year or more old are rarely attacked (Bachy, 1958); (ii) seedlings planted in October and shaded escape attack (Gunn *et al.* 1961); (iii) the severity of the 'short-dry' season of August affects the incidence of Blast, and hence the provision of irrigation water during this season reduces Blast incidence (Robertson, 1959); and (iv) shading a main-season nursery from October to February reduces Blast incidence (Bachy, 1958; Rajagopalan, 1974).

It has been stated that shading the nursery from October to late January was, in West Africa, the only protection against Blast until the role of the Homoptera, *Recilia mica*, in its transmission was discovered (Quencez, 1982). In fact, however, experience in the Ivory Coast cannot be applied to the whole of West Africa since Blast incidence varies widely from territory to territory and it has undoubtedly been a more serious problem in the Ivory Coast than elsewhere. In Nigeria the shading of nurseries was stopped in 1954 as it was believed that any possible reduction of Blast was counterbalanced by the more drawn-up and weakly seedlings produced. Later, improved cultural practices and the provision of irrigation during the short dry season so improved the standard of growth that shading at such a late stage was not reintroduced. But for the special dry-season nursery it was shown that shading the October-planted seedlings for 3 months almost eliminated Blast during that period, and shade can, at this stage, be provided inexpensively by single bent-over palm leaves. These gradually wither, letting more and more light through and when removed in December or January the seedlings make very rapid progress. However, this type of nursery is no longer recommended (Aya, 1979).

The above findings seem to apply as much to polybag as to field nurseries; in the Nigerian experiment referred to on p. 339. Blast incidence in unshaded nurseries was actually less with polybags than in the field nursery (Aya, 1974). For West Africa as a whole the decision on whether to use a systemic insecticide for the control of *Recilia mica* will depend on experience of the locality concerned, the time of planting, the irrigation facilities available and the origin of the planting material. This is further discussed on pp. 589–92.

Pests are more troublesome in nurseries in Asia and America than in Africa and although regular prophylactic spraying is not usual a close watch has to be kept for night-flying beetles (*Apogonia* and *Adoretus* species), crickets and grasshoppers, and leaf-cutting ants and spraying must be started as soon as necessary (see pp. 630 and 656–7).

Costs

Detailed costs of polybag nurseries in terms of man-days have been given by Quencez (1982) for Africa and by Bevan and Gray (1969) for Malaysia for nurseries planted with prenursery seedlings. The former author, using a nursery spacing of 70 cm triangular and allowing for the planting of 18,000 seedlings per hectare of nursery, estimated that these would be sufficient for 100 hectares of field planting at 9 m triangular. Labour usage per hectare of nursery was 1,815 man-days, and this constituted (in 1982) 46 per cent of the total cost of the nursery. Thus for 1 hectare of plantation the labour requirement was 18 man-days, or 126 man-days per 1,000 transplantable seedlings.

With the Malaysian spacing of 90 cm triangular, only about 11,000 seedlings will be planted per hectare of nursery, i.e. about 39 per cent less. Labour usage per hectare of field planting will therefore be higher, and a mean figure for Malaysian nurseries was found to be 24 man-days, or 168 man-days per 1,000 transplantable seedlings.

Prenursery labour costs are very small, amounting at most to 4 man-days per hectare of field planting, or 28 man-days per 1,000 transplantable nursery seedlings, and estimates of less than half this amount have been quoted (Ruer, 1963). Taken together, prenursery and nursery *work* form a very small proportion indeed of the capital cost of bringing a plantation into production. Economies which reduce the standard of the transplanted seedling are therefore misplaced as they are likely to be followed by a substantial reduction in early production.

References

Anon. (1953) Notes on the establishment of an oil palm nursery, *J. W. Afr. Inst. Oil Palm Res.*, **1**(1), 88.

Aya, F. O. (1969, 1971) *NIFOR fourth annual report, 1967–8*, pp. 25–7. *Sixth annual report, 1969–79*, pp. 31–2.

Aya, F. O. (1974) The use of polyethylene bags for raising oil palm seedlings in Nigeria. 1. Relative performance of seedlings grown in ground beds and polyethylene bags in the nursery. *J. Nig. Inst. Oil Palm Res.*, **5**(19), 7.

Aya, F. O. (1979) The use of polyethylene bags for raising oil palm seedlings in Nigeria. 2. The influence of bag size, colour, drainage, and planting date on the performance of the seedlings in the nursery. *J. W. Afr. Inst. Oil Palm Res.*, **5**(20), 15.

Bachy, A. (1958) Le 'Blast' des pépinières de palmier à huile. *Oléagineux*, **13**, 653.

Bevan, J. W. L. and **Gray, B. S.** (1966) Germination and nursery techniques for the oil palm in Malaysia. *Planter, Kuala Lumpur*, **42**, 165.

Bevan, J. W. L. and **Gray, B. S.** (1969) *The organisation and control of field practice for large-scale oil palm plantings in Malaysia.* Incorp. Soc. of Planters, Kuala Lumpur.

Coomans, P. (1971) L'arrosage des pépinières de palmiers à huile en sacs de plastique. *Oléagineux*, **26**, 295.

Dupriez, G. (1956) Prépépinières d'*Elaeis*. Bull. Inf. INEAC, **5**, 141.

Fraisse, A. (1962) Les pépinières de palmiers à huile. *Oléagineux*, **17**, 173.

Gilbert, A. C. (1979) Sprinkler irrigation and practice under Malaysian conditions. *Planter, Kuala Lumpur*, **55**, 170.

Gunn, J. S. (1960) *WAIFOR eighth annual report, 1959–60*, p. 41.

Gunn, J. S. and **Sly, J. M. A.** (1961) *WAIFOR ninth annual report, 1960–1*, p. 34.

Gunn, J. S., Sly, J. M. A. and **Chapas, L. C.** (1961) The development of improved nursery practices for the oil palm in West Africa. *J. W. Afr. Inst. Oil Palm Res.*, **3**, 198.

Hew Choy Kean and **Tam Tai Kin** (1969) A comparison of oil palm nursery techniques. In *Progress in oil palm*, Incorp. Soc. of Planters, Kuala Lumpur.

Klaver, H. (1961) Le transport des plantules de palmier à huile. *Oléagineux*, **16**, 601.

Marynen, T. and **Poels, G.** (1960) Considérations sur les méthodes de mise en place du palmier à huile. *Bull. Inf. INEAC*, **9**, 81.

Ochs, R. (1963) Utilisation du test stomatique pour le contrôle de l'arrosage du palmier à huile en pépinière. *Oléagineux*, **18**, 387.

Quencez, P. (1974, 1975) Arrosage par aspersion des pépinières de palmiers à huiles en sacs de plastique. *Oléagineux*, **29**, 405; **30**, 355, 409.

Quencez, P. (1982) Les pépinières de palmier à huile en sacs de plastique sans ombrière. *Oléagineux*, **37**, 397.

Rajagopalan, K. (1974) Influence of irrigation and shading on the occurrence of Blast disease of oil palm seedlings. *J. Nig. Inst. Oil Palm Res.*, **5**(19), 23.

Robertson, J. S. (1959) Blast disease of the oil palm: its cause, incidence and control in Nigeria. *J. W. Afr. Inst. Oil Palm Res.*, **2**, 310.

Ruer, P. (1963) Les prépépinières de palmiers à huile en sachets de polyéthylène. *Oléagineux*, **18**, 693.

Sly, J. M. A. and **Sheldrick, R. D.** (1961) The practical aspects of irrigation of an oil palm nursery. *J. W. Afr. Inst. Oil Palm Res.*, **3**, 273.

Surre, Ch. (1968) Les pépinières de palmiers à huile en sacs de plastique. *Oléagineux*, **23**, 573.

Taillez, B. (1969) Le désherbage sélectif des prépépinières de palmier à huile. *Oléagineux*, **24**, 541.

Toovey, F. W. (1947) *Oil Palm Research Station, Nigeria, seventh annual report 1946–7*, pp. 17–18.

Turner, P. D. and **Gillbanks, R. A.** (1974) *Oil palm cultivation and management.* Incorp. Soc. Planters, Kuala Lumpur.

Vanderweyen, R. (1950) Pépinières en paniers ou en pleine terre. *INEAC Réunion Congopalm*, Comm. No. 10, 58.

Chapter 8

The preparation of land for oil palm plantations

New oil palm plantations are usually established in areas of primary or high secondary forest and there has been much study of different establishment methods and their costs. Increasingly, however, oil palms are being planted on land under light secondary growths or where strong growing weeds such as lalang (*Imperata cylindrica*) or Siam weed (*Eupatorium odoratum*) have become dominant following food-cropping; under these circumstances rather different problems are encountered and these will be discussed later in this chapter.

Usually, oil palm plantings are on flat or gently undulating land, but on inland soils quite steep slopes may sometimes be encountered. In these cases the normal triangular planting distances are maintained horizontally, but the palms may be planted on platforms. Planters accustomed to rubber cultivation have been anxious in some areas to adopt or maintain rubber terracing methods (Gawthorn, 1967), but the fact that the oil palm obtains water and nutrients predominantly from the topsoil and that interference with the rooting area leads to reduced growth suggests that platforms should be used except where terracing is absolutely necessary for soil conservation (see p. 364).

On flat areas of heavy clay soil and on undulating areas where the permeability of the soil is low and the natural drainage courses are sluggish or subject to flooding, preparations of planting include the design and construction of a drainage system which will prevent the stagnation of water in the upper layers of soil and allow for the rapid removal of flood-water after heavy rain.

By far the most expensive work in the establishment of a plantation is the actual felling of the forest and the clearing of it so that the seedlings may be planted out. The cost of this work may be as much as two-thirds the total cost of establishment, and it is not surprising therefore that attempts have been made, not always successfully, to reduce these costs by: (1) directional felling; (2) the use of chainsaws and hand winches; (3) full mechanization of felling (see p. 369); (4) burning the felled forest; and (5) exploiting forest products.

The practice of burning

Early opinion in Africa was that burning would lower fertility. Falling yields in Zaire and Nigeria after 8 or 9 years of bearing were attributed to depletion of nutrients through the practice; to this was added the loss of soil organic matter and the provision of conditions leading to erosion (Sheffield and Toovey, 1941; Toovey, 1947). Early workers in Zaire (Beirnaert, 1942; Vanderweyen, 1943) and Malaysia (RRIM, 1939) supported these ideas, though in Malaysia it was later realized that, with the increasing employment of tractors on estates for all manner of work, the improved access obtained through burning followed by mechanized or hand clearing might outweigh any advantages of leaving the forest unburnt (RRIM, 1957). The advantages of burning were seen by its advocates to be managerial and economic: a reduction in the costs of establishment and the creation of conditions where early work in the plantation was easier and supervision and maintenance therefore less time-consuming and costly (Plate 8.1).

The theory that burning 'lowers fertility' was further considered in soil studies, and, in Nigeria, field trials comparing burning and non-burning were laid down (Sly and Tinker, 1962). These trials were unique, and though all three of them were on Acid Sands soils (though one was in an area of heavier rainfall some 400 km from the others) their results are likely to be generally applicable. The yields and soil analysis results obtained are given in Table 8.1. No yield differences were found between burning and non-burning except in the first 4 years in one of the experiments at Benin where increased yields were obtained from the burnt plots. Increased early yield might well be expected owing to the larger release of some nutrients, particularly nitrogen, and the lack of competition from natural vegetation following burning; the latter effect was probably exaggerated in this experiment as maintenance during the year after planting was inadequate and this, of course, favoured the burnt plots where natural vegetation takes longer to re-establish itself.

The effects of burning on soil nutrient levels 20, 10 and 9 years after planting in the three experiments were also examined and the differences between burnt and unburnt areas found to be small and unimportant. Burning tends to conserve potassium, this effect showing up particularly in the later years. Where magnesium and calcium are in good supply burning lowers their levels to a small extent, but this does not occur when the levels are already low.

A

B

Pl. 8.1   Before and after the burn at the end of the dry season in Nigeria.

Table 8.1 *Effects of burning on yield and soil analysis in Nigeria*

Fruit bunches per hectare per annum; soil data – depth in cm

Expt	Age at soil sampling	Treatment	Yield		Exchangeable cations (meq per 100 g)								Exch. cap. (meq per 100 g)		C(%)		N(%)	
					K		Na		Mg		Ca							
			First 4 yrs	Adult years†	(cm) 0–15	(cm) 15–30	(cm) 0–15	(cm) 15–30	(cm) 0–15	(cm) 15–30	(cm) 0–15	(cm) 15–30	(cm) 0–15	(cm) 15–30	(cm) 0–15	(cm) 15–30	(cm) 0–15	(cm) 15–30
			(kg)	(kg)														
33–2 Benin	20	Burnt	4,860	9,947	0.045*	0.034	0.070*	0.064*	0.68	0.24	3.15	1.89	6.04	3.55	1.10	0.47	0.095	0.041
		Unburnt	5,242	9,730	0.038	0.021	0.039	0.033	0.63	0.27	3.82	1.89	6.27	3.43	1.20	0.49	0.100	0.041
3–4 Benin	10	Burnt	4,958*	6,549	0.070	0.047	0.048	0.038	0.69	0.43	2.18	1.07	5.87	4.93	1.18	0.73	0.098	0.061
		Unburnt	4,363	6,499	0.073	0.045	0.056	0.041	0.83*	0.47	2.75*	1.39	6.15	4.70	1.24*	0.68	0.106*	0.059
506–2 Abak	9	Burnt	2,617	6,038	0.073	0.049	0.044	0.042	0.15	0.08	0.14	0.10	6.01	5.16	1.14	0.79	—	—
		Unburnt	3,017	6,168	0.073	0.040	0.046	0.028	0.16	0.09	0.13	0.07	6.01	5.10	1.19	0.76	—	—

† Means of 11 years (33–2); 6 years (3–4); 7 years (506–2).

* Significant at $P = 0.05$.

Source: Sly and Tinker (1962).

Burning causes a slight depression of organic matter and nitrogen in the top 15 cm, a rapid nitrate flush occurs immediately after burning and this leads to an earlier loss of nitrogen. Under some conditions, therefore, an increased need for nitrogen fertilizers might be expected in the early years. It should be noted that in the experiments quoted above no leguminous covers were planted after burning. With the rapid establishment of leguminous covers the loss of organic matter following burning might be reduced.

On theoretical grounds, Bocquet and Michaux (1961) postulated that, in soils with a clay content of 20 per cent, if the amount of carbon in the top 20 cm is less than 2.2 per cent, the forest should not be burnt. It would not be possible to use this figure as a standard since, in general, C per cent is positively correlated with clay content (Tinker and Ziboh, 1959); moreover the loss of carbon in the topsoil following oil palm planting appears to vary widely. Bocquet and Michaux quote cases of losses of organic carbon at a steady rate for 8–12 years after planting from primary or secondary forest with 2.5–3.5 per cent C in the top 10 cm of soil. In Nigeria, however, a special study of soil changes under plantation conditions (Kowal and Tinker, 1959) showed no fall of organic carbon in a burnt area during a period of 15 years from 1 year after planting; indeed, under certain cultural conditions there was a tendency for organic carbon to rise. It must be borne in mind that high carbon content is not always a sign of fertility; that poor degraded soils may have a high organic matter content has been shown both in West Africa and Malaysia (Tinker and Ziboh, 1959; Coulter, 1950). It is not possible, therefore, to fix any exact soil criterion for the practice of burning and the evidence available shows that, for all *usual* situations, the practice will not lead to any diminution of crop and will prove less costly than the opening of forest areas without burning.

Nevertheless, there is evidence that poor, sandy areas which have for many years had only a light covering are better left unburnt. One example is an area of palm grove and farmland with a structureless sandy soil in which farming had been practised in the past as part of a 'bush-rotation' (Gunn, 1960). Palms were planted in this area under various cultural treatments. Burnt plots normally maintained with a natural cover showed a lower response to an NKMg fertilizer mixture first applied in 1957 than was shown by unburnt plots:

Mean annual yield of fruit bunches per hectare (Abak, Nigeria, Expt. 506–2)

	1955–7	1958–65	
		No fertilizer	Fertilizer
	(kg)	(kg)	(kg)
Burnt	1,556	4,876	7,142
Not burnt	2,167	4,516	7,673

In many parts of the world oil palms are planted in districts containing primary or very old secondary forest. Quantities of very valuable timber may exist in such forests and proper exploitation may be profitable and provide a substantial contribution to the costs of establishment. Quite apart from this possibility, however, the forest will supply a great deal of useful small timber for fence posts, bridges, etc. and this must be extracted to meet future requirements while the areas are being opened up.

Layout

The general layout of a plantation is determined by the terrain, the position of the mill and the length of 'carry' of bunches by harvesters to the road or rail side. If the area is flat or gently undulating, the collection roads or railway tracks should be laid straight and in an E–W direction and should connect with one or more sub-main N–S roads in turn leading to the main road to the mill. If the country is hilly the collection roads will, at least on parts of the estate, have to follow a winding course to avoid steep gradients. In very steep country the roads must be arranged so that the carry is a good deal less than on flat land since the harvesting work is more arduous. A railway system will not be suitable on a hilly estate. A typical estate layout for hilly land is shown in Fig. 8.1. The interlocking of drainage and transport systems on a flat plantation is shown in Fig. 8.3 (p. 381).

The maximum practicable length of carry is generally considered to be 200 m, so the roads or railway tracks are 400 m apart or less. On some estates, however, rail tracks have been as much as 660 m apart. With the opening of new estates and the increase in operating costs much thought has been given to the type and arrangement of the transport system to be used. Between the wars many railways were laid down and many are still being operated economically. The capital cost of these systems is now considered prohibitive, however, and newly opened estates now construct road systems.

It has been shown that the determination of the optimum frequency of collection roads is by no means a simple matter since it depends on labour costs and construction and maintenance costs, and consideration must also be given to the loss of crop through the excision of road areas from the total possible productive area; moreover the optimum frequency will vary according to whether land or capital is limited (Barrett, 1965; Sankar, 1965). As labour rates rise, the optimum road frequency for maximizing return on capital will tend to rise also, i.e. the distance between roads should fall. Calculations might show that a 200 m carry was optimal but, looking ahead, it might be wiser to accept a 160 m carry with a larger road area, higher road construction and maintenance costs, but easier

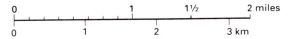

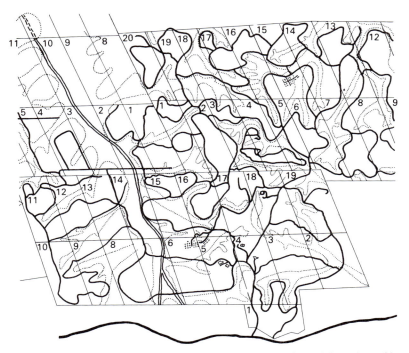

Fig. 8.1   Example of an estate layout (hilly estate): (..............) rough boundary of low-lying land; (——) roads; (====) main drainage course.

harvesting. However, this general thesis might need modification if more extensive use could be made of inter-row transport and mechanical harvesting.

Many new estates are being laid down with collection roads 320 m apart, with intervals as low as 100 m on very hilly land. With 320 m intervals between collection roads the length of road per hectare will be 36 m, assuming sub-main roads every 2 km and not including peripheral roads. With sub-main roads, every 1 km there will be 41 m of road per hectare. A design which has been used in Africa and America consists of a 1 km^2 block (100 hectares) with collection roads at 250 m intervals giving four fields of approximately 25 hectares each; the maximum carry of 125 m is rather less than is usual in Asia and there are 50 m of road per hectare not including peripheral roads. With these systems the roads will occupy between 2 and 4 per cent of the total area.

Methods of land preparation

Various sequences of operations for land preparation have been followed in different parts of the world. The methods chosen often depend on local circumstances and experience. In the days of non-burning it was usual to underbrush (clear the ground vegetation and small saplings) and line out before felling. With the advent of burning, underbrushing was still the first operation though lining was often put off until after felling and burning. It was found in many areas, however, that even a severe burn did not consume the hardwood lining pegs and that there were several advantages in carrying out underbrushing and lining together with holing, where holing was necessary, in the shade of the tall trees.

Various sequences which have been employed after marking out the plantation are given in Table 8.2. It will be seen that the usual sequence is underbrushing, felling, burning and lining, but where attention is given to directional felling as in E, lining and the preparation of the planting rows precede felling and burning. It is important that the newly planted young palms shall be easily accessible and this is assured only if large tree trunks and branches are not allowed to cross the row paths. This can be arranged either by felling the trees along the rows or by moving them into line when once felled. Clearly the latter method is more expensive and skill in felling is therefore important. Method B is designed to concentrate felled material which is not consumed by a light burn into narrow windrows in alternate interlines thus leaving the other interlines and 2 m on each side of each row absolutely clear except for the stumps of the very largest forest trees. This method increases further the ease of inspection and makes some mechanical maintenance possible; it is, however, expensive and if carried out entirely by hand labour would, in terms of man-days, be at least twice as costly as the more usual methods. Even with mechanical saws and mechanical stacking the method may be more expensive in Africa than ordinary hand methods (Surre *et al.*, 1961).

The usual operations carried out in preparing land for planting will now be considered separately, but excluding road building which has already been mentioned, and drainage which will be discussed later.

Table 8.2   *Methods of field preparation*

A. Zaire	B. Ivory Coast	C. Ivory Coast	D. Nigeria
1. Underbrushing 2. Felling forest trees 3. Heaping 4. Burning 5. Lining 6. Clearing paths 7. Opening rides along palm lines 8. Terracing 9. Sow cover crop	1. Underbrushing 2. Marking the position of the lines on the base line and the first stacking line (windrow) 3. Felling directionally along the interlines and marking windrows 4. Stumping the interlines which are to be cleared and stumping along the lines 5. Beating down and cutting the large branches 6. Cutting up trunks 7. Carrying out a light burn of branches and debris 8. Verification and marking of the stacking lines 9. Heaping the logs into windrows in alternate avenues, by tractor 10. Levelling and final clearing 11. Sow cover crop	1. Underbrushing 2. Felling 3. Beating down and cutting up 4. Heaping ⎫ 5. Burning ⎭ 6. Clearing the lines 7. Sow cover crop	1. Underbrushing 2. Felling 3. Beating down 4. Burning 5. Lining 6. Opening paths along rows 7. Sow cover crop

E. Cameroon	F. Malaysia	G. Malaysia	H. Malaysia
1. Underbrushing 2. Lining and clearing paths 3. Holing 4. Felling small trees into interlines* 5. Felling large trees into interlines* 6. Sow cover crop along lines 7. Burning 8. Realign and peg	1. Underbrushing 2. Felling, cutting larger branches (dry out for 3 months) 3. Burning 4. Lining 5. Clear paths along rows 6. Holing 7. Sow covers	1. Rough lining 2. Underbrushing 3. Felling, cutting larger branches 4. Pile branches on top of large timber (dry out 2–4 weeks) 5. Scorch burn 6. Lining 7. Clear paths along rows 8. Holing 9. Sow covers	1. Underbrushing 2. Felling 3. Burning 4. Cutting up, stacking 5. Reburning 6. Lining 7. Clearing paths 8. Weeding 9. Sow covers 10. Holing

* Paths kept clear of debris.

Sources: (A) Vanderweyen (1952); (B) and (C) Surre *et al.* (1961); (F) and (G) Edgar (1958); (H) Bevan and Gray (1969).

Underbrushing

This is done to make access to the area easy and, in the cases mentioned above, to allow lining (and holing) to be carried out. The work is done with cutlasses and axes and will require between 10 and 20 man-days per hectare according to the thickness of the undergrowth; all herbaceous undergrowth, lianas and young saplings 7.5 cm or less in diameter should be cut, the latter as low as possible.

Lining

Many methods of lining have been used and suggested. In the first place it must be realized that the rows are going to run north and south and that with triangular planting the palms in one row will not be opposite the palms in the adjacent row; the line from each palm to the nearest palm in the adjacent row will make an angle of 60° with the row and the distance between the rows will be less than the planting distance. This orientation and planting arrangement ensures that the maximum sunlight falls on the individual palms.

Some lining methods allow for the making of a base line with palms marked off at the planting distance, say 9 m. From this base line, which should of course be N – S, equilateral triangles with sides of 9 m are built up by running lines at 60° to tbe base. On flat land, however, the road will be running E–W and it will be easier to have the base line parallel with the road or E–W boundary. The rows can then be marked off along the base line according to the interline distance which is the perpendicular distance from any of the three points of the 9 m-sided triangle to the opposite side. This is easily calculated by multiplying the planting distance by 0.866, e.g. for 9 m spacing the rows will be 7.8 m apart. Alternatively 9 m triangles can be made from the first two palms of the first row and the first palm of the second row (which will be 4.5 m from the base line) and the rows can then be extended from the base line and the palms marked off at 9 m intervals. The other rows can then be 'placed' by means of similar triangles touching the base line, and every point throughout the field is brought into alignment in three directions. Guide lines can be put in at an angle of 60° from the first line. Lining pegs should be made of some hard wood if they are to survive the burn.

It must be remembered, when lining on hilly land, that the planting distance is a horizontal distance and the lining ropes must therefore be kept horizontal when measuring from one stand to another. With a gradient of 1 in 3 the distance along the ground will be 5.3 per cent greater than the horizontal distance. Contour lining is only recommended where terracing is considered essential since the unevenness of stand produced will lead to some reduction in yield. However, on very hilly land maintenance of harvesting paths and the sowing of covers along the contour may assist soil conservation, and on erodible soils small contour bunds or silt pits may be constructed.

Platforms and terraces

Except in certain inland areas of Malaysia (usually following rubber) the oil palm has been very little planted on steep land. Although terracing has been advocated for Malaysia, there is strong evidence that palms on platforms establish and grow better than those on terraces (Plate 8.2). Platforms are usually circular with a diameter of 3–4 m and have a slope back into the hillside of 7–8°. Terrace construction, apart from its high cost, has certain disadvantages: terrace soil becomes trampled by harvesters and forms a poor rooting medium for the palm, while the backward slope of the terrace often leads to pools of water collecting round the young palms, retarding their growth. In a hilly area of Malaysia where one part was terraced while on the adjoining area platforms of 2.5 × 3.0 m were constructed the terraced area came into bearing nearly a year later than the area with platforms although planted at the same time with the same material. On very steep terraced land it is difficult for harvesters to move between the palm and the bank behind. Advocates of terracing have put forward arguments which are mostly of a managerial and non-agronomic nature (Turner and Gillbanks, 1974). Nevertheless on some very steep land with slopes over 20° terracing may be unavoidable, though it is questionable whether such land should not have been left in forest (Plate 8.3). For Malaysia, Leamy and Panton (1966) define the steep land boundary as the line separating land with slopes greater and less than 20°, and rightly assert that though land below the boundary is suited to tree-crop agriculture, land with slopes in excess of 20° is better suited to permanent forest.

Clearing paths

Where lining follows underbrushing (see E, Table 8.2) it may be possible to carry out the clearing of paths 1.5 m wide at the same time, espcially if these paths are to follow the rows. Clearing paths is usually carried out later, however, but also after lining (e.g. D and H), and their construction in the centre of every other avenue rather than along every line is advocated; these paths will act as

Pl. 8.2   Palm planted on a platform on hilly land in Malaysia.

essential inspection paths in the first few years and then become
harvesters' paths.

Holing

It will be noted that holing is not included in some of the schemes
set out in Table 8.2. This is because on the free-draining sandy soils

Pl. 8.3   Oil palms terraced on steep land in Malaysia.

of West Africa it has been shown that holing prior to planting has no effect on establishment or subsequent growth and yield (Spar-naaij and Gunn, 1959). On heavy or stony land, however, holing is usually undertaken several weeks before planting, the holes being 60 or 90 cm cubes according to the soil type. Before planting, the holes should be refilled with surrounding topsoil and allowed to settle. The smaller hole in which the seedling's ball of earth is placed is not dug until the time of planting. Where holing is not carried out in advance of planting, the planting site must be well weeded so as to prevent any competition between the newly planted seedling and surrounding weeds.

In Malaysia, experiments on the need for holing have not been carried out and several different practices exist. Holing prior to planting as described above is carried out on most soil types but it has been suggested that on heavy coastal clays it is disadvantageous to hole before planting as the holes may fill with rain-water and planting, filling and consolidation become difficult. Under these circumstances a hole sufficiently large to accommodate the ball of earth and to allow for proper firming of the soil around it is all that is needed (Bevan and Gray, 1966). Holes with dimensions 45 × 45 × 40 cm are recommended.

Felling

Costs of clearing planting lines will be considerably reduced if directional felling is practised. On flat or undulating land this felling should be N–S to avoid heavy trunks falling across the planting lines. On steep areas the forest is best felled along the contour. It is sometimes easier to fell the small trees first and later to fell the large trees in a separate operation.

A varying number per hectare of large buttressed trees will have to be felled by axing from a platform 2 m or more above ground level. The stumps of these trees will have to remain in the ground. Felling must be followed by the lopping of any branches or the cutting up and moving of any trunks which fall across the cleared lines. These operations may, however, be left until after burning. There may be some trunks which are too large for cutting up and removing, but with good directional felling very few of these should have fallen across the lines.

With hand axing, felling may take up to 50 man-days per hectare, but if chainsaws are used much less labour is required. On inland soils in Malaysia two chainsaw men with four axemen can fell more than a hectare a day (Balakrishnan and Lim Cheong Leong, 1976). In peat jungle chainsaw felling has been done, after underbrushing, at a rate of less than 3 man-days per hectare, cutting small trees (15–150 cm diameter) at 30 cm and larger ones (30–150 cm diameter) at 60–150 cm above ground level (Rasmussen *et al.*, 1981). In Colombia, where much detailed attention has been given to felling trees of different sizes (Huguenot, 1965), a rate of 5 man-days per hectare can be maintained in forest with 230 trees per hectare of diameter less than 50 cm, 24 of 50–100 cm, 14 of 1–1.5 m, and 10 trees of over 1.5 m diameter.

Felling must be correctly timed. In regions with a distinct dry season the work will be done as early in the dry season as possible, so that burning may be carried out easily as soon as the material has dried out. Beating down and some cutting up of branches is often done before the burn (see A, C and D, Table 8.2), and the burn is so complete that little more than the clearing of planting rows is then needed. In non-seasonal countries such as Malaysia felling is done 8–12 weeks before burning, the latter being timed for one of the two drier periods of the year. If the felled forest is left longer there is danger of regrowth. The burn is often incomplete, so cutting up and stacking for reburning is undertaken after the burn (see H, Table 8.2) and this work may span the next dry period (e.g. June/July) so that planting is unlikely to begin less than 6–7 months after the first burn. In regions of very even rainfall, however, where the seedlings may be planted almost throughout the whole year, felling can proceed at a steady rate regardless of season, burning follows some 3 months later and planting will be done as soon after stacking and reburning as possible.

Burning

In regions with a distinct dry season a good burn is usually achieved without difficulty unless clearing operations have been started too late and the dry weather is brought to an abrupt end by heavy rain. Indeed, the greater danger is usually the fire hazard to neighbouring planted areas where dry debris may be lying about or combustible weeds such as *Imperata* or *Eupatorium* have gained access. For this reason wide cleared fire traces are essential; these may vary from 20 to 40 m in width to prevent scorching, but however wide a trace may be there is always a chance of sparks setting alight adjoining areas and for this reason water carts or drums of water at the roadside should be ready so that any smouldering vegetation may rapidly be quenched. *Fire is a perpetual hazard to planted areas throughout the dry season in West Africa whether large burns are in progress or not; several large planted areas have been totally lost through failure to take elementary precautions.* Some damage to small areas has also resulted during a burn owing to a change of wind and inadequate fire traces. A fire may start in a felled area before it was intended and before adequate fire traces have been cut. For this reason, on any developing plantation, some form of patrol and emergency fire service is needed.

Firing in these areas is simple: a line of men sets fire to the felled forest at the windward side at a time of day when the prevailing wind is blowing moderately. The fire burns rapidly and fiercely through to the other side with the men following it; any portion which may, by chance, be missed as the fire proceeds on its course may then be kindled on the windward side.

Firing in non-seasonal regions such as Malaysia is carried out rather differently (Edgar, 1958). The men are drawn up at 20 m intervals along the side *away* from the prevailing wind. They then advance through the area into the wind setting fire, with firebrands and kerosene, to heaps of dried material at about 20 m intervals, sprinkling kerosene if needed. This method is necessary in a region where the felled forest is not so easily fired. Firing should be carried out at about 1400 hours on a day when there is a breeze. On a still day the burn may be a failure. Firebelts of standing forest are favoured; these are felled, cleared and burnt separately after the main burn.

Post-burning operations

From earlier descriptions it will have been apparent that the amount of work remaining after the burn will depend on whether such operations as lining and path clearing have already been done and on the success of the burn or thoroughness of reburning. If the lining and clearing of the rows have been well done before the burn, the work of realigning and final pegging will be very light and the field will be ready for planting almost immediately. If lining follows burning, the methods already described must now be followed.

In Malaysia the work of cutting up, stacking and reburning is usually very considerable and the stacking is often done round the base of large buttressed tree stumps. On steep or undulating land this work is followed by lining and the construction of platforms or terraces. With slopes of 12–20° platforms are suitable and the normal plantation layout and planting distances can be maintained (Taillez, 1975). Above 20°, however, terracing will probably be necessary although, as already mentioned, it is doubtful if planting should be undertaken at all on such steep land.

Field preparation by mechanical methods

Up to about 1960 the great majority of plantations were opened by hand labour without even the assistance of chainsaws, although some early experience had been gained in Zaire with the use of both winches and tractors with bulldozers for moving axed timbers into the interlines (Dufrane, 1954). These methods were, however, still slow and the amount of labour considerable.

A number of more complete methods of mechanical clearing have how become established, the first closely following method B of the Ivory Coast already referred to in Table 8.2. Heavy-track tractors of at least 235 hp are fitted with special bulldozer blades which include an upward extension for the protection of the tractor and a projection at the left-hand end of the blade (Huguenot, 1963). The blade measures 4.38 × 1.60 m. Dealing with the smaller and medium-sized trees presents no difficulty, but buttressed trees of over 75 cm diameter have to be prepared for felling in two stages, firstly by breaking up the sides of the base of the trees with the bulldozer blade and extension, and secondly by breaking up the buttresses opposite the felling side. Under the conditions in the Ivory Coast two or three trees per hectare will still have to be felled by hand (Fraisse, 1964). Experience has shown that with tree densities varying from 300 to 500 per hectare the time taken to fell by these methods will vary from 3 to 10 hours per hectare. After felling, the larger trunks are cut up by mechanical saws to ease the subsequent work of windrowing. Methods of mechanical forest felling have recently been further advanced by the introduction of the 'tree crusher', a 50 tonne tractor and blade which is capable of clearing a hectare in less than $1\frac{1}{2}$ hours (Martin, 1970).

After the forest is felled the area is marked out with lines whose distance apart is twice the distance between rows. For instance, with

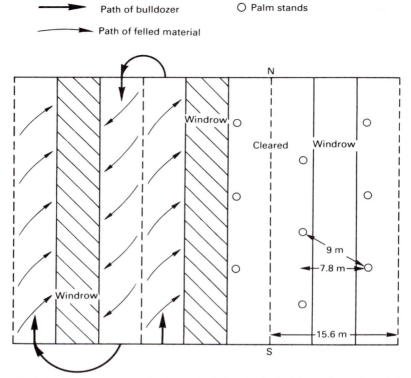

Fig. 8.2 Alternate avenue clearing and windrowing by bulldozer, Ivory Coast (after Huguenot, 1963).

9 m triangular planting the distance between rows being 7.8 m the lines would be 15.6 m apart. As can be seen in Fig. 8.2, the cut-up forest trees and stumps are then worked into the centre of the blocks formed by these lines and windrows of felled material fall into place between each pair of rows, the space within the pairs and along the rows being left absolutely clear. Felling is usually followed by a light burn and the sawing up with mechanical saws of the felled vegetation in order to make the subsequent windrowing easier. Stumping, which is a costly operation, may be done with the aid of projections to bulldozer blades, but the shearing of the stumps with a Rome KG spur blade is preferred. Windrowing is estimated to take 3 tractor-hours per hectare for forest of 300–400 trees per hectare of which only 2–5 are over 150 cm in diameter. Martin (1976) has given estimates of the additional time needed both for felling and windrowing denser forests.

The system outlined above has the advantage of keeping every row clear for ease of inspection and it is also possible, from the start, to maintain every other avenue by mechanical means. While a good burn could greatly reduce the amount of material in the avenues containing the heaped debris, these avenues will inevitably become covered with creepers and all manner of plants will grow up within them. Maintenance will therefore be difficult until considerable rotting has taken place. The use of weed-killers may, however, appreciably reduce this disadvantage. The method relies on the heaviest of tractors and very expensive equipment, and unless the size of the programme justifies high capital outlay or contractors with such equipment are available in the district, methods in which lighter and cheaper tackle are employed are to be preferred. The method also entails burning debris after the cutting up of the trunks rather than the burning of the whole field after felling and drying out.

Another method, which has been used in Nigeria, employs lighter crawler tractors and avoids gross disturbance of the soil, but entails the continued use of hand labour. The sequence is: (1) underbrushing at ground level; (2) lining; (3) holing; (4) felling as many trees as possible at ground level with axes and saws and large buttressed trees above the buttresses; (5) burning; (6) cutting felled trees by chainsaw into 6 m lengths or less; (7) clearing the avenues by moving the logs by tractor and angledozer on to the planting lines but between the holes; and (8) reburning the stacked timber. After this reburning, the lines may be tidied up by a tractor pushing timber at least 1.2 m away from the holes in the line (Sly, 1963). The method has the advantage that all avenues between the palms are cleared and ready for mechanical maintenance while the remaining timber is reduced to a minimum by burning. Moreover, the amount of labour used in planting can be greatly reduced since the plants can be brought by wheeled tractor and trailer into the field and off-loaded at the planting sites.

One of the difficulties of tractor and bulldozer clearing in a seasonal climate is that if, through breakdown, there is an insufficiency of tractors, the forest may not have been felled and dried out in time to obtain a satisfactory burn. The need to carry through all the work during the relatively short period of the dry season means that more equipment per unit area is required. Where the planting season is less restricted, the work can proceed at a slower and more even pace and operations can be under way in one area while planting is proceeding in another.

The question of whether to employ fully mechanical methods or to use largely hand methods aided by mechanical saws has been much debated. It has been argued, for example, that the use of the methods described on p. 370 and Fig. 8.2 enables larger areas to be planted in shorter time thus ensuring the highest return on investment (Martin, 1970). Experience in Malaysia has shown, however, that equally large areas can be opened per annum through the employment of contractors using mechanical saws and hand labour; and in Africa some plantation companies continue to prefer hand methods. Cost comparisons on a per hectare basis will be affected by prevailing labour rates, and in West Africa labour has been cheap enough to render fully mechanical methods much more expensive than hand clearing assisted by mechanical saws. A very complete study of three clearing techniques, viz. (1) manual with chainsaws for felling and cutting up; (2) as (1) but with tractors and bulldozers for stacking; and (3) bulldozers for felling and stacking, was recently carried out in Malaysia. This showed that the capital outlay for tractors and bulldozers for technique 3 was excessive. Techniques 1 and 2 showed little difference in cost and both were much cheaper than technique 3; however, technique 2 was preferred because its cost was not so dependent on a good initial burn, and pruning, stacking and reburning could be carried out without difficulty in all circumstances (Khamaruddin, 1980). One argument used against hand methods has been the need to recruit and house large labour forces at an early stage of development; however, in most countries there is a labour surplus eager for employment and a very high proportion of the labour employed for field preparation can later be employed in maintenance and harvesting. At the present time (1985) fully mechanical methods can only be justified in areas where sufficient manpower is not obtainable for clearing.

In spite of the high cost, mechanical felling and clearing have been used on a large scale in the Ivory Coast and some other countries. Experience has shown that excellent plantations can be obtained where the terrain is suitable and the work is up to time; there have been cases, however, where operations have become seriously behindhand and, with the onset of wet weather, soil puddling has followed and the early growth of the palms has been adversely affected.

In some situations a combination of hand (including chainsaw) and mechanical methods has been found most economical. In Malaysia, if felling becomes seriously behindhand or if the burn has been less than 30 per cent, it has been found best to stack with bulldozers before reburning; but full mechanization is still considerably more expensive than either hand clearing or partial mechanization (Balakrishnan and Lim Cheong Leong, 1976).

Preparation in non-forest areas

There are many areas of the wet tropics where the forest has been felled and the land used for food-cropping and then abandoned. When the intensity of cultivation has not been high and the district is one of low population these food-cropped areas usually revert to forest in stages, shrubs taking hold through the initial ground cover of grass and creepers and the shrubs being succeeded by secondary and later primary forest trees. In these cases preparation would of course be carried out in the same manner as with primary forest or jungle.

Where population density is high or the areas extensive, many factors may prevent a reversion to forest. In Malaysia for instance, areas abandoned to lalang grass (*Imperata cylindrica*) are often regularly, though unintentionally, fired. This tends to kill out the young shrubs and perpetuate the lalang. In West Africa, areas of 'derived savannah', consisting of grasses and fire-resistant trees, are extensive. These are usually to be found in districts with a prolonged dry season where food-cropping has been intensive and been followed by deliberate annual firing of the bush to assist in the capture of game. In other districts (e.g. the 'bad lands' of eastern Nigeria) a poor grassland has even encroached on the areas of palm grove which have become ever-decreasing 'islands' in a sea of grass. Here the palms do not replace themselves and so the grassland obtains entry wherever a patch of palms dies out. The whole succession is often assisted by the roaming of large herds of cattle. In the Ivory Coast, areas of this type are, as in the Far East, dominated by *Imperata*, while in both the Ivory Coast and Nigeria strong growths of Siam weed (*Eupatorium odoratum*) have lately become a menace both in cleared areas and even in recently planted areas. This plant takes fire easily in dry weather.

It was in Malaysia that the recovery of large areas under *Imperata* was first investigated. During the Japanese occupation of 1942–5 large tracts of forest had been felled for tapioca and other food crops while on rubber estates many new plantings were soon swamped by lalang. Hand eradication was not easily carried out owing to high cost and lack of labour. Early experiments showed that disc ploughing followed by several disc harrowings was the most satisfactory method and that the completeness of eradication was much improved by a final cultivation with a straight-tine cultivator (Hartley, 1949). The establishment of legume covers was an essential part of the sequence and very small quantities of lalang had to be dug out in subsequent patrolling. In cocount and oil palm plantations, where light growths of lalang had become established, rotary cultivation behind wheeled tractors was found to be satisfactory.

Costs of mechanical eradication of lalang over large areas were found to be less than mechanical spraying of sodium arsenite, but two sprayings at weekly intervals *before* rotary cultivation effectively aided eradication (Keeping and Matheson, 1949).

More extensive work in the Ivory Coast confirmed the Malaysian methods (Surre *et al.*, 1961); there, ploughing is preceded by the use of the heavy 'cutaway' disc harrow known as the 'Rome-plow' which breaks up the vegetation. Ploughing with a six-furrow disc plough is then carried out to turn over the soil; this is followed by one or two cross-harrowings with the Rome-plow and four to six disc harrowings with a wide and light set of disc harrows at weekly or fortnightly intervals. The whole sequence takes about $10\frac{1}{2}$ tractor hours per hectare or 13 hours if a second cross-harrowing with the Rome-plow is needed.

In derived savannah areas free from trees and with no dominating weed, cultivation may not be required. The planting sites only need to be cleared and covers planted among the sparse vegetation. Such areas are likely to be very poor and their satisfactory usage for the oil palm has not yet been established. There is no reason, however, why intrinsically suitable areas which have become lalang-dominated through mismanagement should not be reclaimed by the planting and proper management of oil palms and, indeed, several such areas have been successfully brought into bearing in the Far East and Africa; the low capital cost of planting has been an encouragement. Fertilizer requirements will need added attention, however. The procedure in Malaysia has been similar to that in Africa, the disc plough being the main implement, preceded or followed by (or sometimes replaced by) the heavy cutaway Rome-plow. The latter will cut down to 20–23 cm but disc ploughing to 30 cm is sometimes required.

No weed-killer has yet approached sodium arsenite in effectiveness and cheapness in the destruction of lalang, but as it is highly poisonous its use has been carefully controlled and, in some countries, avoided altogether. Sodium arsenite is a contact weed-killer and successive sprayings are required to destroy the leaves and weaken the plant so that it can no longer produce shoots from the runners. The quantity and concentration of sodium arsenite to be sprayed depend on the strength of lalang growth, the weather at the time of spraying, and whether a wetting agent is used. On average a $1\frac{1}{2}$ per cent solution obtained by dissolving 10 kg powdered sodium arsenite (80 per cent As_2O_3) with 250 g of a wetting agent in 700 litres of water will be enough for an initial spraying of 1 hectare. Spraying is continued every 8–10 days for seven to ten rounds, the quantity of solution required to wet the leaves decreasing with each round. After the spraying has been completed, sporadic lalang may be eradicated by digging or by 'wiping' with oil. Several proprietary oils such as Shell Lalang Oil W or A-601, Soracide PY or PYD are on the market. An absorbent cloth is simply dipped into the oil, squeezed of excess oil and then wiped along the blades from the base upwards.

As alternatives to sodium arsenite, dalapon (sodium 2,2-

dichloropropionate) and Roundup (isopropylamine salt of glyphosate) have been successfully used. Dalapon is sprayed as a 1.5 per cent solution with three treatments at 3-weekly intervals which results in the use of about 7–8 kg of the commercial product per hectare per treatment (Coomans, 1976). Roundup has been recommended at 4 kg a.e. (acid equivalent) per hectare, with a 'touch-up' spraying 60–90 days later (Wong Phui Weng, 1976).

In recent years Siam weed, *Eupatorium odoratum*, has become a serious weed in Africa, and it is also to be found in the Far East where, however, it does not appear to grow quite so high or luxuriantly. Its control in established plantation is described on p. 437. If planting is to be carried out in an area where patches of *Eupatorium* exist, these patches should first be eradicated. If eradication can be undertaken and completed before planting, then a mixture of 2,4-D and 2,4,5-T at a rate of 5.5 kg a.i. (active ingredient) per hectare may be used. If, however, eradication must continue after planting, it is too dangerous to use these weed-killers since they affect the young palms. The latter are tolerant of substituted triazine and substituted urea herbicides, and of these atrazine at 4.5 kg a.i. per hectare and diuron (DCMU) at 5.5 kg per hectare have been found effective against Siam weed if applied in dry weather (Sheldrick, 1968). Other herbicides which have been successfully used against *Eupatorium* in young palms are Asulox 40, Roundup and diuron/paraquat mixtures (Aya and Fayemi, 1981).

Inter-row planting of covers or crops

Many areas of oil palms have been established without any planting of inter-row cover plants. The natural vegetation was allowed to come up and was kept under control. In many parts of West Africa *Pueraria phaseoloides* becomes dominant between the rows without deliberate seeding, while occasionally patches of *Centrosema pubescens* arise in the same manner. With the advent of burning the early establishment of a legume cover became more urgent, and it is now regarded as an essential part of any planting programme. As, however, the establishment and maintenance of covers continue into the post-planting period, cover planting will be dealt with in Chapter 10. The sowing of a cover is normally done at the end of the land-preparation operations and before planting although there may be exceptional circumstances where covers are sown after planting. In West Africa, for instance, planting can start in March, well before the rains have fully set in, but covers establish better if planted with the first heavy rain in April or May. In Malaysia covers may be sown from 3 to 6 months before planting.

The inter-planting of food crops is rarely an attractive proposition for large estates since they are not organized for such work and the profit to be obtained is very doubtful. To the smallholder, however, a quick-growing cash or food crop while awaiting a return from the palms may be an attractive proposition and sound husbandry. Such cropping should not be confused with a deliberate attempt to combine the cultivation of other crops with the planting of oil palms; such combinations present quite different problems which will be discussed in Chapter 12. *Establishment inter-cropping* is temporary cultivation only and it has no effect on the preparation of the plantation or the spacing of the palms, except of course that a more thorough clearing and burning will be needed.

Several trials of establishment inter-cropping have been carried out in Africa. In Zaire the planting (alone or in succession or combination) of cassava, maize, hill rice and bananas for 1, 2 or 3 years had, in general, a beneficial effect on palm yields and, after 16 years, no deleterious effects had appeared (INEAC, 1955). The area used was in forest, the catch-crops were planted immediately after the palms and the control plots were planted with *Pueraria phaseoloides*. A similar trial planted on forest land near Benin in Nigeria in 1940 included both cropping for 2 years and cropping for as long as crops could be obtained (Sparnaaij, 1957). In practice the latter treatment entailed cropping with maize, yams and cassava until shade made the growing of cocoyams the only possible culture, and this was continued until the twelfth year. In the 2-year cropping plots and the early years of the cropping-to-exhaustion plots, good crops were obtained similar to those in the Zaire trial, and showed that food-cropping for 2 or 3 years may, in these circumstances, be an attractive proposition to a smallholder. In both trials there was a tendency for early bunch yields to be improved by inter-cropping; in the Nigerian trial both leaf production and early yields were significantly increased. In neither trial, after 16 or more years of harvesting, was there any significant fall in yields following the early inter-cropping treatments, though in Nigeria a tendency for yields to fall in later years has been noticed. The cumulative yields over periods of 16 and 19 years' harvesting and the early and late yields are given in Table 8.3 as percentages of the control (*Pueraria* cover).

The early beneficial effect of planting food crops with the palms as a last 'preparatory' treatment in place of a leguminous cover or a natural cover is perhaps unexpected, but soil studies in the Nigerian experiment have helped to provide an understanding of it. Soil analyses were carried out in 1941, 1945, 1951, 1956 and 1961 and the inter-cropping treatments showed the soil effects which one would expect to follow tillage, namely a general reduction of fertility, as judged by total exchangeable cations, below that of the plots having a natural cover. In this respect the *Pueraria* 'control' showed similar, though not such accentuated, soil effects as the

Table 8.3   *Bunch yields of cropped plots as a percentage of control (*Pueraria cover) in Zaire and Nigeria

Inter-palm treatment	Zaire			Nigeria		
	3rd year	16th year	16 years cumulative	1st–3rd year cumulative	14th–19th year cumulative	19 years cumulative
Control (*Pueraria*)	100	100	100	100	100	100
Natural cover	—	—	—	103	94	96
Two years' food cropping	125	105	117	108	88	95
Cropping 'to exhaustion'	—	—	—	120	89	98
Three years' bananas	124	100	113	—	—	—

cropped plots, thus indicating that the cultivation needed to establish the cover may be of significance.

The early beneficial effects of establishment inter-cropping have therefore been obtained in spite of a small reduction in topsoil fertility. It is believed (Kowal and Tinker, 1959) that this is because competition from weeds is reduced considerably by cropping (and this may apply, in the early stages, to cover establishment), and further because cultivation makes some soil nutrients more readily available. Soil differences due to initial cultivation tended to be large after 5 years, but after the palms had been in the field for 16 or 20 years the differences in potassium were insignificant, those of magnesium small, while those of calcium became relatively large. Small differences in exchange capacity and carbon and nitrogen percentages persisted (Tinker, 1963).

The results of the trials so far mentioned were obtained in areas of Zaire and Nigeria opened from heavy forest. In areas degraded by food-crop farming or by annual burning, fertility needs to be rebuilt. In experiments on such land in Nigeria (Sparnaaij, 1957) even moderate yields could not be obtained without heavy fertilizer dressings, the beneficial effect of establishment inter-cropping was short-lived and regular tillage soon became harmful (pp. 424–7).

Some interest has been taken in Asia in establishment inter-cropping and in some smallholders' schemes inter-cropping has been introduced. An experiment in Malaysia on marine alluvial clay soil showed that much depends on the crop used and the amount of tillage entailed. The field was a replanting, and cultivation was carried out to 1.8 m from the young palms. With soya beans planted for 2 years there was no effect, in comparison with a legume cover, on growth or early yield, though there was a suggestion that if deep tillage is undertaken for two annual crops then palm growth is slightly impaired through root damage. Two cassava crops grown in successive years severely retarded growth and reduced early yield by 14 per cent through competition for both light and nutrients, particularly N and P. However, all inter-cropping treatments increased in yield relatively to the legume cover treatment with palm age, and by the fourth bearing year differences were very small (Chew Poh Soon and Khoo Kay Thye, 1976).

It may be concluded that establishment inter-cropping can be practised on fertile, water-retentive soils for 1 or 2 years without substantial loss of crop if the crops are not allowed to compete severely for light, nutrients and water, and provided tillage is not too severe. On light soils in Africa and elsewhere inter-cropping should only be practised on forest land or where an adequate programme of manuring both inter-crops and palms is possible; and the practice should be avoided on very light soil already degraded by bush-rotation farming or periodic burning.

Soil drainage

The oil palm will not grow and fruit in standing water whether the water level be at the soil surface or below it. The majority of plantations are situated either on comparatively light soils of great depth or on undulating land with soils of varying physical composition where the permeability of the soil and the natural drainage courses are adequate. On such plantations drainage problems tend to be localized, i.e. they are confined to small pockets of valley land where the central stream needs to be kept clear and where flood-waters from adjoining rivers may temporarily back up over some planted areas. In these cases the clearing of main drains (canalized streams) and the digging of subsidiary drains from a few low-lying areas is often all that is needed. The bunding of small areas from occasional river floods is expensive and seldom justified. Assistance can sometimes be given to these areas by the digging of 'foothill drains' which lie at the foot of the slopes which rise from a flat valley bottom; these drains are dug to follow the contour and carry seepage water from the slopes directly to the main valley drain at its lower end. They must be deep enough to ensure that the seepage water enters them.

Although the majority of plantings may be on areas which need comparatively little attention to drainage, flat coastal or estuarine alluvial clay soils account for a considerable proportion – and the highest-yielding section – of the plantations of Malaysia. These plantations are flat, are sometimes below high-tide level, and the clay soil may be only slowly permeable. Initially these clays may be overlain by a layer of muck or peat soil the drainage of which presents special problems.

A drainage system for these soils should be 'built' back from the chosen outlet point or points; a natural outlet should be utilized if possible and the main drain extended from it, but no natural outlet may exist and one has then to be constructed across a main river bank or into tidal waters. The main drain is cut as straight as possible through the area to be drained and inter-field drains or collection drains are then cut at right angles to meet it. Finally inter-row or field drains are cut parallel to the main drain to meet the inter-field drains.

On areas of heavy coastal clay, whether overlain by peat or not, field drains are commonly dug every fourth row and sometimes even every other row. On the better structured clays, drains may be less frequent, every sixth row or about 47 m, while on lighter soils intervals of 100 m are possible. Inter-field or collection drains are dug at intervals of about 200–400 m. With every fourth row drainage and the palms spaced at 9 m triangular the length of field drains to be dug will be about 320 m per hectare.

In heavy clay soils the drain sides may be sloped at an angle of 70° from the horizontal but angles of 50–60° are more usual. The main drain should be 3.5–5.0 m wide at ground level and about 2.5 m deep. Collection drains may be 2.0–2.5 m at the top, 1.2–1.8 m deep and 0.6–1.0 m at the bottom. The most usual size is 2.5 × 1.8 × 0.6 m. Inter-row drains are 1.2 × 1.0 × 0.5 m or 1.2 × 1.2 × 0.3 m. All drains are nowadays dug with mechanical equipment, drag line excavators being used for main and inter-field drains and trencher equipment for inter-row drains (Plate 8.4).

The drainage system should be designed to interlock with the transport system so that the minimum of bridging is required. Neither the railway nor the road system may pass over the inter-row drains, but excavators and tractors can be driven into each field from one side of the collection road without passing over a bridge. Access to the rows of palms for inspection can be obtained at both sides of a field either directly or over a footbridge. This is illustrated in Fig. 8.3 (Gray, 1965).

When peat overlies clay, care needs to be taken to see that drainage from the peat is not too rapid. If peat is dried out quickly it is found that reabsorption of water is difficult and slow ('irreversible drying'), and the palms suffer accordingly. In any part of the plantation which is covered with peat, a system of small water gates must be installed in the collection drains so that in dry weather water may be drained very slowly from the area.

Many flat coastal or estuarine areas must be protected from tidal waters by bunding. This is an expensive undertaking since substantial water gates need to be constructed in the main outlets through the bunds. Hinged automatic gates can be installed which will close under the pressure of the tidal water and open when the tide

Pl. 8.4 Constructing an inter-row drain with an excavator on coastal alluvium in Malaysia.

recedes and water flows from the main drains. Soil should, if possible, be taken from the seaward or river sides; bunds should be at least 35 cm above high-tide level and be grassed over. The base of the bund should be double the mean height of the tide, while the top should be about 1 m wide. A 3 m tide thus demands a bund 6 m wide at the base, 3.5 m high and 1 m wide at the top. This gives a slope of about 55° to the horizontal. In the Far East crabs bore holes through the bunds, but they can be kept down by applying to each hole 0.1 litre of diesoline in which 70 g of lindane dispersible powder has been mixed. Where crabs are likely to be troublesome bunds are now constructed by excavator from borrow pits 14 m from the bund on the landward side (Edgar, 1958).

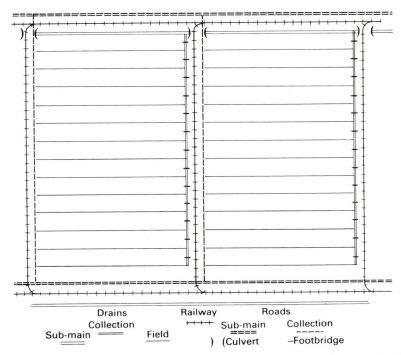

Drains
Collection
Sub-main === Field

Railway
++++ Sub-main

Roads
Collection
Sub-main ====

Collection
– – – –
–Footbridge

) (Culvert

Fig. 8.3 Diagrammatic representation of drainage, rail and road systems for flat, low-lying oil palm plantations.

Irrigation

In Chapter 3 a description was given (p. 104) of the relationships found by French workers between bunch production and water deficit, the latter being calculated for periods of 10 days on the assumption of a soil water reserve of 200 mm and a potential evapotranspiration of 150 mm for dry months and 120 mm for months with more than 10 wet days (IRHO, 1969). At Grand Drewin in the Ivory Coast, where the mean annual water deficit for the 10-year period 1957–66 was 590 mm, an irrigation experiment was conducted on palms planted in 1960 in which each irrigated palm was surrounded by four nozzles each watering a circle of 3 m diameter. The treatments were: A. Sufficient water applied to assure 50 mm rain + irrigation water per 10-day period. B. Water requirements judged by midday stomatal aperture (see p. 146). T. Control. The mean bunch yields for 1966–7 and 1967–8 in tonnes per hectare per annum were: A. 23.6, B. 22.2, T. 10.6 (Desmarest, 1967).

Other irrigation experiments have been conducted at La Mé, also in the Ivory Coast, and in Benin (Dahomey). At La Mé, where the annual deficit averages 254 mm, water was supplied through a system of spray lines laid along the ground (IRHO, 1969/73). The cumulative yield over the first 3 years of bearing in the irrigated plots was 32.7 tonnes per hectare of bunches against 18.5 tonnes in the non-irrigated plots; the increased yield was mainly due to a higher sex ratio and number of bunches produced per palm. At Pobé in Benin (Dahomey) where the mean annual deficit is 520 mm, a drip irrigation system, designed to compensate fully for the dry-season evapotranspiration, has given similar results even though the volume of wetted soil represented only 35–40 per cent of the total rooting volume (de Taffin and Daniel, 1976).

In the even climate of the Far East, calculated water deficits (p. 106) are only occasional, and little benefit is therefore to be expected from irrigation. Midday closure of stomata can occur in dry periods however (Corley, 1973), and an irrigation trial in southern Malaysia showed that small increases in bunch number, largely attributable to reduced abortion, can result from carefully gauged water applications during the drier periods of the year (Corley and Hong Theng Khong, 1981). Positive effects of irrigation on leaf production, sex ratio and oil to mesocarp were also detected in this trial, but the overall increase in bunch yield of only 2.4 per cent, and in palm oil of 5.8 per cent, would not justify irrigation as a commercial practice. The effect on kernel production was not determined.

The remarkable results obtained from irrigation in West Africa indicate that in a climate with several dry months irrigation can raise yields to levels near to those obtained in the Far East. Although the methods described above will not always be economic, large-scale irrigation has been employed in Benin (Dahomey), and in several other regions, e.g. south-western Mexico, northern Colombia and south-western Ecuador, there is now an interest in raising yields by irrigation or in planting in areas where the rainfall is clearly insufficient but irrigation water can be supplied. These areas are largely flat and low-lying, and the provision of water from canals or drains is being attempted. The essential conditions for successful irrigation are: (a) deficits must be made good by frequent and calculated applications; occasional applications will be largely valueless; (b) the method used must supply water to the surface roots; (c) areas must not remain waterlogged for long periods; (d) the expected yield increase must be substantial.

References

Aya, F. O. and **Fayemi, A. A. A.** (1981) The effects of some weed control treatments on young oil palms and the weed spectrum in the field. In *Oil palm in agriculture in the eighties*, ISP, Kuala Lumpur.

Balakrishnan, V. and **Lim Cheong Leong** (1976) Jungle clearing by FELDA for planting oil palm. *International developments in oil palm*, ISP, Kuala Lumpur.

Barrett, R. G. (1965) A method of calculating the optimum 'maximum carry' on oil palm estates. Sabah Planters Association. Oil Palm Seminar. Mimeograph.

Beirnaert, A. (1942) Nienwere in de cultuurtechniek van warme landen. *Bull. agric. Congo belge*, Leopoldville, **33**, 55.

Bevan, J. W. L. and **Gray, B. S.** (1966) Field planting techniques for the oil palm in Malaysia. *Planter, Kuala Lumpur*, **42**, 196.

Bevan, J. W. L. and **Gray, B. S.** (1969) *The organisation and control of field practice for large-scale oil palm plantings in Malaysia.* Incorp. Soc. of Planters, Kuala Lumpur, 166 pp.

Bocquet, M. and **Michaux, P.** (1961) Choix d'une méthode d'ouverture en forêt pour une plantation de palmiers à huile. *Oléagineux*, **16**, 533.

Chew Poh Soon and **Khoo Kay Thye** (1976) Growth and yield of intercropped oil palms on a coastal clay soil in Malaysia. In *International developments in oil palm*, ISP, Kuala Lumpur.

Coomans, P. (1976) Contrôle chimique de l'*Imperata*. *Oléagineux*, **31**, 109.

Corley, R. H. V. (1973) Midday closure of stomata in the oil palm in Malaysia. *MARDI Res. Bull.*, **1**(2), 1.

Corley, R. H. V. and **Hong Theng Khong** (1981) Irrigation of oil palms in Malaysia. In *Oil palm in agriculture in the eighties*, ISP, Kuala Lumpur.

Coulter, J. K. (1950) Organic matter in Malayan soils. *Malay. Forester*, **13**, 189.

Desmarest, J. (1967) Essai d'irrigation sur jeune palmeraie industrielle. *Oléagineux*, **22**, 441.

de Taffin, G. and **Daniel, C.** (1976) Premiers résultats d'un essai d'irrigation lente sur palmiers à huile. *Oléagineux*, **31**, 413.

Dufrane, M. (1954) Contribution à l'étude de la mécanisation de l'aménagement et de l'entretien des plantations de palmiers. *Comptes Rendus des Journée d'Etudes sur la Mécanisation de l'Agriculture au Congo Belge*, Oct. 1954, p. 119.

Edgar, A. T. (1958) *Manual of rubber planting (Malaya).* Incorp. Soc. of Planters, Kuala Lumpur, Malaysia.

Fraisse, A. (1964) Essais de mécanisation pour l'abattage de la forêt à La Mé. *Oléagineux*, **19**, 73.

Gawthorn, D. J. (1967) The planting of *Elaeis* on terraces. *Planter, Kuala Lumpur*, **43**, 502.

Gray, B. S. (1965) Private communication.

Gunn, J. S. (1960) *WAIFOR eighth annual report, 1959–60*, p. 52.

Hartley, C. W. S. (1949) An experiment on mechanical methods of lalang eradication. *Malay. agric. J.*, **32**, 236.

Huguenot, R. (1963) Coup mécanique des souches et andainage sur sol de forêt. *Oléagineux*, **18**, 623.

Huguenot, R. (1965) Utilisation des scies mécaniques pour l'établissement des plantations sur forêt. *Oléagineux*, **20**, 303.

INEAC (1955) Division du palmier à huile. Palmier à huile et plantes vivrières. *Bull. Inf. INEAC*, **4**, 319.

IRHO (1969) Recherches sur l'économie de l'eau à l'IRHO. L'eau et la production du palmier à huile. *Oléagineux*, **24**, 389.

IRHO (1969–73) *Rapport annuel 1968*, p. 57, and *Rapport d'activites 1972–1973*, pp. 67–8.

Keeping, G. S. and **Matheson, H. D.** (1949) Mechanical spraying for the eradication of lalang. *Malay. agric. J.*, **32**, 253.

Khamaruddin, Megat (1980) Mechanical land clearing. *Planter, Kuala Lumpur*, **56**, 559.

Kowal, J. M. L. and **Tinker, P. B. H.** (1959) Soil changes under a plantation established from high secondary forest. *J. W. Afr. Inst. Oil Palm Res.*, **2**, 376.

Leamy, M. L. and **Panton, W. P.** (1966) *Soil survey manual for Malayan conditions.* Min. of Agric. & Cooperatives, Kuala Lumpur.

Martin, G. (1970) Le défrichement mécanique pour la création de palmeraies industrielles. *Oléagineux*, **25**, 575.

Martin, G. (1976) Méthode d'estimation des temps de travaux pour le défrichement et l'andainage mécanique d'un palmerie industrielle. *Oléagineux*, **31**, 59.

Rasmussen, A. N., **Kanapathy, K.**, **Norman Sant Maria** and **Gurmit Singh** (1981) Establishment of oil palms on deep peat from jungle. In *Oil palm in agriculture in the eighties*, ISP, Kuala Lumpur.

RRIM (1939) Planting without burning. *Plts' Bull. Rubb. Res. Inst. Malaya*, No. 6, 1.

RRIM (1957) Jungle clearing. *Plts' Bull. Rubb. Res. Inst. Malaya*, No. 33, 105.

Sankar, N. S. (1965) Fruit collection and evacuation by road. Sabah Planters Association, Oil Palm Seminar. Mimeograph.

Sheffield, A. F. W. and **Toovey, F. W.** (1941) Oil Palm Research Station, Nigeria. Second annual report, Part A, paras 23–27. Mimeograph.

Sheldrick, R. D. (1968) The control of Siam weed (*Eupatorium odoratum* Linn.) in Nigeria. *J. W. Afr. Inst. Oil Palm Res.*, **5**, 7.

Sly, J. M. A. (1963) *WAIFOR eleventh annual report, 1962–3*, p. 47.

Sly, J. M. A. and **Tinker, P. B. H.** (1962) An assessment of burning in the establishment of oil palm plantations in Southern Nigeria. *Trop. Agric. Trin.*, **39**, 271.

Sparnaaij, L. D. (1957) Mixed cropping in oil palm cultivation. *J. W. Afr. Inst. Oil Palm Res.*, **2**, 244.

Sparnaaij, L. D. and **Gunn, J. S.** (1959) The development of transplanting techniques for the oil palm in West Africa. *J. W. Afr. Inst. Oil Palm Res.*, **2**, 281.

Surre, Chr., **Fraisse, A.** and **Boyé, P.** (1961) Plantation de palmier à huile sur le sol de forêt et sur savane de *Imperata*. *Oléagineux*, **16**, 91.

Tailliez, B. (1975) L'aménagemen des terrains vallonnés et accidentés pour la plantation de palmiers à huiles. *Oléagineux*, **30**, 299.

Tinker, P. B. H. (1963) Changes occurring in the sedimentary soils of Southern Nigeria after oil palm plantation establishment. *J. W. Afr. Inst. Oil Palm Res.*, **4**, 66.

Tinker, P. B. H. and **Ziboh, C. O.** (1959) A study of some typical soils supporting oil palms in Southern Nigeria. *J. W. Afr. Inst. Oil Palm Res.*, **3**, 16.

Toovey, F. W. (1947) *Oil Palm Research Station, Nigeria, seventh annual report.* p. 30.

Turner, P. D. and **Gillbanks, R. A.** (1974) *Oil palm cultivation and management.* Incorp. Soc. of Planters, Kuala Lumpur, 672 pp.

Vanderweyen, R. (1943) Quelques directives pour l'établissement d'une palmeraie. *Bull. agric. Congo belge*, Leopoldville, **34**, 80.

Vanderweyen, R. (1952) *Notion de cultures d'Elaeis au Congo Belge.* Brussels.

Wong Phui Weng (1976) Use of Roundup (glyphosate) for lalang control prior to planting oil palm and rubber. In *International developments in oil palm*, ISP, Kuala Lumpur.

Chapter 9

The establishment of oil palms in the field

In planting oil palms in the field, the first object is to bring them into bearing as early as possible and so to reduce the period in which no return on capital outlay is being obtained. Growth to the bearing stage can be influenced at planting time by:
1. The stage of development and the general health of the nursery seedling.
2. The method of transplanting.
3. The time of transplanting.

The second object is so to space the plants in the field that the optimum economic yield will be obtained from the whole period of production.

Planting in the field

Stage of seedling development

In regions with no distinct dry season it is preferable to plant well-developed seedlings which have been growing in the nursery for 10–16 months. Early experiments in Malaysia showed that 12- to 18-month-old seedlings (from the two-leaf stage) come into bearing earlier and give appreciably higher early yields than palms only 6 months old (Gray and Hew Choy Kean, 1963–6), and that polybag seedlings transplanted at 13 months from the germinated seed stage or older give significantly higher yields in the first 3 bearing years than seedlings transplanted at younger ages (Hew Choy Kean and Tam Tai Kin, 1971). Other experiments showed that older seedlings maintain a higher leaf production, bear earlier, and have heavier bunches, a higher F/B ratio and, perhaps most importantly, a higher oil to mesocarp in the first year of harvesting (Khoo Kay Thye and Chew Poh Soon, 1977). The growth of older palms may be checked after planting, the first leaves to be opened being shorter than the last leaves opened in the nursery, but this tendency can be reduced by good transplanting techniques and is reversed soon enough for the older seedlings to maintain a lead over the younger ones.

The right age to transplant is really determined by balancing a number of factors. Small seedlings cost less to transplant and do not show so much check in growth; but they will be more uneven owing to their greater susceptibility to pests and diseases and because they have been subjected to less culling, and they will take longer to come into bearing. Older seedlings on the other hand are more costly to transplant and are subject to a greater initial check; but nevertheless they come into bearing earlier. When these factors are all taken into account it will be found inadvisable to transplant seedlings younger than about 10 months or older than about 20 months from the germinated seed stage.

The age at which seedlings are transplanted may, however, depend on special circumstances. The object may sometimes be to bring an area into bearing as soon as possible, when nurseries have not yet been prepared nor seed purchased. Such circumstances may result from a sudden acquisition of land or a sudden decision, on economic grounds, to change from another crop to the oil palm. The question then is how can the seeds or prenursery seedlings purchased be brought into bearing in the field in the shortest possible time? The answer will be to raise them in polythene bags and to transfer them to the field as soon as possible as young transplants. Where, however, a lengthy planting programme is being undertaken over a number of years or a previous crop is being removed at a steady rate, the question becomes what kind of seedling should be raised to reduce to a minimum the time from transplanting to bearing? The answer in this case will be to prepare nurseries sufficiently far ahead of planting to give large robust seedlings which will come to the bearing stage in the shortest possible time *from transplanting*.

In markedly seasonal climates such as obtain in West Africa, seedlings for planting into the field at the beginning of the rains will have normally been 11–13 months in the nursery, i.e. since the beginning of the previous rains. However, if dry-season nurseries (see p. 346) have been established, seedlings for transplanting from these nurseries will be 7 months old. Experiments have shown that while these seedlings will not come into bearing so soon as the larger seedlings, if seed supplies and the land to be planted are both available and prepared, it is better to plant at 7 months from a dry-season nursery than either to keep the seed for the next season or to hold the seedlings for another year in the nursery (Gunn and Sheldrick, 1963). Seedlings which pass through a full wet season in West Africa after being forced forward in a dry-season nursery will attain an enormous girth and height by the time they are 18 months old; but apart from the difficulties of transplanting such large seedlings, their contemporaries already planted into the field in the previous year rapidly overtake them in growth and flowering.

Seedling selection and roguing in the nursery

The question of which seedlings should be discarded or 'rogued' from a nursery is one that has been given much attention. The obtaining of an even stand in the field is of great importance and this has led to the advocacy of a heavy elimination of 'poor doers'. A very careful definition of a 'poor doer' is needed, however. Obviously it would be dangerous to transfer to the field any seedlings which are deformed or have had their growth seriously retarded by disease, but selected progenies show very different conformations and rates of elongation in the nursery and in a mixed lot of crosses, therefore, many plants will be relatively short because this is a characteristic of their parentage.

Differences in height in the nursery of palms *of the same age* tend to even up when the plants are transferred to the field. In one nursery in Malaysia groups of seedlings averaging from 95 to 150 cm in height (range 55 cm) at planting time showed a group average height range of only 21 cm after 14 months in the field. A group of palms averaging only 67 cm in height in the nursery was still markedly underdeveloped, but in the following 12 months this group made rapid growth, reaching a height within 5 cm of the mean height of the other groups. In the first bearing year, significant yield differences between groups were not clearly associated with height in the range 103–150 cm (Gray and Hew Choy Kean, 1964). On an estate in Nigeria, seedling height groups of 1.1–1.5 m and 1.5–1.8 m (length of youngest, fully opened leaf) gave similar early yields; larger palms, 2.0–2.4 m high, came into bearing earlier but gave slightly reduced early yields, while palms only 0.6–1.1 m in height gave considerably lower yields (Unilever, 1961).

There is therefore no case for heavy selection on the basis of height or size of plants, and roguing should be of plants which are markedly stunted or whose growth shows abnormalities of habit or leaflet form. Roguing can be conveniently done at 8 months and it will then only be necessary to discard very few plants at transplanting time. A normal polybag plant at 8 months has been described as having a height of 0.6–1 m, a girth of 15–22 cm, a width greater than its height, five to eight leaves the middle ones of which form an angle of 45° with the axis, and leaflets spreading at an angle greater than 60° to the leaf rachis (Wuidart and Boutin, 1976).

Palms which show habit abnormalities and should be rogued are: (a) unusually upright and narrow with height greater than width (*dressé*); (b) flat-topped with successively shorter leaves giving a bunched appearance (*ramassé*); (c) unusually spread out with flaccid, curved leaves also giving rise to a flat-topped appearance (*étalé*); (d) maintaining a juvenile type of growth, although large, so that the leaves do not become fully pinnate (*folioles soudées*). Leaflet abnormalities may be classified as: (i) those inserted at an acute angle to the rachis, i.e. 45° or less instead of 60–90°; (ii) unusually narrow leaflets, rolled longitudinally to give a narrow appearance (*étroite*); (iii) unusually short but broad leaflets which come to a point abruptly (*courte*); (iv) leaflets tending to be crowded together and often short and crimped (*rapprochée*); (v) leaflets much wider apart on the rachis than usual (*espacée*). Cases of Collante and Leaf Crinkle, noted in prenurseries (p. 344–5), can also occur in polybags, and, lastly, there may be genetic abnormalities characterized by yellow or white stripes or patches on some of the leaflets.

In all the main oil palm growing regions it should now be possible to plant out in the field at least 80 per cent of the nursery seedlings. Of the remaining 20 per cent, 10 per cent may be casualties or deformed seedlings, while the other 10 per cent may be small seedlings which are less than half the average height of the nursery. With a loss of 5 per cent at the prenursery stage (if used) it should therefore be possible to plant in the field at least 75 per cent, and more usually 80 per cent, of the germinated seed provided. Where Blast makes an annual appearance, however, the supply of transplantable seedlings is less predictable.

Methods of transplanting

Nursery seedlings cannot be transplanted to the field with bare roots unless they have been given special treatment, and even when such treatment is given, results are variable and establishment is comparatively slow. The failure of many early plantings in West Africa was due to the fact that bare-root seedlings had to be carried over long distances (Toovey, 1948). Transplanting is now done almost exclusively from polythene bag nurseries, so the work of transplanting from a field nursery will only be briefly described.

The early plantings in Africa soon demonstrated that it was essential to transplant field-nursery seedlings with a substantial ball of earth (Toovey, 1953). When seedlings are planted with naked roots many of the primaries decay and the plant's survival depends on soil conditions being right for their early substitution by other primaries developing from the base. When preparing a seedling with a ball of earth of 15 cm radius many primary and secondary roots are cut, and these are replaced by thick secondaries growing out from the cut ends as pseudo-primaries or *racines de substitution* (Moreau and Moreau, 1958; Sparnaaij and Gunn, 1959). Young primaries and secondaries within the ball of earth continue to develop. The growth of a transplanted field-nursery seedling can be further improved by prior root cutting. Half the circumference of the circle to be dug round the plant is cut with a sharp spade 6–8 weeks before transplanting, and the other half 4–6 weeks before transplanting. This causes a proliferation of the root system within the ball and helps to hold the soil together so that there is less breakage on transference to the field.

In dry weather a field nursery must be watered before lifting begins. With most soils, and with root cutting, the seedlings can be carried to the field in head pans (West Africa) or square sloping-sided pans (Malaysia) and, if necessary, they can be wrapped in sacking. With very sandy soils, however, special implements may be needed for lifting and carrying the seedling with its ball of earth (Dekester, 1962; Sly *et al.*, 1963).

Seedlings in polythene bags should be well watered the day before transplanting and, if necessary, fungicide and insecticide treatment can be given. The preparation of the planting site has been described in Chapter 8. At the time of planting a hole is dug with a diameter slightly wider than that of the polybag. The base of the bag is then cut away and the whole plant in the bag is placed in the hole. The side of the bag may then be slit and the bag is pulled out leaving the soil intact in the ground. If the soil in the bag is very firm, the polythene may be cut off altogether before planting. The small gap between the ball and the surrounding ground is then filled with topsoil which is firmly pressed in with the feet or some implement. This is a very important operation as the young roots must be able to enter the surrounding soil without delay or hindrance. To make certain that the plant is firm in the ground a wooden soil rammer 1 m long and 7–8 cm in diameter may be used to ram the soil in successive layers round the ball of earth. Unless seedlings are planted firmly they are in danger of being blown over by high winds before their root systems are fully established in the soil. It is also important that the seedling should be planted so that its base is just above the level of the ground. Deep planting has been shown to retard subsequent growth (WAIFOR, 1955).

With very large 12- to 18-month-old seedlings it may be necessary to leaf prune to make handling easier when transplanting. Experiments in Nigeria showed that seedlings 1.5 m high could have their leaves cut back to give a height of about 1 m without affecting subsequent growth (Gunn and Sheldrick, 1963). In Malaysia, leaf pruning of very large seedlings to a diamond shape is favoured, but those below 1.5 m do not need pruning. In the Ivory Coast it is considered that the advent of the polybag nursery has made pruning generally unnecessary, but that some of the older leaves, especially if prematurely withered by Freckle, can be severed at the lowest point of leaflet insertion and the cut painted with Otina-C grease to prevent attack by the weevil, *Temnoschoita* sp. (Barnabe, 1976).

The time of transplanting

In regions where there is no very distinct dry season it is possible to plant all the year round, though it is more usual to choose those months in which the average rainfall is high. In the south of Malaysia, for instance, some estates start to plant during April (a month with a comparatively high average rainfall) and continue right through the drier months of June and July into the second 'wetter' season of September to December. Experience has shown that the weather at planting time is not of the first importance; if rain falls within 10 days of planting, good establishment should follow, and even if rainfall is still further delayed seedlings correctly planted will survive and rapidly establish themselves with the onset of a rainy period. A great deal of the 'waiting for rain' which is a common feature of the planting sequence is unnecessary, and it is better for the plant to be in the ground when rain comes than for it to be planted during or just after a period of torrential rain.

Nevertheless, in most regions near the equator, e.g. Zaire and Malaysia, where there are two 'wetter' seasons, the one in April and May and the other from August to October or November, it is best to choose the *beginning* of one or other or both of these periods for planting out. If a big programme is envisaged the longer season should be chosen with, perhaps, a further small planting at the beginning of the other season. Even in a typical equatorial climate young seedlings may be retarded by periods of drought – expecially in very permeable soils or in those clay soils which tend to bake hard – and it is therefore best to allow for a period of establishment in wet weather before a dry period is expected. The most severely retarded seedlings will be those planted in wet weather just before the drought period sets in.

In the seasonal climate typical of West Africa the time of planting becomes of much greater importance, the main consideration being to enable the plant to establish itself and grow to the maximum extent possible before the dry season sets in. Before the advent of the polybag nursery, the effect of weather on seedlings planted with a good ball of earth and with naked roots was closely examined in Nigeria by planting every day throughout the year except during the early part of the dry season in the months of November, December and January (Sparnaaij and Gunn, 1959). With satisfactory balls of earth no correlation was found between rainfall during planting and survival or subsequent growth of main-nursery seedlings. With naked-root seedlings, however, survival in the early months (before the wet season had fully set in) was largely dependent on the weather at planting time. Such seedlings planted in February or early March only survived if planted within a period of 3 days before rain. Palms planted with balls of earth or with bare roots were of heaviest weight 2 years later when planted in the early months of intermittent rainfall and high number of hours of sunshine. These months were April and May, but ball-of-earth seedlings planted in dry weather in February and March also grew very well. More recent experiments with polybag nurseries have confirmed that April and May are suitable months for transplanting in Nigeria (Aya, 1978), though earlier planting was not tried.

In the Nigerian experiments with field nursery seedlings it was found that plants kept in the nursery through another dry season and planted as 2-year-old nursery seedlings in February to May were heavier after $1-1\frac{1}{2}$ years in the field than seedlings planted *late* in the previous year and which had been $2-2\frac{1}{2}$ years in the field. This shows that under West African conditions it is better to leave main-nursery seedlings for 2 years in the nursery rather than plant them in the middle of, or *late* in, the wet season. These 2-year-old seedlings will, of course, be very large and difficult to handle, and this result is more of a warning against late planting than an advocacy of the deferment of transplanting from an already existing nursery.

In the Ivory Coast, where the main dry season is much less severe than in Nigeria but the 'short-dry' season of August is more pronounced, planting in October has been practised. In a seed block at La Mé, out of 13 groups of progenies (planted in 1962, 1963, 1964, 1965 and 1966) from each of which some had been planted in May and some in October, the May-planted subgroup was superior in yield in 12 out of the 13 groups in the first bearing year, in 9 out of the 13 in the second year, in 11 out of the 13 in the third year and 10 out of the 13 in the fourth bearing year; only by the fifth bearing year had the yields of the two subgroups become similar (IRHO, 1974).

As a general axiom for all countries it may be said that it is better to transplant just *before* a period of intermittent rain and sunshine is normally experienced rather than to wait for more continuous rain to set in, and that planting late in a wet season should be avoided.

Cultural practices at transplanting

Cultivation and mulching

The site of the planting hole should be levelled and cleared of all vegetation to a radius of 1 m before planting. In some regions mulching of this circle has been recommended and practised. In the Ivory Coast, where rather severe leaf pruning is practised, mulching with a circle of cut vegetation or with a square piece of black polythene sheeting is recommended (Dekester, 1962). The latter is about 1.3 m across, has a central hole 20 cm in diameter for the seedling, and is perforated with small holes of 1 cm in diameter and

10 cm apart. In Nigeria two experiments have shown that mulching with cut vegetation and with black polythene improves early growth. The latter appears to have a slightly greater effect in the first year, but if mulching is continued the mulch of cut vegetation seems to gain the advantage (Sly and Sheldrick, 1965).

Elsewhere, the mulching of normal, large transplanted seedlings has not been shown to be advantageous, though if very dry weather were to follow planting the practice might be expected to be of value. Apart from cut vegetation and black polythene, bunch refuse has been used as a mulch.

Hoeing around the plants at the end of the rains showed a small effect on growth in Nigeria but this effect was not as great as that of mulching.

In areas where the soil is very impermeable and young palms suffer severely during periods of heavy rain, improvement through the construction of raised platforms or terraces before planting has been claimed (Poncelet, 1965; Ollagnier and Delvaux, 1966). It should be realized that these measures cannot replace a proper drainage system, though they may be justified in special circumstances. In Colombia, platforms about 1 m in diameter and of varying height have been employed, the small drain around the platform being connected with the interline drains, On the heavy *massape* soil of Brazil, mounded terraces 2.5 m wide and 0.7 m high have been raised by disc harrows but this practice has not proved very successful.

Manuring at planting

Nitrogen is almost always required during the establishment of oil palms in the field, but there is a danger, which was particularly noticeable in West Africa in the days of field nurseries, of damaging the root system if nitrogen fertilizers are wrongly applied. Several cases have been recorded of deaths following application of fertilizers, particularly sulphate of ammonia, in the planting hole before or at planting time, and on other occasions leaves paled and dried off prematurely (Sparnaaij and Gunn, 1959). The severity of the effect is increased if there is a dry period immediately after planting.

It is a common practice in Malaysia, taken over from the rubber industry, to apply 200 g of Christmas Island rock phosphate in the planting hole; there is as yet no evidence that this practice is of value, however. Phosphorus may or may not be required before the palms come into bearing, depending on the soil type; where it is required it is probably best applied after the seedlings have established themselves and in the circle of the feeding roots.

Both potassium and magnesium may be required in young plantings. Where required, applications of K have been shown to bring palms earlier into bearing. Magnesium is required to counteract a deficiency which is found on certain soils and which is accentuated by N and K dressings. Where magnesium is likely to be deficient, it may be added at the rate of 100 g anhydrous magnesium sulphate or 200 g Epsom salts or kieserite per palm.

The most usual need is for sulphate of ammonia and sulphate or muriate of potash, and these may be applied in a ring around the seedling at the rate of 200 g each per palm 4–6 weeks after planting. In areas requiring higher N applications the sulphate of ammonia dose may be repeated after 5 or 6 months. Compound fertilizers containing unnecessary elements should be avoided.

Protection from rodents

In Africa, rodent damage to young transplanted seedlings may be very serious owing to the presence of the large rodent, the 'cutting grass', *Thryonomys swinderianus*. This animal is usually present in primary or secondary forest adjacent to new plantings and will attack seedlings up to 3 years from planting; it eats through the leaf base to get at the heart of the plant. The single apical bud is therefore almost always destroyed and the plant dies. A new planting may be almost totally destroyed by these rodents in one night. The only really effective measure against this destruction is to surround the young plant *immediately after planting* with a wire collar. Although the collar is open at the top the animal will not normally attack the plant from inside the collar; but if the latter is not properly pegged down it may be lifted off and the plant will then be open to attack. Moreover if the collars are too wide, smaller rodents may enter and cause damage. Experiments have shown that surrounding the plants with wire collars or a fence of sticks packed closely together are the only effective measures; painting the base of the seedlings with tar is not effective.

Owing to the rapidity with which young plants grow it may be necessary to provide wire collars of two different sizes, viz:
1. At planting time: height 45 cm, circumference 75 cm, diameter 23 cm.
2. After 12–18 months: height 60 cm, circumference 120 cm, diameter 38 cm.

With the use of large plants at transplanting time, collars 45–60 cm in height and 120 cm in circumference may be found satisfactory for the whole period. They may be removed between 2 and 3 years after planting depending on the speed of growth of the seedlings.

Supplying

Transplanting techniques have now reached such a high standard that transplanted polybag seedlings should seldom require replacement. Moreover, growth in the field is so rapid that a supply planted

2 years after its neighbours will be able to make little headway, and supplying even after only 1 year from planting is of doubtful utility. In large well-regulated plantings the amount of supplying required after 1 year should be only of the order of 0.5 per cent, though in areas with a very severe dry season rather higher supplying, up to 3 per cent, may be needed.

For these reasons it is best to pay attention to supplying shortly after planting rather than to let a year go by before taking any action. In seasonal climates, if really early planting has been undertaken, it is possible to have all stands inspected about a month after planting; any plants which have clearly not established themselves or have been severely attacked by pests can then be replaced before the season is too advanced for planting. In non-seasonal climates, too, it is best to carry out an inspection 1 or 2 months after planting with a view to supplying any dead or doubtful plants before their neighbours have developed so far that supplies will later become permanently retarded. When there are two 'wetter' seasons it may be best to leave this first supplying until just before the next period of heavy rainfall is expected (Lafaille and Daniel, 1965).

In spite of this early supplying, some further supplying may be required 1 year after planting, particularly if a group of palms has suffered insect or other damage. In this case the plants used should be from a new nursery having seedlings of suitable size for transplanting, supplying should be carried out as early in the season as possible, and the plants should be given special care and attention so that they have the best opportunity of catching up with the other palms.

Yeow *et al.* (1981) drew attention to the yield frequency distribution in oil palm fields and to the fact that palms which fail to produce female inflorescences in their first 30 months remain poor bunch yielders both as young palms and through their lives. In the fields they examined, over 7 per cent of the palms produced no female inflorescences before 30 months, and 15 per cent gave the equivalent of 1.5 tonnes bunches per hectare per annum against a field mean of 25.8 tonnes. They therefore suggested that low-yielders could first be detected from their failure to produce female inflorescences and then be replaced by specially raised nursery seedlings of the same age. While the potential production of the replaced palms and the replacements may well be correctly assessed, the thesis overlooks the fact that the replacements will inevitably be checked in their growth on transplanting with the result that their neighbours will be larger and will overshade them. This will cause the supplies to have a reduced sex ratio and probably a higher abortion rate; so whether they will yield more than, or even as much as, those removed remains doubtful. In short, supplying should be *avoided* as far as possible by good cultural practice and control of pests and diseases.

Spacing the plants in the field

In discussing the flowering and fruiting of the palm it was seen that shade had an effect on apparent sex ratio and it would therefore be expected that yield per palm would diminish rapidly with increasing density not only because of increased competition for nutrients but also because of the mutual shading effect of the palm leaves. This has been shown to be the case. Very closely spaced adult palms always produce a preponderance of male inflorescences and Sparnaaij (1960) showed that the pruning of the leaves of surrounding palms in a close-spaced planting has a marked effect on the apparent sex ratio and hence on the yield of unpruned palms. Apart from the decrease in sex ratio, the weight per bunch decreases with closer spacing, and the usual vegetative effects are an increase in height, a smaller trunk girth, a reduced leaf production and an increased leaf length; the overall leaf area per leaf is, however, little affected (see p. 155).

The optimum yield of a hectare of palms over a given period can only be determined by a knowledge of the potential cumulative yield of palms at different spacings. At a certain density the number of palms per hectare multiplied by the cumulative yield per palm becomes optimal. When plantations were first established, spacing was determined by judging the probable spread of the palms' leaves under plantation conditions. Spacings adopted varied from 7.5–10 cm (25–33 ft) and a triangular arrangement was usually employed. Such spacings gave between 115 and 205 palms per hectare. Because the leaves are produced spirally and the palm cannot 'fill in' any gaps in the canopy by production of branches, a very regular planting arrangement with the largest possible number of similarly spaced surrounding palms is desirable; a triangular spacing in which each palm is surrounded by six of its fellows fulfils this requirement best.

Young palms are little affected by spacing since they are not yet in competition for nutrients or light. Where growth is not rapid, the yield of palms may for several years be unaffected by their spacing. In the Ivory Coast, palms planted in 1926 at 8 and 10 m triangular, gave similar yields per palm until 8 years after planting (Prevot and Duchesne, 1955), while in Nigeria palms planted in 1945 at 6.4 m triangular gave higher yields per palm than palms planted at 12.8 m triangular until 9 years after planting (Sly and Chapas, 1963). Close-planted palms may, through their higher proportion of weeding circles and paths, and the higher overall shading of ground vegetation, be better cared for than wider-spaced palms, and so, through reduced competition with surrounding vegetation, be able to give higher yields in the period before mutual shading begins to have the opposing effect. Where growth in the early years is very rapid, however, close spacing effects may be noticeable almost as soon as the palms come into bearing. On both coastal clay in Malaysia and volcanic soil in Papua New Guinea, palms at the high density of 186 per hectare were showing, by their second bearing year, a lower per palm yield than palms at the 'normal' density of 148; and, on the volcanic soil, yield per hectare was already higher at normal density by the third bearing year (Breure, 1977). The high-density palms had, by this time, developed trunks which were appreciably higher than those of normally spaced palms.

Old palms may be markedly affected by a reduction of density (increase of spacing). Thinning out of a plantation from 124 to 99 palms per hectare at 15 years has been done on fertile land without more than a temporary drop in yield per hectare. Similarly, where stands have been reduced by 10–20 per cent by Vascular Wilt disease or *Ganoderma*, yields per hectare have been largely maintained. Heavier bunches per palm are initially obtained and larger numbers of bunches per palm follow. At La Mé in the Ivory Coast there was little difference, from 17 years of age onwards, between the yields of plots with a 'normal' density of 123 palms per hectare and of low-density plots of 100 palms per hectare. In two triangular spacings where the density in one case was over 50 per cent higher than the other, the yield in the less dense plots from 17 years of age onwards was more than 20 per cent *higher* than in the denser plots (Prevot and Duchesne, 1955).

In short, high densities can be expected to give high early yields, while lower densities than normal may not reduce yields per hectare in old fields. What, then, is the optimum density for the whole plantation cycle, and what will be the effect of early high densities on later yields in unthinned and thinned plantations?

The early spacing trials which shed light on these problems were carried out in Africa. Prevot and Duchesne (1955) showed, on the basis of an experiment with two triangular and three square spacings, that it is possible to calculate the density giving optimum production for any given situation and planting material. This is made possible by the fact that, within the limits of planting distances which are likely to be used, there is a linear relationship between the cumulative bunch yield per palm to any age and the density per hectare. This finding was confirmed with data from Zaire (Beirnaert and Vanderweyen, 1940) and Nigeria (Sly and Chapas, 1963) and is illustrated in Fig. 9.1.

The slope of the straight line relationship between cumulative

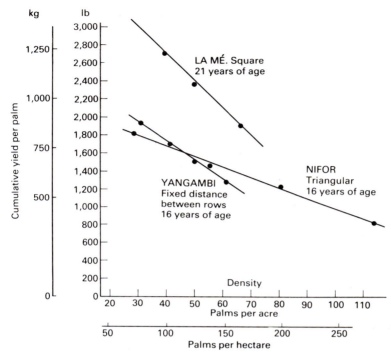

Fig. 9.1 Relationship between cumulative yield per palm and density in the Ivory Coast, Zaire and Nigeria. In each case the relationship is linear.

yield per palm and density gives a competition factor which is calculated as follows:

$$C = \frac{\text{Cumulative yield (wide spacing)} - \text{Cumulative yield (close spacing)}}{\text{Density (close)} - \text{Density (wide)}}$$

This factor, which is an estimate of the amount by which the cumulative yield per palm falls with an increase of one palm in the density, is obtained from the two extreme points of the straight line and is a rough method of calculating the regression coefficient b which is normally calculated from all points available. It should be noted that any particular competition factor applies only to the cumulative yield to the particular number of years for which data have been obtained, and it can be used for finding the optimum density for yield for that period of years. It has already been seen, however, that with older palms yield per hectare does not vary appreciably with density. Provided the data are available for around 12 years, therefore, the optimum density determined is likely to be applicable to a bearing period of 20–25 years or the whole life of the plantation. In the La Mé data presented by Prevot and Duchesne the optimum density became constant when the palms were 18 years old and 13 years' data had been obtained.

Production at any density is calculated from the competition factor as follows:

$$P_x = (p_1 \pm (D_x - D_1)C) \times D_x$$

where P_x is the production per hectare at the density, D_x, concerned, and p_1 is the production per palm at density D_1.

For instance, if the competition factor is 2.14 kg and the cumulative production per palm at the age of 16 years and a density of 70 palms per hectare is 821 kg, then production per hectare at 138 palms per hectare (9.14 m $\triangle$), P_x is expected to be

$$(821 - (138 - 70)2.14) \times 138 = 93,216 \text{ kg}$$

The actual accumulated yield in this case was 91,626 kg, the divergence from the expected figure being 1.7 per cent.

Calculation of the density, D_0, which would give the optimum yield per hectare, P_0, over a given number of years is obtained by the formula

$$P_0 = (p_1 - (D_0 - D_1)C) \times D_0 \text{ (as above)}$$

to give

$$P_0 = p_1D_0 - CD_0^2 + CD_1D_0 = (p_1 + CD_1)D_0 - CD_0^2$$

hence

$$p_1 + CD_1 - 2 \times CD_0 = 0$$

$$D_0 = \frac{p_1 + CD_1}{2C}$$

for maximum yield per hectare, P_0.

Taking the previous example, the density for maximum yield up to 16 years of age will then be:

$$\frac{821 + 2.14 \times 70}{4.28} = 226.8 \text{ palms per hectare.}$$

(Calculated from the regression coefficient b, the number of palms per hectare for optimum yield was found to be 228.)

Using these formulae, Prevot and Duchesne (1955) showed that a higher cumulative yield per hectare can be obtained from triangular than from square spacing, but that the planting density for optimum yield is lower in the triangular spacings than in the square spacings. This is illustrated by the curves in Fig. 9.2. It was also shown that the decrease in yield per palm with increased density is due both to a reduction in the number of bunches and the weight

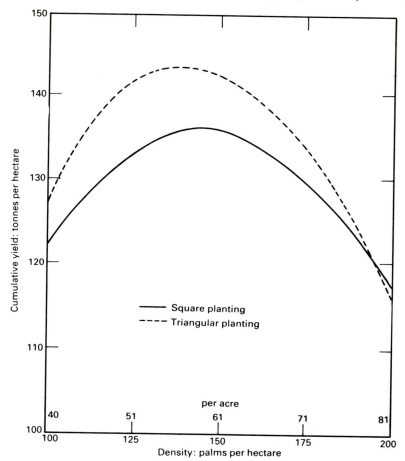

Fig. 9.2 The relationship between cumulative yield per hectare and density in square and triangular plantings (Prevot and Duchesne, 1955).

per bunch, but that the effect on the number of bunches was much the more pronounced.

Linear relationships have also been shown with both square spacing (Ivory Coast) and triangular spacing (Nigeria) between height increment and density, high densities inducing a rapid increase in height. For ease of harvesting, therefore, low densities are advantageous.

The most complete of the early experiments was the one conducted at La Mé in the Ivory Coast and it showed that where the same density is to be maintained throughout the planting cycle, a density of 139 palms per hectare with a triangular spacing of 9.1 m should be used. The Zaire experiment, with 16-year-old palms and using different rectangular spacings but with rows always 8 m apart, gave a calculated optimum density of 203 palms per hectare. With rectangular spacing, however, the optimum is always higher than with triangular spacing, and, moreover, it was clear that a constant optimum density had not yet been reached. In Nigeria a much higher optimum was found (92.3 per acre, 228 per hectare) largely owing to the slowness with which the plants came into bearing. When the calculations were made the palms were the same age as those in the Zaire experiment and 12 years' data were available. But the palms were not competing with each other, i.e. there was no positive competition factor, until the palms were 10 years old. Moreover there were progeny differences which were unfortunately confounded with blocks (see p. 404). Thus several circumstances made it difficult to determine a valid optimum density, though much interesting subsidiary information emerged. However, from all the information available up to 1970 a density of 143 palms per hectare (9 m triangular) seemed suitable for African conditions, and this density and spacing became standard in many parts of the world.

Later, a number of density experiments were undertaken in Malaysia (Plate 9.1). In an experiment on inland soil, Deli *dura* progenies planted at densities of 96, 114, 138, 158 and 183 palms per hectare (triangular) were recorded from the seventh to nineteenth year after planting (Ramachandran *et al.*, 1973). The

Pl. 9.1   An aerial view of a large spacing experiment in Malaysia.

optimum density was calculated, by the formulae already given, to be about 130 palms per hectare. However, the lack of early yields suggests that the real optimum density to 19 years was higher. A rather similar experiment on coastal clay soil (Tan Yap Pau and Ng Siew Kee, 1977) had densities of 111, 136, 161 and 185 palms per hectare (triangular). On the basis of 7 years' data the optimum density for the economic life of the palm was calculated to be 156 palms per hectare. This optimum was also calculated from the same formulae, but a rather lower estimate of 144 palms per hectare was found by using the method of Corley *et al.* (1973) to be described below.

Using the data of four density experiments in Malaysia, Corley *et al.* (1973) found that there was little effect of planting density on mean area per leaf (see p. 155), but that the latter factor was directly related to optimum density for current yield and that this relationship appeared to hold good for any level of production. In Fig. 9.3 mean leaf area is plotted against optimum densities for current yield and against competition factors calculated for current yield using data from experiments on two soils in Malaysia and from the Nigerian experiment already referred to. It should be noted that the competition factor is zero below a certain leaf area or palm

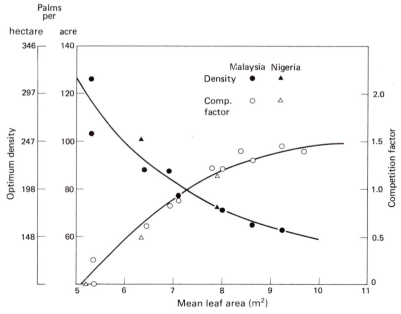

Fig. 9.3 Relationship between mean area per leaf and (1) optimum density for current yield, and (2) the competition factor for current yield (Corley *et al.*, 1973).

density and tends to become constant when the leaf area per leaf has reached about 10 m^2. The current-yield competition factor can be used for calculating the current yield per palm (Y_D) at any density by the formula

$$Y_D = Y_x - \frac{C(D - X)Y_x}{100}$$

where Y_x is the known or expected yield at a standard density, X, and C is the competition factor as a percentage of the expected yield at the standard density.

Using competition factors for current yield derived from leaf data and the yield trends found in experiments conducted under different conditions, Corley *et al.* (1973) calculated cumulative yields for 25 and 30 years for densities varying from about 143 to 180 palms per hectare (58–72 per acre). They found that optimum densities varied from 150 to 170 palms per hectare (60–70 per acre) and that the longer the period and the better the growing conditions the lower the optimum within these limits. However, the curves of cumulative yield plotted against density were so flat-topped that a 10 palm per hectare deviation on either side of the optimum would not reduce cumulative yield by more than 1 per cent. They concluded that about 158 palms per hectare or 64 per acre was a good general compromise, but suggested 150 palms per hectare for good coastal soils, 158 per hectare for good inland soils and 165 per hectare for poorer soils, using current planting material and with adequate manuring.

A rather new light on cumulative yield/density relations has been shed by Goh (1981) who recognized that yield per palm does not in fact decrease linearly with density, but only appears to do so over a middle range of densities. He therefore tried a number of models for fitting to yield/density data and found that the equation of Holliday was most suitable when applied to both a conventional and a systematic or fan-shaped spacing trial. The latter trial (Goh, 1977) included seventeen fertilizer treatments, the results of which are discussed elsewhere (p. 556). The conventional experiment also included fertilizer treatments and, on the basis of 9 years' yields, the Holliday equation gave an estimated density for maximum yield of 143 palms per hectare and the best fit to the yield/density data.

Spacing other than triangular

While the superiority of the triangular over the rectangular arrangement is established beyond doubt, there are some circumstances, such as mixed planting or semi-permanent inter-cropping with food crops, in which avenues of palms, corresponding to rectangular planting, may be employed. It is therefore useful that a study in Nigeria (Sly and Chapas, 1963) has shown the effect on yield of different spacings in the row. Two comparisons were made:

1. Rectangular spacing with constant interline and varying spacings in the row.
2. Various row spacings with constant distance within the row.

The results are given in Table 9.1. In the wide avenues of 19.8 m (65 ft), close spacing in the rows failed to compensate for the heavier and more numerous bunches in the wider spacings. It is of interest to note that with triangular spacing at the same density per hectare as the 19.8 × 3.7 m (65 × 12 ft) planting, the yield was nearly double. The figures in parentheses in the first column of Table 9.1 relate to this triangular spacing. Many of the palms in the 19.8 × 3.7 m planting were spindly and overshaded by their fellows, and there were more dead palms.

The number of bunches and the weight per bunch normally decrease with rising density, but with a constant distance in the row, there is a limit to the advantage to the individual palm of increasing

Table 9.1 *Effect of spacing on yield (cumulative and mean annual yields) on Acid Sands soils in Nigeria*

A. Rectangular spacings with constant interline width

Cumulative yields	19.8 × 3.7 m 138 palms per hectare*		19.8 × 6.4 m 79 palms per hectare	19.8 × 9.1 m 55 palms per hectare
No. of bunches per palm	53.7	(83.0)	82.7	92.7
Wt of bunches per palm (kg)	358	(665)	660	810
Mean wt per bunch (kg)	10.2	(12.0)	12.2	13.8
Mean annual yields (12 years)				
No. of bunches per hectare	618	(954)	544	425
Wt of bunches per hectare (kg)	4,115	(7,631)	4,331	3,717

B. Various spacings with constant distance in the row

Cumulative yields	5.5 × 6.4 m 282 palms per hectare	12.2 × 6.4 m 128 palms per hectare	19.8 × 6.4 m 79 palms per hectare
No. of bunches per palm	60.6	83.1	82.7
Wt of bunches per palm (kg)	369	646	660
Mean wt per bunch (kg)	8.8	11.8	12.2
Mean annual yields (12 years)			
No. of bunches per hectare	1,421	887	544
Wt of bunches per hectare (kg)	8,642	6,893	4,331

* Figures in parentheses relate to same density planting spaced 9.14 m triangular.

the distance between the rows. This can be seen in the number and weight of bunches in the 12.2 × 6.4 m and the 19.8 × 6.4 m spacings where reducing the density by two-fifths reduces the yield by about the same amount, whereas in reducing the density by over half, from a spacing of 5.5 × 6.4 m to 12.2 × 6.4 m, the yield was reduced by less than a quarter.

Two conclusions may therefore be drawn:
1. The oil palm cannot be hedge-planted; if for special reasons it is to be grown in avenue formation then the spacing in the rows must be near to the normal spacing for ordinary field planting.
2. Yield per palm cannot be expected to increase further if the palms are moved more than about 12 m (40 ft) away from each other. Beyond this distance yield per palm or per row will remain constant and the actual spacing of the rows will be determined by other considerations such as the value of inter-crops.

Progeny effects

The general conclusions drawn so far may be said to apply to mixed high-yielding material such as is generally available to the grower. The heterogeneity of planting material is such, however, that it would be unlikely that all progenies would require the same spacing for optimum yield; and, as already described (see pp. 155 and 249), the exploitation of variation in growth factors may provide progenies which maintain a high yield per palm at high density.

In the Nigerian experiment already referred to each block was planted with a different progeny. Although this precluded statistical analysis of progeny differences, the latter were so striking that there was little doubt that they were real. Table 9.2 shows the yield in the twelfth bearing year and the height of six selfed *dura* progenies at four different triangular spacings. It will be noted that in the majority of progenies the yield had by this age fallen off seriously in the close spacing of 6.4 m triangular. However, in two cases (103.426 and 103.151), the yield at the latter spacing was still higher

Table 9.2 *Effect of spacing on selfed* dura *progenies (Nigeria) (mean height per palm and weight of fruit bunches per hectare in twelfth bearing year)*

Spacing	6.4 m △		7.6 m △		9.1 m △		12.8 m △		Diff. 6.4 – 12.8 m	
Parent	Height (m)	Yield (kg)	Height (m)	Yield (kg)	Height (m)	Yield (kg)	Height (m)	Yield (kg)	Height (m)	Yield (kg)
103.149	4.97	7,679	4.42	12,266	4.30	13,329	4.45	10,696	0.52	−3,017
103.90	5.79	7,255	4.33	10,212	3.99	12,119	3.99	9,676	1.80	−2,421
103.426	4.24	11,297	4.05	12,156	3.38	8,019	3.32	6,120	0.92	+5,852
103.114	6.95	6,448	5.49	8,811	5.27	11,055	4.48	10,367	2.47	−3,919
551.256	6.58	6,080	5.82	9,947	6.40	13,299	4.94	10,019	1.64	−3,939
103.151	4.21	8,345	4.33	9,741	3.50	9,331	4.02	5,327	0.19	+3,018

than at the widest spacing of 12.8 m and in one case (103.426) was higher than at the 'normal' spacing of 9.1 m triangular. These two progenies, it will be noted, were shortest and, with the exception of progeny 103.149, were least 'drawn up' by high density. This suggested that progenies showing a short annual height increment may be adapted to high yield at close spacing, and Corley *et al.* (1973) suggested that the progeny differences of optimum density found in this trial might be due to differences in mean leaf area.

In the future, palms showing a higher bunch yield per unit leaf area are likely to be sought (Hardon *et al.*, 1972); and Breure and Corley (1983) have provided evidence that palms having a high bunch index (ratio of bunch dry matter to total dry matter) in their young, non-competitive years yield better in their later years, at both standard and higher densities, than palms selected for other growth parameters including bunch yield itself. Breure (1986) found in Papua New Guinea that above-ground assimilation, measured as total dry matter (vegetative dry matter + bunch yield dry matter), continued to increase for 11 years after planting with palms at fifty-six palms per hectare (i.e. with no inter-palm competition). Through a study of assimilation and light interception of these palms and palms at a range of densities up to 186 per hectare, he concluded that high densities could be exploited for high bunch yield if progenies could be selected which achieved the most rapid crown expansion.

The effects of deaths and thinning

Small reductions of stand following disease outbreaks may result in only temporary and very small effects on yield per hectare. Bachy (1965) presented data from 2,211 square-spaced and 1,467 triangularly spaced stands in fields in the Ivory Coast and, from the yields of palms with different numbers of adjacent palms missing, calculated coefficients for the 'correction' of their yields. Such coefficients are useful in selection work for reducing to normal the yields of palms favourably affected by the absence of one or more of their six neighbours.

In the case of Deli palms at the usual spacing of 9 × 9 m triangular, Bachy found a linear regression of both the weight of bunches and the number of bunches per palm on the number of adjacent palms missing. Calculated coefficients of correction for weight of bunches per palm were as follows:

Number of missing adjacent palms	0	1	2	3	4	5
Coefficient of correction for wt. of bunches/palm	1.00	0.87	0.77	0.70	0.63	0.58

Though it would naturally be desirable to calculate corrections for each area and spacing, these estimates probably have very general application.

In the case of end palms, Bachy suggested that, with triangular planting, half of these must be treated as having three neighbours missing, while the other half, i.e. those at the end of rows which terminate within the field, have only one neighbour missing. It must be remembered, however, that at the side of a field a whole straight row of palms will be concerned and that each of these palms has two neighbours 'missing'. Moreover, it was noticed in Nigeria that the yield of edge rows depended on how they were oriented. Rows unguarded on one side have their yields increased more when they run N–S than when they run E–W and this may be due to the greater reduction of shading from the sun in the case of the former. Coefficients of correction may therefore need to be adjusted to the orientation of the rows in the case of end palms.

Bachy (1965) also drew attention to the relevance of these findings to commercial production. If, for instance, a palm is sterile, its removal will enhance the yield of the field, e.g. if the average yield per palm is k, then the removal of a sterile palm will increase the yield of the six surrounding palms from $6k$ to $6k/0.87 = 6.90k$, an increase of 15 per cent. Thus in fields of very variable material a planter might increase yields considerably by a wise elimination of non-bearing palms.

The optimum economic density

The possibility that the optimum spacing for bunch yield may not be equivalent to the optimum economic density has been considered (Corley *et al.*, 1973; Surre, 1955). Both in the Ivory Coast and, as already mentioned, in Malaysia, changes of density of 10–15 per cent have been shown to lead to only very small changes in cumulative yield over periods of 16–17 years of harvesting. It has been claimed that small reductions of yield resulting from a suboptimal density are amply compensated by much larger reductions in costs of establishment, maintenance and harvesting. It is true that *establishment* costs will be lessened by reduced charges for planting material, nurseries, lining, holing, planting and disease and pest control, and it has been estimated that a density reduction of 10 per cent will reduce the capital cost of all planting operations by 4 per cent. *Maintenance* costs, however, may or may not be reduced; while there will be fewer rings to weed and fewer paths, the wider spacing will, in most fertile areas, induce a heavier growth of interline vegetation which will cost more to control. This may be of particular moment in the early years. If mechanical maintenance is envisaged, however, it is likely that maintenance costs will be reduced because tractors and implements will have easier access to the plantation although cutting a larger area of vegetation outside the rings. Pruning costs will be reduced owing to fewer palms. The relative *harvesting* costs will depend on the method of payment employed. If payment to harvesters is made by number of bunches (or by an allotted number of palms) costs will be reduced as the bunches will be fewer but larger; if however, as is more usual, payment is being made by weight, costs will be unaltered.

In short, a small reduction (up to 10 per cent) from the optimum density will result in a very small reduction in cumulative yield, will reduce the capital required for establishment and may reduce certain of the costs when the plantation is in production. On these grounds it has therefore been claimed that the optimum economic density is always below the optimum density for cumulative yield per hectare. If, however, the yield trend over time is taken into consideration a different conclusion may emerge. Higher density results in higher early yields and under the discounted cash flow method of appraisal these early yields will be worth more than those obtained later on. Although higher densities will produce these higher early yields, they will tend to have somewhat lower yields than those of lower densities in later years. Corley *et al.* (1973) found the optima for discounted cash flow differed very little from the optima for cumulative yield.

Variable densities

It has been seen that if a constant density is to be maintained throughout a planting cycle (i.e. from planting a field to replanting) high densities will tend to give high yields per hectare in the early years and low yields per hectare in the later years, while small reductions in densities in mature plantations will have only small effects on yield per hectare. It is logical, therefore, to search for a method of maintaining a high density in the early years and reducing this to a density which will be a around the optimum for the middle and later years of the planting cycle. Many fields have in the past been planted at double density, i.e. a planting distance of, say, 9 m triangular having been decided upon, plants are first set at 4.5 m apart in the rows with the intention of taking out every other plant at a later date. The evils of hedge planting have already been noted, however, and the tendency has been, more often than not, to leave the field at double density until the production per palm – and, it should be noted, this includes the palms to be left in – has been seriously reduced and the palms have grown abnormally fast in height. In one such area in Nigeria yield per palm was still well below that in an adjoining normal density area 5 years after thinning although weight per bunch had become the same in the two areas. Thus if high densities are to be employed with a view to thinning, the latter operation must be carried out well before a serious reduction of yield per palm has started.

The lower number of bunches per palm produced at higher densities arises from a higher abortion rate and perhaps also a lower sex ratio. Thinning to a 'normal' density might therefore be expected to restore the production of a palm after about 1 year, with any sex ratio effects taking a further $1\frac{1}{2}$ years for adjustment. Malaysian experiments showed that palms thinned after 2 years of production adjusted their yields after little more than a year, thus indicating that a higher abortion rate was the predominant cause of the lower bunch production per palm at high density. There was a suggestion that if thinning is further delayed recovery will take longer than a year (Fig. 9.4). These experiments also showed that the loss, after thinning, in the recovery year was about 30 per cent and that the

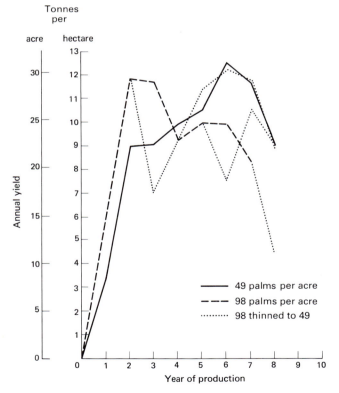

Fig. 9.4   Effects on yields of thinning from high to low density (Corley *et al.*, 1973).

overall gain in yield from high-density planting and thinning was small. When the higher planting costs and the cost of thinning were taken into account the practice was considered unlikely to be profitable under Malaysian conditions (Corley *et al.*, 1973). There is a greater likelihood of finding this practice profitable where growing conditions are slower and double density provides a higher yield for a longer period.

In any policy of thinning, the eventual stand must also be considered. In an area where high densities can be supported for several years, a relatively low eventual stand, perhaps reached in two stages, might be suitable. However, in areas where high densities lead to a rapid reduction in yield per palm, the final density may be imposed at an early stage and should not therefore be too low.

The simplest form of double density stage is obtained, as already described, by doubling the number of palms in the rows to form long rectangles. This arrangement has the advantage of providing a triangular spacing after thinning, but the early stand takes very poor advantage of the available ground before thinning. A better arrangement in this respect is a hexagonal one which, in effect, is the same as double density in the rows except that every other palm, instead of lying *in* the row, is set in the interline so that it is equidistant from the two adjoining palms in the row *and* the palm opposite in the adjacent row (Sly and Chapas, 1963). Every palm is thus surrounded by three palms equidistant from it, and the removal of every other palm in the zig-zag lines leaves a final triangular formation. This arrangement has the further advantage that, if triple density is possible, this can be planted at triangular spacing by placing one palm in the centre of each hexagon. The sequence:

Triple density: triangular
Double density: hexagonal
Final density: triangular

can be seen in Fig. 9.5.

The relative yields of fields with the same density but with hexagonal and triangular spacings are not known, but even coverage of the ground being in the order triangular, square, hexagonal, and higher yields being obtained from triangular than from square planting of the same density, it is logical to expect that hexagonal planting will give yields inferior to those of triangular plantings of the same density. It would not therefore be advantageous to have planting in hexagonal formation for a lengthy period; this formation could perhaps be best regarded as a short stage either between two triangular spacings or between the onset of bearing and the final triangular spacing.

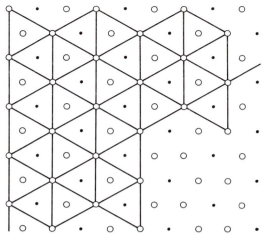

Fig. 9.5   Triangular–hexagonal–triangular planting: thinning from triangular to hexagonal and from hexagonal to triangular spacings. Palms marked ● are in the original small triangles only; (○) palms in small triangles and hexagons; those left in the large triangles are shown in the diagram joined by lines which indicate the eventual spacing.

Thinning schedules should be determined before planting on the basis of a knowledge of the growth rate of palms in the area and any yield figures which may be available per palm for high-density plantings and normal density plantings in the area. When reductions of yield per palm take place from the start of bearing, 2–3 years at the high density is as much as can be expected. Where two stages are possible, the total period from planting to the final thinning may extend to 5 or 6 years.

To summarize, the advantages of high-density initial planting are:
1. Fuller use is made of the ground from the start of bearing.
2. An earlier return, approaching that obtained from mature areas, is made on capital.
3. Earlier and fuller use is made of capital invested in the mill.
4. The period of no and low yield is reduced when replanting, and thus more frequent replanting with higher-yielding material is made possible.

The disadvantages are:
1. Higher initial cost of planting or replanting.
2. Additional field costs, viz. felling half or two-thirds of the stand.
3. Possible reduction in mature yields due to maintaining too high a stand for too long.

Table 9.3 give densities per acre and hectare at different spacings with some conversion factors.

Table 9.3   *Table of planting distances and densities*

Triangular planting distance (ft)	Distance between rows (ft)	Density per acre	Density per hectare	Metre equivalent Planting distance (m)	Metre equivalent Between rows (m)
18	15.6	155.1	383.2	5.49	4.75
19	16.5	138.9	343.2	5.79	5.03
20	17.3	125.8	310.9	6.10	5.27
21	18.2	114.1	282.0	6.40	5.55
22	19.1	103.9	256.7	6.71	5.82
23	19.9	95.1	235.0	7.01	6.07
24	20.8	87.3	215.7	7.32	6.34
25	21.7	80.5	198.9	7.62	6.61
26	22.5	74.5	184.1	7.92	6.86
27	23.4	68.9	170.3	8.23	7.13
28	24.2	64.3	158.9	8.53	7.38
29	25.1	59.8	147.8	8.84	7.65
30	26.0	55.9	138.1	9.14	7.92
31	26.8	52.4	129.5	9.45	8.17
32	27.7	49.1	121.3	9.75	8.44
33	28.6	46.2	114.2	10.06	8.72
34	29.4	43.5	107.5	10.36	8.96
35	30.3	41.1	101.6	10.67	9.24

(m)	(m)	per acre	per hectare	Foot equivalent (ft)	Foot equivalent (ft)
7.0	6.06	95.4	235.7	23.0	19.9
7.5	6.50	83.0	205.1	24.6	21.3
8.0	6.93	73.0	180.4	26.2	22.7
8.5	7.36	64.7	159.8	27.9	24.1
9.0	7.79	57.7	142.6	29.5	25.5
9.5	8.23	51.8	127.9	31.2	27.0
10.0	8.67	46.7	115.5	32.8	28.4

Notes: To obtain the hexagonal density, with the same spacing, multiply the triangular density by two-thirds.
To convert density per acre to density per hectare multiply by 2.4711.
To convert density per hectare to density per acre multiply by 0.4047.
1 kg per hectare = 0.8922 lb per acre. 1 lb per acre = 1.1209 kg per hectare. 1 ton per acre = 2.51 tonnes per hectare.
Distance between the rows in triangular planting is 0.866 × distance between palms.
To obtain density per acre or per hectare of triangular planting, multiply the square of the planting distance in feet or metres by 0.866 and divide this product into 43,560 or 10,000.

Costs of field establishment

The costs of establishing oil palms in the field vary considerably both between and within countries according to terrain, forest

Table 9.4   *Man-day costs per hectare*
Africa

From high forest	Man-days per hectare				
	Nigeria	*Ivory Coast*[1]	*Zaire (1)*[2]	*Zaire (2)*	*Zaire (3)*
Underbrushing	19	10	10 ⎫		44
Felling†	50	50	40 ⎭	58*	55
Beating down	7	30‡ ⎫	90 ⎫		90
Burning	1	30§ ⎭		55 ⎫	90
Opening rides	26 ⎫	20	30 ⎬		
Lining and realigning	10¶ ⎭		4 ⎭		19
Planting‖	21	12	22	14	25
Cover sowing		5	2	5	5
Fixing wire collars	3				
	137	157	198	132	238

* Light to medium forest.
† Axing. With chainsaws and axes the man-days can be reduced to 5–10 per hectare.
‡ Includes cutting up.
§ Includes heaping, etc.
¶ Lining after underbrushing.
‖ Including lifting from nursery.

Sources: [1]Ivory Coast – Surre *et al.* (1961) and Dekester (1962); [2]Zaire (1) – (Vanderweyen, 1952).

Malaysia

From high forest*	Man-days per hectare	
	Manual with chainsaws	*Partial mechanization*
Underbrushing	6.0	6.0
Felling†	5.0	5.0
Burning	0.25	0.25
Cutting up†, stacking and reburning	12.5	1.35‡
Mechanical stacking	–	0.5
Lining and holing	6.4	6.4
Planting (from polybags)	16.5	16.5
Cover sowing	7.2	7.2
	53.9	43.2

* Data from underbrushing to reburning are from FELDA experience (Balakrishnan and Lim Cheong Leong, 1977); lining, holing, planting and cover sowing are from the mean of figures of nine estates.
† Units of one chainsaw man and two axemen.
‡ Assuming 5 per cent of the area cannot be done mechanically. If manual assistance is also required with the mechanically stacked areas, then up to a further 5 man-days may be needed.

density or previous land usage, soil, labour availability, capability and organization, and the degree of mechanization. The man-day figures in Table 9.4, which cover the work described in Chapters 8 and 9, are given only as a guide to what has been accomplished in practice in a number of regions; they are either from published material or are averaged figures from plantation sources.

Notes: Africa. It will be seen that there is not much variation in felling or in planting costs and that the differences lie largely in the amount of effort expended in the various beating down, cutting, clearing and heaping operations which are undertaken between felling and planting. Felling costs are for felling by axe. With chainsaws the man-days should be reduced to 10–20 per cent of the axing figures, thus reducing overall man-day requirements for clearing and planting to around 100–150 man-days per hectare. Mechanical felling and windrowing (see p. 370) reduces labour needs by as much as a further 60 man-days per hectare (Surre, 1961).

Malaysia. Clearing costs in terms of man-days are considerably less in Malaysia than in Africa even when the use of chainsaws is taken into account. However, more man-days appear to be expended on lining, holing, planting and cover sowing.

Road and initial marking-out costs are not included in the above comparisons as they vary widely with the frequency of roads, the terrain and other factors.

References

Aya, F. O. (1978) A preliminary assessment of the influence of age and the time of transplanting on the performance of polybag seedlings in the field. *J. W. Afr. Inst. Oil Palm Res.*, **5** (20), 5.

Bachy, A. (1965) Influence de l'éclaircie naturelle sur la production du palmier à huile. *Oléagineux*, **20**, 575.

Balakrishnan, V. and **Lim Cheong Leong** (1977) Jungle clearing by FELDA for planting oil palm. In *Int. devel. in oil palm*, ISP, Kuala Lumpur.

Barnabe, G. A. H. (1976) La mise en place des jeunes palmiers à huile élevés en sacs de plastique. *Oléagineux*, **31**, 259.

Beirnaert, A. and **Vanderweyen, R.** (1940) Note préliminaire concernant l'influence du dispositif de plantation sur les rendements. *Publs INEAC*, Série Tech., No. 27 bis.

Breure, C. J. (1977) Preliminary results from an oil palm density × fertilizer experiment on young volcanic soils in West New Britain. In *Int. devel. in oil palm*, ISP, Kuala Lumpur.

Breure, C. J. (1986) The effect of different planting densities on the yield trend of oil palm. *Expl. Agric.*, (in the press).

Breure, C. J. and **Corley, R. H. V.** (1983) Selection of oil palms for high density planting. *Euphytica*, **32**, 177.

Corley, R. H. V., Hew, C. K., Tam, T. K. and **Lo, K. K.** (1973) Optimal spacing for oil palms. In *Advances in oil palm cultivation*, 52, Incorp. Soc. of Planters, Kuala Lumpur.

Dekester, A. (1962) La mise en place des jeunes palmiers à huile. *Oléagineux*, **17**, 475.

Goh, K. H. (1977) A systematic design for oil palm spacing trials. In *Int. devel. in oil palm*, ISP, Kuala Lumpur.

Goh, K. H. (1981) Analysis of oil palm spacing experiments, In *The oil palm in agriculture in the eighties*, ISP, Kuala Lumpur.

Gray, B. S. and **Hew Choy Kean** (1963–6) Annual reports, Oil Palm Research Station, Banting, 1962, 1963, 1964 and 1965. Harrisons and Crosfield (Malaysia) Ltd. Mimeographs.

Gunn, J. S. and **Sheldrick, R. D.** (1963) The influence of the age and size of oil palm seedlings at the time of transplanting to the field on their subsequent growth. *J. W. Afr. Inst. Oil Palm Res.*, **4**, 191.

Hardon, J. J., Corley, R. H. V. and **Ooi, S. C.** (1972) Analysis of growth in oil palm. II. Estimation of genetic variances of growth parameters and yield of fruit bunches. *Euphytica*, **21**, 257.

Hew Choy Kean and **Tam Tai Kin** (1971) Annual report, agronomy. Research and advisory scheme. 1970. Harrisons and Crosfield (Malaysia) Sdn. Berhad. Mimeograph.

IRHO (1974) *Rapport d'activités 1972–1973.* Doc. No. 1138.

Khoo Kay Thye and **Chew Poh Soon** (1977) Effect of age of oil palm seedlings at planting out on growth and yield. In *Int. devel. in oil palm*, ISP, Kuala Lumpur.

Lafaille, J.-P. and **Daniel, M.** (1965) Soins aux jeunes palmiers et remplacements. *Oléagineux*, **20**, 667.

Moreau, C. and **Moreau, M.** (1958) Lignification et réactions aux traumatismes de la racine du palmier à huile en pépinières. *Oléagineux*, **13**, 735.

Ollagnier, M. and **Delvaux, R.** (1966) La plantation du Val d'Iguape. *Oléagineux*, **21**, 663.

Poncelet, M. (1965) Plantations de palmiers à huile en terrasses sur terrains humides. *Oléagineux*, **20**, 431.

Prevot, P. and **Duchesne, J.** (1955) Densités de plantation pour le palmier à huile. *Oléagineux*, **10**, 117.

Ramachandran, P., Narayanan, R. and **Knecht, J. C. K.** (1973) A planting distance experiment on *dura* palms. In *Advances in oil palm cultivation*, 72, Incorp. Soc. of Planters, Kuala Lumpur.

Sly, J. M. A. and **Chapas, L. C.** (1963) The effect of various spacings on the first sixteen years of growth and production of the Nigerian oil palm under plantation conditions. *J. W. Afr. Inst. Oil Palm Res.*, **4**, 31.

Sly, J. M. A. and **Sheldrick, R. A.** (1965) *NIFOR first annual report 1964–5*, p. 27.

Sly, J. M. A., Sheldrick, R. D. and **Zeven, A. C.** (1963) *WAIFOR eleventh annual report 1962–3*, p. 25.

Sparnaaij, L. D. (1960) The analysis of bunch production in the oil palm. *J. W. Afr. Inst. Oil Palm Res.*, **3**, 109.

Sparnaaij, L. D. and **Gunn, J. S.** (1959) The development of transplanting techniques for the oil palm in West Africa. *J. W. Afr. Inst. Oil Palm Res.*, **2**, 281.

Surre, C. (1955) Densité économique de plantation pour le palmier à huile. *Oléagineux*, **16**, 411.

Surre, C., Fraisse, A. and **Boyé, P.** (1961) Plantation de palmier à huile sur sol de forêt et sur savane à *Imperata*. *Oléagineux*, **16**, 91.

Tan Yap Pau and **Ng Siew Kee** (1977) Spacing for oil palms on coastal clays in peninsular Malaysia. In *Int. devel. in oil palm*, ISP, Kuala Lumpur.

Toovey, F. W. (1948) *Oil Palm Research Station, Nigeria, seventh annual report, 1946–7*, pp. 19–20.

Toovey, F. W. (1953) Field planting of oil palms. *J. W. Afr. Inst. Oil Palm Res.*, **1**(1), 68.

Unilever Plantations (1961) Annual review of research, p. 36. Mimeograph.

Vanderweyen, R. (1952) *Notions de culture d'Elaeis au Congo Belge.* Brussels.

WAIFOR (1955) *Third annual report 1954–5*, p. 65.

Wuidart, W. and **Boutin, D.** (1976) Choix des plantes de pépinière. *Oléagineux*, **31**, 371.

Yeow, K. H., Tam, T. K., Poon, Y. C. and **Toh, P. Y.** (1981) Recent innovation in agronomic practice: palm replacement during immaturity. In *The oil palm in agriculture in the eighties*, ISP, Kuala Lumpur.

Chapter 10

The care and maintenance of a plantation

Once the plants have been established in the field they need care and protection so that they may come into bearing early and give a high bunch production throughout their bearing life. The palms need protection against competition, soil depletion, diseases and pests, and care must be taken to see that in their pruning and harvesting no damage is done which may affect their future performance. Eventually they must be replaced by another generation of palms by replanting.

Cultivation and soil covers

In Chapter 8 it was shown that in opening new areas from forest by burning the last operation was the planting of covers, except where establishment inter-cropping was to be undertaken. Examples were given of the effects of early cultivation on soil nutrient levels and on subsequent yields. The removal of competition was seen to have early advantageous effects on yields but, on already impoverished soils, to reduce soil fertility to a point where later yields were affected. Two of the West African experiments there mentioned also contained treatments in which, although ring-weeding and paths were maintained, the inter-row natural covers were neglected to the extent that only one round of slashing per year was undertaken as against normal maintenance in which the undergrowth was kept down to about 1 ft (30 cm) by regular slashing rounds. The effects on yields are as shown in Table 10.1.

It will be seen that competition from undergrowth had a severe effect on early yields, this being severest on already impoverished land. In the forest area, however, once the palms had fully established themselves, their shade effected an ascendancy over the cover which resulted in yields no longer being reduced by lax maintenance. This effect was less marked in the other area. It is thus clear that in considering the ground cover policy for optimum yield, the effects of competition and cultivation need as much consideration as the type of cover to be maintained.

Table 10.1 *Yields of annually slashed plots as a percentage of the normal maintenance yield*

	Forest area	Previously cultivated area
First 8 years of bearing		
Normal maintenance	100.0	100.0
Annual slashing only	79.4	69.1
Later bearing years	(11 years)	(2 years)
Normal maintenance	100.0	100.0
Annual slashing only	102.0	85.4

The type of cover to be maintained

There has been much discussion, but little experimentation, on the most desirable type of ground cover to grow in a plantation. Many plants considered manageable as covers are difficult to establish and the cultivation effect of establishing them may readily be confused with an effect of the cover itself. On the other hand an easily established plant may appear deleterious because it enters quickly into competition with the palms through not being properly maintained. It is not unnatural, therefore, that cover policy was for a long time a matter on which opposing opinions were firmly held on very little scientific evidence and that fashions for the use or encouragement of certain plants have come and gone.

Historically, the attitudes to ground cover have been different in Africa and in the Far East. In Africa, particularly in the wet but seasonal parts of West Africa, forest regrowth and the species which follow the felling of forest grow with tremendous rapidity particularly at the beginning and, to a lesser extent, at the end of the rains (April/June and September/October). At the same time the leguminous cover *Pueraria phaseoloides* establishes itself naturally and vigorously and increases the volume of vegetation. The maintenance of the interline covers has therefore consisted of cutting back this adventitious growth to a height of about 1 ft (30 cm). The frequency of cutting depends on the locality, but in areas opened from high forest and subject to a sharp dry season, as many as six 'cutlassing' rounds have been required in the second and third years while ring-cutlassing is needed as or more frequently to prevent the rings around the palms being overrun and the palms themselves covered with fast-growing creepers. As the palms shade the ground, so the frequency of cutlassing and ring-cutlassing decreases. Cutlassing is reduced to two rounds (or even one) per year and ring-cutlassing to two to four rounds.

In the less seasonal climates of Zaire and the Ivory Coast, initial regrowth of natural covers is not so vigorous, but *Pueraria* still establishes itself well. In the Ivory Coast *Pueraria* is encouraged by low cutting of natural regrowths, and cutlassing is only required in young areas twice a year (Surre and Ziller, 1963). In adult areas cutlassing is reduced to one round per year but ring-cutlassing continues four times a year.

By contrast, maintenance in the Far East has been thought of in terms of weeding rather than cutting. The plants held to be desirable, or at least not obnoxious, do not tend to grow rapidly in height, e.g. the leguminous covers and the fern *Nephrolepis biserrata*. A weeding round has therefore largely consisted of chopping out those species not wanted and clean weeding the rings around the palms. One clearly dangerous weed, lalang (*Imperata cylindrica*), must be kept at bay, and this more than anything else has been the cause of maintenance being thought of in terms of keeping out undesirable plants. Lalang is particularly undesirable as it invades the rings and competes directly with the oil palm. Therefore in all countries where lalang is ubiquitous, regular weeding, or oil wiping, patrols are necessary. In Africa, lalang is usually confined to certain locations and seldom presents a problem.

Cover policy has been much influenced in the Far East by current practice in the rubber industry. This has undergone a considerable number of changes in the last 60 years and these changes have been reflected in similar changes in practice on oil palm estates, though fortunately the oil palm was established as a plantation crop late enough to escape the full effect of the clean weeding era which afflicted the rubber industry. In the 1930s, mixtures of two or more of the legumes *Calopogonium mucunoides, Centrosema pubescens, Pueraria phaseoloides* and *Dolichos hosei*, with the non-leguminous creeper *Mikania cordata* were recommended in Malaysia (Bunting *et al.*, 1934). In Sumatra the Carilla gourd, *Momordica charantia*, was favoured as was the shrub, *Leucaena glauca*, which was sometimes used as a green manure. Immediately after the Second World War cover planting fell out of favour and it was reported by research workers of plantation companies (OPRS, Nigeria, 1949) that 'there was a great wave of feeling against the use of cover crops in Malaya'. The idea of 'forestry' covers took the place of planted covers and seemed to suit the non-burning procedure then adopted. The natural cover consisted of shrubs from stumps, and seedlings which grew from the seed of the forest flora. Plants of secondary forest growth such as *Macaranga* species were common. Slashing rounds were undertaken at 4- to 6-month intervals.

Where areas had been burnt, natural covers were of different species. Soft weeds were followed by grass species, typically those of *Paspalum* and *Axonopus*, with small shrubs such as *Trema* and *Lantana* and the creepers *Mikania cordata* and *Passiflora* species. Later, the fern *Nephrolepis biserrata* would become dominant, though mixed with grasses and *Mikania*.

In the new oil palm plantings of tropical America, grass species are the dominant feature of natural covers and present a serious problem. The presence of ranches, big or small, in almost all parts of the region is responsible for this. Even where plantations are being opened from forest, seeds from grasses in adjoining pastures soon gain access. Most of these grasses are upright and grow vigorously throughout the wet season and many, e.g. *Hyparrhenia rufa* (Uribe or Yaragua), produce tall flowering shoots which seed freely and can easily 'submerge' young plants. *Panicum maximum* and *Pennisetum purpureum* are common. A serious weed of the Pacific coast, particularly in Ecuador, is the Camacho (*Xanthosoma* sp.) the leaves of which grow rapidly to enormous size; the plant requires special control measures.

Where areas in America are opened from forest *Pueraria* establishes well, but can quickly become invaded by grass species. Many young areas have been seriously retarded by grass competition, so thorough eradication by a suitable herbicide such as dalapon should be completed before planting. Even when this is done, however, grasses can regain their foothold unless a good stand of *Pueraria* is quickly established and spot-spraying of volunteer grass done regularly both in the interlines and the circles, great care being taken not to allow the spray to fall on the young palms. Cattle have been used in America in an attempt to graze out the grasses, but unless the grazing is carefully managed the land may become seriously puddled and drainage impeded; moreover the grasses can never be completely eradicated by grazing. If ring-weeding is neglected, the grasses soon invade the circles from outside, grow up through the leaves of young palms and enter into competition with them for nutrients and even for light. Once an invasion has taken place, ring-weeding becomes very difficult as the root systems of the palm and the grasses are intertwined. In Honduras a fern cover not dissimilar to the *Nephrolepis* cover of the Far East has been found to establish itself and needs little attention.

With the almost universal practice of burning, and with the replanting of oil palm, coconut and rubber areas with oil palms, the planting of leguminous covers has become the general rule in all oil palm growing regions. Recent reviews have suggested, however, that the controversy of planted legumes versus natural covers is not yet dead. Broughton (1977) reviewed Malaysian experiments and showed that legume plots invariably outyielded plots with natural covers or any other managed covers. Moreover palms in the legume plots showed higher leaf levels of N and P and gave higher responses to fertilizers. He found that, in the early years when the inter-row cover is not yet competing with the root system of the young palm, the legumes grew faster than other covers, and that later, when the palm canopy closed in, the covers began to die and returned large quantities of their nutrients to the soil. For Nigeria on the other hand Aya and Lucas (1977) considered the initial poor establishment of legumes a hindrance to the control of erosion, and they viewed the subsequent vigorous growth as providing strong competition for water and nutrients. They found no increased yields from legume planting in experiments except in some later years in a long-term experiment at Benin; and, although they agreed with Broughton that this later benefit was likely to be due to residues of the legumes when they died out, they did not consider this an overriding advantage. They drew attention firstly to the strong competitive effects of legumes, particularly in the drier regions of West Africa, and secondly to the work entailed in controlling the cover and preventing the invasion of the weeded circles by runners, and suggested that the encouraging of a natural cover is preferable. While it is undoubtedly true that in seasonal climates the prevention of undue competition between covers and palms in the dry season is of special importance, nevertheless in most parts of West Africa it has not been found at all difficult either to establish a legume cover quickly or to control it by normal maintenance methods. Moreover in countries such as Ghana or the Ivory Coast *Pueraria* has been shown to be easily managed, to persist for many years and thus to return large quantities of nutrients to the soil. It seems unlikely therefore that the planting of legumes will be discontinued in any appreciable areas of oil palm cultivation.

The important considerations in determining a cover maintenance policy following the burning of forest or replanting are:

1. The need to prevent, by adequate ring-weeding and cover control, competition for both water and nutrients between the palms and other plants.
2. The need for early establishment of suitable plant species which will cover the intervening ground and, while requiring the minimum of maintenance in the early years, will prevent erosion.
3. The need to maintain in adult areas a cover requiring minimal care and attention and not containing species actually harmful to the palms.
4. The need to maintain soil fertility at all stages.

Ring-weeding and path maintenance

Young palms have a small, developing root system and they will suffer greatly if they have to compete in their development with other plants. It is therefore of great importance that the soil in which the young roots are developing should be kept free of weeds. This is commonly done with a cutlass in Africa and America and with a hoe in Asia. With the latter instrument there is a tendency with successive hoeings for the soil to be drawn away from the palm and for a saucer-like depression, in which rain-water may stand, to be formed. An implement for pulling back a creeping cover to the correct circumference is useful. Ring-weeding should be to 1 m radius for the first year and thereafter to 1.5–2 m radius. In the early years the purpose is predominantly the prevention of competition, but later on the ring is needed for bunch and fruit collection and so that the ripe fruit dropping from a bunch may be seen. In many areas carpet grass, *Axonopus compressus*, establishes itself on rings and paths which are frequently cut. This is favoured by some on the grounds that it holds the soil together, preventing saucering, the hard-baking or puddling of the surface and the creation of runoff channels; it is also held that *Axonopus* does not compete with the main root system of the adult palm. However, with the advent of herbicides *Axonopus* is less often seen on the circles.

With hand weeding, maintenance in the early years must be frequent though the actual frequency will depend on the rainfall and soil fertility. In most areas at least six rounds per year will be required for the first few years and this will then be progressively reduced to two to four rounds a year depending on the locality and the degree of shading of the ground by the palms. In non-seasonal climates monthly ring-weeding rounds are normally necessary in the early years and this is gradually reduced to 6-weekly, then 2-monthly rounds and finally to four rounds per year.

With increasing labour costs the maintenance of the rings or circles by herbicide application is becoming more common. There has been some reluctance to use herbicides in the early years for fear of damaging the young palm or its root system, but while care must be exercised in both the method of application and the choice of herbicides there is no evidence of deleterious effects where such care has been taken. Indeed it has been argued that, with no disturbance of the soil, erosion from the circle will be less and the root system will be undamaged; consequently growth should be improved. Experiments in Nigeria tend to support this conclusion (Sheldrick, 1968a).

With regard to choice of herbicides, it is best to apply contact rather than translocated herbicides and particularly those which are inactivated by soil contact. However, no hard and fast rules can be laid down; paraquat, which has been widely used, is translocated to some extent but is inactivated on contact with the soil. A very full and helpful account of the commoner herbicides and their use in oil palm fields has been provided by Turner and Gillbanks (1974). The herbicides which produce abnormal growth in young palms are the phenoxyacetic acids, 2,4-D and 2,4,5-T, or the halogenated aliphatic acids, dalapon and TCA. 2,4-D produces long-necked and bending symptoms in oil palm seedlings with subsequent death, while dalapon and TCA produce extreme fasciation of the leaves and leaflets and later a flattened rosette appearance (Sheldrick, 1962, 1968b). Lesser damage is caused by aminotriazole (amitrole or ATA), individual leaves showing severe chlorosis, but recovery is rapid. The substituted ureas (often used as pre-emergence herbicides), monuron and diuron, and the triazines, atrazine and simazine, although persistent, are also less dangerous, but they can cause temporary damage and the exact effects on the root system are unknown.

Of the contact herbicides sodium arsenite, which was much employed in the rubber and oil palm industries, has fallen out of use because of its human toxicity. More widely used now is the sodium salt of methane arsonic acid, MSMA, which is effective against certain grasses. The most complete screening tests for the effect of herbicides on the young oil palm are those of Sheldrick (1968b), whose results are summarized in Table 10.2. The effects of MSMA have not been reported.

Table 10.2   *Symptoms in the oil palm associated with particular herbicides*

Herbicide	Type of response				
	Immediate leaf scorch	*Subsequent blotching and leaf necrosis*	*Leaflet chlorosis*	*Epinastic response*	*Leaf fasciation*
Diquat	***				
Paraquat	***				
MCPA	**	***		***	
2,4-D	**	**		**	
Mecoprop	**	***		**	
Dichlorprop	**	***		**	
TCA	*	**			**
Dalapon	***	***			***
Diuron	*	*	**		
Mono-linuron			*		
Linuron			*		
Simazine					
Atrazine			**		
Aminotriazole (Amitrole)	*	**	***		

*** Severe.
 ** Moderate.
  * Slight effects.

Source: Sheldrick (1968).

Herbicide mixtures used for ring-weeding will in practice depend on the above considerations and the current cost of the ingredients. For young palms in Nigeria applications of 2 kg a.i. of paraquat with 3–4 kg atrazine, monuron or diuron per hectare of sprayed ground applied twice a year has been found to give satisfactory weed control (Sheldrick, 1968a). In the Ivory Coast a range of formulations for palms older than 5 years has been evolved (Coomans, 1971). Up to 7 years MSMA at 1.8 kg a.i. per hectare and paraquat at 0.7 litre per hectare have been found effective. The latter is applied at the end of the dry season, the MSMA at the end of the main wet season (August) while MSMA with amitrole at 1 kg a.i. per hectare or 2,4-D at 0.7 kg a.i. per hectare is applied at the end of the short wet season (December). For areas of older palms various mixtures of MSMA, 2,4-D, TCA, sodium chlorate and/or amitrole are recommended. These mixtures contain herbicides dangerous to the palm but with older palms the danger is much reduced.

In Malaysia circle spraying is usually done three or four times a year. In mature areas a mixture of 6–9 litres sodium arsenite liquid + 2–3 kg sodium chlorate per sprayed hectare has been used, but for young palms formulations of paraquat, MSMA and diuron are more common. With young palms it was shown that spraying with mixtures of paraquat, MSMA and diuron can start 3–9 months after planting without adversely affecting growth. A satisfactory mixture in terms of a.i. was paraquat 0.28 kg per hectare, MSMA 1.68 kg per hectare and diuron 0.28 kg per hectare (Seth *et al.*, 1970).

In West Cameroon MSMA replaced paraquat altogether on grounds of cost, but needed to be sprayed more frequently (Smith, 1973). Ametryne (a triazine herbicide) was also found effective. With MSMA alone spraying was undertaken twice in the year of planting, monthly in the second year, six times in the third year and thereafter four times a year.

Circle spraying is most frequently done with knapsack sprayers, a man covering up to 500 palms per day (Smith, 1973). It should be observed that with 143 palms per hectare the proportion of total area to be sprayed is as follows: 1 m circles, one-twenty-second; 1.5 m circles, one-tenth; 2 m circles, one-sixth. Years of spraying with mixtures such as those described above and usually based on paraquat has often resulted in the encouragement of grasses (Turner, 1983). However, with the development of low-volume spraying with controlled droplet application it was found that glyphosate (Roundup), which was previously considered too costly for oil palm plantation use, could be employed at much lower rates than previously recommended; and if sprayed in mixture with 2,4-D amine both the grasses and broad-leaf weeds could be controlled (Turner, 1982; Jollands *et al.*, 1983). In the Ivory Coast, ultra-low-volume spraying of herbicides with a battery-driven knapsack sprayer has been found to be 20 per cent cheaper than ordinary knapsack spraying methods (Hornus, 1983).

The maintaining of paths through the palm fields is desirable in the early years for inspection and in the later years for use by harvesters. They can be maintained by any of the methods described above for ring-weeding. Harvesters' paths, which in adult areas fall naturally along the rows or on hilly land along the contours, tend to become denuded of vegetation or covered with a 'lawn' of carpet grass (*Axonopus compressus*). The position of inspection paths for the early years has been largely a matter of custom, and maintenance is of course more onerous than in later years. In Zaire (Vanderweyen, 1952) paths were maintained between each pair of rows and, where opening by alternate row heaping has been practised, this is convenient. In Nigeria, it was found convenient for inspection to run paths along the palm rows for the first 2 years, but later on, when the palms were still 'on the ground' but the leaves far-spreading, to cut a path between each pair of rows. Later, when the palm crowns were well off the ground, the harvesters' paths again ran along the rows. It is now generally recommended that inspection paths should be maintained from the start between each pair of rows and these later become harvesters' paths. If these are 1.5 m wide they will occupy one-tenth of the area; circles plus paths will then come to occupy about one-quarter of the area.

Cultivation and the early establishment of covers

It was seen in Chapter 8 that cultivation of the soil in the planting of inter-row food crops has a deleterious effect on nutrient status and, if practised on infertile areas, lowers bunch yield. A certain amount of soil cultivation is always required in the establishment of cover plants, the amount usually depending on the ease with which these plants can be established in the locality concerned. Such cultivation consists of the opening of drills or pockets for the planting of seed and the hoeing required for keeping the area free from volunteer competing plants. In some localities the latter cultivation can be very considerable and become very expensive.

Few cultivation experiments with established palms have been carried out. Heavy cultivation, through eliminating competition, usually leads to improved early yields followed by a heavy fall in yield. In Nigeria, for instance, where regular tillage was carried out with a hoe, yields were enhanced by cultivation in the first 3 bearing years in comparison with those of plots maintained normally under a natural cover, but subsequently the yields of the cultivated plots declined as shown below:

Yield of cultivated plots as a percentage of the control plots' yield, Abak (Nigeria)

	First 3 years	4th–7th years	8th–11th years
Burning, normal maintenance	100	100	100
Burning, regular tillage	277	92	82

Similar results were obtained in an early cultivation trial in Malaysia (Lucy, 1941) and, in a more recent trial (Chew and Khoo, 1977), plots in which the soil was maintained bare by herbicide spraying gave yields initially as high as those with a leguminous cover, but by the third bearing year leaf length and area and bunch yield were significantly lower.

Soil data from regularly tilled areas are not available except where tillage was combined with the continuous or temporary growing of food crops between the young palms. These results from an afforested area in Nigeria have already been referred to (p. 376), but it should be repeated here that the soil effects in the inter-cropped plots were those which would be expected from tillage, i.e. a general reduction of fertility as measured by exchangeable cations K, Mg and Ca. Plots prepared for and planted with a leguminous cover – *Pueraria* – showed soil changes intermediate between those of plots left to natural covers and those tilled for food crops. Plots under natural covers showed the minimum nutrient losses. In an early experiment in Malaysia plots in which a selected 'forestry' cover was eventually established showed an initial drop in yield, but after 2 years a large increase in yield was recorded. Kowal and Tinker (1959) believe that the comparatively high nutrient status of plots maintained under such a cover for as much as 10 years after burning is due to the return to the surface of nutrients washed down the profile during the first rains after the burn by the deep-rooted cover which establishes itself.

On the scanty evidence available, therefore, it is reasonable to suppose firstly that, to avoid soil depletion, covers should be planted as early as possible and with the minimum of cultivation, and secondly that there is some merit in a natural cover which includes deep-rooted plants.

In regions with a marked dry season, cultivation has been advocated at the beginning of the dry season to reduce competition between palms and cover for the limited moisture which will be available in the coming months. Such treatment in a trial in Nigeria opened from forest had a marked beneficial effect on yield in the early years (Sly and Sheldrick, 1963). The effect was similar to, but greater than, that of establishment inter-cropping and was still substantial after the plants had been in the ground for 7 years. Data

Table 10.3 *Effect of establishment inter-cropping and dry season soil cultivation in Nigeria (yield per palm (planted 1957))*

Years	Control (kg)	Inter-cropping for 2 years (kg)	Cultivation in the dry season (kg)
1961 + 1962	80	90	103
1963	79	84	89
1964	85	89	90
1961–4	244	263	283

Note: Differences between treatments and control are significant in all years except 1964.

Source: Sly and Sheldrick (1963).

from this trial are presented in Table 10.3 and show that effects last for 3 years of bearing, but then wear off. This may well be due to the fact that when the palms are well up they shade the ground and are able to compete adequately for moisture without the aid of cultivation. Under the extremely dry conditions of Benin (Dahomey) (1,200 mm rainfall) the maintenance of bare soil under young palms to the age of 6 years has been shown to improve vegetative development including the root system and to increase early yields by nearly 100 per cent (Daniel and de Taffin, 1974).

The purpose of sowing an 'artificial' cover or cover mixture after burning is to provide a cover for the soil at the earliest moment and thus prevent erosion and soil nutrient depletion. A secondary purpose is to check the invasion of undesired plants. Because of lack of valid comparisons, the choice of cover plants is usually dictated by experience of their establishment in the locality. Leguminous plants are almost always chosen because they form nodules which contain colonies of nitrogen-fixing *Rhizobium* bacteria which live symbiotically on the host plant. The leguminous creeping covers which have, in mixture, stood the test of time and are still widely planted in all parts of the tropics are *Calopogonium mucunoides*, *Pueraria phaseoloides* and *Centrosema pubescens* (Figs 10.1, 10.2 and 10.3). In Africa and America *Pueraria phaseoloides* is now usually planted as a sole cover because it establishes and spreads so rapidly and can maintain its dominance for many years. Much better establishment of these covers than previously is now obtained by (i) seed treatment with concentrated sulphuric acid or scarification or by soaking for 12–24 hours with subsequent heat treatment for 1 hour at 39–40 °C, (ii) inoculation with the correct *Rhizobium* species for nodulation, (iii) planting with rock phosphate, (iv) dispersing concentrations of wood ash from the planting lines or pockets, (v) (in West Africa) planting when the rains have set in (Sheldrick, 1968c; Aya, 1973).

0   25
mm

Fig. 10.1 *Calopogonium mucunoides* (RRIM).

Calopogonium and *Centrosema* seed should be soaked for 15–20 minutes in concentrated sulphuric acid for satisfactory germination. *Pueraria* requires at least 30 minutes and, in Nigeria, 90 minutes has been found to be the optimum time. The acid must be rapidly washed off by repeated changes of hot water after treatment. Mechanical scarification is used in Malaysia as an alternative. The soaking and heat-treatment method mentioned above has been used for *Pueraria* in Nigeria (Aya, 1973).

Rhisobium inoculum is not always required, and must be purchased from a reliable source. It is only worth obtaining *Rhizobium* culture if there is an absolute assurance that the correct organism is included and that the culture will be effective. The seed rate of the mixture is 5–6 kg per hectare. The recommended

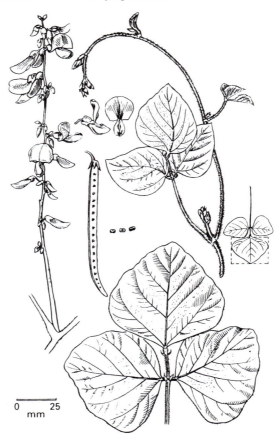

0 ═══ 25
mm

Fig. 10.2   *Pueraria phaseoloides* (RRIM).

proportions in West Africa are *Calopogonium : Pueraria : Centrosema*
– 2 : 2 : 1. In Malaysia a mixture in the proportions 2 : 2 : 3 is
often employed, but a mixture of *Pueraria* and *Centrosema* only in
the proportion 2 : 3 has also been recommended for oil palm estates
(Bevan and Gray, 1966).

The mixture should be sown in three or five drills per interline.
No drill should be nearer than 1.5 m to the line of palms. With 9 m
triangular planting three rows would be about 2.2 m apart, five rows
about 1.1 m apart. Cross-drills may also be sown between the palms
at the same spacing. *Calopogonium* germinates first and grows
vigorously but soon dies out to be replaced firstly by *Pueraria*.
Centrosema is not much in evidence for the first few years, but when
the palms start to shade the ground it may become dominant.

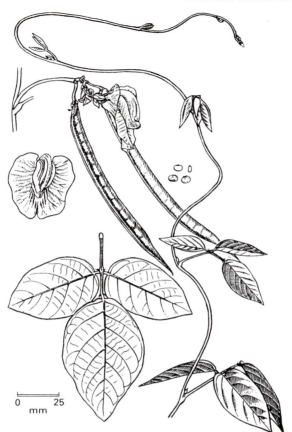

0   25
mm

Fig. 10.3   *Centrosema pubescens* (RRIM).

Failure of establishment of this cover mixture in Africa is rare. Such failures as have occurred have been due to planting before the rains, in too high a concentration of ash or to faulty pretreatment of seed. In the Far East, however, establishment has been often more difficult. Creeping covers may, for no obvious reason, become chlorotic and die off and in some cases this is due to lack of appropriate fertilizers. On most Malaysian soils additional applications of rock phosphate at the rate of 60–120 kg per hectare are recommended at 2, 5, 9 and 12 months after planting. Potash, in burnt areas, is usually in sufficient supply but elsewhere a small initial seed dressing of 1 kg muriate of potash to 1 kg seed is recommended with the phosphate seed dressing. When magnesium deficiency is expected or in evidence 120 kg per hectare of magnesium limestone is applied. Legume failure in Malaysia has also been ascribed to the fungus *Rhizoctonia solani* (as well as to *Fomes lignosus* and *Ganoderma pseudoferreum*) and to an aphid, *Aphis laburni*, the ladybird beetle, *Epilachna indica*, which may be particularly destructive of *Centrosema*, the beetle *Pagria signata* (on *Pueraria* and *Calopogonium*), the bug *Chauliops bisontula*, the caterpillar *Lamprosema diemenalis* and a number of other caterpillars and grasshoppers, slugs and snails (Edgar, 1958). Many cover failures in Malaysia are also due to the nematode, *Meloidogyne javanica*, against which there is as yet no economic control.

It is not surprising, therefore, that in Malaysia there should have been widespread trial of, and interest in, covers other than 'the standard three'. This interest has included bush covers which, besides growing vigorously, have been thought to have the virtues of penetrating and perhaps improving the structure of heavy impervious soils and bringing to the surface nutrients which have been washed down after the forest burning. Among the leguminous plants which have been tried may be mentioned the following:

1. *Creepers.* *Dolichos hosei* is by no means a 'new' creeper, having been mixed with the 'standard three' on many occasions as it has the reputation for persisting under shade. It is planted from cuttings or seed. *Indigofera endecaphylla* has also been in occasional use for a very long time as has *Desmodium ovalifolium*. The latter is shade-tolerant. Neither is very successful as a cover though useful for protecting drain banks. Both can be planted from cuttings. *Stylosanthes gracilis* is a strong grower in pure culture and usually takes full possession of the land. It has a reputation for being 'aggressive', i.e. entering into competition for water and nutrients with the oil palm, and Yeow *et al.* (1981b) have shown that, even when used with extra wide circles round the palms, the latter are adversely affected in the first 2 bearing years. *Stylosanthes sundaica* is similar but less persistent. *Phaseolus calcaratus* (Fig. 10.4) is probably the most promising and widely grown of the more recently employed creepers. It somewhat resembles *Pueraria*, but in the Far East it gives a healthier appearance though it is reported to die back early. There are two varieties, *gracilis*, a slender short-lived climber with very hairy pods, and *glabra*, a more vigorous and persistent creeper with thick hairless leaves. A number of shade-tolerant legumes have recently been tried. Of these, *Calopogonium caeruleum* has tended to dominate on Malaysian coastal clays; *Psophocarpus palustris* has also grown well and has been popular in Sumatra for a long period. Seed rates are usually high: 8 kg per hectare.

2. *Erect covers.* Several species of *Tephrosia*, *Crotalaria* and *Leucaena* have for generations been available in most parts of the tropics as bush covers which can be cut back from time to time to

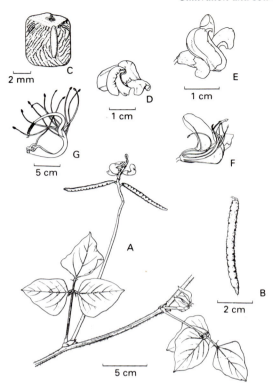

Fig. 10.4 *Phaseolus calcaratus* (RRIM).

provide material for mulching. Several of these (*T. candida, vogelii, C. anagyroidus, usaramoensis, striata, L. glauca*) together with the more recently employed *T. noctiflora* and *Flemingia* sp. (RRIM, 1962) have been tried in Malaysia. *Flemingia congesta* var. *semialata* has trifoliate leaves and shiny black seeds. It requires more cutting back than the variety *latifolia*, which is distinguished by its dense covering of brown silky hairs on young stems and petioles. The other species, *F. strobilifera*, has simple leaves and is only likely to be of value in adverse conditions since it grows on sandy soils especially near the sea.

It has been shown that many of these erect covers have tap-roots which penetrate to great depths (2–4 m) into soils of good structure and that the amounts of loppings obtained from them may reach 25–40 tonnes per hectare per annum. The question which needs answering is whether these factors are of advantage in oil palm cultivation or whether these bush covers may not enter too far into competition with the young palm for either nutrients or light. A further consideration is how the presence of these covers, perhaps – as is sometimes advocated – with creepers growing over them, will affect the development of the natural cover which eventually establishes itself in adult fields. These questions can only be answered by prolonged study of the effects of different covers and cover mixtures grown under varying management systems over a long period of time.

So far *Flemingia congesta* (Plate 10.1) either alone or mixed with *Pueraria* has shown no deleterious effect on the crop when compared with the standard covers, but it must be borne in mind that the maintenance of a pure *Flemingia* stand may entail appreciable cultivation through frequent weeding. Any advantages this cover may have are likely to be seen at a much later date.

The establishment of leguminous covers is sometimes assisted by the use of pre-emergence or other herbicides before or around the time of sowing or by spraying the interlines after the cover is visible. As a general rule legumes are sown on ground which has been cleared of other vegetation either by ploughing, hand cultivation or spraying, and herbicides are used to prolong the weed-free period and so give the legumes a better start. In Nigeria (Sheldrick, 1968c), band-spraying with atraton, simazine, monuron or diuron at 2 kg a.i. per hectare, was successful, though the danger of herbicide-laden soil being washed into the drills cannot be overlooked. In Malaysia, the use of the pre-emergence diphenyl ether herbicide RH 2915 (oxyfluorfen) (Teoh and Chong, 1977), or combinations

Pl. 10.1   The bush cover, *Flemingia congesta*, growing between rows of young palms in Malaysia. (Unilever Ltd.)

of the pre-emergence herbicide alachlor with post-emergence applications of a paraquat/diuron mixture between the drills (Tan *et al.*, 1977), has been effective; but owing to the peculiar difficulty in establishing covers in the Far East frequent hand weeding is still often employed. As the weeds are usually grasses (particularly *Paspalum conjugatum*), glyphosate has been used in developing covers (Pillai, 1978), but more recently fluazifopbutyl has been found less toxic to legumes (Voon *et al.*, 1981) and has been recommended for legume purification at the rate of 0.25 kg per hectare when *P. conjugatum* is the dominant weed.

The maintenance of a suitable cover throughout the palm's bearing life

Attempts to maintain planted covers for a prolonged period raises many problems, both agronomic and economic. In the Far East, difficulties of establishment may necessitate frequent and prolonged weeding or spot-spraying and, later on, grasses, ferns and *Mikania*, along with many woody plants which will be mentioned later, crowd out the legumes. There have been advocates of maintaining the leguminous mixture, by constant weeding rounds, for as long as possible; this is expensive work, however, and other oil palm growers have favoured the controlled natural replacement of legumes by natural covers. Tan and Ng (1981) have advocated the establishment of mixed legume/natural covers from the start. By inoculation of legumes with appropriate *rhizobia*, sowing at the right time, applying sufficient fertilizer and regularly removing 'noxious' species, they found that a much quicker coverage of the ground was obtained than by maintaining a pure stand of legumes. Mixtures containing *Pueraria, Centrosema, Desmodium, Sesbania* and *Mucuna* species and *Calopogonium caeruleum* were found suitable on inland soils. Ground coverage with the mixtures was faster than with either pure legumes or a natural cover, and the mixed cover litter mineralized more slowly than that of the pure legume. Total nutrients in the mixed litter were of the same order as in the pure legume litter but higher than in the natural cover litter. While this is the best type of cover for soil conservation, particularly on sloping land, it requires a high degree of management skill and labour availability for full achievement, and there is a danger of legumes being crowded out altogether or in patches. The system is less applicable to Africa and America where legumes, particularly *Pueraria*, so easily and quickly dominate the interlines provided strong-growing grasses are suppressed.

Early studies of natural covers of rubber plantations in the Far East have influenced practice in oil palm plantations, although the shade under oil palms is much less than under mature rubber and for this reason a much more permanent natural cover becomes established. The study of natural covers was given impetus by the publication by Dr W. B. Haines (1940) of a treatise on *The uses and control of natural undergrowth on rubber estates*. An attempt was made to classify plants into Class A: those to be encouraged, Class B: those probably useful but needing to be controlled, and Class C: those undesirable and needing to be suppressed. It was pointed out that the classification was not hard and fast and that where Class A plants were absent, more Class B plants and even some Class C plants must be tolerated. While certain of the Class C plants (e.g. lalang) were clearly undesirable on the grounds that they quickly entered into close competition with the crop, the whole scheme was tentative and was open to criticism on the grounds that the cover plants present in any situation were a reflection of the soil and other conditions obtaining, and that the performance of the rubber was determined by those conditions rather than by the presence of allegedly 'undesirable' plants.

In considering the natural flora of oil palm plantations, it is important to be aware of this background to the subject. Few plants have been shown, experimentally, to be deleterious to the oil palm though many may be considered undesirable owing to their mode of growth and the high expenditure required for their control or elimination. A plant may be considered desirable if:

1. It is not shown to enter into competition with the palm to the extent that yields are reduced.
2. It does not grow so rapidly that it shades young palms and needs to be constantly slashed.
3. It provides a substantial cover of the soil, but needs little attention.
4. It provides a quantity of herbage which rots rapidly to give a mass of decaying vegetation suitable for the growth of oil palm roots.
5. It has a substantial penetrating root system, but is not so woody that it provides little soil cover and its vegetative parts are resistant to decay.

On almost all plantations in the Far East, the cover which eventually establishes itself is a mixture of the fern, *Nephrolepis biserrata* (Fig. 10.5), with varying quantities of the grasses *Paspalum conjugatum* and *Axonopus compressus* and species of the non-leguminous creeper *Mikania*. As the latter has been shown to be harmful to young palms it should be spot-sprayed in the early years (Gray and Hew, 1968) and is then unlikely to form a substantial part of the undergrowth later on. Grasses are thought to be more harmful in young areas than in older ones, since they are certainly more aggressive where shade is light. *Nephrolepis* has virtues 3 and 4 above, and rarely grows to such a height that it needs slashing; in

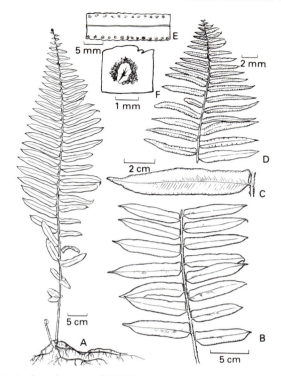

Fig. 10.5   *Nephrolepis biserrata* (RRIM).

situations where it does become bulky it can be given a light slashing and the cut fronds then form a thick decaying mat on the soil surface into which the palm roots grow. Care needs to be taken, however, to see that the slashing is light enough to allow regeneration and does not lead to dying out of the fern. This natural mixture may therefore be considered satisfactory, particularly where the *Nephrolepis* dominates.

Some account may now be given of the plants which may invade the natural association.

Obviously undesirable plants

These are lalang, *Imperata cylindrica*, other strong-growing grasses, *Mikania* sp., Siam weed, *Eupatorium odoratum* and *Asystasia coromandeliana*, a creeping plant which roots freely and tends to work its way into fields from roadsides (Rajaratnam *et al.*, 1977), and can reduce yields if left unchecked. In recent years *M. micrantha* seems to have replaced *M. cordata* as the dominant species of *Mikania* (Teoh *et al.*, 1985). Control of these plants is undertaken by regular patrolling, lalang being dug out or wiped with oil and *Eupatorium* being pulled out, and by spot-spraying with herbicides.

Other ferns

The tropical bracken, *Gleichenia linearis*, and maiden-hair creeper, *Lygodium* sp., are common invaders of the poorer and drier soils which will not be improved by their presence. If there is a reasonable chance of *Nephrolepis* taking their place they may be sprayed out with herbicides. Stagmoss, *Lycopodium cernuum*, may obtain a hold in acid, wet soils. It, too, may be sprayed out, though a need for improvement of soil conditions is indicated by its presence. *Blechnum orientale* somewhat resembles *Nephrolepis* and tends to grow when palm shade is less dense.

Shrubs and young tree seedlings

A number of very common shrubs may become widespread in an oil palm plantation. Certain of them are rated as 'Class C' plants though, in oil palm plantations, there is no evidence that they are harmful. However, they may grow tall and so need constant slashing, and many are indicators of poor soils. Among the latter may be mentioned the Straits rhododendron, *Melastoma polyanthum, Lantana aculeata* and *Clidemia hirta*. These are usually eradicated in the weeding round by hand pulling or chopping below ground level.

Wycherley (1963) has drawn attention to one matter which is fundamental to the consideration of cover policy and is of particular importance in the Far East where various schools of thought have arisen. Referring to cover policy in rubber plantations he writes:

> It is tempting to fall back on general experience, but this can be very misleading because cover plant effects are easily confounded with two other factors. Firstly, many plants are indicators of conditions rather than their cause . . . so the mere removal of these plants will effect no improvement. . . . Secondly, the presence or absence of certain plants may testify only to the skill or otherwise of those required to conform with a cover policy based on preconceived verdicts on the nature of the plants concerned. For example, if all grasses are considered harmful, their successful exclusion may owe much to the availability of trained labour under efficient management, whereas the expression of the latter in other aspects of husbandry such as regular manuring may be responsible for any apparent beneficial results of eliminating grass.

These warnings apply with equal force to a consideration of cover policy with the oil palm.

The natural covers in Africa are not so well documented or so stable as in the Far East. Moreover, it is only recently that any serious weed eradication problem has arisen.

The arrival of Siam weed, *Eupatorium odoratum*, in West Africa presented a serious problem. If left unchecked it will rapidly become dominant and grow to a height of 3–6 m, thus competing with young palms not only for water and nutrients, but also for light. If detected in a plantation early enough it can be kept at bay by regular patrols which pull up the young plants by hand well before they flower and which occasionally spot-spray any older ones that may have survived. This was accomplished for more than 15 years at NIFOR, Nigeria, in an area containing both old and young fields. The expenditure was equivalent to about 2 man-days per hectare per year and control was effective in spite of the whole plantation being surrounded by *Eupatorium*-infested areas. Once the weed gets a hold, however, it can only be tackled by herbicides. Patches can be killed by 2.5 kg a.e. of 2,4-D per hectare, but if spray drift is feared then 4 kg a.i. of atrazine or 5 kg a.i. of diuron per hectare will be effective for about 10 months if spraying is done just before the dry season (Sheldrick, 1968d). Further spraying is then required until the area is sufficiently under control to be taken over by the regular patrol.

Apart from its severe competition with the palm for water, nutrients and light, *Eupatorium* can be dangerous because it easily takes fire in dry weather and young palms may then be destroyed. In this respect it is a particularly serious pest in smallholdings, and in the Ivory Coast the weed presents a peculiarly difficult problem since it has infested village plantings, where there is little control, and from these it invades the large plantations. Here too, 2,4-D has been found the most satisfactory and economical herbicide to use (Dufour and Quencez, 1978).

The succession from leguminous covers to natural covers in Africa is not the same as in the Far East. In Africa *Pueraria* and *Centrosema* tend to persist and only gradually to be replaced by small and retarded specimens of forest species which are shade tolerant (Plates 10.2 and 10.3). Cases are known where *Pueraria* has been maintained in pure stand and without special measures for over 10 years. However, if not dominated by *Pueraria*, the land may become covered with many typical invaders among which may be mentioned *Canna bidentata*, which is often found in a thick stand, *Hibiscus seratensis*, *Solanum* sp., *Ipomoea involucrata*, *Sida acuta*, *Ageratum conyzoides*, *Passiflora foetida*, *Momordica* sp. and other cucurbits, *Aspilia latifolia* and the common grass *Paspalum conjugatum*. Also present in some localities are *Pityrogramma calomelanus* and *Sarcophrynium macrophyllum*.

The succession in the Ivory Coast seems to be rather different from that of Nigeria (Coomans, 1971). After 5 years grasses tend to predominate, chiefly *Axonopus compressus* and *Paspalum conjugatum* with some dicotyledons, viz. *Commelina nudiflora, Borreria*

Pl. 10.2   A *Pueraria phaseoloides* cover still in almost pure stand between 7-year-old palms at the Oil Palm Research Centre, Kusi, Ghana, with good ring-weeding.

Pl. 10.3   A good mixed *Pueraria* and natural cover in a 'closed in' field in Nigeria.

latifolia, Desmodium adscendens and *Talimum* sp. After 12 years the proportions of grasses and dicotyledons vary but of the latter the following are common: *Commelina* sp., *Ageratum conyzoides, Sida carpinifolia, Dissotis rotundifolia, B. latifolia, Urena repens, Phyllanthus amaras, Oldenlandia affinis, Asystasia gangetica, Cyathula prostata* and *Mariscus umbellatus*.

These covers are controlled by regular slashing (cutlassing). In Nigeria this is required four times a year in areas from 5 to 8 years old, twice a year thereafter, but perhaps only once a year in areas over 15 years old where the palm foliage is dense. In the Ivory Coast rather less frequent slashing is required, one or two rounds being sufficient once the palms are above ground.

While slashing is usually resorted to in West Africa and America, in Asia manual weeding of undesirable species with a hoe was practised until spot-spraying with herbicides took its place (Turner and Gillbanks, 1974). The herbicides used must be suited to the vegetation it is desired to eradicate. *Imperata* and *Eupatorium* control has already been discussed. Spot-spraying rounds will usually be monthly in the first year, then 2-monthly, and 3- or 4-monthly by the fourth year after planting. Paraquat is often used for *Mikania* control, but here again the actual herbicide chosen will depend largely on local prices and availability, and new formulations are frequently coming on the market.

One other method of interline cover control, which has been adopted on some estates in Malaysia and is bound up with harvesting methods and mechanization, should be mentioned here since it is entirely different from the standard slashing or weeding methods practised elsewhere. In certain flat coastal estates a strip is cut down the centre of each, or every other, interline by a rotary cutter such as the Servis. This cuts all vegetation near ground level and a lawn of *Axonopus compressus* is formed. Down these lawned paths harvesting carts drawn by bullocks or tractors pass to collect the harvested bunches. The system is clearly suited only to mature plantations on flat land when the crown is well above ground.

Mechanization of cover control

Inter-row cover control may be mechanized by the use of: (1) slashers or mowers, which cut the cover but do not disturb the soil; (2) rollers, which flatten the cover and, when horizontal blades are fitted, also cut up the cover to some extent; (3) cultivators, such as disc harrows, which both cut the cover and disturb the soil.

Experiments on close cutting of the interline vegetation have clearly shown that this practice is deleterious to both growth and yield (Hew and Tan, 1970). In a young area the practice depressed both leaf nitrogen and leaf phosphorus. Even in a mature area there was a small yield decrease, leaf size was reduced and leaf-N and leaf-P depressed. It must be assumed that the close cutting of wide paths between rows will have the same deleterious effect, though the effect may not be as severe as with complete close cutting. The practice of sending vehicles into fields with the harvesters has also been employed in parts of Latin America and not only has the cover suffered but severe puddling of the surface soil has been common. There is little doubt that these practices cause considerable yield reduction.

Experience in the Ivory Coast (IRHO, 1962) has been that any implement which cuts or cultivates tends to destroy a leguminous cover, and only heavy logs of wood rolled behind a tractor or disc harrows set with the discs absolutely straight maintain the cover satisfactorily. In Malaysia a round hardwood log measuring 183 × 30 cm diameter with four narrow horizontal blades (6.4 cm) attached to it was used in mixed Straits rhododendron (*Melastoma*) and legumes. With this light roll, leading to the minimum of cutting, the legume cover re-established itself satisfactorily. Later work with the Holt weed breaker (RRIM, 1957), which has a much wider blade, showed that legume covers suffered while grasses tended to take their place. In an experiment in Nigeria involving all three types of implement (Sly and Sheldrick, 1963), all treatments discouraged the spread of legume covers. Frequent mechanical slashing or cutter-rolling encouraged grass, the cutter-roller also encouraging annual weeds. Disc harrows used frequently left the soil bare, while any mechanical treatment carried out infrequently encouraged bush covers interspersed with grasses, soft weeds or legumes according to the implement used. Another experiment in Nigeria in which different combinations of implements were used in the early, mid and late rains, showed that a Cambridge ring roller assisted the development and maintenance of a strong *Pueraria* cover. The use, additionally, of disc harrows just before the dry season increased early yields, presumably through its effect in reducing competition for soil water (Sheldrick, 1968e).

Where leguminous covers have already disappeared, small or large rollers with cutting bars (*charrues landaises* and *rouleaux débroussailleurs*) have been successfully employed in the Ivory Coast, but it is recognized that the continuous use of one implement tends to select one species of plant, and alternation of implements is recommended (IRHO, 1962).

In the strong-growing grass areas of America it is now the practice to spray out the grasses and replace them by *Pueraria*. Rotary cutters and heavy 'Marden' cutter-rollers were used in the early days of oil palm planting, but the latter often failed to chop up the tougher stems and left a flattened sward which soon grew up again.

Some estates even harvested the grasses with cutter and blower and fed them to penned cattle, but this practice, besides removing quantities of nutrients from the fields, failed to eliminate the competitive grasses.

To summarize: the close mechanical mowing or slashing of oil palm areas has been shown to reduce yields. Other implements which destroy a leguminous cover or which cause excessive disturbance or puddling of the soil will also be deleterious. Mechanical maintenance can indeed only be considered where light rollers are used and the soil structure is unlikely to be disturbed, though very careful use of disc harrows before a dry season may increase yields in seasonal climates. In very wet climates mechanical maintenance is certainly harmful and spraying herbicides is sounder practice. Spraying with tractor-drawn booms has been practised in Latin America, but it is doubtful if the small reduction in labour usage balances the full tractor costs, quite apart from the damage done to the soil and cover and hence the loss of crop.

Pruning

The practice of pruning leaves from the palm is motivated by the desire of the grower to view the production and ripening of bunches in the crown and by the wish of the harvester to have the burden of his work lightened. The latter wish is particularly dangerous and may have been responsible for the practice of 'pruning up to a bunch', i.e. the cutting of all leaves 'below the next bunch' presumably on the assumption that such leaves are functionless, and that therefore it is advantageous to leave the next bunch to ripen where the harvesters can plainly see it and easily cut it down. Early experiments in Sumatra showed that pruning in excess of this practice was deleterious, but *less* severe pruning was not tried (Sly, 1968). This system thus remained unquestioned for many years.

The ripe bunch is found to be subtended by leaf 30 to 32. This, however, is not the 'leaf below the bunch' since the latter leans out from its subtending leaf and lies on one in another spiral on a lower whorl, i.e. on some leaf around number 35 to 37 (Plate 2.12). If pruning is done to leave one whorl of leaves below the bunches then about 35 leaves or just over 4 per spiral will remain; if 2 whorls are left then just over 5 leaves per spiral will remain. But Corley (OPGL, 1971–2) found that on an inland plantation in Malaysia 'minimal pruning' (i.e. pruning only those leaves which harvesters find they need to remove to harvest the bunches) left 35–40 leaves with occasional cases of only 27 leaves. It is clearly difficult therefore to leave more than 40 leaves on the palm unless a deliberate policy of insisting on all green leaves being retained is imposed.

Table 10.4   *The effect of pruning on yield*

Treatment	No. of bunches per hectare per annum	Mean bunch weight (kg)	Bunch yield per annum (tonnes per hectare)	No. of male inflorescences per hectare per annum
Nigeria*				
1. No pruning	803	10.0	8.01	
2. Dead leaves only, annually	813	10.4	8.51	
3. Pruning to one leaf below the bunch, 6 monthly	726	9.9	7.16	
4. Pruning to flowering inflorescence, annually	699	9.7	6.76	
Malaysia				
Coastal[†]				
1. Leaving 7–8 leaves per spiral	1,589	15.3	23.36	1,203
2. Leaving 7 leaves per spiral	1,533	15.0	22.17	1,285
3. Leaving 6 leaves per spiral	1,553	15.0	22.53	1,268
4. Leaving 5 leaves per spiral	1,559	15.0	22.41	1,335
5. Leaving 4 leaves per spiral	1,490	14.1	20.24	1,345
6. Leaving 3 leaves per spiral	1,486	12.5	17.94	1,378
Inland[‡]				*Per palm*
1. Leaving 40 leaves (5/spiral)	1,415	16.5	25.24	9.31
2. Leaving 32 leaves (4/spiral)	1,344	14.5	24.48	10.15
3. Leaving 24 leaves (3/spiral)	1,287	13.9	20.50	9.52
4. Leaving 16 leaves (2/spiral)	1,032	11.2	13.29	10.18
5. Leaving 8 leaves (1/spiral)	212	6.9	4.35	6.00

* Adjusted means for 6 years (pretreatment yields available for 4 years).
† Means for 5 years for yield, 3 years for male inflorescences.
‡ Bunch mean yield for 2 years, other data for the second year.

Sources: Sly (1968); OPRS, Bantang (1971–2) and OPGL, Layang Layang (1971–2) Malaysia.

All experiments indicate that maximum yields are obtained by retaining as many green leaves as possible, but the increment obtained by leaving more than 40 leaves is usually small. Table 10.4 gives the results of three experiments, one in Nigeria and two in Malaysia (Sly, 1968; OPRS, Banting, 1971–2; OPGL, 1971–2). It will be seen from the Malaysian data that a drop of up to 2 tonnes bunches per hectare can be expected from reducing the canopy from 40 to 32 leaves, and this is through reductions of both number of bunches and weight per bunch.

Severe pruning has a rapid effect on production. In the inland experiment of Table 10.4 mean bunch weight was reduced within 4 months and bunch number fell suddenly after 8 months. The latter result indicates an immediate increase in the inflorescence abortion rate. Preferential abortion of female inflorescences is shown by the fact that male inflorescence production is not adversely affected until the most severe pruning treatment is reached. Severe pruning reduces the size of developing leaves but does not alter the rate of leaf production (Calvez, 1976). In the coastal experiment it was shown (Yeow *et al.*, 1981b) that to attain a leaf area index of 6–7 – the optimum for maximizing yield (Corley, 1973, see p. 155) – young palms should be pruned as little as possible, palms aged 4–7 years should retain 6–7 leaves per spiral, those aged 8–14 years 5–6 leaves, and those of 15 years and over 4–5 leaves per spiral.

If no pruning is done it may become difficult for harvesters to see into the crown and to gain access to the bunches; much loose fruit may also be lost. Time of pruning may be important in seasonal climates. Pruning at the beginning of the wet season should be avoided; in an experiment in Nigeria in which intensive pruning methods were used bunch yield was about 10 per cent higher in the first 4 years when pruning was done at the end of the rains than when carried out at the beginning of the rains (Gunn *et al*, 1962). In non-seasonal climates pruning is usually done at times of low crop when labour is available.

On some estates in the Far East it is the practice to carry out a special 'pruning up' when bringing a field into harvest. This is not usual in other parts of the world where the ripe young bunches are simply cut out with a chisel and withered leaves are removed later either during pruning rounds or during later harvests. With palms growing very vigorously, however, the improved access for harvesters is considered to make this practice of early 'pruning up' to a ripe bunch worth while. An experiment in Malaysia showed no effect of this practice on yield when compared with no pruning up (OPGL, 1971–2).

When harvesting begins, pruning is at first confined to senescent leaves lying close to the ground, but when the palms are well up and the canopy closes in harvesters are allowed to cut leaves during harvesting as may be necessary, though two whorls of fronds below the ripe bunch are maintained (Turner and Gillbanks, 1974). This rule is sometimes relaxed for palms over 10 years old, but the criterion of two whorls or a minimum of forty leaves (five per spiral) should be maintained as long as possible. An annual special pruning round to remove excess withered leaves is then all that is additionally necessary.

Pruning may be carried out with cutlasses, axes or chisels. In Africa the cutlass is the usual implement but in Asia the chisel is supplanting the axe for young palms. In both continents the pole and sickle is now being widely used for tall palms. Varying advice is given about the length of the leaf base which should be left on the palm. 'As short as possible' is often advised on the grounds that large leaf bases catch loose fruit, encourage epiphytes (Piggott and Piggott, 1976) and possibly assist in the spread of disease, though for this there is no evidence. There is evidence, however, both in Africa and America that cutting the bases very short may be followed by a serious increase in attack by *Rhynchophorus* species leading either directly to the death of the palm or to Red Ring disease. It is noticeable that attempting to cut the bases very short leads to wounding of underlying leaf bases and this gives a suitable place for the weevil to lay its eggs. The petiole must therefore be cut at a sufficient distance from its base to give a clean cut and no damage to adjoining parts of the palm.

The epiphytes which grow on the trunk of a palm before it has dropped its leaf bases, and which may also grow up into a neglected crown, are usually species of fern. There is no evidence that they have any direct effect on yield. Piggott and Piggott (1976) found that in Malaysia *Nephrolepis biserrata* and *Davalia denticulata* are the earliest colonizers and most widespread, with *Vittaria* sp. common on older palms. Epiphytes may catch falling loose fruit and hide them from view, and in this respect the Bird's Nest fern, *Asplenium nidus*, is a special culprit and large specimens should be removed. Oil palm seeds germinate fairly readily in the debris in cut leaf bases, so the presence of a large crop of oil palm seedlings developing in the leaf bases is an indication that much loose fruit is being lost.

Bunch production and harvesting

Bunch production in the early years may be affected by the practice of ablation (castration) (See p. 448). Until recently it was also influenced in Asia by assisted pollination, a practice which, with the importation of *Elaeidobius kamerunicus*, is no longer needed. Harvesting methods aim at gathering the whole crop without loss of oil or loose fruit, while ensuring that both the palm oil and kernels will be of acceptable quality.

Harvesting implements

In the early days of plantations in Africa, harvesting was carried out almost entirely with the cutlass and tall trees were climbed either by ladder or by means of ropes. The cutlass is not a handy implement for harvesting, however, and climbing with ladders is slow and with ropes very slow. These methods have largely given place to methods developed in the Far East.

The ideal instrument for harvesting is a chisel on the end of a 0.9–1.5 m wooden handle (Plate 10.4). Suitable chisels for

Pl. 10.4   Harvesting bunches: a wooden-handled chisel in use on a young palm.

harvesting and pruning have blades 7–12 cm wide and about 25 cm long with overall length including the shaft hole of about 45 cm. These instruments can be used for cutting the bunch peduncle without cutting the green leaf which subtends the bunch. Chisels are often used for the first 3 years of harvesting only, but they should be employed for as long as possible as other methods lead to the cutting of too many green leaves. A common procedure in the Far East was for short axes to succeed the chisel when the palms are 'off the ground', i.e. have formed a trunk, but the continued use of the chisel is now preferred. As soon as the bunches can no longer be easily reached from the ground harvesting is carried out with a long curved knife lashed to the end of a bamboo pole (Plate 10.5). These poles are gradually increased in length and palms of over 9 m in height can be harvested with them. The curved knife or sickle is about 60 cm long (Plate 10.6).

When harvesting with either an axe or the knife and pole, the leaf or leaves below the bunches are inevitably cut. There is a tend-

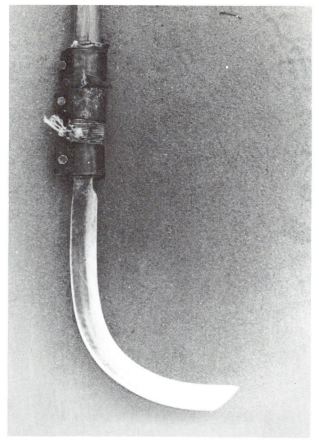

Pl. 10.5   Hooked knife attached to the end of a bamboo pole.

ency, indeed, for more leaves to be cut by the pole-and-knife than by the old method where the harvester climbed with a rope or ladder and used a cutlass. This is because, with a cutlass, the bunch can often be cut out without cutting either the leaf below it or many of the leaves to the side. The pole is, however, much less expensive to use. Moreover a skilled operator can manipulate it so that the minimum of leaves are cut. The blade is first hooked round the peduncle and given a strong downward pull; if the bunch does not fall the blade is applied to the apex of the bunch to ease it out from behind uncut or partially cut leaves.

Recently, many estates have realized the importance of preventing the cutting of green leaves and they have therefore tried

Pl. 10.6 Harvesting bunches: the hook and pole method for tall palms.

to retain the use of the chisel for as long as possible and to eliminate the axe stage altogether. Harvesting is often done with the chisel until the palm is 12 years old and measures perhaps 3.5 m to the crown. The harvester climbs the palm tying a wire rope to a stout leaf base to support him while using the chisel. When chisel harvesting is no longer practically possible pole-and-knife harvesters should be trained to cut bunches with the minimum removal of green leaves. The curved knives on the ends of the long poles may be a danger to other workers. When being carried along roads from field to field, therefore, they should be guarded by the attachment of a piece of wood.

When to start harvesting

The first bunches are small and of poor quality both as regards percentage fruit to bunch and oil to mesocarp. However, well-grown modern material comes very rapidly into bearing giving, in Africa, over 5 tonnes bunches per hectare and double that amount in the Far East in its third or fourth year in the field. It is possible to increase the yield of the first harvesting years by the removal, by hand, of the first female inflorescences appearing on the young palms. This has, curiously, been named 'castration' though ablation is a better term. Obviously, if an area can produce more highly in its first harvesting year, field costs per tonne are reduced and more economical use is made of the oil mill. On the other hand, it must also be the planters' object to reduce to a minimum the period of no crop. All these economic considerations, therefore, must be carefully weighed before a decision is taken on when to start harvesting.

If no ablation is carried out, a delay in harvesting will have no effect on yields in later years (WAIFOR, 1960), but there will be an accumulation of rotting and overripe bunches on the palms and harvesting of any of the latter will affect the quality of the oil extracted.

Ablation

Experiments have covered ablation periods of 6–33 months (Obasola, 1970; Taillez and Olivin, 1971; Hew and Tam, 1972; Chew and Khoo, 1973). The general effects of ablation are: vegetative growth is improved, with an increase in the number and length of leaves and in the palm's girth; there is also evidence that the root system is improved. When ablation ceases a high initial yield is obtained; the early bunches are of increased size and the oil to mesocarp is higher than in the earlier fruit bunches of untreated areas, though there does not appear to be any difference in oil content of contemporaneously harvested bunches. The superior vegetative growth does not persist and in the early bearing period the rate of vegetative development may become inferior to that of untreated palms. After initial high production treated palms suffer a drop in production and there is therefore a tendency for high yield fluctuations in the early years. The drop in yield appears to be due to inflorescence abortion. Male inflorescence production is considerably reduced and this may persist for some time.

In view of the fact that no greater yield is obtained by ablation, it must be shown to have other economic advantages if it is to be employed. The advantages which have been claimed are: (a) that the same yield is concentrated into a shorter period thus reducing harvesting costs; (b) that overall in the early years there will be a higher outturn of oil to bunch; (c) that in practice many small bunches are often imperfectly harvested, are subject to rat damage and, in Asia, may become infected with *Marasmius*; (d) that mill investment can be delayed and that the equipment is put to better initial use by the larger early crops which will be obtained. Against these factors must be set the cost of the ablation itself. There is evidence in Malaysia that additional fertilizer applications may in some circumstances be needed to prevent excessive abortion of bunches (Chan and Mok, 1973).

It is clear from the above that the decision to carry out ablation must depend on local circumstances. Plantation managers tend to favour the practice partly because of the improved cleanliness and general appearance of the palms. In areas of very high initial production or with very precocious material, however, palms may come into early heavy bearing without ablation and the possible period of ablation is then so short that it is hardly worth while. Owners of small or medium-sized plantations are unlikely to favour the practice since they are usually eager to obtain revenue as soon after planting as possible.

In practice ablation usually starts as soon as about 50 per cent of the palms are bearing female inflorescences. This will be at about 14–18 months after planting in Asia and up to 2 years in Africa. Ablation will then continue for 6–18 months with palms being brought into bearing at 3–4 years after planting. The shorter period and the earlier harvesting time are preferable, particularly in areas of rapid growth. Longer ablation would only be undertaken in special circumstances, e.g. where the operation of the only available mill was unexpectedly delayed. It has been suggested that in seasonal climates ablation should be continued until a palm girth of 150 cm has been reached and that the time of discontinuation should be the beginning of the dry season (e.g. November in West Africa). This leads to a delay of harvesting by 1 year if 150 cm girth has not been reached at $2\frac{1}{2}$ years from field planting (IRHO, 1976).

Young inflorescences can be pulled out by hand using a suitable glove. If ablation has to be prolonged beyond 6 months it will probably be necessary to use a narrow-bladed chisel. There is, however, a great danger of wounding the palm and thus allowing entry of *Rhynchophorus* sp. This has to be particularly guarded against in America where there is danger of Red Ring disease, and it is doubtful if ablation should be undertaken at all in that continent

Table 10.5 The results of some ablation experiments

Treatment	Planting date	Ablation period (months)	Harvesting dates	Parameter	Production per palm kg and number				
					1966–7	1967–8	1968–9	1969–70	Total
Ivory Coast[†]									
Control	May 1963	Nil	July 1966	Bunch yield	57	59	81	97	294
				No. of bunches	13.3	9.9	10.0	10.1	43.2
				Wt. per bunch	4.4	6.5	8.4	9.8	
Ablation 1	May 1963	14	Nov. 1966	Bunch yield	70*	65	70*	97	302
				No. of bunches	13.6	10.7	8.2	9.0	41.5
				Wt. per bunch	5.2*	6.4	8.7	10.7	
Ablation 2	May 1963	19	Mar. 1967	Bunch yield	18*	117*	70*	93	298
				No. of bunches	3.1*	18.0*	8.9	9.1	39.1
				Wt. per bunch	6.0*	6.8	8.2	10.4	
					1965	1966	1967	1968	Total 48 mths
Nigeria[‡]									
Control	May 1961	Nil	Jan. 1965	Bunch yield	48.4	71.1	45.1	61.6	226.2
				No. of bunches	12.2	11.3	7.6	7.8	38.9
				Wt. per bunch	4.0	6.7	6.4	7.6	
Ablation 1	May 1961	10	Jan. 1965	Bunch yield	50.0	63.2	46.0	62.7	221.9
				No. of bunches	12.8	9.0*	7.9	8.1	37.8
				Wt. per bunch	3.8	7.7*	6.1	7.7	
Ablation 2	May 1961	19	Jan. 1965	Bunch yield	61.3*	76.4*	42.5	70.4*	250.1*
				No. of bunches	11.6	10.0	7.0*	8.5*	37.1
				Wt. per bunch	5.4*	7.4*	6.4	8.3*	

Table 10.5 (continued)

Treatment	Planting date	Ablation period (months)	Harvesting dates	Parameter	Production per palm kg and number				
					1969–70 (6 mths)	1970 (last 9 mths)	(1971 first 9 mths)	1971–2 (9 mths)	Total
Malaysia§									
Control	Oct. 1967	Nil	Oct. 1969	Bunch yield	9.9	47.3	74.7	113.2	245.1
				No. of bunches	4.4	15.8	16.3	14.3	50.8
				Wt. per bunch	2.3	3.0	4.6	7.9	
Ablation 1	Oct. 1967	12	April 1970	Bunch yield	—	58.7	78.3	111.3	248.3
				No. of bunches		16.0	16.6	14.6	47.2
				Wt. per bunch		3.7	4.7	7.6	
Ablation 2	Oct. 1967	24	Jan. 1971	Bunch yield	—	—	101.2	109.6	210.8
				No. of bunches			14.6	13.1	27.7
				Wt. per bunch			6.9	8.4	

† This experiment covered two types of plantation material. An asterisk (*) indicates that at least the P = 0.05 level of significance between an ablation treatment value and that of control was reached in both halves of the experiment.
‡ In this experiment it should be noted that harvesting started in the same month for each treatment. An asterisk indicates a significant difference at at least the P = 0.05 level, between the ablation treatment and control.
§ Assisted pollination was carried out at ten rounds per month from the cessation of ablation.

Sources: Tailliez and Olivin (1971); Obasola (1970); Chew and Khoo (1973).

if the use of chisels is necessary. Ablation rounds are usually monthly, but 5–6 weekly is permissible in periods of low crop in Africa.

Table 10.5 summarizes the results of three ablation experiments. It will be noted that in the Nigerian experiment harvesting was started 44 months after planting in each treatment whereas in the control in the Ivory Coast experiment harvesting was started after 38 months. From the yields of the former experiment it is fair to assume that a production of some 20 kg per palm could have been harvested in 1964 from the control plots in the Nigerian experiment, in which case the yield to the end of 1968 would not have been significantly less than in treatment 'Ablation 2'. The higher bunch weight induced by ablation was demonstrated in all experiments as was the tendency to greater yield fluctuations. In the Malaysian experiment the earlier and higher yields obtained in Asia can be seen. In the treatment 'Ablation 2' the yield in three successive 6-month periods of harvest was high, low, high.

It has been suggested that the practice of ablation is of more practical utility in regions of low than of high production and particularly so in extremely dry conditions such as obtain in Benin. Experiments in the latter country have shown that early maintenance with clean weeding improves vegetative development, including the root system, and that this is further improved by ablation for 20 months. Though the effect of ablation on overall yield to 6 years of age was small, there was evidence of improved drought resistance and the harvesting costs per tonne of bunches were much reduced. Special timetables for ablation have been devised (Daniel and de Taffin, 1974, 1976).

Corley (1977) has suggested that partial ablation could be used to maintain monthly bunch numbers at or below an estimated maximum thus flattening yield peaks and perhaps filling in yield troughs. He showed that partial ablation besides increasing the weight of the remaining bunches also increases carbohydrate storage in the trunk, though the overall carbohydrate demand, and consequently the rate of supply, decreases.

Time, frequency and criteria of harvesting

Harvesting is carried out with the object of obtaining the maximum quantity of palm oil of a quality, as judged by the free fatty acid content, acceptable to the purchaser. Underripe fruit contains less oil than ripe fruit; overripe fruit provides oil of a higher f.f.a. content. Ripening takes place unevenly over the bunch, and although the period from anthesis to harvest is shortest in a dry season (Broekmans, 1957), the final ripening of fully developed bunches is hastened by wet weather.

Fruit colour has been discussed in Chapter 2, but it may be mentioned here that *nigrescens* fruit is colourless at the base before ripening, and that it ripens from the top downwards to a reddish-orange colour, the exterior fruits retaining their brown or black caps; *virescens* (green) fruit ripens in the same manner to a light reddish-orange, but it does not retain the cap, only a small ring of green remaining around the persistent stigma at the apex of the fruit.

Bunches usually ripen about $5\frac{1}{2}$–6 months after pollination. In Zaire the average maturation period lies, according to age, between 170 and 195 days, though palms are known on which bunches ripen in as little as $4\frac{1}{2}$ or as much as 7 months (Vanderweyen, 1952). Differences due to season may amount to as much as 12 days.

The ripening of the fruit, which becomes detached or easily detachable when ripe, usually takes place from the tip of the bunch downwards and over a varying period of time. Small bunches may ripen their fruit fully in 11 days whereas large bunches from palms of the same age may take up to 20 days to ripen. Bunches of the same weight take longer to ripen on older palms than on young palms. Before ripening, the mesocarp contains a high percentage of water and carbohydrates and has a low oil content. One week before ripening the oil content may have risen to within 80 per cent of its final amount. The final oil production takes place rapidly with the change of colour of the fruit. When oil content is at its maximum the f.f.a. of the oil is about 0.5 per cent. When the fruit has become detached or the bunch harvested, oil production ceases. Moisture then tends to be lost, and if the fruit or bunches are stored a false impression will be obtained of the oil extraction rate. While the oil to fruit, calculated on a moisture-free basis, remains constant, the mesocarp to fruit may, through loss of moisture, fall by 3 per cent in 4 days and the oil to fruit will rise accordingly. During this post-ripening period, the f.f.a. content of the oil will gradually increase.

Harvesting criteria

Though these facts have been known in broad outline for many years, the criterion for harvesting a bunch, i.e. a stated number of loose fruit per bunch, has remained arbitrary and has varied widely from plantation to plantation. Any harvesting arrangement is a compromise: the harvesting round must not be so infrequent that bunches become overripe and most of their fruit drop out and are damaged; nor must it be so frequent that the process is expensive and there is a temptation to harvest underripe.

Dufrane and Berger (1957) in Zaire were the first to provide a logical scheme based on the facts of ripening and f.f.a. formation. They analysed 400 bunches averaging about 5 kg in weight and demonstrated a linear relationship between the number of loose fruit and percentage oil to mesocarp. An increase in the number of loose fruit from 5 to 74 (2–46 per cent) resulted in an increase of oil to mesocarp of just over 5 per cent. At the same time the f.f.a. content of the fruit rose from 0.5 to 2.9 per cent. The f.f.a. content of the loose fruit was approximately constant at 5 per cent, whereas the f.f.a. of oil from unloosened fruit never rose above 1.2 per cent even for bunches with 46 per cent loose fruits.

These authors obtained, with their 5 kg bunches, an equation

$$Y = 45.2 + 0.083x$$

where Y is the percentage oil to mesocarp, x the number of loose fruit (below 50 per cent) and 45.2 the percentage oil to mesocarp calculated to exist in bunches which have not yet loosened any fruit though on the point of doing so. Although linear to 50 per cent loose fruit, this relationship does not appear linear when the number of loose fruit is expressed as a percentage and suggests that a maximum oil percentage of just over 51 per cent was reached at 50 per cent loose fruit (Fig. 10.6). It was suggested that the relationship could be extended to bunches of different average weight by modifying the equation to

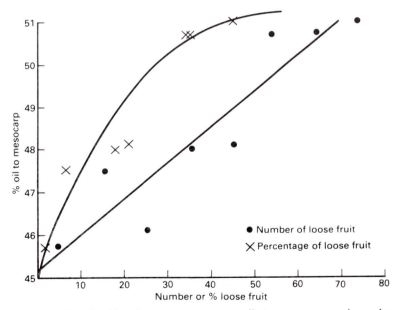

Fig. 10.6 Relationships between percentage oil to mesocarp and number or percentage of loose fruit in bunches averaging 4.8 kg (Dufrane and Berger, 1957).

$$Y = 45.2 + 0.415 \frac{x}{p}$$

where p = the average weight of bunches in kg. Thus for a given standard of ripeness, as judged by the percentage oil to mesocarp, the number of loose fruit would be greater the larger the bunch size. A similar relationship has been obtained in Malaysia with D × D and D × P material where the maximum oil to mesocarp in 5-, 8- and 11-year-old plantings was reached at 30 per cent detached fruit or less (Fig. 10.7). Free fatty acid at 30 per cent detached fruit lay between 2 and 3 per cent (Ng and Southworth, 1973).

Theoretically, if a criterion of x detached fruit per bunch is used and bunches are harvested daily, all bunches harvested might be represented graphically by a vertical straight line at the x point as shown in Fig. 10.8A, i.e. whatever number of bunches are harvested none has more loose fruit than the criterion. Similarly if the interval is extended to a lengthy period, say 20 days, bunches with

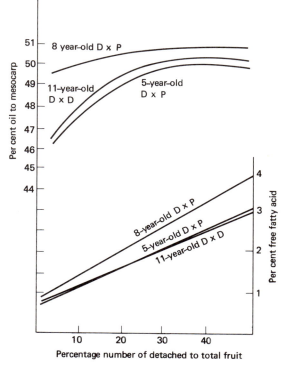

Fig. 10.7 Relationship between percentage oil to mesocarp, free fatty acid and percentage number of detached fruit to total fruit in Malaysia (Ng and Southworth, 1973).

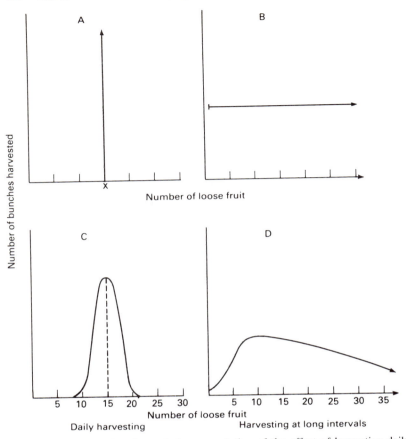

Fig. 10.8 Theoretical and practical representation of the effect of harvesting daily (**A** and **C**) and at long intervals (**B** and **D**) (after Dufrane and Berger, 1957).

all numbers of loose fruit up to the total number of fruit per bunch might be represented by the horizontal straight line shown in Fig. 10.8B. In practice the frequency curves for loose fruit per bunch for 1 day and for 20-day harvesting periods would be as shown in the lower graphs C and D of Fig. 10.8, the asymmetry of the 20-day graph being due to the fact that the curve of bunch maturation is logarithmic (Dufrane and Berger, 1957).

A curve of maturation showing the mean number of fruit becoming detached over a period of days can be constructed for a given area. From this curve can be seen, for any given average number of loose fruit taken as a criterion, the upper and lower limits of loose fruit production for any number of days' harvesting

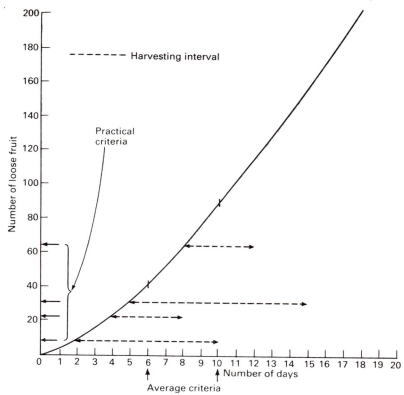

Fig. 10.9 The relation between the harvesting interval, the average number of loose fruit taken as a criterion and the practical criterion (after Dufrane and Berger, 1957).

interval. For instance, if, using Fig. 10.9, the average number of loose fruit taken as the criterion is 40, then a 4-day harvesting interval will entail the harvesting of bunches with between 22 and 64 loose fruit, i.e. the 'practical criterion' will be 'at least 22 loose fruit'. An 8-day interval would reduce the practical criterion to 8 loose fruit, but bunches with about 90 loose fruit would also be being harvested. If the average criterion is set higher, at 90 loose fruit, then 4- and 10-day intervals will give ranges of 62–116 and 30–158 loose fruit, the first number being the practical criterion in each case. From graphs of this kind the effect of the setting of any practical criterion can be seen for different harvesting periods, and the carrying out of this exercise can prevent arbitrary judgements of harvesting standards being made in the field. The increased oil obtained by increasing the practical criterion, in terms of loose fruit per bunch, needs to be balanced against any reduction of price per tonne due to increased f.f.a. When maintaining a given *average* number of loose fruit, reducing the frequency of harvesting may reduce costs but may lead to loss of oil through failure to pick up all loose fruit.

The translating of these considerations into practical field methods designed to give maximum oil at an acceptable level of f.f.a. is not easy and it is therefore not surprising that different systems have been canvassed. A standard that has been widely used is that there should be not less than two loose fruit on the ground per kilogram of bunch, i.e. if a 10 kg bunch has any number of loose fruit on the ground less than twenty, then it will be left until the next harvesting round (Speldewinde, 1968). It is claimed that this system works well with experienced harvesters if used flexibly, e.g. if the round is widened to 10 days or if it is very wet, fifteen fruit on the ground would be acceptable for harvesting a 10 kg bunch.

This system has been criticized on the grounds that it cannot be accurately checked after the bunch has been cut and that the fruit will perhaps be too ripe. Ng and Southworth have shown that, in Malaysia with D × P material, the proportion of detached fruit on the ground to total detached fruit is about 50 per cent when the total of detached fruit is 15 per cent and somewhat more at higher percentages of detachment. A minimum or practical criterion of 20 loose fruit on the ground will give a mean of about 60 with a weekly round (Fig. 10.9), and 60 loose fruit on the ground is equivalent to a total of 120 detached in a 10 kg bunch having 6 kg of fruit. Assuming a fruit weight of 9–12 g, the detached fruit will constitute between 18 and 24 per cent of the total fruit. This will give maximum oil to bunch according to Ivory Coast estimates (Wuidart, 1973) and a little below maximum oil with Malaysian estimates (Ng and Southworth, 1973).

Southworth, (1977), who regards 20 per cent total detached fruit as the aim, recommends two detached fruit per pound of bunch weight (4.4 per kg), defining a detached fruit as 'one which has either fallen from its bunch or can be easily detached by hand'. In practice this is close to the standard of two detached fruit *on the ground* per kilogram of bunch discussed above; he points out, however, that although there is the general relationship already mentioned, there is a wide variation from bunch to bunch in the proportions of fruit on the ground to total detached fruit, so that if harvesters can estimate the total detached fruit this is likely to be the most satisfactory measure. In many parts of the world, however, it may be difficult for harvesters to do this with any accuracy and in this case the criterion of loose fruit on the ground will be the more practical one.

With young palms it is possible to specify the actual number of loose fruit or fruit easily detachable with the finger and in the Ivory

Coast two per kilogram of bunch has been recommended (Gerard *et al.*, 1968). This is low by other standards but it has to be remembered that very small bunches ripen their fruit more rapidly than larger bunches and that if a small bunch with loose fruit is left to the next round it may become overripe.

Rajanaidu *et al.* (1987) claim that enforcement and monitoring of harvesting standards are always more important than the actual criterion adopted. This conclusion was reached from studies of oil to bunch on both a dry weight and wet weight basis using groups of bunches with different numbers of loose fruit (Rajanaidu *et al.*, 1985; Bealing, 1987). Significant differences between groups were found in oil to bunch on a wet basis but not on a dry basis. Similar results were obtained in the analysis of inner and outer fruit and detached fruit. It was concluded that higher extraction ratios obtained with bunches with higher numbers of loose fruit are due to loss of moisture rather than a change in the absolute quantity of oil. It was also found that, from the time of initial fruit loosening onwards, there was no increase in oil to dry mesocarp in either inner or outer fruit. These authors suggested, therefore, that a criterion of one loose fruit per bunch of any weight will suffice, and that the monitoring of harvesting to ensure that there is no cutting of bunches which have yet to loosen any fruit is of first importance. This standard has the advantages of reducing the number of loose fruit which have to be collected and ensuring the production of low f.f.a. oil.

Other factors of importance in harvesting procedure are discussed below.

Yield, payment and harvesting organization

As yields vary so widely from region to region it is not possible to state a general quantity of bunches which a harvester can bring in. In Malaysia a harvester cuts 100–200 bunches per day in mature but not tall areas, this being equivalent to between $1\frac{1}{2}$ and $2\frac{1}{2}$ tonnes of bunches. Payment is per bunch on a variable scale or per unit of weight at a fixed scale (Pratt, 1963). If the harvester also carries, then the amount harvested will be reduced by about one-third; it is common practice, however, for a cutter–carrier pair, often husband and wife, to work as a team and $1\frac{1}{4}$–2 tonnes per day will be a common achievement per team. About 1,100 kg is considered an average task for cutter–carrier pairs in Zaire where a good deal of attention was given to the rate of work of harvesters, i.e. the time taken to cut, walk between palms, carry, collect loose fruit, etc. (Dufrane and Berger, 1957). Similar studies have been undertaken by Boyé and Martin (1969) in the Ivory Coast and by Sankar (1965) and Gillbanks (1967) in Malaysia. While this work will be of interest to those establishing plantations in areas where the oil palm is a new crop, it must be borne in mind that all these components of harvesting time will vary with the age and yield of the field concerned, with the productivity of the region, with the distance between roads and with the ability of workers of different traditions.

The collection of loose fruit is an important part of harvesting. Southworth (1979) has shown that even the loss of one fruit per bunch can be of financial significance. The work is often done by women and children, and supervision is difficult. To avoid contamination the fruit must be picked up individually and not scraped up into a pile.

Frequency of harvesting

Harvesting is usually more frequent in Africa than in the Far East. In Zaire the calculations of Dufrane and Berger (1957) suggest a 6- to 7-day cycle with an increase of frequency to 4–5 days at times of peak crops. In the Ivory Coast 4–5 days is suggested as normal for peak periods, though this may be reduced to 8–10 days during periods of low productivity. In Nigeria weekly rounds have been adopted as standard on some plantations though more frequent harvesting is sometimes practised in the peak period. In Malaysia once-a-month harvesting has been practised for the first 6 months, with twice-a-month for a further year. Thereafter 7- to 10-day rounds are the rule.

Layout of road or rail system; area covered per day

Roads or rails 320–400 m apart keep the carry of bunches to a maximum of 160–200 m; this has been found satisfactory in Malaysia and Zaire. With layouts of this type 4 hectares of mature plantation per day can, under Zaire conditions, be covered by a pair of workers in an area yielding about 12 tonnes bunches per hectare per annum. In high-yielding areas in Malaysia the weight of bunches harvested is much higher but the area covered nevertheless ranges from 3 to 5 hectares per team per day (Bevan and Gray, 1969).

Lastly, mention should be made of the possibility of inducing the bunch, with the aid of growth regulators, to retain its fruit after full ripening. Oil synthesis could then continue in the later-ripening fruit after it would ordinarily have become necessary to harvest the bunch. A number of growth regulators have been tried and of these NAA (d-naphthalene-acetic acid) appears the most promising (Chan *et al.*, 1972). Abscission of the fruit is significantly supressed, but the synthesis of oil seems to proceed at the normal rate so that for any given percentage abscission the oil content per bunch will be higher in treated bunches. Application can be done to the ripening bunches 140 days after anthesis and this avoids the inducement of parthenocarpy. Though this method of reducing the problems of

harvesting is obviously attractive, the economics of it and the techniques to be adopted require further study and elaboration.

Transport of bunches

Bunches are normally carried from the field to collecting points on
the road or railside and then manually loaded into lorries, tractor-
drawn trailers or rail cages. Road transport is now the rule in all
new plantations and special efforts have in recent years been made
to reduce labour costs both in carrying within the field and in
loading at the roadside. In these developments the need to maintain
f.f.a. at a low level through careful handling of the bunches has
been borne in mind.

In South America mules with panniers are used for carrying out
the bunches to the roadside (Plate 10.7), while in Malaysia bullock
carts (Plate 10.8) or buffaloes drawing sledges or carts are employed
in some areas (Wan, 1973). The first of these systems can be much
abused if the mules are allowed to wander uncontrolled through the
fields and much puddling of the soil has resulted on many plantations in high rainfall areas. The system is only permissible under
strict discipline where the mule is confined to a clear harvesters'
path within alternate avenues. Buffaloes and sledges might also be
criticized on the grounds of their soil compacting and puddling
effects, but here again the effect can be minimized by moving down
defined paths in every other avenue. Buffalo transport with sledges
or wheeled carts is now well established in some parts of Malaysia

Pl. 10.7 Bunches carried out of the fields on mules in Ecuador.

Pl. 10.8   Harvest collection from field to railside by bullock cart on a coastal estate in Malaysia. (T. Menendez)

(Anon. 1982); each animal pulls about 1.5 tonnes bunches over a distance of 250 m to the roadside with up to twenty stops for loading (Lau Kok Chin, 1982). In a working day 2.5 tonnes can be brought out from 2.5 hectares, and it is claimed that the productivity and the earnings of harvesters are much increased and the need for maintenance of harvesters' paths is eliminated because the buffaloes graze as they move along. With both manual and animal carrying the productivity of the carrier varies directly with the distance between roads, which determines the length of carry (Muirhead, 1980).

Where animals are employed it is very necessary to remember that they must be kept out of fields where poisonous herbicides have been used, and that the persistence of each such herbicide must be ascertained.

The passage of tractors and trailers through the fields or the use of special small vehicles designed for the job have been tried in many places. The main problem, particularly with tractors, has been the wetness and unevenness of the ground, which leads to much puddling, ponding and sometimes the bogging down of the tractors themselves. In Malaysia a container system of carrying bunches to the factory was combined with the use in the field of a small three-wheeled 2.6 hp machine with hydraulic jack (Jackpak) which could lift the containers and carry them to collecting points (Cunningham, 1969). Other systems have included tractors with trailers, modified transport boxes or rear-mounted fork-lift attachments, and variously modified dump trucks, one of which was actually designed to catch the bunches falling from the palm as well as to transport them to the roadside (Zohadia Bardaie and Wan Ishak, 1982). In general these methods have not taken on because the capital outlay and maintenance costs have been too high. Muirhead (1980) pointed out that the very nature of oil palm cropping works against large machines, since the crop is so scattered and much ground must be covered to obtain even a small load. At present animals are the most widespread aid to collection though in Malaysia some harvesting teams have improved their productivity by adapting bicycles or wheelbarrows or a combination of both. Animals have the advantage of being relatively inexpensive, maintaining themselves, and having value when past working.

In Honduras and Costa Rica harvester-carrying booms mounted on track or high flotation wheeled tractors have been used (Washburn, 1973). These bring the harvester to the bunch on palms up to 12 m high and the fallen bunches are loaded into a trailer drawn behind the tractor. The full cost of this operation in comparison with manual or animal collection and an efficient system of roadside-to-mill carriage is not known, but it must be very high and, moreover, the full effect on yield of the considerable puddling and ponding which result needs to be considered. In general, a system of this kind with large tractors and trailers can only be justified where there is an overriding need to reduce labour.

There have been several recent developments in transport methods from roadside to mill. Tipping lorries or tipping trailers are now the rule so that off-loading at the mill, either onto storage ramps or into sterilizer cages is rapid. Where a rail system is used with the cages (Plate 10.8) no further storage is required and the cages go direct to the sterilizer. These cages can also be used with low trailer frames drawn by tractors. The use of ramps has been given as one cause of increased f.f.a., because of the relatively severe impacts on loading and unloading; however, without a rail system and/or direct loading at the fieldside into sterilizer cages, it is difficult to avoid millside ramps in a large plantation where bunches must at times be accumulating at a faster rate than they can be milled. Tractors sometimes draw up to six trailers from the field to the millside. In this case the tractors must be unhitched to tip each tractor in turn or the trailers must be side opening and run on to the tipping platform.

Bunch collection and transport in nets lifted by cranes have been steadily gaining ground in all regions. This system, known in Malaysia as the 'Kulim system', can be used with lorries or tractors and trailers (Price and Kidd, 1973; Khera, 1974). The former are more likely to be used where the journey to the mill covers an appreciable distance on a public road. With tractors and trailers it is unnecessary to fit all tractors with cranes since some tractors will be used simply for towing full trailers to the mill. Nets, which are designed to carry bunches and loose fruit, are set out on roadsides at the bunch collection points, and the harvested bunches are carried to them. The lorry or tractor-trailer with attached crane lifts the net and the bunches are discharged into the vehicle. The empty nets are carried to the mill or other collecting points and are redistributed to the fields in a sidecar or light vehicle. There is considerable saving in both time, handling and loading labour and oil quality should thus be improved (Plate 10.9).

With the container system already mentioned (p. 462) the plywood containers on steel frames with a capacity of 0.5 tonne bunches are placed along the roadsides and filled by the harvesters. They are loaded by tractor cranes onto trailers and the bunches are eventually off-loaded at the mill into sterilizer cages or into lorries for further transport; this is done by a chain hoist with bridle gear on a gantry (Cunningham, 1969).

Pl. 10.9 Lifting bunches from the roadside in nets by tractor-mounted crane in the so-called 'Kulim system' of bunch collection and transport.

Replanting

A field of oil palms should be replanted when (i) a large number of palms have become so tall that they cannot be conveniently or economically harvested, or (ii) the yield is so low that replacement with high-yielding young palms would give a much higher return. Tallness has usually been considered the more important factor (Ferwerda, 1955), but with the change-over from *dura* to high-yielding *tenera* material, and with the high incidence of disease in some old fields, the second factor is often of equal or greater importance. Pole harvesting can now be continued until palms reach a height of over 10 m, and replanting may well be justified before the palms have passed that height. Clearly every case must be considered on its own merits, account being taken of the present yield, the cost of harvesting, the bunch yield potential of the area and the planting material available.

Growth in height depends both on environment and on genetic constitution. Some seed suppliers can provide material embodying a short-stem factor (see p. 296), and this may help to prolong the generation. In general it may be expected that palms should be replaced when 25 years old (22 years of harvesting) though many fields have been left for 30 years or more before being replanted. However, on volcanic soils where palms tend to grow more rapidly in height, replanting may need to be undertaken after only 15 bearing years.

Early workers were interested in maintaining some crop from the old stand by underplanting (Ferwerda, 1955), but recent workers have given prime attention to the new crop. An unreplicated trial in Nigeria (Gunn and Sly, 1961) in which all old palms were retained for a year but were then subjected to a variety of treatments, suggested that if alternate E–W rows were retained for 3 years, the yield from the old palms would more than compensate for the reduced early yield of the young palms. The retention of N–S rows appeared to suppress the early bearing of young palms presumably because of a greater shading effect. In a trial at Yaligimba in Zaire planted at 9.5 m triangular (128 palms per hectare) replanting was done at the same density and at a density of 160 palms per hectare. Four methods were used, and these are shown in Table 10.6. Where three fellings were undertaken, the least productive palms were chosen in all the rows for the first felling, then all the palms in alternate rows were removed at the second felling, and the remainder were removed at the third felling (Unilever, 1961; Green, 1964).

Table 10.6 *Methods of replanting employed in an experiment at Yaligimba, Zaire (Unilever, 1961)*

Proportion of palms felled in each year

Method of felling	Year of replanting			
	1953	*1954*	*1955*	*1956*
3/3	All palms			
2/3	2/3	1/3		
1/3	1/3	1/3	1/3	
0/3	—	1/3	1/3	1/3

Table 10.7   *Replanting experiment, Yaligimba, Zaire (yield of bunches and estimated yield of oil per hectare, in tonnes)*

Treatment	Bunches per hectare			Oil per hectare		
	Young palms 1956–63	Old palms 1953 to time of felling	Total	Young palms 1956–63	Old palms 1953 to time of felling	Total
3/3	70.9	—	70.9	13.0	—	13.0
2/3	63.5	2.8	66.3	11.7	0.5	12.3
1/3	64.9	9.8	74.7	12.1	1.6	13.7
0/3	57.0	17.2	74.2	10.9	2.8	13.7

Source: Unilever (1961).

The results at the end of 11 years are shown in Table 10.7. The bunch yields are as recorded, while the oil yields, though estimated, raise an important consideration. Replanting was often carried out in *dura* and Deli *dura* fields which were being replanted with *tenera* material. In these cases, total bunch weights do not therefore give the full picture of yields obtained. While oil-to-bunch ratio will remain steady at a comparatively low figure in the old palms, the oil-to-bunch ratio for the young palms will rise year by year from a low figure, say 10 per cent, to a much higher figure. In this case the oil-to-bunch ratio of the old palms was known to be 17 per cent while in the young palms other determinations suggested a progressive increase from 10 per cent in the first year to 22 per cent in the seventh.

The increases in total yield of the 1/3 plots over the clear-felled 3/3 plots were: bunches 5.35 per cent, oil 5.38 per cent, indicating that, over the limited period, the increased oil percentage of the *tenera* material had not yet affected the comparison.

The young palms clearly suffered from the competition of the old palms since their early yields (and their growth measurements) were lower. This effect was not permanent, however, as the data in Table 10.8 show.

The results of this experiment suggested that treatment 1/3, in which one-third of the old palms were removed in successive years,

Table 10.8   *Replanting experiment, Yaligimba, Zaire (bunch yield of young palms per hectare, in tonnes)*

Treatment	1959–60	1960–1	1961–2	1962–3
3/3	11.9	8.9	12.5	13.1
2/3	10.8	9.3	11.4	12.5
1/3	11.1	10.1	12.2	13.2
0/3	9.6	9.6	12.1	13.0

Source: Unilever (1961).

starting in the replanting year, would provide the highest overall yield and the shortest period of no crop (some small yield was obtained from the young palms during the last year of felling) and that the young palms would suffer no permanent ill effect. If the old palms were *tenera*, this treatment would be still more favourable. The method, which came to be known as 'the Congo system', was at one time used in estate practice and incorporated in trials carried out both in Malaysia and in West Africa.

In Nigeria the 'Congo system' was tried against (a) cutting out either alternate N–S or E–W rows at replanting and the remainder 2 years later; (b) pruning old palms to half canopy at replanting, at 3 months and (to the spear) at 1 year, and felling after 2 years; and (c) felling all palms at replanting (Sheldrick, 1968f). Of the underplanting treatments the pruning treatment was the most effective, giving a significantly higher total yield to the fourth bearing year of the young palms. The young palms in the full felling treatment gave significantly higher yields in the first 3 bearing years than in the underplanting treatments, but thereafter there were no differences; so this treatment did not overtake any of the underplanting treatments when the old-palm production was included. To year 8 the mean overall bunch yield of the underplanting treatments was 28 per cent above that of the complete felling treatment, while the pruning treatment bunch yield was 18 per cent higher than that of the other underplanting treatments.

In Asia, although underplanting was tried in the early years, it soon became apparent that the danger of *Ganoderma* infection of the young palms precluded this practice. Even in Africa, and in spite of the results of experiments, it remains doubtful whether the leaving in of any part of the old stand is worth while. Firstly, growth of unshaded young oil palms is now so rapid that they may be brought into bearing at least a year earlier than was previously the case. Experience has shown that underplanting makes it impossible to take advantage of this precocity. Secondly, there are suggestions that species of *Ganoderma* are now attacking earlier in Africa and for this reason underplanting has been considered of doubtful wisdom there too. In Malaysia and Sumatra replanting has, until recently, tended to be dominated by the need to combat *Ganoderma* and to reduce *Oryctes* attack (see pp. 607 and 651).

To minimize *Oryctes* and *Rhynchophorus* attack mechanized systems of replanting have been recommended for the Ivory Coast. The main objective is to obtain rapid growth of a *Pueraria* and *Centrosema* cover at the beginning of the wet season, all cultivation and felling having been done early in or during the dry season. If the trunks are then covered by a leguminous cover within 4–6 months of felling, very little *Oryctes* breeding occurs (Boyé and Aubrey, 1973). In the second method the palms are first shredded into pieces of varying sizes while still standing and then the boles are uprooted (Dupré, 1982).

In Malaysia, where so much replanting is taking place, there has been a recent return to the consideration of methods of shortening the period of no crop. Experiments on both coastal and inland soils of underplanting gave results not dissimilar to those of the 'Congo system' already described (Mohd, Nazeeb, *et al.*, 1987), i.e. treatments with partial retention of the old stand for up to 2 years gave a higher total yield than normal replanting methods. Doubts remain, however, about the later incidence of *Ganoderma*; but it was claimed that, even though the young palms are already in the ground, the risk is reduced if the boles are dug out during the felling of the old stand.

A novel idea has been the production of 'advanced planting material' or 'giant' nursery plants raised at 1.8 m triangular in the nursery in bags measuring 60 × 75 cm and planted out when 24 months old. Special machines have been tried for transplanting these palms, and the importance of the nursery being close to the field to be replanted has been stressed, but the real value of the method was shown by the production of bunches only 6 months after planting out.

Disposal of oil palms

The old palms may be disposed of by felling with axes, felling mechanically or poisoning with chemicals.

Axing of palms

This is heavy and tedious work. The number of mature palms which can be felled per man-day is between five and twelve. Directional felling is necessary, particularly if the area has been underplanted.

Mechanical felling

This requires the use of heavy tractors and tackle and will only be possible where large areas are to be cleared. Two methods were tried in Malaysia (Hartley, 1949), (a) pulling the palms over with a 2.5 cm cable drawn by two heavy tractors, one on each side of a row or pair of rows; (b) bulldozing the palms with an 85 hp tractor and bulldozer blade. The latter method is now generally used and under moist conditions the whole bole of a palm will emerge from the ground. In the Ivory Coast uprooting is considered particularly desirable where Wilt disease is prevalent (Dupré and Toulouse, 1978), and the methods mentioned above are employed.

Poisoning

This is an effective way of killing the palms but where there is a danger of the young palms becoming infected with *Ganoderma* it must be combined with felling and uprooting. In Malaysia if the old palms are poisoned and left to stand while disintegrating, *Ganoderma* infection of the young palms may be serious (Turner and Gillbanks, 1974). In these circumstances the palms are felled as soon as the foliage has turned brown and the crown begun to collapse using a special palm de-stumper which pushes the palms over from a height of 2.5 m (Stimpson and Rasmussen, 1973). More recently a method of shredding the palm trunks with a bulldozer blade, to hasten decay, has been introduced.

A trial in Malaysia with Deli palms showed that sodium arsenite is an effective and rapid killer at doses as low as 28 g per palm (Hartley, 1949). A hole is bored into the trunk by a 2.5 cm auger to as far as half the distance between the circumference and the centre of the palm. The hole may be 30–90 cm above the ground and has a gentle downward slope. The powered sodium arsenite is placed in the hole, water is added and the hole is tamped with clay or stopped up with a wooden plug. More than forty trees per day can be bored by one man. The usual safety precautions in using sodium arsenite are, of course, necessary. All palms show signs of dying within a few days and 40–80 per cent wither within 10 days. Within a month the crowns are dead and the stems begin to collapse.

Trials in Nigeria showed that West African palms could not be poisoned with sodium arsenite quite so readily as Deli palms in Asia. Up to 170 g per palm were required and in some palms the rotting portion of the trunk around the injection hole became sealed off by a hard layer of tissue and the remainder of the trunk remained normal (WAIFOR, 1953). Some exudation of the poison took place and larger holes more firmly plugged were probably needed. In Zaire (Marynen and Gillot, 1957) the introduction of 60–120 g in concentrated suspensions was effective in killing over 95 per cent in 70 days but, as in Nigeria, some resistant palms remained apparently healthy. Also in Zaire it was noted that survivors of a first application were not affected by subsequent efforts to poison them (Ferwerda, 1955). These palms developed cavities and necrotic areas in the vicinity of the auger hole. In general, sodium arsenite has been shown to be the cheapest and most effective substance for poisoning palms and, although 28 g per palm is effective in Malaysia, rather larger doses seem to be required elsewhere.

Attempts have been made to find poisons as or more effective than sodium arsenite and with a lower toxicity to man. In Nigeria thirteen herbicides were screened but only two, the bipyridyl herbicides diquat and paraquat, showed toxic qualities similar to sodium arsenite (Sheldrick, 1963). The effect of diquat is as follows:

within 2 weeks most of the leaves are desiccated; after 6 weeks they have usually collapsed basally and remain cloaking the stem. In a few palms the leaves remain in their normal position but there is some fracturing along their length. The spear remains upright though dried out. This is in contrast to sodium arsenite poisoning where the spear falls out and the trunk collapses. The trunks of diquat-treated palms collapse much later. The internal progression of change has been described in detail by Gunn and Tatham (1961). Although there is extensive drying out of the leaves, portions of the spear remain green for some time. The stem becomes dry and has a pink discoloration and a wet rot eventually develops in the leaf bases and the growing point. Later a rot develops near the point of injection. Further collapse of the palm normally occurs in high winds or heavy rain.

Diquat at the rate of 20 g a.i. per palm is more effective than a similar quantity of paraquat; 35 g a.i. of the latter is, however, rather more rapid in its action than 20 g diquat. Diquat may be introduced into a single sloping hole in the trunk 20 cm deep and punched 60–90 cm from the ground with a crowbar. The hole is then filled up with water.

With the banning of sodium arsenite in Malaysia, other arboricides have been sought there also. Diquat and paraquat were found to be effective at 23 g a.i. per palm. Glyphosate, at 17 g a.e., was also effective though slower and more expensive (Han, 1979).

The importance of replanting methods in the control of *Ganoderma* Trunk Rot is discussed on p. 610.

Field Costs

Table 10.9 shows representative costs of field maintenance work in terms of man-days per hectare per annum; these are costs of manual work only (*vide* Ch. 9, p. 413) and are averaged figures from several plantation sources. Quite different costs will be incurred if, for instance, maintenance is with herbicides or by mechanical means, and for these the portions of this book dealing with these subjects and the references given therein should be consulted. The upkeep of roads entails 2–3 man-days per planted hectare.

Maintenance of adult plantations by hand in the Ivory Coast is stated to provide 20–30 days (IRHO, 1962) work per hectare per annum.

The very much higher expenditure on weeding in the Far East compared with that on ring-weeding and slashing in Africa is due to the difficulty of, and heavy expenditure on, the establishment and maintenance of covers in the first few years. In addition more atten-

Table 10.9 *Field maintenance work – man-days per hectare per annum*

(a) **Africa**

Young palms Years of age:	Nigeria 0–5	Zaire (1)† 0–2	3–4	Zaire (2) 0–3	3–7	Zaire (3) 0–4
Ring-weeding	12	17	17 ⎱	20	20	22*
Paths or 'rides'	12	6	6 ⎰			
Slashing	20	12	6	20	15	12
	44	35	29	40	35	34

Middle years Years of age:	6–11	6–7				4–11
Ring-weeding	7	12 ⎱				10
Paths or 'rides'	10	6 ⎰				
Slashing	15	6				8
	32	24				18

Adult palms Years of age:	11+	9–20		8+		12–25
Ring-weeding	5	10 ⎱		12		6
Pests and diseases	10	6 ⎰				
Slashing	7	4		6		7
	22	20		18		13

(b) **Malaysia**

Year in the field	1st	2nd	3rd	9th (adult field)
Ring-weeding and weeding, including spraying if done	100	72	47	20
Pests and diseases	12	17	12	12
Paths, drains, roads, bridges, etc.	15	12	12	15
Manuring	2	5	2	2
	129	106	73	49

* Including mulching.
† Vanderweyen (1952)

tion has to be given to pests and, on coastal clay estates, to upkeep of drains and bridges.

In West Africa, maintenance costs in the first 4–5 years may be increased by about 40 man-days per hectare per annum by the need to spray every 3 weeks against Freckle (*Cercospora elaeidis*). With the use of motor-operated sprayers the work is reduced by about one-half. If the area has been prepared so that tractors can be driven between the rows then the man-day requirements can be reduced by four-fifths, i.e. to about 8 man-days per hectare, spray lines being used in conjunction with a tractor-mounted sprayer.

Pruning, usually carried out annually though sometimes more frequently, entails between 2 and 5 man-days per hectare per annum in the early harvesting years and 5–12 man-days once the palms are well above ground. Ablation (eight to twelve rounds in a year) entails, in Malaysia, about 10 man-days per hectare.

Harvesting

Harvesting tasks have been discussed briefly in this chapter. Owing to varying yields, labour expenditure per tonne of bunches and per unit area varies widely from region to region and month to month. The figures from Zaire and Malaysia in Table 10.10 for cutting plus carrying to the roadside are means of several estate figures and calculations are based on yields per hectare per annum on the following scales:

Year of harvest	1	2	3	4	5	6+	Old fields
	Tonnes bunches per hectare per annum						
Zaire	2.5	3.7	6.2	10.0	10.0	12.5	7.5
Malaysia (medium)	5.0	7.5	12.5	17.5	20.0	20.0	
(high)	7.5	10.0	15.0	20.0	25.0	30.0	

Harvesting costs per tonne are found to vary greatly even within regions and the figures given in Table 10.10 must therefore be taken as very rough averages.

Table 10.10   *Harvesting: cutting and carrying*

Age	Man-days per tonne of bunches	Man-days per hectare per annum	
Zaire			
Year 5	3.7	14	
Year 14	1.5	19	
Old fields	3.2	23	Low yields
Malaysia			
Young fields (chisel harvesting)	4.5	33	
Medium fields (axe harvesting)	1.5	19	Medium yield
	1.5	30	High yield
Adult fields (pole harvesting)	1.1	22	Medium yield
	1.1	33	High yield

Plantation labour requirements

From the information given above and in Tables 10.9 and 10.10 it can be seen that with manual methods the labour requirement on an adult plantation will vary as shown in Table 10.11. Figures of 3–5 hectares per worker are usually accepted in Malaysia for the total labour force required, including transport, mill and other labour; this suggests that the above man-day figures for Malaysia are rather higher than average, and that, in particular, labour on maintenance can be kept lower by mechanization of drainage and other work. It should be observed that these figures are based on maintenanc of adult fields and that a high proportion of young areas on a plantation raises the labour requirement, though the difference between the upkeep costs of young and adult areas appears to be much greater in the Far East than in Africa.

Table 10.11 *Labour requirements on an adult plantation*

	Man-days per hectare	
	Africa	*Malaysia*
Field maintenance	14–22	49*
Pruning	5–12	4.4
Harvesting	19–23	22–33
Roads	2–3	
Total	40–60	81
	Africa	*Malaysia*
Hectares per worker working 300 days per year	7.5–5.0	3.6

* Including roads.

References

Anon. (1982) Buffalo-assisted harvesting. Editorial, *Planter, Kuala Lumpur*, **58**, 332.

Aya, F. O. (1973) Germination inhibitors in the seeds of *Pueraria phaseoloides* (Ruxb.) Benth. *J. Nig. Inst. Oil Palm Res.*, 5(18), 7.

Aya, F. O. and **Lucas, E. O.** (1977) A critical assessment of the cover policy in oil palm plantations in Nigeria. In *Int. devel. in oil palm*, ISP, Kuala Lumpur.

Bealing, J. (1987) *Annual research report*. Palm Oil Research Institute of Malaysia (PORIM), 1986, Kuala Lumpur.

Bevan, J. W. L. and **Gray, B. S.** (1966) Field planting techniques for the oil palm in Malaysia. *Planter, Kuala Lumpur*, **42**, 196.

Bevan, J. W. L. and **Gray, B. S.** (1969) *The organisation and control of field practice for large-scale oil palm plantings in Malaysia*. Incorp. Soc. of Planters, Kuala Lumpur.

Boyé, P. and **Aubrey, M.** (1973) Replantation des palmeraies industrielles. Méthode de préparation de terrain et de protection contre l'*Oryctes* en Afrique de l'Ouest. *Oléagineux*, **28**, 175.

Boyé, P. and **Martin, G.** (1969) Organisation générale de la récolte en palmeraie industrielle. *Oléagineux*, **24**, 451.

Broekmans, A. F. M. (1957) Growth, flowering and yield of the oil palm in Nigeria. *J. W. Afr. Inst. Oil Palm Res.*, **2**, 187.

Broughton, W. J. (1977) Effect of various covers on the performance of *Elaeis guineensis* Jacq. on different soils. In *Int. devel. in oil palm*, ISP, Kuala Lumpur.

Bunting, B., Georgi, C. D. V. and **Milsum, J. N.** (1934) *The oil palm in Malaya.* Kuala Lumpur.

Calvez, C. (1976) Influence de l'élagage à différents niveaux sur la production du palmier à huile. *Oléagineux*, **31**, 53.

Chan, K. W., Corley, R. H. V. and **Seth, A. K.** (1972) Effects of growth regulators on fruit abscission in oil palm, *Elaeis guineensis*. *Ann. appl. Biol.*, **71**, 243.

Chan, K. W. and **Mok, C. K.** (1973) Castration and manuring in immature oil palms on inland latosols in Malaysia. In *Advances in oil palm cultivation*, p. 147, Incorp. Soc. of Planters, Kuala Lumpur.

Chew, P. S. and **Khoo, K. T.** (1973) Early results from disbudding trials on oil palms. In *Advances in oil palm cultivation*, p. 133, Incorp. Soc. of Planters, Kuala Lumpur.

Chew, P. S. and **Khoo, K. T.** (1977) Growth and yield of intercropped oil palms on a coastal clay soil in Malaysia. In *Int. devel. in oil palm*, ISP, Kuala Lumpur.

Coomans, P. (1971) Entretien chimique des ronds dans les palmeraies adultes de Côte d'Ivoire. *Oléagineux*, **26**, 595.

Corley, R. H. V. (1973) Effects of plant density on growth and yield of the oil palm. *Expl. Agric.*, **9**, 169.

Corley, R. H. V. (1977) Oil palm yield components and yield cycles. In *Int. devel. in oil palm*, ISP, Kuala Lumpur.

Cunningham, W. M. (1969) A container system for the transport of oil palm fruit. In *Progress in oil palms*, p. 287, Incorp. Soc. of Planters, Kuala Lumpur.

Daniel, C. and **de Taffin, G.** (1974) Conduite des jeunes plantations de palmiers à huile en zones sèches au Dahomey. *Oléagineux*, **29**, 227.

Daniel, C. and **de Taffin, G.** (1976) L'ablation des inflorescences de jeunes palmiers. Cas particulier des zones sèches. *Oléagineux*, **31**, 211.

Dufour, F. and **Quencez, P.** (1978) Lutte chimique contre *Eupatorium odoratum* L. sous palmeraie. *3° Symposium sur le désherbage des cultures tropicales*, Dakar, Vol. 2, p. 377.

Dufrane, M. and **Berger, J. L.** (1957) Étude sur la récolte dans les palmeraies. *Bull. agric. Congo belge*, **48**, 581.

Dupré, C. (1982) L'abattage des palmiers à huile agés en vue de la replantation des plantations industrielles: la technique du déchiquetage sur pied. *Oléagineux*, **37**, 283.

Dupré, C. and **Toulouse, J.** (1978) Extirpation des souches de palmiers. *Oléagineux* **33**, 491.

Edgar, A. T. (1958) *Manual of rubber planting (Malaya).* Incorp. Soc. of Planters, Kuala Lumpur.

Ferwerda, J. D. (1955) Questions relevant to replanting in oil palm cultivation. Thesis, Wageningen.

Gerard, P., Renault, P. and **Chaillard, H.** (1968) Critère et normes de maturité pour la récolte des régimes de palmiers à huile. *Oléagineux*, **23**, 299.

Gillbanks, R. A. (1967) Harvesting and fruit transport. A discussion on current practice. *Planter, Kuala Lumpur*, **43**, 322.

Gray, B. S. and **Hew Choy Kean** (1968) Cover crop experiments in oil palms on the West Coast of Malaya. In *Oil palm developments in Malaysia*, p. 56, Incorp. Soc. of Planters, Kuala Lumpur.

Green, A. H. (1964) Private communication.

Gunn, J. S. and **Sly, J. M. A.** (1961) *WAIFOR ninth annual report*, p. 58.

Gunn, J. S., Sly, J. M. A. and **Sheldrick, R. D.** (1962) *WAIFOR tenth annual report*, p. 52.

Gunn, J. S. and **Tatham, P. B.** (1961) Diquat as an arboricide. *Nature, Lond.*, **189**, 808.

Haines, W. B. (1940) *The uses and control of natural undergrowth on rubber estates.* Rubb. Res. Inst. Malaya, Planting Manual No. 6, Kuala Lumpur.

Han, K. J. (1979) Some arboricide investigations on oil palms. *Planter, Kuala Lumpur*, **55**, 371.

Hartley, C. W. S. (1949) The felling and disposal of old oil palms prior to replanting. *Malay, agric. J.*, **32**, 223.

Hew Choy Kean and **Tam Tai Kin** (1970) The effects of maintenance techniques on oil palm yield in coastal clay areas of West Malaysia. In *Crop protection in Malaysia*, p. 62, Incorp. Soc. of Planters, Kuala Lumpur.

Hew, C. K. and **Tam, T. K.** (1972) The effect of removal of the initial inflorescences on the establishment, growth and subsequent yields of the oil palm on the coastal clays in West Malaysia. *Malay Agriculturalist*, **11**, 13.

Hornus, P. (1983) Adaptation de techniques TBV à gouttelettes contrôlées pour les traitements des ronds des palmiers à huile. *Oléagineux*, **38**, 301.

IRHO. (1962) L'entretien des palmeraies adultes. *Oléagineux*, **17**, 777.

IRHO. (1976) Ablation des inflorescences des jeunes palmiers à huile. *Oléagineux*, **31**, 9.

Jollands, P., Turner, P. D., Kartika, D. and **Soebagyo, F. X.** (1983) Use of controlled droplet application technique for herbicide application. Basic considerations, equipment, trials and recommendations. *Planter, Kuala Lumpur*, **59**, 388.

Khera, H. S. (1974) Innovations in Malaysian agriculture – a case study of Kulim System of FFB loading and transporting from field to palm oil mill. *Rev. Agric. Econ. (Malaysia)*, **5**(2), 14.

Kowal, J. M. L. and **Tinker, P. B. H.** (1959) Soil changes under a plantation established from high secondary forest. *J. W. Afr. Inst. Oil Palm Res.*, **2**, 376.

Lau Koh Chin (1982) Buffaloes in Pamol estates. *Planter, Kuala Lumpur*, **58**, 48.

Lucy, A. B. (1941) A comparison between natural covers and clean weeding on yields of oil palms. *Malay. agric. J.*, **29**, 190.

Marynen, T. and **Gillot, J.** (1957) L'élimination des vieux palmiers par empoisonnement. *Bull. Inf. INEAC*, **6**, 167.

Mohd, Nazeeb, Loong, S. G. and **Wood, B. J.** (1987) Trials on reducing the nonproductive period at oil palm replanting. *Int. Oil Palm Conf.*, Kuala Lumpur, June 1987.

Muirhead, P. R. S. (1980) In-field assisted collection of FFB. *Planter, Kuala Lumpur*, **56**, 543.

Ng, K. T. and **Southworth, A.** (1973) Optimum time of harvesting oil palm fruit. In *Advances in oil palm cultivation*, p. 439, Incorp. Soc. of Planters, Kuala Lumpur.

Obasola, C. O. (1970) *NIFOR fifth annual report*, Nigeria, p. 41.

OPGL (Oil Palm Genetics Laboratory), (1971–2), Layang Layang, Malaysia. *Progress report, 1970 and 1971.*

OPRS (Oil Palm Research Station), (1971–2) Banting, Malaysia. *Agronomy annual reports, 1969, 1970, 1971.*

OPRS, Nigeria (1949) Proceedings of the Conference on Oil Palm Research held at the Oil Palm Research Station, near Benin, Nigeria, Dec. 1949. Mimeograph.

Piggot, A. G. and **Piggot, C. J.** (1976) What should we do about oil palm epiphytes? *Planter, Kuala Lumpur*, **52**, 354.

Pillai, K. R. (1978) A review of chemical weed control in rubber and legumes. *Planter, Kuala Lumpur*, **54**, 669.

Pratt, N. S. (1963) *Notes on the cultivation of oil palms.* Incorp. Soc. of Planters, Kuala Lumpur.

Price, J. G. M. and **Kidd, D. D.** (1973) Mechanised loading and transport of oil palm fruit. In *Advances in oil palm cultivation*, p. 415, Incorp. Soc. of Planters, Kuala Lumpur.

Rajanaidu, N., Abdul Aziz Ariffin, Wood, B. J. and **Sarjit Singh** (1987) Ripeness standards and harvesting criteria for oil palm bunches. *Int. Oil Palm Conf.*, Kuala Lumpur, June 1987.

Rajanaidu, N., Tan, Y. P. and **Rao, V.** (1985) Harvesting and estimation of oil in a bunch. *Proc. of the symposium on impact of the pollinating weevil on the Malaysian oil palm industry*, Feb. 1984, pp. 177–86, PORIM, Kuala Lumpur.

Rajaratnam, J. A., Chan Kook Weng and **Ong Hong Tong** (1977) *Asystasia* in oil palm plantations. In *Int. devel. in oil palm*, ISP, Kuala Lumpur.

RRIM (1957) Trial of the Holt Weed Breaker. *Plts' Bull. Rubb. Res. Inst. Malaya*, No. 31, 74.

RRIM (1962) Species and varieties of *Flemingia* in Malaya. *Plts' Bull. Rubb. Res. Inst. Malaya*, No. 61, 78.

Sankar, N. S. (1965) Fruit collection and evacuation by road. Sabah Planters' Association, Oil Palm Seminar. Mimeograph.

Seth, A. K., Abu Bakar and **Sivarajah, S.** (1970) New recommendations for weed control under young palms. *Crop Protection Conference*, 1970, p. 85, Incorp. Soc. of Planters, Kuala Lumpur.

Sheldrick, R. D. (1963) A note on recent investigations into palm poisoning. *J. W. Afr. Inst. Palm Res.*, **4**, 101.

Sheldrick, R. D. (1968a) The control of ground cover in oil palm plantations with herbicides. 3. Development of ring weeding techniques. *J. W. Afr. Inst. Oil Palm Res.* 5(17), 57.

Sheldrick, R. D. (1962 and 1968b) The control of ground cover in oil palm plantations with herbicides. 1. An introduction and some early investigations. *J. W. Afr. Inst. Oil Palm Res.*, **3**, 344. 2. Screening trials for herbicides suitable for ring weeding. *J. Nigerian Inst. Oil Palm Res.*, **4**, 417.

Sheldrick, R. D. (1968c) Weed control with herbicides during legume cover establishment. *J. Nigerian Inst. Oil Palm Res.*, **5**, 67.

Sheldrick, R. D. (1968d) The control of Siam weed (*Eupatorium odoratum* Linn). *J. Nigerian Inst. Oil Palm Res.*, **5**, 7.

Sheldrick, R. D. (1968e) Mechanical maintenance in oil palm plantations. *Nigerian agric. J.*, **5**, 7.

Sheldrick, R. D. (1968f) *NIFOR third annual report, 1966–7*, p. 48.

Sly, J. M. A. (1968) The result of pruning experiments on adult palms in Nigeria. *J. Nigerian Inst. Oil Palm Res.*, **5**, 89; quoting Rutgers, A. A. L. *Investigation on oil palms*. AVROS, Medan, 1922.

Sly, J. M. A. and **Sheldrick, R. D.** (1963) *WAIFOR eleventh annual report*, pp. 28 and 46.

Smith, R. (1973) Notes on chemical weed control in oil palms in Cameroon. *Oil Palm News*, No. 16, 12.

Southworth, A. (1977) Harvesting – a practical approach to the optimization of oil quantity and quality. In *Int. devel. in oil palm*, ISP, Kuala Lumpur.

Southworth, A. (1979) Field factors affecting quality. *Planter, Kuala Lumpur*, **55**, 440.

Speldewinde, H. V. (1968) Harvesting and harvesting methods. In *Oil Palm developments in Malaysia*, Incorp. Soc. of Planters, Kuala Lumpur.

Stimpson, K. M. S. and **Rasmussen, A. N.** (1973) Clearing the old stand and some preparation for replanting coastal oil palms. In *Advances in oil palm cultivation*, p. 116, Incorp. Soc. of Planters, Kuala Lumpur.

Surre, Ch. and **Ziller, R.** (1963) *Le Palmier à huile*. G.-P. Maisonneuve and Larose, Paris.

Taillez, B. and **Olivin, J.** (1971) Nouveaux résultats expérimentaux sur l'ablation des jeunes inflorescences du palmier à huile en Côte d'Ivoire. *Oléagineux*, **26**, 141.

Tan, H. T., Pillai, K. R. and **Fua, J. M.** (1977) Establishment of legume covers using pre- and post-emergence herbicides. In *Int. devel. in oil palm*, ISP, Kuala Lumpur.

Tan, K. S. and **Ng, W. C.** (1981) Preliminary results of nutrient cycling of covers in oil palm on inland soils. In *The oil palm in agriculture in the eighties*, ISP, Kuala Lumpur.

Teoh Cheng Hai and **Chong Choon Fong** (1977) Use of pre-emergence herbicides during establishment of leguminous cover crops. In *Int. devel. in oil palm*, ISP, Kuala Lumpur.

Teoh, C. H. and six others (1985) Prospects for biological control of *Mikania micrantha* HBK in Malaysia. *Planter, Kuala Lumpur*, **61**, 515.

Turner, P. D. (1983) Economies in weed control costs following changes from conventional knapsack spraying in plantation crops. *Symposium on Application and Biology*, BCPC monograph No. 28, 33.

Turner, P. D. and **Gillbanks, R. A.** (1974) *Oil palm cultivation and management.* Incorp. Soc. of Planters, Kuala Lumpur.

Unilever (Plantations) (1961) Annual review of research, p. 32. Mimeograph.

Vanderweyen, R. (1952) *Notions de culture d'Elaeis au Congo Belge.* Brussels.

Voon Ching Hoi, Han Kee Juan and **Lim Jit Kim** (1981) Fluazifopbutyl (PP009) – a selective herbicide for legume purification in oil palms. In *The oil palm in agriculture in the eighties*, ISP, Kuala Lumpur.

WAIFOR (1953) *First annual report, 1952–3*, p. 84.

WAIFOR (1960) When to start harvesting. *J. W. Afr. Inst. Oil Palm Res.*, **3**, 187.

Wan, D. (1973) The use of buffaloes in oil palm fruit collection. In *Advances in oil palm cultivation*, p. 432, Incorp. Soc. of Planters, Kuala Lumpur.

Washburn, R. A. (1973) The African oil palm in Costa Rica. *Oil Palm News*, **16**, 1.

Wuidart, W. (1973) Evolution de la lipogenèse du régime de palmier à huile en fonction du pourcentage de fruits détachés. *Oléagineux*, **28**, 551.

Wycherley, P. R. (1963) The range of cover plants. *Plts' Bull. Rubb. Res. Inst. Malaya*, No. 68, 117.

Yeow, K. H., Mohd. Hashim and **Tam, T. K.** (1981a) Effects of frond pruning on oil palm performance. In *The oil palm in agriculture in the eighties*, ISP, Kuala Lumpur.

Yeow, K. H., Tam, T. K. and **Mohd. Hashim** (1981b) Effects of interline vegetation and management on oil palm performance. In *The oil palm in agriculture in the eighties*, ISP, Kuala Lumpur.

Zohadia Bardaie, M. and **Wan Ishak Wan Ismail** (1982) Collection and transportation of oil palm FFB during harvesting. *Planter, Kuala Lumpur*, **58**, 562.

Chapter 11

The nutrition of the oil palm

Nutrient demand and employment

Nutrient uptake and usage is an important physiological process but it was not dealt with in Chapter 4 as it can more usefully be discussed in direct connection with manuring. Attention has already been given to the cumulative production of dry matter in the palm (p. 150) and to the distribution of dry matter production in the various organs. The immobilization of nutrients taken up in the course of this dry matter production will now be considered.

That nutrient requirements can be indicated by the chemical composition of the plant has not been generally accepted, but it has been claimed that analysis can have some value when the crop is bulky, as in the oil palm, and the long-term nutrient supplying power of the soil is poor, as in so many tropical soils (Tinker and Smilde, 1963a). Results of analyses in the case of the oil palm are of particular interest in a consideration of replanting since by the time one cropping cycle of 25–30 years is completed considerable quantities of nutrients have been removed in the bunches, stored in the trunk, leaves and roots, or returned to the soil in dead leaves and roots. A good idea can therefore be obtained at that time of the proportions in which the various nutrients have been lost in the produce or have been or are to be returned to the soil.

A number of attempts have been made to carry out a complete estimation on these lines. Ferwerda (1955) combined data from African sources to construct a table of plant nutrients immobilized per hectare by a 20-year-old plantation of *tenera* palms in Zaire. Accounted as immobilized were nutrients in (i) the trunk, (ii) the crown with the leaves and the roots existing on the palms at the time of estimation and (iii) the bunches removed from the palms since they started to bear. The total of these sources constitute the cumulative net nutrient uptake to any given age.

In Nigeria, Tinker and Smilde (1963a) analysed palms aged 7, 10, 14, 17, 20 and 22 years taken from a forest-felled area where no clear responses to N, P, Ca or Mg in adjoining fields had been obtained and where K had only become deficient in the later years. Bunch yield throughout the palm's life was known and nutrient losses were thence estimated through analysis of bunch stalk and calculations of bunch composition. Analysis was made of the following above-ground parts of the palm separately: leaflets, rachis, apical tissues and trunk. Roots of two 17-year-old palms were excavated to 90 cm and analysed, and on the basis of Ferwerda's work quantities were assumed to be similar for all age-groups except the 7-year-old palms. The 22-year-old palms, though not showing symptoms, were in an area becoming potassium-deficient, and the study also included a group of three 22-year-old palms showing obvious potassium deficiency and two 21-year-old palms in an area of magnesium deficiency.

Interest in the data of these palms (Table 11.1) lies largely in the changes that occur with age and with the onset of deficiency symptoms and in the estimations that can be made of the total quantity of nutrients removed by the palms in the course of their productive life. There is very little variation with age in percentage nitrogen or phosphorus content, but potassium percentage decreases with age of palm while there is a corresponding increase in magnesium and calcium percentages. Potassium deficiency is reflected by lower quantities in all parts of the K-deficient 22-year-old palms and it may be assumed from the very low trunk K content of the 20-year-old palms that they must have been on the verge of potassium deficiency. Extremely small quantities of magnesium are left in any tissues of the Mg-deficient palms.

The cumulative net nutrient uptake per hectare, which differs from the total uptake by the exclusion of nutrients recycled to the soil in old leaves, male inflorescences and dead roots, was calculated from the analytical data, estimates of live roots and the actual bunch production from the start of bearing. The means of the 20- and the healthy 22-year-old palms were used. These data are shown in Table 11.2 where they are compared with Ferwerda's estimate for 20-year-old Zaire palms. Considering all the approximations inevitable in both sets of data and the composition from various sources of the Zaire estimate, remarkably similar figures emerge.

The cumulative net uptake per palm of phosphorus, potassium and magnesium over lengthening periods of years is illustrated from the Nigerian data in Fig. 11.1. It will be seen that though the P uptake is much smaller than the K uptake, the proportion of the P diverted to the bunches is large. Phosphorus also accumulates in the trunk at a steady rate. Nitrogen uptake follows a similar course to that of phosphorus, but a lower proportion is removed in the bunches. Large quantities of potassium are diverted to the bunches and it is apparent that in these palms the K accumulation in the palm itself is very slow; there is here a suggestion that in the later

Table 11.1   *Dry matter and nutrient content of palms of different ages in Nigeria*

Age	Part	N	P	K	Mg	Ca	Dry matter
A. *Kilograms per palm*							
7	Crown	0.61	0.07	0.63	0.15	0.26	68.5
	Trunk	0.33	0.04	0.30	0.14	0.14	64.4
	Total	0.94	0.11	0.93	0.29	0.40	132.9
10	Crown	0.82	0.07	0.69	0.17	0.31	92.2
	Trunk	0.85	0.08	0.65	0.25	0.27	167.5
	Total	1.67	0.15	1.34	0.42	0.58	259.7
14	Crown	0.80	0.08	0.48	0.21	0.36	85.4
	Trunk	1.31	0.17	0.87	0.42	0.36	238.8
	Total	2.11	0.25	1.35	0.63	0.72	324.2
17	Crown	0.89	0.09	0.76	0.19	0.32	96.9
	Trunk	1.53	0.17	1.51	0.66	0.63	280.0
	Total	2.42	0.26	2.27	0.85	0.95	376.9
20	Crown	1.31	0.17	0.68	0.48	0.87	151.5
	Trunk	1.93	0.30	0.83	1.49	1.12	439.1
	Total	3.24	0.47	1.51	1.97	1.99	590.6
22	Crown	1.06	0.11	0.67	0.41	0.64	117.5
	Trunk	1.17	0.20	1.15	0.90	0.50	343.8
	Total	2.23	0.31	1.82	1.31	1.14	461.3
22 K-deficient	Crown	0.86	0.11	0.31	0.44	0.73	99.1
	Trunk	1.26	0.24	0.50	0.93	0.60	311.6
	Total	2.12	0.35	0.81	1.37	1.33	410.7
21 Mg-deficient	Crown	0.44	0.05	0.37	0.02	0.08	47.0
	Trunk	0.56	0.07	0.45	0.03	0.15	88.0
	Total	1.00	0.12	0.82	0.05	0.23	135.0
17	Roots	0.46	0.03	0.60	0.19	0.09	128.0
B. *Percentage of dry matter*							
7	Leaflets	1.64	0.12	0.95	0.27	0.55	
	Rachis	0.41	0.07	0.84	0.16	0.25	
	Apical tissue	2.20	0.38	3.30	0.80	0.82	
	Trunk	0.52	0.07	0.47	0.21	0.22	
14	Leaflets	1.90	0.13	0.70	0.36	0.68	
	Rachis	0.37	0.06	0.43	0.14	0.28	
	Apical tissue	2.18	0.36	2.55	0.91	0.73	
	Trunk	0.55	0.07	0.36	0.17	0.15	
20	Leaflets	1.94	0.14	0.77	0.38	0.78	
	Rachis	0.33	0.08	0.26	0.28	0.47	
	Apical tissue	2.00	0.40	1.75	1.00	1.00	
	Trunk	0.44	0.07	0.19	0.34	0.26	
22 K-deficient	Leaflets	1.82	0.15	0.48	0.46	0.92	
	Rachis	0.38	0.08	0.20	0.43	0.64	
	Apical tissue	2.00	0.48	1.61	0.99	0.87	
	Trunk	0.40	0.08	0.16	0.30	0.19	
21 Mg-deficient	Leaflets	1.63	0.12	0.99	0.07	0.20	
	Rachis	0.45	0.10	0.60	0.05	0.12	
	Apical tissue	3.00	0.47	3.33	0.53	0.53	
	Trunk	0.64	0.08	0.50	0.03	0.17	

Source: Tinker and Smilde (1963a).

Table 11.2 *Nutrients removed from or immobilized in 20- and 22-year-old palms in Nigeria and Zaire (kg per hectare)*

Nigeria	N	P	K	Mg	Ca
Parts – Above ground	390	55	250	230	220
Roots	70	5	90	30	14
Bunches*	430	90	500	65	76
Total	890	150	840	325	310
Zaire estimate					
Parts – Above ground	713	125	305	174	297
Roots	84	9	86	4	1
Bunches*	564	97	585	82	88
Total	1,361	231	976	260	386

* Based on yields: 1,060 kg per palm, Nigeria, and 1,371 kg per palm, Zaire.

Sources: Tinker and Smilde (1963a); Ferwerda (1955).

years the reserves in the trunk were being drawn upon. In contrast, magnesium (where not deficient) accumulates in the trunk most rapidly in the later years and a low proportion is removed in the bunches. This pattern is also followed by calcium.

In drawing any general conclusions from these data notice must be taken of the depletion of potassium which was clearly taking place in this soil under palms, the known antagonisms which exist between the nutrient elements, and the low level of bunch yield – though typical of the region concerned – compared with other producing regions. It is reasonable to suppose that if K dressings had been applied or the soil had had a higher K content, accumulation in the trunk would have been greater and that the K graph in Fig. 11.1 might have approximated more closely to that of P; bunch yield would have been higher and the proportion of K diverted to bunches might have remained the same.

Phosphorus immobilization is on a much smaller scale, but owing to the high proportionate diversion to bunches it is hardly surprising that a phosphorus need often arises when bunch production has been raised by the supply of primarily deficient nutrients.

Owing to the potassium–magnesium antagonism, Mg contents, both absolute and as a percentage of dry matter, tend to be high in the prevailing state of K deficiency of the Nigerian palms. As Table 11.1 shows, however, Mg contents can fall to very low levels indeed.

An extensive study of nutrient uptake was undertaken in Malaysia by Ng Siew Kee (1967, 1968). It is not possible to compare the results directly with those from Nigeria since gross annual uptakes were computed in place of net nutrient uptake over the life

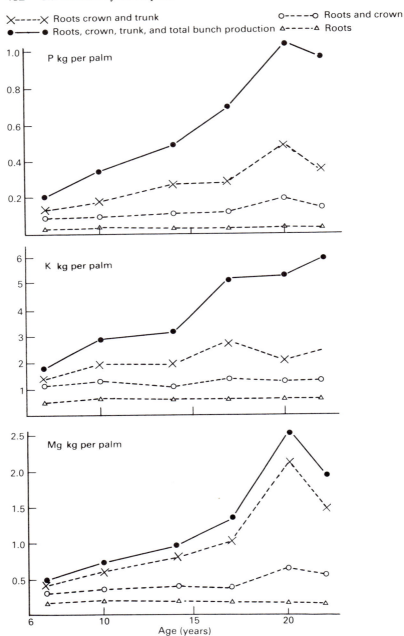

Fig. 11.1 Cumulative net uptake of phosphorus, potassium and magnesium in a field in Nigeria (Tinker and Smilde, 1963a).

of a palm. The gross annual nutrient uptake of a group of adult palms growing on marine clay and yielding 25 tonnes of bunches per hectare is given, together with the gross uptake less nutrients in leaves and male inflorescences recycled to the soil, and the uptake in the bunches, as follows:

	Kilograms per hectare per annum				
	N	P	K	Mg	Ca
1. Gross nutrient uptake (GNU)	193	26	251	61	89
2. GNU, less leaves and male inflor.	144	15	149	32	23
3. Bunches	73	12	93	21	20

From these figures and from Table 11.2 it may be deduced that cumulative net nutrient uptake is much higher under Malaysian conditions than under conditions obtaining in Nigeria since, for example, the potassium taken up by the palm less K returned to the soil in leaves and male inflorescences in 6 adult years is more than that removed from and immobilized in the Nigerian palms in their whole 20–22 years of life. The Malaysian analyses showed figures comparable to those of Nigeria for nutrient concentration in the various organs except in the case of potassium, the percentage of which in the trunk was much higher in Malaysia. In short, the conditions of growth on the Malaysian coastal soils were, in comparison with conditions in Nigeria, conducive to much higher uptake of nutrients for bunch production and storage in the trunk, a greater development of the trunk both in height increment and width, a higher bunch production, but only a small difference in leaf production. From the bunch analyses, the mean quantities of nutrients per tonne of bunches were estimated as follows:

Per tonne of bunches	N	P	K	Mg	Ca	Mn	Fe	B	Cu	Zn	Mo
Kilograms	2.89	0.43	3.65	0.76	0.80						
Grams						1.49	2.43	2.12	4.69	4.85	0.0083

Methods of detecting nutrient need

Field experiments with fertilizers are the primary means of detecting and determining nutrient need, but these may be greatly assisted by (a) methods using plants and (b) soil analysis methods. Wallace (1957) has stated that the main methods of estimating nutrient needs are qualitative rather than quantitative, are empirical in character and some are obviously suited only for preliminary diagnosis. This is especially the case in the methods depending on the behaviour of the palm itself and these will be considered first.

Deficiency symptoms

Work on deficiency symptoms dates from a classic experiment at Nkwelle in Nigeria in 1940 (Toovey, 1948) and the discovery by Hale (1947) of the connection between leaf potassium content and 'bronzing' or orange spotting of the leaves. In the Nkwelle experiment an application of 17 tonnes of wood ash per hectare applied over a 3-year period restored affected palms to health and increased their yield fivefold. Later, dressings of sulphate of potash in the same area showed that the responses to ash were attributable to its potash content.

These startling results stimulated an interest in deficiency symptoms and in the following 15 years a considerable amount of knowledge was gained both in Africa and Asia. This knowledge was obtained by three methods: (i) leaf injection and spraying of chlorotic leaves, (ii) the correlation of symptoms with leaf nutrient contents and with yield and (iii) the inducement of symptoms in sand culture. In the following descriptions the symptoms induced on seedlings are those described by Broeshart (1955) and Bull (1961a, b), while the adult palm symptoms are attributable to general observations in several parts of the world and to accounts by several authors.

Nitrogen deficiency

In the very young seedling the first symptom is the development of a uniform pale green colour over the leaves. Paling is followed by yellowing, green gradually disappears and the production of a deeper yellow colour precedes necrosis. In larger seedlings and in young palms in the field paling and yellowing follow a similar course and, in addition to the necrosis, the leaflet midribs and the rachis take on a bright yellow colour and the laminae tend to be narrow and to roll inwards. Thus N-deficient plants are quite distinctive (Plate VI) and when the deficiency is corrected by applications of sulphate of ammonia the plants very rapidly turn a dark green colour.

Nitrogen deficiency symptoms are often associated with waterlogged conditions and they are soon counteracted by drainage. The leaves of severely waterlogged seedlings, however, often have a dull, olive green and flaccid appearance. Severe grass competition also induces nitrogen deficiency symptoms.

Phosphorus deficiency

Symptoms of phosphorus deficiency can be obtained in pot culture under rigorous systems of sand and water purification (Bull, 1961a).

In small seedlings the oldest leaves become dull and assume a pale olive green colour. The chlorotic condition increases in severity but the seedlings do not become fully yellow before necrosis of the tips sets in. The necrotic areas are usually of a dark brown colour and in transmitted light the tissues near to these areas are seen to be pale and water-soaked, suggesting rapid cell collapse. The leaves are much reduced in size.

Though responses to phosphorus fertilizers are often obtained, clear symptoms of phosphorus deficiency have not appeared in the field. In an experiment in southern Zaire, however, it was noted that palms not receiving phosphorus showed a high incidence of premature desiccation of the older leaves, a symptom usually found in areas of potassium deficiency.

Potassium deficiency

In very young seedlings in sand culture, leaves begin to show a pale green to white interveinal mottling with minute white or yellowish rectangular spots. Later, the leaves become pale olive or ochre in colour and the tips and margins become necrosed, the necrosis being typically pale grey or silvery. In larger seedlings the chlorosis is similar, being interveinal with the veins and adjacent tissues remaining a normal green colour. Necrosis of the leaflet tips and margins also proceeds in the same manner as in smaller seedlings, the transition zone between chlorosed and necrotic tissue being very thin and pale brown. Minute clear spots also appear scattered over the laminae. These symptoms are accompanied or followed by marked shortening of the rachis and of the leaflets and this gives the leaf a 'bunchy' appearance, but these symptoms do not necessarily persist in later-formed leaves.

A variety of symptoms are associated with potassium deficiency in the mature oil palm. These variations are thought to be caused by contemporaneous variations in the concentration of other ions which may in turn be caused by environmental or genetic factors (Bull, 1957a). Those symptoms which have unquestionably been associated with potassium deficiency as shown by fertilizer responses or leaf analysis are described below.

Confluent Orange Spotting

This name, first used by Waterston, was adopted by Bull (1954a) to describe a condition in which chlorotic spots, changing from pale green through yellow to orange, develop and enlarge both between and across the leaflet veins and fuse to form compound lesions of a bright orange colour. Under a lens a pale yellow halo can be seen around the spot, but to the naked eye the boundary between spot and green leaf appears sharp. Necrosis within spots is common but irregular, and may be accompanied by fungus invasion which is secondary. Orange Spotting described in Malaysia and Sumatra is virtually identical except that in Malaysia the orange spots tend to be more elongated (Anon., 1952), a tendency not unknown in Africa.

Bull (1954a) has listed the large number of alternative names which have been employed for this condition. Chloroses in the oil palm, when not meticulously examined, were often simply termed 'yellowing'. One of the earliest cases of Confluent Orange Spotting was termed 'Nkwelle Yellows' from the place in Nigeria where it was widespread, but the spotted nature of the chlorosis was early recognized in such Nigerian plantation terms as 'speckled yellows', 'speckled bronzing', etc. Significantly, the term 'bronzing' came to be widely used in both Nigerian and Malaysian areas of Confluent Orange Spotting because the leaves with the most dense production of coalescing spots take on a bronzed appearance when viewed from a distance. This is particularly pronounced in certain coastal areas in Malaysia. Coulter and Rosenquist (1955) showed that bronzing, as the term was used in Malaysia, included two distinct conditions, firstly the orange spotting similar to that of Nigeria, and secondly a yellowing which was more pronounced on the upper ranks of leaflets. Chapman and Gray (1949) distinguished, but did not describe, bronzing and yellowing associated with low leaf K and Mg respectively.

Hale (1947) analysed upper, middle and lower leaves from healthy, mildly and severely affected palms growing at Nkwelle and on two estates. The Nkwelle material is of most interest as all palms were shown to be K-deficient though some did not show the spotting symptom; those with the symptom were, however, more deficient than those without. Middle leaves gave the following K and Mg percentages of dry matter and indicate the extent to which K deficiency had proceeded:

| | Per cent of dry matter | |
	K	Mg
Healthy	0.37	0.49
Mildly affected	0.17	0.69
Severely affected	0.16	0.63

The outstandingly large responses to K manures at Nkwelle have already been referred to. Equally distinct effects of K applications on Confluent Orange Spotting symptoms were later seen at Umudike and Mbawsi (Bull, 1957b) in eastern Nigeria. At the latter place palms exhibiting the symptoms were shown to contain considerably less potassium in leaves 9, 17, 25 and 33 than a palm growing in the same field, but showing no symptoms (see Table 11.4).

The fields of palms in Africa in which these Orange Spot palms appeared were of mixed genetical origin and the symptom, though widespread, was not found on every individual. In recent years it has been shown that individual palms and progenies exhibiting a high degree of Confluent Orange Spotting may have a high leaf potassium content. Six Orange-spotted palms found in a field of generally healthy palms in Nigeria gave a leaf-K percentage of 0.94 against 0.87 per cent for adjacent healthy palms; one selfed progeny showing a high incidence of the condition had a K percentage of 1.05 while adjacent healthy progenies, whether selfed or crossed, gave similar K contents. The palms were not in the areas of severe K deficiency but progenies showing symptoms gave a significantly lower yield than healthy progenies (Smilde, 1962, 1963). Forde and Leyritz (1968) made a detailed study of the palms and divided their symptoms into three types. They showed that these symptoms could not be attributed to K deficiency and appeared to be genetic in origin. On the basis of this study it was affirmed that Confluent Orange Spotting is not a deficiency symptom of the major elements; however, in view of all the evidence from elsewhere, it is not poss-ible to accept such an unconditional statement.

In Benin (Dahomey), also, individual palms with Confluent Orange Spotting were found which could not be associated with a low leaf-K content (Ochs, 1965).

In Malaysia both leaf-K differences and progeny differences were studied by Coulter and Rosenquist (1955) in a field of palms showing Confluent Orange Spotting. In this study the youngest fully opened leaf, which was counted as leaf 3 (not leaf 1) and leaf 17 on this scale were taken. A very significant correlation was found between the Orange-Spotting scores and the potassium in the ash and in dry matter of leaf 3, but no correlation was found in the case of leaf 17. A high calcium content was associated with Orange Spot-ting, and this condition was, as might be expected, associated with a decrease in yield. Marked differences in Orange Spotting inci-dence were found between progenies. In leaf 3, the less affected progenies had a significantly higher potassium content in the ash than the most affected progenies, but this difference did not appear in leaf 17. There were also indications that some susceptible pro-genies might have a higher percentage of K in the ash of affected than of unaffected palms.

Various attempts have been made to find an association between Confluent Orange Spotting and an excess or deficiency of other elements or to explain its connection with K deficiency under special conditions of nutrient balance. No definite and consistent trends have come from this work.

It may be affirmed that where Confluent Orange Spotting is found as a widespread condition in a population of mixed genetic origin it is symptomatic of acute potassium deficiency. However, certain individual palms or progenies may be prone to this condition when potassium need is slight, or even absent, and correlations between the symptom severity and potassium leaf content will not then be obtained.

Mid-Crown Yellowing

This symptom was first described by Chapas and Bull (1956) from highly K-deficient areas in eastern Nigeria, but has since been recognized in old palms in less deficient fields. Leaves around the tenth position on the phyllotaxis became pale in colour and terminal and marginal necrosis follows. A band along the midrib usually remains green. There is a tendency for later-formed leaves to be shorter and the palm has an unthrifty appearance with much premature withering, sometimes referred to as 'grey withering', of older leaves. Leaf analysis of seventy-one palms in a field near Benin where this symptom was prevalent lends support to the belief that this is a primary potassium deficiency symptom, but there is a suggestion that nitrogen deficiency may be a contributing factor (Table 11.3). The different effects of reduced leaf K on the Ca and Mg contents is of particular interest and will be referred to later.

Table 11.3 *Leaf analysis of the seventeenth leaf of palms in a field showing Mid-Crown Yellowing: Benin, Nigeria*

Mid-Crown Yellowing:	Percentage of dry matter				
	N (%)	P (%)	K (%)	Ca (%)	Mg (%)
Absent	2.55	0.17	0.63	0.85	0.40
Slight to medium severity	2.36	0.17	0.39	0.87	0.54
Medium to severe	2.17	0.17	0.30	0.90	0.64
	Percentage of total cations				
Absent			33	45	22
Slight to medium			22	48	30
Medium to severe			16	49	35

Source: Leyritz (1964).

Symptoms described from Benin (Dahomey) as *décoloration diffuse* also agree with those of Mid-Crown Yellowing and it has been shown that this symptom does not usually appear until the K content of leaf 17 has fallen to about 0.3 per cent of dry matter (Ochs, 1965). The 'shading effect' characteristic of magnesium deficiency is also found in this symptom, i.e. where one leaflet is covered by another the covered portion does not show chlorosis.

Mbawsi Symptom

This distinctive symptom takes its name from a highly K-deficient plot in eastern Nigeria. Large yellow or orange patches appear on affected leaflets. The midrib and a narrow strip on either side of the midrib remain green. The orange patches usually contain a mass of minute orange spots showing little or no necrosis.

Table 11.4 shows the analyses of leaves of a palm without symptoms in a K-deficient area at Mbawsi together with analyses of palms with Confluent Orange Spotting only and the Mbawsi Symptom with some Orange Spotting (WAIFOR, 1956). The latter two palms also suffered some tip and edge necrosis from the seventeenth leaf.

It will be noticed that the only striking differences are the markedly lower K content of both the 'COS' and 'MS' palms in comparison with the symptomless palm which itself has a rather low K percentage.

Table 11.4 *Leaf analysis of single palms showing Orange Spotting and Mbawsi symptoms: Mbawsi, Nigeria (percentage of dry weight)*

	N			P			K			Mg			Ca		
Conditions of palm Leaf No.	H (%)	COS (%)	MS (%)	H (%)	COS (%)	MS (%)	H (%)	COS (%)	MS (%)	H (%)	COS (%)	MS (%)	H (%)	COS (%)	MS (%)
1	2.3	2.5	2.5	0.19	0.20	0.18	0.9	0.9	0.7	0.2	0.4	0.2	0.5	0.3	0.4
9	2.2	2.4	1.9	0.15	0.18	0.14	0.5	0.3	0.2	0.6	0.6	0.3	0.7	0.8	0.6
17	2.2	2.1	1.8	0.14	0.16	0.13	0.7	0.4	0.4	0.2	0.9	0.4	0.7	1.0	0.6
25	2.1	2.1	2.0	0.14	0.17	0.14	0.6	0.4	0.3	0.2	0.7	0.6	0.9	1.0	0.9
33	1.9	2.0	1.7	0.13	0.16	0.12	0.5	0.2	0.2	0.1	0.4	0.2	0.9	0.9	0.9

H = Healthy in appearance. COS = Confluent Orange Spotting. MS = Mbawsi symptoms with some Confluent Orange Spotting

Source: Bull in WAIFOR (1956)

Magnesium deficiency

In small seedlings in sand culture the older leaves first develop an ochre colour changing later to pale to bright yellow, the deep orange tints of magnesium deficiency in adult palms being absent. Older seedlings develop a similar chlorosis on the distal leaflets of the older leaves and the appearance of magnesium deficiency in a nursery, often induced by unbalanced manuring, is quite characteristic. The symptoms are always most strong on the older leaves which, in a nursery, are still entire or bifurcate. Here the chlorosis is most pronounced in the central part of the leaflet, the paling only gradually proceeding to the edge. Eventually, necrosis sets in at the tip of the older leaves and is reddish or chocolate brown in colour.

On close inspection it can be seen that the apparently clear chlorotic parts of the leaflets contain a mass of small orange spots (Plate VIIA).

In adult palms and on large seedlings in the field severe magnesium deficiency symptoms are most striking and have been named Orange Frond (Plates VIIB and C, between pp. 230 and 231). While the lower leaves will be dead, those above them show a graduation of colouring from bright orange on the lower leaves to a faint yellow on leaves of a young of intermediate age. The youngest leaves show no discoloration. In young palms in the field there is often a much smaller proportion of bright orange leaves. The chlorosis proceeds as follows: discoloration begins about 10–12 cm from the leaflet tip with an ochre-coloured strip which extends first between and then across the veins until the whole leaflet, except small areas at the tip and the base, becomes first yellow then deep orange. Typically the chlorosis spreads right across the midrib and there is no green band. Later the leaflets are infected by fungi and umber and purplish-coloured areas appear at the tip and extend down the edges. This necrosis is quite distinct from the brown-grey withering of orange-spotted leaves. Most typical of magnesium deficiency symptoms is the strong shading effect of one leaflet lying over another; the shaded portion of the lower leaflet, particularly when in proximity to the upper one, will be found to be dark green.

The fungus *Pestalotiopsis gracilis* occurs regularly in the purplish-brown lesions just described and the organism can be parasitic on moribund Orange Frond leaflets (Bull, 1954b; Coulter and Rosenquist, 1955).

Thompson (1941) was the first to induce magnesium deficiency symptoms in water culture with seedlings, and Chapman and Gray (1949) reported 'yellowing' of older leaves where leaf levels of magnesium were low; but it remained for Bull (1954b) to establish adult-palm symptoms as magnesium deficiency through leaf-tip injection and spraying trials.

Calcium deficiency

This has only been seen in small seedlings raised in sand culture where rigorous conditions of sand and water purification have been adopted (Bull, 1961a). It has been evident in all pot culture work that oil palms can develop normally even when only minute quantities of calcium are available. Absence of calcium first produces abnormally short and narrow leaves with prominent veins. On older, bifurcate leaves one-half of the blade shows poor development, with apical splitting and necrosis. Later leaves are progressively smaller, terminal necrosis increases, and eventually after a progression of malformed blades the youngest leaves merely comprise the basal part of the petiole. There is no chlorosis.

Sulphur deficiency

In pot culture with small seedlings the symptoms in the early stages are not unlike those of nitrogen deficiency. Leaves are small and pale green or almost white in colour. Some interveinal streaking occurs. Later, brown necrotic spots appear on the older leaves followed by terminal necrosis. In older seedlings the interveinal chlorosis which follows the general paling is mottled or spotted. These chlorotic areas of spots become orange and brown and then coalesce and become necrotic.

Broeshart *et al.* (1957) reported sulphur deficiency symptoms in a field trial in Zaire though it is possible that these were confused with those of boron deficiency. More recently Turner *et al.* (1983) described an interveinal chlorosis on large nursery seedlings which gradually became more intense until the youngest leaves were entirely pale yellow. The condition was cured by sulphur powder applications.

Chlorine deficiency

Although chlorine has been considered to be an essential and important element in oil palm nutrition, no visual deficiency symptoms have yet been reported.

Minor element deficiencies

Boron

Boron deficiency is perhaps the most elusive of the deficiencies of the oil palm. Seedling deficiency symptoms have been determined, field symptoms have been described and experiments in areas of supposed deficiency have been conducted, but much of the latter work has lacked the final conclusiveness which has been so widely sought.

Symptoms produced in Nigeria with small seedlings in pot culture under rigorous conditions of sand and water purification were as follows (Bull, 1961b): newly emerging bifurcated leaves were much smaller than normal, having a shortened petiole and a truncated leaf blade. Affected leaves were less than one-tenth the size of leaves of the same age in control pots and the continued production of these small leaves produced congestion in the centre of the seedling. Hooking of the apex or corrugations of the blade were not seen. However, in similar trials in Malaysia Rajaratnam (1972a) reported some cases of hooks and flaps on the leaf margins and puckering of the laminae. Some chlorosis of the sixth (bifurcate) leaf was also reported with characteristic wide angles between the two lamina sections. Later leaves, besides being very small, were erect, compact and had necrotic tips. With 8-month-old healthy seedlings transplanted into sand culture minus boron chlorotic streaking was seen after the sixth leaf to be produced; this was followed by the production of small leaves with no reduction in the number of leaflets but a marked reduction in their size. After fifteen of these leaves both 'incipient little leaves' and, subsequently, 'fish-bone' leaves were found. In the former, the leaflets were much bunched on the shortened rachis and the number of leaflets reduced. The 'fish-bone' leaves were abnormally stiff with leaflets reduced to projections. This symptom was in turn followed by the production of shorter stump leaves and eventual rotting.

Rajaratnam (1972b) has also reported other symptoms in the recovery of boron-deficient seedlings after application of the nutrient. These have been described as 'Fish-tail' Leaf in which the terminal leaflets are fused and basal leaflets absent, and 'Hook Leaf', a symptom often reported on oil palms of all ages in which the end of the leaflet is bent over in the form of a hook (Plate 11.1).

In adult palms the symptoms of boron deficiency have been difficult to substantiate. Ferwerda (1954) found 93 per cent of the palms in plots receiving all elements except boron to be showing symptoms of Hook Leaf and a narrowing of the distal leaflets with some leaflets corrugated, reduced in size or even absent. The stunted rachis of an abnormal leaf was often calloused but rotting was limited to some leaflets of the younger spear leaf. There was no chlorosis.

Pl. 11.1 The 'Hook-Leaf' condition, which is of widespread occurrence, seen in Ecuador. (P. F. Arens)

The Fish-bone symptom has been frequently seen in Malaysia in areas, particularly young plantations, with low boron leaf content. Following the work of Rajaratnam it is reasonable to conclude that the main symptoms of boron deficiency in the field are a reduction of leaf area in certain leaves producing either incipient 'Little Leaf' (defined in this case as leaves smaller than normal with leaflets only

Pl. 11.2 Palm showing shortened leaves devoid of leaflets, probably through deficiency of boron (Malaysia).

half normal width and three-quarters normal length, without reduction in number of leaflets), advanced Little Leaf with extreme reduction of leaf area and bunching and reduction in number of leaflets, and Fish-bone Leaf (Plate 11.2). Much of the confusion that remains over boron deficiency symptoms is due to the uncritical use of the term 'Little Leaf'. The leaf reductions already described produce little leaves, but these must not be confused with Little Leaf as produced in Spear Rot-Little Leaf disease and which Robertson (1960) showed by simple surgery to be caused by a pathogen whose point of entry is the base of the spear leaf (p. 616 and Plate 13.5). Dufour and Quencez (1979) claim that in young palms the (unspecified) Little Leaf symptom is preceded by irregularly longitudinal, raised transparent spots on the leaflets.

In Colombia Ollagnier and Valverde (1968) found that symptoms including reduction in leaf size, malformations of the leaflets and the temporary halting of leaf production were arrested by boron applications. The leaf malformations included Hook Leaf and a more extreme form of Hook Leaf (*en baionette*); however, Hook Leaf is so commonly though sporadically seen that it is undoubtedly an error to attribute it to boron deficiency without other supporting evidence. Chlorosis, if it is a symptom at all, appears to be present only at certain stages of symptom development (Rajaratnam, 1972a) and there is no evidence that White Stripe (p. 596) is a primary symptom of boron deficiency as has been claimed (Tollenaar, 1969).

Iron

Cases of iron deficiency in the field have not been detected. In pot sand culture (Bull, 1961b) lack of iron is shown by a very marked chlorosis from pale green to pale yellow, spreading uniformly over the whole surface of the seedling's leaves. Necrosis then sets in at the tips, the leaves become yellowish-brown, growth is arrested and the plants die.

Manganese

Seedlings grown in sand culture without manganese assume a flattened appearance and the leaves become dull and pale yellowish-green in colour (Bull, 1961b). Later, a paler longitudinal chlorotic striping appears between the veins. These stripes necrose and become greyish-red. Apical splitting occurs and necrotic areas expand. According to Eschbach (1980), the hybrid *E. guineensis* × *E. oleifera* is more sensitive in hydroponic culture to lack of Mn than is *E. guineensis* itself.

Copper, zinc and molybdenum

Visual symptoms of molybdenum and copper deficiencies have not been obtained with seedlings in pot culture although plants grown without copper show a marked reduction in dry weight (Eschbach, 1980).

The condition known as Peat Yellows in Malaysia has been associated with low leaf levels of potassium, copper and zinc (Plate XV). In a study in Malaysia this condition was divided into Mid-Crown chlorosis, Peat Yellows and a combination of the two (Ng and Tan, 1974). In this case the Mid-Crown chlorosis was the symptom associated with very low levels of leaf copper. This symptom (not to be confused with Mid-Crown Yellowing of K deficiency) is characterized by a chlorosis on the tips of the leaflets of the younger leaves and is followed by necrosis or desiccation from the tip downwards. The chlorosis starts with the appearance of small yellowish green specks along the leaflets and these gradually coalesce. Very young palms can be affected.

Seedlings grown without zinc show no characteristic visual symptoms but growth is markedly reduced (Eschbach, 1980). No effects of zinc deficiency on growth and yield have ever been seen in the field, expect on peat soils in Malaysia (see p. 546).

Multiple deficiencies

In pot culture, where two elements are lacking, the deficiency symptoms of one of them is usually dominant (Smilde, 1961). Thus where nitrogen is deficient in combination with a deficiency of any other major element, the foliar symptoms are identical with those of N deficiency. Phosphorus deficiency symptoms are dominant when the deficiency is in combination with that of K, Ca or S. Magnesium deficiency symptoms dominate when in combination with lack of P and K, but very severe necrosis occurs in the latter case. Exceptionally, calcium produces its deficiency symptoms – short, narrow leaves with prominent veins – in combination with those of K and Mg and, to a lesser degree, of P.

Toxicity

Under plantation conditions oil palms have not shown symptoms of any general toxicity due to one element. Fertilizers, however, can be applied in toxic quantities if care is not taken. Sulphate of ammonia applied within 6 weeks of transplanting may damage seedlings seriously. The effect is usually seen on the youngest fully opened leaf which blackens off, but other leaves become dried up and the palm may die. It is clear that the initial effect in these cases has been on the root system, since similar symptoms follow the transplanting in dry weather of large seedlings with damaged root systems.

Cases of toxic effects following applications of boron to the soil have been known. Growth was severely checked and flowering inhibited. Toxicity symptoms have been induced in 18-month-old seedlings growing in polythene bags (Rajaratnam, 1973); they appeared when the boron level exceeded 250 ppm of dry weight in leaf 3. These symptoms were a terminal chlorosis progressing along the leaflet tips and margins, necrotic spotting on the chlorotic tissue and a drying out and shattering of the necrosed tissue.

The oil palm leaf is extremely sensitive to copper and the common copper fungicides cannot be used. Copper toxicity through the soil has not been encountered, however.

Leaf analysis

Leaf analysis for diagnostic purposes has been particularly useful for perennial crops which are relatively slow growing, can provide easily defined and standard leaf material for analysis, and whose responses to fertilizers will in any case take a longer time than will those of annual crops. One of the dangers of relying exclusively on leaf analysis is that it may fail to show in time the soil depletion that is taking place through poor cultural practices or other causes, and the detection of soil changes, with a crop such as the oil palm, may be an important factor in assuring satisfactory growth and yield in future crop cycles.

Leaf analysis has come to the fore in the last 40 years largely through the work of Lundegardh, and the oil palm has been among the handful of perennial crops on which much of the important work has been undertaken. Following the work of Hale (1947), already mentioned, and of Chapman and Gray (1949) in Malaysia, great progress was made through the investigations of Prevot and his co-workers using material of the Ivory Coast, Benin (Dahomey) and elsewhere (Prevot and Ollagnier, 1954, 1957, 1961). In much of the early work, attention was necessarily given to the leaf contents of the separate nutrients. It has been found with other crops that the levels of leaf nutrient content can be zoned into a sucession of ranges according to their effects. This can be seen in Fig. 11.2 where a growth or yield curve is constructed to show the changes of concentration of a mineral element through a gradual increase in the supply. These zones are as follows:

Zone A. Decrease in concentration with increased growth through initial increase in supply (the Steenbjerg effect).

Zone B. Growth response with *no change* in mineral concentration.

Zone C. Normal growth response with *increasing* mineral concentration reaching the 'critical level' or zone.

Zone D. No growth response, but continued *increasing* mineral concentration. The work of Foster and Chang suggests, however,

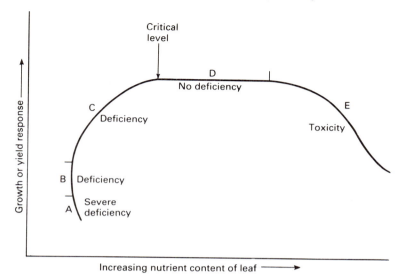

Fig. 11.2 Diagram of deficiency and toxicity in relation to leaf nutrient content and growth and yield (adapted from Drosdoff Prevot and Ollagnier 1954, and Smith, 1962). According to Foster and Chang (1977) the luxury uptake represented by D does not usually occur with the oil palm.

that this further increase does not usually occur with the oil palm; they found on inland soils in Malaysia that maximum yield and leaf nutrient levels were simultaneously reached.

Zone E. Increase of concentration with toxic effect.

Chapman and Gray (1949), who were the first to show correlations between both leaf-P and leaf-K and yield in oil palm fertilizer experiments, did not concern themselves with critical levels, but considered that the K_2O/P_2O_5 ratio in the leaf ash was of primary importance. They showed that where this ratio was low potassium increased yields, but where it was high, phosphorus increased yields. An important further observsation was that where the ratio was very low and potassium therefore extremely deficient, additions of phosphorus, in disturbing the balance still further, decreased yield. Moreover potassium applications were found to assist the uptake of nitrogen.

Prevot and Ollagnier (1954), defined the critical level as 'the concentration of the element in the leaf (dry matter basis) above which a yield response from the element in the fertilizer is unlikely to occur' and proposed the following critical levels for N, P and K in leaf 17: N – 2.70 (later revised to 2.50); P – 0.15; K – 1.00 per cent of dry weight. Later, Ca and Mg levels were added, viz. Ca – 0.60, Mg – 0.24 per cent, the basis for Ca level being only that a total of about 2 per cent for the sum of K + Ca + Mg had been observed. Correlations, both positive and negative, between the leaf contents of the element were shown to exist, and the relative constancy of the total leaf K + Ca + Mg content was noted.

It was soon to be seen that the critical levels given above did not have universal application since they could be modified by many other factors (Smith, 1962); and, in their later work, the French workers were at pains to point out that the critical level was useful as a first approximation and should not be used in a mechanical manner (Prevot and Ollagnier, 1957). The critical level is that at which the supply of the element is barely, but just, above the point of limiting growth or yield; when the critical level for element A has been reached, some other factor may then become limiting for growth or yield. This factor may be element B or it may be any environmental factor such as climatic conditions, or a cultural factor, e.g. drainage. Prevot and Ollagnier (1961) have described this factorial method of foliar diagnosis as being based on a modification of Liebig's Law of the Minimum, and they provided data which showed, for instance, that only when the leaf level of N had reached 2.70 per cent was a positive correlation to be found between K levels and yield, i.e.

$$N < 2.70\%. \quad \text{K: Yield correlation } r = +0.239 \text{ Insig.}$$
$$N > 2.70\%. \quad \text{K: Yield correlation } r = +0.643^{***}$$
$$\text{conversely,} \quad K < 1.1\%. \quad \text{N: Yield correlation } r = -0.042 \text{ Insig.}$$
$$K > 1.1\%. \quad \text{N: Yield correlation } r = +0.655^{***}$$

This, then, explains why responses to fertilizers applied to palms with a nutrient content below the supposed critical level are sometimes not found. In particular, little is known of the critical levels of minor elements, and where one of them is deficient, then normal responses to major elements may not occur.

Factors likely to affect critical levels even within narrow geographical and soil boundaries are the progenies used and the previous cropping. Significant differences have been found between progenies in West Cameroon and between fruit forms (*dura* and *tenera*) in Malaysia (Poon Yew Chin *et al.*, 1970a). In the latter case, however, it was pointed out that the differences might be the result of productivity differences varying the demand on nutrient reserves. The number of bunches produced by the *tenera* palms was 20 per cent higher than by the *dura* palms.

In areas where the environmental circumstances encourage high production, critical levels may easily be set too low. Initial experimentation may show a response to one element alone following an obviously low leaf content of that element, and the levels of the other elements may be thought to be at or above the critical level.

When the deficiency is satisfied, however, either the level of another element may fall below the believed critical level or it may remain stationary or rise, but nevertheless be deficient for the new production level. The usual effects of applications of the major elements on leaf composition are now known (Table 11.5). An effect is described as antagonistic when the leaf content of an element is reduced by the application of another element and synergistic when application increases the leaf content of the element concerned. The latter phenomenon is not common in the oil palm, though Chapman and Gray (1949) and Chan (1981) reported an increase of leaf-N associated with K applications, and increases in P following N applications have also been reported. Antagonisms are common, the most important being between K and Mg. Ng (1972) has pointed out that effects are often not easy to interpret because fertilizers may contain another anion or cation which also has an effect. Thus the presence of calcium in phosphorus fertilizers suppresses potassium, while a rather longer-term effect would be the increased leaching of potassium or magnesium, and hence their reduced uptake, following application of ammonium sulphate or fertilizers in the form of nitrate or chloride salts.

Table 11.5 *Usual effects of applied elements on the mineral composition of the oil palm leaf*

Element added	Element measured in the leaves				
	N	P	K	Ca	Mg
N	+	+	—	0	—
P	0[†]	+	—	+	0
K	+	0*	+	—	—
Ca	0	0[†]	—	+	0
Mg	0	—	—	—	+

* Both increases and decreases have been reported.
[†] Synergism has been reported.
+ = Increase. — = Decrease. 0 = No marked effect.

In recent years the relationships between the elements in the nutrient composition of leaves has been examined statistically by computing partial correlation coefficients (calculated for each pair of nutrients, all others being held constant) and by principal component analysis. Using the latter method Holland (1967) showed, from the data of a factorial fertilizer experiment, that some 80 per cent of the total variation could be attributed to antagonism between K on the one hand and Ca and Mg on the other, and synergism between N and P. In a foliar uniformity trial in Malaysia Poon *et al.* (1970b) showed the same relationships through the computation of partial correlation coefficients from the data of a

uniformity trial, although additionally there was a significant synergism between P and K. It is thus possible that the latter is masked
in fertilizer experiments by the opposing effect of calcium in the
phosphorus fertilizers.

Macronutrients

Under each set of environmental circumstances the critical levels for
leaf nutrients should be related to the results of fertilizer experiments. Deductions made from leaf analysis data not so based may
be misleading. French workers estimated their critical levels from
analyses of palms in fertilizer experiments and in areas of obvious
deficiency or health, as well as by indirect methods such as the
comparison of effects, under different conditions, of the supply of
one element on the level of another (IRHO, 1962). Special attention was given to potassium deficiency and the relation of K levels
to the supply of phosphorus and nitrogen.

It was shown that where the leaf-P content in leaf 17 was less
than 0.15 per cent of dry matter there was no correlation between
K content and yield, but that when phosphorus deficiency was

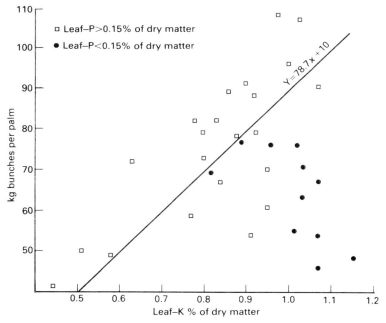

Fig. 11.3 The relation between yield and leaf-K levels with leaf-P content above
and below 0.15 per cent of dry matter. Correlations: yield/leaf-K with P<0.15 per
cent, insignificant; yield/leaf-K with P>0.15 per cent, $r = 0.739***$ (Ochs, 1965).

rectified and the leaf-P content brought above 0.15 per cent, a strong correlation existed and the highest yields were obtained in the leaf-K range 0.9–1.1 per cent (Fig. 11.3). Similarly, where K was below 1.0 per cent there was no leaf-P/yield correlation but where leaf-K was above 1.0 per cent there was a strong correlation between leaf-P and yield (Ochs, 1965). The N–K relationship has already been mentioned (p. 498).

In considering the leaf levels of N and P, Ollagnier and Ochs (1981) examined the results of fertilizer experiments in all three continents and concluded that, for phosphorus, there is a critical line which is a function of the N level, while for nitrogen critical levels depend very largely on palm age. The combination of these two concepts is illustrated in Fig. 11.4.

No satisfactory critical level or line for magnesium has been determined. The early suggestion of 0.24 per cent of dry matter for leaf 17 is quite inapplicable on young volcanic soils such as those of Ecuador or Papua New Guinea. In Nigeria the strong antagonism between K and Mg led to very high leaf-K values being recorded on Calabar soils of very low magnesium status, with figures of 1.62 per cent for leaf-K and 0.07 per cent for leaf-Mg in unfertilized plots. By contrast, on magnesium-deficient Acid Sands soils where yield responses to both K and Mg were obtained, leaf-Mg could be raised to 0.30 per cent with leaf-K in the range 0.73–0.90 per cent (Smilde, 1963). In Papua New Guinea, on volcanic ash soils, leaf

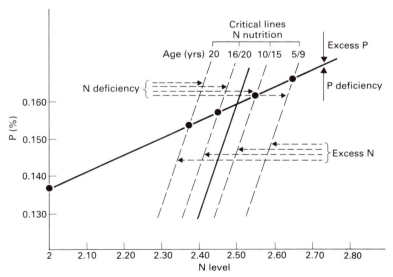

Fig. 11.4 Critical standards for nitrogen and phosphorus nutrition (Ollagnier and Ochs, 1981).

levels of both Mg and K are low, but applications of Mg have had little effect on yield (Breure and Rosenquist, 1977); and in Ecuador, in spite of deficiency symptoms and very low leaf-Mg levels (0.13–0.14 per cent), high yields have been recorded without responses to Mg. Ollagnier and Ochs (1981) consider that in these circumstances the low leaf content of both Mg and K is due to the high level of exchangeable calcium in the soil which is reflected in unusually high Ca leaf levels.

A comprehensive study in Malaysia of leaf levels in relation to yield was undertaken by Foster and Chang (1977) working on the data of ten coastal and ten inland-soil fertilizer experiments. They found that the leaf levels for optimum yield of any specific nutrient differed with the presence or absence of fertilizers providing other elements. For instance, a bunch yield of 24 tonnes per hectare was obtained in one experiment in the following circumstances:

At zero levels of P and Mg: Leaf-N 2.82–2.92% Leaf-K 1.07–1.36%
At optimum levels of P and Mg: Leaf-N 2.72–2.89% Leaf-K 0.83–1.07%

Thus yields can only be related to the leaf level of a specified nutrient if other nutrients are kept constant, and any critical levels proposed must at least assume that the optimum requirement for other elements has been met. Using response functions fitted to bunch yield and leaf nutrient plot data, Foster and Chang found that predicted leaf nutrient levels at optimum fertilizer rates for yield were not the same for inland and for coastal soils. Table 11.6 shows the mean of these predicted levels and the critical ranges they deduced from this work. The question of why there should be such markedly differing levels for the two groups of soils has not been fully resolved. Ng (1970) suggested that this was because of differences of moisture status since coastal soils normally have a more regular water supply and this might result in more efficient utiliz-

Table 11.6　*Means of predicted nutrient levels (per cent of dry matter) at optimum rates of all fertilizers, and the range of critical levels for two groups of soils in Malaysia, Leaf 17*

Leaf nutrient	N	P	K	Mg	Ca
Coastal soil trials (10)	2.59	0.169	0.94	0.29	0.45
Inland soil trials (10)	2.87	0.177	1.11	0.26	0.74
Range of critical levels					
Coastal soils	2.55–2.65	0.165–0.175	0.85–1.05	—	—
Inland soils	2.85–2.95	0.175–0.185	1.10–1.15	—	—

Source: From Foster and Chang (1977).

ation of nutrients which would require lower levels. Foster and Chang (1977), however, have pointed to the marked difference of soil calcium status and suggest that as leaf-Ca is lower on coastal soils, requirements for K and Mg will also be less, since a balance between the nutrients is likely to be more important than the actual levels. Later work in Malaysia (Foster *et al.*, 1987b) showed that on all soils the optimum levels of leaf-N and P increased with annual rainfall, while on inland soils leaf-N optimum levels increased with the silt content of the soil. Age was also found to influence N levels while planting density influenced P levels.

The claim that chlorine is an essential element in the nutrition of the oil palm is based largely on the interpretation of leaf analysis data (Ollagnier and Ochs, 1971; Ollagnier, 1973a). In Colombia applications of KCl, while giving small increases in bunch yield, actually decreased leaf-K and increased leaf-Cl from 0.18 to 0.54 per cent of dry matter. Following a world survey of leaf chlorine levels it was suggested that 0.1–0.3 per cent must be considered the acute deficiency range, with 0.5–0.6 per cent being regarded as satisfactory. Applications of either KCl or MgCl to palms of high Cl status is held to increase leaf-Ca as well as leaf-Cl and, through antagonism, to reduce K levels. There has been support for this thesis in Papua New Guinea where applications of KCl increased leaf-Cl but reduced leaf-K in an area where leaf-Cl was below 0.20 per cent (Breure and Rosenquist, 1977). Ollagnier (1973b) has suggested a critical leaf-Cl level of 0.50 per cent. Levels below 0.40 per cent are commonly found in Peru, Colombia, Ecuador and northern Sumatra, and it has been recommended that in these areas the chlorides of K and Mg should be employed.

Some evidence of sulphur deficiencies has come, through leaf analysis, from the Ivory Coast and Papua New Guinea. In an experiment in the Ivory Coast with young palms the length of leaf 4 was found to be correlated with the S level in the leaves though not with the Mg level in spite of the fact that both the S levels and Mg levels were raised by kieserite application (Ollagnier and Ochs, 1972; Ollagnier, 1973a). On the basis of this finding a critical range of 0.20–0.23 per cent of dry matter in leaf 17 was proposed.

Evidence that in degraded Tertiary sands in the Ivory Coast sulphur deficiency is of importance has been provided by Calvez *et al.* (1976). They found that when the same quantities of N were supplied by urea and sulphate of ammonia, applications of the latter resulted in higher leaf-S levels, improved early growth and dry matter production, and a reduced incidence of *Cercospora* Leaf Spot. Leaf-S levels reached 0.208 per cent of dry matter in the palms treated with sulphate of ammonia compared with 0.181 and 0.185 per cent in the control and urea palms.

In Papua New Guinea, where leaf-17 S levels were initially

0.161–0.168 per cent, industrial sulphur applications had only a small effect on leaf levels but significantly increased leaf production when supplied with manganese (Breure and Rosenquist, 1977).

Micronutrients

In the Ivory Coast no deficiency symptoms were detected with leaf boron levels of 6 ppm of dry matter or less; in Colombia, however, with more rapid growth, Little Leaf symptoms were common in young palms at this level and were reduced by soil application in rings of 30 g borax per palm (Ollagnier and Valverde, 1968). Application in the axils of three leaves above the zone of developing bunches was shown to have larger immediate effects on leaf-B, though this effect was short-lived. In Sumatra symptoms have been reported with leaf-B levels as low as 4–5 ppm but it was pointed out that such levels could also be encountered on the palms not showing foliar symptoms (Purba and Turner, 1973). In most areas leaf-B levels of 10–20 ppm are encountered.

In a study of the mobility of boron within the palm Rajaratnam (1972c) has shown that it does not move from leaves to the shoot but tends to move to the extremity of the leaflet and is not retranslocated. Some boron is lost by guttation. This work indicates that soil application is likely to be the best method since axillary application will not allow direct translocation to the shoot or younger leaves.

Manganese leaf-content levels vary greatly and, though leaf levels can be raised, experiments have not shown any yield responses to this element. Concentrations in leaf dry matter vary from below 100 to around 1,000 ppm (Eschbach, 1980).

Molybdenum levels are low, usually less than 1 ppm. Attempts to raise the N level of the oil palm by application of Mo to the legume cover have failed (Eschbach, 1980). Iron levels range from around 60 to over 400 ppm. Soil applications have never had any effect on leaf levels or yield.

Leaf-copper is usually about 6–15 ppm of dry matter. Some indications of the leaf levels required for satisfactory growth and for the elimination of deficiency symptoms (Mid-Crown chlorosis) were obtained from experiments on peat in Malaysia (Ng *et al.*, 1974). With mature palms soil applications of copper sulphate raised leaf 17 levels from 1.4–2.1 ppm to 2–5 ppm and increased leaf length; with young palms foliar sprays of copper sulphate in 200 ppm Cu solution improved the colour of the leaves, increased their length in comparision with untreated palms and gave leaf-3 copper levels of 3.7 and 2.3 ppm at 3 and 6 months after treatment against levels of 2.8 and 1.8 ppm for untreated palms. It is not possible to gauge a critical level for copper from these results because there is a gap between the normal levels of 6–15 ppm prevailing on non-peat soils and the highest level to which Cu content has been raised by Cu application on peat soils. Cheong and Ng (1977) showed that, on peat, Cu deficiency should be rectified at a very young stage in the field. Although correction is still possible up to 6 years of age, older palms showing severe chronic symptoms cannot usually be cured.

The leaf levels of zinc lie between 9 and 40 ppm for leaf 17. Applications of zinc sulphate can raise leaf levels but, except on peat, have no effect on growth or yield (Eschbach, 1980). Application of both Zn and B increase leaf-P (Ataga *et al.*, 1981).

Leaf-nutrient content variation

Early workers made a deliberate choice of either the first fully opened leaf (usually designated leaf 1), leaf 9, leaf 17 or a leaf making an angle of 45° with the horizontal. The choice of leaf 17 was based on convenience and reasoning but not on any exhaustive sampling trials. It was argued that this leaf lay conveniently on the easily recognized spiral in the succession 1, 9, 17, 25, of the phyllotaxis and was in the middle of the foliage, the lower part of the rachis making an angle of about 45°. This leaf was fully developed, not yet senescent and carried an inflorescence in its axil. In some, usually older, palms however, leaf 17 is hard to identify or is senescent, and in such a case a leaf round about the leaf 17 position, with its lower rachis making an angle of 45°, was often chosen; or recourse was had to the easily identified leaf 9.

Chapman and Gray (1949) sampled the central 10 cm of the central two pairs of leaflets. With this material, taken from a fertilizer experiment, they found that there was a downward trend in the percentage P_2O_5 and K_2O of leaf ash with age of leaf, but that with N there was a rise to leaf 5 followed by a decrease with age. From the control plot and the plot receiving the highest PK dressings it appeared that malnutrition showed its greatest effect on the ash composition of leaves of medium age; moreover, although there were no increases in yield due to fertilizers in the experiment except from PK in combination, correlations were found between the nutrient contents of the leaf ash and palm yields. These correlations were high and significant for all four leaves chosen (leaves 1, 9, 17 and 25) for N per cent, but for P_2O_5 and K_2O the correlations were high and significant for leaves 17 and 25 but not significant in the younger leaves. It was largely on this work that the use of leaf 17 was based.

Calculation of leaf nutrient content on a dry matter rather than an ash basis has now become accepted practice. Prevot found decreasing concentrations of N, P and K and increasing concentrations of Ca and Mg with increased age of leaf, though N increased slightly in the younger leaves (Scheidecker and Prevot, 1954; Prevot and Montbreton, 1958). The variations in Mg were very small. Small significant differences were found between morning and evening sampling. Many workers followed Chapman and Gray in selecting two pairs of leaflets in the centre of the leaf and taking the centre 10 cm of these leaflets for analysis. Gradients of composition within the leaf have been variable, however, and no consistent trend has emerged. Hale took leaflets on alternate sides of the rachis at intervals of ten leaflets and used the whole lamina.

The first comprehensive investigations of the errors of leaf sampling were those of Smilde and Chapas (1963) and Smilde and Leyritz (1965). Using leaves 1, 17 and 25 of thirty-two 15-year-old palms in Nigeria they showed that there were significant differences between palms, age of leaf and month of sampling in leaf N, P, K, Ca and Mg. In some cases there were significant interactions between palms and months and months and age of leaf, though not between age of leaf and palms. The results showed that different weight must be attached to the results of analyses of different elements when the same number of palms are used for the analysis of each element. Table 11.7 gives estimates from 2 years' data of the minimum number of palms required in a uniform area of palms such as that investigated in Nigeria to make it possible to detect differences exceeding 5, 10 and 20 per cent of the mean nutrient level at the 5 per cent level of significance; to obtain a proper comparison, the formula $n = 2s^2t^2/D$, where s is the coefficient of variation, t is the reading in the t table for $P = 0.05$, and D is the difference (per cent) of the mean nutrient level to be detected, has been applied to the data of both years.

The main conclusions drawn from this work were that the necessary degree of sampling may be expected to be about the same from one year to another and that, while the seventeenth leaf is the

Table 11.7 *Estimation from two sets of data from thirty-two palms of the extent of leaf sampling necessary to detect significant differences exceeding 5, 10 and 20 per cent of the mean nutrient level (P = 0.05) (Nigeria) (number of palms to sample)*

Nutrient measured	N		P		K		Mg		Ca	
Difference of the mean nutrient level to be detected:	1960–1	1962–3	1960–1	1962–3	1960–1	1962–3	1960–1	1962–3	1960–1	19
17th leaf sample										
20%	—	1	—	1	—	14	—	9	—	
10%	4	3	4	3	56	54	40	33	16	3
5%	12	10	18	11	224	212	156	131	60	11
1st leaf sample										
10%	10	—	8	—	18	—	28	—	28	—
5%	40	—	32	—	68	—	108	—	110	—

most satisfactory for N and P determinations and has no disadvantages for Ca, the first leaf gives less variation for Mg and, more especially, for K. Most workers will feel that sampling must be standardized on one leaf and in this case it could be argued that the first leaf should be used since a lesser number of leaves overall would be required. On the other hand the seventeenth leaf has such considerable precision for N and P that its use is justified except where specially precise investigations on K, and to a lesser degree Mg, are being carried out. It was also concluded from this work that in fertilizer trials all palms in the plots must be sampled; moreover, as plots rarely contain more than sixteen palms an accuracy of 10 per cent of mean values will not be achieved for K, Mg and Ca and differences of less than about 20 per cent of the mean values for these elements cannot be regarded as significant, even when all palms in each plot are sampled. For advisory or confirmatory purposes Smilde and Leyritz recommended the taking of composite leaf-17 samples of 10 per cent (30) of the palms in 2 hectare homogeneous blocks to give results accurate to within 20 per cent of mean values for K, 10–20 per cent for Mg and Ca, and about 5 per cent for N and P.

Other studies of leaf nutrient variation were made on plantations in both Nigeria and Malaysia. In the former (Ward, 1966a), the first fully opened leaves of 512 palms in an area of about sixteen times that number of palms were sampled and analysed for N, P, K, Ca, Mg, Mn and B. Thus a much larger number of palms and a different leaf (leaf 1) was used than in the sampling of Smilde and Leyritz. The investigation did not cover leaf age or seasonal factors. The results (Ward, 1966b) are compared in Table 11.8. Variation of leaf-

Table 11.8 *Comparison of deductions from the results of three leaf samplings*
Minimum number of palms to sample in order to detect differences of 5 and 10 per cent of the mean nutrient level with a 5 per cent level of significance (confidence interval)

Locality	Nigeria*		Malaysia[†]				Nigeria[‡]	
Leaf	17		17		3		1	
Differences of mean	10%	5%	10%	5%	10%	5%	10%	5%
N	3	10	4	14	5	17	6	24
P	3	11	4	13	5	17	9	35
K	54	212	22	85	10	39	8	31
Mg	33	131	26	104	22	88	24	94
Ca	30	117	34	137	41	161	615	2,462

* From Smilde and Leyritz (1965)
[†] From Poon *et al.* (1970b)
[‡] From Ward (1966b)

N and P was low in both studies, while leaf-K variation was lower in the study using leaf 1 and leaf-Ca variation much higher. In spite of the fact that the overall results (except for Ca) of the two studies were not markedly different the conclusions drawn were not the same, Ward considering that a 1 per cent intensity of random sampling would give sufficiently accurate results for all major nutrients except calcium. The results of Poon *et al.* (1970) in Malaysia were remarkably similar for leaf 17 to those of Smilde and Leyritz (1965) except for potassium, the variation of which appears to be considerably less in Malaysia (Table 11.8). Ward's leaf 1 corresponds to the Malaysian leaf 3 and again the results are not dissimilar except for the anomalous calcium.

Ng and Walters (1969) have compared coefficients of variation from three sites on coastal clays in Malaysia and found these to be as follows:

	Coefficients of variation (per cent)				
	N	P	K	Mg	Ca
Site J*	12.9	14.3	14.1	27.1	22.3
Site B	6.6	7.5	16.1	19.5	21.4
Site SD	4.9	7.0	13.5	20.1	24.1

* Estimated from composite fifteen palm samples.

Using these coefficients they calculated the sample sizes needed for the different nutrients to detect differences of 5 and 10 per cent using three different sampling methods: (1) a compact block within the area, (2) random sampling, (3) random groups of nine palms. The lowest number of palms is required under the completely random system (2), but system (3) only increased the number of palms required by some 50 per cent and the number of points (number of palms divided by 9) was very much reduced. This system was considered to combine practical convenience with satisfactory precision.

The optimal size of the area to be sampled clearly depends on the terrain and soil variations and may lie between 4 and 5 hectares. Poon *et al.* (1970b) recommended the use of 4-hectare sampling units for a start, with the combining of these areas into larger units if the analysis results indicate that this is feasible. Ochs and Olivin (1977) recommended the combining of pairs of their standard 25-hectare fields into 50-hectare sampling units, and combining these into 100-hectare sampling blocks if the latter are found to be homogeneous. This is a very low sampling intensity and the statistical basis for the recommendation has not been given. Assuming a sampling area of 10 hectares and using Ng and Walter's system (3) above, the number of palms required for leaf-17 sampling would vary according to locality from 2 to 5 per cent of all palms to achieve for potassium a precision within 10 per cent of the mean. Five random points of nine palms, as recommended by Ng (1972), is 3.1 per cent of 10 hectares. It is important to realize that with sampling of this order differences of less than 10 per cent are unlikely to be signficant with K and Mg though they may be signficant with N and P and the viewing of analytical results without these statistical considerations in mind may well lead to unjustified conclusions.

A number of sampling methods are currently in use on plantations. The system of taking a block within an area is obviously not to be recommended. The sampling of the tenth palm in every tenth row, giving a 1 per cent sample, is quite common. For this system the unit area must be at least 18 hectares, since not less than twenty-five palms per unit should be sampled. In West Africa, a twenty-five-palm sample, taken from every other palm in a selected pair of rows per 50 or 100 hectares, is used (Ochs and Olivin, 1977). For many areas such low percentage sampling will be inadequate, however, and the nine-point random group system of Ng (1972) described above is more suitable and can be used for areas of 10 hectares.

Season and time of day

The detailed study in Nigeria indicated, as might be expected, that the beginning of the wet season is a bad time for sampling particularly on account of the nitrogen flush at that time. Leaving this period out of account there is no time of lowest palm-to-palm variability for nitrogen (Smilde and Chapas, 1963). Differences in the variability of other elements during the year were marginal, though there was some tendency for leaf-K to be less variable in leaf 1 in April to September than in other months and less variable in leaf 17 in October to January. Magnesium levels may vary rather widely between months, but no consistent pattern has emerged from year to year. It was concluded on the basis of 2 years' results that sampling for foliar analysis can be carried out in West Africa at any time of year except during the April to June period of nitrogen flush. Ochs and Olivin (1977), however, prefer to use the dry season only as, they claim, leaf levels are more stable and K deficiency is best exhibited at that time.

It is usually recommended in Malaysia that the periods of drought or very heavy rain should be avoided and that 6 months, or an absolute minimum of 3 months, should elapse between fertilizer application and leaf sampling (Ng, 1972). For each area it is also best to keep to the same sampling period each year if annual analysis is to be undertaken. Foster and Chang (1977) found low coefficients of variation for seasonal fluctuations varying from 3 to 4 per cent for N and P and 6 to 11 per cent for K, Mg and Ca. They also detected and influence of soil moisture in the month before sampling: in coastal areas P and K leaf levels were as much as 5 and 9 per cent higher after a very wet month than after a dry month. This relationship was not, however, maintained for K on inland soils. Teoh *et al.* (1981) also found seasonal fluctuations to be small but recommended May–July sampling for inland Malaysian soils and May–August for coastal alluvial soils.

It is sometimes recommended that sampling should only be done between 6.30 a.m. and noon. Smilde and Chapas (1963) considered this to have little justification since the small differences found, besides being insignificant, are unlikely to be of a greater order than day-to-day differences, and the sampling programme always covers many days.

Leaf number (age of leaf)

The leaf concentration of N, P and K decreases with the age of the leaf, with leaf-Mg there is no clear-cut trend, but leaf-Ca increases with age. Thus the K + Mg + Ca sum is likely to be constant at about 2 per cent unless there are any marked deficiencies. Leaves younger than No. 17 are now rarely used since a leaf-17 sample can usually be obtained in areas of satisfactory growth by the age of 3 years.

Age of palm

Leaf analysis is likely to be most useful with bearing palms at the commencement or in the middle of their bearing life. Broeshart (1955) compared analyses of very young seedling palms with those of older palms growing in the same field and which had reached the bearing stage; he found little difference with leaves 1 and 9 except in the case of magnesium of which the young palm has a lower leaf content. It has been general experience, however, that leaf levels of certain nutrients tend to decline with age even where liberal quantities of fertilizers have been supplied. Such a decline is common with K and N, usual with Mg, but not common with P or Cl; some increases of Ca have been reported (Knecht *et al.*, 1977). Declines of N and Mg can occur at quite an early stage; K declines have been more pronounced on some soils than others. In the twenty trials examined by Foster and Chang (1977) declines with age were only found with N, P, Mg and Ca on coastal soils. On inland soils there was some tendency for P to increase; and there was no decline in K levels. These results suggest that the decline of K with age which is often seen may be due to insufficient application rather than age itself. It is also clear that in Asia changes of nutrient leaf levels with age depend to a great extent on soil type.

In West Africa Bachy (1964) found declines in K contents with age in all three localities investigated; N and Mg declines were found in one area and an increase of P also in one area only.

The leaf analysis of very old palms may show some peculiar results. For instance, in Nigeria, 38-year-old palms giving a 264 per cent increase in yield as a result of K dressings applied 12 years previously showed leaf analyses as follows (per cent of dry matter):

Treatment	Yield 1960–1 (%)	K (%)	Mg (%)
K_0	100	0.37	0.35
K_1	218	0.62	0.26
K_2	264	0.77	0.25

Sulphate of ammonia depressed yields, though N contents were as follows: N_0 – 2.08 per cent, N_2 – 2.03 per cent (Smilde, 1963).

Environment in relation to yield and critical levels

Such results as those quoted above and anomalous results from other parts of the world have long suggested that critical levels may be influenced by environmental factors and not simply by palm and leaf age, the supply of other nutrients, etc. Ruer (1966) has used the effective sunshine factor of Sparnaaij *et al.* (1963) (see p. 181) to demonstrate that when climatic conditions are unfavourable the correlation between the leaf-K content and yield becomes insignificant. It was shown, for instance, at La Mé in the Ivory Coast that when effective sunshine over a period of 12 months from 1 March to 28 February was low (below 1,620 hours) there were no signficant correlations between the yield in the 12-month period 28 months later and the K or N leaf contents determined 2, 1 or 0 years previously; whereas with effective sunshine above 1,620 hours significant correlations were usually found, particularly between yield and the K or N leaf contents 2 years previously. Through the fitting of yield/leaf-K curves to the data obtained for years of high and average effective sunshine, it was also estimated that much larger yield responses to high leaf-K contents would be obtained in yield years which corresponded to the years of high effective sunshine.

From these results Ruer (1966) argued that in climates of even rainfall with regularly high effective sunshine, the critical levels will be higher than where these conditions do not obtain, and he suggested leaf levels of 0.7–0.8 per cent K for a dry climate such as Benin (Dahomey), with levels of 1.0 per cent for the Ivory Coast and 1.2–1.3 per cent for even rainfall climates. Later work has shown, however, that no such simple relationship exists through the whole range of water deficits or effective sunshine hours (Ollagnier *et al.*, 1987). While a reduction of water deficit from a very high level of 600 to 250 mm does indeed lead to an increase in the potassium critical level in a dry country like Benin, below 250 mm the critical level tends to fall. Even this relationship cannot, however, be directly transferred to countries where low water deficits are usual and where leaf-K levels vary considerably between soil types.

Table 11.9 gives examples of leaf analysis data from a number of regions and illustrates the leaf-content percentages typical of certain deficiencies. As will be seen from the examples of Ecuador and Papua New Guinea, leaf levels on volcanic soils may be markedly anomalous. Quite apart from the very low Mg levels already mentioned (p. 502) and thought to be due to the high exchangeable Ca status of the soils, it is often difficult to raise the leaf dry matter content of either P or K to levels normally encountered elsewhere.

Relation of yield to leaf nutrient levels

A greater study has been made of leaf-potassium levels than of those of any other nutrient and some attempts have been made to judge, in the case of seriously deficient palms, the yield increments that may be expected to result from the raising of the leaf-K level by a given amount. In the study of nine sets of experimental data Ochs (1965) found that yield was raised by various proportions from 4 to 12 per cent for every gain of 0.1 per cent in the leaf-K level; he therefore concluded that other uncontrolled factors made it impossible to relate yield with increases in leaf-K with any precision, but he put forward an approximate average increment of 10 per cent in the yield for a rise of 0.1 per cent in leaf-K. This corresponds with a correlation found in a field in Nigeria between individual palm yields and their leaf-K content (Forde *et al.*, 1965). In this case the relationship appeared to hold good with palms showing leaf-K contents up to 1.10 per cent. Correlations between yield and leaf-N and leaf-P have already been referred to (pp. 498 and 501); however, no attempt was made to attribute any generally applicable percentage rise in yield with any given rises in the leaf levels of these nutrients. Prevot and Ollagnier (1961) found overall significant positive correlations between N, P and K and yield, but showed that these needed special interpretation owing to the existence of interactions, i.e. in all cases where elements x or y or both x and y are below certain levels, no correlation is found between the level of z and yield. This is illustrated for K from Ochs's work in Fig. 11.3.

From their examination of twenty fertilizer trials in Malaysia Foster and Chang (1977) have concluded that an increase in bunch yield of 1 tonne per hectare is accompanied by increases in leaf nutrient levels of 0.03–0.05 per cent of dry matter for N, 0.003–0.005 per cent for P and 0.05–0.08 per cent for K. In Asia

Table 11.9 *Some leaf analysis data of the oil palm (percentage of dry matter in leaf 17)*

Country	Locality or soil	Leaf nutrient content						Notes
		N (%)	P (%)	K (%)	Ca (%)	Mg (%)	Cl (%)	
1. Malaysia	Coastal alluvium	2.79	0.185	1.13	0.54	0.33		No apparent deficiency
	Muck soil	2.47	0.113	1.87	0.31	0.12		P and Mg deficiencies. K high through antagonism. N marginal
	Rengam series	2.72	0.172	1.20	0.88	0.21		K fertilizer applied
		2.66	0.172	1.08	0.84	0.27		K Mg fertilizer applied
		2.72	0.177	0.90	0.97	0.25		No fertilizer
2. Benin (Dahomey)	Pobé	1.93	0.149	0.43	0.81	0.68		Severe K and N deficiency
3. Nigeria	Benin. Acid Sands	2.82	0.20	1.24	0.72	0.44		Leaf 1 $\}$ K
	Sands	2.55	0.15	0.69	0.92	0.40		Leaf 17 $\}$ deficient
	Calabar	–	–	1.62	0.50	0.07		Severe Mg deficiency
	Benin	2.29	0.13	0.81	1.20	0.31		P_0 plots $\}$ P effect
		2.28	0.14	0.68	1.33	0.31		P_2 plots $\}$ on K
4. Colombia	Alluvium	1.68	0.13	0.67	0.69	0.31		Extreme NPK deficiency: grass competition
5. Ecuador	Pacific Coastal plain: Volcanic	2.50	0.145	0.72	0.98	0.12	0.337	Mg deficiency symptoms
6. Papua New Guinea	West New Britain: andosols	2.49	0.158	0.81	0.92	0.19	0.34	K_0 plots
		2.47	0.156	0.75	1.05	0.18	0.56	K_3 plots

1 tonne may represent a 4–5 per cent yield increase, so that in the case of potassium a yield increase of 8–10 per cent might be accompanied by an increase of 0.1–0.16 per cent in the leaf-K level; this agrees closely with the suggestion of Ochs referred to above. Foster has concluded that, provided factors affecting critical levels are taken into account, leaf analysis is a useful guide to the nutrient status of the palm, and that research directed towards improving the knowledge of critical-level variations will be most productive.

The precise value of leaf analysis with the oil palm has been much debated in recent years. Its ability to show gross deficiencies is not disputed, and as a tool for diagnosing these deficiencies or obvious imbalances it is of undoubted utility. Many plantations, applying substantial annual fertilizer dressings, use annual leaf analysis for monitoring the effect of these dressings on leaf nutrient changes. While this may lead to the detection of some gross errors of fertilizer practice leading to serious leaf nutrient imbalance and consequent lowering of yield, there is a dangerous tendency to deduce more from small changes than can possibly be justified when regard is had for the variability of the material and the statistical considerations which should be taken into account (p. 506). Moreover many growers expect that annual leaf-nutrient reviews can, with expert examination, provided the correct types and quantities of fertilizers to be applied in that year for maximum production. Green (1974) has rightly tried to dispel the idea that these experts exist and concludes that 'the unpalatable but inescapable truth is that in the present state of our knowledge once nutrient levels are somewhere around the believed optima, leaf analysis of itself offers no guide to the quantities of fertilizer that ought to be applied' and 'can only tell us, and then retrospectively, when we have gone wrong again and a faulty policy has re-created imbalances or induced new deficiencies'. It is important to note the words 'of itself' in the above sentence since it is in conjunction with the results of fertilizer experiments that leaf analysis data can be most intelligently viewed; and Green's dictum that 'interpretation by experienced and competent advisers is almost always based on their wide familiarity with the subject than on any quantitative interpretation of the data' may be taken to embrace familiarity with relevant fertilizer experiment results.

Sampling practice and analysis

Two methods are commonly used for sampling leaf 17. In one, after the leaf has been cut down, it is cut into three equal parts and six leaflets are taken from each side of the middle section; of these twelve leaflets six will by upper-rank and six lower-rank leaflets (Poon, 1969). Upper-rank leaflets nearly always contain more K and less Mg than lower-rank leaflets (Bull, 1960). The leaflets are bulked from each sampling unit and placed in a polythene bag. Another method is to take leaflets from the entire length of the leaf on alternate sides of the rachis at intervals of about ten leaflets (Smilde and Chapas, 1963).

Leaflet samples should be prepared within 24 hours. The leaflets are first thoroughly cleaned with damp cotton wool. The midribs are then removed and the marginal 2 mm of the laminae cut off (Martin, 1977). The taking of the middle 20–30 cm has been advocated but there is slightly less between-palm variation when the whole lamina is used. The laminae are cut into small pieces and dried for about 5 hours at 65–70 °C in an air-draught oven. This does not remove all moisture but dries the samples sufficiently for storage or despatch but not so far that alternative methods of analysis cannot still be employed. The samples are placed in polythene bags and hand-crushed. The separate retention of one side of each leaflet sample has been advocated so as to guard against loss in transit or anomalous analysis results (Martin, 1977).

There are now many analytical laboratories dealing with oil palm leaf material and an outline of the modern methods most commonly used for oil palm leaves has been provided by Poon (1969).

Soil analysis

There is a persistent demand by plantation companies, development organizations and others for laboratories to decide from the results of the analyses of soil samples whether an area is suitable for oil palms or not, and what fertilizers should be applied. Soil classification in feasibility studies has been discussed in Chapter 3 (pp. 114 and 134) where it was seen that it is primarily the physical properties of the soil which determine suitability. Within a particular climatic environment it will be these properties also which will to a great extent determine the general level of yield. Here, however, we are concerned with the determination of the nutrient-supplying power of the soil and its fertilizer needs; and it must be admitted that soil analysis has rarely been of primary value in this respect. There is no set of quantitative standards of soil nutrients comparable to the admittedly imperfect and fluctuating critical levels in the leaf. Some typical profile analyses were given in Chapter 3 and it is difficult to discern from them even the most general relationship between soil nutrient levels and fertility. Soil chemists themselves are not slow to point out their inability to relate fertility to analysis. In the case of the Sabah soils now under oil palm cultivation, Paton (1963) has written:

> If the analytical figures alone were available the conclusion would be reached that in the soils of Group A there are a number of families of considerable agricultural potential while the soils of the B Group would

be considered of very little value. The situation is profoundly altered by the fact that the Table family of soils, in Group B, is actually extremely productive. . . .

The great difficulty of correlating soil analysis with 'fertility' is due to the fact that (i) roots extract nutrients from three sources (Russel, 1961), i.e. the soil solution, the exchangeable ions and the readily decomposable minerals of the soil, which are interrelated in a complex manner, (ii) crops differ considerably in their needs and in their power to extract nutrients from these three sources, (iii) a measure of any of these sources may not indicate the real availability of nutrients for any particular crop, since soils differ in the rate at which they release non-exchangeable ions, (iv) the rooting volume available to the plant varies considerably and (v) the physical condition of the soil affects the power of the root system to exploit it. It should not be surprising therefore that in areas where there are abrupt changes of soil type with marked changes of underlying parent rock, limited use has been made of analytical data, but that in regions where large areas are covered with soil of more or less the same origin correlations were soon found.

Initially, the soil scientist provides the best service by recognizing, through soil surveys, distinct soil series and, through experience, soil analyses and fertilizer experiments, providing information as to their suitability and requirements. In Malaysia considerable progress was made on these lines and many plantations have undertaken detailed soil surveys and the mapping of soil series; soil and leaf analysis are carried out using these maps as a basis (Ng and Selvadurai, 1967).

Soil analysis for the detection of fertilizer needs made most headway in West Africa where Tinker and Ziboh (1959) established relationships both on the wide areas of Acid Sands soils and the basement complex soils of Nigeria between certain functions and responses to potassium and magnesium fertilizers. In these areas potassium is usually the dominating need, but no direct relationship was found between exchangeable or total K and yield response. Through the main areas of oil palms in Nigeria a relationship was found, however, between the mole fraction K – exchangeable K/cation exchange capacity – and yield. There were some exceptions, these being either in areas with unusually high clay content in the subsoil or on soils of very different origin from that of the Acid Sands. However, in later work on equilibrium ionic activity ratios, Tinker (1964) showed a better relationship of yield with a function AR_u,[*] the unified activity ratio, in which account was taken

---

[*] $AR_u = \dfrac{K^+}{\sqrt{[(Ca^{++}) + (Mg^{++})]} + P^3\sqrt{(Al^{+++})}}$ P is a constant.

of the activities of the cations K, Mg and Ca and of aluminium. From this work it was then found that potassium response was closely related to a function of the exchangeable cations, including aluminium, as follows:

$$ ER = \frac{K_e}{Ca_e + Mg_e + 2.5\ Al_e} $$

where $_e$ = exchangeable. Since $Ca_e + Mg_e + 2.5\ Al_e$ is correlated with the exchange capacity, the previously found relationship with the mole fraction K (above) becomes clear (Tinker, 1961). This relationship was found to cover palms on Acid Sands, basement complex and a soil probably of volcanic origin, being particularly suited to soils of high acidity with low or varying amounts of calcium.

Correlations have also been found within a field on Acid Sands soil between various functions indicative of the potassium status of the soil and both leaf-K content and individual palm yields (Forde *et al.*, 1965, 1968). In this case correlations between yield and both exchangeable K and mole fraction K were found in addition to a relationship with the ionic activity ratio without inclusion of aluminium, as the soil was only weakly acid. Such within-field correlations are of interest since they show that oil palm yields vary with relatively small variations in the nutrient status of the soil. In these Acid Sand soils, where exchangeable K falls below 0.10 m eq per 100 g of soil, response to potassium is certain provided no other deficiency or climatic circumstance is inhibiting a response. Responses in soils with exchangeable K in the range 0.10–0.20 m eq per 100 g are probable. There are strong positive correlations between leaf-K and soil-K. Leaf symptoms of K deficiency may be expected with leaf-K in the range 0.3–0.5 per cent of dry matter which will correspond with soil exchangeable K of below 0.10 m eq per 100 g (Forde *et al.*, 1968).

While this work has hardly led to the determination of critical levels for soil potassium parallel with those for leaf-K, there are indications that, on the Acid Sands soils of Nigeria, soils with a mole fraction of below 0.015 are likely to give responses, while values of 0.015–0.020 may be marginal. On the West African soils studied by Tinker an AR_u value in (moles per litre)$^{\frac{1}{2}}$ of less than 0.006 is also likely to indicate a potassium response (see Fig. 11.5).

Ataga (1974) studied the availability and fixation of potassium in the Acid Sands and basement complex soils of Nigeria using the b-index (number of successive extractions with 50 ml of 0.005 N $CaCl_2$) and other closely correlated extraction methods. Available reserves were found to be high in basement complex soils with a high proportion in exchangeable form. Reserves in depleted topsoils on the Acid Sands were low, with considerable quantities from non-

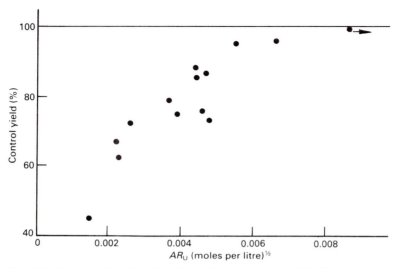

Fig. 11.5 Relationship of the function AR_u in the topsoil of fertilizer experiments with the yield responses obtained to K fertilizer applications (Tinker, 1964).

exchangeable sources. When K is added to these soils at rates equivalent to about 1 kg per palm, 30–65 per cent is fixed, though later released gradually to the palm. The clay fraction contributes the bulk of the non-exchangeable K released by these soils, thus providing a further reason for preferring those Acid Sands which contain a higher proportion of clay. Studies of the potassium reserves and fixation in Malaysian soils (Ng, 1965) gave results rather similar to those in Nigeria. Greater fixation was found in montmorillonitic and micaceous clays than in kaolin clay fractions, but the eventual supplying power of the former soils is greater.

Soil deficiencies of magnesium are widespread in Africa but owing to the strong potassium/magnesium antagonism magnesium need tends to be exhibited only where potassium is in good supply, or where the levels of both elements are exceptionally low. In fertilizer experiments, symptom development (Orange Frond) has been common and yield response to magnesium applications occasional. Prevot and Ziller (1958) showed a relationship in Congo (Brazzaville) between water-soluble magnesium in the topsoil and both deficiency symptoms and leaf-Mg. Tinker and Ziboh (1959), working with twenty-six Nigerian soils and ten samples from Etoumbi in Congo (Brazzaville), found that symptoms were usually, but not always, related to the level of exchangeable magnesium in the soil. In the worst areas the level of exchangeable Mg was below 0.1 m eq per 100 g but in the Etoumbi soil and some soils in Nigeria symptoms were found where the Mg levels were well above 0.1 m eq per 100 g and in these cases exchangeable potassium would be about 0.2 m eq per 100 g. It was found that a good relationship existed between the topsoil exchangeable Mg : K ratio and symptoms, but the steepness of the curve indicated a very critical ratio level of about 2 below which symptoms appeared.

Later experiments both in the field and with soil in pots showed how easily magnesium deficiency could be induced by potassium applications to the soil, particularly in areas where potassium had, in the course of previous cultivation, been less depleted than magnesium (Tinker and Smilde, 1963b). It was shown that young palms were particularly prone to magnesium deficiency symptoms when K + N fertilizers were applied. This condition appeared to be due not only to the K/Mg antagonism but also to the leaching effect of sulphate of ammonia on Ca and Mg in the soil. At the same time it became clear that, with young palms, symptoms of Mg deficiency could be induced where the topsoil exchangeable Mg/K ratio sank below 4; and the critical level in this case seemed to be about 3.5.

Foster *et al.* (1985a, b) have attempted to assess N and K needs on Malaysian soils by new methods. In the first place they charac-terized seventeen trial sites on alluvial soils and seventeen sites on sedentary soils by their soil structure and consistency, drainage and water-table depth, soil slope, root growth and impedance (gravel and pan), available water-holding capacity, annual rainfall and number of dry months, as well as by physical and chemical analysis in which total extractable bases (Ca + Mg + K) and total extract-able cations (Ca + Mg + K + Al + H) were taken particularly into account. Trials on soils considered likely to be anomalous, e.g. organic, volcanic, very sandy and highly lateritic soils, were excluded.

Using yields from the zero levels of N and K in these trials, and a stepwise multiple regression procedure, these workers were able to predict the yields obtainable under various environmental combinations in the absence of N and K fertilizer and at non-limiting levels of P and Mg. It was shown that, on alluvial soils, yields without either N or K fertilizer are largely determined by soil drainage conditions and the level of rainfall, though without K the yields are also influenced by the percentage soil-extractable K. On sedentary soils yields without N fertilizer are most influenced by palm age and density, soil consistency and rainfall, while both firm soil consistency (e.g. Batu Anam series) and excess rainfall reduce N uptake. Yields without K on the other hand were most affected by the ratio of soil-extractable K to other extractable bases and by soil organic matter content and drainage conditions.

This basic work made it possible to assess the environmental conditions under which economic responses to N and K could be expected and so, eventually, to establish the fertilizer needs of all soil series over a range of climatic and soil conditions taking palm age and density into account. Using equations which allowed the prediction of N and K yield responses for a range of site characteristics, Foster *et al.* (1985b) found that N responses on alluvial soils depended on the receipt of adequate but not excessive rainfall, inadequate rainfall inhibiting N responses particularly at sites which had high yields without fertilizers. Nitrogen responses were also found to increase with clay content but were depressed by poor drainage. Similar results were found on sedentary soils although the most important factor was soil slope; severe slopes, through fertilizer runoff and, perhaps, the development of an inadequate root system, markedly reduce responses to N.

Potassium responses on alluvial soils were much affected by soil properties, being negatively correlated with silt content and with total extractable cations. These effects were thought to be caused by increased K buffering which resulted in a reduced proportion of fertilizer K passing into soil solution. Excess rainfall in the 2–3 months after application also depressed K responses. On sedentary soils the K buffering capacity is usually much less, and K responses increase with annual rainfall though they are reduced by soil slope. On sedentary soils with high silt contents, however, responses are small.

A particularly interesting deduction from this work was that, though yields are high without fertilizers on alluvial soils, it is unlikely to be economic to push them higher than 26 tonnes bunches per hectare, whereas sedentary soils, which have a much lower yield without fertilizers, can give economic responses up to 30 tonnes per hectare.

The difficulties of assessing phosphorus need from soil total phosphorus or various measures of 'available' phosphorus are notorious. In Malaysia no clear differences can be seen in these measures between soils on which the oil palm responds to phosphorus and on soils on which the palm does not respond. There is perhaps more hope of finding a relationship where the soils have similar genetic properties, and in West Africa some progress has been made with such work. Forde (1965) used a number of reference soils from typical areas growing oil palms in West Africa in an experiment in which oil palm seedlings were planted in pots treated with complete nutrients with or without phosphorus. With the majority of soils there was a response to added phosphorus in seedling dry matter production and, with some notable exceptions, those soils showing greatest response had given responses in growth or yield in the field. Additionally, certain methods of estimating 'available' phosphorus showed a significant negative correlation between the log percentage response in dry matter of the pot seedlings and the soil values of available phosphorus, the Olsen and Saunders methods giving the highest significant correlations. Work on phosphorus release from these soils suggested that they could be grouped into three categories: (A) those giving no response with the oil palm to P fertilizers owing to initial high levels of P; (B) those having a low level of P and responding to fertilizer and (C) those having a low level of P but showing no response owing to high fixation of phosphorus. Ataga (1978) pursued this work a stage further and found that Bray and Kurtz's extraction method showed soils of category (A) had P > 10 ppm, those of category (B) had 5–10 ppm, and those of category (C) 5 ppm.

The areas of sandy loams around Belém near the mouth of the Amazon in Brazil provide the outstanding case of phosphorus deficiency, palms growing without P dressings showing very poor growth and low yield but giving marked and prolonged responses to all forms of P fertilizers. These soils have the lowest total P (80 ppm) of any soil supporting oil palms and their low available P (Saunders) of 22 ppm is matched only by a very P-deficient soil in Sumatra (Ollagnier and Ochs, 1981).

Other analytical methods

The difficulties in interpretation of both soil and leaf analysis have stimulated interest in possible other methods. Growth analysis has been discussed in Chapter 4 and its value in relation to nutrition is considered later in conjunction with fertilizer experiments (p. 534).

In an area where K was highly deficient but P was not, interesting relationships have been found between leaf contents of K and P and the contents of these elements in mature primary roots of the same palms (IRHO, 1974). With potassium it was shown that there was a good correlation between leaf and root levels particularly where K was supplied as a fertilizer. Although there was a sufficiency of phosphorus, there was a large difference in root content between palms which had received and had not received a dressing of dicalcium phosphate 3 years previously, while the leaf levels were only marginally, though significantly, different as shown below:

Per cent of dry matter	Without P	With P	Increase (%)	Without K	With K	Increase (%)
Leaf content	0.170	0.177	4	0.578	0.907	57
Root content	0.051	0.114	124	0.346	0.739	114

These results indicate that the root can act as a storage organ for P, providing a reserve against possible later deficiency, whereas with a deficiency of K the leaves and roots appear to be reacting similarly to fertilizer. The large percentage differences in the roots also suggest that this analysis method merits much further investigation. Root analysis does not, however, appear likely to be so promising for N or Mg. With the former, the values are low compared with leaf values, while with Mg there is little variation even when leaf-Mg varies considerably (IRHO, 1975). The range of root nutrient values (per cent of dry matter) is approximately as follows: N, 0.30–0.45; P, 0.03–0.16; K, 0.20–0.95; Ca, 0.05–0.15; Mg, 0.10–0.15.

Soil changes during the life of a plantation

Most plantations of oil palms are planted from forest, pass through 25–35 years of life under one stand of palms and are then replanted. This cycle of growth and replanting is then likely to be repeated. Long-term soil changes during a plantation's life have only been reported from Nigeria, fortunately from an oil palm field containing a cultural experiment (Kowal and Tinker, 1959). The field was planted in 1940 on Acid Sands soil, Benin fasc (see p. 127) and sampled in 1941, 1945, 1951, 1956 and 1961. Shorter studies were made of an experiment planted in 1951 on the more leached Calabar fasc of the Acid Sands, and of another planting on the Benin fasc.

The results of these studies showed that where there was no burning there was a rapid release of K from the vegetation followed by substantial increases in Mg and Ca in the topsoil. In burnt areas there was a larger early build-up of K, but by the fifth year these differences had largely disappeared and there was then, to a soil depth of 46 cm, a gradual reduction of all three cations over the next 15 years. Organic matter is surprisingly little affected unless there has been tillage or inter-cropping, and the changes in carbon and nitrogen are small. The most serious steady losses during the life of the palms were found to be in potassium, but magnesium and calcium losses were also substantial.

Estimating that the loss of exchangeable K from the top 3 m of soil in 15 years would be around 335 kg per hectare, Tinker (1963) pointed out that, to fulfil a requirement of 670 kg uptake, the other 335 kg would be provided by only a very small proportionate release of K from the non-exchangeable sources, and a manuring recommendation of 335 kg KCl or K_2SO_4 per hectare every 3 years thus seemed an adequate supply.

Long-term studies of this nature in other oil palm regions would be of great interest and value. One of the practical difficulties, however, is the finding of areas which have not been subjected to fluctuating fertilizer policies.

Fertilizer experiments and application

Some account must now be given of the knowledge that has been gained of nutrient requirements through fertilizer experiments. It will already have been noted how important such experiments are in relation to the correct interpretation of both leaf and soil analysis, but they have, of course, a particular value in themselves since not only do they show which nutrients are needed but, if properly designed, they can assist in determining the most economic rate and frequency of application.

Early experimentation

Although some fertilizer experiments were conducted in the between-wars period, extensive trials were not begun in most producing areas until after 1945. Reviews of experiments up to 1955 have been provided by May (1956a, b) and the early knowledge of manurial needs is summarized below.

In most areas opened from undisturbed or very old secondary forest little or no responses were initially obtained to fertilizers. In Zaire many experiments were conducted without positive results and in Malaysia the same lack of response was experienced even on the poorer soils. In Nigeria, areas opened from old secondary forest showed no responses until the palms were well into bearing, though when improved planting techniques and selection began to induce more rapid growth in the early years responses in growth and early yield to nitrogen fertilizers appeared. An exception to this general experience was the Belém area of Brazil where P dressings were found to be essential from the time the forest was felled. In all areas the period of satisfactory growth tended to be short-lived, and on all but exceptionally suitable soils some nutrients began to be required soon after bearing started.

Where an area had a history of any kind of cropping, fertilizers were soon shown to be required from the start. The early experiments on bearing palms showed responses typically to potassium in Africa and on some soils in the Far East; to nitrogen in the early years, but less commonly with bearing palms; to phosphorus on certain soils in the Far East; to magnesium on very poor soils in West Africa, Congo (Brazzaville), Zaire and Malaysia; to organic manures and wood and incinerator ash in West Africa and Zaire. The most startling responses were obtained to potassium, particularly in the case of old plantations which had never received fertilizers and where yields were declining and leaf chlorosis appearing. Lack of K responses in early experiments in Malaysia was due to the fact that experiments were largely confined to soils derived from sedimentary rocks. Soil applications of magnesium fertilizers in Nigeria were spectacular in their elimination of the Orange Frond chlorosis.

Rates of application were largely guesswork, being based on some experience with other permanent crops, and applications were either annual or once-and-for-all. Thus dressings of potassium sulphate, for instance, varied from 1 to 9 kg per palm for bearing palms, this being the equivalent of about 150–1,250 kg per hectare. Young seedling palms in the field might receive anything between 0.25 and 1 kg per palm. Some of the negative results may have been due to the very small dressings, although in other experiments results were obtained from applications which were two to four times as large as were required. Comparatively few of the early experiments allowed for varying rates or different methods of application. In one set of experiments in Nigeria arrangements were made for increasing the dosage by 0.45 kg of fertilizer per palm for each year of age until a ceiling dosage was seen to have been reached. As early responses were not obtained in forested areas, however, there was a tendency to allow these applications to reach very large and unprofitable amounts before they were discontinued.

However, this period saw the carrying out of three important sets of experiments, the Unilever 'Crowther Experiments' and the early WAIFOR and IRHO experiments which threw some light on fertilizer rates and frequencies applicable to African soils.

The Crowther experiments, designed by Dr E. M. Crowther, were NPK, NPKMg and NPKCaMg factorial experiments started in 1940 on young 1932–40 plantings and they covered the Benin and Calabar fascs of the Acid Sands of Nigeria, lateritic clay basalt-derived soil and sandy gneiss basement complex soil in Cameroon, and a sandy latosol in Zaire (Haines and Benzian, 1956). They showed K as a primary need on the Acid Sands, with P and N enhancing yield further when a sufficiency of K had been applied. The Cameroon basalt-derived soil and the Zaire latosol also gave K responses, but the basement complex soil had a primary P requirement. Fertilizer rates were between 1 and 3 kg of each fertilizer per palm per annum.

The early experiments of the West African Institute for Oil Palm Research in Nigeria covered the Benin, Calabar and intergrade soils of the Acid Sands. At Nkwelle, where there were advanced symptoms of Confluent Orange Spotting (see p. 485), much higher rates of potassium sulphate, up to 9 kg per palm, were used, and yield responses of nearly 200 per cent were obtained on mature palms planted in the 1930s. An even more remarkable response was had at Umudike in the heart of the palm belt of eastern Nigeria where the broadcasting of a single application of KCl under 23-year-old palms at rates equivalent to 3.65 and 7.3 kg per palm gave the

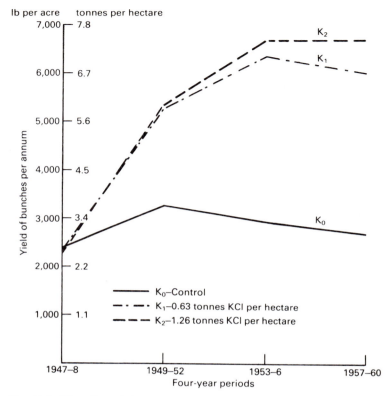

Fig. 11.6 The effect of single applications of potassium chloride on the yield of 23-year-old palms at Umudike, Nigeria.

results illustrated in Fig. 11.6 (Chapas and Bull, 1956). Soil analysis 10 years after application showed that differences in exchangeable potassium in the topsoil between treatments had completely disappeared, but that the fertilized plots had maintained a higher *total* potassium at all depths down to 1.2 m, thus indicating that part of the original applications had been fixed in a form which was not immediately available but may be assumed to be released slowly over a long period (Tinker and Ziboh, 1959; Ataga, 1974). It is also probable that the original application so improved the general health and hence the root system of the palms that the dry season effect on growth and yield was not so severe.

Similar results were obtained at the IRHO stations at Pobé in Benin (Dahomey) (Prevot and Ziller, 1957) and at Dabou in the Ivory Coast (Ochs, 1965). At the latter station palms receiving only 1.5 kg KCl per palm still showed an increased yield of over 100 per cent of control 5 years after application, though the leaf-K content of the treated palms had by that time reverted to the same level as that of the control palms. These experiments did not, however, allow for different levels of application, and few responses other than to K were obtained. In view of these results and those at Umudike it is perhaps not surprising that an experiment at Akwete in south-eastern Nigeria showed that there was no advantage in distributing the potassium sulphate in small doses through the season, and that annual applications were no more productive than those made every 3 years.

The predominant effect of potassium on West African soils generated an interest in supplying this element through bunch refuse or ash. Comparisons of fertilizers with farmyard manure were also made. On the Benin fasc Acid Sands soil of the WAIFOR Main Station in Nigeria (now NIFOR), one of the first fertilizer trials started from the time of planting showed N, P and K requirements with N showing its effect at an early stage and K and P at a later stage (Table 11.10). Farmyard manure could, at the rates used, supply the same needs as the fertilizers. Another trial compared bunch refuse and complete fertilizer applications and, apart from indicating the value of the refuse as a source of K, showed the importance, in Nigeria, of manuring at the beginning rather than

Table 11.10 *Responses to NPK fertilizers and farmyard manure on Acid Sands soil, Benin fasc., Nigeria (Expt 34–1) (yield of fruit bunches per hectare per annum)*

Fertilizer and manure	Rate of application Per palm per year of age (kg)	1952–5 (kg)	1956–9 (kg)	1960–3 (kg)
N_0	—	2,446	7,024	—
N_1	0.30	2,718	7,759	—
N_2	0.61	2,798	7,553	—
P_0	—	2,694	7,008	6,587
P_1	0.30	2,618	7,642	7,530
P_2	0.61	2,651	7,687	7,599
K_0	—	2,734	7,109	6,787
K_1	0.30	2,565	7,428	7,297
K_2	0.61	2,663	7,800	7,634
FYM_0	—	2,283	6,600	6,478
FYM_1	19	2,731	7,871	7,656
FYM_2	38	2,949	7,807	7,581
LSD, $P = 0.05$		394	654	621

N =Sulphate of ammonia. P = Superphosphate. K = Sulphate of potash.

at the end of the wet season. Pre-war trials in Malaysia demonstrated that similar yield increases are obtained from bunch refuse applied in equivalent amounts as raw refuse, composted refuse or bunch ash and that, for reasons of costs, application in the form of ash is to be recommended (Arokiasamy, 1969). The ash obtained constitutes 2.4 per cent of the refuse or 0.6 per cent of the original bunches and provides 30–35 per cent K_2O which is about half to two-thirds the K_2O equivalent of muriate of potash (Uribe and Bernal, 1973). Ash applications should therefore be about double the quantities that would be applied as KCl. The ash has an MgO content of 3–5 per cent.

If an area is yielding 20 tonnes bunches per hectare about 120 kg ash per hectare or 0.84 kg per palm will be produced in a year, equivalent to 0.42 kg KCl per palm. If dressings of, say, 3 kg KCl per palm were required, ash would provide one-seventh of requirements.

The information which emerged from the early work in Africa may be summarized as follows:

1. The need for potassium is widespread; where old plantings are very deficient, dressings up to 4 kg per palm of KCl on K_2SO_4 may be needed for full responses, but the latter will be maintained for many years. In areas opened from good secondary forest there may not be a potassium requirement for several years and annual dressings of the order of 1 kg per palm will be sufficient. Potassium applications induce magnesium deficiency on poor Acid Sands soils.
2. Nitrogen dressings are of importance in the early years and may be of the order of 225 g per palm sulphate of ammonia in the planting year and thereafter 450 g per palm per year of age for 3 or 4 years.
3. Phosphorus is a primary need on certain soils such as the basement complex soils of West Africa, while on other soils it tends to be a secondary need when the K and N requirements have been satisfied.
4. Applications of fertilizer early in the rains are most advantageous and there is no advantage in dividing the dressing into several small doses.
5. Yield responses can be obtained to organic manures in the form of farmyard manure or bunch refuse, but refuse incinerated to form ash is just as effective, less expensive to apply and can supply a small but significant portion of potassium needs.

Recent experiments and fertilizer practice

The foregoing may be said to represent the published knowledge of the manuring of oil palms up to about 1955, by which date very few results had been published in the Far East though many experiments were under way. At this time, however, the common need for magnesium, the dependence of responses of one nutrient on the adequacy of supply of another, and the possibilities of minor element requirements were beginning to be seen. From about 1955 a very large number of factorial experiments covering different rates of application were laid down both in the Far East and in Africa by both governmental and commercial research organizations. It is clearly impossible here to deal fully with the extensive data which have now accumulated. In the following pages the interpretation of experiments and advances in fertilizer practice will first be discussed and this will be followed by some selected examples of results from various parts of the tropics.

Factorial experiments

Factorial designs have been employed for the majority of fertilizer experiments in all parts of the tropics. Green (1972), in a stimulating review of a number of such experiments, has drawn attention to two important aspects of their interpretation. Firstly, these experiments usually rely on the pooling of second- and third-order interactions as error, and there is a great need to distinguish between those interactions which are genuinely small or non-existent, and so can be legitimately pooled, and those which should be taken into account.

Secondly, each of the main effects of N, P, K, etc. only show the response to these elements in the presence of the mean level of all the other elements (or other treatments) in the experiment; the effect of each nutrient *alone* or in *specific combinations* is not shown in the summarized data or two-way tables; but it is exactly this information which is required for an economic appraisal of the results, since fertilizer recommendations must depend on the gains in bunch weight, oil and kernels believed to be obtainable from specific formulations applied alone and not along with varying quantities of additional nutrients. Green has now provided a method of estimating the expected yield value of all eighty-one treatment combinations in 3^4 factorial experiments and has applied this to his experiments. Using these values, he calculated the profit or loss expected to result from the application of each single nutrient or specific combination.

From three experiments in West Cameroon and one each in Nigeria, Zaire and Malaysia it was shown that (a) maximum yield was usually obtained from a balanced fertilizer mixture of three or four nutrients; (b) applications of the 'wrong' fertilizers or fertilizers in inappropriate proportions could lead to substantial reductions in yield and considerable financial loss; (c) the increased yield was usually accountable predominantly to one or two nutrients and the highest profit was thus obtained from applying one or two nutrients only, the addition of others resulting in financial loss; (d) only in one instance, in Malaysia, did application of all four major nutrients give a maximum profit, and in this case the profit was not significantly different from that obtained from two or three nutrients only; (e) foliar analysis data from unfertilized plots indicated that these data could not have shown the requirements for maximizing yield let alone the mixtures for highest profit. Gross deficiency was only apparent in two cases. In spite of substantial yield increases due to P and K, leaf-P and leaf-K were only increased significantly in one case each by the optimum treatment. Leaf-Mg followed the manurial applications more closely.

In general 3^4 factorial experiments using plots of twelve to sixteen palms have given very useful results in many countries, but more recently experiments with four or more levels of certain nutrients have been used. It has been claimed that for obtaining a sound production function for the economic assessment of results four to five levels of nutrients are needed and, to reduce the number of plots, incomplete factorial designs have been advocated (Lo and Goh, 1973).

The economics of fertilizer application

When manuring results in increased bunch yield and hence in increased oil and kernel outturn, there are no increases in general overheads, and calculations of profitability are relatively simple (Gunn, 1962). If the cost of harvesting, transporting and milling a tonne of bunches and the extraction rate of oil and kernels are known, then the net value of any additional tonne of bunches produced will be

$$V_{nt} = a + b - c$$

where a and b = value of oil and kernels extracted respectively from a tonne of bunches, and c = the cost of harvesting, transporting and milling a tonne of bunches. In this latter figure should be included, of course, charges for maintenance, etc. of vehicles and milling overheads, but it can be noted that where a mill is running below capacity the milling of a larger crop may increase costs very little and so reduce the overall milling cost per tonne.

The cost of fertilizer and its application in the field now has to be set against the expected bunch weight increase at this net value per tonne, V_{nt}. This is best done by first calculating the weight of bunches of equivalent net value to the cost of fertilizer and its application:

$$xV_{nt} = C + A,$$

where C = cost of fertilizer per hectare and A = cost of application per hectare. Thus if C and A are respectively 8 and 0.5 monetary units and V_{nt} is 8 units, then $x = 8.5/8 = 1.0625$ tonnes. If the actual response is expected to be greater than x, or in this case greater than 1.0625 tonnes per hectare, then the fertilizer application will be profitable.

While these calculations are very simple for adult palms responding directly in bunch yield to applied fertilizers, it is more difficult to assess the profitability of fertilizing young palms. Any assessment must be somewhat arbitrary since the object of manuring at this early stage is to establish strong healthy plants which will come into bearing early and, if thereafter supplied with nutrients sufficient for optimum yield, have a productive bearing life. It is probable that palms insufficiently fertilized when young will never be able to produce at the high rate of others properly manured. To a certain degree, therefore, early manuring is an establishment cost. Moreover the quantities required during the early years are so small on a per hectare basis that yield increases of 1–5 per cent over the first 3 years of bearing would amply cover the costs of pre-bearing applications on areas of fair production.

Another important point to note is that the quantity, x, given above for the weight of additional bunches required to balance the cost of fertilizers and their application, is an absolute amount; therefore, if the *basic yield*, i.e. the yield expected without the use of fertilizers, is low, then the *percentage increase* in yield must be high to be profitable. On the other hand if high yields can be obtained *without fertilizers* then very small percentage responses are required for their profitable use. The proposition, therefore, that a high-yielding area requires no fertilizer needs careful examination; increases as low as 5 per cent can often be profitable, though such increases can only rarely be shown to be statistically significant in a fertilizer experiment. As Gunn (1962) has pointed out, the corollary of this *for high-yielding areas*, is that any response shown in a fertilizer experiment to be significant is almost bound to be profitable.

Studies of the profitability of fertilizer applications have also been carried out in Malaysia. Hew *et al.* (1973) presented the discounted accumulated gross profit to be obtained from N, P, K and Mg applications with yields similar to those obtained in five experiments on different soils. The calculations allowed for the time-lag of response by a discounted cash flow procedure; certain constants were used for costs with three levels of oil and kernel prices. Three- or four-year yields from the start of bearing were employed. As might be expected, results differed widely between soils. The only two consistent results were that nitrogen was uneconomical to apply and that the higher rates, e.g. over 3 kg per palm, of muriate of potash or rock phosphate were not justified. The lower rate of muriate of potash was the most profitable treatment on Munchong, Serdang, Rengam and Briah series (see pp. 116–19), but was unprofitable on the Selangor series marine clay; the lower rate of rock phosphate was the most profitable treatment on Munchong soil only. Magnesium, as kieserite, was uneconomical to apply on all the three soils where it was used. The main effects in this study appear to have been taken straight from the experiments and not adjusted to 'expected values' as advocated by Green (1972), nevertheless the results do indicate that fertilizers applied before bearing or in the early bearing years will only give economic returns if used with discrimination through knowledge obtained from fertilizer experiments.

Lo and Goh (1973) applied a response surface analysis to an experiment on Rengam series (p. 117) with adult palms using varying fertilizer and oil price levels. They concluded that optimum fertilizer rates depend on yield response, price of produce and price of fertilizer in that order of importance, but that there are difficulties of adjustment to prices in view of the time-lag between application and response. Rather than making continuous changes to fertilizer rates with prices, managements are recommended to estimate average prices of both produce and fertilizers for several years and to adhere to these estimates. Another factor is the variation in the estimates of yield response. If the profit margin is in any case small and the 95 per cent confidence limits of the experimental results are wide (i.e. the error is large) management may be inclined to reduce costs to conserve the cash flow. All these considerations point to the conclusion that the economics of manuring the oil palm is not a simple or straightforward matter, that decisions can only be taken after a painstaking appraisal of all the factors involved and that the oil palm grower should be on his guard against short cuts.

Leaf analysis in fertilizer experiments

Much has already been said about the uses, and danger of misinterpretation, of leaf analysis and of the need to adjust critical levels following appraisals of the results of experiments. Numerous examples could be given of leaf analysis data from experiments, but two examples will suffice to show that where deficiencies are pronounced good correlations between leaf data and yield responses can be found.

With adult Deli *dura* palms in Malaysia on Kulai series soil derived from volcanic tuff a 3^4 NPKMg experiment showed, over a 7-year period, yield responses to N, P and K (Warriar and Piggott, 1973). Figure 11.7 shows the effects of these three elements and Mg on both leaf and soil analyses, and Table 11.11 gives the correlation coefficients between these functions and yield. It will be seen that where the nutrients had not been applied the leaf contents of N,

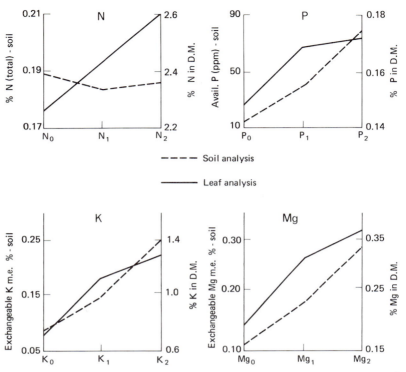

Fig. 11.7 The effect of main nutrient applications on leaf and soil levels in a fertil-izer experiment on Kulai series soil in Malaysia (Warriar and Piggott, 1973).

K and Mg were very low, and of P rather low. Palms without fertil-izers were yielding less than 7 tonnes bunches per hectare in 1969, whereas those with complete dressings at the higher rates were producing over 25 tonnes. In some plots leaf levels as low as N – 2.00 per cent, P – 0.129 per cent, K – 0.319 per cent and Mg – 0.094 per cent could be found and K and Mg deficiency symptoms were visible. A curious feature of the results was that, although soil and leaf analysis were well correlated, yield and soil analysis showed little relationship, and under these circumstances of extreme de-ficiency leaf analysis therefore seemed to be a better guide. The negative correlation between Mg leaf levels and yield is of interest here and simply indicates the strong K deficiency.

A very different situation existed in an experiment at La Mé in the Ivory Coast where potassium was the principal deficient element (Bachy, 1969). Here 2 kg per palm per year each of sulphate of ammonia, superphosphate, muriate of potash and magnesium sulphate were applied to clayey-sand soil derived from Tertiary

Table 11.11 *Total and partial correlation coefficients between analysis data and bunch yield*[†]

	Total correlations			Partial correlations[‡]			Bunch yield, 1969–70, high levels as per cent of nil levels[§]
	Soil and leaf	Leaf and yield	Soil and yield	Soil and leaf	Leaf and yield	Soil and yield	
N	−0.119	0.640***	−0.240*	0.046	0.634***	−0.215	155
P	0.561***	0.492***	0.279*	0.507***	0.421***	0.004	123
K	0.648***	0.449***	0.056	0.698***	0.543***	−0.345***	140
Mg	0.531***	−0.395***	−0.153	0.518***	−0.375***	0.073	101

[†] Yield of year 1970; analyses carried out in Feb. 1970.
[‡] The partial correlations are calculated by keeping the third variable fixed.
[§] Kilograms of fertilizer applied over 7 years: N, 5.54; P, 4.81; K, 11.53; Mg, 2.45.

Source: Warriar and Piggott (1973).

sandstone in a 2^5 experiment. On an area planted in 1946 applications were made from 1954 with the regular dressings of 2 kg from 1956 (N, K and Mg) and 1959 (P). The effect of K became significant in 1957–8 and over 11 years the plots without K gave 62 kg bunches per palm per annum and the K plots 98 kg. This 58 per cent increase was believed to be exaggerated by some 20 per cent owing to unusually low yields from some of the P without K plots which were highly deficient at the beginning of the experiment. Leaf analysis data computed as the mean of the 1961, 1963 and 1969 determinations are shown in Table 11.12.

It will be noted that the effect of K on both leaf content and yield were very great and that K altered the leaf content of the other nutrients, though not sufficiently to bring Mg into the range of deficiency; N and P had only minor effects on leaf contents and no effect on yield, and the P effects on K and Ca were no doubt due to the Ca content of the fertilizer. In the Ivory Coast the yield-

Table 11.12 *Leaf analyses from an NPKMg experiment at La Mé, Ivory Coast (per cent of dry matter – mean of 1961, 1963, 1967 data)*

Plots	With N	Without N	With P	Without P	With K	Without K
N	2.64*[1]	2.60	2.61	2.64	2.66*[3]	2.58
P	0.173*[1]	0.170	0.175*[3]	0.168	0.170*[2]	0.173
K	0.644*[1]	0.679	0.634*[1]	0.689	0.849*[3]	0.474
Ca	0.868	0.876	0.919*[3]	0.824	0.791*[3]	0.953
Mg	0.402	0.396	0.401	0.397	0.338*[3]	0.461

Significant differences in 1 year only (*[1]), 2 years(*[2]), or in all 3 years (*[3]), at $P = 0.01$.

Source: Bachy (1969).

limiting factor is water and it is not improbable that if this impediment were removed nutrients other than K might have an effect in spite of their present leaf levels.

While these experiments show useful correlations between leaf nutrient contents and yield, such correlations may, as already mentioned, be entirely absent. Among the experiments of Green (1972) (see p. 528) for instance, the leaf analysis data for the experiment in Nigeria were rather similar to those of La Mé; maximum yield was obtained by K applications with the addition of N and P but significant differences were not seen in the N and P leaf analyses between fertilized and unfertilized plots. In the Cameroon and Zaire experiments N and P leaf differences were also absent although these nutrients featured in the optimum treatment. Potassium leaf values were anomalous, showing reductions (two significant) with K applications, while Mg, although featuring in all optimum dressings, showed a significant leaf increase in only one case. In the Malaysian experiment only a very small leaf-P increase was shown although the NPKMg mixture increased yields by 32 per cent.

Growth and density in fertilizer experiments

The anomalous and variable results of leaf analysis in fertilizer experiments stimulated interest in growth factors in their relationship to yield. Corley and Mok (1972) examined the net assimilation rate (E), crop growth rate (CGR), leaf area index (L), vegetative dry matter production (VDM) and bunch index (BI) in relation to yield in the plots of two 3^4 NPKMg experiments on Rengam series (p. 117) in Malaysia. On this soil the main response is to K, but responses can additionally be obtained to N, P and Mg (Rosenquist, 1962). VDM, L and CGR were increased by N, P and K with the increases in yield. E was increased only by N, which also increased the number of leaves per palm. In one of the experiments the responses to N were dependent on the application of K and vice versa.

The main point of interest in these findings was that BI, which is the ratio of bunch yield to bunch yield plus VDM, was unaltered by manurial treatment, suggesting firstly that this is an inherited character (see p. 249), and secondly that VDM might be used as a tool for diagnosing nutrient need. The highest values of VDM found in the two experiments (139 and 135 kg per palm per annum) were similar to those found on the highly productive marine clays in Malaysia and Corley and Mok (1972) suggested that only where annual values of VDM fall below 120 kg per palm might leaf analysis be undertaken. Lo et al. (1973) have compared the correlation coefficients between VDM and bunch yield in the same two experiments with those of N, P, K and Mg and yield and found that the former were of a greater magnitude and less fluctuating.

There are a number of factors which complicate this approach. Firstly, it appears likely that environmental factors are of the greatest importance in the determination of *BI* and that available assimilates are first used for vegetative growth (Corley, 1973a), and that therefore variation in availability will have a greater effect on yield than on *VDM* (see p. 153). Secondly, density has an effect on the growth parameters and complicates the interpretation of data. For instance, the two experiments cited above were planted at 114 and 138 palms per hectare respectively and Corley and Mok believed that density accounted for the lower *E*, higher *L* and the lower *BI* in the denser planting. Breure (1977) showed that bunch yield responses to fertilizers may be markedly affected by density. He obtained a 40 per cent yield increase at a density of 110 palms per hectare compared with only 6 per cent at normal density for the same medium rates of application, and no response at all at higher density.

Effect of fertilizers on bunch composition

Fertilizer effects are usually measured only in terms of bunch yield. Corley (1973), however, described three experiments carried out in southern Malaysia in which the oil-to-bunch percentage was found to decrease with increasing doses of potassium fertilizers, as follows:

Percentage oil to bunch

K *level*	0	1	2
Expt. PF 73 *Dura*	20.0	18.6	18.2
Expt. PF 73 *Tenera*	23.5	21.8	20.8
Expt. PF 75 *Dura*	19.0	17.4	17.0
Expt. PF 78A *Tenera*	23.5	22.2	22.3

This decrease in oil percentage was more than compensated for by increased bunch yield. Further work in Malaysia (Foster *et al.*, 1987a) confirmed that KCl depresses oil content on inland soils, but in two experiments on coastal soils an increase was found. The decrease in oil to bunch on the former soils was due to a reduction of dry mesocarp content and this was partly compensated by an increase in kernel-to-bunch ratios; fertilizer recommendations were therefore not thought to need alteration. The K increase on coastal soils was mainly due to an increase in dry mesocarp to fruit and the profitability of KCl applications was thereby enhanced; N and P fertilizers were not found to have any appreciable effects on bunch composition.

Rates, frequencies and methods of application

Rates of application have featured in the majority of experiments though rarely with more than two actual levels of the fertilizers concerned. 3^n experiments have been most common and the quantities used for adult palms have varied within the following limits: nitrogen as sulphate of ammonia or urea, 0.5–2 kg; triple super-phosphate, 0.5–4.5 kg; rock phosphate, 1–14 kg; potassium chloride or sulphate, 0.5–7 kg; and kieserite, 0.5–5 kg per palm per annum. Rates of fertilizers tried in Malaysia have generally been higher than in Africa, but the lower rates, usually about 1.5–2.5 kg per palm, have often been sufficient for maximum yield. Rates of magnesium fertilizers tried tend to be lower than with the other nutrients. In plantation practice also, the Malaysian rates tend to be higher than those in Africa (Green 1974), and this reflects, though it is not necessarily justified by, the much higher general level of yields resulting from climatic factors.

Very few trials have been carried out on *frequency* of application on adult palms. The long-lasting effect of the potassium application in the Umudike experiment in Nigeria and in Dabou experiment DA–CP$_2$ (pp. 524–26) suggested that applications on a triennial basis might be as satisfactory and cheaper than annual applications and several fertilizer experiments have been laid down on this assumption in Nigeria; it has been contended, however, that large triennial dressings may induce magnesium deficiency when the smaller annual ones would not (Forde *et al.*, 1968), and, in general, annual applications are more convenient for management.

In Malaysia the tendency has been to apply nutrients more frequently rather than less frequently than once a year. The effect of nitrogen on growth and colour in the first few years of the palm's life has often been so rapid but short-lived that frequent N applications have been strongly indicated. Application of 200 g sulphate of ammonia per palm 4–6 weeks after planting, with a similar amount of muriate of potash, has been successfully employed in Nigeria, with one further similar dressing in the same year towards the end of the rains. In Malaysia dressings of N at 3, 6 and 12 months have been advocated with P and K or PKMg at the two latter times (Turner and Gillbanks, 1974). Magnesium is often required in young plantings and can be applied as 200 g kieserite per palm at the same time as the N or NK. Application twice a year has also been advocated until bearing or even later, and where there are two wetter seasons this might be justified. However, no evidence has been provided to support this practice.

As already seen (p. 527), in seasonal climates fertilizer applications should be made shortly before the onset of the wet season and not before or during the dry season or in the middle of the rains. Where rainfall is better distributed the drier periods and periods of very heavy rainfall should also be avoided and, if this is possible, those months should be selected in which a fair amount of light or medium rainfall occurs, preferably preceding a rainy season rather than a dry season.

The *method of application* of fertilizers has been a subject of some controversy but few experiments. With adult palms in Nigeria the concentrated placement of fertilizers in sectors of trenches showed less yield response than broadcasting treatments (Forde *et al.*, 1968). Experiments in the Ivory Coast and Benin (Dahomey) have failed to show differences between broadcasting and application in circles round the palms (Ollagnier *et al.*, 1970). Ng (1977) quotes one case in Malaysia where there were no differences in uptake of N and K, as judged by leaf analysis, between broadcasting and application in the weeded circle and another case in which there was a slightly greater uptake of N from broadcasting. The relationship of the mode of development to nutrient uptake has been discussed by Ollagnier *et al.* (1970) who pointed out that since absorption is through the quaternaries and the tips only of the larger roots it is reasonable to supply nutrients to all areas where quaternaries are found in quantity. Although the density of primary and secondary roots tends to decrease with distance, the total quantity of absorbing roots increases in successive surrounding rings at least to 3.5–4.5 m (Ruer, 1967). Furthermore, roots show a positive tropism to areas where water and nutrient supply is good. The argument that fertilizer should be applied only within the weeded circle because there is a concentration of roots of all kinds near the base of the palm is unsound since it ignores the facts that there are larger quantities of absorbing roots at a distance from the palm, that the roots are deeper and more lignified in the circle, and that better conditions are provided in the interline for the encouragement of positive tropism.

Ruer (1967) recommended increasing the circle of application from 1.5 m at 1 year to as far as 4 m at 5 years of age provided the doses being applied are not too small (i.e. giving less than about 40 g per m^2); and broadcasting on strips in the interline or in wide circles to 3.9 m (halfway across the interline), for adult palms. However, the fertilizer should not be thrown on the bare ground near the base of the palm, and application in a broad band to a little beyond the extremity of the leaves corresponds approximately with Ruer's recommendations for young palms. Once the palms have closed in the fertilizers should be broadcast in the interline or in a very broad band around the palms.

It has also been shown that development of the root system is much greater under a good cover of *Pueraria* than under grass (Taillez, 1971) or under the paths along the rows. Therefore when broadcasting between or around adult palms small isolated areas of grass should be avoided if possible and the fertilizer should not be scattered on the trampled ground of paths.

Choice of fertilizers

Nitrogen

Nitrogen is commonly required for the rapid growth of young palms in the field. Sulphate of ammonia should be used unless, on the basis of nitrogen content, urea is found to be appreciably cheaper. This is because up to 25 per cent of the nitrogen of urea can be lost through volatilization as ammonia. Chan (1981) found sulphate of ammonia to be more efficient than urea on inland soils in Malaysia. The need for nitrogen in the production years is not very common but it varies according to the situation; for instance, at La Mé in the Ivory Coast the effects of annual dressings of sulphate of ammonia were positive and significant with young palms, but there were no effects with adult palms in the presence of adequate K supplies. Where K was not supplied, N could actually reduce yields (IRHO, 1974). By contrast, the Malaysian experiment quoted on p. 531 showed a 55 per cent yield response to N by adult palms.

Nitrogenous fertilizers containing calcium must be avoided because of the antagonistic effect of calcium on potassium uptake. Ammonium phosphate can be used where there is a need for both N and P and this fertilizer has been found particularly valuable in prenurseries and nurseries.

Phosphorus

Rock phosphate (from Christmas Island, 36 per cent P_2O_5) has been the traditional source of P in the Far East while the superphosphates, single, double or triple (18, 38 and 48 per cent P_2O_5), have been the most available forms of P fertilizer in Africa and tropical America. On a very P-deficient soil in northern Brazil a large initial application of rock phosphate spread over the whole ground had a superior effect to annual doses of triplesuperphosphate (IRHO, 1974); rock phosphate may therefore be a more suitable source for oil palms than the superphosphates.

Phosphorus is required on specific soils in all three continents and in these cases often produces very large yield increases. On other soils it may provide a further yield increment when K requirements are satisfied, but on its own it may actually reduce yields, possibly through the antagonistic effect of the calcium content of the phosphatic fertilizers. The indiscriminate application of phosphorus 'maintenance dressings' is therefore inadvisable.

Potassium

Potassium is the most commonly required element for adult palms and with nursery or young palms it may reduce the incidence of leaf diseases such as *Cercospora elaeidis*. It is usually applied as the chloride (containing the equivalent to 60–62 per cent K_2O) or sulphate (48–52 per cent K_2O) according to availability and price; but it has been shown (see p. 503) that the effects of some KCl applications can be attributed to Cl rather than K (Ollagnier and Ochs, 1971). Potassium is sometimes not required where there is a strong P or Mg deficiency and in the latter case its application can produce or increase intense Orange Frond. Potassium has also been applied as Patent Kali, sulphate of potash magnesia, containing 26–30 per cent K_2O and 9–12 per cent MgO, and as wood ash or bunch ash (see p. 527). The latter is particularly useful for very acid soils such as the acid sulphate soils of Malaysia.

Magnesium

The need for Mg is commonly exhibited through its deficiency symptoms which are often induced by potassium manuring. Magnesium is required on the poorer soils of West Africa and on many of the American soils derived from recent volcanic material whether sedentary or alluvial. However, its application rarely results in very large increases in yield. Magnesium is usually applied as hydrated magnesium sulphate (Epsom salts, 46–48 per cent $MgSO_4$ or 16 per cent MgO), crude magnesium sulphate (kieserite, 26 per cent MgO) or anhydrous magnesium sulphate (96–98 per cent $MgSO_4$, 30–32 per cent MgO). Magnesium limestone is not usually suitable owing to the K/Ca and Ca/Mg antagonisms, but may be useful where magnesium is required in very acid conditions as on the Malaysian acid sulphate soils. Magnesium chloride has been used quite extensively in Sumatra and Ecuador and has been especially advocated where a chlorine deficiency is also suspected. It is not uncommon for Mg deficiency to be exhibited by young palms in areas where it is not seen on adult palms and where dressings of Mg fertilizers are not needed to maximize the crop. In such cases a small dressing will correct the symptoms but they will also disappear on their own without fertilizers as the palm develops. Magnesium deficiency is much commoner in replantings than new plantings (Chan and Rajaratnam, 1977).

The use of compound fertilizers of fixed formulae is not advisable with the oil palm except perhaps where very small quantities are needed in prenurseries or nurseries. Most compound fertilizers contain, for the oil palm, a low amount of Mg which, when Mg is not needed, will be detrimental or simply wasted, or, when required, will be insufficient. Oil palm requirements with young or adult palms are usually so specific that it would be very rare for a compound to meet these requirements, and some ingredients are almost always wasted or actually harmful. Costs per unit of the nutrients supplied are also higher with compounds.

Asia

Indonesia

Some information on past work and fertilizer practice has been provided by Werkhoven (1965). Rock phosphate became widely used in Sumatra owing particularly to its effect when applied to 'liparitic' latosols and to some soils derived from sedimentary rocks, but Dell and Arens (1957) showed that the indiscriminate use of P might be dangerous. In areas shown by leaf analysis to be deficient in K, rock phosphate applications either depressed yields or had no effect, while on the liparitic soils where leaf analysis showed K to be in good supply a yield response to P of 24 per cent over 4 years was obtained. The yield depression due to P was most marked on marine soils where already low leaf-K levels were further depressed.

Recent experiments (Umar Akbar *et al.*, 1977; Ollagnier and Ochs, 1981; Taniputra and Panjaitan, 1981) have been mainly on the liparitic soils referred to as yellow or yellowish-red podzolic soils. In young plantations on previously cultivated land deficiencies of N, P, K and Mg soon become apparent and Taillez (1982) has shown that balanced dressings of moderate amounts of fertilizers are needed; additionally, large dressings of rock phosphate to the legume cover and 500 g per plant in the planting hole are recommended.

On adult plantations experiments demonstrated a positive interaction of N and P; 4 kg ammonium sulphate per palm with 3 kg triplesuperphosphate nearly doubled yields while application of one of these fertilizers alone only gave a 40 per cent increase (Table 11.3). A distinctive feature of these soils are the very low Mg levels. In one experiment a correlation was found between leaf-Mg and yield responses to kieserite or magnesium chloride and substantial yield increases were had only where leaf-Mg fell below 0.10 per cent of dry matter. Responses to K were not found on these soils in spite of leaf levels falling to between 0.8 and 0.9 per cent.

Table 11.13 *Effect of sulphate of ammonia and triple superphosphate on adult palms on liparitic latosols (yellow podzolic) in Sumatra*

P levels	Fertilizer rates (per palm per annum)	N_0 0 kg	N_1 2 kg	N_2 4 kg
		kg of bunches per palm per annum, mean of 3 years		
P_0	0 kg	86	98	127
P_1	1.5 kg	97	134	142
P_2	3.0 kg	121	124	167

Source: Umar Akbar *et al.* (1977).

On latosols derived from Tertiary sandstones experiments have shown strong P and K requirements with the need for corrective dressings of Mg.

Malaysia

A series of experiments on soils of the Rengam series (derived from granite) and the Pamol series (derived from shale) gave results which may be taken as a first guide to the fertilizer requirements of inland soils derived from acid igneous and sedimentary rocks (Rosenquist, 1962). The experiments allowed for three levels of each nutrient, but unfortunately at each site one replicate of a 3^3 or $3^3 \times 2$ experiment had the fertilizers applied only once, in 1949, while the other received the fertilizers annually from 1949 to 1953 on the Pamol soil and from 1949 to 1956 on the Rengam soil, after which, in the latter case, annual dressings were given to both replicates. The results were presented for the 1956–9 period for the Rengam soil and the 1950–3 period for the Pamol soil. The experiments are therefore not fully annual-rate experiments, though nearly so. Rates of application were as follows, in kg per palm:

Sulphate of ammonia		Rock phosphate		Muriate of potash		Kieserite	
N_0	0	P_0	0	K_0	0	Mg_0	0
N_1	0.91	P_1	1.36	K_1	0.91	Mg_1	0.91
N_2	1.82	P_2	2.72	K_2	1.82		

The results on the sedimentary derived soil, where Mg was not included, were straightforward: K gave no significant response even at the high rate of P; the higher rate of N was required for a response to that element and this was only obtained in the presence of added P; and there was a response to P but no significant advantage in the higher rate.

On the Rengam soils the primary deficiency is potassium, and the experiments showed that other elements are of little value, or may be actually harmful, if a primary deficiency remains undiscovered (Tables 11.4 and 11.5). In this case there were significant negative responses to P in the absence of K and to N where P was high and K low.

The experiments covered two sites of rather dissimilar Rengam series soil, site A being on a typical sandy clay while site B had a coarser soil of lower fertility as judged by yield. The negative response to P in the absence of K was found on the typical soil of higher general fertility (Table 11.4). Positive responses to P in the presence of K were more marked on the coarser soil. Responses to

Table 11.14 *Responses to N, P, K and Mg on Rengam soil series (granite derived), sites A and B, Malaysia (weight of bunches per palm per annum, 1956–9 (kg))*

Major response Both sites	K_0	K_1	K_2
	66	89	94

Secondary responses – both sites

	Nitrogen			Phosphorus			Magnesium	
	K_0	K_1/K_2		K_0	K_1/K_2		K_0	K_1/K_2
N_0	62	88	P_0	69	86	Mg_0	69	89
N_1	72	89	P_1	66	91	Mg_1	64	94
N_2	65	96	P_2	63	96			
Linear effects	Insig.	Sig.*		Insig.	Sig.**		Insig.	Sig.*

Site A				Site B		
	K_0	K_1/K_2			K_0	K_1/K_2
P_0	77	94		P_0	61	79
P_1	69	97		P_1	63	86
P_2	62	101		P_2	64	90
Linear effects	Sig.*	Insig.			Insig.	Sig.*

For rates of application see text.

Source: Rosenquist (1962).

Mg were only obtained in the presence of K and the positive response was similar at both levels of K.

Responses to N were also dependent on an adequate supply of K but the presence or absence of P fertilizer also had an effect. When K was not supplied or, as in one replicate, was inadequate owing to discontinuity of applications, high P applications caused the N effect to be negative, although in the absence of P a positive N effect was obtained. Where K was sufficient, however, N responses were positive at all levels of P and the highest yields were obtained from N_2P_1 and N_1P_2 with K_1, and from N_2P_2 with K_2 (Table 11.15). In experiments with this type of response pattern a simple response to K may show a high degree of curvature, indicating that a low level of application is all that is required; however, when the right balance of other elements is supplied the response to K may be linear, and it is then difficult to say whether the higher rate is sufficient or not (Fig. 11.8). Experiments of the design 4^n with really high top rates of application may therefore be needed.

These experiments indicated the importance firstly of determining the primarily deficient element, secondly of determining the other elements needed for maximizing responses to the primary element, and thirdly of providing rates of application sufficient to discover the optimum level of application for any combination used.

Table 11.15 *Interactions between* N *and* P *at different rates of* K *on Rengam series soil derived from granite, site B (annual dressings), Malaysia (weight of bunches per palm per annum, 1956–9 (kg))*

Potassium	Nitrogen	Phosphorus			
		P_0	P_1	P_2	Mean
K_0	N_0	50	75	73	66
	N_1	71	73	65	70
	N_2	66	67	58	64
	Mean	63	72	65	67
K_1	N_0	85	99	92	92
(Muriate of potash	N_1	73	97	107	92
0.9 kg per palm per annum)	N_2	96	104	99	101
	Mean	84	101	99	95
K_2	N_0	86	77	96	86
(Muriate of potash	N_1	90	99	103	98
1.8 kg per palm per annum)	N_2	99	101	113	104
	Mean	92	93	104	96

LSD between means, P = 0.05 − 14

Annual rates of application N_1, N_2 − 0.9, 1.8 kg S/ammonia; P_1, P_2 − 1.4, 2.7 kg rock phosphate.

Source: Rosenquist (1962).

On both soils it was seen that the application of certain elements without satisfying the primary requirement could actually reduce yields below a 'control' level which was itself already low. In certain circumstances N applications may be dangerous in the absence of K or P, P applications may be dangerous in the absence of K, while K applications may be dangerous in the absence of Mg or vice versa.

Other experiments on soils derived from sedimentary rock have confirmed the need for N and P but have suggested that in some circumstances K is additionally required for maximum yields and Mg may be needed for early growth. The mean effects over 3 years in an experiment on Munchong series soil were as follows (Hew *et al.*, 1973):

	N_0	N_1	N_2	P_0	P_1	P_2	K_0	K_1	K_2
Tonnes bunches per hectare per annum	24.8	27.7	28.5	23.1	28.6	29.3	25.3	27.6	28.0
Leaf 17 N, P or K per cent of d.m.	2.73	2.94	3.06	0.137	0.152	0.156	1.17	1.26	1.32

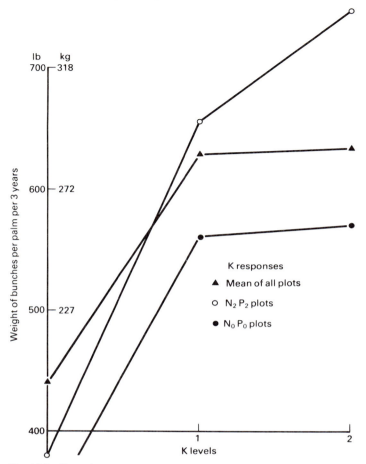

Fig. 11.8   Potassium responses on Rengam series soil, Malaysia (Data, 1962: Rosenquist, 1962).

It will be seen that P gives the highest responses and shows the most obvious deficiency; N and K leaf levels are well above the supposed critical levels but these elements gave substantial yield increases. An NK experiment on Serdang series (Tan, 1973), where all the plots had received applications of P and Mg, also showed substantial responses to K.

Several further experiments have been carried out on the common Rengam series soil derived from granite. The mean effects in two of these experiments were as shown in Table 11.16 (Lo *et al.*, 1973).

Lack of P responses may have been due to the lasting effect of

Table 11.16 *Responses to* N, P *and* K *in bunch yield and leaf nutrients in two experiments on Rengam series soil in Malaysia*

	N_0	N_1	N_2	P_0	P_1	P_2	K_0	K_1	K_2
1. Tonnes bunches per hectare per annum, 5 years	19.04	22.9	22.9	21.1	22.2	21.5	18.6	22.7	23.5
Leaf 17, N, P or K per cent of d.m.	2.60	2.79	2.86	0.175	0.179	0.184	0.766	1.090	1.169
2. Tonnes bunches per hectare per annum, 4 years	18.9	23.4	25.1	22.0	22.0	23.3	19.7	23.6	24.2
Leaf 17, N, P or K, per cent of d.m.	2.53	2.83	2.92	0.168	0.176	0.178	0.629	1.017	1.096

Source: Lo et al. (1973).

considerable quantities of rock phosphate applied in pretreatment dressings. There were no responses to Mg. The lower levels of N and K appeared to be sufficient. These levels were 3.6 kg per palm per annum of ammonium sulphate and potassium chloride in the first experiment and 4.5 kg in the second. On another area of Rengam series soil where all plots had received P and Mg, N responses did not appear until the third year of bearing, while levels of K higher than 2.7 kg KCl per palm per annum gave no further return.

Using data of a long series of experiments carried out from 1930 onwards, studies have been made of the effects of magnesium and nitrogen on both Rengam series and a range of inland soils derived from sedimentary rock. Previous land use was found to be important for magnesium (Chan and Rajaratnam, 1977), no response being obtained until after 23 years with palms growing on land opened from primary forest. With replants, however, responses to Mg were common and were magnified by heavy applications of N and K. Without the latter Mg gave little yield response, but both N and K effects were enhanced by quite moderate dressings of Mg. Once deficiency symptoms were corrected and leaf levels reached 0.20–0.22 per cent of dry matter, no further yield responses were obtained; dressings of around 1 kg of magnesium sulphate per palm per year were usually sufficient.

For nitrogen (Chan, 1981), it was again shown that responses on land opened from primary forest were rare, but with replants N was needed both for early vegetative growth and for yield. The higher rates tried, up to 4 kg per palm per annum, gave yield responses even though there was no corresponding increase in leaf-N; and it was shown that for full expression of N effects the strong synergistic effect with K needs to be exploited. The responses to N dressings were short-lived, so a frequency of at least once a year is required.

On the fertile marine clays (Selangor series) in Malaysia early trials gave responses only to K and even this was not achieved until after several years of bearing and did not amount to more than about 5 per cent or 1–1.5 tonnes of bunches. Dressings of under 3 kg per palm per annum of KCl are usually sufficient. More acid phases of the marine clays such as Sedu series or the acid sulphate soils respond more sharply to potassium particularly when applied as bunch ash; in an experiment on Sedu series 2.7 kg KCl per palm per annum gave nearly a tonne of bunches less than an equivalent rate (K content) of bunch ash (Hew and Poon, 1973). Magnesium also appears to be deficient on acid sulphate soils and leaf levels are raised by application but yield responses have not always been obtained.

The Briah series is derived from mixed marine and riverine alluvium and oil palms normally respond to both K and N (Hew and Poon, 1973; Hew *et al.*, 1973), dressings of about 2.5 kg each of sulphate of ammonia and KCl being sufficient. The effect of P is often masked in Malaysian experiments by the practice of giving overall dressings of rock phosphate, but clear responses to P have also been obtained on this soil together with a positive interaction with K (Mollegaard, 1971; United Plantations, 1985).

The peat soils of Malaysia present special problems, and the Peat Yellows symptoms have already been described (p. 495). Very low leaf levels of copper (1.4–2.5 ppm) and zinc (6–7 ppm) are usual, with low leaf levels of N and K. Ng *et al.* (1974) obtained considerable changes in the growth of palms suffering from Peat Yellows or Mid-Crown chlorosis. Foliar sprays with 200 ppm Cu solution of copper sulphate resulted in large increases in leaf and leaflet lengths and raised leaf-Cu levels. Soil applications of 2.5 kg per palm copper sulphate increased leaf length in adult palms after a 3-month delay. Spear drenching with copper sulphate had a quicker effect. Although small yield responses have been obtained (Cheong and Ng, 1977), it has proved very difficult to improve the production of older palms suffering from acute symptoms. However, if proper treatment is given to young palms early enough, satisfactory growth and yields can be obtained on peat. The most recent experiments (United Plantations, 1985) have shown that even on the deepest peat good yields can be obtained by applying bunch ash (6–9 kg per palm), KCl (3 kg), rock phosphate (1 kg) and urea (1 kg), and spraying the foliage with about 4.5 litres per palm of copper and zinc sulphates at 150 ppm Cu and 1,000 ppm Zn two or three times a year. Gurmit Singh (1987) has claimed that zinc is the dominant element in curing or preventing Peat Yellows; his experiments on young replants on deep peat in southern Perak showed that zinc alone (1,000 ppm at 4.5 litres per palm) not only cured the condition and increased bunch yields, but also raised the leaf levels of N, P, K and Cu. Peat soils vary in depth and age and the variations of ground water level are always likely to be of first importance. Whether copper or zinc or both are needed will need determining for each area.

Foster and Goh (1977) fitted response functions to yield data of twenty fertilizer experiments, ten on coastal and ten on inland soils. They found that the yield without fertilizers was appreciably higher on the coastal than on inland soils and that responses to individual fertilizers, in the absence of other limiting factors, declined with increasing yield level. Thus responses to P and K were less on coastal soils and this was considered to be due not only to higher soil supplies but also to their higher buffer capacities. Nevertheless P application was found to be profitable on all soils except river alluvium (Briah series) and peat; and, provided P requirements had been satisfied, K and N fertilizers gave profitable responses on all inland soils and on Briah series. Profitable responses to Mg were rare.

Foster and Goh (1977) also found, as so frequently indicated earlier in this chapter, that responses to individual fertilizers were dependent on the adequacy of other fertilizers. There were also indications that responses could be inhibited on coastal soils by limited rainfall and on inland soils, in the case of N, by leaching. These authors charted responses on isoquant, or yield contour, maps showing both the range of dressings of two fertilizers giving positive responses and the most profitable combination at different yield levels; for this interesting method their paper should be consulted.

Africa

Nigeria

A new phase of fertilizer investigations started in West Africa around 1959 on rates of application, on the effect of the supply of one nutrient on the need for others, and on frequency of application. Experiments in Nigeria and Sierra Leone were, in general, of a 4^n design though four levels were not always retained for each nutrient. The different levels allowed for rising total quantities of fertilizer applied, over the 8 or 9 years, annually in the case of nitrogen, but triennially with other elements (Gunn, 1960). The nutrients and the number of levels to be employed were chosen after a consideration of both the soil type and the earlier, less exact, information on nutrient needs. For instance, on an area of K-deficient Acid Sands soil where N responses had sometimes been obtained, Mg responses were probable and P responses in the pres-

Table 11.17   *A $4^3 \times 2$ fertilizer experiment undertaken in Nigeria (kg per palm)*

Expt. 508–1 Fertilizer	Level	Year									Total (kg)
		0	1	2	3	4	5	6	7	8	
Urea	N_0										0
	N_1	0.11	0.11	0.23	0.34	0.34	0.34	0.34	0.34	0.34	2.50
	N_2	0.23	0.23	0.45	0.68	0.68	0.68	0.68	0.68	0.68	5.00
	N_3	0.34	0.34	0.68	1.02	1.02	1.02	1.02	1.02	1.02	7.49
Super-	P_0										0
phosphate	P_1	0.23	0.45	0.91	3.41	—	—	4.09	—	—	9.08
Sulphate of	K_0										0
potash	K_1	0.11	0.23	0.57	0.91	—	—	1.14	—	—	2.95
	K_2	0.23	0.45	1.14	1.82	—	—	2.27	—	—	5.90
	K_3	0.34	0.68	1.70	2.72	—	—	3.41	—	—	8.85
Magnesium	Mg_0										0
sulphate	Mg_1	0.06	0.18	0.34	1.02	—	—	1.14	—	—	2.72
(anhydrous)	Mg_2	0.11	0.34	0.68	2.04	—	—	2.27	—	—	5.45
	Mg_3	0.18	0.51	1.02	3.06	—	—	3.41	—	—	8.17

Note: Year 0 is the year of planting.

ence of K were not unlikely, the $4^3 \times 2$ design shown in Table 11.17 was arranged. Micronutrients were included in some experiments.

On the *Acid Sands* soils of Nigeria interest centred on K and Mg needs with N and P as subsidiary requirements. On young palms N has usually been needed. Applications of NK can sometimes induce magnesium deficiency on these soils and dressings of anhydrous magnesium sulphate at about half the rate of the minimum potash dressings are then advisable.

On the *Benin fasc* of the Acid Sands an experiment of the same general design as that shown in Table 11.17, but with only two rates (0 and 1) of N and Mg, showed that responses to K are obtained up to the K_2 rate, that P also increases yield and that there is a possible boron requirement, though the latter is only shown in the presence of Mg; N and Mg responses are absent, at least in the early years. On the *Calabar fasc*, with treatments as in Table 11.17, the first 4 years' yields showed that responses to K were significant up to the K_3 level and Mg responses to the Mg_1 level (Table 11.18). Here also there was a suggestion of a boron requirement but only in the presence of P (Aya, 1972). In the early years there were severe Mg deficiency symptoms in the Mg_0 plots and K applications increased these symptoms though improving growth (Sly *et al.*, 1963). The presence of P also increased Mg-deficiency symptoms though slightly improving initial growth. Nitrogen dressings enhanced early growth on both Benin and Calabar fasc soils, but this effect was not reflected in enhanced early yields; in fact on the Benin fasc there was a small initial negative response. It is a curious feature of some experiments that in spite of the better growth that

Table 11.18 *Responses to potassium, magnesium, phosphorus and boron on Acid Sands soils, Nigeria (NIFOR) (kg per palm per annum in first 4 or 5 years of bearing)*

Expt.	K_0	K_1	K_2	K_3	P_0	P_1	B_0	B_1	LSD P = 0.05
Benin fasc. [*] (2–15)	44.2	50.1	52.4	52.5	48.2	51.4	44.3	46.2	
					Mg_0	Mg_1	Mg_2	Mg_3	
Calabar fasc. [†] (508–1)	29.4	39.2	42.1	46.1	36.6	40.4	39.3	40.5	3.1

[*] First 5 years for K and P, 4 years for B.
[†] First 4 years of bearing.

is obtained from early applications of N fertilizers, this effect does not always persist, either in later development or in bunch yields.

Experiments on widely separated fields of older palms on the Calabar fasc showed yield responses to K and Mg. Many of the experiments were on old fields of poorly grown palms and, although Mg-deficiency symptoms were largely abated, yield responses to Mg were small and irregular. At Itu, where duplicate experiments were sited on adjoining areas showing, and not showing, Orange Frond symptoms, even combined KMg dressings did not raise the yield of old palms to levels normally expected in Nigeria, and it was evident that the Mg requirement was as high where there were no symptoms as where symptoms were shown (Sly and Sheldrick, 1965). Applications of Mg to old palms on very poor sandy soil at Akwete gave similar positive responses, but here also yields remained generally low. The fact is that many soils which are extremely Mg-deficient and are bearing old palms in south-eastern Nigeria are intrinsically very poor sands with a history of long usage for annual crops. Besides the exceptionally low exchangeable Mg level they have a much lower level of exchangeable K than, for instance, the Mg-deficient area at Etoumbi in Congo (Brazzaville) and their clay plus silt content is exceptionally low (only 12 per cent at Akwete) even at 60 cm depth.

Information on the fertilizer requirements of palms growing on *basement complex soils* remains meagre. In the Kwa Falls area, where soils derived from mica and quartzose schists predominate, responses to P have not been as great as expected. Highest yields were obtained in one experiment from ammonium phosphate with K applications, each fertilizer being applied at 2.27 kg per palm every 3 years. In the Calaro area young palms on soil derived from granite gneiss showed a 24 per cent bunch yield increase to half the P_1 levels of phosphorus shown in Table 11.17 and a 7.5–13.5 per cent increase to the higher rates of magnesium in the first 4 years of bearing. In the same area, on soils derived from mica schist, early yields suggested P and N requirements (Sheldrick, 1967).

Ghana

Fertilizer experiments laid down by NIFOR were extensively reviewed by Van der Vossen (1970). These experiments, of 2^4 or 2^5 design, were on soils derived from Tertiary sands contiguous with those of the Ivory Coast and on soils derived from granites and phyllites (Lower Birrimian formation). All fertilizers were applied at the rate of 0.23 kg per palm in the planting year and then, for 6 years, 0.45 kg per palm per year of age, i.e. reaching 2.72 kg per palm. Thereafter, P, K, Mg (and Ca if included) were applied triennially at 4.54, 2.27 and 2.72 kg of single superphosphate, muriate of potash and magnesium sulphate per palm respectively (and 3.18 kg lime if included), with sulphate of ammonia applied annually at 1.82 kg per palm.

On the silty clay soil of the Lower Birrimian phyllites, in Ghana's main cocoa area, very good yields were obtained through a 12-year period even on unfertilized plots. Apart from an early indication that the highest yields might be obtained from KMg applications, leaf analysis indicated that, later on, N, P and K may all be required. On the sandy and silty clays with some concretions overlying granites there was a P and K requirement with a positive PK interaction at Pretsea (Table 11.19 (2)) and a negative KMg interaction at Assin-Foso (Table 11.19 (1)). Nitrogen responses were also recorded at the latter place, but Mg was clearly not required on these soils.

The results of the experiment on Tertiary sands soil are of particular interest since they were in marked contrast to those obtained in the Ivory Coast on soils of the same derivation. At Aiyinasi high early yields were obtained from the superphosphate dressings with a later positive PK interaction giving up to 16 tonnes bunches per hectare in certain years. These yields fell away rapidly from the seventh year of bearing some 4 years after the triennial dressings started, and this decline continued until yields were only one-third of what they had been in the palms' early adult years. Before this time foliar analysis had indicated low P even in the P_1 plots and low N. Van der Vossen (1970) has considered the possible causes of this unusually severe decline and concludes that, although in part it could be accounted for by a nearly continuous fall from 1962 to 1968 in effective sunshine (see p. 179), the main and later factor was a depletion of organic matter and available nutrients, particularly P, in the soil, through root uptake, leaching and P fixation. He suggested a reversion to annual dressings, with at least 2–2.5 kg superphosphate per palm with addition of 1–1.5 kg of

Table 11.19 *Responses to fertilizers on soils derived from Pre-Cambrian rocks in Ghana (tonnes bunches per hectare per annum)*

	N_0	N_1	P_0	P_1	K_0	K_1	Mg_0	Mg_1
Birrimian phyllites								
First 3 years	9.44	9.80	9.46	9.78	9.27	9.71	9.51	9.73
Next 9 years	12.29	12.85	12.59	12.54	12.30	12.84	12.41	12.73
Leaf 17 per cent of dry matter	N 2.41	2.37	P 0.142	0.145	K 0.83	0.93	Mg 0.39	0.41
Granite								
(1) First 3 years	2.83	3.48	2.71	3.60*	2.81	3.51*	3.17	3.14
Next 6 years	8.61	9.96*	8.06	10.51*	8.80	9.77	9.30	9.26
Leaf 17 per cent of dry matter	N 2.68	2.71	P 0.134	0.155*	K 1.25	1.34	Mg 0.30	0.32
(2) First 3 years	8.76	8.53	8.21	9.09**	8.56	8.74	8.80	8.49
Next 6 years	9.02	9.26	8.74	9.54	8.57	9.70*	9.40	8.87
Leaf 17 per cent of dry matter	N 2.35	2.42	P 0.151	0.159	K 0.61	0.74	Mg 0.48	0.51
Interaction								
Last 6 years	P_0K_0 8.83	P_0K_1 8.66	P_1K_0 8.32	P_1K_1 10.76*				

For fertilizer dressings see text. * Significance at least at P = 0.05. ** Significance at P = 0.1.

Source: Van der Vossen (1970).

muriate of potash, and applications in August instead of May. From the leaf analysis it was seen that K content was severely reduced by P applications and the K applications were insufficient to bring the leaf level in the P_1 plots above 0.9 per cent of dry matter. Soils developed over Tertiary sands in the adjoining Ivory Coast are predominantly K-deficient with little or no demand for P and Van der Vossen suggested that the Aiyinasi soils are not typical of the Tertiary sands soils as a whole, being comparatively thin and underlain by granite-derived material.

Ivory Coast

In the Ivory Coast experiments up to 1968 reviewed by Bachy (1968) showed K to be the predominant need on the Tertiary sands soils. In areas of derived savannah plots yielding 17 kg per palm were raised to a mean production of 82 kg per palm (11.7 tonnes per hectare) by the application of 1 kg KCl per palm per annum; four years were needed to establish the new level of yield. Also in the savannah region a small increase in yield following magnesium sulphate applications was attributed to sulphur on the grounds that there was a small positive correlation between yield and leaf sulphur (IRHO, 1972). On the forest soils of the Tertiary sands the onset of potassium deficiency can be delayed until well after the palms come into bearing. In an experiment using a K_1 level of 3 kg per palm of potassium chloride or sulphate the leaf-K level of K_0 plots did not sink below 0.9 per cent until 8 years after planting (Fig. 11.9). Until that point was reached the yield increase due to K was not more than 5 per cent, but thereafter it rose to over 10 per cent with even greater increases in mean weight per bunch (Ollagnier and Ochs, 1981).

Phosphorus has only given small yield increases on typical Tertiary sands, though at Grand-Drewin in the Sassandra region, where soil phosphorus is reported as unusually low, responses to 2 kg per palm per annum of dicalcium phosphate were obtained with high rates of K (1.6 kg per palm).

Nitrogen has been found to depress yields on the Tertiary sands through, it is believed, its depressing effect on leaf-K; in the presence of K there is slight increase in yield due to N (IRHO, 1974).

There has been some anxiety that the effect of repeated KCl applications could be deleterious through the raising of leaf-Cl levels above the believed optima (0.5–0.6 per cent). These levels are attained in the Ivory Coast without K application. Some depressive effects of KCl have been recorded, and a change is being made to potassium sulphate in some experiments. However, it is also believed that positive N effects, where K is sufficient or excessive, may be due to an N-Cl antagonism and consequent reduction of the leaf-Cl content (IRHO, 1974). In one experiment (LM–CP14), in

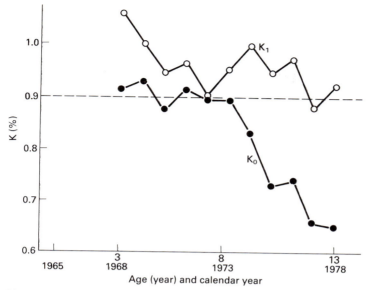

Fig. 11.9 Leaf level responses to potassium according to palm age on soils of the Ivory Coast Tertiary sands (Ollagnier and Ochs, 1981).

which there was a reduced yield with higher K dressings in the absence of N, application of sulphate of ammonia restored the yield of the K_2 plots to the control plot level.

Cameroon

In East Cameroon, oil palms respond to both K and Mg on Tertiary sands soil, experiments indicating an application of 1 kg KCl and 1 kg kieserite per palm per annum as probably optimal (IRHO, 1972, 1974). In West Cameroon on poor sandy soil a strong K requirement has also been demonstrated. In an experiment on poor lateritic soil derived from basement complex there were P and K responses with a positive PK interaction and a further yield increment provided by N and Mg. Rates giving optimum yield appeared to be 3 kg each of sulphate of ammonia and rock phosphate and 1 kg each of KCl and kieserite (Green, 1972). In another experiment, on soils developed over basaltic flows, an early P response was later supplemented by responses to K and Mg.

Benin (Dahomey)

On the *terre de barre* soils (p. 129), K deficiency is common, but responses are limited by the water deficits experienced in this dry part of the West African coast. In one experiment significant responses were only obtained when the water deficit was below

400 mm (IRHO, 1974). Leaf-K levels for optimum yield appear to lie between 0.75 and 0.85 per cent according to the current water deficit. An average of 0.75 kg KCl per palm per annum has been recommended (de Taffin and Ochs, 1973). Nitrogen is not recommended for young palms beyond their second year. Low production in some areas supplied with KCl suggested a possible excess of chlorine and in certain experiments leaf-Cl levels have reached more than 0.9 per cent of dry matter.

Sierra Leone

In Sierra Leone NIFOR experiments were planted on river alluvium and on lateritic soils. The sedimentary soils are similar to the Acid Sands of Nigeria. Symptoms of magnesium deficiency are widespread in Sierra Leone and it was expected that this deficiency would be found on both lateritic and inland alluvial soils. In the experiment on the former soil, with treatments as in Table 11.17 but including four levels of P and two of K instead of the reverse, there were early Mg-deficiency symptoms on K and NK plots, but the main requirements for yield have been shown to be P and N; yield levels are poor on this soil in which both rooting volume and water-holding capacity are low (Aya, 1970/72).

On river alluvium yields are considerably higher and the experiment allowed for six rates of magnesium sulphate and two rates of sulphate of potash (0, 1.36, 2.72, 4.09, 5.45 and 6.81 kg per palm of Mg sulphate and 0 and 4.09 kg of sulphate of potash over the first 6 years). Magnesium increased yields up to the Mg_2 level while there were also responses to K and a positive KMg interaction.

Zaire

Many prewar and postwar experiments at the Yangambi station of INEAC and elsewhere showed no responses to fertilizers. Examination of the treatments of these experiments (Vanderweyen, 1952) suggested that there had been an expectation, perhaps as the result of early Sumatran experience, that phosphorus would be the primary requirement on Zaire latosols. Most experimental treatments allowed for fairly liberal doses of phosphorus fertilizers with none or only small applications of potassium; even when included, dressings of the latter rarely rose above 1 kg of chloride or sulphate of potassium. In none of the earlier experiments reported were substantial potassium applications made either alone or with small supplements of other nutrients or as part of a factorial experiment. Indeed, in one experiment, dressings of 1.8 and 3.2 kg per palm were considered to be 'shock' applications.

A strong indication that palms in northern Zaire would in fact readily respond to fertilizers was later given by an experiment at Binga where application of 12 kg per palm of an NPKMg plus Zn, Cu, B, Mn fertilizer mixture distributed over 4 years gave rise to a 34 per cent yield increase over a 5-year period (Anon., 1962). Later, small responses to N were reported from northern Zaire, to P from both north ad south, and to K in northern Zaire (Green, 1972). At Yaligimba on the northern latosols responses were primarily to P and K but with additional yield increments from N and Mg, so that relatively high rates of all four elements (2 kg per palm per annum of Calnitro, Fertiphos and KCl and 1 kg of kieserite) maximized yields. However, owing to the high cost of fertilizers to inland Zaire plantations only 1 kg per palm of Fertiphos appeared to be economic, this treatment by itself giving a calculated yield increase of 12 per cent.

America

In *Brazil*, experiments have shown that on the soils developed over Tertiary sands around Belém phosphorus is the main requirement. A single application of 1 tonne of rock phosphate per hectare gave 100 per cent yield increase over a period of 6 years. Annual applications of superphosphate were almost as effective, and KCl at 0.25 kg increasing to 2 kg per palm gave a small additional yield increase.

In *Colombia*, on the alluvial soils of the Rio San Alberto, yield increases following KCl applications were attributed to chlorine rather than to potassium. Leaf-Cl showed a marked increase while leaf-K levels were reduced through the increase of Ca promoted by chlorine (Ollagnier and Ochs, 1981).

On the andosolic soils of *Ecuador* there are serious nutritional problems. Although deficiency symptoms of Mg and very low leaf levels are common, neither the latter nor yields are appreciably increased by Mg fertilizers. Leaf analyses often indicate what Ollagnier and Ochs (1981) consider to be P deficiency with N excess (see Fig. 11.4), but there have been no published experimental results confirming this. Potassium levels are sometimes low, but this can be partly attributed to the very high Ca levels. Throughout the oil palm areas of the Costa (Pacific plain), but particularly near to the Andes, palms have suffered from yellowing and leaf withering with consequently reduced yields. How far nutritional causes may be concerned in these symptoms is not known. The Costa is subject to marked variations in the length and intensity of the July to November drier period (see p. 97), and there is no doubt that water deficit plays a major part in the variations of both leaf symptoms and yield.

In *Peru*, on riverine alluvium east of the Andes, leaf levels of chlorine are exceptionally low and experiments have indicated that where applications of KCl or MgCl have given yield increases these have been due to chlorine (Daniel and Ochs, 1975).

Much of the nutritional experience gained on andosolic soils in

Latin America, particularly with regard to K, Mg and Cl, has been paralleled on andosolic soils of New Britain (Papua New Guinea – Breure and Rosenquist, 1977).

Planting density and fertilizer responses

Several density experiments have indicated that responses to fertilizers are much more pronounced at low planting densities and may be absent altogether at high densities. In New Britain (Breure, 1977) NKMg fertilizers gave over 30 per cent increase in bunch yield at 110 palms per hectare, less than 10 per cent increase at 148 palms per hectare and had no effect at all at 186 palms per hectare. Increased density augments mutual shading and increases the leaf area index; net assimilation rate is, however, reduced, and the decrease in total dry matter production per palm is entirely accounted for by a lower production of bunches (Corley, 1973b). Application of fertilizers may therefore be entirely or partially wasted if a high density, or in some circumstances even a 'normal' density, is employed.

Fertilizer effects should therefore be considered in conjunction with growth factors. Different combinations of soil and climate produce palms differing in their vegetative growth. Palms which are grown under conditions conducive to rapid and extended vegetative development may, if planted at high or normal density, not only fail to respond to fertilizers, but also show a rapid decline in bunch yield when their canopies interlock.

Goh (1981) concluded from Malaysian experiments that the optimum density tends to be lowest in the best environment, and that this is especially the case where high rates of fertilizers are employed. Two questions therefore arise. Firstly (and this is especially important where fertilizer prices are high or rising), will it be advantageous to increase the density and use less fertilizer? Alternatively, having regard to the flat-topped nature of cumulative yield/density curves (Corley *et al.*, 1972), might it be better to reduce the density below the calculated optimum and so take fullest advantage of fertilizer effects? Without being able to predict future costs it is of course impossible to give simple answers to these questions, but Goh believed it would be prudent on good Malaysian inland soils to accept a density of not more than 136 palms per hectare.

References

Anon. (1952) Notes on current investigations, Oct. to Dec. 1951: oil palm. *Malay. agric. J.*, **35**, 41.

Anon. (1962) Modalités de replantation du palmier à huile avec ou sans apport d'engrais. *Bull. Inf. INEAC*, **11**, 113.

Arokiasamy, M. (1969) Investigation on the best method of using the oil palm bunch waste as a fertilizer. *Commun. (Agron) Chemara Res.*, **7**, 1–7.

Ataga, D. O. (1974) Release and fixation of potassium in some soils supporting the oil palm (*Elaeis guineensis* Jacq.) in Nigeria. In *Potassium in tropical crops and soils*, Proc. 10th Colloquium Int. Potash Inst., p. 131.

Ataga, D. O. (1978) Soil phosphorus status and responses of the oil palm on some acid soils. *J. of NIFOR*, **5**(20), 25.

Ataga, D. O., Omoti, U. and **Ojuederie, B. M.** (1981) Effect of micronutrients on the yield and mineral nutrition of the oil palm on an Acid Sand soil in Nigeria. In *Oil palm in agriculture in the eighties*, Kuala Lumpur.

Aya, F. O. (1970/72) *NIFOR fourth and fifth annual reports (1967–8 and 1968–9)*, pp. 32 and 31 respectively.

Aya, F. O. (1972) *NIFOR fifth annual report (1968–9)*, pp. 28–33.

Bachy, A. (1964) Diagnostic foliaire de palmier à huile. Niveaux critiques chez les arbres jeunes. *Oléagineux*, **19**, 253.

Bachy, A. (1968) Principaux résultats acquis par l'IRHO sur la fertilisation du palmier à huile. *Oléagineux*, **23**, 9.

Bachy, A. (1969) A propos d'un cas typique de carence potassique du palmier à huile en Côte d'Ivoire *Oléagineux*, **24**, 533.

Breure, G. J. (1977) Preliminary results from an oil palm density fertilizer experiment on young volcanic soils in West New Britain. In *International developments in oil palm*, Kuala Lumpur.

Breure, C. J. and **Rosenquist, E. A.** (1977) An oil palm fertilizer experiment on volcanic soil in Papua New Guinea. In *International developments in oil palm*, Kuala Lumpur.

Broeshart, H. (1955) The application of foliar analysis in oil palm cultivation. Thesis presented to the Agricultural University of Wageningen, Holland.

Broeshart, H., Ferwerda, J. D. and **Kovachich, W. G.** (1957) Mineral deficiency symptoms of the oil palm. *Pl. Soil*, **8**, 289.

Bull, R. A. (1954a) A preliminary list of the oil palm diseases encountered in Nigeria. *J. W. Afr. Oil Palm Res.*, **1**(2), 53.

Bull, R. A. (1954b) Studies on the deficiency diseases of the oil palm. 1. Orange Frond disease caused by magnesium deficiency. *J. W. Afr., Inst. Oil Palm Res.*, **1**(2), 94.

Bull, R. A. (1957a) Techniques for visual diagnosis of mineral disorders and their application to the oil palm. Proceedings of the Anglo-French conference on the Oil Palm, Jan. 1956, *Bull. agron. Minist. Fr. outre mer*, No. 14, p. 137.

Bull, R. A. (1957b) *WAIFOR fifth annual report 1956–7*, p. 98 and Tinker, P.B.H. ibid., p. 79.

Bull, R. A. (1960) *WAIFOR eighth annual report, 1959–60*, p. 104.

Bull, R. A. (1961a) Studies on the deficiency diseases of the oil palm. 2. Macronutrient deficiency symptoms in oil palm seedlings grown in sand culture. *J. W. Afr. Inst. Oil Palm Res.*, **3**, 254.

Bull, R. A. (1961b) Studies on the deficiency diseases of the oil palm. 3. Micronutrient deficiency symptoms in oil palm seedlings grown in sand culture. *J. W. Afr. Inst. Oil Palm Res.*, **3**, 265.

Calvez, J., Olivin, J. and **Renard, J. L.** (1976) Study of a sulphur deficiency on young oil palms in the Ivory Coast. *Oléagineux*, **31**, 251.

Chan Kook Weng (1981) Nitrogen requirements of oil palms in Malaysia: fifty years of experimental results. In *Oil palm in agriculture in the eighties*, Kuala Lumpur.

Chan Kook Weng and **Rajaratnam, J. A.** (1977) Magnesium requirement of oil palms in Malaysia: 45 years of experimental results. In *International developments in oil palm*, Kuala Lumpur.

Chapas, L. C. and **Bull, R. A.** (1956) Effects of soil application of nitrogen, phosphorus, potassium and calcium on yields and deficiency symptoms in mature oil palms at Umudike. *J. W. Afr. Inst. Oil Palm Res.*, **2**, 74.

Chapman, G. W. and **Gray, H. M.** (1949) Leaf analysis and the nutrition of the oil palm. *Ann. Bot.*, NS, **13**, 415.

Cheong Siew Park and **Ng Siew Kee** (1977) Copper deficiency of oil palms on peat. In *International Developments in oil palm*, Kuala Lumpur.

Corley, R. H. V. (1973a) Oil palm physiology: a review. In *Advances in oil palm cultivation*, p. 324, Incorp. Soc. of Planters, Kuala Lumpur.

Corley, R. H. V. (1973b) Effects of plant density on growth and yield of oil palms. *Expl. Agric.*, **9**, 169.

Corley, R. H. V., Hew, C. K., Tam, T. K. and **Lo, K. K.** (1972) Optimal spacing for oil palms. *Advances in oil palm cultivation*, Kuala Lumpur, 52.

Corley, R. H. V. and **Mok, C. K.** (1972) The effects of nitrogen, phosphorus, potassium and magnesium on growth of the oil palm. *Expl. Agric.*, **8**, 347.

Coulter, J. K. and **Rosenquist, E. A.** (1955) Mineral nutrition of the oil palm. *Malay. agric. J.*, **38**, 214.

Daniels, G. and **Ochs, R.** (1975) Amélioration de la production des jeunes palmiers à huile du Pérou par l'emploi d'engrais Chlore. *Oléagineux*, **30**, 295.

Dell, W. and **Arens, P. L.** (1957) Inefficacité du posphate naturel pour le palmier à huile sur certains sols de Sumatra. *Oléagineux*, **12**, 675.

De Taffin, G. and **Ochs, R.** (1973) La fumure potassique du palmier à huile au Dahomey, *Oléagineux*, **28**, 269.

Dufour, F. and **Quencez, P.** (1979) Etude de la nutrition des oligo-éléments du palmier à huile et cocotier cultivés sur solutions nutritives. *Oléagineux*, **34**, 323.

Eschbach, J. M. (1980) Les oligo-éléments dans la nutrition du palmier à huile. *Oléagineux*, **35**, 281.

Ferwerda, J. D. (1954) Boron deficiency on oil palms in the Kasai region of the Belgian Congo. *Nature, Lond.*, **173**, 1097.

Ferwerda, J. D. (1955) Questions relevant to replanting in oil palm cultivation. Thesis presented to the Agric. University of Wageningen, Holland.

Forde, St C. M. (1965) The phosphorus status of some soils of West Africa. Nigeria Institute for Oil Palm Research, Conference Paper.

Forde, St C. M., Leyritz, M. J-P. and **Sly, J. M. A.** (1965) The importance of potassium in the nutrition of the oil palm in Nigeria. *Potash Review*, Subject 27, 46th suite, June 1966.

Forde, St C. M. and **Leyritz, M. J-P.** (1968) A study of Confluent Orange Spotting of the oil palm in Nigeria. *J. Nigerian Inst., Oil Palm Res.*, **4**, 372.

Forde, St C. M., Leyritz, M. J-P. and **Sly, J. M. A.** (1968) The role of potassium in the nutrition of the oil palm in Nigeria. *J. Nig. Inst. Oil Palm Res.*, **4**, 331.

Foster, H. L. and **Chang, K. C.** (1977) The diagnosis of the nutrient status of oil palms in West Malaysia. In *International developments in oil palm*, Kuala Lumpur.

Foster, H. L., Chang K. C., Mohd. Tayeb Hj. Dolmat, Ahmad Tarmizi Mohammed and **Zin Z. Zakaria** (1985a) Oil palm yield responses to N and K fertilizers in different environments in peninsular Malaysia. *PORIM Occ. paper*, 16.

Foster, H. L., Mohd. Tayeb Hj. Dolmat and **Gurmit Singh** (1987a) The effect of fertilizers on oil palm bunch components in peninsular Malaysia. *Int. Oil Palm. Conf.*, Kuala Lumpur, June 1987.

Foster, H. L., Mohd. Tayeb Hj. Dolmat and **Zin Z. Zakaria** (1985b) Oil palm yields in the absence of N and K fertilizers in different environments in peninsular Malaysia. *PORIM Occ. paper*, 15.

Foster, H. L. and **Goh, H. S.** (1977) Fertilizer requirements· of oil palm in West Malaysia. In *International developments in oil palm*, Kuala Lumpur.

Foster, H. L., Tarmizi Mohammed, A. and **Zakariah, Zin Z.** (1987b) Foliar diagnosis of oil palm in peninsular Malaysia. *Int. Oil Palm Conf.*, Kuala Lumpur ·June 1987.

Goh Khen Hing (1981) Analyses of oil palm spacing experiments. In *Oil palm in agriculture in the eighties*, Kuala Lumpur.

Green, A. H. (1972) Annual Review of Research, 1970. Unilever Plantation group, Unilever, London. 140 pp. Mimeograph.

Green, A. H. (1974) Book review. *Planter, Kuala Lumpur*, **50**, 242.

Gunn, J. S. (1960) *WAIFOR eighth annual report 1959–60*, p. 57.

Gunn, J. S. (1962) The economics of manuring the oil palm. *J. W. Afr. Inst. Oil Palm Res.*, **3**, 302.

Gurmit Singh (1987) Zinc nutrition of oil palms on peat soils. *Int. Oil Palm Conf.*, Kuala Lumpur, June 1987.

Haines, W. B. and **Benzian, B.** (1956) Some manuring experiments on oil palm in Africa. *Emp. J. exp. Agric.*, **24**, 137.

Hale, J. B. (1947) Mineral composition of leaflets in relation to the chlorosis and bronzing of oil palm in West Africa. *J. agric. Sci., Camb.*, **37**, 236.

Hew, C. K., Ng, S. K. and **Lim, K. P.** (1973) The rationalisation of manuring oil palms and its economics in Malaysia. In *Advances in oil palm cultivation*, p. 306, Incorp. Soc. of Planters, Kuala Lumpur.

Hew, C. K. and **Poon, Y. C.** (1973) The effects of muriate of potash and bunch ash on uptake of potassium and chlorine in oil palms on coastal soils. In *Advances in oil palm cultivation*, p. 239, Incorp. Soc. of Planters, Kuala Lumpur.

Holland, D. A. (1967) The interpretation of the chemical composition of some tropical crops by the method of component analysis. *Oléagineux*, **22**, 307.

IRHO (1962) Vingt ans d'activité. *Oléagineux*, **17**, 249.

IRHO (1972) *Rapport d'activités 1971*, pp. 56–7.

IRHO (1974) *Rapport d'activités, 1972–73*. Document 1138, Paris.

IRHO (1975) Private communication.

Knecht, J. C. K., Ramachandran, R. and **Narayanan, R.** (1977) Variation of leaf nutrient contents with age of palms in oil palm leaf sampling. *Oléagineux*, **32**, 139.

Kowal, J. M. L. and **Tinker, P. B. H.** (1959) Soil changes under a plantation established from high secondary forest. *J. W. Afr. Inst. Oil Palm Res.*, **2**, 376.

Leyritz, M. J-P. (1964) *WAIFOR twelfth annual report 1963–4*, p. 79.

Lo, K. K., Chan, K. W., Goh, K. H. and **Hardon, J. J.** (1973) Effect of manuring on yield, vegetative growth and leaf nutrient level of the oil palm. In *Advances in oil palm cultivation*, p. 324, Incorp. Soc. of Planters, Kuala Lumpur.

Lo, K. K. and **Goh, K. H.** (1973) The analysis of experiments on the economics of fertilizer application on oil palms. In *Advances in oil palm cultivation*, p. 338, Incorp. Soc. of Planters, Kuala Lumpur.

Martin, G. (1977) Préparation et conditionnement des échantillons pour le diagnostic foliaire du palmier à huile et de cocotier. *Oléagineux*, **32**, 95.

May, E. B. (1956a) The manuring of oil palms – a review. *J. W. Afr. Inst. Oil Palm Res.*, **2**, 6.

May, E. B. (1956b) Early manuring experiments on oil palms in Nigeria. *J. W. Afr. Inst. Oil Palm Res.*, **2**, 47.

Mollegaard, H. (1971) Results of a fertilizer trial on a mixed colluvial alluvial soil at Ulu Bernam in West Malaysia. *Oléagineux*, **26**, 449.

Ng Siew Kee (1965) The potassium status of some Malaysian soils. *Malaysian Agric. J.*, **45**, 143.

Ng Siew Kee (1967 and 1968) Nutrient contents of oil palms in Malaya. I. Nutrients required for reproduction: fruit bunches and male inflorescences. II. Nutrients in vegetative tissues. *Malay. agric. J.*, **46**, 3 and 332.

Ng Siew Kee (1970) Greater productivity of the oil palm (*Elaeis guineensis* Jacq.) with efficient fertilizer practices. In *Role of fertilization in the intensification of agricultural production*, Proc. 9th Congress Int. Potash Inst., p. 357.

Ng Siew Kee (1972) *The oil palm, its culture, manuring and utilization*. International Potash Institute, Berne, Switzerland.

Ng Siew Kee (1977) Review of oil palm nutrition and manuring – scope for greater economy in fertilizer use. In *International developments in oil palm*, Kuala Lumpur.

Ng, S. K. and **Selvadurai, K.** (1967) Scope for using detailed soil maps in the planting industry in Malaysia. *Malaysian Agric. J.*, **46**, 158.

Ng, S. K. and **Tan, Y. P.** (1974) Nutritional complexes of oil palms planted on peat in Malaysia. I. Foliar symptoms, nutrient compositions and yield. *Oléagineux*, **29**, 1.

Ng Siew Kee, Chan, E., Tan Yap Pau and **Cheong Siew Park** (1974) Nutritional complexes of oil palms planted on peat soil in Malaysia. I. Foliar symptoms, nutrient composition and yield. II. Preliminary results of copper sulphate treatment. *Oléagineux*, **29**, 1–14, 445–56.

Ng Siew Kee and **Walters, E.** (1969) Field sampling studies for foliar analysis in oil palms. In *Progress in oil palm*, p. 67, Incorp. Soc. of Planters, Kuala Lumpur.

Ochs, R. (1965) La fumure potassique du palmier à huile. *Oil Palm Conference*, London, 1965, and *Oléagineux*, **20**, 365, 433 and 497.

Ochs, R. and **Olivin, J.** (1977) Le diagnostic foliaire pour le contrôle de la nutrition des plantations de palmiers à huile. *Oléagineux*, **32**, 211.

Ollagnier, M. (1973a) La nutrition anionique du palmier à huile. Application à la détermination d'une politique de fumure minérale à Sumatra. *Oléagineux*, **28**, 1.

Ollagnier, M. (1973b) Anionic nutrition of the oil palm: application to fertilizer policy in northern Sumatra. In *Advances in oil palm cultivation*, Kuala Lumpur, p. 227.

Ollagnier, M., Daniel, C., Fallavier, P. and **Ochs, R.** (1987) The influence of climate and soil on potassium critical levels. *Int. Oil Palm Conf.*, Kuala Lumpur, June 1987.

Ollagnier, M. and **Ochs, R.** (1971) Le chlore, nouvel élément essentiel dans la nutrition du palmier à huile; *and* La nutrition en chlore du palmier à huile et du cocotier. *Oléagineux*, **26**, 1–15 and 367–72.

Ollagnier, M. and **Ochs, R.** (1972) Les déficiences en soufre du palmier à huile et du cocotier. *Oléagineux*, **27**, 193.

Ollagnier, M. and **Ochs, R.** (1981) Management of mineral nutrition on industrial oil palm plantations – fertilizer savings. In *Oil palm in agriculture in the eighties*, Kuala Lumpur.

Ollagnier, M., Ochs., R. and **Martin, G.** (1970) The manuring of the oil palm in the world. *Fertilité*, **36**, March–April 1970, 63 pp.

Ollagnier, M. and **Valverde, G.** (1968) Contribution à l'étude de la carence en bore du palmier à huile. *Oléagineux*, **23**, 359.

Paton, T. R. (1963) *A reconnaissance soil survey of soils of the Semporna Peninsula, North Borneo.* Dept. Tech. Cooperation. Colonial Research Studies, No. 36, HMSO.

Poon Yew Chin (1969) An outline of the technique of oil palm foliar analysis. *Planter, Kuala Lumpur*, **45**, 452.

Poon Yew Chin, Varley, J. A. and **Ward, J. B.** (1970a) The foliar composition of oil palms in West Malaysia. III. Differences in leaf composition between fruit types. *Expl. Agric.*, **6**, 335.

Poon Yew Chin, Varley, J. A. and **Ward, J. B.** (1970b) Foliar composition of the oil palm in West Malaysia. I. Variation in leaf nutrient levels in relation to sampling intensity. II. The relationships between nutrient contents. *Expl. Agric.*, **6**, 113–21, 191–6.

Prevot, P. and **Montbreton, C. Peyre de** (1958) Étude des gradients en divers éléments minéraux selon le rang de la feuille chez le palmier à huile. *Oléagineux*, **13**, 317.

Prevot, P. and **Ollagnier, M.** (1954) Peanut and oil palm foliar diagnosis interrelations of N, P, K, Ca and Mg. *Pl. Physiol.*, **29**, 26.

Prevot, P. and **Ollagnier, M.** (1957) Méthode d'utilisation du diagnostic foliaire. In *Plant analysis and fertiliser problems*, IRHO, Paris.

Prevot, P. and **Ollagnier, M.** (1961) Law of the minimum and balanced mineral nutrition. In *Plant analysis and fertiliser problems*, Am. Inst. Biol. Sci., p. 257.

Prevot, P. and **Ziller, R.** (1957) Étude d'une carence en potasse et en azote sur palmier à huile au Dahomey. *Oléagineux*, **12**, 369.

Prevot, P. and **Ziller, R.** (1958) Relation entre le magnésium du sol et de la feuille de palmier. *Oléagineux*, **13**, 667.

Purba, A. Y. L. and **Turner, P. D.** (1973) Severe boron deficiency in young oil palms in Sumatra. *Planter, Kuala Lumpur*, **49**, 10.

Rajaratnam, J. A. (1972a) Observations on boron-deficient oil palms (*Elaeis guineensis*). *Expl. Agric.*, **8**, 339.

Rajaratnam, J. A. (1972b) 'Hook Leaf' and 'Fish-tail Leaf': boron deficiency symptoms of the oil palm. *Planter, Kuala Lumpur*, **48**, 120.

Rajaratnam, J. A. (1972c) The distribution and mobility of boron within the oil palm, *Elaeis guineensis*. I. Natural distribution. II. The fate of applied boron. *Ann. Bot.*, **36**, 289–97, 299–305.

Rajaratnam, J. A. (1973) Boron toxicity in oil palms (*Elaeis guineensis*). *Malaysian Agr. Res.*, **2**, 95.

Robertson, J. S. (1960) *WAIFOR eighth annual report 1959–60*, p. 112.

Rosenquist, A. E. (1962) Fertiliser experiments on oil palms in Malaya. Part 1. Yield data. *J. W. Afr. Inst. Oil Palm Res.*, **3**, 291.

Ruer, P. (1966) Relations entre facteurs climatiques et nutrition minérale chez le palmier à huile. *Oléagineux*, **21**, 143.

Ruer, P. (1967) Répartition en surface du système radiculaire du palmier à huile. *Oléagineux*, **22**, 535.

Russell, E. W. (1961) *Soil conditions and plant growth*, 9th edn. Longman, London.

Scheidecker, D. and **Prevot, P.** (1954) Nutrition minérale du palmier à huile à Pobé (Dahomey). *Oléagineux*, **9**, 13.

Sheldrick, R. D. (1967) *NIFOR Third Annual Report (1966–67)*, pp. 38–9.

Sly, J. M. A. and **Sheldrick, R. D.** (1965) *NIFOR first annual Report*, pp. 45–6.

Sly, J. M. A. *et al.* (1963) *WAIFOR eleventh annual report, 1962–3*, p. 35.

Smilde, K. W. (1961) *WAIFOR ninth annual report 1960–1*, p. 82.

Smilde, K. W. (1962/63) *WAIFOR tenth and eleventh annual reports 1961–2, 1962–3*, pp. 74 and 75 respectively.

Smilde, K. W. (1963) *WAIFOR eleventh annual report 1962–3*, pp. 70–2.

Smilde, K. W. and **Chapas, L. C.** (1963) The determination of nutrient status by leaf sampling of oil palms. *J. W. Afr. Inst. Oil Palm Res.*, **4**, 8.

Smilde, K. W. and **Leyritz, M. J-P.** (1965) A further investigation on the errors involved in leaf sampling of oil palms. *J. Nigerian Inst. Oil Palm Res.*, **4**, 251.

Smith, P. F. (1962) Mineral analysis of plant tissues. *A. Rev. Pl. Physiol.*, **13**, 81.

Sparnaaij, L. D., Rees, A. R. and **Chapas, L. C.** (1963) Annual yield variation in the oil palm. *J. W. Afr. Inst. Oil Palm Res.*, **4**, 111.

Taillez, B. (1971) Le système racinaire du palmier à huile sur la plantation de San Alberto (Colombie). *Oléagineux*, **26**, 435.

Taillez, B. (1982) Importance des fumures équilibrées sur jeunes palmeraies au Nord-Sumatra. *Oléagineux*, **37**, 271.

Tan, K. S. (1973) Fertilizer trials on oil palms on inland soils on Dunlop estates. In *Advances in oil palm cultivation*, p. 248, Incorp. Soc. of Planters, Kuala Lumpur.

Taniputra, B. and **Panjaitan, A.** (1981) An oil palm fertilizer experiment on yellowish-red podsolic soil in north Sumatra. In *Oil palm in agriculture in the eighties*, Kuala Lumpur.

Teoh, K. C., Chew, P. S. Soh, A. C. and **Chow, C. S.** (1981) A study of the seasonal fluctuations in leaf nutrient level in oil palm in peninsular Malaysia. In *Oil palm in agriculture in the eighties*, Kuala Lumpur.

Thompson, A. (1941) Notes on plant diseases. *Malay. agric. J.*, **29**, 241.

Tinker, P. B. H. (1961) *WAIFOR ninth annual report 1960–1*, p. 105.

Tinker, P. B. H. (1963) Changes occurring in the sedimentary soils of Southern Nigeria after oil palm plantation establishment. *J. W. Afr. Inst. Oil Res.*, **4**, 66.

Tinker, P. B. H. (1964) Studies on soil potassium. IV. Equilibrium cation activity ratios and responses to potassium fertiliser of Nigerian oil palms. *J. Soil Sci.*, **15**, 35.

Tinker, P. B. H. and **Smilde, K. W.** (1963a) Dry-matter production and nutrient content of plantation oil palms. II. Nutrient content. *Pl. Soil*, **19**, 19.

Tinker, P. B. H. and **Smilde, K. W.** (1963b) Cation relationships and magnesium deficiency in the oil palm. *J. W. Afr. Inst. Oil Palm Res.*, **4**, 82.

Tinker, P. B. H. and **Ziboh, C. O.** (1959) Soil analysis and fertiliser response. *J. W. Afr. Inst. Oil Palm Res.*, **3**, 52.

Tollenaar, D. (1969) Boron deficiency in sugar cane, oil palm and other monocotyledons on volcanic soil in Ecuador. *Neth. J. Agric. Sci.*, **17**, 81.

Toovey, F. W. (1948) *Seventh annual report, Oil Palm Research Station, Nigeria, 1946–7.*

Turner, P. D. and **Gillbanks, R. A.** (1974) *Oil palm cultivation and management.* Incorp. Soc. of Planters, Kuala Lumpur.

Turner, P. D., Soekoyo, P. and **Pani, H. A.** (1983) Carence en soufre dans une pépinière de palmier à huile au nord de Sumatra. *Oléagineux*, **38**, 7.

Umar Akbar, Tampubolon, F. H., Amiruddin, D. and **Ollagnier, M.** (1977) Fertilizer experimentation on oil palm in north Sumatra. In *International developments in oil palm*, Kuala Lumpur.

United Plantations Research Department (1985) *20th annual report, 1984.*

Uribe, A. and **Bernal, G.** (1973) Incinerateur de rafles des régimes de palmier à huile. Utilisation des cendres. *Oléagineux*, **28**, 147.

Van der Vossen, H. A. M. (1970) Nutrient status and fertilizer responses of oil palms on different soils in the forest zone of Ghana. *Ghana J. Agric. Sci.*, **3**, 109.

Vanderweyen, R. (1952) *Notions de culture d'Elaeis au Congo Belge.* Brussels.

WAIFOR (1956) *Fourth annual report, 1955–6*, p. 82.

Wallace, T. (1957) Methods of diagnosing the mineral status of plants. In *Plant analysis and fertiliser problems*, p. 13, IRHO, Paris.

Ward, J. B. (1966a) Sampling oil palms for foliar diagnosis. *Oléagineux*, **21**, 277.

Ward, J. B. (1966b) Private communication.

Warriar, S. M. and **Piggott, C. J.** (1973) Rehabilitation of oil palms by corrective manuring based on leaf analysis. In *Advances in oil palm cultivation*, p. 289, Incorp. Soc. of Planters, Kuala Lumpur.

Werkhoven, J. (1965) *The manuring of the oil palm.* Verlagsgesellschaft für Ackerbau GmbH, 51 pp. Hanover.

Chapter 12

Mixed cropping, rearing livestock among oil palms and tapping for wine

The desire to combine either the cultivation of a subsidiary crop or the raising of livestock with the growing of a plantation crop does not usually arise where the latter is being cultivated on a large scale. Those in control normally wish to concentrate all their efforts and resources upon the opening of new land and the successive establishment of large fields, well planted and designed solely for the crop concerned. It is to the small farmer that a subsidiary means of livelihood is more usually attractive since in the first place he often needs an income for the few years before the main crop comes into bearing, and secondly his acreage may be so small that he is forced to consider a more intensive use of the land throughout the life of the crop.

With the oil palm these latter considerations are further influenced by the small stand per hectare, though this is counteracted in some measure by the rapidity with which the palm comes into bearing. A farmer depending largely on his own resources is likely to wish not only to engage in establishment inter-cropping, which was discussed in Chapter 8, but to use his land to provide himself each year with food crops and perhaps livestock. Furthermore, he may wish to use some of his palms for the production of palm wine. In Africa, particularly where population pressure is high, the growing of food crops among oil palms is already standard practice, and any small farmer who planted an area with palms would almost certainly wish to cultivate the interlines with such crops in rotation with a natural cover (Waterston, 1953; Sparnaaij, 1957).

The low stand per hectare has also had some influence on the large grower, and the inter-planting of such crops as cocoa or coffee has been tried or advocated on many occasions and has sometimes met with success.

There is very little tradition for cattle raising in the wet tropical lowlands, except in Latin America, but improvements in the control of cattle diseases and an increasing demand for animal products has encouraged an interest in cattle where this only existed previously on a very small and primitive scale. On some of the new American plantations the grazing of cattle or sheep among the palms is almost taken for granted.

Inter-cultivation of food crops: planting palms and farming

In those parts of Africa where the oil palm is the cash crop of the peasant population, the farmers are also engaged in subsistence farming. The relationship of the palm groves to farming has already been discussed in Chapter 1. In these regions palms and food crops have to share the same areas permanently and the question arises as to how they should be arranged. The majority of food crops, e.g. maize, cassava, yams (*Dioscorea* spp.), are not suited to growing under shade, though cocoyams (*Colocasia* and *Xanthosoma* spp.) are adapted to shade conditions. Where food crops are grown haphazardly under palms, yields suffer through competition both for nutrients and light, and if both cultures are to be organized on the same holding they must either have separate areas allotted to them or the spacing of the oil palms must be such that competition between the crops is reduced to a minimum. As the growing of food crops usually demands a resting period, the farmer will normally wish to cultivate them over as wide an area as possible, and the allotment of separate areas for food crops and palms may not be to his taste.

These considerations gave rise to several experiments in Nigeria where this problem is most acute. In an early spacing experiment, already described in Chapter 9, rows were spaced 65 ft (19.8 m) apart to allow sufficient room for food-crop cultivation or for grazing. It was shown that the palms could not be crowded in the rows to compensate for the wide spacing. Palms spaced at 19.8 × 3.7 m (65 × 12 ft) gave a lower yield than palms spaced at 19.8 × 6.4 m (65 × 21 ft). Provided both the palms and the food crops received satisfactory manurial treatments, individual palms spaced normally *within* the widespread rows yielded slightly more than palms at normal spacing both within and between the rows. Food crops kept 3.7 m away from the palm rows had no effect on the palms' yield. These results suggested that the combination of planting palms with farming should be studied in more detail and the effects of fertilizers, establishment inter-cropping and cultivation should be taken into account (Gunn and Sparnaaij, 1957; Sly *et al.*, 1963). Experiments laid down on forest and old grove areas in Nigeria and on previously cultivated land in Sierra Leone had the treatments shown in Table 12.1.

In view of the poor growth of palms crowded in the rows, all palms were planted at the basic spacing of 8.83 m (29 ft) triangular;

Table 12.1   *The treatments of farming and planting experiments in Nigeria and Sierra Leone*

Factor	Level 0	Level 1	Level 2
A. Fertilizer (to palms)	None	In April (beginning of rains)	In October (end of rains)
B. Establishment inter-cropping or cultivation	None	Establishment inter-cropping for 2 years	Dry season cultivation for 2 years*
C. System of planting and farming	Normal density; no farming	Half density; rotational farming in wide interlines	Two-thirds density; rotational farming in wide interlines[†]

* Omitted in the old grove area in Nigeria, and in Sierra Leone.
[†] In Sierra Leone this was replaced by half density with *grazing* in the wide interlines.

the half density plots represented areas in which every other pair of rows had been omitted, and the two-thirds density plots represented areas with one row omitted in every three, these two arrangements giving a wide interline for farming of 15.24 m (50 ft) in the first case and 7.62 m (25 ft) in the second, no farming being permitted within 3.8 m (12½ ft) of the rows. Establishment inter-cropping (Plate 12.1) covered the whole area except the weeded circles of the palms and produced the expected results: increased early yields of the palms. Dry-season cultivation also increased yield per palm in the first 4 years (Sly and Sheldrick, 1965). Fertilizers were immediately shown to be necessary in the areas not opened from forest.

In these experiments the palm rows were set N–S since it was expected that the high incidence of sunlight thus obtained might increase the sex ratio and thus induce a higher yield per palm; a reduction of stand by one-third or one-half to enable food-cropping to take place would not then result in reductions of crop of the same magnitude. This expectation was fulfilled and after 15 years the half density plots were still giving a yield per hectare of more than two-thirds that of the normal density plots (Aya, 1974). Dry-season cultivation increased yields in the previously forested area up to the eleventh year from planting, while the beneficial effect of establishment inter-cropping lasted for 7 years in the forest area and for at least 13 years in the old grove area (Aya, 1975). Satisfactory food crops were obtained in the wider interlines, but the experiments were unfortunately concluded after 16 years whereas experience of both cultures over a complete oil palm crop cycle would have been valuable. Leaving out one row in three did not provide enough space for satisfactory food-crop cultivation and the best means of combining the two cultures at present seems to be the alternating of normally spaced pairs of rows with food-crop cultivation in the lanes which would have been occupied by other pairs of rows, the

Pl. 12.1   Establishment inter-cropping: young palms growing between lines of yams (*Dioscorea* sp.) in Nigeria.

food-crop cultivation being kept about 4 m away from the palm rows.

Experience of replanting palm groves has shown that there is often the greatest difficulty in persuading farmers not to plant food crops right up to the palms and to damage the palms in so doing. Moreover cultivation tends to continue in the vicinity of the palms for very many years after the latter are planted. As a result, the palms come late into bearing and their subsequent yield is reduced. It is therefore best to confine even establishment inter-cropping to strips of 4–5 m width down the avenues between the rows and not to allow the whole area to be covered by annual crops. Considerably more interest is now being shown in mixed cropping with food crops in response to both population pressure and de-forestation (Watson, 1983), but there is no doubt whatever that if mixed cropping of palms and food crops is to be successful, the work must be carried out under disciplined control; if this is not practicable, then palm areas and food-crop areas are better separated entirely one from the other.

In Malaysia it appears that the effect on the palms is likely to be a lesser problem in inter-cropping; there the main effort has been directed towards improving the standard of cultivation of the food crops themselves (Cheng, 1970).

Mixed cropping with other perennial crops

Wherever a main crop is a large tree leaving plenty of space between individuals, there has been a desire to make fuller use of the land by planting some smaller economic crop in the intervals. In considering this question from the viewpoint of the rubber industry, Allen (1955) stated that the secondary crop should be shade tolerant, exploit a different soil horizon, be less susceptible to diseases the two crops may have in common, be compatible with, and not damaging to, the main crop in its cultural and harvesting requirements, and should have a similar or shorter economic life. To these characteristics may be added the more positive ones that the soil shall be suitable for both crops, that the combined yield of the two crops shall be greater in monetary terms than that of the main crop when grown alone, and that, when the subsidiary crop comes to the end of its bearing life, the yield of the main crop shall continue at an economic level unaffected by the previous presence of the subsidiary crop.

Rubber and oil palms were planted together in the early years of oil palm cultivation in Indonesia. These crops are obviously incompatible, however, since rubber will overtop the oil palm and its canopy will spread out and shade the palms. Competition for nutrients will be fierce, and while the rubber must be widely spaced and so yield poorly per hectare, oil palm fruiting and yields will be seriously suppressed. Harvesting and transporting costs will be much increased.

Coffee and cocoa are small trees which can be planted among oil palms. Both these crops, but particularly cocoa, are more exacting in their soil requirements than is the oil palm. In West Africa satisfactory yields of these crops cannot be obtained on the great areas of Acid Sands soils where oil palms are found in greatest number. On the cocoa soils of Ghana and western Nigeria, cocoa is likely to be more profitable as a sole crop than when mixed with palms, even though the latter will yield well on these soils. Robusta coffee will produce quite well on the soils of the Congo river basin, on the Tertiary sands and other soils of the wet coastal belt of south-western Ghana, the Ivory Coast and Liberia, and in some parts of Sierra Leone. In the Far East cocoa has proved particularly suitable on volcanic soils in East Malaysia and on coastal clays in West Malaysia. Robusta coffee is also a fair yielder in many parts of the Far East, but the return to be obtained does not make it such an attractive proposition as cocoa.

Work in Zaire confirmed that considerations of soil type were the most important for inter-planting with cocoa and that, though cocoa is shade tolerant, shading with the oil palm presented certain diffi-culties; while cocoa benefits from greater shade when young, the palm provides increasing shade as the plantation matures. Reducing shade by leaf pruning is impracticable because of the mode of growth of the oil palm and the adverse effect of pruning on yield, while planting at abnormally wide spacing might reduce oil palm yields to a point where the cocoa yields were insufficient compensation. Vanderweyen (1952) recommended the planting of cocoa when the palm trunks are about 1.8 m in height (7–8 years old) either in two rows between palms normally spaced at 9 m triangular, or in three rows between pairs of palm rows approximately 10.5 m apart, the palms being spaced 7.5 m apart in triangular formation in the pairs of rows.

In an experiment in Zaire on the inter-planting of Robusta coffee among oil palms, the cultivation of coffee had no overall effect on oil palm yields, but in one treatment it appeared that the cultivation applied to the coffee may have had an effect similar to that of cultivation for establishment inter-cropping with food crops, since the bunch yield was slightly enhanced. Coffee yields soon declined, however, and it was concluded that it was better to plant the two crops separately. In this experiment all treatments allowed for an almost full stand of both palms and coffee bushes per hectare, so the result was not unexpected. Coffee can, of course, be planted at the same time as the palms; there is no need, as has appeared to be the case with cocoa, to await the palm's shade. Marynen (1960) suggested that coffee can be grown with oil palms in Zaire for 6 or 7 years after which it must be cut out.

A more informative experiment has given interesting results on an area of river alluvium in Sierra Leone. The treatments (Table 12.2) were:

(A) Control. Pure stands of coffee and oil palms, the plots being divided between the two crops. Oil palms spaced 30 ft (9.1 m) triangular, coffee 10 ft (3 m) square.

(B) Oil palms at 30 ft (9.1 m) triangular but with one row in three omitted. In the wide interline thus formed, coffee planted at 10 ft (3 m) square.

(C) Oil palms 30 × 40 ft (9.1 × 12.2 m) rectangular; coffee at 10 ft (3 m) square.

(D) Oil palms 30 ft (9.1 m) square, but alternate plants in alternate rows omitted thus leaving squares into which coffee at 10 ft (3 m) square was planted.

It should be noted that in treatments (C) and (D) in which the crops are intermingled there is a higher number of palms plus coffee bushes per unit area than there is in the treatment (A) where half the area is planted with palms and half with coffee. Early results showed that a higher total of produce may be obtained in the first years by an arrangement such as (C) than by having separate areas of palms and coffee. The arrangement in treatment (B) had too low

Table 12.2 *Inter-planting oil palms and coffee: Sierra Leone*

Treatment	Stand		Yield of coffee, 1962–6 (kg per hectare per annum)*	Yield of oil palm bunches, 1962–6 (kg per hectare per annum)*	Total production (kg per hectare per annum)
	Per acre	Per hectare			
A. Separate stands within the plots	OP 55.9 (27.9)†	138 (69)	991	3,771	4,762
	C 435.6 (217.8)†	1,076 (538)			
B. Coffee in wide interlines	OP 37.3	92	551	5,315	5,866
	C 130.1	321			
C. Wide oil palm spacing, coffee interplanted	OP 36.3	90	975	5,506	6,481
	C 286.7	708			
D. Coffee in hollow squares	OP 36.3	90	1,111	3,932	5,043
	C 289.2	714			

* The figures for treatment A are 'per hectare of plot' since it is this that must be compared with the other per hectare figures.
† Stands per acre in treatment A were 55.9 and 435.6 respectively but, for comparative purposes, the 'stands per acre of plot' are taken as half these figures, i.e. 27.9 and 217.8 (69 and 538 per hectare).

569

a stand of coffee to compete with treatment (C). The 'hollow square' system of treatment (D) does not appear to be advantageous. Higher economic returns from mixed cropping depend on the relative value of the two crops and the ability of the subsidiary crop to continue production at a good level. In this case there was evidence that coffee yields were in fact declining; and with low prices of coffee it would clearly be better for the *whole* area to be under palms.

Both cocoa and Robusta coffee have been grown commercially with oil palms on estates in the Congo basin. Coffee is grown as a catch crop and planted at the same time as the palms, and three crops are taken. A total yield of about 3 tonnes per hectare is expected and this provides revenue during the mainly unproductive years of the palms' life. Cocoa, on the other hand, is planted under the shade of mature oil palms. With two rows of cocoa trees between the rows of palms a yield of around 135 kg of dry cocoa is obtained and, provided there is no reduction in the palms' yield, this forms a useful cash increment when prices are satisfactory.

On the coastal alluvium of Malaysia coconuts underplanted with cocoa has become an industry of its own, and the mixing of cocoa with oil palms bids fair to follow suit. Oil palms throw a heavier shade than coconuts and the inter-planting of cocoa with oil palms is therefore more difficult and needs special study. One experiment has shown that the yield of cocoa planted within wide avenues of oil palms at 99 palms per hectare is lower than the yield of a similar number of cocoa trees planted under coconuts at 119 palms per hectare (United Plantations, 1985). On the fertile Selangor series soil there is little doubt that oil palms with cocoa make profitable use of the land; but comprehensive trials are needed, in both adult fields and on newly planted areas, to show exactly what yields of the two crops will be obtained from a range of spacings. Normal oil palm spacing (9 m triangular) throws too heavy a shade in old fields so, for cocoa planting, some thinning is necessary. Reducing the stand by one-seventh, one-quarter or one-third leaves hexagons in which cocoa can be conveniently planted. When a quarter of the stand is removed by taking out every other palm in every other row a hexagonal stand of 107 palms per hectare is obtained, and one experiment has shown (United Plantations, 1985) that with this number of palms and 648 cocoa trees per hectare a yield of between 500 and 700 kg dry cocoa per hectare per annum can be obtained. Oil palm yields were vitiated in this trial by *Ganoderma*, but it should be noted that a stand reduction of one-quarter does not lead to a similar yield reduction; if Bachy's formula is applied (p. 405), the expected drop in yield would be less than 10 per cent.

With cocoa grown among oil palms from the time the latter are planted it is probably best to maintain a triangular stand since this gives the most even ground coverage and hence the optimum palm yields, as well as providing the most evenly distributed shade for the cocoa. The best arrangement can only be determined by experiment, however, and Lee and Kasbi (1980) have claimed that avenue planting of oil palms at 10 × 7 m (which gives a stand equivalent to 9 × 9 m triangular planting) with a single row of cocoa between the rows is most promising.

A novel method of mixing oil palms and cocoa is being tested in Malaysia: cocoa is planted among already established palms until the yield of the latter falls, through *Ganoderma* losses, below a certain figure and the cocoa trees are providing mutual shade. The palms are then poisoned and cocoa remains in pure stand for 5 years after which young palms are planted in cleared spaces at normal density. The object is to have continuous cropping from the land and to suppress *Ganoderma*, but whether the latter can really be suppressed remains to be seen.

The grazing of livestock under oil palms

Oil palm fields provide shade which is a good deal lighter than that of crops such as rubber and cocoa, and even in fully mature fields there remains a considerable quantity of undergrowth which has to be kept under control. This undergrowth is, of course, much increased by the use of slightly wider spacings. It is not unnatural therefore that attention has been given to the raising and rearing of livestock on oil palm plantations as a subsidiary source of income.

In Africa, oil palm plantations lie in areas where livestock are subject to trypanosomiasis, but immune breeds of cattle exist and sheep and goats are also to be found. In the Far East it is not uncommon for herds of cattle (and sometimes buffaloes) to roam through plantations, but these animals are rarely owned by the estate and are usually regarded by managers as a nuisance.

In Latin America, many of the new oil palm plantings are in areas where ranching is already an established occupation. These plantings may be laid down in old pasture land or may be adjacent to pastures so that strong-growing grasses soon gain access (Hartley, 1965). There has been some grazing in young plantations, but the palms in these cases usually suffered damage. On heavy soils in high rainfall areas grazing caused puddling of the soil and adversely affected the root system of the palms which demand a good surface structure. In many plantations in Colombia and Ecuador *Panicum maximum, Pennisetum purpureum* or *Hyparrhenia rufa*, often mixed with other grasses, have invaded young plantations, severely retarding the growth of the palms. They then have to be eradicated by herbicides and replaced by *Pueraria*.

Neither sown leguminous covers nor natural covers are suitable for intensive grazing; the former tend to be grazed out and the latter to provide an insufficiency of material and the flora changes in composition. The main problem with combining oil palms with the production of animal products is that the best vegetation for live-stock, grass, is the worst for the palms. If one recognizes that the palm is the main crop and must take precedence in its requirements, then the stocking rate must be such that minimum harm comes to the palms while the livestock make the best use possible of an undergrowth which is suitable for oil palm cultivation (Samuel, 1974). Some early experiments were, nevertheless, directed at supplying suitable grasses in the interlines. In Nigeria elephant grass (*Pennisetum purpureum*) grows vigorously and if it is allowed to become too luxuriant and encroaches on the circles reduced early yields may be expected. Later, the palms tend to shade out the grass in their immediate vicinity and the effect on yield disappears. This is shown by the following figures of bunch yield from a spacing–inter-cropping–grazing experiment:

9.14 m (30 × 30 ft) spacing:	*Pueraria plots*		*Elephant grass plots*	
Bunch yield:	*Tonnes per hectare per annum*	*per cent*	*Tonnes per hectare per annum*	*per cent*
1st–3rd year of bearing	3.18	100	1.78	56
4th–6th year of bearing	7.85	100	8.89	113
7th–9th year of bearing	10.00	100	10.89	109

If elephant grass is to be planted at the same time as the palms, therefore, it must be kept out of the circles. Up till about the eighth year it will grow fairly satisfactorily among normally spaced palms in West Africa, but experiments have shown that after 10 or 12 years it is completely shaded out (Gunn *et al.*, 1962).

In the spacing–inter-cropping–grazing experiment in Nigeria (pp. 403 and 564), plots were planted with elephant grass or *Pueraria* or allowed to grow natural covers. Grazing was started 8 years after planting (the palms had established slowly) in plots of the following spacings:

	Triangular			*Rectangular*		
Metres:	*9.14*	*12.8*	*9.14 × 19.8*	*6.4 ×19.8*	*3.66 × 19.8*	*6.4 × 12.8*
Feet:	*30*	*42*	*30 × 65*	*21 × 65*	*12 × 65*	*21 × 40*
Density						
(per acre)	55.9	28.5	22.3	31.9	55.8	51.9
(per hectare)	138	70	55	79	138	128

The rows were arranged E–W so that the interline obtained a minimum of shading and this assisted the growth of elephant grass. As already mentioned, the elephant grass was soon completely shaded out in the 9.14 m triangular plots but in other spacings growth was satisfactory. From the point of view of palm yields, however, the wide 19.8 m lines and the very low density plots were quite unsatisfactory, the overall yield for both elephant grass and other cover plots being as follows:

	Fruit bunches per hectare per annum (12 years)					
Spacing:	*Triangular*			*Rectangular*		
Metres:	*9.14*	*12.8*	*9.14 × 19.8*	*6.4 ×19.8*	*3.66 × 19.8*	*6.4 × 12.8*
Feet:	*30*	*42*	*30 × 65*	*21 × 65*	*12 × 65*	*21 × 40*
Density (per hectare)	138	70	55	79	138	128
Yield (tonnes)	7.63	4.80	3.72	4.33	4.12	6.89

It will be noted that the adoption of the rectangular spacing 21 × 40 ft (6.4 × 12.8 m), where the elephant grass grew quite satisfactorily in the wide interlines, resulted in a yield only slightly lower than that of normal triangular spacing. Half of each elephant grass plot was used for grazing while the remaining half was normally maintained by regular slashing. The yield of bunches was unaffected by the grazing and the carrying capacity of the plots growing elephant grass was found to be as high as one beast per 0.8 hectares (2 acres) in a herd of small upgraded Ndama cattle. Strict control of the grazing was required to maintain a steady growth of the grass. The general conclusion was that with a very small reduction in yield owing to wider spacing, a considerable contribution might be made to meat supplies in palm-growing areas where the diet of the people is deficient in protein, and that at the same time the cost of maintaining the palm fields would be reduced (Gunn, *et al.*, 1961). Nevertheless, for such a system to be successful very strict management is needed and the early adverse effect of strong-growing grasses on the establishment of young palms cannot be ignored. It is difficult to imagine that on a plantation of any size management could afford to plant large areas of elephant grass and rotate them efficiently without the main crop suffering to some degree. Experience in Asia, Africa and America gives support to these doubts. From a study in Malaysia, Samuel (1974) concluded that as the oil palm is the main crop it should be planted at optimum density. Consequently, it should be accepted that normal covers be maintained and stocking densities kept low; and he found that, under Malaysian conditions, a stocking density of not more than one beast per hectare, with movement from one 3-hectare field to another every 7–10 days, could be maintained on mixed leguminous covers with *Axonopus compressus* without significant damage to the palms or the cover.

In the Ivory Coast Rombaut (1974) found that, at a stocking rate of 125 kg cattle per hectare (equivalent to one head of the Baoulés breed to 2 hectares), cattle could be carefully rotated by herdsmen during the day over the whole plantation and confined at night by electric fencing in enclosures at the rate of four head per hectare. A live weight gain of about 500 g per day was obtained. Under this relatively light stocking rate it is claimed that the vegetation is not in general degraded nor is the soil poached and that the cost of herd management is met by the saving in field maintenance expenditure which the system makes possible.

In Colombia it was found that a similar stocking rate was possible with *Pueraria* and natural covers provided the grazing was carefully rotated so that the regeneration of good covers was encouraged (Van den Hove, 1966). If overgrazing was allowed to take place the cattle soon began to eat the leaves of the young palms and to damage the drains. Many plantation owners in Latin America do not yet have staff available to exercise proper control of such grazing and, with heavy soils and high rainfall, the use of cattle in this manner cannot yet be generally recommended.

The production of palm wine

Mention was made in Chapter 1 of the tapping of oil palms in the vicinity of African towns and villages for the production of palm wine. The industry is of considerable economic and nutritional importance in Nigeria and other parts of West Africa. Tuley (1965a, b) studied production from palms in eastern Nigeria and described tapping methods; Okereke (1982) described the industry and its organization. Bassir (1962) investigated the composition and fermentation of palm sap from the male inflorescence stalk. From samples taken from one palm he found the chemical constituents of fresh palm sap to be as follows:

	g per 100 ml palm sap
Sucrose	4.29 ± 1.4
Glucose	3.31 ± 0.95
Ammonia (NH_3)	0.038 ± 0.015
Lactic acid	Present
Amino acids	Present

Many yeast cells and two main types of bacteria were found to be present after 8 hours' fermentation in the open air at about 25 °C. Experiments suggested that fermentation proceeded in two stages: in the first, bacterial action was responsible for the production of organic acids; in the second, sucrose was inverted by yeast and ethyl alcohol and more organic acids were produced.

The production of the so-called 'down-wine' from the terminal 'cabbage' of a felled palm is entirely destructive of the crop. It is said to be 'preferred' in Ghana (Sodar Ayernor and Mathews, 1971), but this is of course simply because the rural economy of the country has not forced the people to preserve their palms as has been the case in eastern Nigeria. The flow of sap is induced from the stem apex by an incision in the region of the unopened leaf bases and it continues for about a month. The fresh wine appears to be of different composition from the inflorescence wine as it is reported to contain glucose, sucrose, fructose, maltose and raffinose. However, the presence of the latter three sugars might be due to post-felling changes since even with inflorescence wine they are found after 24 hours' fermentation (Bassir, 1968). Another difference, however, is in the alcohol content. Fermented inflorescence palm wine is reported to contain only ethanol whereas 'down-wine' contains some methanol and propanol and this is another undesirable feature of 'down-wine' production.

Stem tapping (Tuley, 1965b) of standing palms is damaging to the soft tissues around the growing point and may kill the palm or provide entry for injurious insects, bacteria or other organisms. The only acceptable and manageable form of tapping is through excising the male inflorescence and drawing off the sugary solution which will exude from the cut surface.

The leaf subtending an immature male inflorescence is removed to obtain access to the inflorescence enclosed in its spathes. An incision is made near the apex of the inflorescence and the top of the tissue inside the spathes is removed. A piece of the front spathe is removed and the main stem of the spadix is cut horizontally to form a 'tapping panel'. The cut is covered with a piece of 'felt' composed of the fibrous leaf sheath fabric, and a new slice is taken daily until the wine begins to flow. A funnel of bamboo is inserted in the felt cover which is then set in position and the wine allowed to flow into calabashes or bottles. It is collected morning and evening and a new slice taken from the tapping panel at each collection. The spathes have been described by Tuley (1965b) as forming a cylindrical casing for the collection of the wine (Plate 12.2).

Wine production depends on the number of male inflorescences available and on the quantity produced per inflorescence. In the

Pl. 12.2   Palm wine flowing from the tapped immature male inflorescence.
(P. Tuley)

tapping of sixty palms over a 4-year period in eastern Nigeria it was
found that yields were highest at the beginning and end of the rains
in the periods March to April and October to November (Tuley,
1965a). It appeared that high production in the latter period was
due to the large number of male inflorescences available for
tapping, whereas the high yield in March to April was due to high
production per inflorescence, this presumably being associated with
the increased rate of development of the organs of the palm at the
beginning of the rains. The mean monthly production and the mean
number of palms in tapping are shown in Fig. 12.1. The mean
annual yield per palm was 26.3 litres of wine, which, at Nigerian
internal prices, was estimated to have a value more than double that
of the oil and kernels from similar palms and some 60 per cent
higher than oil and kernels from high-yielding palms in eastern
Nigeria.

Oil palm wine has a milky appearance due to high concentrations
of yeast. It has a slightly sulphurous smell because of its sulphur-
protein content. As it is not collected under sterile conditions
fermentation is quite rapid, but if this has not proceeded too far
palm wine forms a nutritious drink which provides an important
source of the vitamin B complex.

The pH of fresh wine is 7.4; it falls to 6.8 in the first stage of
fermentation and then to 4.0 in the second stage when fermentation
virtually comes to an end (Bassir, 1968). The wine is usually drunk
at pH 5.5–6.5 after about 12 hours' fermentation. In Nigeria the
organic acid content includes acetic, lactic and tartaric acids and
thirteen amino acids have been identified. Vitamins B_1, B_2, B_6 and
vitamin C are present. In Zaire inflorescence wine was found to
contain seven organic acids, twenty-five amino acids and vitamin B_{12}

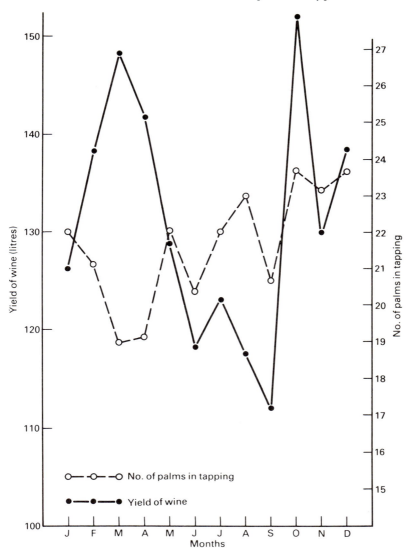

Fig. 12.1   Mean monthly production of palm wine and number of palms in tapping, 1959–62, at Umudike, Nigeria (Tuley, 1965a).

(Van Pee and Swings, 1971). The microbial flora causing fermentation appeared to differ from that found in Nigeria.

Sealed bottling of palm wine is undertaken in West Africa (Bassir,

1968). In Cameroon a process of double pasteurization and controlled fermentation has been used with *Raphia* palm wine (Fyot, 1973). the distilling of a spirit or 'gin' from palm wine has long been practised in crude distilleries. The raw spirit was later used in Ghana for the industrial production of branded spirits (Sodah Ayernor and Mathews, 1971).

References

Allen, E. F. (1955) Cultivating other crops with rubber. *Plts' Bull. Rubb. Res. Inst. Malaya*, 16.

Aya, F. O. (1974) *NIFOR ninth annual report, 1972–3*, p. 30.

Aya, F. O. (1975) *NIFOR tenth annual report, 1973–4*, p. 26.

Bassir, Olumbe (1962) Observations on the fermentation of palm wine. *W. Afr. J. biol. chem.*, **6**, 2, 20.

Bassir, Olumbe (1968) Some Nigerian wines. *W. Afr. J. biol. chem.*, **10**, 2, 42.

Cheng Yu Wei (1970) Improving the performance of catch crops in Malaysia. In *Crops diversification in Malaysia*, p. 66, Incorp. Soc. of Planters, Kuala Lumpur.

Fyot, R. (1973) Industrial processing of a traditional beverage: pasteurized palm wine. *Techniques and Développement*, No. 8, 10.

Gunn, J. S. and **Sparnaaij, L. D.** (1957) *WAIFOR fifth annual report 1956–7*, p. 55.

Gunn, J. S. and **Sly, J. M. A.** (1961) *WAIFOR ninth annual report 1960–1*, p. 44.

Gunn, J. S., Sly, J. M. A. and **Sheldrick, R. D.** (1962) *WAIFOR tenth annual report 1961–2*, p. 39.

Hartley, C. W. S. (1965) Some notes on the oil palm in Latin America. *Oléagineux*, **20**, 359.

Lee, A. K. and **Kasbi, H.** (1980) Intercropping cocoa and oil palms. *Proceedings of the Int. Conf. on cocoa and coconuts*, Kuala Lumpur, Malaysia, 1978, p. 158.

Marynen, T. (1960) Précis de phytotechnie des principales cultures industrielles. *Publs INEAC Hors Série*.

Okereke, O. (1982) The traditional system of oil palm wine production in Igbo Eze local government area of Anambra State of Nigeria. *Agric. Systems*, **9**(4), 239.

Rombaut, D. (1974) Étude sur l'élevage bovin dans les palmeraies de Côte d'Ivoire. *Oléagineux*, **29**, 121.

Samuel, C. (1974) Cattle in oil palm. 1. The effects of an integrated grazing system. *Planter, Kuala Lumpur*, **50**, 201.

Sly, J. M. A. and **Sheldrick, R. A.** (1965) *NIFOR first annual report 1964–5*, p. 30.

Sly, J. M. A., Sheldrick, R. D. and **Zeven, A. C.** (1963) *WAIFOR eleventh annual report 1962–3*, p. 28.

Sodah Ayernor, G. K. and **Mathews, J. S.** (1971) The sap of the palm *Elaeis guineensis* Jacq. as raw material for alcoholic fermentation in Ghana. *Trop. Sci.*, **13**, 71.

Sparnaaij, L. D. (1957) Mixed cropping in oil palm cultivation. *J. W. Afr. Inst. Oil Palm Res.*, **2**, 244.

Tuley, P. (1965a) Studies on the production of wine from the oil palm. *J. Nigerian Inst. Oil Palm Res.*, **4**, 284.

Tuley, P. (1965b) How to tap an oil palm. *The Niger. Fld.*, **30**, 28.

United Plantations (1985) *Twentieth annual report*. Res. Dept, Teluk Anson, Malaysia.

Van den Hove (1966) Utilisation du bétail pour la lutte contre les graminées dans les plantations de palmiers à huile en Colombie. *Oléagineux*, **21**, 207.

Vanderweyen, R. (1952) *Notions de culture d'Elaeis au Congo Belge*. Brussels.

Van Pee, W. and **Swings, J. G.** (1971) Chemical and micro-biological studies on Congolese palm wines (*Elaeis guineensis*). *E. Afr. agric. J.*, **36**, 311.

Waterston, J. M. (1953) Observations on the influence of some ecological factors on the incidence of oil palm diseases in Nigeria. *J. W. Afr. Inst. Oil Palm Res.*, **1**(1), 24.

Watson, G. A. (1983) Development of mixed tree and food crop systems in the humid tropics: a response to population pressure and de-forestation. *Expl. Agr.*, **19**(4), 311.

Chapter 13

Diseases and pests of the oil palm

Until the time of the Second World War it was true to say that the oil palm was largely free from serious diseases and pests; since that time, however, there have been serious, and at times devastating, outbreaks of disease in several parts of the world. Of greatest importance have been the devastation caused by *Fusarium* Wilt and a bacterial Bud Rot in southern Zaire, the considerable losses sustained through Dry Basal Rot (*Ceratocystis*) in Nigeria and through *Ganoderma* Trunk Rot in old and replanted areas in Asia, and the sudden and devastating attacks of bud rots and Sudden Wither on new plantations in Colombia, Peru and Central America.

Diagnosis and cure of the more recent of these diseases have proved extremely difficult, firstly because the size and manner of growth of the palm makes investigation difficult and time-consuming, and secondly because there has been a lack of plant pathologists to apply themselves single-mindedly to the diseases in question. Partly for these reasons there has been a tendency to avoid the usual plant pathological approach and to search for resistance to, or tolerance of, the diseases both within *E. guineensis* material and in interspecific hybrids with *E. oleifera*, and some progress has been made with this work. Arnaud and Rabechault (1972) have compared the roots of the two species and postulate that resistance to pathogens gaining access through the roots will be greater in *E. oleifera* and the interspecific hybrid because of the greater lignification of the hypodermis and the external cortical parenchyma and the presence of tannins in the endoderm and phloem. The exact relevance of these differences to specific diseases has yet to be worked out, however; the most outstanding resistance yet shown by *E. oleifera* and its hybrids has been to undiagnosed bud rots in Colombia and it is not thought that the pathogen in this case enters through the roots.

Attacks by one pest, the Hispid *Coelaenomenodera elaeidis*, have been growing more serious in West Africa, while sporadic defoliations have also been caused by caterpillars and bagworms of various species in Malaysia and South America. With the greatly increased areas under oil palms there has been a general increase, particularly in Asia and America, in the incidence of pests of several natural orders, but particularly of the Lepidoptera and Coleoptera. The prediction that owing to the vast assembly of palm species in that continent oil palms in America would soon be troubled by many species has been amply fulfilled.

A comprehensive work on diseases has been provided by Turner (1981). For pests, Wood's *Pests of oil palms in Malaysia* (1968) gives much information of general application on ecology and control, while much detailed information on African and American pests is to be found in special issues of the journal *Oléagineux* (IRHO, 1981; Genty *et al*, 1978). Descriptions in this chapter must necessarily be condensed, but reference is made to original papers where greater detail can be found. Nutritional disorders have been described in Chapter 11 and this chapter will therefore deal with conditions caused by pathogenic organisms, with important diseases of unknown cause, and with insect and other animal pests causing more or less serious damage to the palm.

Diseases

It will be most convenient to deal with diseases according to the stage of growth at which the palm is attacked and the organs affected.

Germinating seed diseases

Brown Germ

Precise cause unknown. Turner (1981) has listed twenty-seven fungi associated with the disease of which *Aspergillus* spp. and *Penicillium* spp. are most frequent. Many are secondary invaders, as are bacteria spp.

Distribution
This disease is now universal.

Symptoms
Brown spots appear on the emerging 'button'. These spread and coalesce as the embryo develops, and the tissues become slimy and rotten.

Control
The disease develops most readily under moist conditions at a temperature of 38–40 °C; use of the wet heat treatment for germi-

nation therefore encourages its spread. Though sanitary measures in the germinator may reduce incidence, the best method of control is to adopt the dry heat treatment method of germination (p. 322), since the seeds are dry when being heated at 39.5 °C, and when germinating they are at around 27 °C, a temperature which does not encourage the growth of the organism.

Other seed infections

Where removal of the mesocarp in cleaning the seed has been incomplete a species of *Schizophyllum* may invade the shell. Infection can spread to the inner shell surface and kernel. The condition is not of great importance since it is unlikely to occur where removal of the mesocarp is complete and the seeds have been properly dried.

Seedling diseases

There are certain leaf diseases which may attack seedlings in the prenursery, nursery or when they are young palms in the field. They will be dealt with together here even though some of them are characteristic of very small and some of larger seedlings.

Leaf diseases

Anthracnose

Cause: Species of *Botryodiplodia, Melanconium, Glomerella.*

Distribution

The disease has a wide distribution and several causal organisms; the appearance of the necrotic area depends on the organism concerned (Plates IX and X, between pp. 230 and 231).

Symptoms

Anthracnose is a disease characterized by limited lesions, necrosis and arrested development typically caused by one of the Melanconiales, e.g. *Melanconium, Pestalotiopsis*, but also by other fungi. It appears as dark necrotic lesions on the leaves of seedlings, usually at the prenursery or young nursery stage. The pathogenicity of species of three Anthracnose genera has been confirmed by tests in Nigeria (Robertson, 1956). These are:

1. *Botryodiplodia palmarum*. Small translucent spots, typically near the top or edge of the leaf or where the leaf is damaged, change to dark brown and are surrounded by a yellow halo or transition zone. The lesions enlarge and their centres turn grey while drying

out; pycnidia develop here and liberated spores give rise to further lesions.

2. *Melanconium* sp. (probably *M. elaeidis*, also found in Malaysia). Development rather similar to that of *B. palmarum*, but the lesion is lighter brown with a pale yellow halo; dries out more rapidly and therefore has a proportionally larger grey and dry area. Acervuli develop with spherical spores which cause further infection.

3. *Glomerella cingulata* produces long lesions between, but not crossing, the veins. The necrosed tissue is brown or black and at first has a water-soaked appearance. The transition zone is yellow. The fungus produces acervuli with conidia and, on the older dried-out lesions, flask-shaped perithecia containing asci each with eight ascospores.

Control

The fungi concerned appear to be weakly parasitic since the disease rarely appears if agronomic practices are sound, i.e. sufficient space between plants in the prenursery, regular watering to prevent wilting and fertilizing to induce steady growth, transferring plants to the nursery or bag with a block of soil to prevent leaf desiccation, and careful planting. As the disease may be carried from the prenursery to the nursery prophylactic spraying with a fungicide weekly in the prenursery from the two-leaf stage and for 6 weeks after transplanting to the nursery is advocated in West Africa. Captan (Orthocide M50) and ziram at 1 and 0.5 kg respectively per 500 litres water have given good protection but thiram and Dithane M45 are likely to be equally effective. In Malaysia spraying is thought to be unnecessary unless the disease appears, and Thibenzole (80 per cent a.i. at 0.1 per cent) has given best control (Turner, 1981).

Freckle, or Cercospora Leaf Spot (Cercosporiose)

Cause: *Cercospora elaeidis*.

Distribution

Cercospora Leaf Spot, is widespread throughout Africa but has not been reported in Asia or America (Plate XI, between pp. 230 and 231).

Symptoms

Freckle is a disease of nursery seedlings which sometimes starts in the prenursery and is frequently carried to field plantings where it survives for many years.

The youngest leaves of nursery seedlings become infected and minute translucent spots surrounded by yellowish-green haloes enlarge and become dark brown. Conidiophores emerging through the stomata in the centre of the spots, mainly on the under surface of the leaf, produce conidia which give rise to further, surrounding spots. This results in a freckly appearance, but later the lesions coalesce and the tissue dries out to become greyish-brown and brittle. The disease tends to become aggressive as the leaves age, and the process described above may proceed very rapidly at certain periods of the year. In West Africa this is usually the middle or end of the wet season, and in the following dry season the drying out of the older leaves is much hastened by Cercospora incidence. Proof of pathogenicity was obtained by Kovachich (1954) in Zaire and Robertson in Nigeria (1956). For details of growth and reproduction of *Cercospora* in the host, the papers of these authors and of Weir (1968) should be consulted.

Control

The gradual increase in the severity of Cercospora invasions, particularly in the field, has caused anxiety. Even moderate attacks materially reduce the green leaf area and are therefore likely to affect early bunch production. The obvious course is to try to eradicate the disease in the nursery and to prevent reinfection of the young seedlings in the field. Copper-based fungicides will control the infection, but unfortunately they have a toxic effect on palm leaves and so cannot be used. Many organic fungicides have been tested but those now generally used are Dithane M45 and benomyl (Benlate). The former has been shown to be considerably more effective than captan even when applied at only 0.5 kg per 500 litres of water (Rajagopalan, 1973). Benomyl is a systemic fungicide and is taken up by oil palm seedlings both through soil application and by spraying on the leaves (Renard, 1973). Penetration is higher when application is to the under-surface of the leaf. If resistance by the pathogen to benomyl is suspected, either 0.15 per cent thiophanate (Pelt), 0.15 per cent carbendazim or 0.2 per cent mancozeb can be used (Renard and Quillec, 1977; Quillec and Renard, 1977).

Prophylactic spraying is recommended in the first year after transplanting to the field. Benlate, Pelt and Dithane M45 are effective (Renard *et al.*, 1977).

In the nursery, pruning rounds are organized for the purpose of removing all old dry leaves and any others that may be badly infected. In the field, however, pruning becomes an anxiety since on the one hand removal of green leaf will reduce growth, delay flowering and induce cycles of male inflorescences; on the other hand, failure to prune and destroy prunings may increase the severity and prolong the incidence of the disease. The practical planter has to find some middle way which does not affect growth, even though this may not substantially reduce the incidence of

Cercospora. A pruning standard suggested is: any leaf which shows dead or badly necrosed areas over more than one-third of its total surface should be cut off, removed and burned.

Nitrogen manuring may cause a small increase in the incidence of Freckle in the nursery, but potassium substantially reduces it. Small favourable effects of phosphorus have also been noted (Robertson, 1960a).

There are significant differences between *E. guineensis* progenies in Cercospora susceptibility (Robertson, 1963a) and, since serious loss of crop through Cercospora attack in field plantings has been demonstrated, Duff (1970) suggested that breeding for tolerance would be worth while. With better control being obtained from the newer fungicides however, it is doubtful if such work will be included in breeding programmes. *Elaeis oleifera* progenies planted in Africa have shown a marked susceptibility to the disease and frequent spraying for 18 months may be needed. Interspecific hybrids are rather less susceptible.

Seedling Blight or Curvularia Leaf Spot

Cause: *Curvularia* spp. Turner (1981) has grouped Curvularia Leaf Spot with leaf spots attributed to species of *Helminthosporium, Drechslera*, and *Cochliobolus* under the term 'Nursery Leaf Spot'.

Distribution

This disease has been recognized in Malaysia since 1952 and in 1959 was reported to have spread to seedlings in several states in that country (Johnston, 1959). It is not present in Africa.

Symptoms and control

Like other Leaf Spot diseases, this disease originates in a small, translucent yellow spot, but its subsequent development is distinct. The spots tend to become irregularly elongated along or between the veins. They consist of a well-defined yellow halo with a narrow raised greenish-brown rim within it; inside is a reddish-brown region with concentric ridges, while the centre of the spot is thin and light brown in colour. Spots may reach 7–8 mm in length.

An attack may be light, with just a few spots, or heavy; in this case the leaves dry out, curl downwards and may disintegrate. The disease progresses inwards and in severe cases all the leaves are affected and the plant dies. The most serious attacks occur when the plants are a few months old in the nursery. Serious attacks do not usually occur in the field, though they have been reported. Pathogenicity tests confirmed that a *Curvularia*, resembling *C. eragrostidis*, was causing the disease. Early experiments showed the value of organic fungicides both on the incidence of the disease and on seedling growth, which was considerably improved. Treatment similar to that for *Cercospora* may therefore be applied. Thiram is favoured in Malaysia.

Other seedling leaf spots

Leaf spots caused by *Helminthosporium* spp., *Drechslera* spp. and *Cochliobolus* spp. have been reported from Malaysia and Zaire, and for their distinctive symptoms Turner (1981) should be consulted.

Control is similar to that of Curvularia, though Thiobenzole has been suggested for *Cochliobolus* spp. *Leptosphaeria elaeidis* (the imperfect stage of *Pestalotiopsis*) has been isolated from seedlings in Nigeria (Booth and Robertson, 1961) and Sabah (Williams, 1965). The dull orange spots surrounded by a yellow halo enlarge and elongate and necrosis of the tissue follows.

Other leaf conditions of seedlings

In the prenursery malformed seedlings are not uncommon and are variously attributed to the after-effects of Brown Germ or incorrect orientation of the germinated seed at planting. Some nursery diseases, for which many causes (including virus infection) have been canvassed but none established, have been constant enough in their symptoms to acquire distinctive names.

Infectious chlorosis

There is no evidence that this condition is infectious, and its name is therefore misleading. The tips of the apical leaflets become chlorotic interveinally and later turn dull or bronze. The condition was first recognized in Zaire, but is also found in Malaysia.

Bronze Streak

This is a very curious condition which was first reported in Nigeria but is also found in Malaysia. Typically, the condition is found on nursery seedlings shortly before transplanting time when they are seen to have a diffused orange or bronze streaking along the leaflets. This originates in a white chlorotic streaking or spotting at the tips of newly opening spear leaves. The bronze streaks on the upper surface of the leaves sometimes appear water-soaked and they may cover as much as 80 per cent of the leaf surface. The condition may appear alarming when seen to affect a large group of seedlings but, curiously, the affected seedlings are often large ones eminently suited, in other respects, for transplanting, and almost as soon as they are transplanted to the field the condition disappears.

Ringspot

A condition in which yellow or white rings with dark green centres are found on the leaflets has been a feature of nurseries in Nigeria

(Robertson *et al.*, 1968). The rings may coalesce in severe cases and cause some stunting of the seedling but in general the disease has not been serious. Similar symptoms have been described in Malaysia.

Cylindrocladium leaf disease

This has only been recorded in the Ivory Coast. *Cylindrocladium macrosporum* is the pathogen responsible. Brown speckling of the spear leaf is followed by brown lesions with whitish or light brown centres. These give rise to secondary lesions; other tissue becomes chlorotic. Benlate spraying has been suggested.

Freak conditions

These occur in young prenursery and nursery seedlings at the bifur-cate leaf stage and have been described as (i) *Leaf Crinkle*, in which the lamina between the veins is folded in lines across the leaf; (ii) *Leaf Roll*, in which the lamina is rolled under the leaf giving it a spiky appearance; (iii) *Collante*, in which the lamina between the veins becomes laterally compressed at a band about halfway along the leaf so as to form a constriction there (Gunn *et al.*, 1961). (Plate 7.7, p. 345). Other abnormal conditions which necessitate roguing in the nursery have been described in Chapter 9, p. 387.

Spear and bud rots

Phytophthora Spear Rot

Cause: *Phytophthora* sp.

Distribution

This disease has only been reported by Kovachich (1957) from Zaire where pathogenicity of the *Phytophthora* was demonstrated. A nursery spear rot has occasionally been seen in Malaysia but has not been diagnosed.

Symptoms

The median leaflets of the spear leaf are affected by a rotting which varies greatly in intensity. When the leaf opens the rotted portions become desiccated though bounded by a well-defined orange-brown transition zone. Adjacent tissue is chlorotic. Sporangiophores of the *Phytophthora* species are produced on the diseased leaflets under moist conditions. Attention has been drawn to the similarity between this disease and Crown disease of older palms.

Control

The disease has not yet been sufficiently virulent to require special control measures.

Corticium Leaf Rot

Cause: *Corticium solani*. It is not clear whether this disease should be classified as a leaf disease or a spear rot since Kovachich (1957), while mentioning that in nursery palms the leaflets rot in the spear leaf, also states that its attack on prenursery seedlings is similar to Anthracnose as described by Bull (1954). Pathogenicity was established, but the disease, though affecting many thousands of seedlings in southern Zaire, was not thought to be of great economic importance. The fungus has also been reported in Malaysia and in high-rainfall areas of Sabah and New Britain. Here seedlings are attacked at a young age. The rot appears at the base of unexpanded leaves which, when opened, show transverse rows of lesions. These are dark brown, become grey and brittle and may fragment leaving holes in the leaves. Thiobenzole (80 per cent a.i. used at 0.1 per cent) has been recommended where the disease is severe.

A *Nursery Bud Rot* has been reported from Zaire as causing about 6 per cent deaths in some nurseries through a rot at the base of the spear leaf proceeding into the bud (Kovachich, 1957). The roots are healthy in the early stages of the disease. No pathogen has been discovered.

Root diseases

Blast disease

Cause: *Rhizoctonia lamellifera* and *Pythium* sp.

Distribution

Blast has been a serious nursery disease throughout West Africa. It was particularly severe in the Ivory Coast and of considerable importance in Nigeria and Cameroon. Most of the research on Blast was done in the days of field nurseries, but it can be prevalent in polybag nurseries. The disease has also been recorded from Malaysia (Turner, 1966b), Indonesia, Brazil (Cardoso, 1961) and Colombia, and Turner (1981) considers that it could occur in any country where climatic conditions and nursery techniques are likely to favour its development.

Symptoms and causes

The symptoms of the disease have been described in great detail by Bull (1954) and Robertson (1959a). Affected seedlings lose their normal gloss and become dull and flaccid, the leaf colour changing successively to olive green, dull yellow, purple or umber (at the tips) and finally, with full necrosis and drying out, to a brittle dark brown and grey. Necrosis of the central spear is usual and death occurs in a few days. Incipient cases of Blast can be picked out in a nursery by a practised eye and, after a few days, these plants will be seen to be dead (Plate XII, between pp. 230 and 231).

Necrosis of the spear leaf may proceed from the tip or the base and, in the latter case, though the exposed part of the spear may remain green, the basal rot can destroy the growing point and affect the whole of the spear. In about 5 per cent of all cases the rot may not reach the growing point nor may the leaves become completely withered; the seedling then survives to make a partial recovery as a weakly and unacceptable plant.

The roots of diseased plants will be found to include large numbers whose parenchymatous tissue within the hypodermis has been rapidly destroyed from the tip towards the 'bulb', the stele remaining loose within the hollow cylinder. When the rate of cortical rotting becomes greater than the rate of production of new absorbing roots desiccation and death follow very rapidly.

In 1954–5 *Rhizoctonia lamellifera*, and a *Pythium*, probably *P. splendens*, were isolated (Robertson, 1959b). The former was present in decaying cortical tissue behind the transition zone and its small black sclerotia were to be found on the naked stele. The *Pythium* species was only isolated from primary infections of the root tips where there had been no secondary infection by saprophytes or *R. lamellifera*; under these circumstances it was shown to penetrate the cells and cause their collapse. Sporangia were found within infected cells just behind the transition zone.

In laboratory experiments Robertson (1959b) showed that the *Pythium* may be parasitized by *R. lamellifera*. In inoculation experiments, mixed inoculum of the *Pythium* sp. and *R. lamellifera* produced more extensive root rotting and the leaf symptoms were more pronounced than with individual inoculations. Inoculations with *R. lamellifera* alone were successful only when the roots had been artificially damaged. In *Pythium* sp. inoculations, damage was confined in the root tips. In all these cases pathogenicity was established by reisolation of the organisms. Typical leaf symptoms rarely appeared in any of the experiments which were carried out on seedlings which were younger than the normally recognized susceptible stage. One set of experiments was carried out with seedlings at this stage and in this case typical symptoms appeared. It was concluded that *R. lamellifera* plays a very important part in Blast disease in the destruction of cortical tissues and that it gains access either through a prior invasion of *Pythium* sp. which it parasitizes or through root damage from some other cause. The *Pythium* species is thought to be of importance through its role as a primary invader and its ability to penetrate the parenchyma cells and develop within them. For a further discussion of these relationships Robertson's paper (1959a) should be consulted.

Entirely new light has been shed on the etiology of the disease in the Ivory Coast where the Blast problem has always been peculiarly severe. It was noted in 1973 that plants grown in metal cages covered with mosquito netting showed only 0.75 per cent Blast in comparison with 15.5 per cent outside in unshaded areas. In 1974 a polythene bag nursery trial compared a completely closed cage with very fine netting (to give the minimum shading effect), an open-top cage, plots treated twice weekly with parathion (40 g a.i. per hectolitre), and unshaded control plots which had natural grass between the bags (Renard *et al.*, 1975). The results were as follows:

Treatment	1. Completely caged	2. Open-top cage	3. Parathion	4. Control
Per cent Blast by				
end of Dec.	0	6	27	46
end of Jan.	2	9	35	63

These results gave rise to the hypothesis that an insect vector is concerned in promoting the disease. Further trials showed that plants covered during the whole night had a low Blast incidence (6 per cent) as against 48 per cent for uncovered plants, and that plants covered between 6 a.m. and 10 a.m. suffered an intermediate attack. Covering from 4 p.m. to 6 p.m. was less effective. It was noted that night covering prevents dew deposition and it was postulated firstly that this could account for the effect of normal shading practice where dew, which may well attract insects, does not form, and secondly that the lesser effect of covering during a morning period would be due to the fact that this would be the daylight period when dew would normally be on the plants.

Later, it was established that the insect involved was a Jassid, *Recilia mica*, of which *Paspalum* spp. and *Pennisetum* spp. were the host plants (Julia, 1979). The insect moved to the palm nursery only in October and November. Desmier de Chenon (1979) found that the removal of grasses in the vicinity of the nursery reduced Blast incidence; the application of aldicarb monthly from the start of the nursery was also effective and made it possible to eliminate the shade which had always been found necessary in the Ivory Coast (Quencez, 1982). The exact connection between *R. mica* and Blast disease has not, however, been determined. Reproduction of the disease by *Pythium* sp. and *R. lamellifera* inoculation has not been accomplished in the Ivory Coast, as it was in Nigeria, and Turner (1981) has discussed the possibility of there being either several causes of Blast or of two apparently different causes being linked. The possibility of transmission of a microplasm has been mooted.

Control

One of the most interesting features of Blast disease is the import-ance of time of attack. It has been shown both in the Ivory Coast (Bachy, 1958) and Nigeria (Robertson, 1959a) that a relationship exists between Blast incidence and the age of the seedlings at the time of attack. If the seedlings are either very young (1–4 months) or old (11 months or over) at the beginning of the Blast season, the casualties are very small. Results from the Ivory Coast were as follows:

Number of months in the nursery to 1 November	1	2	3	4	5	6	7	8	9	10	11	12
Blast (%)	0	1	3	5	18	22	23	20	21	21	12	5

Plants left in a nursery beyond 12 months or young plants in the field are only very rarely attacked by Blast.

The precautions which are taken in nurseries against high Blast incidence were briefly discussed in Chapter 7 (p. 350) as they form an integral part of the agronomy of the crop at this stage. The effect of shade in reducing Blast incidence has been established, but the provision of shade for large plants nearing the end of their nursery life has disadvantages, and in Nigeria generally proved unnecessary. A significant negative correlation was found between Blast inci-dence and rainfall for August and September (Robertson, 1959a). This period covers the 'short-dry' season, which is very variable in severity and which precedes the onset of the Blast season which normally extends from October to January. Experiments confirmed that the provision of irrigation during the short-dry season in August, and the extension of this as needed into September, substantially and significantly reduced Blast incidence. Later exper-iments, while confirming the effect of short-dry season irrigation except in years of high rainfall, also showed that for shade to have its maximum effect it must be maintained from planting time (Rajagopalan, 1968–74). This of course leads to considerable etio-lation and cannot be generally recommended.

It must be emphasized that for satisfactory control of Blast a high standard of nursery cultural practice and the observance of correct planting dates are essential. The control measures for Blast in West Africa may be summarized as follows:
1. Plant well-developed prenursery seedlings early in the rainy season and ensure their rapid growth.
2. Pay particular attention to irrigation during the short-dry season and make sure that polythene bags have a sufficient though not excessive water supply throughout the nursery period.

3. Where *Recilia mica* is prevalent spray out host grasses in the vicinity and apply aldicarb as Temik granules monthly at 2 g per seedling.

Dry Bud Rot

A disease of coconuts, this has been reported only from nurseries in the Ivory Coast (Julia, 1979).

Diseases of the adult palm

In this section will be included all those diseases known to attack the palm after it has been established in the field. Some of the conditions mentioned are characteristic of the palm's early life, e.g. Freckle or Cercospora Leaf Spot, which is primarily a nursery disease, some are characteristic of early bearing life, e.g. Dry Basal Rot, and some are characteristic of senescence, e.g. Ganoderma Trunk Rot; but there is much overlapping and it is not therefore practicable to allot the diseases to particular stages of growth. Moreover some diseases, e.g. Ganoderma Trunk Rot, may be characteristic of one age in some regions and of a different age in other regions, while other diseases may have different manifestations at different ages.

Leaf diseases

Patch Yellows (*Déchiqueture*)

Cause: *Fusarium oxysporum.*

Distribution

This disease appears to be confined to Africa where it is widely distributed, though sporadic (Plate XIII, between pp. 230 and 231).

Symptoms

Wardlaw (1946a) reported that following the discovery of *Fusarium oxysporum* associated with Vascular Wilt disease, a second strain of *F. oxysporum* which closely resembled the first had been shown to be associated with a leaf disease in Zaire known as Patch Yellows. Kovachich (1956a) later proved the pathogenicity of the organism. Infection takes place in the unopened spear leaf and for this reason the lesions at the sites of infection appear opposite each other on the leaflets when the leaf opens. When this stage is reached the lesions are circular or oval with rings of pale yellow surrounding straw-coloured or brown centres where conidiophores can be found; some of the lesions, however, are simply chlorotic with no brown centres. The patches may appear all along the lamina. Later the centres of the patches dry out and drop away giving rise to the typical 'shot-hole' appearance, or, if the patches are towards the edge of the leaflets, to a raggedly indented appearance. The purely yellow patches persist and darken and can be seen to have small orange spots within them (Bull, 1954).

Susceptibility and control

The disease affects between 0.2 and 1.8 per cent of the palms in areas where it is found, and evidence for genetic susceptibility was provided in Kovachich's pathogenicity tests. He suggested that seed palms might be sprayed with inoculum to eliminate very susceptible lines which might otherwise be multiplied, but the disease does not seem to have reached sufficient proportions for this precaution to be taken, and increased incidence has not been reported.

A condition in Malaysia known as *Wither Tip*, from which both *F. cocysporum* and *F. solani* have been isolated, has been described by Turner (1981) who suggests that it is allied to Patch Yellows both in symptoms and cause.

Crown disease (*Arcure défoliée*)

Cause: Unknown.

Distribution

The disease is found in all oil palm areas but is largely confined to palms of Deli origin.

Symptoms

A young 2- to 4-year-old palm suffering severely from Crown disease has many of its leaves bent downwards in the middle of the rachis; at this point the leaflets are absent or small and ragged. These symptoms originate in the spear leaf where the folded leaflets begin to show a rot of their edges or centre (Kovachich, 1957). This rot is brown and spreads throughout the central portion of the leaf so that when the leaf unfolds the leaflets of this section are disintegrating or already missing. The bend of the leaf at the point where the leaflets are absent is a curious symptom and the real cause of it has not been discovered, though Thompson (1934) stated emphatically that it had been established that the 'decreased rigidity' was due to insufficient lignification of the parenchymatous tissue. The leaves, however, tend to be quite rigid, though bent. In severe cases all the leaves surrounding the spear may be bent down, and the spear itself may have a rot of its terminal portion which turns brown and hangs down. Under these extreme circumstances Crown disease may have a severe effect on early development and yields. The disease normally affects palms in the second to fourth year in the field, but instances have been reported in the nursery and up to 10 years of age.

Causes and control

The disease was most prevalent in the Far East, particularly in the early Deli plantations, and in the absence of a pathogen it was assumed that the disorder was physiological and might be inherited. This latter assumption proved correct, and with regard to the former it has been suggested that palms suffering from the disease have low leaf-Mg contents and that the incidence of the disease may be affected by Mg and K manuring (Hasselo, 1959). However, the evidence for this is weak.

The most valuable work on Crown disease is that of de Berchoux and Gascon (1963) who showed that pure Deli progenies in the Ivory Coast were highly susceptible, that La Mé material, free of Crown disease, gave crosses with Delis which were also free of Crown disease, but that Zaire material, which showed several cases of the disease, gave Deli × Zaire crosses with a quarter to a half of the palms showing the disease. The authors postulated that susceptibility to Crown disease is due to a monofactorial recessive character. In twelve La Mé × Deli crosses the disease was either absent or amounted to less than 2 per cent. In one Deli × Deli cross the disease was present in 99.5 per cent of the progeny. In Zaire × Deli crosses, or (La Mé × Deli) × Deli crosses, 3 : 1 ratios were expected and often nearly obtained except with the presumed homozygous recessive Deli, where 1 : 1 was expected. Some examples from de Berchoux and Gascon's data are given in Table 13.1. Crown disease susceptibility is assumed to be genetically aa, and from the results obtained by these workers it seems practicable to select palms which will not throw susceptible individuals in their progeny; in particular it would be valuable to have *pisifera* shown to be homozygous for absence of Crown disease (AA), as the Zaire (Sibiti) palm S 127 P appears to be.

De Berchoux and Gascon consider the fact of susceptibility in some Zaire palms at La Mé as supporting evidence for a Zaire origin of the Deli palm. The immunity of African material in the Far East was early noted by Thompson (1934), but in Nigeria cases have occurred in individual palms not thought to have any Deli parentage. In Sierra Leone the disease has also been reported on palms not of Deli origin (Bull, 1954).

These apparent anomalies may perhaps be explained by the work of Blaak (1970) in West Cameroon. He found that with four palms in crosses and selfs the expected inheritance occurred, assuming that each was Aa. However, the results of crossing and selfing three other palms and crossing them with the first four gave segregations which could only be explained by the presence of an inhibitor gene

Table 13.1   *The incidence of Crown disease in Deli, African and crossed progenies in the Ivory Coast*

Presumed type of cross	Cross	Progenies No. of palms		Expected segregation	
		Without Crown disease	With Crown disease	AA or Aa (%)	aa (%)
aa × aa	Deli D 115 D Selfed	1	205	0	100
AA × aa	L 10 T (La Mé) × D 115 D	130	0	100	0
Aa × aa	L 219 T (La Mé × Deli) × D 115 D	116	87	50	50
Aa × Aa	L 219 T (La Mé × Deli) × D 10 D (Deli)	104	25	75	25
aa × AA	D 115 D × S 127 P (Zaire)	217	0	100	0
Aa × aa	S 7 T (Zaire) × D 115 D	68	62	50	50
Aa × Aa	L 236 T (Zaire) × L 269 D (Deli)	105	25	75	25
Aa × Aa	L 239 T (Zaire) × D 128 D (Deli)	149	61	75	25

Source: de Berchoux and Gascon, 1963

which, in homozygous condition, prevents the expression of the disease in the aa genotypes. Blaak points out that the presence of an inhibitor gene complicates selection for absence of Crown disease since detection of a palm of aa genotype (susceptibility) is only possible by test crossing with a palm which is known not to have the inhibiting gene, and this will require 3 years.

Leaf Wither

Cause: *Pestalotiopsis* sp.

Distribution

A virulent type of leaf withering has been troublesome in parts of Colombia, Ecuador and Honduras and has caused much defoliation. It is also commonly seen in Colombia on *Elaeis oleifera* palms. The disease has been described as Pestalotiopsis Leaf Spot and Grey Leaf Blight in Malaysia, but there the fungus is only associated with old and near-moribund leaves and is not considered of economic importance (Turner, 1981). The virulence of the attacks in Latin America seem to be due to the easy access given to the young leaves by the feeding activities of insects, but it is also possible that the strains of the *Pestalotiopsis* species involved have a greater pathogenicity.

Symptoms

The first symptom is the appearance of small brown spots with yellowish halos. These spots soon coalesce into brown necrotic areas which spread over the leaflet tissue and later become grey and brittle. There is a sharp line between the brown and grey areas and in the latter a species of *Pestalotiopsis* is found, black specks indicating the location of spore-bearing acervuli (Hartley, 1974) (Plate XIV, between pp. 230 and 231).

The disease has caused considerable defoliation on certain plantations and, as would be expected, this has been followed by serious yield decline; falls in bunch production from 18–20 to 12–15 tonnes per hectare in adult areas, and from 11 to 7–8 tonnes per hectare in young plantings have been reported.

Cause and control

The attacks on moribund tissue in Africa and Malaysia are often associated with magnesium deficiency symptoms (Bull, 1961), but in Colombia leaf analysis does not suggest that this deficiency is an underlying cause. Genty *et al.* (1975) showed that a Tingid, *Leptopharsa gibbicarina* (see p. 632), was the principal means of infecting young leaves. This insect punctures the leaflets alongside the midribs, producing whitish spots with their surrounds stained with excrement (Genty *et al.*, 1983). Two species of *Pestalotiopsis* are the usual entrants, but species of *Helminthosporium, Curvularia* and other genera may also gain access to the leaflets. Control of the insect by aerial spraying of propoxur, fenitrothion or phosphamidon has had considerable success in checking the disease, but more recently the injection of monocrotophos at 8 g a.i. per palm has been advocated and investigations of biological control are under way (Genty *et al.*, 1983; Guerrero, 1985). Monocrotophos injection has also been successful in Honduras (Vessey, 1981). In Ecuador *Peleopoda arcanella* has been implicated in the provision of access to the leaves by *Pestalotiopsis* sp. (Turner, 1981).

White Stripe

Cause: Unknown.

Distribution

This condition is sporadic and, in Asia, is said to be more common on alluvial soils, particularly organic clays or mucks.

Symptoms

Narrow white stripes are found on each side of the leaflet midrib and extend its whole length. The stripes are at any point from the midrib to the margin and are sharply divided from the adjoining green (often dark green) tissue. Many affected palms recover, and Rajaratnam (1972) reports that the chlorotic tissue may turn green after about 7 months and that the symptom is more prominent in young leaves than in old. He also showed that chlorosis was due to failure of the palisade mesophyll cells to elongate and that apparent recovery was through an increase in the chlorophyll content of the spongy mesophyll and not through development of the palisade cells. Turner (1981) states that symptoms are more severe in Malaysia than elsewhere and that typically they appear at 2–3 years of age, becoming more severe at 3–5 years and then becoming chronic. Moderately affected palms are thought to suffer a 20–30 per cent reduction in yield.

Causes and control

The cause of the disease has been thought to be nutritional: either boron deficiency or a high leaf N/K ratio. These claims have not been substantiated. It has also been suggested that the disorder is of genetic origin. A certain *tenera* × *dura* cross showed similar percentages of White Stripe when planted in the Ivory Coast and in East Cameroon; also certain Deli selfs in the Ivory Coast showed the symptoms while others did not (Ollagnier and Valverde, 1968; Gascon and Meunier, 1979). While the cause of White Stripe remains unknown no control measures can be confidently recommended though Turner (1981) has suggested substantial applications of potash with reduction of nitrogen applications.

Minor leaf diseases

The oil palm leaf is unusually susceptible to patchy discoloration and necrosis from minor pathogens and to surface covering by epiphytic and saprophytic organisms. These often cause the older leaves to appear far from healthy and the area of actively photosynthesizing leaf may be seriously reduced.

Necrotic Spot

A disease of the older leaves long distinguished in Africa (Robertson *et al.*, 1968) and caused by *Cercospora elaeidis*, though of a different strain from that causing Freckle. Brown necrotic spots on the leaves are surrounded by a narrow greenish yellow halo which later expands and becomes bright orange. The disease sometimes becomes aggressive and causes much premature withering, leaflets tending to die off from the tip and the margins. Proof of pathogenicity was obtained in Zaire by Kovachich (1956b). The fungus *Oplothecium arecae* is found in the old spots but there is no evidence that it is a pathogen.

Crusty Spot (Croûtes noires)

This is a disease of senescent leaves in which circular or oval orange spots appear on the leaflets with a blackened crust at the centre of each. Commonly the spots are at the bases of the leaflets. Around the solid crust the tissue becomes grey and is bounded by an orange halo which merges into the green of the leaf. Only rarely does the disease become aggressive, in which case large areas of the leaflets die off (Robertson *et al.*, 1968). The crusts have been found to contain the stromatic tissues of the Ascomycete *Parodiella circum-data*, and pathogenicity has been established.

Orange Leaf Blotch (Bigarrure)

This is a disease of older leaves in which large irregularly-shaped orange patches appear on the lamina; in advanced cases necrosis occurs in the centre of the patches giving rise to dark brown mottling. The cause has not been established though *Pestalotiopsis* sp. and *Helminthosporium* sp. (Moreau, 1952) have been suggested; Turner (1981) considers it to be a variant of leaf wither.

Algal Leaf Spot

Leaf damage caused by an alga, *Cephaleuros virescens*, is now common. Pin-point yellow spots develop on the upper surface of the leaflets and on the rachis. Later, the production of sporangiophores gives the lesions an orange tinge. In Zaire the organism has been regarded as mildly parasitic on senescent leaves, but Weir (1968) has shown that even on vigorously growing palms up to 10 years old leaves which are only entering the second half of their life span may be infected and thereafter the infection may increase so that there may be as many as twenty algal spots per centimetre length of the entire leaflet. It has proved difficult to assess the real effect of the alga in hastening senescence, since it is often found in company with other conditions of the ageing leaf. Profuse development of *C. virescens* has been referred to as Red Rust, though Turner (1981) has pointed out that this term is misleading owing to possible confusion with a rust fungus.

Epiphytic and saprophytic moulds and lichens

Black 'Sooty Mould' is often found to grow on the older leaves of adult palms and occasionally spreads over a large proportion of the leaf surface giving the palms a blackish-grey appearance. Although mainly epiphytic, the flora concerned may harm the plant by blocking stomata and screening the leaves from light.

Several of the commonest fungi to be found in Africa as constituents of the epiphytic flora appear in Turner's 'Micro-organisms associated with oil palm' (1971). Among these the Ascomycetes *Apiospora* sp., *Meliolinella elaeidis* and *Meliola elaeis* may be mentioned. *Meliolinella elaeidis* is recorded as also being found in America (Costa Rica) on *Elaeis oleifera*. Epiphytic flora may appear on the upper or lower surface of the leaves. In West Africa the black mould usually found on the upper surface consists of discrete circles of about 5 mm diameter; on the lower surface the black mould is in irregular patches of less dense material. In Malaysia, sooty moulds of *Brooksia, Ceramothyrium,* and *Chaetothyrium* sp. develop on insect secretions on the leaves (Williams, 1965; Turner, 1981). *Brooksia tropicalis* is common in Africa.

Lichens are often found among the epiphytic flora on oil palm leaves, forming small grey-green incrustations on the upper surface of the leaflets (Turner, 1971).

Root and stem diseases

Root and stem diseases are characterized by fracture and drying out of fully developed leaves, leaving the spear leaf and some surrounding leaves standing erect until the disease has proceeded much further. These early symptoms may be accompanied by a change of colour, drying out or wilting of one of the more erect younger leaves. Bud and spear rots on the other hand tend to be characterized by symptoms in the centre of the crown. The spear leaf may be directly affected or the surrounding leaves show a sudden chlorosis. Successive spear leaves may be shortened, have peculiar 'little leaf' formations, or cease to develop, leaving a palm with an empty centre.

These general symptom differences between the two groups of diseases give a first rough guide when deaths occur or alarming disease symptoms appear; but dissection of the palm must follow to see exactly where the site of *destruction* is. While the root and stem diseases kill the palm by a process of denial of water and nutrients to the crown, the bud rots eventually kill the palm by growing towards and reaching the single growing point. The site of decay in the first case may be expected therefore to be in the bole, trunk or roots, but with bud and spear rots the changes will be found in the 'funnel' or 'cabbage' of the crown.

Dry Basal Rot

Cause: *Ceratocystis paradoxa* (imperfect stage = *Thielaviopsis paradoxa*).

Distribution

This disease appears to be confined to West Africa and, though the pathogen is a common soil inhabitant, the disease was not discovered in epidemic form until 1960. One estate in Nigeria was devastated and thereafter minor outbreaks occurred in several parts of Nigeria, West Cameroon and Ghana. In the first epidemic deaths were common, but recovery has now become more usual and further serious outbreaks have not been reported.

Symptoms

The foliar symptoms are preceded by extensive bunch and inflorescence rot. The rachis of certain leaves then becomes fractured submedianly, though the leaflets remain green for a considerable period before they eventually die. Occasionally a young leaf high up in the centre of the crown becomes necrotic and dries out, and this precedes the necrosis of the older leaves. It is quite common for a complete ring of leaves to exhibit the submedian fracture while the upper leaves are still erect, and this gives the newly affected palm its characteristic appearance. Later, the upper leaves and the spear will be similarly affected and the palm dies, or it may make a recovery at any stage. A recovered palm will take several years to come back into bearing (Plate 13.1).

The characteristic internal symptom of the disease is a dry rot at the base of the trunk. This rot is well established by the time the primary leaf symptoms are apparent. In the transition zone between rotted and healthy material many vascular bundles are necrotic, and it is possible to trace infection from an infected root or leaf base into the base of the trunk.

The majority of palms attacked have been palms which have recently come into bearing, but 10-year-old palms have also been affected. Symptoms usually appear at the end of the dry season.

The cause, spread and control of the disease

The cause of Dry Basal Rot was shown in pathogenicity tests conducted by Robertson (1962a, b) to be due to the Ascomycete *Ceratocystis paradoxa* of which the imperfect stage is known as *Thielaviopsis paradoxa*. *Ceratocystis paradoxa* is a soil inhabitant widely distributed throughout the tropics of Africa and Asia and causes diseases of several other crops. Its sudden appearance in West Africa as the cause of a serious condition was unexpected and has given rise to investigations on conditions conducive to its spread. An epidemic at Akwukwu in Nigeria occurred on Acid Sands soils with an unusually low clay content at depth (15–17 per cent at 2 m), and minor outbreaks at the NIFOR Main Station also occurred on fields with less clay in the profile. This led to the belief that incidence might be connected with soil–climate relationships. A further outbreak at NIFOR in 1967 followed a severe dry season. Incidence varied between fields from 0.1 to 10.0 per cent.

Two features of the spread of the disease are important. In the outbreak at Akwukwu many deaths occurred in the first 2 years, amounting to about 30 per cent in one area. Thereafter very few deaths occurred and there was considerable recovery; though new infections occurred, these did not give rise to many further deaths (Robertson, 1963). All the palms which recovered were bearing bunches by three years after the last survey (Rajagopalan, 1965).

The other feature of the spread of this disease was mentioned in Chapter 5 (p. 293). At the NIFOR Main Station incidence in one

Pl. 13.1   Dry Basal Rot, *Ceratocystis paradoxa*: **A**, a severely infected palm showing sub-median fracture of the lower leaves; **B**, a palm showing external symptoms of the disease, dissected to expose the dry rot at the base of the trunk.

field of 9-year-old palms was mainly confined to progenies having the same female parent. In Robertson's pathogenicity tests he found he could infect all seedlings through a root dipping technique; nevertheless inoculated progeny lines planted in the nursery showed marked differences of disease incidence (Robertson, 1962c). Selection for resistance is therefore a promising line for the future should the disease once more become important.

Vascular Wilt Disease (Fusariose or Tracheomycose)

Cause: *Fusarium oxysporum*. Schl. f. sp. *elaeidis* Toovey.

Distribution

Since its description by Wardlaw (1946b) in Zaire, this disease has always been considered the most menacing of all oil palm diseases. This view has been somewhat mitigated in course of time by the absence of serious epidemics in West Africa and elsewhere, but its earlier reputation has died hard.

The early history of the disease was briefly stated by Wardlaw (1950) as follows:

> During a visit to the Belgian Congo in 1946 I observed a wilt disease of the oil palm (*Elaeis guineensis*), and isolated *Fusarium oxysporum* from the necrosed vascular strands. In 1947, Messrs. S. de Blank and F. Ferguson, in a private report, announced the presence of this disease in Nigeria and submitted cultures to me for identification; and in 1948 I was able to confirm their diagnosis during a visit to the affected plantations. Substantial proof of the pathogenicity of *F. oxysporum* is now being obtained by workers in the Belgian Congo.

Pathogenicity was confirmed by Fraselle (1951) in 1948. Thereafter Vascular Wilt was found on several plantations in Nigeria and West Cameroon and in the Ivory Coast and elsewhere in West and West Central Africa. The greatest devastation occurred on replantings on estates in southern Zaire. In West Africa the disease was largely confined to plantations, particularly replantings, and for a long time was not much noticed in the groves. However, Aderungboye (1981) found that it was widespread in the drier Ogun and Ondo states of Nigeria though infrequent or absent in the high-rainfall areas of the south-east.

There have been occasional isolations of *Fusarium oxysporum* from diseased palms in America (e.g. Sanchez Potes, 1964), but the only report that suggests the disease itself has come from Brazil (van de Lande, 1984), though pathogenicity remains to be proved.

Symptoms

In the plantation areas of southern Zaire Wilt is commonly found in young palms which have recently come into bearing; this is also so in Nigeria in replantings. However, in West Africa the disease has attacked older palms which have been in production for 10 years or more (Prendergast, 1957).

In the most usual, chronic form of the disease in mature palms the older leaves become desiccated and the rachis breaks near the base or at some distance from the base, the ends of the leaves hanging downwards. This feature has been used to distinguish the disease from Ganoderma Trunk Rot in which the leaves collapse at the base and closely cloak the stem. The disease usually proceeds gradually along several leaf spirals with younger leaves becoming successively affected. The erect and still green leaves in the crown are now much reduced in size and are often chlorotic and the palm may stay in this state for several years before the crown eventually collapses.

Occasionally a mature palm suffers a rapid death through an acute attack. The leaves dry out and die rapidly while still in an erect position and then snap off about a metre or more from the trunk, usually during strong winds. The remaining leaves die quickly. All stages between the acute and chronic forms are encountered (Plates 13.2 and 13.3).

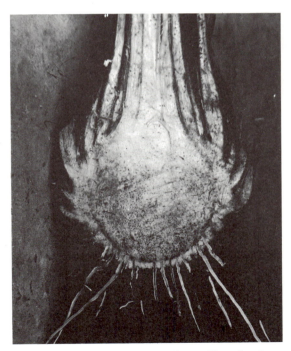

Pl. 13.2 Vascular Wilt – longitudinal section of a seedling, showing continuation of necrosed vascular strands from roots to stem. (A. G. Prendergast)

Pl. 13.3 Vascular Wilt disease, caused by *Fusarium oxysporum*, of a 16-year-old palm. (A. G. Prendergast)

Symptoms of the disease in young palms of up to about 6 years of age in which no trunk has yet been formed are somewhat different. In these palms the 'Lemon Frond' symptom is frequent; a leaf somewhere in the upper middle part of the crown (fourth to fifteenth leaf) develops a bright lemon yellow colour before drying out from the tip to the base. Leaves at about the same level then turn yellow and dry out to be followed by some of the younger leaves which will die while many of the older ones remain green. Newly developed leaves become successively smaller, and death of the whole palm usually takes less than a year. It should be noted that the striking 'Lemon Frond' symptom is not always to be seen, and in southern Zaire a general yellowing of the leaves before death was more usual.

Nursery plants suffering from Wilt often recover. Shortening of the leaves, causing a bunched appearance, is followed by browning and drying off of the older leaves and the condition progresses inwards.

The pathogen is soil-borne and usually enters the palm through the roots, growing along the stele which becomes blackened. Infection can take place through wounds in the stem base and through uninjured roots (Kovachich, 1948). Renard (1970) considered that entry of the mycelium was much impeded by lignification even with wounding and that rapid infection was mainly through the transmission of spores in the vascular system. Locke (1972), also working with seedlings, showed that the pathogen is confined to the conducting elements of the xylem and can reach the stele from the tip of a lateral root or the damaged cortical tissue of a pneumathode. He considered that the plant had little defence against serious infection in spite of resin formation and tyloses.

From the roots the mycelium penetrates into the wood vessels of the vascular strands where conidia and chlamydospores are also found. The vascular bundles are normally pale yellow or whitish, but when diseased they become brownish-grey or black, and a cross-section of the trunk therefore shows a speckled appearance. Discoloration, which is associated with the presence of gum, is confined to the wood vessels; blackened fibre strands do not indicate Vascular Wilt. Such blackening often occurs in older palms and sometimes in other conditions and the inexperienced observer can therefore be misled into a wrong field diagnosis. Moreau (1952) pointed out that though discoloration of the vessels is normal in palms over 20 years old, in these palms the blackening decreases towards the top of the palm instead of becoming accentuated as in the case of Wilt. Drying-up of the leaves and death of the palm are caused partly by the destruction of the roots and partly by the blocking of the wood vessels by gum. Diseased vessels may at first occur in only one section of the stem base and this probably accounts for leaf symptoms being confined at first to certain spirals. Vessels in the centre and at the top of the stem then become diseased and the symptoms spread across the stem so that a large proportion of the vessels at the top are affected. In young palms up to 6 years old diseased vessels are usually widely dispersed throughout the base.

Incidence, susceptibility and spread

Prendergast (1957) claimed that healthy vigorous palms in good soil suffer little from the disease and he showed that in areas of K deficiency incidence is substantially reduced by applications of potassium fertilizers. This finding has been substantiated in potassium experiments in areas of Vascular Wilt in both the Ivory Coast and Benin (Dahomey) (Ollagnier and Renard, 1976).

It has been shown that although, where prevalent, the outbreaks are often sporadic, new cases do tend to develop next to old ones. Young palms in a replanting tend to become infected when they are near to the sites of previous cases of Vascular Wilt. On one estate in southern Nigeria incidence reached 23 per cent in one-half of the area and 12 per cent in the other half with 17-year-old palms. In fertilizer experiments incidence was over 30 per cent in plots not receiving potassium but usually below 20 per cent when potassium had been applied.

Control

In Zaire the destruction by fire of all diseased palms and their neighbours was recommended and the replanting of areas where Vascular Wilt had been prevalent was discouraged (Moreau 1952). In the Ivory Coast, Renard and Quillec (1983) recommend the monitoring of first-generation palms and removing any infected palms not producing more than a bunch a year. After replanting with tolerant material the ground is either kept bare for 3 years or planted with grass species of *Brachiaria* instead of *Pueraria*. It is claimed that competition for nitrogen between the grass and the palm discourages *Fusarium* infection. There have been some suggestions for the use of fungicides (e.g. Moreau and Moreau, 1960; Renard, 1976), and in particular the systemic fungicide benomyl, but Turner (1981) considers that the effect of any such treatment is likely to be short-lived.

The most promising method of control is by the breeding of resistance lines or the screening of selections for tolerance to the disease. Prendergast (1963) was the first to develop a technique for testing seedling resistance at the nursery stage and his methods were adopted with very little modification by Renard *et al.* (1972). For detailed descriptions these authors' papers should be consulted, but in essence the method consists of applying a mixed inoculum of isolates of the causal organism from different sources to nursery seedlings by pouring 10 ml of the inoculum on to the bulb of the seedlings or the exposed roots around the collar. Later, prenursery seedlings were used. The work showed clearly that no seedlings are resistant to invasion and that differences are really ones of tolerance to the fungus.

Different and empirical methods of division of progenies into 'resistant' or 'susceptible' have been used. Prendergast divided progenies into (1) good, (2) probably resistant, (3) possibly resistant and (4) poor. (1) and (2) had a lower percentage loss in a given test than the mean of that test; (1) had a lower percentage loss than the mean of (1) and (2). Renard *et al.* (1972) calculated a wilt index which is the percentage of wilt-infected plants in a progeny as a percentage of the wilt percentage of all the plants in a trial. Another index, Ig was attributed to a parent by taking the mean of the wilt indices of the progenies in which it figured. Those with indices above 100 were called 'susceptible', those below 100 'resistant'. Special attention was given to *pisifera* parents.

The methods referred to above do not ensure that the dose of inoculation per seedling is uniform. Locke and Colhoun (1973a) developed a method (of which full details are given in their papers) of inoculating very young seedlings grown in compost with two known levels of inoculum, and then comparing their growth with that of seedlings grown in uncontaminated compost. Determinations were made of the number of propagules in the soil so that subsequent inoculations could bear a relation to normal soil levels. A technique was devised for providing talc-chlamydospore mixtures for storing and for use in inoculation experiments. The fungus was recovered from progenies showing both large and small reductions either in weight per plant or in 'leaf area product'. No progenies were immune. It was demonstrated that some progenies were tolerant of infection in lightly contaminated compost only, some in both lightly and heavily contaminated compost; others showed high susceptibility at both levels. A high level of repeatability was attained.

Evidence has been presented to show that tolerant seedlings do give rise to palms which have a low incidence of Wilt in the field (Renard *et al.*, 1972, 1980; Green and Ward, 1973). In Nigeria, Rajagopalan *et al.* (1978) found that among 336 progenies none was immune but 149 showed sufficient tolerance to be considered valuable for breeding; and certain *pisifera* showed their worth by providing consistent tolerance in crosses with a range of dura. Meunier *et al.* (1979) also found parents which consistently gave crosses with a high degree of tolerance and they attributed most of the variability to genes with additive effects.

Locke and Colhoun (1973b) have shown that, in spite of some routine disinfection methods, *Fusarium oxysporum* can easily be transmitted by seed from one country to another.

Ganoderma Trunk Rot (Basal Stem Rot)

Cause: *Ganoderma* sp.

Distribution

Species of *Ganoderma* are soil-borne fungi of widespread occurrence, causing disease in a great many economic crops. In Africa the species attacking the oil palm are found throughout the palm groves, killing the old palms and occasionally younger ones which have not yet dropped their leaf bases. In planted areas it is also a disease of senescence, being rarely found except in fields which should already have been replanted. In Asia, though it is also a disease of old palms, it is of much greater economic importance since in certain areas it has attacked palms of 5 years old and upwards, particularly in replantings. Disease caused by *Ganoderma* spp. has not yet been reported from America.

Symptoms

In old palms the onset of the disease is shown by the collapse of one or more of the lower leaves which do not snap, but hang vertically downwards complete with their petioles. This is followed by the drooping of younger leaves which turn a pale olive green or yellowish colour and die back from the tip. The leaflets roll back around the rachis and the top of the stem becomes heavily cloaked by the desiccated leaves (Plate 13.4).

Later, the base of the stem blackens, gum may be exuded and the well-known fructifications of *Ganoderma* sp. appear. The whole head of the palm may then fall off, or the trunk collapse.

Bull (1954) has described and illustrated the internal symptoms of old palms exhibiting Ganoderma Trunk Rot. Briefly, it is found that the peripheral tissues are hard and unaffected by the rot, the

Pl. 13.4   *Ganoderma* Trunk Rot of an old palm on coastal alluvium in Malaysia.

black fibres in this zone being normal. Within the stem at the base of the palm there is usually a large peg of dark brown infected material; between the peg and the outer part of the stem are narrow dark brown bands. In this zone the majority of the tissue is yellow-coloured and breaks up easily; mycelium can be found extending through the tissue.

Roots are also found to be infected, the cortex being brown and decaying, the stele black. In Malaysia, white mycelium of *Ganoderma* spp. has been found surrounding the cortex which becomes spongy. Large numbers of fructifications, the sporophores, may be formed, the early ones being small and rounded, the later ones being typical brackets. The upper surface of the sporophore is reddish-brown and rather uneven, though shining. There is a clear white band at the outer edge; the underside is pale yellow-brown and covered with minute pores.

The symptoms of young palm infection in Asia are variable but a common first symptom is that of drought conditions, i.e. a failure of the young leaves to open so that a number of fully elongated but unopened 'spears' are seen in the centre of the crown (Turner, 1966). The internal symptoms are similar to those of infected old palms, but the sporophores often appear at an early stage, thus confirming that the failure of leaf opening is a disease-induced symptom.

Cause, spread and control

Turner (1981) has listed fifteen species of *Ganoderma* which have been recorded as likely pathogens, and he considers that a single species is unlikely to be the sole cause of the disease in any particular area. Species pathogenic to the oil palm also attack a wide range of other hosts.

It is in areas such as Malaysia, where Ganoderma Trunk Rot is not simply a disease of senescence but has caused serious damage in young replantings, that investigations of its spread and control have proceeded furthest. Navaratnam (1961) obtained successful inoculations, with mycelium, of both roots and stems of 40-year-old palms and from this and the general pattern of the disease it seemed likely that infection under natural conditions is mainly by root contact. The disease is much more prevalent on the coastal clay soils than on inland soils and it is on the former that the serious attacks on young palms have occurred. Turner (1965a), examining attacks on young palms, showed that incidence on areas where the preceding crop was coconuts was much higher than where planting followed forest or rubber. Cases were quoted where fields of 15-year-old oil palms after rubber had 4 and 2 per cent Ganoderma attack while adjoining areas which followed coconuts had incidences of 39 and 35 per cent. Where oil palms follow oil palms there is a similar build-up of Ganoderma on old stumps and trunks, but Turner (1981) considers oil palm tissue a less conducive medium for the fungus than coconut tissue. However, old oil palm stumps or trunks, especially when poisoned and left to rot *in situ*, produce many fructifications, and replanting without treatment of the old material will result in a high incidence of the disease in the replant. Devastating attacks have been experienced in replanted coastal areas of Sumatra where underplanting was practised or the old stand not treated.

For a full discussion of methods of treatment and prevention of Ganoderma Trunk Rot Turner's (1981) detailed account should be consulted. Briefly, once an attack has started in a field little can be done to combat it by cultural methods other than by removal of palms actually seen to be diseased. Attempts to control the disease by systemic fungicides have not yet been successful, probably owing to the massive lesions involved, though work is continuing (e.g. Loh, 1977; Jollands, 1983). Breeding for resistance has been considered and differences of incidence between West African and Deli material, and between *dura* and *tenera*, have been quoted (Umar Akbar *et al.*, 1971; Akbar and Kusnadi, 1976), but Turner considers that any natural resistance is likely to be over-whelmed by the volume of inoculum present. Biological control by inoculating lesions with cultures of various micro-organisms has been suggested (Varghese *et al.*, 1975) but not pursued. Surgery of large discrete lesions has been practised (Turner, 1968), sometimes successfully, but is expensive, and treated palms may later collapse and will, of course, need constant inspection. All in all, the only method of dealing with the disease in an existing plantation is by regular inspection rounds and removing individual cases by poisoning, felling, cutting up the trunk and excavating the bole tissue to hasten decay of the whole palm.

In the long term, the best method of reducing Ganoderma incidence is to deal with it at replanting time by thoroughly ridding the fields of as much oil palm tissue as possible. Mechanical methods have been outlined by Turner (1981), and Stimpson and Rasmussen (1973) have given an account of a system used on the coastal clays of Malaysia which entails burning or, if this is not possible, cutting up, splitting the boles and windrowing the old oil palms so that they rot rapidly. The method adopted includes prior poisoning of the palms and subsequent root raking and ploughing to bring up and dispose of pieces of palm base and other material which may form a focus for *Ganoderma* spp. The operations are costly but are considered essential in coastal areas. In inland areas where the hazard of Ganoderma is not so great, poisoning and felling may be followed by any method which encourages the rapid rotting of the old palms.

Some soil and leaf analysis studies have been made in relation to Ganoderma incidence, and these indicate that nitrogen and magnesium may have special roles in combating the disease (Umar Akbar *et al.*, 1971). The difference in incidence in fields on coastal soils in Asia and inland soils remains unexplained. One curious feature of the disease is its low incidence on clay areas where there is a covering of peat.

Armillaria Trunk Rot

Cause: *Armillariella* (*Armillaria*) *mellea*.

Distribution

This disease became conspicuous in the northern oil palm areas of Zaire in the late 1940s. Although *A. mellea* was recorded from the oil palm in Ghana in 1927, the disease as such has not been recorded as present in West Africa nor in any other oil palm region of the world.

Symptoms

The symptoms have been described by Wardlaw (1950b) and Moreau (1952). Palms 4–12 years of age are usually affected. External symptoms are similar to Vascular Wilt disease but, in addition, the lower leaf bases become rotten and easily fall from the trunk. Internally, the trunk shows regions of mass infection in a soft rot which may progress across the base of the trunk or in an upward direction. The disintegration becomes so complete that the trunk collapses. Infected tissue has, at first, a yellowish-brown colour and the fibre strands, not the wood as in Vascular Wilt, become dark. Occasionally, the palm seals off the rotted portion of the trunk, produces new roots and survives. The fungus produces light brown mushroom-like sporophores near the base of the trunk during the wetter parts of the year.

Invasion takes place through the roots, only a few of which may be attacked; the cortex is destroyed by the matted white mycelium. Rhizomorphs spreading round the base of the palm cause the loosening of the leaf bases.

Incidence and control

Although the incidence of *Armillaria mellea* has varied considerably it can become locally important. The fungus may sometimes be found in the palm in conjunction with *Fusarium oxysporum* and this makes it difficult to measure its incidence and to determine its pathogenicity.

A number of control measures such as the regular inspection and disposal of infected palms, removal of infected leaf bases and the removal of infected forest stumps have been suggested, but the disease has not proved sufficiently widespread for investigation to proceed very far, and its incidence seems to have decreased considerably since the 1950s.

Sudden Wither (Marchitez Sorpresiva)

Cause: Uncertain.

Distribution

This disease has been serious on one plantation in Colombia and has been reported on other plantations in that country and in Ecuador and Peru. In Surinam the disease has been described as Hartrot (Van Slobbe *et al.*, 1978), and a similar condition has been prevalent in Bahia, Brazil.

Symptoms

The disease is characterized by a sudden rotting of all developing bunches, a reddish discoloration of the top of the petioles and a rapid drying out of the leaves from the oldest ones upwards. This drying out is preceded by the appearance of reddish-brown streaks at the ends and centres of the lowest leaflets. The leaf then becomes successively pale green (like nitrogen deficiency), yellow, reddish-brown and ash-grey. The palm dies in 2–3 weeks and as soon as the external symptoms appear the root system will be found to have rotted and to a large extent dried out. Similar symptoms though proceeding at a slower rate are sometimes seen, and in Colombia this has been referred to as 'Marchitez progresiva'. However, this description may be due to confusion with palms suffering from Lethal Bud Rot (Corrado, 1970). In the typical Marchitez symptoms the spear is initially unaffected.

The root rot is cortical. The cortex liquefies in wet weather but in the dry season tends to become necrosed and to detach itself from the stele. The rot starts to develop from the extremities and moves towards the trunk and towards the lower roots. The trunk itself usually remains healthy but cases are reported where the base is rotted sufficiently to form a cavity (Martin, 1970; Van den Hove, 1971).

Palms have been attacked by Sudden Wither from the age of 4 years onwards.

Cause and control

Sudden Wither has the appearance of an adult Blast attack and it is clear that, as in Blast, the destruction of the roots quite suddenly exceeds their replacement to such a degree that the palm no longer absorbs sufficient water for survival.

In some areas the disease has been found mainly near to rivers or on the periphery of plantations near forest. At first it was attributed to an effect of heavy, compacted, poorly drained soils; on the plantation in Colombia suffering greatest devastation the soils had been compacted by cattle and tractors. However, the appearance of the disease in other situations suggested that soil conditions could not be a cause. Evidence has been accumulating that the prime cause is infection by protozoan flagellates of the genus *Phytomonas* which have been found in several countries in association with the disease in the phloem of roots, meristem zone, spear base and inflorescence stalks (Dollet *et al.*, 1977; Dollet and Lopez, 1978; Dzido *et al.*, 1978). Little is yet known, however, of exactly how the flagellates cause the disease symptoms.

In certain cases a connection between infection and insect attack has been thought likely, though here again the role of the associated organisms has not been clear. Some workers (López *et al.*, 1975; Genty, 1976) have considered that the root miner, *Sagalassa valida* (Genty, 1973a, b, see p. 641) may be a carrier, though the miner is often present in areas where the disease is absent. In Colombia, where the disease devastated an area where palms were growing in a heavy stand of *Panicum maximum*, the bug *Haplaxius pallidus* (see p. 632) was found on palm leaves while the nymphs were present on the roots of *Panicum*. The use of herbicides and insecticides reduced the incidence of Marchitez (Mena Tascon *et al.*, 1975), while inoculation experiments (Mena Tascon and Martinez-Lopez, 1977) also suggested that *H. pallidus* might be playing a part in the transfer of the disease.

The likelihood that *Phytomonas* has a major role in the disease has increased through its discovery in the sieve tubes at the start of leaf symptoms (Genty, 1981) and its presence in certain weed plants (Dollet, 1982). Most recently, Desmier de Chenon *et al.* (1983) and Perthuis *et al.* (1985) have claimed that the bug *Lincus lethifer* and other species of *Lincus*, which live in the axils of the leaves (Dollet *et al.*, 1987), are the vectors of the flagellate.

In view of the uncertainty of diagnosis it is difficult to recommend definite control measures. On the grounds that *Sagalassa valida* is likely to be playing a part in the transmission of the disease, applications of endrin around the base of the palms at rates of 2 litres of solutions of between 0.75 and 1.5 per cent endrin (19.5 per cent a.i.) have been used to suppress the insect and have been strongly recommended in Colombia, Ecuador and Peru (Genty, 1977b; López *et al.*, 1975).

Cases of Sudden Wither in *E. oleifera* hybrid × *E. guineensis* have not until recently been recorded and it is possible that the planting of hybrids may be a method of avoiding the disease. In Surinam, however, some hybrids have been subject to 'Hartrot', though wild *E. oleifera* palms with the disease have not been observed (Alexander and Kastelein, 1983). Certain *E. guineensis* palms on a plantation devastated by this disease have remained healthy; it may therefore be possible to select resistant progenies within the species.

Upper Stem Rot

Cause: *Phellinus (Fomes) noxius.*

Distribution

Thompson (1937) described a lethal trunk rot, attributable to *Fomes noxius*, which was serious only on deep peat and inland valley soils. This disease has, however, appeared on other soils in both Malaysia and Indonesia.

Symptoms

Fructifications of *Phellinus noxius* only appear on palms where the leaf bases are extensively decayed. The brown decay appears to proceed slowly inwards from the leaf bases and in many cases a typical collapse of the stem at one point occurs, this usually following high winds. Pathogenicity was proved by inoculation experiments.

Investigations of this condition remained largely in abeyance for nearly 30 years, but Navaratnam and Chee Kee Leong (1965) and Turner (1969, 1981) have now given extensive accounts of its symptoms, incidence and control. Examination of affected palms showed that the disease was confined to the stem and did not enter the roots. Typically, the lower leaves first become yellow and this symptom gradually extends to the middle leaves and then to the spear. It was evident that spore infection of leaf bases takes place and that from these the fungus gains entry to the peripheral tissues of the stem. The rot thence spreads upwards and downwards in the stem, eventually killing the palm by invading the crown. Two forms of fruiting bodies (normal and resupinate) appear later; these are small greyish-brown bodies with velvety-brown margins and are inconspicuous among the leaf bases.

Control

As there is usually much penetration of the stem by the time sporophores appear it is desirable to detect the disease at an earlier stage. This can be done on palms of 10 years or older by a sonic method, i.e. striking the leaf bases with a wooden pole to detect the dull sound of an infected base. Incidence is insufficient to justify surveying palms below 10 years. When the diseased leaf bases are cut away the extent of the infection can be explored. The lesion is exised from the stem with a harvesting chisel and the cut surfaces are treated with a preservative (Turner, 1969). Coal tar has been reported to give the best overall results. Treated palms give as high a yield as untreated palms, so the measures are considered well worthwhile wherever incidence is likely to be significant. If palms are allowed to collapse with this disease they become a focus for *Ganoderma* spp.

In a fertilizer experiment containing different progenies, there was evidence firstly that fertilizers containing potassium reduced incidence, and secondly that progeny resistance and susceptibility existed. (Navaratnam and Chee Kee Leong, 1965).

Other stem or trunk rots

Basal Decay

Known in Malaysia as Stem Wet Rot, it is found sporadically in Africa. In both regions palms 3–8 years old are affected and very occasionally older palms. For a stem rot the disease is unusual in that the centre of the crown is first affected, the spears being shorter than normal so that the palm shows a central depression. Soon, however, the lower leaves are affected and then the younger ones. Some of them are subject to curious twisting and all of them eventually collapse and the palm dies. The deaths tend to be so sporadic and occasional that they are often not noticed until all characteristic symptoms have passed; if noticed early enough, however, the base of the young palm's trunk will usually be found to be destroyed internally by a wet, putrid rot which, on dispersion, leaves a large cavity with a surrounding zone in which yellow-brown fibres are found although the cortex has been destroyed. Incidence is not often more than 2 per cent. In Malaysia the disease is more sudden in its onset, all unexpanded leaves and a few expanded ones dying suddenly at the same time and complete death of the crown taking place within 2 weeks (Turner, 1981).

The cause of Basal Decay is unknown. Turner (1981) considers that the symptoms suggest infection through the roots and that a bacterium may be involved. No control measures can be recommended; fortunately incidence is low and is confined to young palms.

Charcoal Base Rot

A minor disease of Malaysia attributed to the fungus *Ustulina zonata* or *U. deusta* (Thompson 1936; Turner, 1981), but its pathogenicity is uncertain. A black dry rot is formed at the base of palms whose foliage, particularly the older leaves, have become chlorotic. After the rot has advanced across the base the palm may fall over. Incidence is low and the disease so far unimportant.

Diseases of the bud or stem apex

Under this heading must be grouped any disease condition occurring in the emerging spear and younger leaves inside the crown. Such diseases normally move towards the growing point through the enclosed, developing leaves of the 'cabbage' and when they reach it the palm is killed. Bud and spear rots have been of wide occurrence in all three continents and provide, perhaps, the most difficult problems of oil palm pathology.

Investigation is difficult owing to the position of the transition zone, often in the heart of the palm, the rapid entry of secondary organisms into any rot within the cabbage and the multiplicity of confusing symptoms, some of which may be similar to those of deficiencies or genetic abnormalities.

Turner (1981) has suggested that the term 'spear rot' should apply to diseases in which the primary rotting affects the spear, while 'bud rot' should be used only for diseases first destroying the unemerged leaves and the hidden base of the spear and also, usually, the apical meristem. The latter diseases are usually fatal, the former frequently not.

Spear Rot-Little Leaf Disease

Cause: Bacterium of the genus *Erwinia*.

Distribution

This disease, previously called Bud Rot-Little Leaf, has caused serious losses in the oil palm areas of southern Zaire where deaths exceeding 30 per cent have been common. In northern Zaire it is of occasional occurrence while in West Africa cases rarely exceed a few per cent and are often confined to certain progenies; deaths occur but are rare. Symptoms would suggest that the disease in West Africa is the same as that in Zaire, but this has not been proved. Spear rots in the Far East have been insufficiently described and studied; it is therefore not possible to say if this specific disease exists in that region. Spear and bud rots in America have different and varying symptoms.

Symptoms

Many causes have been assigned to the 'little leaf' sympton and its place in the symptomatology of bud and spear rots and possible nutritional disorders was not made clear until reviewed by Bull and Robertson (1959); for the early history of investigations of the little leaf symptom their paper should be consulted.

The first sign of attack is a wet, brown rot on the lower part of the unopened spear leaf. Robertson (1960), working in Nigeria on palms of a susceptible progeny having regular cycles of infection, showed that the cause of Spear Rot-Little Leaf disease was an active pathogen since the appearance of little leaves and bud rot could, he found, be prevented by cutting off the rotting spear below the rotted portion. Although prior insect attack is often suspected it is not known for certain how spear infection takes place. However, spear rotting is the primary symptom and Duff (1963) has described how in very mild cases only the leaflets may be affected and the leaflet rot is passed from spear to spear until it either develops further or the palm grows out of the attack. Normally, however, the rachis becomes infected and the spear collapses and hangs down; it is not uncommon to find a spear leaf, in which the infected portion has rotted away altogether, lying on the ground where it has fallen.

The spear rot grows downwards and becomes a bud rot, but it is only if this reaches the growing point that the palm dies; in other cases a callus layer is formed and the palm produces a varying number of 'little leaves' according to how many unemerged, undeveloped leaves have been partially destroyed by the rot. The first leaves to emerge after the spear rot are stumps consisting of the malformed basal portion of the rachis. Subsequent leaves are very short with a few corrugated shortened leaflets, but each successive leaf will be longer, and the leaflets less abnormal, until fully normal leaves are again produced. Little leaf is therefore a recovery symptom and does not precede rotting (Plate 13.5).

Cause and susceptibility

A bacterium of the genus *Erwinia*, similar to *E. lathyri*, was consistently isolated in Zaire by Duff (1963) from young lesions and from tissue in advance of visible rotting. This bacterium has been found on the surfaces of spears and opened leaves of healthy palms, and inoculation experiments have shown that Spear Rot-Little Leaf symptoms can be induced by it.

Susceptibility seems to be genetic, physiological and seasonal. In a field in Nigeria the disease was confined to one progeny. Genetic differences were also found in Zaire where there was an association between rate of growth and disease incidence. The former was judged by the rate of elongation of spears and in susceptible palms the growth rate fell below normal levels 2 or 3 weeks before an attack of the disease. It was believed that these circumstances, encountered in 'unhealthy' palms, allowed susceptible tissues to be exposed to infection for longer periods than normal. Palms whose growth rate was artificially reduced by root or leaf cutting showed greater than normal susceptibility. Seasonal differences have ranged from there being no seasonal influence to high incidences either at the beginning or end of the rains (Turner, 1981).

Control

Duff has provided growth and health records showing that the more vigorous progenies suffer less from the disease and he infers from this that anything interfering with vigorous growth increases susceptibility. While, therefore, the disease is not likely to be serious

Pl. 13.5 A 'Little Leaf' with pronounced lamina contortion.

enough for control measures to be taken in areas where growth conditions, particularly those of water and nutrient supply, are good, in marginal areas the breeding of particularly vigorous progenies will be necessary.

Lethal Bud Rot (Pudrición del Cogollo)
Cause: Unknown.

Distribution

A lethal bud rot with variable symptoms, but not usually including the typical 'little leaf' progression, has caused serious damage on plantations in Central and South America. Some plantations have been totally devastated while others have suffered serious losses with many palms remaining in a moribund, unproductive condition for long periods. Turner (1981) has called the disease Fatal Yellowing from a characteristic symptom found in Colombia.

Symptoms

This description is taken from a plantation in northern Colombia where the disease killed almost all palms in many of the fields (Sanchez Potes, 1972; Turner, 1970a, b). In many cases about four to six young elongated leaves remain unopened and stuck together; this is the 'baton' effect. On separation these spears are found to be sticky. Whether this symptom is seen or not, spear decay starts some distance from its base. The rot, which is reddish-brown at first, spreads downwards and the spear eventually collapses, this process taking from 1 to 9 weeks. The youngest leaves become yellowish, typically with yellow stripes on each side of the leaflet midribs, and this is often the first symptom noticed. Some of these leaves later collapse and hang down in the same manner as the spear. Fruit rot does not appear at this stage. In some cases the rot may affect the surrounding leaves before it affects the spear. The early stages of the disease have also been accompanied by leaf splitting.

The later stages of the disease are more rapid. After spear collapse the bud tissue and developing leaves show various stages of decomposition into a wet, putrid mass. The rot may advance in a broad zone towards the growing point or it may be found to be in a narrow strip between the spear and a younger leaf running down towards the bud. This rot is of a light orange-brown colour and wet. The spear leaflets also rot and become dark grey. When the rot reaches the growing point the palm collapses and dies. Death may be rapid or the palm may remain moribund for some time before death. Spontaneous recovery occasionally occurs and in some cases 'little leaves' may be produced, but the regular succession found in Spear Rot-Little Leaf is not usually encountered.

In Panama the pattern of Lethal Bud Rot in young palms has been a little different; chlorosis is often absent and spear rotting takes place in the 'funnel' well above the growing point. The rot

Pl. 13.6.   Young palm suffering from Spear Rot, with no central leaves, Panama.

spreads to adjoining rachises and further spears do not emerge though some rotted tips can be seen; this gives the palm a spreading appearance with an empty centre (Hartley, 1965) and the existing leaves remain green for a long time (Plate 13.6). Internally, the rot can be found in the funnel where it continues to attack any new spears starting to elongate. The palm can remain in this state for perhaps a year without the rot penetrating far towards the growing point. Little leaves have occasionally been seen, but complete recovery without little leaf has been noted in other cases (Plate 13.6).

Lethal Bud Rot can attack palms of any age, but in Colombia cases were noted at 30 months and incidence was at its highest on palms which had been in bearing for 2–5 years. However, on replanting devastated areas the disease appeared on palms as young as 3–8 months (Anon., 1974).

In Nicaragua, Bud Rot has occurred as a lethal disease on tall adult palms. The disease starts as a rot of successive spear leaves leading to the production of smaller, usually chlorotic, leaves and finally to the rotting away of all leaves as they emerge; the centre of the crown is thus markedly reduced in size although there are no little leaves. On dissection, the rot is found to be in the funnel, and even when the external symptoms are far advanced rotting has only reached within about 30 cm of the growing point. Sudden recoveries occur, but the disease tends to be recurrent and eventually a damp rot reaches the growing point and kills the palm (Hartley, 1965).

Cause and incidence

The cause of this syndrome remains obscure, and it should be noted that as symptoms vary from country to country it is not certain that they are of the same disease or have the same cause. Insects, fungi and bacteria have been suspected, while unfavourable environmental circumstances such as poor drainage, compacted soils and unbalanced nutrition have been put forward as predisposing factors. No direct evidence of infectious spread was found, and it was thought that the development of large infected areas might be related to local growth factors. In Colombia a relation was found between former land usage and incidence, areas with compacted pasture soils suffering the highest casualties; however, this did not prevent the disease spreading over most of the plantation irrespective of the prior vegetation. A low potassium/magnesium ratio was also suspected, but corrective manuring did not stem the spread of the disease.

Fungi which have been isolated from diseased spears include *Fusarium oxysporum, F. moniliforme* and *Botryodiplodia* sp. in Colombia and *F. solani* and *Sclerophoma* sp. in Ecuador. Invasion of bud tissue by many species of bacteria follows the basal spear rot. Pathogenicity of these organisms has not yet been established. Turner (1981) considers that, while many symptoms suggest a strong bacterial association, they are quite unlike those normally associated with *Fusarium* attack.

It is claimed that bud rots are initiated by insect attacks, and in Colombia a *Cephaloleia* sp. was found to induce symptoms similar to those of the early stages of the disease, and Urueta (1975) studied a range of other insects in diseased material. In Ecuador, Dzido *et al.* (1978) found the larvae of *Alurnus humeralis* and several other insects on diseased palms, but no definite connection between these insects and the disease has yet been established.

Control

Without a knowledge of the causal organism it has not proved possible to initiate direct control measures. Speculative prophylactic applications of mixed fungicides and insecticides to the palm's crown have not been successful.

By far the most promising measure is the introduction of resistant progenies. It is possible that, as with Spear Rot-Little Leaf disease, there may be resistant lines within *E. guineensis*, but the fortunate discovery on La Arenosa (Coldesa) plantation in Colombia that specimens of *E. oleifera* and the interspecific hybrid were immune or resistant to the disease led to the replanting of devastated areas with the hybrid and the establishment of the first commercial plantation of this cross in the world. On this plantation hybrids raised from *E. oleifera* parents growing in Surinam and crossed with six African *E. guineensis pisifera* were laid down in 1963 in a number of fields where the *E. guineensis* population was subsequently devastated (Anon., 1974). From 1968, crosses of Colombia *E. Oleifera* with *pisifera* were also planted in small plots. All these hybrids survived while the surrounding *E. guineensis* palms died off in large numbers. In 1970 *E. guineensis* and hybrids were planted in a field in alternate rows; the *E. guineensis* died in large numbers while the hybrids were unaffected; there seems little doubt, therefore, that in areas subject to Lethal Bud Rot the planting of this cross is the best known method of combating the disease.

Diseases of the bunches and fruit

The occasional bunch and fruit rots which are encountered have not been extensively studied. Bunch-End Rot has been associated with the Deli palm, particularly in Malaysia (Thompson, 1934). Where neither lack of pollen nor insect attack are implicated, both this condition and complete Bunch Failure (see p. 173) are attributed to 'over-bearing' (Turner and Bull, 1967), i.e. the full development of the bunch in cases of heavy female inflorescence production is thought to be more than can be sustained by the palm's processes of assimilation.

A Bunch Stalk Rot has been connected with an undiagnosed condition in West Africa known as Leaf Base Wilt (Bull, 1954). The leaves bend down towards the ground and the stalks of bunches in the leaf axils also bend and may then begin to rot. The disease seems to be of purely mechanical origin and provided the rot is not so extensive that the bunch falls, the majority of fruit will develop. The small splits that appear in the stalk are invaded by a variety of saprophytic bacteria and fungi.

Marasmius Fruit Rot

Cause: Marasmius palmivorus

Distribution

A number of species of *Marasmius* have been recorded on the oil palm in Africa and Asia (Turner, 1971) but only one, *M. palmivorus*, has been associated with rotting of fruit in the bunch and this condition has been confined to the Far East.

Symptoms

Marasmius palmivorus is common on the cut petioles and on the decaying debris between these and the trunk. The white or pinkish strands or rhizomorphs can easily be seen in association with the fructifications. The latter have white caps, 5–8 cm in diameter, which are upturned when fully developed; on the under-surface are the spore-producing white gills. The fructifications are produced in greatest abundance in wet weather, and in drier weather they tend to be smaller and pinker (Turner, 1965b). Both the rhizomorphs and fructifications extend from the leaf axils to rotting or apparently healthy bunches under conditions which will be discussed below.

Susceptibility and parasitism

Marasmius palmivorus is a saprophyte on the leaf bases and in the leaf axils of the oil palm and its standing as a parasite has been much debated. The fungus has been stated to 'grow up over mature fruit bunches and render them useless for oil production' and to 'invade living tissue' and it has thus been accounted a facultative parasite (Turner, 1965b). Pathogenicity has never been proved, but where a large mass of rotting material, e.g. unpollinated bunches and other debris, are provided under moist conditions sufficient inoculum potential appears to be built up for healthy bunches to be invaded from the infection sources. For a full discussion of the factors involved in the spread of the disease, Turner (1981) should be consulted.

Control

The most obvious means of control is to reduce as far as economically possible, through sanitary measures, the media on which the fungus grows on the palm. Where bunches are rotting and *Marasmius* is seen to be prevalent, infected bunches, dead male flowers and as much of the axil debris as possible should be periodically removed; burying or burning of the debris are of doubtful value. The severing of leaves as near to the trunk as possible has been advocated, thus reducing the space where debris may collect; but this practice should certainly not be followed where there is danger of attack by *Rhynchophorus* species. Prophylactic spraying against *Marasmius* is not generally recommended, but Turner (1981) considers that on acid sulphate soils in Malaysia spraying at lengthy intervals may be economically justified: alternatively, partially infected bunches may be sprayed with Antimucin WBR (0.12 per cent) or cycloheximide (Acti-dione).

Other abnormal conditions of the palm

The oil palm is subject to many abnormal conditions of growth and development the causes of which are not known. Usually, though not always, these abnormalities are encountered where conditions are in some way adverse: impoverished sandy soils, long dry seasons, excessively wet conditions, intermittent waterlogging, grass competition, pockets of unusual valley-bottom soils, etc. In the more severe conditions bunch yield is usually negligible (Courtois, 1968). Only a few can be mentioned here.

Plant failure (Boyomi)

This term was used by Wardlaw for palms which almost ceased to grow. Owing to a marked lowering in the rate of root and spear production the number of green leaves decreases and those that remain are erect and crowded through lack of heavy leaflets and bunches in their axils. This in turn leads to a tapering of the trunk and progressive deterioration of the leaves which are subject to various kinds of chloroses, dry out prematurely and become brittle. There has been much speculation on the reasons for such palms being found dotted about among normal ones. The condition rarely occurs in Asia. In Africa it is considered either to be of genetic origin or may be associated with severe potassium and magnesium deficiency.

Dwarfed Crown or Choke

This condition has been encountered in fields in America suffering from Red Ring Disease (see p. 626), but does not appear to have the same cause (Malaguti, 1953). It has been referred to as *hoja pequeña* (little leaf) but this is unsuitable as it is important that the term 'little leaf', in oil palm symptomatology, should be reserved for the recovery symptom of Spear Rot-Little Leaf. In this condition all the leaves are smaller than normal, green, erect, bunched together and twisted with varying amounts of atrophy or corrugation of the leaflets. A sudden recovery from the condition is frequent, a tall bunch of normal new leaves being produced in the centre of the deformed ones giving the palm a two-tier appearance. This type of deformity is not unknown elsewhere and the term 'choke' has been used in Malaysia to describe a similar condition (Plate 13.7).

Genetic chloroses

Some forms of chlorosis are genetic. An orange spotting has been confined to certain progenies in Africa and Asia. It can be distinguished from Confluent Orange Spotting (p. 485) firstly by the fact that it is not general in the field but is confined to a progeny and secondly by the absence of marginal necroses on the affected leaflets of older leaves.

Another chlorotic condition thought to be genetic consists of leaves which may be wholly or partially pale or bright yellow and which continue to be produced on the same palm.

Pl. 13.7   Sudden recovery from an abnormal 'choked' condition on coastal alluvium of low pH in Malaysia.

Rigidoporus (Fomes) lignosus and *Porogramme (Poria) ravenalae*

Fructifications of these fungi are occasionally seen at the base of young palms in Asia, the latter invading the leaf bases. Neither fungus appears to be pathenogenic on the oil palm.

Leaf Base Wilt

The leaves appear prematurely to lose their normal full attachment to the stem and therefore are set at a more obtuse angle than is normal for their age. With older palms they often bend over sufficiently to touch the ground or, if the palm is tall, to hang down to cloak the trunk. The effect on the bunch has already been mentioned (p. 622). The cause of this condition is unknown.

Leaf Mottle

This condition has been reported from Ecuador and Peru and described at length by Turner (1981). When the spears open they fail to become fully green, linear spots of pale tissue remaining. This leaf symptom is followed by the rotting of the root system and spear; death usually follows, though there are cases of recovery. Diseased palms tend to be scattered throughout a field and are much more frequent in grass areas than where a *Pueraria* cover is dominant. A good leguminous cover should therefore be maintained in areas subject to this condition.

Lightning

The oil palm is occasionally killed by lightning strike. Young palms can collapse rapidly and wither, but a sub-lethal condition known previously as Rachis Internal Browning, is now also believed to be caused by lightning (Turner, 1981). In older palms the trunk base is often charred. Surrounding palms show scorching on the side facing towards the strike.

Pests

The pests of the oil palm include nematodes, slugs and snails, mites, insects, birds and mammals.

Nematodes

Nematodes have often been thought responsible for damage to the oil palm, but only one has been responsible for a serious diseased condition, though others have been found in rotting tissue, e.g. the roots of seedlings suffering from Blast and the spears of cases of Spear Rot-Little Leaf.

Red Ring Disease (Anneau rouge, Anillo rojo)

Cause: *Rhadinaphelenchus cocophilus*.

Distribution

'Red ring' of oil palms has been found in Venezuela, Surinam, Brazil and Colombia where the similar disease of coconuts is prevalent. It has been studied on an estate in Venezuela where it has done very considerable damage (Schuiling and Dinther, 1982).

Symptoms

The symptoms of this disease have been described by Malaguti (1953). The centre of the crown takes on a dwarfed appearance and the newly opened leaves become bundled together into an erect compact mass, the leaflets being corrugated, twisted and sometimes adhering to the rachis. Gum is exuded. Later this crown of leaves turns slowly yellow and dries out, the rachis being a light brown colour with yellow spots. One or two of the intermediate leaves become bronzed and after 2–5 months all the leaves gradually become yellow or bronzed, though remaining erect. Developing bunches rot away and inflorescences fail to set fruit.

The most striking interior symptom is the brown (not really red) cylindrical ring found in the trunk, 7–8 cm from the periphery and 1–2 cm broad. This ring is most distinct towards the base of the palm but the infection proceeds upwards into the petioles and rachis of the leaves in the crown in which, on cross-sectioning, necrosed areas or spots can be found (Plate 13.8). This infection does not, however, invade the tissues of the stem apex or surrounding very young leaves. The effect of the invasion of the stem and rachis therefore appears to be the prevention of normal water and nutrient supply to the foliage.

Incidence and cause

It has been shown that in unprotected areas incidence can become high with some rapidity; Malaguti cites a group of 100 palms showing only 16 doubtful cases in January which by August had 22 deaths, 9 doubtful or affected cases and only 69 palms remaining healthy. On this Venezuelan estate about one-third of the original stand is believed to have been affected.

The coconut nematode, then named *Aphelenchus cocophilus*, seems first to have been recorded on oil palms by Freeman (1925) in Trinidad. Proof that the nematode was the cause of the condition was obtained by Malaguti, who undertook inoculations with inoculum both from the oil palm and the coconut. The disease appeared in inoculated palms 2–10 months after inoculation. The vector of the nematode is usually, but perhaps not always, the weevil *Rhynchophorus palmarum*. Maas (1970), who found six cases of Red Ring in coconuts and four cases in oil palms in Surinam, noted that about 7 per cent of weevils were contaminated with the nematode.

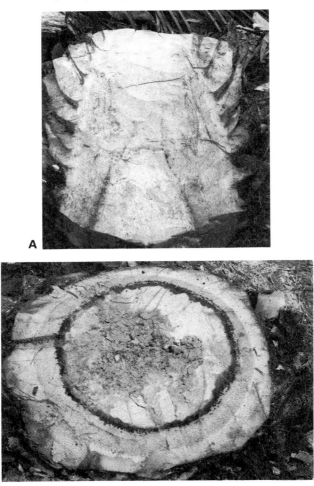

Pl. 13.8   Longitudinal (**A**) and transverse (**B**) sections of an oil palm suffering from 'Red Ring' in Venezuela.

The low incidence of Red Ring appeared to be connected with the fact that there were many recently felled trunks of many palms in which both the nematode and its vector could breed in preference to live palms.

Control

Incidence on the affected estate in Venezuela has been greatly reduced by the taking of regular sanitary measures. Any diseased palm is poisoned with sodium arsenite and, if possible, felled and burned later. The whole estate is searched every 2 months for diseased palms. Most important is the protection of the palm against the type of wounding which will provide sites for *R. palmarum* to lay its eggs. A sudden attack of Red Ring in Brazil was preceded by very close leaf pruning which had resulted in underlying petioles being wounded. Care must be taken when removing leaves to make a clean cut sufficiently far up the petiole to avoid wounding other parts of the palm. Ring-weeding with herbicides instead of with hand tools may also help to prevent wounding. Regular disinfection of tools has been suggested together with treatment of the cut leaf and bunch-stalk surfaces. Providing wounding is avoided, however, it is doubtful if the latter precautions, which are expensive, are necessary.

Arachnids

The arachnida include both spiders and mites, but only mites have been pests of the oil palm.

Germinator mites

The embryo and part of the kernel may be destroyed by mites in the germinator. In Zaire they have been described as being translucent and like elongated drops of water (Frazelle and Buyckx, 1962); those in Malaysia are 'small white specks' which concentrate on the micropyle and fibre tufts (Wood, 1968). No identifications have been made.

Germinators may be fumigated with methyl bromide, but with the dry heat treatment there is little chance of damage from these mites. Seed dressings containing BHC must not be used.

Red spider mite

The true spider mites belong to the family Tetranychidae and the spider mite prevalent in Malaysia has been identified as a species of *Oligonychus*. These are large enough (0.5 mm) to be seen moving on the leaf which they cause to turn a bronze colour. Webbing and cast skins of eggs and young stages can also be seen as white flecks. Badly infected leaves die prematurely and a widespread attack, which is favoured by dry weather, may have severe effects on a nursery.

Control

Tetradifon (Tedion) will control the pest without killing the natural enemies, and its use can be combined with dimethoate (Rogor 40) which has systemic properties. The latter as a 0.1 per cent high-volume spray is followed 10 days later by Tedion in either a 2 per cent emulsion or 1.0 per cent wettable powder (Turner and Gill-banks, 1974). Boron applications have been found to reduce spider mite injury (Rajaratnam and Law, 1975).

Orange spotting mite

A new species of mite, *Retracrus elaeis*, has caused a widespread orange spotting in Colombia characterized by brown greasy spots which turn yellow with time (Genty and Reyes, 1977; Genty, 1981). Heavy attacks can cause a 50 per cent loss of crop. Good control is obtained with sulphur in wettable powder form, but the palms only recover with the growth of new leaves.

Insects

A large number of insects have been recorded on oil palms, but only a few have become major pests. Those orders of insects from which only minor damage has been reported will be briefly mentioned, while those species which have at some time and place caused serious harm to the palm will be described more fully.

Many oil palm pests have been reported from the newly developing areas of America. This region is a vast museum of palm species all with their insect fauna. It is only to be expected that many insect species will turn their attention to the oil palm and that some may for a time become troublesome pests (Rojas-Cruz, 1977).

Orthoptera

Acridoidea (Grasshoppers, locusts and crickets)

Grasshoppers can be a nuisance in prenurseries and nurseries in all oil palm regions and they are normally controlled by hand collection. The stink locust, *Zonoceros variegatus*, has damaged young and adult palms in West Africa following the slashing of overgrown covers. In Malaysia, the grasshopper, *Valanga nigricornis*, eats large pieces of the leaf of young palms and in the last decade has been increasingly damaging over quite large areas. Outbreaks usually occurred on areas previously in primary forest, on palms 2–3 years of age and in regions where several months of drought had just been experienced (Han and Chew, 1978). Treatment with monocrotophos by mist blowers or aerial spraying (200 c.c. in 17 litres water per hectare) was successful; more recently aerial spraying with Tamaron 600LS at 0.62 litres of the product in 17 litres water per hectare gave good control on an area of over 7,000 hectares (Ng, 1980).

Isoptera (Termites)

While termites of the genus *Coptotermes* are known to have damaged the oil palm in Malaysia, particularly in peat areas (Wood, 1968), they cannot be regarded as potentially serious pests.

Hemiptera

Aphididae (Aphids)

Aphids are occasionally found on nursery seedlings. In Zaire they have been reported as being found near the base of the leaves under earth coverings constructed by black ants (Frazelle and Buyckx, 1962). Here the dark blue aphids of less than 2 mm width puncture the epidermis and may cause a slowing down of growth rate. Several species are found in Malaysia both on nursery plants and on the leaves of mature plants (Wood, 1968). The severity of the attack does not usually justify the use of insecticides.

Coccidae (Mealybugs, scale insects)

The species *Planococcus (Pseudococcus) citri* is reported as damaging nursery seedlings in Zaire by attacking the plant just under the surface of the soil. The damage to the epidermis leads to the formation of a small 'canker'. The insect, with its white mealy covering, measures about 4 mm. Control measures are rarely needed where prenursery or nursery plants are growing vigorously, but if necessary the systemic insecticide parathion may be added to the irrigation water at the concentration of 0.5 to 10,000 (Frazelle and Buyckx, 1962).

Smaller Coccids are sometimes found on the leaves and more often on the developing fruit of bearing palms. In Zaire *Aspidiotus destructor* and *A. elaeidis* cover leaves and fruit while *Pinnaspis marchali* is not uncommon on bunches in Africa, causing discoloration of fruit and inhibiting ripening. In Malaysia a *Dysmicoccus* species is found on leaves, unopened spears and fruit, and a species of *Geococcus* has been recorded on seedling and young palm roots causing some root distortion and weakening of the palm (Wood, 1968). *Dysmicoccus brevipes* occurs on fruit in Guyana (Rai, 1977).

Mealybugs and aphids are naturally controlled by many predators of which the ladybird beetles are the most important. Although control measures are rarely required, control of the tending ants is sometimes helpful. Dieldrin may be used as a spray on ant nests or as granules on the ground, or the Mirex bait may be employed.

Other Hemiptera

The presence of *Haplaxius pallidus* (Cixiidae) in the foliage of oil palms and nearby grass cover was mentioned in connection with Sudden Wither on p. 623, and control measures were suggested (Mena Tascon and Martinez-Lopez, 1977). *Leptopharsa gibbicarina* (Tingidae) was mentioned as the vector of *Pestalotiopsis* in Leaf Wither on p. 596. The insect spends its whole life on the under-surface of the leaf and pierces the tissue on each side of the midrib of the leaflets. The piercing depth is so great that the effect can be seen on the upper surface of the leaflets as bleached necrotic spots. The underside of the leaflet is spotted with excrement. The highest populations are found in the middle leaves (17–25) but there are also considerable numbers in the younger leaves (Genty *et al.*, 1975).

The Jassid, *Recilia mica*, was mentioned on p. 590 in connection with the transmission of Blast disease.

Lepidoptera (Moths and butterflies)

The caterpillars which do damage to oil palm leaves are mainly those called slug or nettle caterpillars belonging to the Cochlidiidae or Limacodidae family, or bagworms of the Psychidae family. *Pimelephila ghesquierei* belongs to the allied family Pyralidae. Caterpillars found in Latin America belong to several families including the above. A feature of recent serious attacks has been the mixing of many leaf-eating species on the same palms and leaves.

Pyralidae

Pimelephila ghesquierei – *The African spear borer*

Distribution.   This pest is found damaging leaves of young palms in all African territories. Damage is most common between the second and fifth year in the field, but both nursery seedlings and older palms are sometimes affected. The moth was first described by Tams (1930) from specimens obtained in 1929 by Ghesquière in Zaire.

Incidence, life cycle and damage.   This moth cannot be described as a common pest, but on occasions the damage done has been severe. It has perhaps been more troublesome in Central than in West Africa.

The eggs are laid by the moth at the base of the spear leaves and even one larva hatching can do considerable harm. Usually two or three are found on young palms or up to a dozen on older ones and, typically, they penetrate the rachis and leaflets of the growing, unopened spears, forming galleries through them. The attack may proceed downwards towards the growing point and the rachis may be so damaged that later, in a strong wind, several young leaves may snap near the base. When unbroken spears open, the holes left by the caterpillar are seen to be symmetrically placed on either side of the rachis. The caterpillar does not kill the palm, but may be followed by weevil larvae, e.g. those of *Temnoschoita*, or a bacterial rot which may finally prove lethal (Frazelle and Buyckx, 1962).

The caterpillars reach a length of 3–4 cm before pupating in a cocoon of fibrous debris. The colour of the caterpillar changes from dark red to yellowish as it develops. The olive to brown moths are not long in emerging from the pupae and the whole life cycle takes 35–45 days; attacks can therefore be made at frequent intervals.

Light attacks can be dealt with by removal of infected leaves and collection of the caterpillars and pupae. In the case of more serious infestation, or in the vicinity of infested areas, sprays of 0.02 per cent parathion or 0.1 per cent endrin were found effective in Zaire when directed at the base of the spears and allowed to penetrate. In the Ivory Coast spraying has been carried out at intervals of 2–3 weeks in nurseries and the first 2 years in the field but older palms were found to be less vulnerable to attack and are not usually treated (Mariau and Morin, 1971).

Tirathaba rufivena (mundella) – *The oil palm bunch moth*

Distribution.   Occasionally found on other palms, *T. rufivena* is widespread in Malaysia and Indonesia and can reach epidemic proportions especially in young areas.

Life cycle and damage.   Eggs are laid in the bunches, especially those overripe or rotten, and in inflorescences or bunches lying on the ground. Caterpillars bore into developing fruit or feed on the surface of ripening fruit. They are sometimes found tunnelling into the base of a spear leaf. They are light to dark brown and grow to 4.0 cm before pupating as dark brown pupae inside the bunch.

Control.   Although trichlorphon (Dipterex) at 0.55 kg in 370 litres water per hectare, with a wetter and white oil at 1 per cent, has been commonly used, endosulphan at 0.02–0.08 per cent a.i. has proved more effective (Wood and Ng, 1974). Two larval parasites have recently been identified: a Chalcid wasp, *Antrocephalus* sp., and an Ichneumon wasp, *Venturia palmaris*, of which the latter is more common and appears to be the more promising for possible biological control (Ng Kwang Yew, 1981).

Sufetula *species – The oil palm aerial root caterpillar*

Distribution.   Species of Sufetula feed on the aerial roots in all three continents. *Sufetula sunidesalis* is found in Asia, *S. nigrescens* in West Africa and *S. diminutalis* in Colombia (Genty and Mariau, 1977).

Life cycle and damage.   Eggs are laid on palms having a supply of roots at the base of the trunk; they are inserted among the roots or at their base. *Sufetula diminutalis* (Genty and Mariau, 1975) lays fifty to eighty eggs. The young caterpillars feed on the tips of the growing aerial roots and the older ones can hollow out these roots for a few centimetres. Although attack is normally above ground it can continue when the root has already penetrated a short distance into the soil. The caterpillars grow in five larval stages from 1–1.5 mm to 1.2–2.0 cm. From the fourth larval stage they acquire both dorsal and lateral plates. Pupation takes places in the soil at a depth of a few centimetres and about 50 cm from the trunk, or among the aerial roots. The pupa, which is straw-coloured, is protected by a silk cocoon. The whole life cycle lasts only 28–31 days.

   Experiments in Colombia suggest that aerial root development is severely curtailed by *S. diminutalis* attack. Palms treated with endrin have shown a tremendous development of these roots, which would otherwise be systematically destroyed. It remains to be seen, however, how far the maintenance of a supply of aerial roots by control of the pests will enhance growth and yield.

Other Pyralids

In Colombia a species of *Tiquadra* has been found invading the spear and rotting bunches. This latter pest appears to be the American counterpart of *Tirathaba*.

Limacodidae

Parasa viridissima *and other species – West African slug caterpillars*

Distribution.   Species of *Parasa* defoliating oil palms have been found in West Africa but have not been serious pests in Zaire, although Lespesme recorded the oil palm as a host plant for *Parasa carnapi* in that country. *Parasa* species have also been found on the palm coconut in West Africa. Specimens from Cameroon, Nigeria, Liberia and Uganda have been identified as *P. viridissima*, while *P. pallida* has been a pest of oil palms in the Ivory Coast (Mariau and Julia, 1973).

Incidence, life cycle and damage. *Parasa* attacks in West Africa have been infrequent and localized on planted areas. In Nigeria attacks were mostly on adult plams and the larvae spread rapidly from palm to palm (Allen and Bull, 1954).

In heavily infected palms the caterpillars start feeding from the tip of the leaflets and may consume the whole of the lamina, or they may transfer to another leaf consuming the terminal half of the laminae only. The spear and youngest leaves are not consumed when in that stage. Feeding continues for about a month. The youngest larvae of *P. viridissima* do not consume the vascular network and upper epidermis of the leaflets.

Eggs are laid in groups on the under-surface of the leaflets. The young caterpillars, which are brown, then feed in colonies. After a time the caterpillars disperse and gradually change to a green colour as they grow to a final length of 3–4 cm. There are rows of orange-red bristle tufts along the back. These insects pupate on the under-surface of the leaves and on the rachis. *Parasa viridissima* cocoons are spherical, 1.0–1.3 cm in diameter, and are gummed to the leaf and covered with a whitish secretion and upright brown stinging bristles; those of *P. pallida* are oval 1.5–2.0 cm in length and covered over and held to the leaf by a network of stinging bristles. The adult moth of *P. viridissima* is green and hairy, *P. pallida* is white with black markings except for the posterior wings and the abdomen which are pale yellow.

An unidentified species, probably of *Parasa*, which caused a serious attack on adult palms at Awka, Nigeria, in 1953, had green caterpillars turning yellow and reaching 3.0–3.5 cm. Pupation took place under the soil surface around the palms; the cocoons were dark brown, hard and brittle. Adults have not been bred (Allen and Bull, 1954). (Plate XVII, between pp. 230 and 231).

It appears that these insects have a fairly short life cycle and serious attacks may succeed each other at intervals of 4–9 months. Normally, fungi and natural predators keep the populations in check by attacking the larvae and pupae. For serious attacks in the Ivory Coast trichlorphon (Dipterex) at 0.7–1.0 kg a.i., and carbaryl at 0.8–1.2 kg a.i. per hectare have been recommended for application 3 weeks after the appearance of the first caterpillars (Mariau and Julia, 1973). As the later larval instars do the greatest damage, treatment should not be delayed.

Setora nitens

Distribution.   This nettle caterpillar is found throughout South-east Asia. It also feeds on the coconut and nipa palms.

Incidence, life cycle and damage.   Severe infestations of *S. nitens* and other nettle caterpillars may occur very rapidly as the life cycle is about 6 weeks and reproduction rate high. The eggs of *Setora* are deposited on the under-surface of the leaflets near the tip. The caterpillars usually feed on the under-surface and eat away the whole lamina leaving the midrib. They then drop to the ground to pupate in cracks in the soil, the pupal stage lasting about 25 days.
    The caterpillars are yellowish-green when young, becoming green. They have a longitudinal purple band and four prominent tufts of spines at the corners of their rectangular bodies (Plate XVIA, between pp. 230 and 231). The caterpillar reaches about 3.5 cm in length. The moth is brown with a wing span of 3.5 cm.

Control.   Outbreaks of *Setora* damage have been attributed to the prior use of contact insecticides which kill the parasites. Numerous parasites and predators of *Setora* have been recorded by Wood (1965, 1968) including five species of wasp, four parasitic flies and a bug. Lead arsenite at 4 kg per hectare has been preferred because of its minimal effect on the natural enemies, but trichlorphon (Dipterex) at 2 kg per hectare may also be used. However, results with these insecticides have been variable and Wood *et al.* (1977) have tested a range of chemicals and *Bacillus thuringiensis* insecticides. In general organochlorines and organophosphates gave good kills but are not sufficiently selective, though monocrotophos by trunk injection was promising. The carbamate, aminocarb, was effective and specific.

Darna *species*

Distribution.   *Darna* (formally *Othocraspeda*) *trima* (Plate XVIB, between pp. 230 and 231) is a common pest of South-east Asia. *Darna metaleuca* is troublesome on plantations in Colombia.

Incidence, life cycle and damage.   *Darna trima* is widespread in Malaysia where it consumes the margins of leaves giving a serrated appearance; but total defoliation also occurs. *Darna metaleuca* is among the more serious of the now numerous South America caterpillar pests. The biology of *D. metaleuca* has been studied by Genty (1976) who showed that the life cycle was 44–56 days. The life cycle of *D. trima* is 51–60 days (Tiong and Munroe, 1977). Pupation takes place on the under-surface of the leaves. The caterpillars of *D. trima* are light brown with orange markings and smaller than *Setora*. The moth is dark brown with a wing span of 1.8 cm.

Control.   In East Malaysia (Sarawak) spectacular progress has been made with the control of *D. trima*. A virus inoculum was prepared from diseased late-instar larvae and from healthy larvae confined with them. The larvae were naturally infected by the virus and the inoculum was prepared by simple maceration, straining and dilution to 0.1 per cent with untreated clean soft water. This was sprayed with mist blowers repeatedly until the larval census showed that resurgence was not taking place. There was a high mortality within 8 days in comparison with unsprayed areas, and resurgence such as is common after 5 weeks with chemical insecticide spraying did not take place (Tiong and Munroe, 1977).

Darna metaleuca has several important parasites, including a wasp, *Casinaria* sp., which is itself parasitized. Genty (1976) recommends that the parasite population be carefully examined before control measures are decided upon. However, good results have been obtained with a mixture of Dipterex and carbaryl applied twice at the very young larval stage, and with chlorfen amidine.

Sibine fusca *and other species*

Distribution.   *Sibine* species have a prominent place in the destructive caterpillar populations on oil palms in South America. *Sibine fusca* is the most common, but *S. apicalis*, *S. modesta* and other unidentified species have also been recorded.

Life cycle and damage.   Genty (1972) has described ten stages of development of the caterpillars of *S. fusca* from 1.6 mm in length when they are pale yellow to 2.5–3.5 cm when they are green. They tend to feed in groups until reaching the last stages. The pupa is oval, 1.2–1.6 cm long, brown and fixed to the base of the rachis. The young caterpillars eat the under-surface of the leaflets, but later the whole lamina is consumed. Young and developed caterpillars and cocoons can be found on the palm simultaneously. The moth is brown, the female having a wing span of 4.0–4.5 cm, the male 3.0–3.5 cm.

Control.   *Sibine fusca* is attacked by several bugs, and Genty (1981) records a wasp, *Apanteles* sp., and two flies, (Palpexorista coccyx and *Systropus nitidus*, which not only parasitize the larvae but also transmit a viral disease (Meynadier *et al.*, 1977). A supply of 20 g of infected larvae ground down and immersed in 220 ml of water and applied at 50 ml per hectare has spread the disease over the whole population within 18 days. If this treatment is not possible then carbaryl at 1.0–1.5 kg a.i. per hectare gives satisfactory results. Treatment should be at least 3 weeks before pupation because the three last caterpillar stages are responsible for 95 per cent of the damage; and treatment is considered necessary with more than an average of fifteen to twenty larvae per palm (Genty and Mariau, 1973/75).

Thosea spp.

Distribution.   Outbreaks of *Thosea asigna* have occurred in West and East Malaysia, while *T. bisura* caused heavy damage on one plantation in Johore state (Hertslet and Duckett, 1971). *Thosea vetusta* is occasionally encountered.

Life cycle and damage.   The larva of *Thosea bisura* is green with a straight narrow bluish and; its larval stage lasts 28–35 days. The caterpillar then moves down the palm and pupates in the soil and plant debris. The cocoon is ovoid, 10–1.2 cm long, dark brown to black. The moth is brown, the female larger than the male. *Thosea asigna* is green with a longitudinal purple-grey band of uneven width. The larval stage is longer, lasting 63–75 days, and pupation lasts from 35 to 45 days (Tiong, 1981). Larval infestation begins on the middle or lower leaves and progresses upwards until a large part of the laminae have been consumed.

Control.   *Thosea* spp. are attacked by a virus pathogen, a fungus (*Cordyceps* sp.), a wasp, and at least three species of assassin bug (Tiong, 1979). Tiong (1981) claims that control can be effected by an integrated programme in which chemical intervention (with such insecticides as aminocarb, dieldrin, lead arsenate, monocrotophos and *Bacillus thurigiensis* (Dipel)) is confined to quelling high-density infestation and combined with the encouragement of the natural fungus and insect enemies.

Other Limacodidae

In Malaysia outbreaks of *Ploneta diducta* sometimes occur (Leitch, 1966). Other occasional leaf eaters are *Susica pallida, Cania robusta, Birthamula chara, Cheromettia sumatrensis* and *Trichogyia* sp. All these Limacodidae pests of oil palms in Asia have been well illustrated by Wood (1968). In Colombia the Limacodidae are further represented by *Natada pucara, Phobetron hipparquia* and other species. These may form part of the general leaf-eating population but are not yet individually serious.

Wood *et al.* (1977) described a series of experiments on the control of nettle caterpillars in Malaysia and concluded that the general strategy must be the same as that described above for

Thoseu asigna (Tiong, 1981). The trials confirmed the need for chemical intervention on occasion, choosing insecticides on grounds of good kill, selectivity, low cost and low toxicity to man. The utilization of diseased individuals by spraying suspensions of crushed bodies was in several cases very successful, but this method should be subject to human safety tests.

Psychidae

Bagworms: Species of Metisa, Cremastopsyche *and* Mahasena

Distribution. Bagworms have been pests of the oil palm in Asia since the start of the plantation industry, but the prevalent species have changed. In the between-war period *Mahasena corbetti* was extensively studied, but since 1945 first *Cremastopsyche pendula* and latterly *Metisa plana* have been the common species in Malaysia. In Indonesia *Mahasena corbetti* is still reported to be the principal species (Plate 13.9).

Incidence, life cycle and damage. The larvae are encased in bags constructed of pieces of leaf bound with silk. *Metisa plana* and *Cremastopsyche pendula* feed on the upper surface of the leaf, the scraped portion first becoming dried out and then forming a hole. Further damage is done by the removal of pieces of leaf to make the case. Badly damaged leaves soon dry up and this gives the lower

Pl. 13.9 Bagworms in Malaysia: (*left*) *Metisa plana*; (*right*) *Crematopsyche pendula*. (B. J. Wood)

and middle part of the crown its characteristic grey and upstretched appearance, the only green leaves being the younger ones. *Mahasena corbetti* feeds on the under-surface of the leaf. Surviving caterpillars pupate in their cases on the underside of the leaves. The size and form of the bags and the manner in which they are attached to the leaf help to distinguish the species. *Metisa plana* has a short hooked attachment and the bag is about 13 mm long. The case of *C. pendula* is about 6 mm long, is rather rough and hangs on the end of a long vertical thread. *Mahasena corbetti* is much larger and more ragged. The winged male moths fly from their cases, but the wingless females remain in them and each lay 100–300 eggs. On hatching, the caterpillars acquire their own cases and feed in groups. The life cycles take between $2\frac{1}{2}$ and $4\frac{1}{2}$ months (Syed, 1978). The caterpillars die in large numbers from parasitic and predatory attacks and other causes, but explosions of population do occur locally from time to time when the natural balance is disturbed for one reason or another. It is believed that the increase in bagworm incidence in Malaysia was caused by the use of contact insecticides sprayed both against other minor pests and against the bagworms themselves (Conway and Wood, 1964) thus destroying their natural enemies. This hypothesis was substantiated in experiments done by Wood (1972b), who sprayed a block of about a hectare recurrently with dieldrin and showed that *Metisa plana* gradually increased in and spread from the sprayed block.

Control.   Owing to the danger of contact insecticides, hand-picking was for long recommended for very small attacks and stomach poisons such as lead arsenate (4 kg per hectare) or trichlorphon (Dipterex, $1–1\frac{1}{2}$ kg per hectare) for larger outbreaks. Aerial spraying over wide acreages was successful with Dipterex at 1–2 kg in 12–25 litres water per hectare (Wood, 1968). Later, the systemic insecticide, monocrotophos, applied at 6 g a.i. per palm, was found to be effective (Wood *et al.*, 1974) Trunk injection by pouring into holes made with modified chainsaw drills was used in Malaysia though the method was somewhat slow and costly. On some estates it is now superseded by tractor-mounted generators with electric drills and special equipment for immediate injection following drilling (Sarjit Singh, 1986).

Other bagworms

Wood (1968) mentions species of *Pteroma, Clania* and *Amatissa* as showing limited increases in Malaysia from time to time. In Colombia and Ecuador, *Stenoma cecropia*, which belongs to the Stenomidae, has occasionally caused extensive damage and is remarkable for having a fixed bag and eating by journeys from it while remaining attached by fibres (Genty, 1978). There are some natural enemies but control by them is reported to be weak and, in serious outbreaks, effective control has been obtained by aerial spraying with trichlorphon.

Glyphipterigidae

Sagalassa valida. *Oil palm root miner*

Distribution. The caterpillar of this moth has been found mining in the roots of oil palms in several South American countries including Colombia, Ecuador, Peru and Brazil.

Life cycle and damage. The female moth has a wing span of 2.1 cm, the male 1.8 cm. They live in the undergrowth and among the cut palm leaves in the interline and their dull colour blends with that of the withered material. The moths are frequently captured on the wing by dragonflies. The position of egg-laying has not been observed, but it is presumed that it is in humid material such as lichens, mosses or humus at the base of the palm. The egg is 0.55 × 0.35 mm and the young larvae are at first 0.8–0.9 mm long growing eventually to 2 cm. The larvae penetrate the primary roots immediately after hatching but can also move through the soil to attack roots some distance from the point of hatching. The young caterpillars at first eat the external part of the root leaving the central cylinder intact; this partial destruction stimulates the production of new roots. Larger larvae cause complete destruction of the tissues of the roots in which they mine (Genty, 1973a).

In areas of attack it has been noted that the number of caterpillars increases with the age of the palm, and it is stated that there have been palms in which 50–80 per cent of the root system has been destroyed, including old and recent damage. Some attacked palms fall over (Genty, 1981). The amount of damage is highly variable but tends to be greater on the edge of a plantation near the forest or near to rivers and streams.

Effect of damage, and control. The effect of this caterpillar damage in relation of Sudden Wither (Marchitez sorpresiva) has already been discussed (p. 613). Genty (1977) considers that the extent of the damage done to the roots in itself justifies treatment, and states that a generalized yellowing of the leaves may be due solely to *Sagalassa*. He recommends routine checks by examination of one hole, 40 × 40 × 50 cm deep at the foot of one palm per 20 hectares every 6 months. If more than 20 per cent of the primary roots are attacked more intensive checks are done, and if 20 per cent attack is still found then 2 litres per palm of a 0.4 per cent solution of commercial endrin is applied around the base of young palms and a 1 per cent solution to adult palms. This treatment is always followed by a cessation of the attacks and a quick regeneration of the root system.

Castnidae

Castnia daedalus. *Oil palm bunch miner*

Distribution. This pest has done serious damage to bunches in Guyana, Surinam and Peru.

Life cycle and damage. The butterfly, which has a wing span of 17–21 cm, lays its eggs on unripe bunches. The larvae grow to 13 cm in a period of about 8 months, passing through fourteen larval stages. The insect then pupates in the leaf bases for a period of 30 days. The larvae bore into the peduncles and bunches, causing rotting, and also into the stem. There is a high mortality from wasp and fly parasitism (Korytkowski and Ruiz, 1980).

Palms are attacked as soon as they start bearing, and provided harvesting is complete the larvae will be detected in the bunches and a measure of control obtained (Huguenot and Vera, 1981). Mariau and Huguenot (1983) have described methods of estimating populations of larvae of different stages with the object of initiating control measures before the dangerous later larval stages are reached.

Control. Various control methods have been tried. Stobbe (1983) found injection of monocrotophos and carbofuran ineffective, but applications of 5 per cent granular carbofuran in the spear region was successful. Huguenot and Vera (1981) recommend trichlorphon or carbaryl at 2.7 kg per hectare.

Other caterpillars

Damage by caterpillars of other families is also reported from time to time. In Africa, the Zygaenid caterpillar *Chalconycles catori*, which as a life cycle of about 30 days, has been reported as doing serious local damage in the Ivory Coast (Genty, 1968). Less common are *Pteroteidon laufella* and *Dasychira* sp. *Pyrrhochaleia iphis* (Hesperiidae) and *Epimorius adustalis* (Psychidae) have been reported from Zaire (Frazella and Buyckx, 1962).

In Malaysia, Wood (1968) has recorded a large number of leaf-eating caterpillars, few of which, however, have done any significant damage. These include *Odites* sp. (Xylorictidae); *Elymnias hypermnestra*, and *Amathusia phidippus*, butterfly caterpillars of the Nymphalidae; *Cephrenes chrysozona* and *Erionota thrax* (Hesperiidae); *Turnaca* sp. (Notodontidae); woolly bear caterpillars of *Asota, Asura, Diacrisia* and *Amsacta* genera (Arctiidae); cutworms

(Noctuidae), usually *Agrotis*, sp., which can do damage in prenur-series and *Spodoptera litura* which strips the leaf epidermis in nurseries; and tussock moths (Lymentriidae) *Dasychira mendosa, Laelia venosa* and *Orgyia turbata* which occasionally consume significant quantities of leaf in nurseries.

In America, colonies of mixed species of Lepidoptera have characteristically developed on some plantations and the method and timing of control has influenced the balance between species. *Opsiphanes cassina* (Brassolidae) did much damage as a leaf eater and was reported to be encouraged by carbaryl spraying but reduced by using lead arsenate or *Bacillus thuringiensis* (Rojas-Cruz, 1977). In the mixed colonies there have been species of *Megalopyge* (Megalopygidae), species of Dalceridae and Hesperiidae genera, and *Herminodes insula* (Noctuidae) in the spear leaf, and some Psychidae.

Coleoptera (Beetles)

The beetles which damage the oil palm belong to three distinct groups: (i) the Chrysomeloidea, of which the Hispid leaf miners are of the greatest importance; (ii) the Curculionidae or true weevils; and (iii) the Scarabeoidea of which the Dynastid 'Rhinoceros' beetles form the most damaging group.

Chrysomeloidea: Hispidae

Coelaenomenodera elaeidis – *The West African oil palm leaf miner*

Distribution.   This pest is found on oil palms and, to a much lesser extent, on the coconut and *Borassus* palms throughout West and Central Africa; but the more serious attacks, causing widespread defoliation, have taken place in the drier, more marginal areas of West Africa, e.g. parts of Ghana, Benin (Dahomey) and the western side of western Nigeria. Recently, however, the pest seems to have become of greater importance in more favoured areas and serious attacks have been reported from the Ivory Coast and West Cameroon.

Incidence, life cycle and damage.   For a long period this pest was only reported from Ghana where its method of feeding and life history were recorded by Cotterell (1925). Very detailed studies have now been made by Morin and Mariau (1970–4) of the biology of this insect and of its control. The life history in days is as follows: eggs, 20; larvae, 44; pupae, 12; adult to egg laying, 18; total, 94. The adults continue to live on the under-surface of the leaf for 3–4 months during and after laying eggs. The length of the life cycle accounts for the pest damage reappearing in some cases every 3 or 4 months.

The larvae, which grow to about 6.8 mm in length, are brown and their heads are squeezed into the thorax. Their flattened bodies are transversely divided by deep furrows and they have no feet. They mine under the upper epidermis of the leaflets of palms of all ages except, normally, those below 3 years old in the field. The paths of the miners are longitudinal, and in a severe attack the greater part of the leaf tissue will be destroyed. A single passage or gallery mined by a larva to attain its full development measures about 15 cm in length and is 1 cm broad. Severely attacked palms have a typical appearance; the young leaves are green, being little attacked, while the remainder are grey-brown and withered with desiccated rolled-in leaflets. Later, the withered laminae shatter, leaving the leaflet midribs only.

The pupae are found in the dead tissue of the leaves, and the adults, which are 4–5 mm long, emerge after about 12 days. The pupae are mobile and are found in the centre of the galleries. The adult emerges through the upper epidermis and shows a preference for migrating to the higher leaves. These adults are pale yellow with reddish wing cases; they make grooves about 1 cm long on the leaflets and the female lays her eggs in a small cavity on the underside of the leaf.

It has been observed that natural parasites normally reduce the severity of successive attacks and Cotterell (1925) reported Hymenoptera parasites of both the eggs and the larvae as well as fungal parasitism. In the drier parts of the West African palm belt where leaf miner damage has in some years been serious, the attacks seem ordinarily to have been controlled by the natural predators, and resurgence has not occurred again until, for some reason, the parasite population has fallen below normal.

Control.   The parasitism of *Coelaenomenodera* has been studied in detail by Morin and Mariau (1970–4). The eggs are parasitized by the Chalcid fly, *Achrysocharis leptocerus*, and by *Oligosita longiclavata* (Trichogrammatidae). Three larval parasites attack towards the end of development; these are Eulophid flies *Dimmockia aburiana* and *Pediobius setigerus*, and, less frequently, *Cotterellia podagrica*. A rare attacker of young larvae is another Eulophid fly, *Closterocerus africanus*, and another, unidentified, species.

In the Ivory Coast a method of assessing the level of attack through an *intensity index* has been adopted as follows (Mariau and Bescombes, 1972). Counting of adults and larvae is done on a leaf between the twenty-fifth and thirtieth, small and large larvae, nymphs and adults being recorded separately. As counting is done on one palm per hectare the index on an area basis is obtained by dividing the total insects counted by the number of hectares. The palms selected for counting are changed on each count. Counting is done 3-monthly when the larval index is below 10 and the adult index below 1; monthly when the indices are 10–20 and 1–3; and weekly when more than 20 and 3. When the latter stage is reached treatment is considered necessary; penetrating insecticides then have to be used (Bescombes, 1968).

In mature plantations the most effective insecticide against adults was found to be 25 per cent BHC applied as a dust at the rate of 15 kg per hectare and directed to the under-surfaces of the leaves. Application at 8-day intervals in dry, still weather is advocated. It was suggested that if this treatment, regarded as preventative, is carried out efficiently, then measures against larvae may be unnecessary (Ruer, 1964).

Control at the larval stage is more expensive and more difficult. The spray used must reach the upper surface of the leaves. Lindane (gamma BHC) in vegetable oil emulsion, applied as 3 litres of emulsion containing 12 per cent active ingredient in 650 litres of water to cover 1 hectare of palms, was previously recommended in the Ivory Coast, with a succession of two or three sprayings, and trials in West Cameroon (Shearing, 1964), confirmed the value of this treatment. Lately, however, dusting with BHC (HCH) (Phillippe and Diarrassouba, 1980) has been employed for large plantations; 70–80 hectares can be treated per day and three consecutive dustings are needed for a 95 per cent kill.

A search has been made for possible parasites to introduce into the Ivory Coast for control of the pest. The Eulophid wasp, *Chrysonotomyia* sp., was successfully introduced from Madagascar, but it failed to parasitize *Coelaenomenodera elaeidis* (Lecoustre *et al.*, 1980).

Alurnus humeralis – *The Alurnus beetle*

Distribution. This large Hispid was first reported on the oil palm in Ecuador but is to be found in other parts of tropical America (Merino and Vasquez, 1963).

Description, incidence and damage. The eggs are laid in typical rows of seven to ten joined by mucilage and adhering to the leaflets or rachis. They hatch after 29–43 days. The light brown larvae are 7–8 mm long on emergence but reach 40 mm before pupating. In a severe attack the laminae are almost entirely eaten away leaving the rachis and midribs bare, and this gives the palm its characteristic appearance. In the centre of the crown of young palms, larvae in all stages of development can be found between the leaflets of the unopened spears. The larval period lasts 7–8 months and there are seven stages. The pupae are to be found on the surface of the petioles of the younger leaves, and the pupal stage lasts 44 days. Although larval damage is more severe, the adults also damage the leaflets by consuming longitudinal strips several centimetres long and about 1 mm wide. The adult has an average life of 113 days (Santos, 1968).

Incidence can be heavy; in one group of 410 palms 198 were found to be infested, and the larval foliar damage over a whole plantation was estimated to be about 30 per cent.

Control. The natural predators are mites and wasps and it is to be supposed that severe attacks will occur when the populations of these are temporarily suppressed. Several insecticides have been tried against the larvae in the crown of the palm and heptachlor (0.1 per cent) and toxaphene (0.5 per cent) applied at 4-monthly intervals are reported to have given fair control, reducing damage by 70–82 per cent. Diazinon and dieldrin were also found to cause a high mortality of the larvae when sprayed at strengths of 0.16 and 0.1 per cent respectively (Merino and Vasquez, 1963).

Cassididae

Pseudimatidium neivai *and* P. elaeicola

Distribution. *Pseudimatidium neivai* was discovered in Brazil and named in Bondar's (1940) large work on the insects of Bahia as *Himatidium neivai*. Bondar transferred the genus to the Hispidae. Following the extended planting of the oil palm in Colombia, a species of *Pseudimatidium*, possibly *P. neivai*, was reported from the Magdalena valley, and it has become a pest of the oil palm in all parts of the continent. A new species, *P. elaeicola*, was discovered on the Pacific coastal plain near Calima damaging oil palm fruit, and the genus reverted to the Cassididae (Aslam, 1965) though some authors still refer to the insect under the name *Himatidium neivai* in the Hispidae.

Description and damage. The adult of *P. neivai*, which is of flattened shape, measures 5 × 3 mm and is at first white but rapidly becomes a shiny brown with fine longitudinal lines along the wing cases. Single eggs are laid. The larvae are more flattened than the adult and their feet are short and withdrawn; they are at first translucent, later turn dull red and reach 7 mm in length and 4 mm in breadth. The pupae are brown and otherwise resemble the larvae.

The insect is found on the under-surface of leaves, but the main point of attack is the fruit (Figueroa and Van den Hove, 1967). Most of the damage is from the larvae which nibble the exocarp beginning at the apex. A fungus then develops at the point of attack and the exocarp becomes lignified and grey.

Extent of damage, and control. It has been estimated that a heavy attack leads to a loss of 7–9 per cent oil and that losses from the more usual lesser attacks are, in spite of the alarming appearance of the bunches, negligible (Genty and Mariau, 1973). Young plantations are more vulnerable. *Pseudimatidium elaeicola* is reported to do similar damage to developing fruit so that the latter becomes dry and hardened (Arens, 1965).

The pupae of *P. neivai* are parasitized by two flies, *Terrastichus* sp. and *Psychidosmiera* sp., but the amount of control exercised by them appears small. Ant species are considered to play a more important role in limiting the population. Only if the attack becomes severe (a general attack of more than 70 per cent of the palms, or more than 10 per cent attacked heavily) is insecticidal treatment thought necessary. Endrin in 0.15 per cent solution of active ingredient has been found effective and, once applied, treatment is not necessary for a year (Genty and Mariau, 1973). The toxicity of endrin must be taken into account and time of application should not be near to harvesting rounds.

Other Chrysomeloidea

Hispoleptis elaeidis, the South American oil palm leaf miner (Aslam, 1965), is a minor pest of the oil palm in Ecuador. The adults are slightly larger than those of *Coelaenomenodera* but the larval damage is similar. In Malaysia, *Promecotheca cumingi*, a coconut pest, can also attack oil palms (Wood, 1968). It, too, is closely related to *Coelaenomenodera* but rather larger. A little Cassid, *Calyptocephala marginipennis*, has been found doing minor damage to oil palm leaves in Honduras.

Curculionidae (The weevils)

Temnoschoita *species*

Distribution. These weevils are to be found damaging the oil palm throughout Africa but appear to be more commonly encountered in Zaire than in West Africa. The commonest species is *T. quadripustulata (quadrimaculata); T. delumbata* is less common.

Life cycle and damage. The adults of *T. quadripustulata*, which are 8–10 mm long, are dark brown with the thorax spotted with indentations. The light brown wing cases have four reddish blotches and do not fully cover the abdomen. The females lay their eggs on cuts and wounds on the leaf petioles of young palms, both those recently

Pl. 13.10   *Temnoschoita* damage to leaves in West Cameroon. Note typical 'windows'.

transplanted and palms in early bearing. Nursery plants may also be infested. The young larvae tunnel their way through both dead and living tissue towards the heart of the palm or, in the case of older palms, they move into the inflorescence and cause rotting; they pupate in the tunnels so formed. In bearing palms the adults are attracted to the inflorescences where eggs are also laid. The damage is sometimes severe, and young palms may be killed through penetration of the crown and growing point (Plate 13.10).

Control.   The following measures have been suggested (Fraselle and Buyckx, 1962). In areas where the weevil has been noted, care should be taken to avoid wounding the palms by excessive leaf pruning, particularly just before transplanting. Unfortunately this injunction may conflict with control measures against *Cercospora* Leaf Spot. With bearing palms the collection and destruction of rotted bunches and scattered fruit are also recommended as these may contain eggs, larvae and pupae. When harvesting begins it may be advantageous to undertake a general cleaning of the crown followed by dusting twice with an insecticide such as BHC at 3-weekly intervals, the dust being applied in the crown from the centre to the base, not on the leaves. Traps for the adults have also been constructed from recently cut and split petioles or banana trunks. Banana plants are an attractive host and should not be grown near nurseries or young plantations where infection with *Temnoschoita* is feared. For nursery attacks a 0.3 per cent solution of dieldrin has been recommended (Dubois and Gerard, 1968).

Rhynchophorus *sp.* – *Palm weevils*

Distribution.   Species of the large *Rhynchophorus* weevil are to be found attacking palms in all parts of the tropics. As pests of the oil palm the distribution of the more important species is as follows:

R. phoenicis	Africa	
R. palmarum	America	The Gru-gru beetle
R. ferrugineus	} Asia	{ The red palm weevil
R. schach		{ The red-stripe weevil

(*R. papuanus* is found in the Celebes and New Guinea.)

Description.   The larvae attain a length of some 5 cm and are ovoid or rounded, legless, yellowish-white sacks with small brown heads. The last abdominal segment is flattened and has brown edges carrying bristles (Plate 13.11). The cocoons of the pupae,

Pl. 13.11 *Rhynchophorus palmarum* larva found in a 'spear rot' palm in Nicaragua. (B. J. Wood)

constructed of concentrically placed fibres, extend to 8 cm in length and 3.5 cm in breadth.

The adults show distinct specific differences but are usually about 4–5 cm long and 2 cm broad. *Rhynchophorus phoenicis* is black but on the thorax there are two narrow longitudinal dark brown bands. The wing cases have about a dozen longitudinal grooves. The underside of the body is light brown with diffuse black spots. *Rhynchophorus ferrugineus* is the common red palm weevil of the Far East, while *R. schach*, the red-stripe weevil, is the more dangerous for the oil palm, since it is the species most commonly found in Sumatra and Malaysia. The oil palm has, however, proved far less liable to attack than the coconut palm. *Rhynchophorus ferrugineus* is rather variable in length (2–5 cm) and is red-brown with a few irregular black spots on the thorax. *Rhynchophorus schach*, previously considered a variety, is black with a longitudinal red-brown line down the centre of the thorax (Dammerman, 1929). The American species, *R. palmarum*, is entirely black with velvety thorax slightly prolonged at the base and shiny grooved wing cases.

Life cycle and damage.   *Rhynchophorus* weevils are wound parasites laying their eggs, which are 2–3 mm long, in the cut or damaged surfaces of many palms. The eggs hatch in 3 days and the larvae tunnel into the crown and trunk. The tissues around the growing point then begin to decay and the palm may be killed. The external symptoms of attack have been described as similar to those of *Fusarium* Wilt, i.e. the leaves show a gradually increasing chlorosis, and fracture in strong winds.

The larval stage lasts about 2 months and pupation then occupies about 25 days, the larvae moving towards the periphery of the trunk to pupate. The whole life cycle lasts less than 3 months. The weevil more commonly breeds in the stumps of a felled palm field, newly cut stumps being preferred. *Oryctes* (see p. 653) and *Rhynchophorus* species are often present in a plantation at the same time, wounds made by *Oryctes* giving a means of *Rhynchophorus* infection, while *Rhynchophorus* damage will provide conditions suitable for *Oryctes*. The connection between *Rhynchophorus* attack and Red Ring disease was mentioned on p. 629.

Through its lethal effect, *Rhynchophorus* is a potentially serious pest, but in Asia and Africa it cannot be said that its incidence on the oil palm is very high. Deaths have, however, been noted in Africa where leaves have been cut abnormally short and wounding of adjacent leaf bases has resulted. In America, incidence may well be higher. Deaths from *R. palmarum* attack have been noted in young plantings within the grove areas in Bahia, Brazil, and the pest in quite frequently encountered on oil palms in other parts of the continent.

Control.   Effective control of *Rhynchophorus* attack is not easy. In the first place, however, the wounding of the palm must be avoided and the petioles must not be cut close to the trunk. Secondly, all dead or felled palms should be destroyed within the period of the life cycle. In some areas collection of adult beetles has been resorted to. Measures for the control of *Oryctes* and other large beetles will help to reduce the incidence of *Rhynchophorus*.

In areas of infection, cut portions of the petioles should be inspected, and treated if found to be infested. Though the labour involved is considerable, it is possible to hook out the larvae from their tunnels with the aid of a wire. The tunnels can then be treated with disinfectant and stopped up with clay or putty. These and other preventative and curative measures have been described in detail by Mariau (1968). Several parasites of *Rhynchophorus* species are recorded.

Other weevils

Several other weevils have been reported as doing damage similar to that of *Rhynchophorus*. *Rhinostomus (Rhina) barbirostris* (the bearded weevil) is found in company with *Rhynchophorus palmarum* in Brazil and elsewhere in America while *Rhinostomus (Rhina) afzelii* has been described in Sierra Leone.

Two species of *Prosoestus*, *P. sculptilus* and *P. minor*, are found in Zaire and live in the female inflorescences; they are only some 6 mm long (Fraselle and Buyckx, 1962). The flower stigmas are punctured and become prematurely brown. The eggs are laid and the young larvae tunnel in the stile working their way towards the ovary which naturally fails to develop. The amount of damage is small, however, the adjoining undamaged fruit tend to grow larger to compensate for the smaller number of fruit developing.

The presence of small weevils in the debris around cut leaf bases and bunch stalks is not uncommon. In Malaysia *Diocalandra frumenti*, which is 8 mm long, is common; the larvae develop in tissue around wounds but do not penetrate deeply (Wood, 1968). In Brazil and other parts of South America *Metamasius hemipterus* is also not uncommon and develops in rotting leaf bases. *Leurostenus elaeidis* and *Pseudostenotrupis filum* have been reported on oil palms in Zaire, and the former species also in Sierra Leone. There is no record of damage.

Scarabeoidea: Dynastidae

Oryctes species – The Rhinoceros beetles

Distribution.   *Oryctes* species are to be found throughout the palm-growing areas of Africa, Asia and the Pacific. Goonewardena (1958) states that there are forty-two species of which four are found in South-east Asia. The following species may be mentioned:

O. rhinoceros. The common Rhinoceros beetle of the Far East and which has spread to the Pacific islands (Plate 13.12).

O. gnu (trituberculatus). Asia, less common.

O. boas
O. monoceros } Africa. *O. boas* is probably the most common.
O. owariensis

The Asian species are primarily pests of the coconut palm, but they are to be found attacking many of the other palms of the continent, whether cultivated or wild. The African species attack the coconut and *Borassus* palms but, owing to its ubiquity, the largest population is to be found on the oil palm. Three other species are recorded on oil palms by Jerath (1968).

Description.   The male adult has the characteristic rhinoceros horn; in the female the horn is smaller or, in the African species, is reduced to a triangular protuberance. The beetle is black and measures 4–6 cm long and 2–3 cm broad according to species, *O. trituberculatus* being larger than *O. rhinoceros*, and the African species *O. owariensis* being larger and *O. boas* being smaller than *O. monoceros*. The horn of *O. boas* is particularly long and curved.

The eggs are white, 3–4 mm in diameter and easily observed on the breeding grounds. The young larva is white at first but its head soon becomes brown and its body blue-grey, then yellowish or greenish-white; it reaches a length of 4–10 cm (Plate 13.12).

Life cycle, incidence and damage.   The eggs are laid on rotting vegetable matter, logs or cow dung and compost. On an oil palm estate decaying palm trunks and bunch refuse are common breeding grounds. About twenty eggs are usually laid but higher numbers have been recorded; they hatch after 11–13 days. The length of the larval stage varies considerably, ranging from around 100 to 200 days. Similarly, the insect's life as an adult may last for a few months or extend to over half a year. Before pupation there is a short prepupal stage of a week; the adults emerge after a further 3 weeks.

The oil palm is damaged by the adult beetle which burrows into the cluster of developing spears in the crown and bores its way through the petioles into the softer tissues of the younger unopened leaves. The effect can be seen when these leaves develop and open, but the regularity of wedge-shaped cuts so characteristic with the coconut palm is not always so clearly seen in the oil palm. Where the rachis has been penetrated, leaves may later snap off. Previous attacks may be detected by the presence of holes in the petioles of

A

B

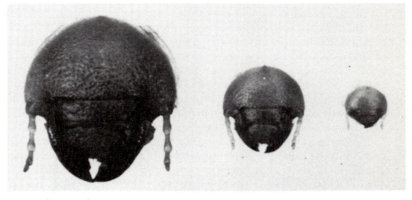

C

Pl. 13.12 *Oryctes rhinoceros* in Malaysia: **A**, adults, male (*right*), female (*left*); **B**, larval instars, 1st, 2nd, 3rd (early, late, prepupal); **C**, head capsules of larval instars; (*left to right*) 3rd, 2nd, 1st.

older leaves. An attack is most dangerous in young palms since the growing point may occasionally be reached or a Bud Rot may develop which will kill the palm. As already mentioned the bore holes also give *Rhynchophorus* weevils access to the palm. In other cases the attack may be so severe that recovery is slow and the new leaves formed are small and contorted. Wood (1968) has described and illustrated the kinds of damage that the palm may suffer from the Asian species.

Control.   The majority of work on the control of the *Oryctes* beetle has been done in connection with coconut cultivation, but these measures are largely applicable to the oil palm. Overriding all other means is the institution of thorough-going sanitary measures for the elimination of the larval stage. All rotting vegetable matter, dung heaps and composts should be dispersed and rotting palm trunks and timbers should be disposed of. This is often more easily said than done, but with proper heaping and firing it is possible to burn both old coconut and oil palm trunks and stumps. In certain Asian countries the disposal of breeding grounds for *Oryctes* can be made compulsory by orders issued under the law.

In replanting oil palms in Asia the measures to be taken against Ganoderma Trunk Rot (see p. 609) are likely also to reduce the incidence of *Oryctes*. Unless the palm trunks or other timbers such as rubber have been fully disposed of by fire, it will be necessary in a replanting to carry out regular inspection by workers who will break up the rotting material and collect the grubs. Thorough inspection by workers armed with large hoes and axes is advocated in Malaysia.

Measures taken against the adult, whether by trapping or treating the palms with insecticides, were often unsuccessful or too costly. In very severe infestations of young palms hand collection may be resorted to, the beetles being extracted from holes by means of a hook. However, larval searches and destruction are considered preferable (Barlow and Chew, 1971). It has also been found that ethyl chrysanthemumate is a strong attractant of *Oryctes* and traps, at the rate of twenty-five per hectare, using 0.2 ml per trap of this substance have been suggested (Turner, 1973).

It has been noticed in Malaysia that palms along or near roadsides may be heavily attacked while those within the field escape injury, and it is thought that inter-row vegetation may form a barrier (Wood, 1968) and, in young areas, may blur the palm silhouette which is believed to attract the beetle. In areas of young palms either kept bare, sown with a mixture of erect and creeping covers *Oryctes* breeding and damage were consideraly higher on the bare areas (Wood, 1981). It is now considered that rapid covering of timbers with a dense ground cover, usually leguminous, is by far the most effective way of suppressing *Oryctes* attack and the encouragement of an early cover is therefore an important part of planting or replanting schedules. Where risk of attack is thought to be high, these measures may be supplemented by application of the systemic insecticide, carbofuran (Toh and Brown, 1978).

Many parasites and predators of *Oryctes* have been known for a long time and larvae in particular are kept in check by rodents, birds, lizards, ants and termites. Certain mites feed on the eggs. Much attention has been given, again in connection with coconut cultivation, to the introduction and spread of insect parasites or fungi. Of the former, the Scoliid wasp, *Scolia procer*, parasitizes the larvae. In Malaysia there is a regular, though small, larval mortality from the virus, *Rhabdionvirus oryctes*, and the fungus, *Meterrhizium anisopliae* (Barlow and Chew, 1971), and Wood (1972, 1981) records that releases of the virus in territories where it was absent (but where *Oryctes* attack was more severe) have raised the virus incidence to the level existing in the endemic areas.

In the large new plantings in West Africa the incidence of *Oryctes* is relatively low, and rapid covering of the felled forest timbers or old palms by leguminous covers is the most effective measure (Boyé and Aubry, 1973; Julia and Mariau, 1976).

Strategus aloeus – The Strategus beetle

Distribution.   This beetle somewhat resembles *Oryctes* and is distributed throughout tropical America where it has been troublesome in several oil palm plantations.

Incidence and damage.   The adults attack young palms in the field or nursery by digging a hole in the ground near the palm from which they bore their way into the plant just above the roots. Often, in a young palm, the growing point is reached and the plant killed. Eggs may be laid in the palm which is then consumed by the developing larvae. However, eggs may also be laid in rotting stumps, trunks and vegetation, and the measures to be taken against the larval stage are therefore the same as for *Oryctes* species.

Control.   In view of the lethal attack on young palms by the adult, additional control measures are required in areas where the beetle is common. During wet weather fortnightly inspections are recommended. Application of 2 litres of a 0.2 per cent solution of endrin to each hole has been found effective in Colombia (Bachy, 1963). In dry weather *Strategus* attack is rare.

Other large Dynastid Beetles

Certain other Dynastid beetles are occasionally troublesome in oil palm plantations. In Africa the large horned beetle *Dynastes (Augosoma) centaurus* can attack and sometimes kill young nursery or field plants. Large numbers may be captured in light traps in the dry season. In Malaysia the rather similar Gideon beetle, *Xylotrupes*

(Dynastes) gideon, is found on coastal estates feeding on the leaf rachis, sometimes causing it to break (Wood, 1968). The males only of these beetles have large thoracic horns. The breeding habits are similar to *Oryctes* and the grubs can be confused with *Oryctes* larvae, though they are much more hairy. The Atlas beetle, *Chalcosoma atlas*, is also found with *Oryctes*.

Cetoniidae

Platygenia barbata

Distribution. This African beetle has long been reported as damaging the oil palm in Zaire (Fraselle and Buyckx, 1962).

Life cycle, incidence and damage. The eggs are laid at the base of the petioles and the larvae develop in the debris in the leaf axils. They then bore holes 2–3 cm deep into the petioles which weaken them so that in a wind the leaf will snap low down and hang parallel with the trunk; this gives an attacked palm a characteristic appearance. The larvae also tunnel into the peduncles and bunches thus fail to ripen. The holes made allow other insects such as *Rhynchophorus* and *Temnoschoita* to enter. The adult is a black beetle, about 30 mm long, with rounded wing cases. The larvae reach 50 mm in length, have a brown head and are distinctly hairy. The pupae are enclosed in a cocoon of palm fibres.

Control. Control measures have been confined to hand collection and the clearing of debris from the leaf axils.

Scarabaeidae (Cockchafers or night-flying beetles)

Adoretus and Apogonia species (Plate 13.13)

Distribution. Cockchafer beetles of these genera can be a considerable nuisance in nurseries in Malaysia by chewing holes in the leaves. Such beetles do not appear to have done significant damage elsewhere. *Adoretus* is in the Rutelinae sub-family and *Apogonia* in the Melolonthionae.

Incidence and damage. The four common species in Malaysia are *Apogonia expeditionis* and *A. cribricollis*, which are black, and the brownish *Adoretus borneensis* and *A. compressus* which are slightly larger and about 12 mm long. They make holes in the leaf at night, *Apogonia* being responsible for the larger incursions (Wood, 1965). During the day they rest in the soil at the base of the seedlings.

Control. The night-feeding habit makes control by collection

Pl. 13.13 Common Malaysian cockchafers which damage oil palm seedlings: (*left to right*) *Apogonia expeditiones, Adoretus borneensis, Adoretus compressus.* (B. J. Wood)

difficult. Dipterex sprayed as 0.1 per cent (high volume) has given good control (Wood, 1968). Dicrotophos (Bidrin) is also effective.

Leucopholis rorida

The larvae of this beetle (Melolonthinae) have fed on the roots of nursery seedlings in Malaysia and more recently cockchafer grub damage has been reported in young field areas (Wood, 1972). In nurseries control by digging aldrin into the soil is recommended.

Another cockchafer, *Psilopholis vestita*, has been reported as damaging young palms and the inter-row vegetation by root feeding by the larvae. Attacks are nearly always confined to forest boundary areas (Wood and Ng, 1969).

Other Coleoptera

Many other beetles are encountered on oil palms but few do serious damage. In Malaysia many of them are co-inhabitants of *Oryctes* breeding sites and they have been listed and illustrated by Wood (1968).

Hymenoptera

Formicidae (Ants)

Leaf-cutting ants have been reported in Zaire, but do much more damage in Latin America than elsewhere. In Costa Rica and Ecuador they have been seen to remove considerable quantities of the leaf laminae from young palms in the field. In Malaysia ants are regarded mainly as a nuisance, particularly the aggressive Kerengga (*Oecophylla smaragdina*). The little ant *Crematogaster dohrni* is also common. These ants attend honeydew-producing scale insects. If the nuisance of these ants becomes excessive then treatment may be given (see p. 632).

Apidae

Bees are reported seriously to have reduced the supplies of pollen on some estates in Malaysia (Gray and Turner, 1966). Control has been effected by tracing out the nests and burning them.

Insect control on oil palms

It will have been noted from the previous pages that more and more attention is being given to what has been called 'integrated control', that is the timing and use of insecticides in a manner that helps to preserve the natural enemies of a pest. This subject has been dealt with very fully by Wood (1968, 1972b) and cannot be expanded here. It is worth mentioning, however, that the evil of indiscriminate application of contact insecticides has been clearly seen in Malaysia with such pests as bagworms and nettle caterpillars (Wood, 1971a), and there is considerable scope for biological control through the spreading of natural enemies whether they be other insects, fungi or viruses. Many of the chemicals which have been and in many cases still are effectively used to combat oil palm pests are highly toxic to humans and must only be handled with great care and according to the safety instructions of the producers. In some oil palm countries pesticides commonly used and found effective have now been banned and oil palm growers must make themselves acquainted with their local regulations.

Methods of insecticide application are also described in some detail by Wood (1968) and much attention has been given to the use of tractor-drawn powered sprayers and to aerial application over adult plantations. For use in nurseries or on young palms in the field shoulder-mounted knapsack sprayers, manually operated, are normally adequate and may be used with taller palms with an extension lance. Motorized knapsack sprayers may be used in areas where palms are large but tractors cannot pass easily through the fields. In the Ivory Coast very efficient use has been made of tractor-drawn dusters and sprayers for control of the leaf miner, *Coelaenomenodera elaeidis*, on both medium-aged and tall adult palms (Bescombes, 1968, 1972). Where large areas have to be covered aerial application has been successful. The Piper Pawnee aircraft has been employed both in Asia and Africa for oil palm pest control and 400 hectares per day can be treated. Detailed instructions for the layout, marking and organization of an aerial spray programme have been given by Wood (1968) and Genty (1977).

Vertebrates (birds and mammals)

Vertebrates, especially birds and rats, have assumed much greater importance in recent decades, particularly in Asia and America.

Birds

Parrots

The long-tailed parakeet (*Psittacula longicanda*), the blue-rumped parrot (*Psittinus cyanurus*) and the Malay lorikeet (*Loriculus galgulus*) have all been troublesome in Malaysia. Most destructive is the long-tailed parakeet, which feeds in flocks of up to thirty birds and carries away ripe fruit from the bunch and tends to scatter it about half eaten. Such damage can be distinguished from rodent damage by the single beak groove in the fruit. The other species feed close to the bunch and do not scatter the fruit. Shooting is the only control known and with the long-tailed parakeet this does not seem to have been very effective (Wood, 1968).

Vultures

The American black vulture (*Coragyps atratus*) has become a serious pest. These birds are scavengers and eaters of carrion and flock around slaughter houses. In Brazil, Colombia, Honduras and elsewhere in America they have migrated to nearby oil palm plantations and established themselves among the palms, gorging themselves on the fruit as it becomes ripe. Much localized loss is suffered; and, once established, the birds tend to distribute themselves through the plantations and losses then become more general. In most countries the birds are protected by law as useful scavengers, and special permission must be obtained to shoot them. This course has been adopted in Colombia. In Africa, the palm-nut vulture, *Gypohierax angolensis*, does some damage in the groves.

Crows

In Malaysia, both the house crow (*Corvus splendens*) and the large-billed crow (*C. macrorhynchos*) have, in recent years, moved from the vicinity of abattoirs and refuse dumps to adjoining oil palm fields in much the same manner as the American black vulture (Siew *et al.*, 1979, 1980; Chua *et al.*, 1980).

Other birds

In West Africa the weaver bird (*Ploceus cuculatus*) may be locally troublesome through stripping the leaflet laminae from a wide area to make nests in adjoining trees. It is usually necessary to fell the nesting trees to disperse the birds. Birds reported as causing local damage to fruit in Asia include the common mina, *Acridotheres tristis* and the Philippine glossy starling, *Aplonis panayensis* (Wood, 1969a).

Rodents

The 'cutting grass' – *Thryonomys swinderianus*

This large rodent is common in Africa and young areas planted near to the forest are particularly at risk from its devastating attacks. Protection against this pest is described under transplanting practices on p. 393.

Rats – *Rattus* sp. and other genera

Studies in Asia have shown that populations of rats of 250–500 per hectare are not uncommon and may rise to 1,000 (Gillbanks *et al.*, 1967; Wood, 1969b). Early infestations by the rice field rat, *R. argentiventer*, give place to *R. tiomanicus*, the Malayan wood rat, which is found in virtually all oil palm plantations. Baiting experiments have shown that the black, house or roof rat (*R. rattus diardii*) can also become an oil palm pest in parts of Malaysia (Wood *et al.*, 1987). Although young palms are sometimes attacked (and these can be protected with wire-netting collars as against the 'cutting grass', though the collar must be turned in at the top (Wood, 1968), the main damage is to the fruit, both ripe and unripe. Occasionally inflorescences may also be damaged or even destroyed. Poison baiting is the main method of systematic control, using anticoagulant chronic poisons. Very detailed instructions for baiting have been published by Turner and Gillbanks (1974) and Krishnan (1980). Paper-wrapped baits of broken rice, prawn powder and 1.5 per cent a.i. coumachlor (Tomorin or Ratafin) in the ratio 16:2:1 have been widely used. This mixture is used in 15 g lots in greaseproof paper. Wax baits are also commonly employed and consist of maize or rice bran, 7 kg; fish heads 0.7 kg; palm oil 2.5 litre; paraffin wax 0.7 kg; coumachlor 0.4 kg. If the anticoagulant used is warfarin (0.5 per cent a.i.) then about double the quantities given for coumachlor are needed. Timing of baiting depends on population and damage assessment (Wood, 1971b).

The increasing populations of the barn owl, *Tyto alba*, have assisted in controlling the size of rat populations (Lenton, 1980). Duckett (1984) showed that these owls are unaffected by the residual warfarin ingested from poisoned, but still living, rats, though the 'second generation' anticoagulants such as brodifacoum are likely to cause owl deaths following secondary ingestion.

Other rat species found in Asian oil palm fields are *R. exulans* and the bamboo rat, *Rhyzomis sumatrensis*.

Other rodents

In West Africa populations of *Dasymys incomtus, Lemniscomys striatus, Lophuromys sikapusi* and *Uranomys ruddi* were investigated on Ivory Coast plantations by Bellier (1965) and Bredas *et al.* (1968). Recommendations for control have been given by the IRHO (1976).

Squirrels (*Callosciurus* sp.) are occasionally troublesome in Asia, eating the mesocarp and sometimes attacking nursery plants. Porcupines (*Hystrix brachyura*) attack young palms on forest margins, gnawing through to the bud. Wire collars are not effective and zinc phosphide baits with palm oil in cassava roots (Wood, 1968) and chemical repellants (Chandrasekharan and Edmunds, 1976) have been employed.

Large mammals

Elephants have done great damage to young plantings in Sabah, systematically uprooting rows of newly planted seedlings. Barriers formed by ditches, electric fences, tangle-felled forest trees and scaring devices have been tried, though not very successfully. Automatically firing carbide guns firing through the night at irregular intervals are valuable (Wood, 1968).

Wild pigs damage or kill young palms, and monkeys occasionally pull up seedlings. Liaw (1983) has described trapping methods currently used in Sabah for the control of these pests and porcupines.

Monitoring disease and pest incidence

With the establishment of large plantations in new areas the need to monitor disease and pest incidence, i.e. to warn oneself of approaching danger before it is too late, has been increasingly apparent. Various census systems have been proposed sometimes with particular pests in mind (Wood, 1968; Dufour and Philippe, 1980). One of the best methods in use is one devised in Sabah (Syed and Speldewinde, 1974) and which can easily be adapted for diseases as well as pests. The system consists of two steps, detection and enumeration. Monthly or bi-monthly *detection* rounds by experienced detecting staff are recommended. The intensity of the detecting work can be varied by taking either every harvesting (or inspection) path or alternate or every third path. The presence of pests or a disease symptom is recorded on special cards, and the work is summarized at the end of each day. Rough levels of infestation (high, medium or low) are recorded. *Enumeration* follows detection and is carried out within a day or two with the object of determining the density of pest attack and so deciding on the control measures to be taken. Diseases detected are of course examined if necessary by plant pathologists and appropriate action taken. In the enumeration of pests about twelve palms are taken per hectare of the affected area and six leaves per palm are used for counting larvae, pupae, etc.

This system has been described as an 'extensive method' in contrast to a 'sampling intensive method' in which one point is chosen per 3 hectares and pest enumeration is done on six leaves, two from each of three adjacent palms (Wood, 1968). As a general system for disease and pest monitoring in a new area the Sabah system is to be preferred since the whole area is covered. It is relatively inexpensive, and a detection worker can usually cover 10 hectares a day.

References

Aderungboye, F. O. (1981) Significance of Vascular Wilt in oil palm plantations in Nigeria. In *The oil palm in agriculture in the eighties*, ISP, Kuala Lumpur.

Akbar, U., Kusnadi, M. and **Ollagnier, M.** (1971). Influence de la nature du matériel végétal et de la nutrition minérale sur la pourriture sèche du tronc du palmier à huile due à *Ganoderma*. *Oléagineux*, **26**, 527.

Akbar, U. and **Kusnadi, T. T.** (1976) Relationship between tolerance to *Ganoderma* basal stem rot and the origin of oil palm planting material. Mimeograph, Medan.

Alexander, V. T. and **Kastelein, P.** (1983) Hartrot or 'sudden wilt' disease in hybrids of *Elaeis oleifera* × *Elaeis guineensis*. *Surinaamse Landbouw (Suriname)*, **31**(1), 18.

Allen, J. D. and **Bull, R. A.** (1954) Recent severe attacks on oil palms by two caterpillar pests belonging to the Limacodidae. *J. W. Afr. Inst. Oil Palm Res.*, **1**(2), 130.

Anon. (1974) Replanting oil palm areas with *Elaeis oleifera* × *Elaeis guineensis* hybrids. *Oil Palm News*, No. 18, 1.

Arens, F. P. (1965) Private communication.

Arnaud, F. and **Rabechault, H.** (1972) Premières observations sur les caractères cytohistochimiques de la résistance du palmier à huile au 'dépérissement brutal'. *Oléagineux*, **27**, 525.

Aslam, N. A. (1965) On *Hispoleptis* Baly (Coleoptera, Hispidae) and *Imatidium* F. (Coleoptera, Cassididae). *Ann. Mag. nat. Hist.*, Ser. 13, **8**, 687.

Bachy, A. (1958) Le 'Blast' des pépinières de palmier à huile. *Oléagineux*, **13**, 653.

Bachy, A. (1963) Insectes et animaux nuisibles au palmier à huile. *Oléagineux*, **18**, 15–18 and 173–6.

Barlow, H. S. and **Chew Poh Soon** (1971) The Rhinoceros beetle, *Oryctes rhinoceros*, in young oil palms planted after rubber on some estates in West Malaysia. In *Crop protection in Malaysia*, p. 133, Incorp. Soc. of Planters, Kuala Lumpur.

Bellier, L. (1965) Evolution du peuplement des rongeurs dans les plantations industrielles de palmier à huile. *Oléagineux*, **20**, 735.

Berchoux, C. de, and **Gascon, J. P.** (1963) L'arcure défoliée du palmier à huile. *Oléagineux*, **18**, 713.

Bescombes, J. P. (1968) Essais de traitements des palmeraies adultes avec le B.S.E. Bangui Special. *Oléagineux*, **23**, 715.

Bescombes, J. P. (1972) Le matériel de traitements insecticides en plantation de palmier à huile. *Oléagineux*, **27**, 479.

Blaak, G. (1970) Epistasis for Crown disease in the oil palm (*Elaeis guineensis*, Jacq.). *Euphytica*, **19**, 22.

Bondar, G. (1940) Notas entomologicas da Bahia. Parts V, VI, VII. *Revta Ent., Rio de J.*, **11**, 199 and 842; (1941) **12**, 268. Also (1941) abstract in *Rev. appl. Ent.*, **29**, 9 and 468; and (1942) **30**, 425.

Booth, C. and **Robertson, J. S.** (1961) *Leptosphaeria elaeidis* sp. nov. isolated from anthracnosed tissue of oil palm seedlings. *Trans. Br. mycol. Soc.*, **44**, 24.

Boyé, P. and **Aubry, M.** (1973) Replantation des palmeraies industrielles. Méthode de préparation de terrain et de protection contre l'*Oryctes* en Afrique de l'Ouest. *Oléagineux*, **28**, 175.

Bredas, J., Stessels, L. and **Gérard, Ph.** (1968) La lutte chimique contre les petits rongeurs en jeune palmeraie. *Oléagineux*, **23**, 15.

Bull, R. A. (1954) A preliminary list of oil palm diseases encountered in Nigeria. *J. W. Afr. Inst. Oil Palm Res.*, **1**(2), 53.

Bull, R. A. (1961) Studies on the deficiency diseases of the oil palm. 2. Macronutrient deficiency symptoms in oil palm seedlings grown in sand culture. *J. W. Afr. Inst. Oil Palm Res.*, **3**(11), 254.

Bull, R. A. and **Robertson, J. S.** (1959) The problems of 'Little Leaf' of oil palms – a review. *J. W. Afr. Inst. Oil Palm Res.*, **2**, 355.

Cardoso, R. M. E. (1961) Podridao de raizes em dendezeiro. *Biologico*, **27**, 246.

Chandrasekharan, K. and **Edmunds, G. C.** (1976). Porcupine – a major pest in oil palm clearing from jungle in Central Johore. *Planter, Kuala Lumpur*, **52**, 216.

Chua, T. H., Siew, Y. C., Ng, S. M., Fong, F. W. and **Lee, T. H.** (1980) Oil palm, a source of food for some large-billed crows. *Planter, Kuala Lumpur*, **56**, 44.

Conway, G. R. and **Wood, B. J.** (1964) Pesticide chemicals – help or hindrance in Malaysian agriculture? *Malay. Nat. J.*, **18**, 111.

Corrado, F. (1970) La maladie du palmier à huile dans les Llanos de Colombie. *Oléagineux*, **25**, 383.

Cotterell, G. S. (1925) The Hispid leaf miner (*Coelaenomenodera elaeidis* Maul) of oil palms (*Elaeis guineensis* Jacq.) on the Gold Coast. *Bull. ent. Res.*, **16**, 77.

Courtois, G. (1968) Arbres anormaux chez *Elaeis guineensis*. Leur production comparée à celle des arbres normaux. *Oléagineux*, **23**, 641.

Dammerman, K. W. (1929) *The agricultural zoology of the Malay Archipelago.* Amsterdam.

Desmier de Chenon, R. (1979) Mise en évidence du rôle de *Recilia mica* Kramer dans la maladie du Blast des pépinières de palmiers à huile en Côte-d'Ivoire. *Oléagineux*, **34**, 107.

Desmier de Chenon, R., Merlan, E., Genty, P., Morin, J. P. and **Dollet, M.** (1983) Research on the genus *Lincus* and its possible role in the transmission of Marchitez in oil palm and Hartrot of coconut. *4th Reun. Com. Tecn. Reg. San. Veg.*, SARH-IICA, Cancun, Mexico.

Dollet, M. (1982) Les maladies de palmiers et cocotiers à protozoaires flagellés intra-phloemiques en Amérique latine (*Phytomonas* sp., Trypanosomatidae). *Oléagineux*, **37**, 9.

Dollet, M., Giannotti, J. and **Ollagnier, M.** (1977) Observations de protozoaires flagellés dans les tubes criblés de palmiers à huiles malades. *C. R. Acad. Sci., Paris*, **284**, Sér. D, 643.

Dollet, M. and **Lopez, G.** (1978) Etude sur l'association de protozoaires flagellés à la marchitez sorpresiva du palmier à huile en Amérique du Sud. *Oléagineux*, **33**, 209.

Dollet, M., Mariau, D. and **Renard, J. L.** (1987) Research needed in Latin America on oil palm diseases. *Int Oil Palm Conf.*, Kuala Lumpur, June 1987.

Dubois, J. and **Gerard, Ph.** (1968) Temnoschoites dans les pépinières de palmier à huile. *Oléagineux*, **23**, 571.

Duckett, J. E. (1984) Barn owls (*Tyto alba*) and the second generation rat-baits utilised in oil palm plantations in Peninsular Malaysia. *Planter, Kuala Lumpur*, **60**, 3.

Duff, A. D. S. (1963) The Bud Rot Little leaf disease of the oil palm. *J. W. Afr. Inst. Oil Palm Res.*, **4**, 176.

Duff, A. D. S. (1970) *Cercospora elaeidis*, Stey., and the oil palm. *Oléagineux*, **25**, 329.

Dufour, F. and Philippe, R. (1980) Surveillance sanitaire des jeunes cultures en Afrique de l'Ouest. *Oléagineux*, **35**, 85.

Dzido, J. L., Genty, P. and Ollagnier, M. (1978) Les principales maladies du palmier à huile en Equateur. *Oléagineux*, **33**, 55.

Figueroa, A. and Van den Hove, J. (1967) Contributión al estudio de la 'escoriación' de los fontos de la palmera de aceite en Colombia, el *Himatidium neivai*. *Oléagineux*, **22**, 15.

Fraselle, J. V. (1951) Experimental evidence of the pathogenicity of *Fusarium oxysporum* Schl. to the oil palm. *Nature, Lond.*, **167**, 447.

Frazelle, J. and Buyckx, E. J. E. (1962) Maladies et animaux nuisibles du palmier à huile. *In Publs INEAC*, Hors Série.

Freeman, W. G. (1925) Report of the Department of Agriculture, Trinidad and Tobago, 1925. Abstract in *Rev. appl. Ent.*, **14**, 546.

Gascon, J. P. and Meunier, J. (1979) Anomalies d'origine génétique chez le palmier à huile *Elaeis*, Description et résultats. *Oléagineux*, **34**, 437.

Genty, Ph. (1968) Deux lépidoptères nuisibles au palmier à huile. *Oléagineux*, **23**, 645.

Genty, Ph. (1972) Morphologie et biologie de *Sibine fusca*, Stoll, lépidoptère défoliateur du palmier à huile en Colombie. *Oléagineux*, **27**, 65.

Genty, Ph. (1973a) Observations préliminaries du lépidoptère mineur des racines du palmier à huile, *Sagalass valida*, Walker. *Oléagineux*, **28**, 59.

Genty, Ph. (1973b) Informe de misión sobre una enfermedad similar a la Marchitez sorpresiva' en varias plantaciones de palma Africana en El Ecuador. Ass. Nac. de Cultivadores de Palma Africana, Quito, Ecuador. Mimeograph.

Genty, Ph. (1976) Etude morphologique et biologique d'un lépidoptère défoliateur du palmier à huile en Amérique latine *Darna metaleuca* Walker. *Oléagineux*, **31**, 99.

Genty, P. (1977a) Traitements aériens des plantations industrielles de palmier à huile. 1. Le matériel et l'organisation. 2. La réalisation. *Oléagineux*, **32**, 5 and 51.

Genty, P. (1977b) Les ravageurs et les maladies du palmier à huile et du cocotier. Les lépidoptères mineurs de racines: *Sagalassa valida* W. *Oléagineux*, **32**, 311.

Genty, P. (1978) Morphologie et biologie d'un lépidoptère défoliateur du palmier à huile en Amérique latine, *Stenoma cecropia* Meyrick. *Oléagineux*, **33**, 421.

Genty, P. (1981) Entomological research on the oil palm in Latin America. *Oil Palm News*, **25**, 17.

Genty, P., Desmier de Chenon, R. and Morin, J. P. (1978) Les ravageurs du palmier à huile en Amérique latine. *Oléagineux*, **33**, 325.

Genty, P., Garzón, M. A. and Garcia, R. (1983) Dégâts et contrôle du complexe *Leptopharsa-Pestalotiopsis* chez le palmier à huile. *Oléagineux*, **38**, 291.

Genty, Ph., Gildardo Lopez, J. and Mariau, D. (1975) Dégâts de *Pestalotiopsis* induits par des attaques de *Gargaphia* en Colombie. *Oléagineux*, **30**, 199.

Genty, Ph. and Mariau, D. (1973/75) Les Limacodidae du genre Sibine. *Oléagineux*, **28**, 225; Utilisation d'un germe entomopathogène dans la lutte contre *Sibine fusca* (Limacodidae). *Oléagineux*, **30**, 349.

Genty, Ph. and Mariau, D. (1973) Le genre *Himatidium*. *Oléagineux*, **38**, 513.

Genty, P. and Mariau, D. (1975) Morphologie et biologie du Pyralidae des racines de l'Elaeis, *Sufetula diminutalis*. *Oléagineux*, **30**, 147.

Genty, P. and Reyes, R. (1977) Un nouvel acarien du palmier à huile: l'Eriiphyidae 'Retracrus elaeis' Keifer'. *Oléagineux*, **32**, 255.

Gillbanks, R. A., Turner, P. D. and Wood, B. J. (1967) Rat control in Malaysian oil palm estates. *Planter, Kuala Lumpur*, **43**, 297.

Goonewardena, H. F. (1958) The Rhinoceros beetle (*Oryctes rhinoceros* L.) in Ceylon. *Trop. Agric. Mag. Ceylon agric. Soc.*, **114**, 39.

Gray, B. S. and **Turner, P. D.** (1966) Removal of pollen from oil palms by *Megaspis dorsata*. *Pl. Prot. Bull. FAO*, **14**, 37.

Green, A. H. and **Ward, J. B.** (1973) Unilever Plantations Group annual review of research, 1971. London. Mimeograph.

Guerrero, A. V. (1985) Cria de *Chrysopa* spp. en laboratorio para control del chinche de encaje *Leptopharsa gibbicarina* (Froech). *Palmas*, **6**(3), 25.

Gunn, J. S. *et al.* (1961) The development of improved nursery practices for the oil palm in West Africa. *J. W. Afr. Inst. Oil Palm Res.*, **3**, 198.

Han, K. J. and **Chew, P. S.** (1978) Control measures and observations in some *Valanga nigricornis* outbreaks in oil palms in Peninsular Malaysia. *Planter, Kuala Lumpur*, **54**, 682.

Hartley, C. W. S. (1965) Some notes on the oil palm in Latin America. *Oléagineux*, **20**, 359.

Hartley, C. W.S. (1974) Oil palm research and development in Colombia. Min. Overseas Dev., London. Mimeographed report.

Hasselo, H. N. (1959) Fertilising of young oil palms in the Cameroons. *Pl. Soil*, **11**, 113.

Hertslet, L. R. and **Duckett, J. E.** (1971) *Thosea bisura* – a new pest of oil palms. *Planter, Kuala Lumpur*, **47**, 398.

Huguenot, R. and **Vera, J.** (1981) Description et lutte contre *Castnia daedalus* Cr. (Lép. Castnidae) ravageur du palmier à huile en Amérique du Sud. *Oléagineux*, **36**, 543.

IRHO (1976) Protection des jeunes palmiers contre les rats. *Oleágineux*, **31**, 165.

IRHO (1981) Les ravageurs du palmier à huile et du cocotier en Afrique occidentale. *Oléagineux*, **36**, 168.

Jerath, M. L. (1968) A list of insects found on palms in Nigeria and their known parasites and predators. *J. Nig. Inst. Oil Palm Res.*, **4**, 411.

Johnston, A. (1959) Oil palm seedling blight. *Malay. agric. J.*, **42**, 14.

Jollands, P. (1983) Laboratory investigations on fungicides and biological agents to control three diseases of rubber and oil palms and their potential applications. *Trop. Pest Management*, **29**, 33.

Julia, J. F. (1979) Mise en évidence et identification des insectes responsables des maladies juveniles du cocotier et du palmier à huile en Côte-d'Ivoire. *Oléagineux*, **34**, 385.

Julia, J. F. and **Mariau, D.** (1976) Recherches sur l'*Oryctes monoceros* Ol. en Côte-d'Ivoire. Parts I, II and III. *Oléagineux*, **31**, 63, 113 and 263.

Korytkowski, C. A. and **Ruiz, A.** (1980) The oil palm bunch miner, *Castnis daedalus* (Cramer) in the Tocachi plantation (Peru). *Oléagineux*, **35**, 1.

Kovachich, W. G. (1948) A preliminary anatomical note on vascular wilt disease of the oil palm. *Ann. Bot.*, **12**, 327.

Kovachich, W. G. (1954) *Cercospora elaeidis* leaf spot of the oil palm. *Trans. Br. mycol. Soc.*, **37**, 209.

Kovachich, W. G. (1956a) Patch Yellow disease of the oil palm. *Trans. Br. mycol. Soc.*, **39**, 427.

Kovachich, W. G. (1956b) Necrotic spotting of the oil palm by *Cercospora elaeidis*, Steyaert. *Trans. Br. mycol. Soc.*, **39**, 297.

Kovachich, W. G. (1957) Some diseases of the oil palm in the Belgian Congo. *J. W. Afr. Inst. Oil Palm Res.*, **2**, 221.

Krishnan, R. (1980) Prebaiting for rat control in oil palms. *Planter, Kualà Lumpur*, **56**, 137.

Lecoustre, R., Mariau, D., Philippe, R. and **Desmier de Chenon, R.** (1980) Contribution à la mise au point d'une lutte biologique contre *Coelaenomenodera*. 2. Introduction en Côte-d'Ivoire d'un hymenoptère Eulphidae du genre *Chrysonotomyia* Ashmead, de Madagascar. *Oléagineux*, **35**, 177.

Leitch, T. T. (1966) *Ploneta diducta* – a pest of oil palms. *Planter, Kuala Lumpur*, **42**, 433.

Lenton, G. M. (1980) Biological control of rats in oil palm by owls. In *Tropical*

Ecology and Development: 5th Int. Symp. Trop. Ecology, Int. Soc. Trop. Ecology, Kuala Lumpur.

Liaw, A. M. K. (1983) Control of some mammalian pests through trapping. *Planter, Kuala Lumpur*, **59**, 215.

Locke, T. (1972) A study of vascular wilt disease of oil palm seedlings. Thesis, University of Manchester.

Locke, T. and **Colhoun, J.** (1973a) Contributions to a method of testing oil palm seedlings for resistance to *Fusarium oxysporum* Sch. f. sp. *elaeidis* Toovey. *Phytopath Z.*, **79**, 77.

Locke, T. and **Colhoun, J.** (1973b) *Fusarium oxysporum* P. sp. *elaeidis* as a seed-borne pathogen. *Trans. Br. mycol. Soc.*, **60**, 3, 594.

Loh, C. F. (1977) Preliminary evaluation of some systemic fungicides for *Ganoderma* control and phytotoxicity to oil palm. In *International developments in oil palm*, p. 555, ISP, Kuala Lumpur.

López, G., Genty, Ph. and **Ollagnier, M.** (1975). Contrôle préventif de la 'Marchitez sorpresiva' de l'*Elaeis guineensis* en Amérique latine. *Oléagineux*, **30**, 243.

Maas, P. W. Th. (1970) Contamination of the palm weevil (*Rhynchophorus palmarum*) with the Red Ring nematode (*Rhadinaphelenchus cocophilus*) in Surinam. *Oléagineux*, **25**, 653.

Malaguti, G. (1953) 'Pudrición de Cogolla' de la palmera de aciete africana (*Elaeis guineensis*, Jacq.) en Venezuela. *Agronomia Tropical*, **3**, 13.

Mariau, D. (1968) Méthodes de lutte contre le Rhynchophore. *Oléagineux*, **23**, 443.

Mariau, D. and **Bescombes, J. P.** (1972) Méthode de contrôle des niveaux de population de *Coelaenomenodera elaeidis*. *Oléagineux*, **27**, 425.

Mariau, D. and **Huguenot, R.** (1983) Méthode d'estimation des populations de *Castnia daedalus* (Lepidoptère Castnidae) sur le palmier à huile. *Oléagineux*, **38**, 227.

Mariau, D. and **Julia, J. F.** (1973) Les Parasa. *Oléagineux*, **28**, 129.

Mariau, D. and **Morin, J. P.** (1971) La Pyrale du palmier à huile. *Oléagineux*, **26**, 379.

Martin, G. (1970) Le dessèchement des feuilles. Maladi du palmier à huile dans la région du Nord-Santander de Colombie. *Oléagineux*, **26**, 22.

Mena Tascon, E., Cardona Mejia, C., Martinez-Lopez, G. and **Dario Jimenez, O.** (1975) Efecto del uso de insecticidas y control de malezas en la incidencia de la marchitez sorpresiva de la palma africana (*Elaeis guineensis*, Jacq.) *Rev. Colombiana Ent.*, **1**, 1.

Mena Tascon, E. and **Martinez-Lopez, G.** (1977) Identificacion del insecto vector de la marchitez de la palma africana (*Elaeis guineensis*, Jacq.). *Fitipatologia Colombiana*, **6**, 2.

Merino, M. G. and **Vasquez, A. V.** (1963) El 'Gusano cogollero' *Alurnis humeralis* (Rosenberg) como plago de la palma Africana aceite y su combate químico en Ecuador. *Turrialba*, **13**, 6.

Meunier, J., Renard, J. L. and **Quillec, G.** (1979) Hérédité de la résistance à la fusariose chez le palmier à huile *Elaeis guineensis* Jacq. *Oléagineux*, **34**, 555.

Meynadier, G., Amargier, A. and **Genty, P.** (1977) Une virose de type densonucléose chez le Lépidoptère *Sibine fusca* St. *Oléagineux*, **32**, 357.

Moreau, C. (1952) in *Notions de culture de l'Elaeis au Congo Belge*, by Vanderweyen, R., Brussels.

Moreau, C. and **Moreau, M.** (1960) Inhibition de la croissance du *Fusarium oxysporum* par divers fungicides organiques. *Revue Mycol.*, **25**, 307.

Morin, J. P. and **Mariau, D.** (1970–4) La biologie de *Coelaenomenodera elaeidis* Mlk. Parts I, II, III, IV and V. *Oléagineux*, **25**, 11; **26**, 83 and 373; **27**, 496; **29**, 233 and 549.

Navaratnam, S. J. (1961) Successful inoculation of oil palms with a pure culture of *Ganoderma lucidum*. *Malay. agric. J.*, **43**, 233.

Navaratnam, S. J. and **Chee Kee Leong** (1965) Stem Rot of oil palms in the Federal

Experiment Station, Serdang. *Malay. agric. J.*, **45**, 175.

Ng, K. Y. (1980) Grasshopper (*Valanga nigricornis*) control by aerial application in plantation crops in Malaysia. *Planter, Kuala Lumpur*, **56**, 362.

Ng, K. Y. (1981) *Venturia palmaris* Wilkinson (Hymenoptera: Ichneumonidae) – a parasite of oil palm bunch moth *Tirathaba rufivena* Walker (Lepidoptera: Pyralidae). In *The oil palm in agriculture in the eighties*, ISP, Kuala Lumpur.

Ollagnier, M. and **Renard, J. L.** (1976) Influence du potassium sur la résistance du palmier à huile à la fusariose. *Oléagineux*, **31**, 203.

Ollagnier, M. and **Valverde, G.** (1968) Contribution à l'étude de la carence en bore du palmier à huile. *Oléagineux*, **23**, 359.

Perthuis, B., Desmier de Chenon, R. and **Merlan, E.** (1985) Mise en évidence du vecteur de la Marchitez sorpresiva du palmier à huile, la punaise *Lincus lethifer* Dolling (Hemiptera Pentatomidae Discocephalinae). *Oléagineux*, **40**, 473.

Prendergast, A. G. (1957) Observations on the epidemiology of Vascular Wilt disease of the oil palm (*Elaeis guineensis*, Jacq.). *J. W. Afr. Inst. Oil Palm Res.*, **2**, 148.

Prendergast, A. G. (1963) A method of testing oil palm progenies at the nursery stage of resistance to Vascular Wilt disease caused by *Fusarium oxysporum*, Schl. *J. W. Afr. Inst. Oil Palm Res.*, **4**, 156.

Philippe, R. and **Diarrassouba, S.** (1980) Méthode de lutte contre *Coelaenomenodera* par poudrage aérien de HCH. *Oléagineux*, **35**, 187.

Quencez, P. (1982) Les pépinières de palmiers à huile en sacs de plastique sans ombrière. *Oléagineux*, **37**, 397.

Quillec, G. and **Renard, J. L.** (1977) Protection contre la cercosporiose du palmier à huile. *Oléagineux*, **32**, 363.

Rai, B. K. (1977) Pests of the oil palm and their control. Annual report, 1977, Central Agric. St., Mon Repos, Guyana, p. 76. Mimeograph.

Rajagopalan, K. (1965) *NIFOR first annual report 1964–5*, pp. 88–9.

Rajagopalan, K. (1968–74) *NIFOR third, fourth and fifth annual reports*, p. 105, pp. 87–8 and 78–9 respectively; and Influences of irrigation and shading on the occurrence of Blast disease of oil palm seedlings. *J. Nigerian. Inst. Oil Palm Res.*, **5**(19), 23.

Rajagopalan, K. (1973) Effectiveness of certain fungicides in controlling *Cercospora* leaf spot of oil palm (*Elaeis guineensis*, Jacq.) in Nigeria. *J. Nigerian. Inst. Oil Palm Res.*, **5**(18), 23.

Rajagopalan, K., Aderungboye, F. O. and **Obasola, C. O.** (1978) Evaluation of oil palm progenies for reaction to the Vascular Wilt disease. *J. Nigerian Inst. Oil Palm Res.*, **5**, 87.

Rajaratnam, J. A. (1972) White stripe disorder of oil palm (*Elaeis guineensis*) in Malaysia. *Expl. Agric.*, **8**, 161.

Rajaratnam, J. A. and **Law I. H.** (1975) Effect of boron nutrition on intensity of spider mite attack on oil palm seedlings. *Expl. Agric.*, **11**, 59.

Renard, J. L. (1970) La Fusariose du palmier à huile. Rôle des blessures des racines dans le processus d'infection. *Oléagineux*, **25**, 581.

Renard, J. L. (1973) Transport et distribution du bénomyl dans les palmiers à huile au stade de la pépinière. *Oléagineux*, **28**, 557.

Renard, J. L. (1976) Diseases in Africa and South America. In *Oil palm research*, eds R. H. V. Corley, J. J. Hardon and J. B. Wood, p. 447, Amsterdam.

Renard, J. L., Gascon, J. P. and **Bachy, A.** (1972) Recherches sur la fusariose du palmier à huile. *Oléagineux*, **27**, 581.

Renard, J. L., Mariau, D. and **Quencez, P.** (1975) Le Blast du palmier à huile: rôle des insectes dans la maladie. Résultats préliminaires. *Oléagineux*, **30**, 497.

Renard, J. L., Noiret, J. M. and **Meunier, J.** (1980) Sources et gammes de résistance à la fusariose chez les palmiers à huile *Elaeis guineensis* et *Elaeis melanococca*. *Oléagineux*, **35**, 387.

Renard, J. L. and **Quillec, G.** (1977) Lutte contre la cercosporiose du palmier à huile. 1. En pépinière. *Oléagineux*, **32**, 43.

Renard, J. L. and **Quillec, G.** (1983) Fusariose et replantation. Eléments à prendre en considération pour les replantations de palmiers à huile en zone fusariée en Afrique de l'Ouest. *Oléagineux*, **38**, 421.

Renard, J. L., Quillec, G. and **Hornus, P.** (1977) Lutte contre la cercosporiose du palmier à huile. II. En plantation. *Oléagineux*, **32**, 89.

Robertson, J. S. (1956) Leaf diseases of oil palm seedlings, *J. W. Afr. Inst. Oil Palm Res.*, **1**(4), 110.

Robertson, J. S. (1959a) Blast disease of the oil palm; its cause, incidence and control in Nigeria. *J. W. Afr. Inst. Oil Palm Res.*, **2**, 310.

Robertson, J. S. (1959b) Coinfection by a species of *Pythium* and *Rhizoctonia lamellifera* Small in Blast disease of oil palm seedlings. *Trans. Br. mycol. Soc.*, **42**, 401.

Robertson, J. S. (1960a) *WAIFOR eighth annual report 1959–60*, p. 107.

Robertson, J. S. (1960b) *WAIFOR eighth annual report 1959–60*, p. 112.

Robertson, J. S. (1962a) Dry Basal Rot, a new disease of oil palms caused by *Ceratocystis paradoxa* (Dade) Moreau. *Trans. Br. mycol Soc.*, **45**, 475.

Robertson, J. S. (1962b) Investigations into an outbreak of Dry Basal Rot at Akwukwu in Western Nigeria. *J. W. Afr. Inst. Oil Palm Res.*, **3** 339.

Robertson, J. S. (1962c) *WAIFOR tenth annual report 1961–2*, p. 82.

Robertson, J. S. (1963a) *WAIFOR eleventh annual report*, pp. 77–8.

Robertson, J. S. (1963b) *WAIFOR eleventh annual report 1962–3*, p. 79.

Robertson, J. S., Prendergast, A. G. and **Sly, J. M. A.** (1968) Diseases, disorders and deficiency symptoms of the oil palm in West Africa. *J. Nigerian Inst. Oil Palm Res.*, **4**, 381.

Rojas-Cruz, L. A. (1977) Insect pests of oil palms in Colombia. In *International developments in oil palm*, ISP, Kuala Lumpur.

Ruer, P. (1964) Les conditions de lutte contre un prédateur du palmier á huile. *Oléagineux*, **19**, 387.

Sanchez Potes, A. (1964) Enfermedades del Algodonero, concetero y palma africana en Colombia. *Boln. Notic. Inst. Fom. algod., Bogotá*, **4**(12), 6.

Sanchez Potes, A. (1972) Dos enfermedades de importancia económica que afectas la palma Africana de aceite en Colombia. Inst. Col. Agropecuaria. Typescript.

Santos, J. V. (1968) Algunas caracteristicas biologicas y etologicas del *Alurnus humeralis* Rosenburg 'Gusano chato o cogollero' de la palma Africana. *Oléagineux*, **23**, 159.

Sarjit Singh (1968) The use of electrical drills for more efficient trunk injection against bagworms in oil palms. *Planter, Kuala Lumpur*, **62**, 58.

Schuiling, M. and **Dinther, J. B. M.** (1982) La maladie de l'anneau rouge à la plantation de palmiers à huile de Paricatuba, Para (Brésil): une étude de cas. *Oléagineux*, **37**, 555.

Shearing, C. H. (1964) A serious attack of Hispid leaf miner (*Coelaenomenodera elaeidis* Maul.) at Mpundu Palms Estate. Cameroons Development Corporation, Ekona Research Unit. Mimeograph.

Siew, Y. C., Ng, S. M., Fong, F. W. and **Lee, T. H.** (1980) Oil palm, a source of food for some large-billed crows. *Planter, Kuala Lumpur*, **56**, 44.

Siew, Y. C., Ng, S. M., Poon, S. K. and **Fong, F. W.** (1979) The house crow – a pest of oil palm? *Planter, Kuala Lumpur*, **55**, 185.

Stimpson, K. M. S. and **Rasmussen, A. N.** (1973) Clearing the old stand and some preparations for replanting coastal oil palms. In *Advances in oil palm cultivation*, p. 116, Incorp. Soc. of Planters, Kuala Lumpur.

Stobbe, W. G. van (1983) Control of *Castnia daedalus*, a major pest of the oil palm in Surinam. *Trop. Agric. (Trin.)*, **60**, 172.

Syed, R. A. (1978) Bionomics of the three important species of bagworms on oil palm. *Malaysian Agric. J.*, **51**, 392.

Syed, R. A. and **Speldewinde, H. V.** (1974) Pest detection and census on oil palms. *Planter, Kuala Lumpur*, **50**, 230.

Tams, W. H. T. (1930) A new moth damaging oil-palm in the Belgian Congo. *Bull. ent. Res.*, **21**, 75; (1930) abstract in *Rev. appl. Ent.*, **18**, 426.

Thompson, A. (1934) in *The oil palm in Malaya*, by Bunting B., Georgi, C. D. V. and Milsum, J. N., Chapter 7, Kuala Lumpur.

Thompson, A. (1936) *Ustulina zonata* on the oil palm. *Malay. agric. J.*, **34**, 222.

Thompson, A. (1937) *Observations on Stem Rot of the oil palm.* Dept. of Agriculture, Malaya, Scientific Series No. 21.

Tiong, R. H. C. (1979) Some predators and parasites of *Mahasena corbetti* (Tams) and *Thosea asigna* (Moore) in Sarawak. *Planter, Kuala Lumpur*, **55**, 279.

Tiong, R. H. C. (1981) Study of some aspects of biology and control of *Thosea asigna* (Moore). In *The oil palm in agriculture in the eighties*, ISP, Kuala Lumpur.

Tiong, R. H. C. and **Munroe, D. D.** (1977) Microbial control of an outbreak of *Darna trima* (Moore) on oil palm (*Elaeis guineensis*, Jacq.) in Sarawak (Malaysian Borneo). In *International Developments in oil palm*, ISP, Kuala Lumpur.

Toh, P. Y. and **Brown, T. P.** (1978) Evaluation of carbofuran as a chemical prophylactic control measure for *Oryctes rhinoceros* in young oil palms. *Planter, Kuala Lumpur*, **54**, 3.

Turner, P. D. (1965a) The incidence of *Ganoderma* disease of oil palms in Malaya and its relation to previous crop. *Ann. appl. Biol.*, **55**, 417.

Turner, P. D. (1965b) *Marasmius* infection of oil palms in Malaya – a review. *Planter, Kuala Lumpur*, **41**, 387.

Turner, P. D. (1966a) Infection of oil palms by *Ganoderma* in Malaya. *Oléagineux*, **21**, 73.

Turner, P. D. (1966b) Blast disease in oil palm nurseries. *Planter, Kuala Lumpur*, **42**, 103.

Turner, P. D. (1968) The use of surgery as a method of treating Basal Stem Rot in oil palm. *Planter, Kuala Lumpur*, **44**, 302.

Turner, P. D. (1969) Observations on the incidence, effects and control of Upper Stem Rot in oil palms. In *Progress in oil palm*, Incorp. Soc. of Planters, Kuala Lumpur.

Turner, P. D. (1970a) Spear rot disease of Plantación 'La Arenosa'. Mimeographed report. Harrison Fleming Advisory Services.

Turner, P. D. (1970b) Oil palm diseases on Plantación 'La Arenosa'. Mimeographed report. Harrison Fleming Advisory Services.

Turner, P. D. (1971) Microorganisms associated with oil palm (*Elaeis guineensis*, Jacq.) Commonwealth Mycological Institute, Phytopathological Papers, No. 14, 58 pp.

Turner, P. D. (1973) An effective trap for *Oryctes* beetle in oil palms. *Planter, Kuala Lumpur*, **49**, 488.

Turner, P. D. (1981) *Oil palm diseases and disorders.* Oxford Univ. Press, 280 pp.

Turner, P. D. and **Bull, R. A.** (1967) *Diseases and disorders of the oil palm in Malaysia.* Incorp. Soc. of Planters. Kuala Lumpur, 247 pp.

Turner, P. D. and **Gillbanks, R. A.** (1974) *Oil palm cultivation and management.* Incorp. Soc. of Planters, Kuala Lumpur, Malaysia.

Urueta, E. J. (1975) Insectos asociados con el cultivo de palma africana en Uraba (Antioquia) y estudio de su relación con la pudrición de la flecha, pudrición del cogollo. *Rev. Colomb. Entom.*, **1**, 15.

Van den Hove, J. (1971) Un type de pourriture des racines du palmier à huile. *Oléagineux*, **26**, 153.

Van de Lande, H. L. (1984) Vascular Wilt disease of oil palm (*Elaeis guineensis*, Jacq.) in Para, Brazil. *Oil Palm News*, **28**, 6.

Van Slobbe, W. G., Parthasarathy, M. V. and **Hesen, J. A. J.** (1978) Hartrot or fatal wilt of palms. II. Oil palm (*Elaeis guineensis*) and other palms. *Principes*, **22**, 15.

Varghese, G., Chew, P. S. and **Lim, J. K.** (1975) Biology and chemically assisted biological control of *Ganoderma*. *Rubber Res. Inst. of Malaysia*, 15 pp.

Vessey, J. C. (1981) Lutte contre une maladie foliaire du palmier à huile au Honduras à l'aide d'insecticides. *Oléagineux,* **36**, 229.

Wardlaw, C. W. (1946a) *Fusarium oxysporum* on the oil palm. *Nature, Lond.,* **158**, 712.

Wardlaw, C. W. (1946b) A Wilt disease of the oil palm. *Nature, Lond.,* **158**, 56.

Wardlaw, C. W. (1950a) Vascular Wilt disease of the oil palm caused by *Fusarium oxysporum* Schl. *Trop. Agric., Trin.,* **27**, 42.

Wardlaw, C. W. (1950b) *Armillaria* Root and Trunk Rot of oil palms in the Belgian Congo. *Trop. Agric., Trin.,* **27**, 95.

Weir, G. M. (1968) Leaf Spot of the oil palm caused by *Cercospora elaeidis* Stey. Aspects of atmospheric humidity and temperature. *J. Nigerian Inst. Oil Palm Res.,* **5**, 41.

Weir, G. M. (1968) Algal spot of the oil palm. *Cephaleuros virescens* Kunze, in Robertson, J. S., Prendergast, A. G. and Sly, J. M. A. (1968).

Williams, T. H. (1965) Diseases of the oil palm in Sabah. Sabah Planters' Association. Oil Palm Seminar. Mimeograph.

Wood, B. J. (1965) *Insect pests of oil palms in Malaya.* Annual report, Johore Planters' Association.

Wood, B. J. (1968) *Pests of oil palms in Malaysia and their control.* Incorp. Soc. of Planters, Kuala Lumpur, 204 pp.

Wood, B. J. (1969a) The extent of vertebrate attacks on the oil palm in Malaysia. In *Progress in oil palm,* Incorp. Soc. of Planters, Kuala Lumpur.

Wood, B. J. (1969b) Population studies on the Malaysian wood rat (*Rattus tiomanicus*) in oil palms, demonstrating an effective new control method and assessing some old ones. *Planter, Kuala Lumpur,* **45**, 510.

Wood, B. J. (1971a) The importance of ecological studies to pest control in Malaysian plantations. In *Crop protection in Malaysia,* Incorp. Soc. of Planters, Kuala Lumpur.

Wood, B. J. (1971b) sources of reinfestation of oil palms by the wood rat (*Rattus tiomanicus*). In *Crop protection in Malaysia,* Incorp. Soc. of Planters, Kuala Lumpur.

Wood, B. J. (1972a) Developments in oil palm pest management. *Planter, Kuala Lumpur,* **48**, 93.

Wood, B. J. (1972b) Integrated control: critical assessment of case histories in developing economies. *Planter, Kuala Lumpur,* **49**, 367.

Wood, B. J. (1981) The present status of pests on oil palm estates in South East Asia. In *The oil palm in agriculture in the eighties,* ISP, Kuala Lumpur.

Wood, B. J., Chung, G. F. and **Sim, S. C.** (1987) The rise of *Rattus rattus diardii* as an oil palm pest. *Int. Oil Palm Conf.,* Kuala Lumpur, June 1987.

Wood, B. J., Hutauruk, Ch. and **Liau, S. S.** (1977) Studies on the chemical and integrated control of the nettle caterpillars (Lepidoptera: Limacodidae). In *International developments in oil palm,* ISP, Kuala Lumpur.

Wood, B. J., Liau, S. S. and **Knecht, J. C. X.** (1974) Lutte contre la chenille à fourreau, *Metisa plana,* par injection d'insecticides systémiques dans les stipes des palmiers à huile. *Oléagineux,* **29**, 499.

Wood, B. J. and **Ng, K. Y.** (1969) The cockchafer, *Psilopholis vestita,* a new pest of oil palms in West Malaysia. *Planter, Kuala Lumpur,* **45**, 577.

Wood, B. J. and **Ng, K. Y.** (1974) Studies on the biology and control of the oil palm bunch moth, *Tirathaba mundella* (Walker) (Lepidoptera: Paralididae). *Malaysian Agric. J.,* **49**, 310.

Chapter 14

The products of the oil palm and their extraction

When the oil palm bunch* arrives at the place of extraction, some of the fruit will already be loose. Complete loosening of fruit from a bunch may take more than a week, and if natural loosening is awaited the quality of oil, as indicated by the higher free fatty acid (f.f.a.) content, will deteriorate, since a proportion of both the loose fruit and fruit on the bunch will inevitably be bruised and lipolysis, which is the splitting of fat molecules by hydrolysis into glycerol and fatty acids through enzyme action, will proceed. Rapid extraction of fruit from the bunch is therefore an important part of the product extraction process and provision must be made to arrest the deterioration of palm oil quality even before the fruit has been removed from the bunch.

Palm oil is solid at ambient temperatures in a temperate climate. At tropical temperatures it is a fluid with certain fractions held in crystalline form. On settling, there is a clear liquid section and a crystalline fluid base. The prerequisites for the release of palm oil from the fruit are a physical breakdown of the mesocarp sufficient to rupture the cells, and a temperature sufficient to aid in this rupturing and fully to homogenize the fat constituents.

For the satisfactory release of the kernels from the fruit the requirements are that the oil-bearing mesocarp shall be removed and the shells cracked without damage to the kernels.

Palm kernel oil is not usually extracted on the plantations, though occasionally mills contain presses designed for this purpose. The conditions for the release of palm kernel oil, which is liquid at tropical day temperatures, are different from those of palm oil, but

---

* As some mill engineers use the word bunch to indicate that part of the bunch which is left over after stripping, it is necessary to define the terms used in this chapter as follows:
Bunch = the whole ripe bunch with loose fruit as cut from the palm (often referred to as FFB – fresh fruit bunches).
Fruit = the palm fruit as detached from the bunch before or during stripping and sometimes containing a proportion of perianth segments ('the calyx leaves').
Bunch refuse = that part of the bunch which is discarded in the stripper.

similar to those of copra and hard oil-bearing seeds. As the kernel matrix is solid and hard it must be crushed to a meal before the oil can be extracted under pressure. Roller mills are usually employed to grind the material so fine that a high proportion of the oil-containing cells are ruptured. 'Cooking' in stack cookers under steam pressure releases the oil still further and expression of the oil is then undertaken in screw-press expellers. Alternatively, solvent extraction may be carried out after prepressing in an expeller to give a cake with about 15 per cent oil; or, under the Direx system, the oil can be extracted without the need for prepressing (Cornelius, 1983).

Oil palm products, their formation and characteristics

The three commercial products of oil palm fruit are palm oil, palm kernel oil and palm kernel cake. Chemically, the word 'fat' is used to cover vegetable oils and fats whether they are in the solid or liquid state, and fats have been defined as the esters of fatty acids with the trihydric alcohol glycerol. The triglyceride fats, which predominate in plant and animal fats, have the following general formula:

$$
\begin{array}{c}
\text{H} \qquad \text{O} \\
| \qquad\quad \| \\
\text{HC}-\text{O}-\text{C}-\text{R1} \\
| \\
\qquad\quad\ \text{O} \\
\qquad\quad\ \| \\
\text{HC}-\text{O}-\text{C}-\text{R2} \\
| \\
\qquad\quad\ \text{O} \\
\qquad\quad\ \| \\
\text{HC}-\text{O}-\text{C}-\text{R3} \\
| \\
\text{H}
\end{array}
$$

where R1, R2 and R3 represent the hydrocarbon chains of fatty acid radicals.

Naturally occurring vegetable fats are mixtures of fats and their characters are taken largely from the fatty acids which predominate in them and from the arrangement of these fatty acids in the tri-glycerides. The fatty acids are hydrocarbon chains in which two hydrogen atoms are attached to all or the majority of carbon atoms within the chain. The carbon atom at one end of the chain has three hydrogen atoms attached to it and the one at the other end is attached to a carboxyl group to give the general structure:

$$\begin{array}{cccccc} \text{H} & \text{H} & \text{H} & \text{H} & \text{O} & \\ | & | & | & | & || & \\ \text{HC} - \text{C} - \text{C}.&..&\text{C} - \text{C} - \text{O} - \text{H} \\ | & | & | & | & \\ \text{H} & \text{H} & \text{H} & \text{H} & \end{array}$$

Fatty acids of this general formula are saturated fatty acids as they have the full number of hydrogen atoms attached to the carbon atoms of the chain. In unsaturated fatty acids there are one, two or three double bonds between carbon atoms which then have only single hydrogen atoms attached to them, in the following manner:

$$\begin{array}{cccc} \text{H} & \text{H} & \text{H} & \text{H} \\ | & | & | & | \\ \text{C} - \text{C} = \text{C} - \text{C} - \\ | & & & | \\ \text{H} & & & \text{H} \end{array}$$

Double bonds can occupy different positions in the chain, thus giving rise to different isomers. Furthermore, there are also geometric isomers (the *cis-* or the *trans-* forms) according to whether portions of a molecule joined by a double bond extend in the same or opposite directions.

Chemically, the most important reaction of the fats from the producer's point of view is *hydrolysis*, i.e. the formation of free glycerol and free fatty acid through a splitting of the fat molecule and the addition of the elements of water which may be partly represented by the equation

$$\text{CH}_2\text{OOCR} + \text{H}_2\text{O} = \text{CH}_2\text{OH} + \text{HOOCR}$$
$$\phantom{\text{CH}_2\text{OOCR}}|\phantom{+ \text{H}_2\text{O} = } |$$

where CH_2 represents one-third part of the glycerol radical and R the hydrocarbon chain of the fatty acid radical. Hydrolysis can be either autocatalytic in the presence of water, be catalysed by metals, or be brought about by the action of the enzyme lipase. The latter is, of course, the fat-splitting enzyme of animal digestion, but it is also found in palm fruit and in fungi and other organisms which gain access to fats. One of the most important tasks of the palm oil producer is to prevent hydrolysis by reducing to a minimum the amount of water and impurities present in the oil and by the destruction of the enzyme. Hydrolysis by alkalis is distinguished as *saponification* and gives rise to soaps and glycerol.

The second important reaction of fats is *oxidation*. Unsaturated fats are commonly oxidized at the double bonds and the oxidation products, the first of which are hydroperoxides, lead to rancidity with the loss of palatibility due to obnoxious flavours and odours, and may affect the bleachability of the oil. In the oil extraction process, the substances most likely to promote oxidation (pro-oxidants) are free atmospheric oxygen and traces of metals; the reaction is accelerated by light. Naturally occurring antioxidants suppress oxidation, which is measured as the *peroxide value* of a fat; this represents the reactive oxygen content, and is estimated through the liberation of iodine from potassium iodide in glacial acetic acid and recorded in terms of milliequivalents of peroxide-oxygen per 100 g fat.

The third important reaction of the fats is *hydrogenation* or 'hardening', which is a process of manufacture and is generally acknowledged to have contributed more to the interchangeability of fats and fatty oils than any other process. Hydrogenation processes add hydrogen atoms at the double bonds of unsaturated fats converting these into saturated fats with higher melting points.

Fourthly, the reaction which is used as a measure of the proportion of unsaturated constituents present in a fat is *halogen addition* to the double bonds of the unsaturated fatty acids, and the quantity of halogen taken up is expressed in terms of iodine as the *iodine value*, which is the number of grams of iodine absorbed per 100 g fat.

Lastly, with much larger quantities of palm oil coming on the market, there has been an increasing interest not only in providing standard products of very high quality but also in improving overall income through *fractionation* of palm oil in the countries of production (Martinenchi, 1972). The liquid fraction can then be marketed as cooking oil for local consumption while the solid fraction is sold as a component of frying fats, margarine and other products. Fractionation methods have been described and compared by Cornelius (1983).

Vegetable fats such as palm oil melt gradually over a range of temperatures not only because they contain a mixture of fats, but also because any solid fat of the same chemical composition may have several crystalline forms or *cis-* and *trans-*isomers. It is outside the scope of this book to describe the techniques used for the identification and estimation of individual fats and fatty acids, their isomers and their geometric isomers. The methods now in use rely mainly on spectroscopy, thin-layer and gas-liquid chromatography and X-ray techniques. Other physical properties of fats which sometimes need to be known for transportation and usage are the smoke point (the temperature at which the fat begins to smoke), the flash point (the temperature at which volatile products are produced rapidly enough to allow instantaneous ignition at the surface), and the fire point (temperature at which evolution of volatiles is sufficiently rapid to support continuous combustion). These values depend on the chain length of the fatty acids and the f.f.a. content (Rossell, 1986; PORIM, 1987).

Fat formation in the ripening fruit

The crude physical changes accompanying ripening have been referred to in Chapters 2 and 10. Development of fats in the kernel precedes that of the mesocarp. At 8 weeks from pollination the content of the seed is liquid. At 10 weeks it becomes semi-gelatinous, and it does not become really hard until the fifteenth week. At 10 weeks from pollination the amount of fat is very small and it is believed that at this stage the oil is present only as basal protoplasmic fat. Unsaturated fatty acids preponderate and this is indicated by the high iodine value of about 85 (Crombie, 1956).

From this stage there is a slow accumulation of fats until about the twelfth to the thirteenth week when fat formation becomes more rapid; the fats laid down are largely saturated, principally lauric (C12) which rises to 46–50 per cent and myristic (C14) which reaches 18–20 per cent by the twentieth week. The biggest accumulation occurs around the fourteenth to sixteenth week.

Fat formation in the mesocarp takes place very late in fruit development (Crombie and Hardman, 1958; Thomas, *et al.*, 1971). From the eighth to the sixteenth week after pollination fats constitute less than 2 per cent of the dry weight. There is in fact very little addition of any kind to the dry weight of the mesocarp from the eighth to the nineteenth week when, just prior to ripening, dry weight increases by 300–500 per cent and fats rather suddenly come to constitute 70–75 per cent of dry matter. During the long period of low oil content palmitic and linoleic acid esters predominate. During the final week of ripening all the fatty acids in combination increase, though oleic acid increases in greatest proportion and becomes second only to palmitic acid in quantity. Some data from the work of Crombie (1956) and Crombie and Hardman (1958) are given in Table 14.1.

The composition of palm oil and palm kernel oil

Palm oil

Although palm oil has a high proportion of the saturated palmitic acid (C16) it also contains a high quantity of unsaturated fats, principally those derived from oleic acid. About three-quarters of the glycerides are mixed saturated and unsaturated triglycerides. The oil melts over a range of temperatures from 25 to 50 °C.

Eckey (1954) quoted the fatty acid composition of twenty-one samples of commercial palm oils, seven from plantations and fourteen from countries producing oil mainly from the groves. The

Table 14.1 *Changes in weight and composition of developing fruit of Palm No. 6–173 in Nigeria*

Weeks after pollination	Dry weight (g per nut)	Oil extracted* (g per nut)	Saturated acids						Unsaturated acids		
			C6 + C8 (%)	C10 (%)	C12 (%)	C14 (%)	C16 (%)	C18 (%)	Oleic (%)	Linoleic (%)	Linolenic (%)
A. Kernel											
10	0.09	0.01	1.6		8.5	4.3	28.2	7.8	35.2	14.4	
11	0.12	0.03	6.4		9.2	4.3	20.0	4.6	48.0	7.5	
12	0.30	0.05	2.3	2.0	34.4	13.4	13.7	5.0	28.1	1.1	
14	0.54	0.20	3.3	0.3	48.4	14.2	7.6	2.9	18.7	4.6	
16	0.93	0.44	3.3	3.6	48.3	20.7	9.6	1.5	12.4	1.0	
19	1.11	0.51									
20	1.46	0.57	2.8	2.5	48.3	19.8	6.8	2.4	16.5	0.7	
B. Mesocarp	g per fruit	Per cent of dry weight						+C20			
8	0.33	1.3									
12	0.26	1.4				tr.	57.5	7.0	0.9	29.6	5.2
16	0.33	1.3				2.8	67.0	8.6	0.0	20.3	1.5
19	0.34	22.4				0.2	55.7	6.1	28.2	9.7	0.0
20	1.70	70.5				0.4	45.5	7.8	34.0	11.8	0.0

* **A.** In the kernel analysis, weights were on a per nut basis. Oil was estimated as light petroleum extract. **B.** In mesocarp, oil was estimated as liquid extract as per cent of dry weight.

Sources: Crombie (1956); Crombie and Hardman (1958).

ranges and the means of the percentages of the constituent acids are given in Table 14.2 together with analyses from other sources. Loncin and Jacobsberg (1963, 1965) found traces of linolenic (C18:3) and lauric (C12) acids in their Zaire oils and stated that oil from grove palms had almost the same composition as plantation oil. Eckey's Sierra Leone, Liberia and Ivory Coast samples had unusually high oleic and low palmitic contents.

Bienaymé and Servant (1958) examined oil from twenty places, the majority in the African palm belt, but some in Malaysia and Sumatra. They confirmed, through iodine value determinations, that from the Ivory Coast westwards there was a higher proportion of unsaturated acids. At the same time they found that when plantations were planted with Deli material the iodine value was slightly lower than for Benin (Dahomey), Nigeria and Zaire. With the considerable range in fatty acid composition it is obvious that progenies will differ significantly. Some significant differences were found in Malaysia between fruit forms and types, but as the ranges of composition overlapped the differences were probably ones of progeny within the forms and types (OPGL, 1972; Ng *et al.*, 1976). Loncin and Jacobsberg (1963) showed that in Zaire the triglycerides were to be found in the following proportions:

Triglycerides S-S-S about 6 per cent
Triglycerides S-S-U about 48 per cent
Triglycerides S-U-U about 43 per cent
Triglycerides U-U-U about 3 per cent

where S = saturated, U = unsaturated fatty acids.

Thus palm oil is firstly a fat containing a very high proportion of palmitic acid (to which may be attributed its value in soap-making) and secondly, the high quantities of oleic and linoleic acids give the fat a much higher unsaturated acid content than that of coconut or palm kernel oils which are essentially lauric oils giving a hard soap with greater lather.

The most important minor constituents of palm oil are the carotenoids; among these the carotenes are so conspicuous that the fat was termed 'red palm oil' in the Far East. Both the total carotenoid content of the oil and the proportions of the constituents vary, and they include α, β, γ and ζ carotene together with lycopene and lutein (xanthophyll). Vitamin A is derived from the carotenes, β-carotene having twice the 'vitamin A activity' of either α-carotene or γ-carotene. The carotenes are hydrocarbons with long chains of conjugated double bonds and have the formula $C_{40}H_{56}$. When the carotene molecule is split with the addition of the elements of water to both halves, vitamin A is produced. In β-carotene the two halves are identical, but in the other carotenes they are not, only one-half being of the constitution necessary for vitamin A formation (Booth, 1957).

Table 14.2 *Fatty acid composition of palm oil (E. guineensis) (per cent: means and ranges)*

	Lauric C12	Myristic C14	Palmitic C16	Stearic C18	Palmitoleic C16:1	Oleic C18:1	Linoleic C18:2	Linolenic C18:3
Eckey (1954)								
Range	—	0.6–5.9	32.3–45.1	2.2–6.4	0.8–1.4*	38.6–52.4	5.0–11.3	—
Sierra Leone, Liberia and Ivory Coast samples	—	1.8	34.8	5.4	—	50.4	7.4	—
Far East, Zaire, Nigeria, Cameroon and other samples (plantations)	—	2.3	41.2	4.3	—	42.5	9.6	—
Loncin and Jacobsberg (1963)								
Zaire plantations	0.2	1.2	43.7	6.2	—	36.5	11.8	—
Malaysia (OPGL 1972)								
Plantations†	—	0.6	49.9	2.8	—	40.6	6.2	—
Range	—	0.1–1.0	45.2–58.5	0.6–5.4	—	34.6–44.0	4.3–9.2	—

* In three samples only.
† Oil from four *dura* and four *tenera* bunches.

Differences in fruit colour have already been noted (p. 81), and apart from the *albescens* type which contains very little carotene, there are big differences in the carotene content of the mesocarp. For instance, two fruit samples which were 'red' and 'orange' when ripe gave the following carotene content (Purvis, 1957):

Fruit	Red	Orange
Carotene in dry mesocarp (mg per 100 g):	207	89
Total carotene in oil (ppm):	2,560	1,100

The content of the red sample is exceptionally high; oils taken direct from individual palms rarely contain more than 2,000 ppm. The carotene content of the orange sample is quite usual for palms in West Africa. Samples of Sumatran oil, however, were found to have carotene contents of between 320 and 475 ppm and commercial Malaysian oils, in the days when the Deli palm predominated had, between 300 and 500 ppm (Ames *et al.*, 1960), showing that, on average, Deli palms produce a lighter coloured oil than the general run of African palms. Carotene contents of progenies in Nigeria were shown to vary widely from bunch to bunch (Nwanze, 1961–4). In one progeny, thirty-eight out of fifty-one bunches had carotene contents in the low range 250–500 ppm, and ten were in the range 501–700. In another progeny, in which sixty-two bunches were examined, there was a much greater scatter, viz:

Carotene range (ppm)	250–500	501–600	601–700	701–800	801–900	901–1,000	over 1,000
Number of bunches	12	7	3	9	11	9	11

Bienaymé and Servant (1958) suggested that carotenoids are in greater concentration in oils from Togo, Benin (Dahomey) and Guinea than elsewhere in Africa and that in the palms of Angola carotene contents are as low as in the Deli palm.

The proportions in which the different carotenoids are found are also variable. In the 'red' and 'orange' samples of Purvis (1957) the proportions were as follows:

	Red (%)	Orange (%)
α-carotene	28	34
β-carotene	54	50
Lycopene	<3	<3
Lutein and ζ-carotene	<1	<1

Of the other minor constituents of palm oil, the tocopherols are found in a number of forms. They are antioxidants and may be found in quantities as high as 800 ppm in well-prepared plantation oil, though in oil coming from the groves the quantity is usually around 500 ppm. Palm oil contains only very small quantities of phospholipids and sterols.

Palm oil melts over a range of temperatures from 25 to 50 °C, and its slip melting point varies from 31 to 39 °C. Its density at 50 °C is 0.89. The solid fat content is 60–61 per cent at 5 °C and this decreases to 4 per cent at 40 °C (Tan and Flingoh, 1981). The smoke, flash and fire points (see p. 674) of palm oil or palm kernel oil have not been published, but the values applicable to fats in which C16 and C18 fatty acids predominate (e.g. palm oil, soya bean oil and sunflower oil) have been given (PORIM, 1987) as follows:

Oil with	5% f.f.a.	1% f.f.a.	0.1% f.f.a. (refined oil)
Smoke point	275	320	390
Flash point	500	560	600
Fire point	625	660	670

Palm kernel oil

Palm kernel oil resembles coconut oil with which it is readily interchangeable. Both fats have a preponderance of saturated fatty acids, but palm kernel oil has a lower quantity of the low molecular weight acids, caprylic and capric. The usual ranges of fatty acid constituents, per cent, are as follows:

	Saturated						Unsaturated	
	Caprylic C8	Capric C10	Lauric C12	Myristic C14	Palmitic C16	Stearic C18	Oleic C18:1	Linoleic C18:2
Range	3–4	3–7	45–52	14–19	6–10	1–3.5	11–19	0.5–2
Typical sample	3.0	6.0	50.0	16.0	6.5	1.0	16.5	1.0

Traces of the saturated caproic and arachidic acids, C6 and C20, and of the unsaturated palmitoleic and linolenic acids, C16:1 and C18:3, are also found (Bezard 1971).

With such a high proportion of saturated acids it is not surprising that saturated triglycerides constitute over 60 per cent and monooleic disaturated triglycerides more than 25 per cent of the total glycerides. The iodine value is about 17. The melting range of palm kernel oil is 27–30 °C (Cornelius, 1977b) and the mean slip melting point 27.3 °C (Siew and Berger, 1981). The relative density is 0.90–0.91.

Oils of *Elaeis oleifera*

The mesocarp oil of the American oil palm, *Elaeis oleifera*, has an iodine value of 78–88, much higher than that of the oil of *E. guineensis*, showing that it contains higher percentages of oleic and linoleic acids and lower percentages of palmitic and other saturated acids. Similarly, the kernel oil of *E. oleifera* has an iodine value of 25–32, indicating that saturated acids do not predominate to such an extent as in palm kernel oil.

Table 14.3 shows analyses of the mesocarp and kernel oils of *E. oleifera* and of its hybrids with *E. guineensis*. The unsaturated fatty acid content of the mesocarp oil of the hybrid is intermediate between those of the parent species. It has been postulated that if the hybrid is back-crossed with either of the parents then half the progeny will have hybrid-type fatty acid analysis and half will have the analysis of the parent; support for this was obtained from two back-cross analyses on Malaysian material which showed both a hybrid-type and an *E. guineensis*-type analysis (Macfarlane *et al.*, 1975).

The fatty acid content of the hybrid kernel oil approximates more to that of the *E. guineensis* parent.

Palm kernel cake

Palm kernel cake contains around 18–19 per cent of protein and is thus the lowest of the oil cakes in protein value. The amino acid composition has been given by Babatunde *et al.* (1975) and shows high proportions of arginine and glutamic acid. An average analysis is as follows:

	Carbohydrates	Oil	Proteins	Fibre	Ash	Water
Per cent	48	5	19	13	4	11

The protein content of *E. oleifera* and hybrid meals is lower than that of *E. guineensis* though the amino acid composition is similar (Quraishi and Macfarlane, 1975).

The quality of palm oil and kernels

Pritchard (1969) has indicated the user quality requirements of palm oil products which the producer, or more particularly the exporter, must aim to satisfy.

Palm Oil

A poor quality palm oil may be defined as having any or all of the following characteristics: (1) a high f.f.a. content; (2) contamination with water and other impurities; (3) poor bleachability. High water

Table 14.3 *Fatty acid composition of the oils of* Elaeis oleifera *and of the hybrid* E. guineensis × E. oleifera *(per cent)*

	Caproic C6	Caprylic C8	Capric C10	Lauric C12	Myristic C14	Palmitic C16	Stearic C18	Palmitoleic C16:1	Oleic C18:1	Linoleic C18:2	Linolenic C18:3	Iodine value
Mesocarp oil												
E. oleifera												
(1)*				0.1	0.3	25.0	1.2	1.4	68.6	2.1	0.9	81.3
(2)				0.5	0.2	22.5	0.6	2.4	55.4	18.0	0.4	81.8
(3)				—	tr.	22.9	1.0	1.3	54.8	20.0	—	81.5
(4)				tr.	1.2	24.6	0.6	1.0	56.6	15.9	0.1	—
Hybrid, E. oleifera × E. guineensis												
(1H)*				0.1	0.5	31.7	4.1	0.1	49.5	13.4	0.5	70.7
(2H)*				0.1	0.9	32.5	3.4	0.2	48.0	13.8	0.4	62.0
(3H)					0.8	27.3	6.1	0.5	52.5	11.4	1.3	69.8
(4H)				tr.	0.6	33.4	3.4	—	51.8	10.9	—	63.0
(5H)					0.4	35.8	1.4	—	55.0	7.5	—	63.0
(6H)					0.7	30.8	3.6	—	51.8	13.1	—	67.0
(7H)				tr.	0.2	31.5	3.6	0.3	55.1	9.2	0.1	—
Kernel oil												
E. oleifera												
(1K)	2.9	0.9	0.8	29.4	23.8	9.9	2.1	—	25.1	5.1	—	30.3
(2K)	0.3	0.4	0.5	25.1	27.8	12.7	2.2	—	24.9	6.1	—	31.9
Hybrid, E. oleifera × E. guineensis												
(1HK)	tr.	2.8	3.0	48.9	17.7	8.2	1.7	—	16.0	1.6	—	16.6
(2HK)	tr.	1.8	2.1	43.1	21.3	9.4	2.2	—	16.3	3.7	—	20.4

* Traces also of arachidic acid, C20.

Sources: Colombia – (1), (3H) Hardon (1969); (6H), (2K), (2HK) Macfarlane *et al.* (1975); Nigeria – (2), (4H), (1K), (1HK) Macfarlane *et al.* (1975); Zaire – (1H) Hardon (1969); Malaysia – (2H) Hardon (1969); (5H) Macfarlane *et al.* (1975); France (Source of material not stated) – (4), (7H) Naudet and Faulkner (1975).

Note: (4H), (5H), (1HK) and (2HK) are means of determinations of *E. oleifera* × *E. guineensis, dura, tenera* and *pisifera.*

and dirt contents are contributory causes of high f.f.a., and factors causing high f.f.a. may also contribute to poor bleachability.

The formation of free fatty acid

The fatty acids of palm oil may be formed by autocatalytic action, by the action of the lipolytic enzyme lipase from the palm fruit, or by microbial lipases. In practice the main 'set-up' of f.f.a. is caused by the action of lipase before processing.

Palm fruit contain a very active lipase which rapidly effects the breakdown of the fats into fatty acids and glycerol when the cellular structure of the fruit has been disturbed. The lipase is inactivated by high temperatures but is active even when the temperature has been reduced below 15 °C. To a small degree degradation of the oil takes place in the intact fruit (Jacobsberg, 1983), though the bulk of it is believed to be protected from the lipase by the membranes of the vacuoles (Loncin and Jacobsberg, 1965).

Oil from fresh ripe fruit contains as little as 0.1 per cent fatty acid (estimated as palmitic acid), but in bruised and crushed fruit the f.f.a. will increase very rapidly. Desassis (1957) showed that a temperature of 55 °C was sufficient to prevent enzyme action, although at ambient temperatures oil from crushed mesocarp can attain over 30 per cent f.f.a. in 5 minutes (Fig. 14.1). With careful harvesting and carriage, however, the bruising of fruit need not be

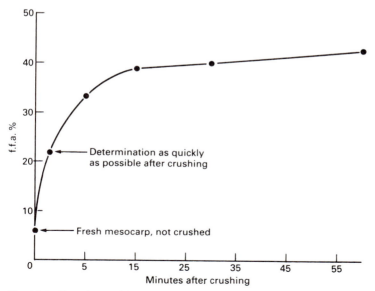

Fig. 14.1 Free fatty acid set up in mesocarp following crushing (determinations by Van Heurn, quoted by Desassis, 1957).

very extensive and on well-regulated estates low f.f.a. oil can often be obtained from bunches kept overnight.

Fruit which has been kept for several days before processing, or which has been allowed to become overripe on the palms, may be covered and invaded by a number of moulds. Usually these fungi invade the base of detached fruit or wounds on the fruit surface. Coursey (1963) identified lipolytic species of *Rhizopus, Aspergillus, Penicillium, Trichoderma, Circinella, Cunninghamella, Fusarium* and *Phoma* in Nigeria while in Malaysia Turner (1969) drew attention to the invasion of fruit by *Marasmius palmivorus* and species of *Sclerotium, Diplodia* and *Glomerella* before the bunch is removed from the palm, and to invasion of loose fruit by seventeen different fungal species, fourteen of which showed lipolytic activity, together with yeasts, bacteria and nematodes. Of the fungi, *Aspergillus* species were particularly abundant. In both countries, however, oil degradation through these agents is likely to be very slight in comparison with degradation originating in the fruit itself.

Turner (1969) also examined the possibilities of microbial degradation of oil during processing and showed that lipolytic fungi can be found in all parts of a mill both in the atmosphere and on the floors, walls and apparatus. Sterilized fruit if left for more than 24 hours before further processing is invaded by micro-organisms, particularly *Neurospora sitophila* which grows rapidly over both bunches and fruit. However, temperatures are normally so high at all stages of processing that the potentially lipolytic organisms have little chance of significant activity except when a forced close-down of a mill takes place.

From the above it will be clear that for the *production* of low f.f.a. oil the major requirements are: (i) minimal bruising of the fruit during harvesting, carriage and movement at the mill side; and (ii) minimal time between harvesting and sterilization. To these may be added (iii) the processing system must be such that the fruit or extracted oil does not cool down and come into contact with apparatus or materials which could cause a recommencement of lipolysis.

After enzyme destruction and when the oil is in *storage* some f.f.a. formation can still take place through autocatalytic hydrolysis (Vanneck and Loncin, 1951; Loncin, 1952; Loncin and Jacobsberg, 1963). This hydrolysis stops almost completely when the moisture concentration is kept below 0.1 per cent, but above that level water concentration has little influence except at high temperatures when higher water concentrations increase the rate of f.f.a. formation (Vanneck and Loncin, 1951). Some residual moisture has been shown to give protection against oxidation (Berger, 1983).

There has been increasing evidence of the action of lipolytic micro-organisms in stored oil. Loncin and Jacobsberg (1965) showed that *Geotrichium candidum*, which was found in many mills in Zaire, was responsible for lipolysis, and Coursey (1963) demonstrated increased hydrolysis brought about by infection with other lipolytic micro-organisms. Examination of samples of oil in Nigeria showed that the more important identified lipolytic fungi were species of *Paecilomyces, Aspergillus, Rhizopus* and *Torula*. Inoculation experiments showed that *additional* rises of f.f.a. of the order of 1–3 per cent could be induced during 8 weeks' storage. The sources of infection are palm fruit, bunch refuse, oily films on drums and other receptacles. Rises in f.f.a. due to lipolytic fungi are likely to occur wherever oil is produced under generally dirty conditions and, with drum storage, where the means of drum sterilization are inadequate. With bulk tank storage at high temperatures the biochemical factor is likely to be much less important.

To provide an oil which will maintain a low f.f.a. content, the final milling operations must therefore be directed firstly to attaining a water content of less than 0.1 per cent, secondly, the dirt content must be reduced to a minimum and clean and sterile conditions must be maintained in order to avoid the invasion of the oil by lipolytic micro-organisms.

The bleachability of palm oil

Although some of the materials for which palm oil is used, e.g. margarine, are coloured, in manufacturing practice oils must be bleached to definite specifications for the various uses to which they are put. The composition, soft texture and plastic range of palm oil make it satisfactory for blending in fair quantities in margarine and shortening fats and for use in commercial baking and biscuit manufactures. But the oil must first be transformed into a bland, colourless, stable, edible product, and colours in the range 5.0 red to less than 1.0 red with the $5\frac{1}{4}$ inch (13.3 cm) Lovibond cell are needed (Jasperson and Pritchard, 1965). Any difficulty in bleaching palm oil and any additional expenditure involved must, therefore, militate against its use in manufacture.

Manufacturers found that the bleachability of oil received from plantations in the Far East was superior to oil received from the west coast of Africa, but later it became apparent that easily bleached oil could also be obtained from plantations in Zaire and other parts of Africa. It was at first thought that the carotene content of the oil would be the primary factor in bleachability, but later there was evidence that oxidation of the oil and of the carotenoids was of more importance than the absolute carotene content.

Oil from the Deli palm estates of Malaysia and Indonesia is generally found to have a fairly uniform carotene content of around 500 ppm. Wide variations exist in Africa, however, both in carotene contents and bleachability. One of the first surveys carried out of oils from fruit and from small-scale and industrial extraction plants gave some remarkably diverse figures (Bienaymé, 1954). Certain fruits in the extreme west of Africa were found to contain over 3,000 ppm carotene while Deli palms growing in the same region gave, as expected, oil with only 400–600 ppm. Some African fruits, however, also gave low carotene contents and mills provided oils varying in carotene content from 600 to 1,600 ppm. Of particular interest was the discovery that oil from the residual fibre was much higher in carotene content than oil as normally expressed, suggesting that the oil first squeezed out from digested fruit is unlikely to show the full carotene content of all the available oil. Subsequent work showed that hand-squeezed oil has a lower carotene content than press-extracted oil. It is thought that the exocarp fibres, which lie in that part of the fruit having the highest pigmentation, are less completely separated in the course of pounding or digestion than those of the remainder of the pulp (i.e. the true mesocarp).

In a survey of Nigerian oils expressed in a laboratory hand press from fruit samples in different producing areas, carotene contents from 400 to 1,800 ppm were obtained (Nwanze, 1961–4). All samples with carotene contents below 1,200 ppm gave a low residual colour after bleaching with 5 per cent Fuller's earth at 105 °C for 1 hour. Progeny differences in carotene content were also found to exist and it is clear that where less colour and superior bleachability are claimed in Africa for 'plantation' oils as against 'grove' oils extracted by the same processes, such differences are due to the chance selection for plantation use of grove material with a lower than average carotene value.

While, therefore, it is recognized that Deli palm estates produce oils of both low carotene content and good bleachability and certain Zaire estates produce oils with almost similar properties, it is not thought that, in general, carotene content is a major cause of poor bleachability, and investigations have been directed to handling and processing methods (Hartley and Nwanze, 1965). Ames *et al.*, (1960) suggested that poor bleachability of West African oil, which was largely produced by small-scale processing methods, was due to: (i) oxidation of lipoxidases in bruised fruit stored for various periods before processing; (ii) oxidation, catalysed by iron, during processing and bulking; and (iii) mixing of oils of good and poor bleachability. It was also suggested that high carotene oils are more likely to deteriorate than oils of low carotene content.

Fruit storage for a few days was common in West Africa but provided care is taken no rise in peroxide values or deterioration of the bleachability of the oil occurs (Nwanze, 1961–4). If fermentation is allowed to take place, as in one of the traditional extraction processes, bleachability will be affected, since oxidation of the unsaturated fatty acids gives rise to compounds responsible for colour fixation in fats. However, these oils, which are also characterized by a high f.f.a. content, do not normally enter into international trade and fruit storage cannot therefore be considered as an important cause of poor bleachability.

In a well-regulated mill of standard design there should be very little oxidation. However, some oxidation will occur if there is too prolonged heating at any stage in the presence of air, or if copper and its alloys, or to a lesser extent iron, are allowed to play a catalytic role (Abdul Gapor and Ong, 1982). Berger (1983) states that unless stainless steel is used in all those parts of the mill subject to attrition, the crude palm oil produced may contain 5 ppm of iron and this tends to be concentrated in the stearin fraction, reaching as much as 20 ppm. Oils with more than 5 ppm iron are more difficult to bleach. Copper parts must be entirely eliminated from the mill.

The manner in which mill processes are operated has also been shown to affect bleachability (Jacobsberg, 1971, 1983). In the sterilization process both failure to eliminate air and the application of too high a pressure (see p. 712) impair bleachability. Digestion with steam injection and clarification with centrifugal separation help to maintain the bleachability of the oil.

Poor handling and storage of oil are nowadays likely to be the main causes of poor bleachability. In village processes metallic containers and drums, and exposure to air, are the chief culprits. At bulk oil establishments the steaming out of drums for reuse is rarely fully effective and drum residues have been found to have high peroxide values and be unbleachable.

Lastly, Jacobsberg *et al.* (1978) have shown the importance of maintaining the initial tocopherol content of the oil throughout extraction and processing, since the initial content appears to represent the optimum antioxidant concentration; while Abdul Gapor and Ong (1982) have shown the additional protection against oxidation that can be had by use of the antioxidant TBHQ in combination with citric acid.

Quality standards

Palm oil is traditionally bought on a 5 per cent f.f.a. basis by importing countries, with penalties for exceeding this figure. To keep within these standards producers must achieve an oil of about 3.5 per cent f.f.a. and many mills in the Far East and elsewhere have been satisfied with the production of oils with f.f.a. ranging from 2.5 to 4.0 per cent. In Nigeria, from whence the bulk of the exported oil produced by small operators was previously derived, the internal requirement for edible oil purchased by marketing authorities was 3.5 per cent of f.f.a. Oil with higher f.f.a. contents was bought at substantial discounts and this policy, together with the production of relatively low f.f.a. oil by Pioneer mills, was responsible for the remarkable improvement in the quality of oil exported from Nigeria during the 1950s and 1960s.

Water and dirt levels have been variable. In Nigeria the small producer was permitted to sell oil containing as much as 0.4 per cent water and, in the non-sterile conditions obtaining, this was responsible for later rises in f.f.a. and perhaps in reduced bleachability. Most mills in the Far East, Zaire and elsewhere achieve moisture contents of 0.1 per cent or below, though in order to obtain some antioxidant effect Berger (1983) suggests storage at 0.15 per cent. Dirt contents of 0.1 per cent were permitted in Nigeria, but in the Far East are often as low as 0.005 per cent.

Arnott (1963) categorized palm oil quality as follows:

Factor	Very low	Low	Medium	High	Very high
f.f.a.	<2.0	2.0–2.7	2.8–3.7	3.8–5.0	>5.0
Moisture	<0.1	0.1–0.19	0.2–0.39	0.4–0.6	>0.6
Dirt	<0.005	0.005–0.01	0.011–0.025	0.026–0.05	>0.05

Certain large producers have for many years been marketing oil of very low f.f.a. content, sometimes as low as 1.5 per cent and usually around or under 2 per cent. Investigation of the chemistry and biochemistry of palm oil extraction led to the production of a special high quality oil in Zaire (Jacobsberg, 1971). The characteristics of this special prime bleach (SPB) in comparison with what may be termed 'ordinary' plantation oil are given in Table 14.4.

Table 14.4 *Characteristics of SPB and ordinary plantation oils in Zaire (Loncin and Jacobsberg, 1965)*

	SPB	Ordinary
f.f.a., as palmitic acid (%)	1–2	3–5
Moisture (%)	<0.1	>0.1
Dirt (%)	<0.002	0.01
Iron (ppm)	<10	>10
Copper (ppm)	<0.5	>0.5
Iodine value	53 ± 1.5	45–56
Carotene (ppm)	500	500–700
Tocopherol (ppm)	800	400–600
Bleachability (Bleaching Standard: Lovibond 5¼ in) (13.3 cm)	2.0R, 20Y	3.5R, 35Y

While commending these standards, there is an inherent danger in producing oil of around 1.5 per cent f.f.a. which should not go unnoticed. The handling of fruit before milling, and in particular the loose fruit, plays an important part in the f.f.a. 'set-up' achieved. Attention has been drawn in this chapter to the very rapid increase in oil content of the mesocarp during ripening, and the relation of harvesting criteria to oil yield and f.f.a. set up has been discussed in Chapter 10. If a very low f.f.a. oil is insisted upon, managers and harvesters may be forced into the cutting of underripe bunches instead of concentrating on the careful handling of fully ripe bunches, and the loss of production may then be far greater than any advantage to be gained by a 1 per cent lower f.f.a. content. This tendency has, indeed, often been noted in the field.

As poor bleachability may be caused by the formation of oxidation products or by a decrease in the natural antioxidants or by the presence of oxidation promoters, it has been difficult to provide a standard bleachability test. The *peroxide value* does not account for the full extent of oxidation (Jacobsberg, 1983). The *Totox value* (2 × peroxide value + anisidine) is now frequently used but still does not account for the full effect of pro- and antioxidants. Various direct bleaching tests have been suggested, and these and other standards have been summarized by Cornelius (1973, 1977b).

This book is not concerned with the treatment of palm oil and kernels beyond the plantation mill, but it should be mentioned here that in the Far East increasing quantities of these products are now directly exported in more refined forms (Anon., 1986). The bulk of the oil palm products exported from Malaysia are now semi or fully refined and leave the country as RBD palm oil (refined, bleached and deodorized), RBD palm olein, RBD palm stearin, RBD palm kernel oil, crude palm olein and crude palm stearin. The storage, handling and transportation of these products have been outlined by Leong and Berger (1982) and Berger (1983), and the standardization of Malaysian palm oil itself has been described by Chin (1979). Bek-Nielsen and Krishnan (1979) have described the beginnings of the palm oil refining industry which has grown up in Malaysia.

Palm kernels, palm kernel oil and palm kernel cake

The quality of palm kernel oil and cake depends primarily on the quality of the kernels. Palm kernel oil is required to be of a low f.f.a. content and a light yellow colour, easily bleached, while palm kernel cake should be relatively light-coloured and its nutritive value, particularly the constituent amino acids, must not be impaired.

To obtain these high-quality products palm kernels must themselves be judged mainly on the f.f.a. content of the oil they contain, on the amount of external and internal discoloration, and on factors such as moisture content and presence or absence of mould which are likely to affect the eventual f.f.a. content and colour. An average sample of palm kernels in a producing country has a composition as follows:

	Oil	Moisture	Protein	Extractable non-nitrogen	Cellulose	Ash
Per cent	47–52	6–8	7.5–9.0	23–24	5	2

There is some variation in the composition of the non-oily and non-protein solids. The portion given as 'extractable non-nitrogen' has been found to contain variable quantities of sucrose, reducing sugar and starch, but in some samples no starch has been detected though mannose has been present. As to their moisture content, kernels eventually come to an equilibrium according to the prevailing relative humidity (Cornelius, 1965). The latter reaching an average of at least 80 per cent in producing countries, the equilibrium moisture content is around 7 per cent, though in very wet periods this might rise to 8 or 9 per cent. It is believed that a safe moisture content for seed storage is 14 per cent of the non-oily portion of the seed, and it would therefore appear that the equilibrium moisture content is on the borderline of what has been termed 'safe'. Therefore, until the moisture has been reduced to its equilibrium level kernels are liable to reactions for which moisture is needed.

These reactions are similar to those of palm oil, namely autocatalytic hydrolysis and lipolysis by fat-splitting enzymes in the kernels and from lipolytic moulds. At ordinary tropical temperatures the more important of the latter are four species of *Aspergillus* and species of *Paecilomyces*, *Syncephalastrum* and *Penicillium* identified in Nigeria (Coursy *et al.*, 1963), and *Rhizopus cohnii* and a lypolytic yeast, *Petasospora rhodanensis*, found in Zaire (Loncin and Jacobsberg, 1964). In stacks of palm kernels which heat up to 50–60 °C a different fungal flora, which is thermophilic, develops.The most prominent species are a *Thermomyces, Chaetomium thermophile*, and a *Penicillium*, probably *P. duponti*. All are actively lipolytic and their optimum temperature for growth lies between 42 and 52 °C. Only *C. thermophile* grows below 30 °C (Eggins and Coursey, 1964; Oso, 1979). Turner (1969) has detected lipolytic micro-organisms at various stages in the milling process but particularly in hydrocyclones and clay baths during shell and kernel separation. Disturbing features of infection with these organisms are their ready carriage by insects and the continued activity of their lipases even when the moisture content of the kernels has been brought well below the equilibrium value after several hours' heating at 100 °C. Lypolysis may be reduced by sterilization and a reduction in the number of cracked and broken kernels, but overheating produces discoloration of the kernels.

For the provision of kernels giving a low f.f.a. oil, therefore, low moisture contents and low breakage are required. This was well illustrated in Sumatra for whole and broken kernels of average and high moisture content and the results obtained are shown in Fig. 14.2 (Stork, 1963). Even though precautions may be taken to reduce f.f.a. production by proper drying and minimum breakage, there still remains, however, the problem of the heat-resistant enzymes already mentioned. In Zaire kernels are found to have in their oil an f.f.a. content, as lauric acid, of 0.5 per cent after extraction from the nut and this rises to 1.0–1.5 per cent after drying (Loncin and Jacobsberg, 1965). During storage the f.f.a. may rise to any point between 2 and 10 per cent and this has been shown to be due to the variable action of microbial lipases, undestroyed by drying or high temperature. Free fatty acid contents of 3–7 per cent or more are common in the oil of kernels arriving in importing countries, and it is not possible to prevent this except by preventing the development of micro-organisms before and during storage. This has been effected by successive injections of steam and cold air, thus sterilizing and drying the surfaces of the kernels before introducing them into the drier. By this method the f.f.a. of the oil

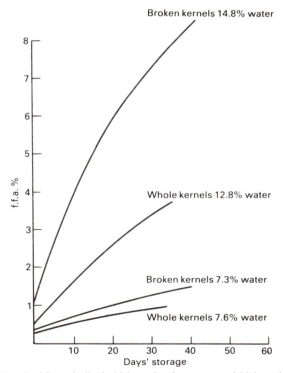

Fig. 14.2 Rises in f.f.a. of oil of old kernels of average and high moisture content stored at about 28 °C (Stork, 1963).

in the kernels has been maintained below 1.5 per cent for 6 months of storage.

In Malaysia steam sterilization at temperatures exceeding 90 °C for 5–6 minutes has been shown to inhibit rises in f.f.a. and to give rise to kernels which are stable during storage (Bek-Nielsen, 1969a; Clegg and Teh, 1972).

It has long been realized that the browning of the endosperm and darkening of the testa of palm kernels extracted in mills fitted with steam-pressure sterilizers were to some degree inevitable. Further discoloration in the digester is normally slight and in the nut bin and kernel drier negligible (Thieme and Olie, 1969). Heating, with consequent discoloration, can also take place in heaps of uncracked nuts (Coursey, 1961) and in kernels stored under unsuitable conditions. Kernel browning affects the colour of the cake as well as of the oil.

The chemical reasons for browning are not fully understood but are likely to be: (i) the reaction of free amino groups of proteins with aldoses to give brown polymers and co-polymers; and (ii) the reaction of mannose (among the carbohydrates) with amino groups to form brown compounds (Cornelius, 1965).

The proportions of white, discoloured and broken kernels, and of shell, in samples of commercial palm kernels received in the United Kingdom, are shown in Table 14.5. These figures give a general idea of the extent of the browning problem. The moisture content of the samples varied very little, being between 4.3 and 5.9 per cent. Oil contents lay between 47 and 54 per cent, leaving broken kernels out of account.

Table 14.5 *Colour of samples of palm kernels received in the United Kingdom (percentage by weight)*

Sample	White	Off-white	Brown	Brown and mouldy	Broken	Shell
Malaysia	62.6	12.7	11.7	4.7	7.3	1.0
Ghana	31.8	24.7	24.0	4.8	12.6	2.1
Sierra Leone	65.1	13.1	9.3	6.2	4.7	1.6
Nigeria (1)	69.4	11.1	4.9	4.9	6.7	3.0
(2)	54.7	22.0	8.3	7.1	5.5	2.4
(3)	62.2	13.0	9.9	5.5	6.8	2.6

Source: Cornelius (1965).

Non-mechanical traditional methods of extraction of palm oil and kernels

The satisfactory extraction of palm oil requires specially designed machinery, whether hand or machine operated, and the provision of ancillary equipment of correctly calculated capacity for the prior preparation of the fruit and for the subsequent preparation of the products for sale.

Before the advent of machinery, palm oil was extracted in Africa by crude means to give a product of generally poor quality; and kernels were extracted from the nuts one by one by hand cracking. These processes still continue to be used in many parts of Africa, but only the kernels still enter world trade. Though they vary considerably in their details, the essentials of these processes are briefly described below (Gray, 1922; Faulkner and Lewin, 1923; Manlove and Watson, 1931).

'Soft oil' production

In the great areas of eastern Nigeria from which the bulk of the early export supplies were brought, the process employed gave rise to the so-called 'soft oil', thus named because the greater part of the oil was liquid at tropical temperatures. The harvested bunches are cut up into sections and kept in heaps for 2–4 days. The heaps are sprinkled with water and covered with leaves.

1. *Boiling.*   The fruit is picked from the bunch sections and boiled in large pots for about 4 hours; 44 gallon (200 l) petrol drums are commonly used. The fruit may be left in the vessel for up to 3 days.

2. *Pounding.*   The boiled fruit is pounded in a wooden mortar with a wooden pestle until a mixture of nuts and crushed pulp of more or less even consistency is obtained.

3. *Separation.*   The oil is separated from this mass of pulp by immersing the latter in water. The initial stage may be carried out either in a special pit with its sides coated with cement or mud, or in some large vessel. The whole mass is stirred and first the crude oil which has risen to the surface is skimmed off into another vessel, then the fibre is sifted out of the water and finally the nuts, now largely free of fibre, are picked out and laid out to dry. The crude oil thus obtained is boiled in smaller vessels where any fibre it contains sinks to the bottom. The now purer oil is again skimmed off and is then 'fried' in a shallow pot to get rid of the last traces of water. When this condition is reached, drops of water sprinkled on the oil will rapidly evaporate with a crackling noise – hence the use of the term 'frying'.

The amount of oil extracted depends largely on how far heat has been maintained throughout the process and how assiduous the women are in skimming and in teasing out the fibre. Usually the mass of pounded pulp is allowed to cool off and extraction rates are low. Such figures as exist for extraction rates in the soft oil process, i.e. 6–10 per cent oil to fruit of low mesocarp content, suggest that a normal efficiency (extracted oil to total oil in the fruit) would be 40–45 per cent, occasionally rising to 50 per cent, but sometimes falling to as low as 30 per cent. Average f.f.a. content used to be about 7–12 per cent, but lower f.f.a. oils can be produced by this method.

'Hard oil' production

In parts of Africa the people are not accustomed to pounding, and a less efficient method, in which the fruit is trodden, is the rule. This is the typical method of the people of the Niger delta. The sequence of events is:

1. *Fermentation.*   After hand picking from the stored, chopped-up bunches the fruit is placed in a pit or in a long wooden canoe and covered over with leaves. Fermentation caused by microbial and enzyme action takes place with the generation of heat and the fruit thus becomes softened.

2. *Treading and separation.*   After some days the fruit is in a condition where it can be vigorously trodden in a canoe. After the first treading the oil is allowed to drain for 3 days from the lower end of the canoe. Water is then added and a second treading is done. When the work is considered to have proceeded far enough and the remaining separated oil has risen to the surface it is skimmed off and boiled in vessels for final preparation as in the soft oil process, though usually with less care. The oil thus prepared solidifies rapidly. Considerable hydrolysis and oxidation take place during the process and the proportion of low-melting-point constituents is reduced. Extraction rates are very low, about 4–6 per cent oil to low mesocarp *dura* fruit, with a correspondingly low efficiency of 20–30 per cent; and the f.f.a. content, owing to the high initial set-up during fermentation, is usually between 30 and 50 per cent.

Hand-operated presses

Before the First World War two hand-operated machines were designed in which the fruit was placed in cylinders with hot water and submitted to the action of beaters, the oil and water subsequently being run off through a sieve. The 'Gwira' machine was the better known and was given serious trial on the west coast, but never made much headway (Anon., 1917). After 1918 attempts were made to introduce pressing into the soft oil process which already provided for fruit boiling and mashing in a mortar. The process was for some time known as the cooker-press system (Barnes, 1926) because special steam or water cookers were introduced. These never became popular, and boiling in a 44 gallon (200 l) drum became the general practice. 'Depericarping' before pressing was at first thought to be necessary and various types of hand depericarpers were tried in combination with a 'Cully-Ducolson' hand press of the 'bridge' type which consisted of a cylindrical steel press-cage and vertical ram screwed down manually from the bridge. A hand-operated centrifuge was also tried (Barnes, 1925), and fair numbers supplied. The system included a steam-pressure fruit sterilizer, a hand-operated digester and the hand-operated centrifuge itself.

The curb press

The press which was eventually found most satisfactory and captured the market was a curb type of press similar to those used for the extraction of juice from soft pulp fruit in the wine and cider industries. The first machines to be used were modified wine presses of the type developed by André Duchscher, founder of Duchscher & Cie of Luxembourg (later Usine De Wecker), but others were later produced by a number of firms and known by various trade names. Adapted for use as oil palm extractors they gave efficiencies of 55–65 per cent with a maximum efficiency of around 70 per cent which was achieved only under very favourable conditions of fruit digestion. These machines were tested both in Nigeria and Malaysia and gradually came to be widely used in the eastern part of Nigeria where soft-oil production had been the rule. It is believed that there were at one time as many as 10,000 operating in eastern Nigeria.

The press consists of a screwed steel shaft, fixed in the centre of a base plate, and a cage, composed of strips of stout wood set vertically about 3 mm apart, and looped externally with two iron bands (Fig. 14.3). The cage, which is in two halves, can be opened and lifted off so that the pressed fibre and nut mixture can be easily removed. Pressure is applied by a ram which is worked downwards on the shaft by a crosshead turned manually by two long iron bars on which two to four men work. The bars, which are detachable, are fitted into the screw crosshead which lies above the ram. The base of the press is wooden and is surrounded by a metal trough fitted with one spout. The oil is squeezed out of the digested material between the wooden strips into the trough and runs through the spout into any collecting vessel. The press is often known as the cage or screw press. The first term is unobjectionable, but the latter may cause confusion with other types of screw press.

In a trial in Malaysia, Deli fruit was boiled in a 40-gallon (181 litres) drum for 5–6 hours and immediately pounded with a wooden

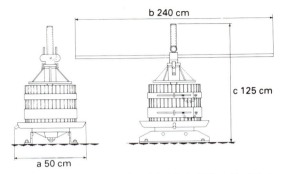

Fig. 14.3   The Duchscher curb press (model NG 1, Usine De Wecker).

pestle in smaller drums for 10–15 minutes and transferred hot to the press (Milsum and Georgi, 1938). A recovery of 72 per cent of the oil contained in the fruit was obtained by double pressing which involved removing the whole charge after the first pressing, separating the nuts and repressing the fibre from the first charges; this fibre had been kept warm and partially dried by placing it on iron sheets over a small fire. Of the 72 per cent oil extracted about 5 per cent was attributable to the second pressing; a maximum of 67 per cent efficiency can therefore be obtained from one pressing. In Africa double pressing is done by adding fresh material to a charge already pressed. Geddes (1976) modified the curb press in Sierra Leone to enable higher pressures to be applied and claimed that 80 per cent of the available oil could be extracted.

The average rate of production of a curb press varies considerably as it is usually operated on a cooperative or family basis, sometimes with employed labour, and production depends on availability of fruit and many other factors. Trials in Malaysia suggested that seven charges could be handled in a day of 10 hours, by two men doing all the ancillary work (Milsum and Georgi, 1938). The quantity of fruit treated would be around 450 kg, equivalent to 770 kg of bunches.

The average charge of the Duchscher NG 1 press is about 70 kg fruit, equivalent to about 115 kg bunches.

Hydraulic hand presses

While the use of the curb press was a substantial advance on traditional methods of extraction and it was soon shown that curb-press operators were able to produce low f.f.a. oil, the loss of extractable oil was still considerable and it was therefore not surprising that sooner or later a hand-operated press of much greater efficiency would come on the market. In 1959 the Dutch firm of Gebr. Stork began to manufacture, and to test in cooperation with the West African Institute for Oil Palm Research (now NIFOR), a hydraulic hand press having a cage capacity of 45–55 kg fruit, equivalent to some 90 kg of bunches, and capable of a higher extraction efficiency than the curb press. The extraction of the crude oil from a single charge was rapid and the problems of its introduction were mainly those of providing sufficient labour and suitable inexpensive ancillary apparatus to keep pace with its demands.

The press is shown in Plate 14.1, and its development has been described by Nwanze (1965a). A ram, which is the cylinder of the

Pl. 14.1 The Stork hydraulic hand press.

hydraulic mechanism, moves downward into a perforated press cage when hydraulic fluid pressure is increased by the hand operation of a two-piston pump; both pistons are operated until considerable force is required and thereafter the small piston only is operated until full pressure is reached. On the release of pressure the press ram is withdrawn upwards by springs.

The cages are filled and emptied on a table in front of the press ram and with skilled operation one pressing, including the insertion and withdrawal of charges, takes 6–10 minutes. As two cages are provided, and one can be emptied and filled with hot digested material during the pressing of the charge in the other, it is possible to complete six to ten pressings per hour and thus to press 270–450 kg fruit per hour, equivalent to 0.45–0.75 tonne bunches per hour as against an equivalent amount per day with the curb press.

The cages are surrounded by covers or oil guards while pressing is proceeding. The crude oil, which is heavily laden with sludge, is channelled into a spout and runs off into buckets. As there is seepage of oil on to the press table this should be perforated with weep holes and be fitted underneath with an oil catch. Nwanze (1965a) showed that the efficiency of the press alone, which must not be confused with full process efficiency, can exceed 95 per cent. The efficiency of this machine led to the construction of hydraulic hand presses by a number of other manufacturers. In some cases pressure was exerted from below, but the mechanism was generally similar. Later, small motors were used by some operators to activate the press, and presses of this type with motor drive can now be obtained.

Most early operators of the hydraulic hand press were surprised at its rate of operation and were unable to supply sufficient equipment and men to deal with the amount of fruit required by the press.

However, Nwanze (1965a) designed for West Africa cheap ancillary equipment which can be made locally, and this, with the quantities needed, is described below.

Sterilization

Bunch sterilizers for use with hand presses were designed to hold about 1 tonne of quartered bunches. These are locally constructed from cut-up oil drums welded together to form a cylindrical vessel 2.5 m tall by 1.1 m diameter and fitted below with two boiler compartments surrounding a fire space. Two or three steam pipes rise from the boiler compartments to distribute the heat. Where the traditional method is followed of picking fruit from the bunches after cutting up and heaping for 2–4 days, then a fruit sterilizer can be used. This consists of a 44-gallon (200 litres) sterilizing drum set

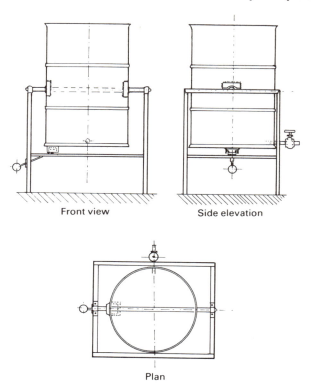

Front view   Side elevation

Plan

Fig. 14.4 Fruit sterilizer for use with the hydraulic hand press.

in a special tipping frame (Fig. 14.4). If sufficient loose fruit is being supplied to keep the press in operation during the daytime then about four drums will be needed. Fruit boiling is carried out for 1 hour after steam appears at the top of the drum.

Stripping

Stripping in a hand-operated mill can be by beating out the fruit from the sterilized bunches by means of wooden batons or pronged forks, or by the use of a hand-turned octagonal slatted drum. The latter has proved very efficient (Plate 14.2).

Digestion

Digestion is by pounding stripped fruit in a mortar. As the fruit will get cold while waiting its turn for pounding two *fruit reheating drums* of similar design to the bunch sterilizers are used for storing the stripped fruit until ready for pounding. The pounding is best done in a large concrete mortar of 90 cm diameter and 50 cm deep;

Pl. 14.2   Bunch stripper or thresher for the hand press mill.

six to eight men using wooden pestles pound the fruit until a 'mash' of uniform consistency is achieved (Plate 14.3).

The pounded material is then placed in special *pounded-fruit reheating drums* which are fed at the top, and from which the material can be continuously removed through a door near the bottom for transfer to the press. This drum has a water compartment with perforations for steam to rise into the drum. At least two preheating drums are needed.

Clarification

With occasional operation of the press, 44-gallon (200 l) drums may be used for clarification. For continuous milling, however, two large drums measuring 90 × 100 cm are employed. These are fitted with a 44-gallon (200 l) drum as an inner compartment, and a funnel of at least 25 cm diameter attached to a tube of 5 cm or more which forms a feed pipe and extends downwards near the side of the large drum to within 7.5 cm of the bottom. The inner 200-litre drum is supported in the centre of the large drum by bars or an attachment to the feed pipe. After the drum has been one-quarter filled with water which is brought to the boil, the oil-and-water mixture obtained from the press is poured through the funnel. The clean oil rises through the water and the dirt tends to fall to the bottom. As more oil or water is added the level rises and eventually the oil overflows into the inner drum and can be drawn off through a pipe leading from the bottom of the inner drum to the exterior. Various arrangements may be made for linking two drums together so that oil partly purified in the first drum may pass over into a second drum for final separation and clarification (Fig. 14.5 and Plate 14.4).

Pl. 14.3 Pounding sterilized fruit.

The operation of this equipment and the number of vessels required for continuous operation have been extensively studied and described by Nwanze (1963–4, 1965a, b), who showed that with one press, three large sterilizing drums, one stripper, two fruit preheaters, two fruit reheating drums and two large clarifiers, throughput could be raised to nearly 100 tonnes of bunches per month on double shift working (3.8 tonnes per working day) and over 130 tonnes if a three-shift system was introduced. The pressing rate was about 0.5 tonne per hour *pressing time*, but when working with two shifts in weekly stretches (i.e starting operations at the beginning of a week and closing down at the end) overall working time was still two or three times the pressing time though daily throughput could be gradually raised to 4 tonnes. The extraction efficiency of the whole process varied from 75 to 85 per cent; it is believed that this could be raised to 93 per cent if pounding were as efficient as conventional digestion (Nwanze, 1956). Over a period of 7 months the following results were obtained:

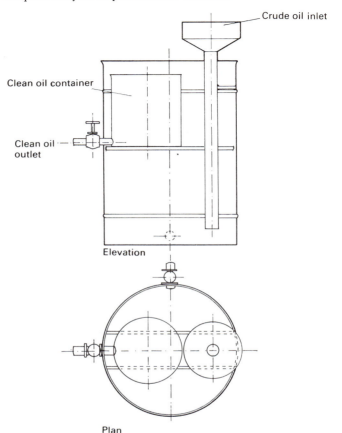

Crude oil inlet

Clean oil container

Clean oil outlet

Elevation

Plan

Fig. 14.5 Clarification drum for use with the hydraulic hand press.

No. of working days	Weight of bunches milled (tonnes)	Weight of bunches per milling day (tonnes)	Weight of oil produced (tonnes)	Extraction rate (%)	Extraction efficiency (%)
183	648.4	3.53	88.9	13.7	80.4

With one press and the equipment described it is clearly possible to achieve annual throughputs of well over 1,000 tonnes of bunches. Thus with West African yield levels, areas of over 80 hectares would be served and the hydraulic hand press is therefore best suited to groups of small grove or replanted holdings milling in cooperation. In spite of its potential and its superiority to the curb press, hydraulic

Pl. 14.4 Clarification drums for the hand press mill.

hand press systems have not, however, taken on widely in West Africa. The reasons for this are controversial and have been discussed by Kilby (1967–8), Purvis (1968) and Blaak (1979).

Kernel extraction

In Africa it has been common practice in hand-operated mills to give back the nuts to the womenfolk for hand-cracking. There are on the market, however, a number of nutcrackers driven by small petrol engines and these can with advantage be introduced into an otherwise hand-operated mill. Motors of around 3 hp are required to turn the cracker at between 2,000 and 2,500 rpm and about 0.75 tonne of nuts may be cracked per hour.

Power-operated palm oil mills

The mechanical milling of oil palm fruit started in Africa before the First World War. Though small-scale extraction was carried out in

Sumatra and, later, in Malaysia as soon as estate planting began, oil was not produced from a large power-operated mill in Sumatra until 1919.* The first power-driven machinery for cracking nuts is said to have been introduced into West Africa in 1877 by Mr A. C. Moore of Liverpool, being devised by Mather and Platt Ltd of Salford (Anon., 1917).

An incentive to the production of palm oil extracting machinery was given by the offer of a prize in 1901 by the German colonial Wirtschaftlichen Committee; this was awarded to the firm of Haake of Berlin for a set of small machines including depericarpers. Progress was slow, but in 1909 Haake erected plants at Mamfe and Victoria in the West Cameroon. A French firm, Fournier, erected a small plant at Cotonou at about the same time and this firm was the first to develop the idea of pressing the fruit whole.

Two pressing methods were tried. In the first, the fruit was sterilized, after removal from the bunch, by steam-heating for 15–30 minutes and was then transferred to hydraulic presses. This was the principle of the Cotonou mill of French design. Two pressings were carried out, with heating in between, and the nuts were then separated from the fibre in a rotating drum and cracked in a centrifugal cracker. The process sounds thoroughly modern except for the omission of digestion. The design of the presses is attributed to Paulmier, and the nutcracker was by Poisson, while the firms of Louis Labarre and Fournier of Marseilles constructed the plant. Digestion was first undertaken in a stamping mill in a plant of German design in Togoland. The entire spikelets seem to have been cut off the bunches and subjected to digestion, so that fruit-loosening and pulp-crushing were simultaneous. The whole mass was treated in hydraulic presses, the nuts were separated in a rotating drum and cracked in a Haake centrifugal cracker.

The earliest machinery of Haake of Berlin followed the second method of press extraction in which the depericarping of fruit before pressing persisted. Apart from Haake, machines were supplied by French companies, by Culley Expressors Ltd, a firm much concerned in the early development of small processing, and by Manlove, Alliott of Nottingham. No really satisfactory machine for continuous working was invented, however (Blommendaal, 1927).

During this period experiments were also being carried out by Lever Brothers in their Zaire and other concessions in Africa, but little information became available until the First World War. Apart from the Lever Brothers mills the only prewar mill of any size was that built at Maka in Cameroon for the Syndikat für Oelpalmen

---

* Palm oil extracting plants were termed mills in Africa from the beginning of their development. In the Far East they were and still are called factories, possibly because of the common use of the word for latex-processing plants on rubber plantations. The earlier and perhaps more accurate word is used in this book.

Kultur by German firms. The mill consisted of six hydraulic presses for pressing before depulping, four depulpers and two presses for pressing after depulping. Stripping fruit from the bunches was done by hand. About 525 tonnes of grove bunches could be processed in a week with day and night work and the extraction rate was 18.5 per cent oil to fruit and 6.9 per cent kernel to fruit (Van Hewin, 1921). Krupp presses were used in the mill.

Centrifugal extraction was first used in Zaire in 1916 (Dyke, 1939) and was the main feature of the Lever Brothers mills. This innovation was taken up by Manlove, Alliott and Co. who, through Nigerian Products Ltd, exhibited a complete mill of that type at the British Empire Exhibition at Wembley in 1924. Centrifuges proved very suited to the African grove fruit with its thin layer of mesocarp, but they also became the standard extracting equipment in Malaysia during the first 10–15 years of the industry's life in that country. The first Manlove mill was erected at Tennamaram estate in 1925. In Sumatra the first mills followed the Maka pattern and the hydraulic press, developed by Konrad Loens, Krupp and later by Gebr. Stork of Amsterdam, underwent further improvement for palm oil extraction. The first large mill was erected at Pulu Radja estate in 1921 and owed its design to Dr Fickendey and its construction and installation to Krupp. Soon after the First World War depericarping had fallen out of favour and both centrifuges and hydraulic presses became efficient oil extractors in combination with rotary-arm digestors. In Africa much of the early machinery in the inter-war period was installed by Lever Brothers, and their successor, Unilever Ltd, was responsible for the design of the small mechanical Pioneer mills which used centrifugal extractors (Plate 14.5) and were introduced into the palm grove areas of Nigeria during and after the Second World War.

While vertical hydraulic presses were to replace the centrifuge in the late inter-war period, so the screw press has gradually replaced the hydraulic press. 'Nowadays', Jacobsberg (1983) has stated, 'throughout the world, milling of palm oil is performed according to one process: sterilization, extraction by screw presses and physical separation of the oil phase from cell debris and water.'

Oil mills are normally powered by steam engines and the boilers use fibre and shell in their furnaces. Transmission in small and medium-sized mills was, from early days, by transmission shaft and belting; this method has now been largely superseded by the generation of electricity for driving electric motors of suitable sizes.

The choice of a plantation mill

In their essential features oil mills are standardized in that they consist of sections, sometimes grouped as 'stations', for: (i) *sterili-*

Pl. 14.5   The centrifuge of a Pioneer mill.

zation of bunches; (ii) *stripping* of bunches; (iii) *digestion* and mashing of fruit; (iv) *extraction* of mesocarp oil; (v) *clarifying* the oil; (vi) *separation of fibre* from the nuts; (vii) *nut drying*; (viii) *nut grading and cracking*; (ix) *kernel separation* and discarding of the shell; and (x) *kernel drying* and bagging.

The central machines of the mill are the extracting machines. It is the rate of operation of these machines, in terms of weight of bunches entering the mill, that determines the capacity of the mill. All other machinery is, or should be, adjusted to the rate of operation of the extracting machines. When choosing a mill it is first necessary or desirable to know the area to be planted, the rate of planting per annum and the expected yield at maturity. From these data estimates can be made of the probable crop for each year until the plantation is fully mature. If there are to be possible extensions, then it must be decided whether to allow for these extensions or to adjust the mill to the original area and to erect further mills for any extensions which may eventuate.

Secondly, it is necessary to estimate the probable distribution of the mature crop throughout the year. In countries with an even climate it is usually reckoned that in the peak month about 12.5 per cent of the total annual crop will be harvested. In certain countries, however, as much as 18–19 per cent of the crop may be harvested in 1 month. The mill's capacity must allow for this.

An example is given in Table 14.6 of the calculation of the capacities required for an area of 4,000 hectares expected to give an eventual yield of 20 tonnes of bunches per hectare. The milling capacity is based on the assumption that three shifts will be worked in the peak months with a total operation of 20 hours per day giving 500 hours per 25 working-day month. Obviously if conditions are such that three-shift work cannot be done then the calculations must be based on a lesser number of hours (including overtime) but this will lead to a larger capital outlay.

Usually, after calculations of this kind have been done, decisions have to be taken on: (i) whether to have one mill or several smaller mills; (ii) the stages of installation of the equipment, particularly the extracting equipment, so as gradually to expand the mill or mills to their maximum capacity; and (iii) the method of dealing with the small crop of the first few years.

There are various methods of dealing with the latter problem. In the first place harvest may be delayed and a larger first harvest obtained by the practice of ablation, i.e. the removal of young inflorescences during the first few months of flowering. Alternatively, a hydraulic hand press or presses may be used and these may be mechanized. This must, of course, be combined with the provision of sterilizing and digesting equipment and, if considered worth while, a simple nutcracker. With a press capacity of over half a tonne of bunches per hour it will be seen that four mechanized hand presses could deal with production in the Table 14.6 scheme for 2–3 years and that the heavy outlay of capital on the large mill could be delayed. Such a scheme may be suitable in areas which are expected to come slowly into production, but, in general, attempts are now made to plant large acreages per year so that full capacity may be reached as early as possible and the capital be fully employed.

Milling machinery can be supplied and installed in stages; these will depend on the rate of plantation development, on the machinery chosen and on the number of lines of machinery to be installed. In the days when the hand-operated hydraulic press was the standard extraction apparatus, the small capacity of individual presses (2–3 tonnes per hour) made it possible to increase the mill capacity very gradually, but with modern screw presses full capacity in a 20-tonne per hour mill would be reached in only two stages.

In the development of very large areas for oil palms, e.g. 8,000–12,000 hectares, it is sometimes questionable whether it is not better to have two mills rather than one very large installation. With the single large mill there are, of course, all the usual economies of large-scale operation: reduced overheads, lower capital outlay

Table 14.6 *Example of calculation of milling requirements*

	1988	1989	1990	1991	1992	1993	1994	1995	1996	1997	1998
Year of planting	1985	1986	1987	1988	1989	1990	Total				
Area planted (hectares)	400	400	800	800	800	800	4,000				
Year of production	1st	2nd	3rd	4th	5th	6th	7th	8th	9th	10th	11th
Estimated tonnes bunches per hectare	2.5	5	10	14	18	20	20	20	20	20	20
Estimated production (tonnes)	1988	1989	1990	1991	1992	1993	1994	1995	1996	1997	1998
1st year's planting	1,000	2,000	4,000	5,600	7,200	8,000	8,000	8,000	8,000	8,000	8,000
2nd year's planting		1,000	2,000	4,000	5,600	7,200	8,000	8,000	8,000	8,000	8,000
3rd year's planting			2,000	4,000	8,000	11,200	14,400	16,000	16,000	16,000	16,000
4th year's planting				2,000	4,000	8,000	11,200	14,400	16,000	16,000	16,000
5th year's planting					2,000	4,000	8,000	11,200	14,400	16,000	16,000
6th year's planting						2,000	4,000	8,000	11,200	14,400	16,000
Total production	1,000	3,000	8,000	15,600	26,800	40,400	53,600	65,600	73,600	78,400	80,000
Tonnage in peak month*	125	375	1,000	1,950	3,350	5,050	6,700	8,200	9,200	9,800	10,000
Capacity needed in peak month†	0.25	0.75	2.00	3.90	6.70	10.10	13.40	16.40	18.40	19.60	20.00

* $12\frac{1}{2}$ per cent of annual crop.
† Capacity in tonnes hourly throughput, assuming 500 hours' operation in the peak month.

(though this economy varies in magnitude according to the size comparisons being made), lower labour costs, lower fuel requirements, etc. per unit of throughput. Against this must be set certain factors which, owing to the multiplicity of circumstances, cannot be dealt with in any great detail here. In the first place if the area is composed of discrete portions of land at a distance one from another, or if it is very long and thin, the extra transport costs may outweigh the advantages of a single central mill. Secondly, the water supply at any one point may be insufficient for a very large mill. Thirdly, if many years are to elapse between the first and last plantings and the details of full development are uncertain, then there is the possibility of technological advance in the interim period; advantage cannot be taken of this if one large mill of a certain design has already been constructed or is in course of development.

In Malaysia and other areas of large-scale development the installation of single large mills (Plate 14.6) serving several plantations has been shown to be more economic than two or more smaller mills even where the plantations are as much as 60 km apart (Cooper and Bevan, 1968). The determining factors in this case were the high standard of the existing road systems, the lower capital outlay per tonne of throughput, lower staff and other running costs and the provision of sufficient storage space on a ramp at the mill side to ensure continuous factory operation (with the

Pl. 14.6. A large mill of 60 tonnes per hour capacity serving 8,000 hectares in Malaysia.

addition of field storage ramps if necessary). However, in recent years the cost of larger mills involving the installation of all the most modern equipment has shown an unprecedented rise and this has cast doubt on the value of large-scale operation in all circumstances. In South America, where road systems are often inferior and the financing of large mills more difficult, the installation of small, less sophisticated, locally manufactured mills with only a few imported parts has become popular. These mills have hourly capacities of 0.75–6 tonnes per hour.

The siting of a mill

Factors which should be considered in siting a mill are (Cornelius, 1983):

1. An adequate and suitable water supply should be within pumping distance. As a rough guide a mill of about 18–20 tonnes per hour capacity will require about 40,000 litres or 40 m^3 per hour. The exact water requirements will be specified by the manufacturers, and if necessary a filtering and purifying plant must be installed.
2. The load-bearing capacity of the soil must be considered, as the choice of a site requiring deep and extensive piling will increase costs considerably.
3. The position should be as central as possible but if there is no suitable central area a location convenient for the shipment of produce and not subject to flooding should be chosen. For an estate of 1,000–2,000 hectares an area of at least 4 hectares will be required for the mill and the ancillary buildings.
4. It must be possible to discharge effluent water and sludge in an acceptable manner (see p. 721).
5. It may be possible to take advantage of topography to reduce expenditure on ramps or, with small mills, to make use of gravity feeds instead of elevators.

The milling process

It is beyond the scope of this book to describe the milling process in large factories in their full engineering detail. The component processes will be described briefly to draw attention to their main features and their effects on the products.

Sterilization

Vertical bunch sterilizers

These are only suitable for small mills. They are usually of 2 or 3 tonnes capacity, but can be constructed to take up to 6 tonnes. In one type the bunches are forked at ground level into a bucket elevator which drops them into the sterilizer through a circular door at the top. After sterilization at pressure the bunches are discharged through a rectangular hinged door at the foot of the sterilizer. The discharge, though manual, is aided by the presence of a sloping perforated plate at the bottom of the sterilizer. Nevertheless, discharge may take 30–40 minutes.

An alternative type of vertical sterilizer containing an immersion basket involves less labour and time in charging and discharging. The bunches are tipped from lorries into the basket which is then lifted by monorail and lowered into the sterilizer which is closed by a specially locked lid. After sterilization, the basket is hoisted out of the sterilizer and moved to the platform above the stripper; there the load is discharged automatically on lowering.

Horizontal sterilizers

These were previously employed only in large mills, but in recent years many small mills in South America have been supplied with short horizontal sterilizers and cages on rails (Plate 14.7). The immediate advantage of the horizontal sterilizer is that it does not require hand emptying and that the bunches are sterilized in cages in which they have been packed and are therefore not subject to further bruising before sterilization. These sterilizers are long cylinders placed at ground level into which cages each holding 1.5–2.5

Pl. 14.7 Small cages, rails and weighing scales for short horizontal bunch sterilizers in a small locally designed mill in Colombia.

tonnes can be pushed on rails. Their capacity depends mainly on their length. A large sterilizer would hold between 13.5 and 22.5 tonnes of bunches in six tonne cages. Several of these would be required for mills of high capacity. A sterilizer with a 'bayonet-type' door is shown in Plate 14.8.

Mention has already been made of the importance of sterilization in reducing the f.f.a. of the oil and the danger, during sterilization, of causing the kernels to become brown. One of the primary purposes of *bunch* sterilization, however, is to loosen the fruit on the bunch so that stripping is not difficult. 'Hard bunches' are often a problem, particularly in Africa, and much attention has been given to methods of injecting and 'blowing off' the steam and raising and lowering the pressure (Stork, 1960a; Nwanze, 1961a). Steam is introduced slowly from a high point to allow the air to flow out from the bottom, and this continues for a short time after steam is seen to be coming from the outlet. Sterilization continues for 60–75 minutes of which 5 minutes is taken up by replacing the air and 15–20 minutes in working up to full pressure (2 kg per cm^2) which is maintained for 40–55 minutes. The longer time is most usual as reduction of sterilizing time has been shown to increase the number of hard bunches. With horizontal sterilizers 5 minutes is occupied in blowing off and 10–12 minutes in discharging and reloading so that the whole cycle occupies about 96 minutes. Some steam may be saved by a system of blowing over from one sterilizer to another. Venting systems are sometimes used: pressure can be maintained

Pl. 14.8   Horizontal bunch sterilizers with bayonet-type doors. (Gebr. Stork & Co.)

for an initial 5 minutes and is followed by venting and recharging. This can be followed by a second 'blow off' after another 5–10 minutes (Nwanze, 1961a). Increasing steam pressure beyond 2 kg per cm^2 (which corresponds to a temperature of about 130 °C) results in kernel discoloration and may also affect the colour of the oil. If a harvest consists solely of bunches from 4- or 5-year-old palms, of only very ripe bunches with a high proportion of loose fruit, sterilizing time can be reduced.

Sterilization usually results in a loss of weight of about 10 per cent of the weight of *fruit*, due to evaporation. With *tenera* fruit the dehydration reduces the volume of digested fruit in the presses and improves oil extraction.

In small mills in which fruit, not bunches, are sterilized, the sterilizing time is much shorter; 12 minutes has been found sufficient in an autoclave designed to hold 200 kg of fruit.

Stripping of fruit from bunches

There are two kinds of stripper (or thresher), the beater-arm type and the rotary-drum type (Stork, 1960b). The former is smaller and cheaper and is suitable for small mills with a capacity of up to 5 tonnes bunches per hour. It consists of an inclined cradle of curved bars between which the beater arms attached to the stripper shaft can pass. Bunches falling into the top of the cradle incline are knocked and turned by the tips of the beater arms and pass gradually down the cradle. Bunches occasionally become jammed in the stripper and this may increase the loss of oil in the bunch stalk.

The rotary-drum stripper has a diameter of about 1.8 m and a length of 2.7–5 m. The longer the drum the more complete the stripping. Bunches falling on the platform above the stripper are usually pushed manually into a shute carrying them into the drum, and loose fruit fall through a grill on the platform to be carried straight to the digester without entering the stripper. Automatic feeders can, however, be employed.

The stripper drum is composed of horizontal metal bars with sufficient space between them to allow the stripped fruit to fall through on to a conveyer. The centrifugal force exerted on the bunches is sufficient to raise them nearly to the top of the revolving drum whence they fall back on to the bottom and scatter their fruit. The bunches, while repeating this motion many times, gradually work their way along the drum aided by guide strips attached to the inside of the drum, and they fall out at the end and may be carried by a conveyer to a special bunch refuse incinerator outside the mill. The speed of the drum is important; if too fast the bunches will be simply carried round and round, if too slow they will remain rolling gently at the bottom. Speeds suitable for large heavy bunches will be too high for small light ones. Speeds of between 21 and 23 rpm are usual with drums of 1.8 m diameter. The maximum capacity of a rotary drum stripper is about 20 tonnes of bunches per hour.

In the use of any stripper a check has to be kept visually on the completeness of fruit removal from the bunches and, in the laboratory, on both the fruit retention and the oil loss in the bunch refuse. Bunches seen to be 'hard', i.e. to have retained their fruit through the stripping process, must be returned for resterilization.

Incinerators are designed to operate by autocombustion, after initial firing, with a natural draught. The resulting ash is a useful fertilizer (see p. 527).

Digestion

Digestion precedes oil extraction and its purpose is to break up the pulp physically and to liberate the oil from the cells in which it is contained. The fault of simple pounding, and of some small-mill digesters, is that heat is not also provided, for heat assists in the loosening of the cells from the fibre.

Modern digesters are steam-jacketed cylindrical vessels with a vertical rotating shaft in the centre driven from the top or, less commonly, from the bottom. The fruit is mashed by pairs of stirring arms attached to the shaft. Various devices are used for preventing the mash going round with the arms. Baffles may be set vertically between liner plates on the wall of the digester or bars may be inserted across the vessel. Sufficient steam pressure is maintained in the jacket to ensure that the mash leaves the digester at about 90 °C. Most digesters have perforations at the bottom to allow seepage of up to 50 per cent of the oil direct to the crude oil tank.

The important points to be checked in the digestion process are: (i) the temperature must be maintained at around 95 °C but the water in the mash must not be allowed to boil; (ii) the mash as produced must be homogeneous without any undigested fruit; (iii) the perforations must be kept cleared so that crude oil may escape at the bottom; (iv) the arms must be replaced when they show wear; and (v) the digester must be kept at least three-quarters full at all times (Stork, 1960–1; Bek-Nielsen, 1969b).

Nowadays digesters are normally supplied as an integral part of screw-press extraction and their size is adjusted to the particular screw press to be used.

Extraction of oil

Except in a few 'Pioneer' mills still in operation, the centrifuge has ceased to be used for oil extraction. An early comparison with a small hydraulic press (Georgi, 1932, 1933) showed that the centrifuge provides a clearer, more easily clarified crude oil but gives a higher oil content in the residual cake. This finding was confirmed on a larger scale by Bek-Nielsen (1969b), who also showed that overall efficiency of extraction was lower with the centrifuge though the percentage of unbroken nuts obtained in the centrifuge process was higher.

The centrifuge was well suited to *dura* fruit since pressing is not efficient where the nut percentage in the digester material exceeds 45 per cent. The latter material is spun in a centrifuge at 950–1,250 rpm in a perforated basket of 91 or 125 cm diameter with capacities of 1–2 tonnes per hour. The overall time for one charge is 20 minutes with 10 minutes' spinning time including 2 minutes of steam injection.

In many medium-sized or 'mini' mills the hydraulic press is still employed, but in mills of over 3 tonnes bunches per hour it is becoming obsolete (Cornelius, 1983). Hand-operated hydraulic presses in large or medium-sized mills have capacities of from 1.5 to 3 tonnes and operate at a maximum cake pressure of 75 kg per cm^2 (Stork, 1961b). The material is introduced into the perforated press cage from the digester shute through a hole (of the same diameter as the press) in a table-like sliding door which is operated by a hand-wheel. The door is closed on the interior of the press by sliding it forward so that the hole no longer lies above the press cage; six or seven pressings can be done in an hour. Hydraulic pressure is applied at the base of the press to a ram which moves upwards to press the mash and returns by its own weight The mash is introduced in portions with up to six circular steel plates inserted between the portions. This assists the operator by enabling him to push off the press cake in manageable slices after pressing. If prepressing or double pressing is adopted a plate is held back to cover the material already in the press before material is added for the second pressing. A typical schedule for pressing at the rate of six pressings per hour would be:

Filling	2 minutes – from digester
Pressing	6 minutes, including double pressing if employed
Emptying	2 minutes – pushing press cake into conveyor

Smaller hydraulic presses with capacities below 1 tonne bunches per hour are also available, and for many years Stork-Amsterdam marketed a fully automatic hydraulic press with a maximum capacity of 6.5 tonnes bunches per hour. This has now been superseded by the continuous screw press.

It has been shown that the performance of a hydraulic press depends to a great extent on the composition of the press cake. Trials conducted by press manufacturers suggest that there is an optimum proportion of nuts to fruit for maximum oil extraction, i.e. there is an increased loss of oil in the fibre when the proportion of nuts to fruit exceeds or falls short of a certain percentage. Where there is an excess of nuts, they bear on each other and full pressure is not exerted on the oil-bearing fibre (Stork, 1961a). No very clear reason has emerged for the low extraction obtained when there is a shortage of nuts but it is thought that the addition of nuts to *tenera* digested fruit may reduce friction resistance within the press cake and help to ensure a better distribution of pressure through the press cake.

Another measure which improves distribution of pressure is the insertion of the metal plates between sections of the introduced digested material. If this is not done the press cake will tend to be more compact and drier in the middle than at the circumference and more fully extracted at the bottom than at the top. The insertion of plates reduces this tendency by redistributing the forces bearing on the cake. It has also been shown that the presence in the cake of small pieces of trash such as the 'calyx leaves' (perianth segments) from the bunch, far from hindering pressing, actually improves oil extraction from the cake.

To obtain the most satisfactory expression of oil in a hydraulic press, therefore: (i) fibre should be added to the sterilized fruit if the fruit has a high nut/mesocarp ratio, e.g. above 50/50; (ii) nuts should be added if the nut/mesocarp ratio is low, e.g. below 25/75; and (iii) no attempts should be made to remove calyx leaves. The actual quantities of nuts or fibre to be added usually need to be found by trial.

Over the last decade screw presses have gradually superseded hydraulic presses for large mills because of their higher throughput, lower capital outlay per tonne of throughput, lower power consumption and lower labour requirement. These advantages outweigh the higher kernel breakage and loss in the fibre and the rather more difficult and costly clarification of the extracted oil. Maintenance costs are now no higher than with hydraulic presses (Bek-Nielsen, 1974; Olie, 1973). The first screw press to be marketed for large mills was of the single-shaft, opposing screw type (Wolversperges, 1963). This press had a gear box and a pressing section consisting of two screws – a feed screw and a press screw. The screws are placed end to end on the same shaft but are of opposite thread and rotate in opposite directions. The feed screw in large presses of this type rotates at 10 rpm while the opposing press screw rotates at 6 rpm. The digested fruit is fed to the press screw by the action of the feed screw and the tapering design of the former gradually reduces the volume available around the screw, thus increasing the pressure until the matte is expelled around the end cone, which is adjustable. Oil is expelled outwards through the perforated cage around the screws and inwards through a short cage around the shaft at the end of the press screw. Presses of this type have a capacity of up to 13 tonnes bunches per hour, but they have now been superseded by double-shafted presses with screws turning on twin shafts or by single-screw presses. In the latter, which have mean capacities up to 7.5 tonnes per hour, a short feed screw in a perforated cylinder connects the digester to the longer main screw and extracts part of the oil. The remainder is extracted by the action of the main screw through its own perforated cage and against the cake-breaker cone at the end.

Twin-screw presses have counter-rotating shafts turning at the same speed but in opposite directions. In one type the digested fruit enters a hopper below which lie the screw and a strainer through which some of the oil will pass. The material then passes into the perforated press cylinder through which oil is expelled; finally the matte is screwed further along the perforated cylinder where the shaft tubes are also provided with holes through which further expelled oil passes. The outlet at the end of the cylinder is regulated by adjusting the cones by means of a hydraulic or absorbed current control system (Bek-Nielsen, 1969b) (Fig. 14.6). Presses of this type are of various sizes with capacities of 1–4, 1.5–7, 7–15 and 10–20 tonnes bunches per hour. The capacity in each case is adjusted by exchangeable V-belt pulleys or a variable speed drive. The usual shaft speed is 10 rpm.

Another make of double-shafted screw press expels the oil through the press cage only and not through the shaft tubes. A feed screw passes the digested material to the twin screws and it is mainly through the variable speed of the feed screw that throughput is adjusted (Olie, 1973; Stork, 1983). The press has a capacity of 10–15 tonnes bunches per hour.

Screw presses have been compared with hydraulic presses by Bek-Nielsen (1969b) who showed that oil losses in the fibre were lower with screw presses but losses in the sludge were higher. This resulted in the overall extraction efficiency being only marginally in favour of screw presses, and the view was expressed that efficiency has now reached a level which, with any kind of pressing, is unlikely to be exceeded.

Although screw presses have been successfully employed with mixtures of *dura* and *tenera* material, they are certainly best adapted to plantations of pure *tenera* material (Maycock, 1975). Another advantage claimed for screw presses is flexibility of throughput. Presses with a nominal throughput of 10 tonnes bunches per hour can be run down to 6 tonnes or up to 15 tonnes when necessary.

Solvent extraction has often been suggested and tried on a small scale, but the technical and economic difficulties have not been entirely overcome. In the early days of the Sumatran industry solvent extraction of the residual fibre after the first pressing (containing over 20 per cent oil to dry matter) was put into practice

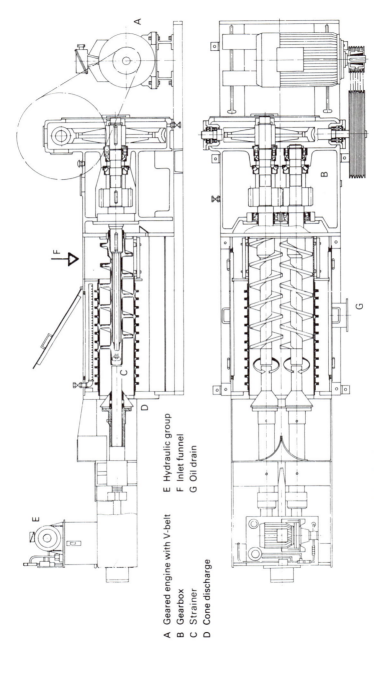

A Geared engine with V-belt   E Hydraulic group
B Gearbox   F Inlet funnel
C Strainer   G Oil drain
D Cone discharge

Fig. 14.6 The double-shafted De Wecker screw press.

on a number of estates (Blommendaal, 1927). The solvent used was naphtha (boiling point 85–105 °C) from local mineral oil refineries. The solvent extraction of residual oils in fibre after normal pressing or centrifuging has also been suggested and much discussed. The quantity of solvent required would be large and the loss, under tropical conditions, considerable. Oil extracted from the residual fibre is known to contain a high proportion of carotenoids and is contaminated with an unacceptable quantity of non-oil components such as waxes (Olie and Tjeng, 1974) and this would introduce bleaching problems.

Several attempts have been made to dispense with mechanical methods altogether and to extract oil after digestion by processes variously described as 'water displacement', 'wet', or 'hot water washing' (Cornelius, 1983). Lever Brothers tried such a process as early as 1911, and again in the late 1950s a system was developed in which a second digester was used for washing out the oil by a hot water spray (Maycock, 1975). The method was further developed for small mills (see p. 732) by Vandekerckhove of Belgium who claimed 78–80 per cent extraction of technical grade oil using a fruit cooker and digester only, and 88–90 per cent extraction using a low-pressure piston press (cake pressure 30 kg per cm^2) followed by warm water washing in a counter-current wet digester (Cornelius, 1983). These extraction processes have capacities of 0.75–2 tonnes bunches per hour.

Clarification

The crude oil coming from the presses or centrifuges is a mixture of oil, water, dirt and cellular matter from the mesocarp. The amount of water accompanying the oil varies widely, but Maycock (1975) has given the following figures as typical of the different extraction processes (in per cent):

	Oil	*Water*	*Non-oily solids*
Wet process	16	79	5
Centrifuge	80	17	3
Hydraulic press	75	20	5
Screw press	66	24	10

Crude oil is largely an oil–water mixture in various phases of oil dispersion in the water. The majority of the oil settles out on top of the water easily enough but, if care is not taken, emulsions may be formed. Impurities tend to prolong emulsification and temperatures of over 100 °C tend to produce emulsions.

Crude oil flows by gravity straight to the crude oil tank where it passes through screens to get rid of the larger impurities. In a small plant the screens may be stationary and set in the crude oil tank, but in larger plants vibrating screens of stretched gauge are used above the crude oil tank and the material which does not pass through is automatically returned to the digester. Sand sinks to the bottom of the tank. Crude oil pumps direct the screened crude oil to the clarification tanks.

In small mills the oil is separated from the water and dirt (the sludge) in simple settling tanks containing water and heated to about 95 °C by steam coils or steam injection. Pure oil is drawn off from the top and sludge passes to another settling tank for further separation after addition of more water. The quantity and kind of clarification equipment depend on the size of the mill and the market requirements. In larger mills continuous clarification tanks are employed. In these tanks (see Fig. 14.7) the crude oil is passed via a heat exchanger (A) at 85–90 °C to near the bottom of the tank, and the oil collects and rises in a thick layer (X) until it reaches the top of a pipe (B) through which it will be automatically or intermittently drawn off. The sludge water syphons off at a

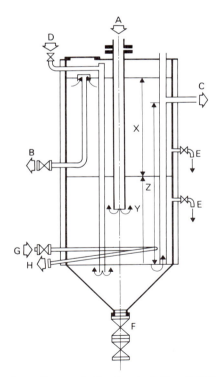

Fig. 14.7   The Stork continuous clarification tank (for explanation see text).

slightly lower level since the specific gravity of the oil plus sludge is lower than the specific gravity of sludge alone which rises in the side pipe (C). The liquid fractions X + Y in the tank and Z in the pipe are in balance. Temperature is maintained by a steam coil in the tank (G–H) or a steam jacket. The quantity of crude oil being fed and the volume of the oil layer together determine the residence time of the oil in the tank and if this becomes too short for satis-factory clarification then an adjustment must be made to the height of the oil exit pipe. A slow supply leads to too long residence and tends to increase the f.f.a. set up (Nwanze, 1962). When supply ceases, the separated oil must be drawn off by allowing entry of water into the sludge layer through pipe D, and the sludge water and the heavy material must be run off by manipulation of the various external taps (E and F).

The remaining clarifying equipment depends on the scale of operations and the results required. Further settling tanks may be employed and vacuum driers or centrifugal separators. A number of systems have been advocated. In one, a centrifugal separator, designed to prevent air entry with consequent oxidization, is used, and the oil is then passed to a vacuum drier which reduces moisture to below 0.05 per cent (Zachariassen, 1969). Several types of centrifugal separator are available (Stork, 1961b), and in another system the discharged water from the separator is recycled to the press, while the purified oil is sent direct, or through a drier, to the storage tanks (Wohlfahrt, 1969).

The old method of dealing with the sludge is to pass it through a series of fat pits. These are arranged in series: the heavier water and dirt go to the bottom of the first and succeeding pits, and pass to the next pit in order under a partition which itself holds back the lighter oil. The partitions can usually be let into the pits in slots and are easily removed.

An alternative method of treating sludge is to conduct it to a tank for boiling and then to a special sludge centrifuge of which several types are available. In one type a star-shaped rotor has four or more nipples with 2 mm orifices. Sludge water enters through the hollow half-shaft and is thrown by centrifugal force towards the four nipples at the extremities of the star. The heavier particles and water pass out, but the oil builds up in the centre and is forced back through a small-diameter outlet shaft inside the inlet shaft and returns to the continuous clarifier.

Effluent disposal

However effective the system of recapturing oil from the sludge may be, the effluent discharged from an oil mill is still objectionable and will pollute streams, rivers or surrounding land. While mills were comparatively few and mostly on large fast-flowing rivers, the problem was not a serious one, but the situation in many countries is now quite different and much attention has recently been given to the subject of effective disposal. Apart from the sludge water itself, which amounts to about 300 kg per tonne of bunches milled (or about 1.5 tonnes per tonne of palm oil), there are also about 175 kg of sterilizer condensate and between 40 and 140 kg of effluent from the hydrocyclone or clay bath separators per tonne of bunches (Olie and Tjeng, 1972; Ma *et al.*, 1983). The total amount of effluent is therefore more than half a tonne per tonne of bunches or 2.5 tonnes per tonne of oil produced. In milling 20 tonnes of bunches per hour, more than 200 tonnes of effluent may be discharged over 24 hours and this may contain up to a tonne of oil and 9 tonnes of dissolved or suspended solids. This effluent has a biochemical oxygen demand (BOD) of about 20,000 mg per litre at 20 °C for 5 days which is extremely high (Singh and Ng, 1968).

Malaysia has taken a more serious view of this pollution than any other country since the BOD of the total effluent produced exceeds that of the sewage of the country's entire population (Cornelius, 1983). Discharge standards were gradually tightened by law from (in mg per litre) BOD 5,000, oily material 150, total N 200, in July 1981 to BOD 50, oily material 50, total N 50, in 1984.

Many methods of attaining the required standards and of utilizing the effluent to advantage were suggested, reviewed and tried during the 1970s and later (e.g. Davis, 1978; Davis and Reilly, 1980; Hemming, 1977; Wood, 1977), and a survey carried out by Ma *et al.* (1983) showed that the successful methods could be grouped as: (1) tank digestion and mechanical aeration; (2) tank digestion and facultative ponds; (3) decanter and facultative ponds; (4) aerobic and facultative ponds; and (5) physical-chemical-biological treatment. The great majority of mills followed method (4) in which the palm oil mill effluent (POME), after passing through a de-oiling tank to trap oil remnants, passes into acidification buffering ponds where it remains for 2–3 days; here acid bacteria convert organic compounds into volatile fatty acids (Chan and Chooi, 1982). The effluent then passes into anaerobic ponds (usually two), remaining for 80 days, and then into facultative ponds before discharge. This effluent now meets the analysis demands, but dealing with the remaining sludge is the main problem. It can, however, be dried in sand beds and the resulting cake can be utilized as a fertilizer.

In the *tank digestion and mechanical aeration* system the effluent first enters cooling/acidification ponds for 1 or 2 days and is then pumped to open-topped digester tanks for residence of about 20 days (Whiting and Lim, 1983). These tank are periodically desludged and the sludge, which has a high N content, is used as a fertilizer. After sedimentation the discharge from the digesters is pumped to an aeration pond and the discharge from the latter passes through a sedimentation tank and the settled sludge is recycled or can be added to the tank sludge for use as a fertilizer. The nutrient contents of these sludges are as follows (Tam *et al.*, 1983):

	Solids (%)	N	P	K (ppm)	Mg	Ca
Digester sludge	7.2	3,552	1,180	2,387	1,509	1,249
Aerobic sludge	3.5	1,495	461	2,378	1,004	1,190

In the *tank digestion and facultative ponds* system the effluent passes through an oil trap into enclosed tanks of up to 3,700 m³ capacity for a residence of 10 days (Quah *et al.*, 1983). The biogas produced (45–70 per cent methane) has been used to generate electricity, or it can be flared off. The digested effluent passes into a holding tank before being used on the land. The production of biogas in plastic tanks or tubes by a 20-day anaerobic fermentation process was devised by Petitpierre (1982). Digester control and the biochemistry of anaerobic digestion have been described by Ma and Ong (1982).

In the *decanter-drier* system the sterilizer and hydrocyclone effluents are discharged separately into ponds. The clarification system itself is modified by allowing the viscous crude oil to pass the crude oil vibrating screen and by inserting a concurrent centrifugal decanter between the screen and the continuous clarification tank (Jorgensen, 1982). The decanter removes 90 per cent of suspended and 20 per cent of dissolved solids. The sludge from the continuous clarification tank is treated in the usual way in a sludge centrifuge, with oil returning to the tank and water with high dissolved solids returning to the presses, while the solid phase from the decanter is treated in a rotary drier using flue gas from the boiler. The palm oil meal produced can be used as a fertilizer (Jorgensen and Gurmit Singh, 1981) or an animal feed (Devendra *et al.*, 1981); it has a moisture content of 5–15 per cent, crude protein 11–13 per cent, ether extract 11–13 per cent, crude fibre 11–14 per cent and ash 15–22 per cent.

Another system (*Antara*) also uses a decanter and drier but treats the whole effluent together and employs coagulants and flocculents to remove solids (Ma *et al.*, 1983).

All these systems, when properly operated, are capable of reducing the BOD of the original effluent by over 99 per cent, and hence achieving the 50 mg per litre requirement. It is also claimed with some processes that the eventual sludge can in certain circumstances be used as a complete replacement of fertilizers (Whiting and Lim, 1983).

Kernel extraction

The matte coming from the presses consists of nuts and moist fibre with some residual oil. To extract the kernels it is necessary to: (i) extract the nuts from the fibre; (ii) crack the nuts; and (iii) separate the kernels from the cracked shells. In large mills built in the 1950s and early 1960s the kernel extraction plant occupied a great deal of space, and in comparison with the oil extraction plant the kernel recovery equipment was very expensive. It was estimated that while the investment and operating costs of the oil extraction equipment were only twice as great as that of the kernel equipment, the value produced in the oil extraction section of the mill was nearly nine times as great as the kernel value produced by the kernel extraction plant. This costly and space-occupying equipment can be simplified by using air only to separate the fibre and to blow the nuts first to the silo and later to the hydrocyclones. A pneumatic transport system can also be used for conveying kernels to the kernel silo and shell to the boiler platform (Olie, 1969).

Separation of nuts from fibre

Fibre separation may be pneumatic, mechanical or hydraulic. The last is no longer used. Pneumatic fibre separators (Stork, 1962a) which are partly mechanical have been the rule in nearly all large modern mills. The press cake has a high temperature as it comes from the press and if it is immediately pushed into a 'breaker-conveyer' – i.e. an open-topped, but steam-jacketed conveyer with a shaft carrying blades which cut up the press cake as well as forcing it along the conveyer – it will dry out appreciably under its own heat. This assists the loosening of the fibre from the nuts. The effect may be enhanced by arranging for the air which is to pass through the separator to be heated by a steam-operated air-heater.

In the most commonly used type of fibre separator the fibre–nut mixture passes into a large drum rotating at about 15 rpm. This drum has baffles mounted on the inside which carry the mixture upward, and allow it to drop. The current of air passing through the drum is sufficient to carry the partially dried fibre to the exit tube and blow it to the boiler platform. The nut and fibre mixture is gradually moved by the rotating motion towards the end of the drum and in the course of this motion is further dried by the hot air current. The nuts will fall into a smaller, lower rotating drum where they are polished by friction. Some air passes through this drum also, carrying any light particles upwards to join the main flow. Various releases can be arranged under the second drum to allow the exit of polished nuts and the passing over to the end of the drum of heavy pieces of bunch stalk and other debris. The separator is efficient but, besides being noisy, it makes comparatively heavy calls on the power supply of the mill.

Purely mechanical fibre separators have been used for a long time, particularly in small mills such as the Pioneer, where a low capital cost was imperative. One type consists of a screened drum which is rotated and allows the separated fibre to fall through the screen. A second type is a modification of a cotton ginning machine; a revolving shaft is fitted with studs which tease off the fibre from the fibre–nut mixture fed into the machine. A third type has a rotary cage bounded by rollers revolving in opposite directions in pairs which remove the fibre to the outside of the cage but retain the nuts inside.

Rotating drum separators are now being replaced by stationary direct air separation columns in which the velocity of the upward current of unheated air which removes the fibre is adjustable. Separators of this type occupy less space and reduce power consumption (Olie, 1969; Unilever, 1967). They can be combined with simplified cracking and kernel separating systems in which pneumatic transport replaces chain elevators and screw conveyers.

Nut screening and cracking

The clean nuts may be dried in a nut silo or, if the drying in the breaker-conveyer and in the fibre separator has been sufficient for cracking, they may be conveyed straight to screens for grading according to size before cracking. It is generally considered that in a mill with a large throughput the additional drying in the silo is necessary, and the opinion has been voiced that the moisture content of the kernels must be less than 16 per cent if they are to be sufficiently shrunk away from the shells for easy cracking. Nevertheless, quite satisfactory cracking is often obtained in mills without nut silos, and if heating in the latter is too high kernel quality will suffer.

The cracking section usually consists of revolving screens to grade the nuts, the crackers themselves, and screens or columns to separate uncracked nuts and/or dust and small shell particles from the mixture. The larger the number of nut grades used, the less will be the number of kernels returned to the nutcracker. This is because the frequency curve of nut size usually overlaps the frequency curve of kernel size (Stork, 1962b). Therefore, if no grading is done and unless a special self-sorting nutcracker is used, *some* large kernels will pass over the cracked mixture screen and be returned to be cracker where they may be broken and partially lost as dust. If the perforations of the cracked mixture screen are made wider, then the return of kernels to the cracker will be less, but more small nuts will pass through the screen and appear in the cracked mixture for separation into kernels and shell. Cracking in at least two fractions is usual and in three fractions if the size of the plant warrants it. This reduces the overlap of kernel and nut size; with *tenera* nuts, which are often small, the overlap of nut and kernel size is inevitably larger than with *dura*, particularly Deli, nuts and it is partly for this reason that kernel recovery is less satisfactory with the former fruit (Olie, 1969; Unilever, 1967).

Modern nutcrackers are centrifugal, the nuts fed into the inlet falling into and then being thrown out of slots on the face of a shaft rotating at high speed and being hurled against a cracking ring. The vertical crackers (horizontal shaft) previously employed are suitable for large *dura* or Deli nuts, but in predominantly *tenera* plantations cracking is best done with horizontal crackers (vertical shaft) which have a larger diameter. The theory behind the use of the latter crackers is that *tenera* nuts, having a distinct rounded head and tapering fibre-covered tail, are enabled to take up a 'head-foremost' position in the extra distance covered and so be cleanly cracked on hitting the plate. If the fibre-covered tail hits the plate the nut may not be cracked. A further development has been the self-sorting nutcracker which incorporates grading slots and pitching blades which grade the nuts and give them the correct speed for cracking; this makes prior nut screening unnecessary (Olie and Tjeng, 1974). The speed of nutcrackers varies from 800 to 2,500 rpm according to the diameter of the rotor.

Not all the nuts will be cracked when they are first fed into the cracker. A 90 per cent cracking is usually achieved and this may rise to 98 per cent with efficient crackers. Cracked mixture screens were previously used to separate dust and small shell particles and to return the uncracked nuts to the cracker. These screens have now been largely replaced by vertical separating columns in which dust and small pieces of shell are blown away; uncracked nuts are recovered *after* kernel and shell separation (Bek-Nielsen, 1969b). Fragments of light shell and small nuts with long fibres have a specific gravity differing little from that of kernels, and if they can first be blown away a more perfect separation of kernels and shell can then be effected.

Kernel and shell separation

In the early days of milling the mixed shell and kernels were placed in a salt bath of such a specific gravity that the shells sank and the kernels floated. The kernels were then skimmed off and dried for bagging. Later a mixture of water and clay to give a specific gravity of 1.17 was found more suitable. The shells have a specific gravity of 1.3–1.4 and sink to the bottom; the kernels, which have a specific gravity of about 1.07, are skimmed off, washed and dried. However, shell of *tenera* material has a lower specific gravity than *dura* shell, and a figure of 1.17 has been quoted (Weko and Sutiardjo, 1969). This makes separation based on specific gravity differences more difficult.

A simple clay bath has a perforated tank in which the mixture is placed; after the kernels have been skimmed off, the tank is raised manually or on hinges and the shell tipped out. The commonest type of mechanical clay bath has a shaft running longitudinally above or at the surface of the clay–water mixture and carrying two discs or wheels with lifting buckets and, on either end, washing drums. The bath tank consists of two compartments separated by an inclined baffle. From one compartment clay water is continuously scooped by the buckets of one wheel. This clay water is tipped out of the buckets and runs into the other compartment so that the latter is continually overflowing back into the first compartment. The cracked mixture is introduced into this overflowing compartment and the floating kernels pass out with the overflow into one of the washing drums while the clay water is running back into the first compartment. At the other end of the bath the shells fall along an inclined baffle to the bottom of the overflowing compartment and are picked up by the second set of lifting buckets and dropped into the other drum where they are washed and pass out of the machine. The efficiency of these machines is limited owing to the tendency for floating kernels to get picked up by the shell buckets.

The advent of the hydrocyclone has reduced the use of clay baths. The hydrocyclone's main advantage is that it is compact and requires only water; however, it has a higher capital cost than the clay bath and its efficiency may not be higher.

A hydrocyclone unit consists of two cyclones, two tanks, two washing drums and two pumps with connecting pipes (Stork, 1963). The mixture is introduced into one tank (A – see Fig. 14.8) which is filled with water, and is forced with the water by a pump (C′) to

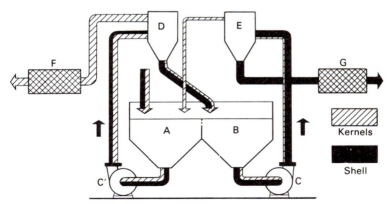

Fig. 14.8 A representation of the operation of the hydrocyclone – see text (Stork, 1963).

a cyclone (D). Here the water is subjected to two forces; firstly, a centrifugal force, and secondly, a non-tangential force directing the water to the vortex in the centre of the cyclone through which the majority of the water flows (Fig. 14.9). A particle lighter than water and/or with high flow resistance tends to leave the cyclone by the vortex with large quantities of water. Heavy particles and/or particles with a low flow resistance follow the centrifugal force and are discharged at the apex at the lower end of the cyclone with the smaller quantity of water. It appears that with *dura* nuts flow resistance plays little part and thus kernels are discharged from the vortex and shell at the apex; but with *tenera* material, separation, as in the clay bath, is less efficient. In practice the greater part of the kernels pass out through the washing drum (F) from which the water is allowed to flow back into the tank (A); the shell, with some kernels, passes into the other water-filled tank (B) and is pumped to another cyclone (E) where separation continues. In this case the kernels going out at the vortex re-enter the first tank (A) while the shell from the apex goes to the other washing drum (G) and so leaves the separator.

Kernel drying

The washed shell passing out of the clay bath or hydrocyclone is carried by conveyor or blower to the boiler house where it serves

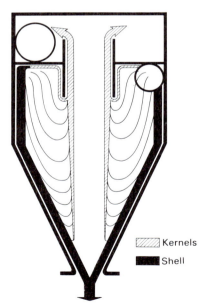

Kernels
Shell

Fig. 14.9   The movement of kernels and shell in the hydrocyclone.

as fuel. The kernels have to be dried before they are fit for bagging and if the separation is not sufficiently complete they must be picked over by hand. This is usually done by passing the kernels slowly in front of a picker on a conveyor belt either before or after the kernels have been dried.

Kernels can be dried on trays passing slowly over hot air, but the kernel silo is the most common method of drying. The wet kernels are admitted at the top and move down the silo in a continuous flow as kernels at the bottom are discharged through a shaking grid and bagged. The silo incorporates a drier which blows heated air through the kernels at different levels and is thermostatically controlled.

Mill layout

In small or medium mills all the equipment is usually under one roof, but in very large mills the boilers, engines and electricity generating equipment may occupy a separate building. A separate power house assists cleanliness and is made possible when all the plant is electrically driven.

When shaft and belt drive is used the boiler room is partitioned off from the rest of the building and the furnaces face outwards; the steam engine is then best positioned in an adjoining room for belt drive on to the middle of the long shaft which drives all the milling equipment in the main part of the building. Next to the engine room will come the clarification plant and next to that a small laboratory and office. Bunches are brought to one end of the building for loading into vertical or horizontal sterilizers while kernels are bagged at the other end. Oil may be drummed at the side entrance to the clarification room or pumped to storage tanks.

Figure 14.10 is a diagrammatic scheme of a large modern mill with horizontal sterilizers, screw presses and hydrocyclones. In a large installation of this kind the reception area must be arranged so that lorries may tip bunches onto a loading ramp into transfer hoppers from which they are discharged into the cages for steriliz- ation. An electric hoisting crane is required for hoisting the cages of sterilized bunches on to the stripper platform. The bunch refuse usually leaves the mill by conveyer, either going to an incinerator or to a cutter and thence to the fields. Incineration is now the usual practice.

A layout for a large electrically driven factory of over 20 tonnes per hour capacity usually has a separate power house with boilers, steam engines and alternators. Clarification is carried out in the main building of the mill though well separated from the presses. The laboratory, offices and workshops are separated from the main building.

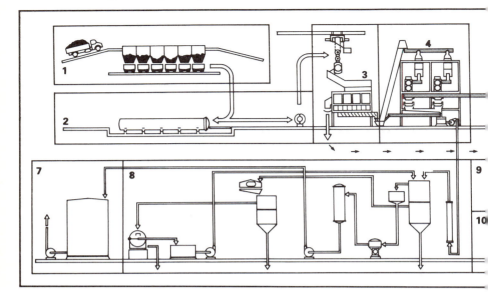

Fig. 14.10 Diagrammatic layout of a large palm oil mill (Stork-Amsterdam): (1) Bunch reception ramp; (2) Bunch sterilizer; (3) Bunch stripper; (4) Digesters, presses and breaker-conveyor; (5) Fibre separator and cyclone; (6) Nut silo, hydrocyclone, kernel silo, dust cyclones and shell storage; (7) Palm oil tank; (8) Oil clarification; (9) Bunch refuse incinerator; (10) Piping system; (11) Power house; (12) Electricity distribution; (13) Boiler house; (14) Water supply.

Small-scale mechanical milling

For a long time palm oil production was considered either as a very large-scale business or as a small-scale peasant undertaking. For this reason there was much technical progress in the installation of very large mills and, at the same time, attention was given to providing for the non-mechanical needs of the small producer. However, in some parts of the world, notably America, the planting of small or medium-sized holdings of a few hundred hectares or less became common and is still increasing, and this has engendered a special interest in small plants of high efficiency. Such plants need to have a capacity of between 1 and 3 tonnes of bunches per hour and a low installation cost; their capacity may sometimes be increased if, as in Ecuador, the owners can gather in bunches from surrounding small properties as well as their own.

The first small-scale or mini-mill was erected at the Central Experiment Station at Serdang, Malaysia, in the 1920s and had both a small centrifuge and a small hydraulic press, but the first widely

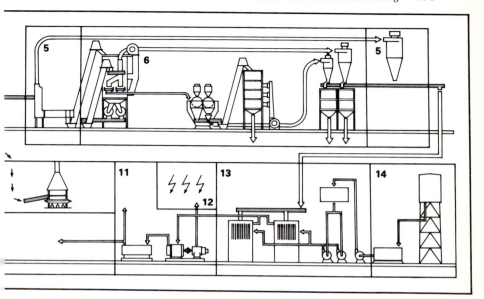

used such mill was the 'Pioneer' which was introduced in fairly large numbers into Nigeria and other African countries in the 1940s. This mill used a centrifuge (Plate 14.5) for oil extraction and was intended mainly to press fruit from the groves, using a relatively large number of low-paid workers. Bunch stripping was done manually in the villages and the fruit was bought in. Vertical boilers and steam engines were installed with shafts and belting, but elevators were not used, so there was much heavy manual work. Sterilization was by fruit autoclave and the digester was open-topped and unheated. The capacity was 1 tonne of bunches per hour (600 kg fruit), and oil extraction efficiency only 85 per cent. When wages rose steeply these Pioneer mills became uneconomic in comparison with small village curb-press businesses, and ingenious schemes of reorganization and machinery improvement came too late to be put into general practice. Many of the alterations could not be effected without considerable injection of capital when costs were already much inflated (Nwanze, 1961b).

Small-scale mills are of two kinds: either they may be based on the hand press and aim to use the very cheapest locally available material for ancillary equipment, or they may be fully mechanized but of low throughput and simple design, with well-built but uncomplicated components.

Mills of the first kind, originally developed by Nwanze (1965a), have already been described (p. 698). Blaak (1979) found that the main design reasons for the lack of support for the system were the method of bunch preparation, the need for manual pounding and reheating the mashed fruit, the high labour requirement, and the rapid wear of the drum plate used. He constructed a village mill in Cameroon which allowed for the mechanization of heavy work though retaining low-cost, simple apparatus. His mill did not provide for bunch sterilization and stripping, and relied on fruit stripped from the bunches in the village. A small boiler provided steam for fruit sterilization and nut drying, while small diesel engines operated the digester, nutcracker and water pump. Production could reach 1 tonne of palm oil in an 8-hour shift, which is equivalent to 0.625 tonne bunches per hour for *tenera*. This type of mill is clearly most suited to the traditional oil palm peasant countries of Africa and is unlikely to attract growers elsewhere. Some of its features were adopted by workers at the NIFOR/FAO Engineering and Development Unit in Nigeria (Hadcock, 1983) with the same objects in view.

Mills of the second kind with a capacity of 0.75–3 tonnes per hour have been installed in several parts of Africa and America. In the latter continent many of these mills have been locally designed and incorporate short horizontal bunch sterilizers with a compact rail system which enables the sterilized bunches to be moved easily from the sterilizers to the stripper (Plate 14.7). In many cases the mills have been constructed on the side of a hillslope so that gravity can be utilized for moving the bunches to the sterilizers and to the strippers and the fruit to the digesters.

Cornelius (1983) has described two interesting types of mini-mill based on the wet or water displacement process which were erected in several parts of West and Central Africa. The VDK (Vanderkerckhove) rural extraction kit consisted of a bunch sterilizer, stripper, cooker, digester and clarifying vessels together with a boiler, nutcracker, oil heater and dehydrator. The VDK compact mill included vertical pressure sterilizers; cooked and digested fruit was first pre-extracted in a low-pressure piston press, after which it was warm-water washed in a counter-current wet digester. These mills have now been superseded by screw-press mills.

Small-scale mills manufactured in Europe are now built around both mechanized hydraulic hand presses and screw presses. They are usually capable of dealing with between 300 and 1,000 hectares according to the level of yield of the region concerned. A plan of a mill incorporating a hydraulic press is shown in Fig. 14.11. The equipment is housed in one open-sided building and space is conserved by employing vertical boilers and bunch sterilizers (though short horizontal sterilizers can be used). The simple vertical boiler provides steam for sterilization, pressing, clarification, etc. and can be used with a small steam engine. However, and especially

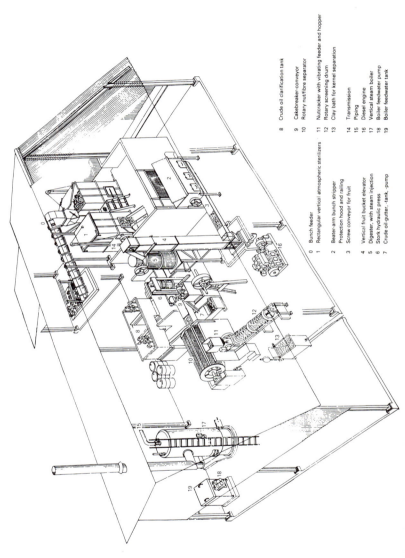

0 Bunch feeder
1 Rectangular vertical atmospheric sterilizers
2 Beater-arm bunch stripper
  Protection hood and railing
3 Screw conveyor for fruit

4 Vertical fruit bucket elevator
5 Digester, with steam injection
6 Stork hydraulic press
7 Crude oil-gutter, -tank, -pump

8 Crude oil clarification tank

9 Cakebreaker-conveyor
10 Rotary nut/fibre separator

11 Nutcracker with vibrating feeder and hopper
12 Rotary screening drum
13 Clay bath for kernel separation

14 Transmission
15 Piping
16 Diesel engine
17 Vertical steam boiler
18 Boiler feedwater pump
19 Boiler feedwater tank

Fig. 14.11 Plan of a 0.75 tonne per hour capacity mill incorporating diesel engine, belt transmission and hydraulic press (Stork-Amsterdam).

in countries where fuel is cheap, most mills now employ small diesel engines for driving either a central pulley shaft or an alternator connected with electric motors. Strippers are usually of the beater-arm type, the digester is fitted with steam injection and clarification is simplified by the use of a single compartmented tank with syphoning arrangements. Kernel extraction is of the simplest kind with a vertical nutcracker and screening drum following the rotary fibre separator, the kernels being separated from the shell in a clay bath.

The mills are usually easily extensible from 0.75 to 1.5 or 1.5 to 3 tonnes per hour by doubling the pressing unit employed. They have tended to become more sophisticated and so more expensive and this applies particularly to those with small screw presses. The main problem with mini-mills is assuring adequate supervision and recording of both quantities and quality. The worst feature of small-scale milling in America has been the lack of provision for milling control. Weighbridges or other means of bunch weighing are often not provided as standard equipment and the owners do not know either the extraction rates or efficiencies being obtained.

Milling control

The processing of bunches for the extraction of oil and kernels cannot be properly controlled unless what is going out of the mill is known in relation to what is entering it.

Coming into the mill are:
1. Bunches
2. Water
3. Steam
4. Dirt

Going out of the mill are:
1. Bunch refuse
2. Palm oil, containing varying quantities of free fatty acids, water and dirt
3. Palm kernels, containing varying quantities of shell, dirt and moisture
4. Sludge water
5. Sterilizer condensate

In addition press fibre and shell are going to the boilers to be used as fuel.

Control of quantity

The calculation of the efficiency of oil and kernel extraction would be much simplified if the composition of bunches was relatively constant or if sufficient of them could be sampled and analysed to give a representative analysis of the daily or weekly intake. Figures of efficiency for oil and kernel extraction could then be obtained by dividing the quantity going out of the mill by the calculated quantity coming in. The bunches vary so much in weight, ripeness and composition, however, that direct sampling as they enter the mill has proved an insufficiently accurate method of estimating the intake of palm oil and kernels (Velayatham, 1975). For this reason the majority of large mills are 'controlled' by a system of analyses for *losses*. All possible sources of loss are examined and analyses for oil on the one hand and kernel on the other are done. The system also necessitates the installation of means of weighing the total and separate outgoings of the mill such as bunch refuse and sludge water, and Southworth (1977) has concluded from his studies of milling control that the expense of these measurements, some-times opposed by mill managers, is fully justified. The losses method is admittedly imperfect and the total of oil produced plus oil losses are rightly recorded in some mill control forms as 'Total Accounted for' but wrongly recorded in others as 'Total Oil present in Bunch'. With oil there is inevitably some loss on machinery and some leakage or spillage which cannot or does not get recorded. Some of the kernel present may disappear as dust. The method does not therefore constitute a complete check on events and the need for supervision of all possible points of leakage remains.

In the quantitative control of the mill the absolute weights obtained are those of bunches entering the mill, clarified oil produced, and kernels bagged. From these weights the true *extraction rates* of the mill as a whole are obtained, e.g.

Month	January
Bunches milled	3,120.65 tonnes
Oil produced	576.03 tonnes
Kernels bagged	151.76 tonnes
Extraction rate: oil	18.45 per cent oil to bunch
Extraction rate: kernels	4.86 per cent kernels to bunch

Extraction rates depend on type of fruit, ripeness, age of palms, season, etc. so although they are exact and very useful for esti-mation of returns, they do not in themselves give any information on how well the mill is operating.

Most of the other data relating to quantity analysis are obtained by sampling and attention should therefore be given to its adequacy. There is still a great need to submit mill sampling methods to statisti-cal examination since most sampling methods, though sensible and possibly adequate, are empirical (Southworth, 1977). Some of the methods commonly used are mentioned below.

Oil losses

Sterilizer waste or 'condensate'

The quantity of sterilizer waste is measured, and samples taken at regular intervals from the discharge line are analysed for oil and hence the total quantity of oil lost over a period is estimated. In many mills this loss is considered too small to be included (Velayutham, 1975).

Bunch refuse

This (sometimes termed empty bunches, though they are not entirely empty of fruit) must be weighed *in toto* or sampled and analysed for (a) *unreleased fruit* and (b) *oil absorbed in the stalk and empty spikelets*. The weight of bunch refuse is commonly obtained by weighing daily about 10 per cent of the cages of refuse going to the incinerator or field. Alternatively a weighing machine is installed in the conveyer line to the incinerator. Samples of bunches are taken from the trucks or conveyer and the unreleased fruit is weighed and analysed for oil and kernel content. Fixed percentages may be used for unreleased fruit if the crop is reasonably uniform, but they must be checked from time to time. For oil absorbed in the refuse stripped bunches can be taken hourly at random from the stripper outlet conveyer and weighed. Chan (1977) found there was more oil in the stalks of overripe than underripe bunches and that more oil was absorbed by the empty spikelets than by the stalk, though there was a positive and significant correlation between the oil contents of stalk and spikelets. He recommended the sampling of three bunches every half-hour and quartering longitudinally to take one-quarter of each sampled bunch. Bunch refuse contains about 0.35 per cent of extractable waxes and in some laboratories this amount is deducted from the palm oil percentage figures obtained (Velayutham, 1975).

Fibre

The oil loss in fibre leaving the mill must be obtained from estimates of the mesocarp fibre entering the mill in the bunches and the oil to dry fibre leaving the extraction plant. The oil to dry fibre estimation is simple and may be done at any stage after the matter has left the press, but in view of the difference in oil content of fibre from different parts of the press cake the sample is perhaps best taken at the fibre cyclone. Then the quantity of nuts and kernels being loose in the fibre may also be analysed and estimated. Samples of about 200 g are taken throughout the day or shift and mixed and quartered down to duplicate 100 g samples.

The simplest method of estimating the amount of mesocarp fibre entering the mill is to determine by sampling the ratio of dry fibre to nuts in the press cake; the total fibre is then calculated from the known tonnages of nuts produced and bunches processed (Velayutham, 1975).

In sampling for oil to dry fibre, Southworth (1977) showed that variations between samples of a bulk sample were small compared with hourly variations, and he found that duplicate tests from a bulk sample made of hourly samples gave a result within 0.6 per cent oil to dry matter of the true figure in 95 out of 100 cases. He recommended bulking hourly samples to give a daily sample and duplicating the analysis, but he admitted that further work on sampling technique is required.

The figure of oil to dry fibre being the largest among the accounted losses and this loss obvious to the eye and touch, a great deal of notice has been taken of it. Moreover gradual improvements in digestion and extraction methods have tended to reduce the average figures over the years from a little under 20 per cent to well under 6 per cent. There has been a tendency even for the oil to dry fibre figure to be taken as a kind of measure of efficiency; and this may lead to a neglect of other and more easily rectified sources of loss. Two points should be borne in mind concerning the oil to dry fibre estimations. In the first place they will depend on the solvent used in the Soxhlet apparatus and will tend to decrease if the fibre is stored for any length of time before analysis. It is necessary, therefore, to adhere to the use of one solvent and to carry out the analysis as soon as possible. Secondly, it should be realized that a difference of 1 per cent in the oil to dry fibre results in a very small difference in estimated milling efficiency. For instance, in a mill where the total accounted losses were estimated to be 1.3 per cent of the weight of bunches entering the mill and the total oil produced by the mill was 17.2 per cent of the weight of bunches, an increase of oil to dry fibre from 9 per cent to 10.5 per cent only led to a reduction of estimated efficiency from 93.0 to 92.4 per cent. In general, with other losses unchanged, a decrease of 1 per cent in oil to dry fibre will only increase efficiency by about 0.4 per cent. Thus, while this loss should clearly not be treated as of no account, it is important that those in charge of mills should not concentrate their attention on oil to dry fibre to the exclusion of other losses which may be occurring, though less obviously, in other parts of the mill.

Sludge

The tonnage of sludge emitted for disposal and treatment is determined by using a basculator at the exit of the final sludge centrifuge or, if tanks are used, by counting the number filled and discharged. The ratio of sludge to bunches processed is then determined. Analysis of samples for oil and for solids not fat are carried out.

Regular samples are taken and mixed and the analysis is by day or by shift.

Nuts

There are various ways of estimating the small amount of oil lost on nuts. In some mills the nuts are weighed automatically entering the nut drier so that samples are taken at that point and analysed for shell, kernel, moisture and oil on shell and these estimations are related to the input of bunches. If the nuts are not weighed, however, then samples must be taken of nuts coming out of the fibre separator and, after cracking, the oil and moisture content of the shell must be estimated in the laboratory. The calculation of the oil loss will then depend on measurement of the quantity of shells produced and determination of the quantity of shell per unit of bunches entering the mill, and on estimation of the percentage of oil and solids not oil in the shell.

Kernel losses

Losses of kernels are incurred: (1) in bunch refuse; (2) among the shells going to the boiler; (3) sticking to shell; (4) in uncracked nuts among the shell; and (5) in nuts in fibre.

(1) and (5) have already been mentioned when oil losses in bunches and fibre were discussed; (2), (3) and (4) can be estimated by sampling the shell going to the boiler house and separately they give an indication of the working of the cracking section of the mill.

In relating these losses to the total weight of bunches entering the mill, account is taken of the quantity of nuts, if this is measured, and of the quantity of shell, fibre and bunch refuse.

Efficiencies

The efficiency of *oil extraction* is obtained from the estimations described, as follows:

$$\frac{\text{Oil produced} \times 100}{\text{Oil produced} + \text{oil losses accounted for}} = \text{estimated oil extraction efficiency}$$

The efficiency of *kernel extraction* is similarly obtained:

$$\frac{\text{Kernels produced} \times 100}{\text{Kernels produced} + \text{losses accounted for}} = \text{estimated kernel extraction efficiency}$$

or, if nuts are automatically weighed entering the nut drier and the dry weight of kernels estimated in nut samples, the efficiency of cracking and separating may be estimated as follows:

$$\frac{\text{Dry weight of kernels produced} \times 100}{\text{Dry weight of kernels entering nut drier}}$$

This, however, leaves out of account losses in bunch refuse and in fibre which take place before the nut drier stage.

The control of quality

The quality analyses required to be done in the mill laboratory are:

(*a*) *Oil.*
1. Moisture content. The sample is weighed to constant weight in a drying oven at 105 °C.
2. Dirt content. This is determined by processes of filtration.
3. Free fatty acid content. This is normally estimated by titration against caustic soda.

(*b*) *Kernels.* This is entirely a physical process, pieces of shell, dirt, and broken kernels being separated from weighed samples of bagged kernels and weighed separately. Moisture content is determined by grinding and drying to constant weight in an oven.

Determination of the amount of discoloration is a somewhat subjective process. Usually it is considered sufficient to divide the sampled kernels, which have been bisected with a sharp knife, into white, slightly discoloured, badly discoloured or mouldy.

(*c*) *Other determinations.* The majority of other determinations in the mill laboratory are either moisture or oil determinations. The former are carried out by drying in ovens to constant weight while the latter, e.g. of fibre, bunch refuse, is usually done in the Soxhlet apparatus.

Several publications are available on laboratory methods for milling control (Arnoth, 1963; IRHO, 1967).

In some mills measurements of the degree of ripeness of the bunches are attempted. This is in the nature of an overall check on the harvesting. The figures are open to misinterpretation since the amount of fruit loosening will depend, apart from ripeness, on the number of hours between harvest and the recording of the number of detached fruit, and on whether the bunches are dry or have become wet in the rain. In some mills the bunches are sampled on receipt and are classified according to the number of detached exterior fruit. In other mills the quantity of loose fruit delivered is measured and related to the total bunch plus loose fruit weight. In spite of the difficulties and admitted inaccuracies of these measurements it is certainly advantageous for the mill engineer to obtain some picture of the state of the produce he is receiving from the field so that he may be able to note any variations in his supply which may affect the quality of his products. In one scheme devised for a group of estates bunches are sampled each day and divided into three ripeness classes: (1) Ripe – more than 1 loose fruit per pound of bunch weight (2 per kg); (2) Overripe – less than one-eighth of the fruit retained in the bunch; (3) Underripe – bunch with less than 1 loose fruit per pound weight. Bunches are further classified and enumerated if found unsatisfactory in the following respects: (1) empty bunches; (2) poor fruit set; (3) diseased; (4) damaged by pests; (5) dirty; (6) bruised; (7) long stalks.

Mill maintenance

As the crop is being harvested throughout the year, overhaul or replacement of machinery often presents a serious problem and thorough routine maintenance is therefore very important. In seasonal climates such as Nigeria, where the harvest falls to a very low level at one period of the year, it is possible to close down the mill for overhaul for 2–3 weeks. Elsewhere, estates may either have sufficient standby capacity to take out some presses or a full line of machinery for overhaul and still continue processing, or bunches may by mutual agreement be sent to another mill for a short period.

References

Abdul Gapor and **Ong, A. S. H.** (1982) Some aspects of trace metals in palm oil; *and* Protection of oils and fats against oxidative deterioration with special reference to palm oil. *PORIM Bull*, No. 4, 19; No. 5, 39.

Ames, G. R., Raymond, W. D. and **Ward, F. B.** (1960) The bleachability of Nigerian palm oil. *J. Sci. Fd. Agric.*, **2**, 194.

Anon. (1917) The African oil palm industry. II. Machinery. *Bull. imp. Inst., Lond.*, **15**, 57.

Anon. (1986) Review of the Malaysian palm oil industry. *Planter, Kuala Lumpur*, **62**, 175.

Arnott, G. W. (1963) The Malayan oil palm and the analysis of its products. Min. Agriculture and Cooperation, Fed. of Malaya, *Bull*. 113.

Babatunde, G. M., Fetuga, B. L., Odumosu, O. and **Oyenuga, A.** (1975) Palm kernel meal as the major protein concentrate in the diets of pigs in the tropics. *J. Sci. Fd. Agric.*, **26**, 1279.

Barnes, A. C. (1925) Mechanical processes for the extraction of palm oil. *Second Special Bull. Dep. Agric. Nigeria*.

Barnes, A. C. (1926) An improved process for the extraction of palm oil by natives. The Cooker-Press process. *Fifth A. Bull. Dep. Agric. Nigeria*, p. 33.

Bek-Nielson, B. (1969a) Quality aspects of oil palm kernel production. In *The quality and marketing of oil palm products*, p. 161, Inc. Soc. of Planters, Kuala Lumpur.

Bek-Nielson, B. (1969b) Palm oil and kernel extraction plants in relation to quality. In *The quality and marketing of oil palm products*, p. 169, Incorp. Soc. of Planters, Kuala Lumpur, and *Oléagineux*, **26**, 483 and 635 (1971).

Bek-Nielson, B. (1974) Technical and economic aspects of the oil palm fruit processing industry. *United Nations publications*, ID/123, 40 pp.

Bek-Nielson, B. and **Krishnan, S.** (1979) Refining palm oil. *Planter, Kuala Lumpur*, **55**, 809.

Berger, K. G. (1983) Problems of palm oil handling and storage. In *Proceedings of* *regional workshop on palm oil mill technology and effluent treatment*, PORIM, Malaysia.

Bezard, J. A. (1971) The component tryglycerides of palm-kernel oil. *Lipids*, **6** (9), 630.

Bienaymé, A. (1954) Les huiles de palme et leur richesse en carotène. *Oléagineux*, **9**, 603.

Bienaymé, A. and **Servant, M.** (1958) Variation des caractéristiques des huiles de palme et notamment de leur caroténoides. *Qualitas Pl. Mater. vegé.*, **1**, 3/4, 336.

Blaak, G. (1979) A village palm oil mill. *Oil Palm News*, No. 23, 5.

Blommendaal, H. N. (1927) *De Fabricage van Palmolie*. AVROS Algemeene Serie No. 33.

Booth, V. H. (1957) *Carotene, its determination in biological material*. W. Heffer, Cambridge.

Chan, K. S. (1977) Sampling of oil palm bunch stalk refuse to determine oil loss. Int. Symposium on Oil Palm Processing and Marketing, Kuala Lumpur, 1976.

Chan, K. C. and **Chooi, C. F.** (1983) Ponding system for palm oil mill effluent treatment. *Proc. regional workshop palm oil technology and effluent treatment*, PORIM, Kuala Lumpur, p. 185.

Chin, A. H. G. (1979) Palm oil standards in relation to marketing and refining behaviour. *Planter Kuala Lumpur*, **55**, 414.

Clegg, A. J. and **Teh, Y. C.** (1972) Production de palmistes hydrolytiquement stables. *Oléagineux*, **27**, 101.

Cooper, I. N. and **Bevan, J. W. L.** (1968) Some factors to be considered when planning the organisation of processing oil palm products. In *Oil palm developments in Malaysia*, p. 118, Incorp. Soc. of Planters, Kuala Lumpur.

Coursey, D. G. (1961) Quelques observations sur l'altération de la couleur des palmistes. *Oléagineux*, **16**, 385.

Coursey, D. G. (1963) The deterioration of palm oil during storage. *J. W. Afr. Sci. Ass.*, **7**, 101.

Coursey, D. E., Simmons, E. A. and **Sheridan, A.** (1963) Studies on the quality of Nigerian palm kernels. *J. W. Afr. Sci. Ass.*, **8**, 18.

Cornelius, J. A. (1965) *Some technical aspects influencing the quality of palm kernels.* Paper presented at the Tropical Products Institute Oil Palm Conference, London, 1965, p. 105, Min. of Overseas Development.

Cornelius, J. A. (1973) The assessment of palm oil quality – international collaborative work. *Oil Palm News*, **15**, 1.

Cornelius, J. A. (1977a) Progress towards international standards for crude palm oil. Int. Symposium on Oil Palm Processing and Marketing, Kuala Lumpur, 1976.

Cornelius, J. A. (1977b) Palm oil and palm kernel oil. *Prog. Chem. Fats other lipids*, **15**, 5.

Cornelius, J. A. (1983) *Processing of oil palm fruit and its products*. Tropical Products Institute, London, 95 pp.

Crombie, W. M. (1956) Fat metabolism in the West African oil palm (*Elaeis guineensis*) Part I. Fatty acid formation in the maturing kernel. *J. exp. Bot.*, **7**, 181.

Crombie, W. M. and **Hardman, E. E.** (1958) Fat metabolism in the West African oil palm (*Elaeis guineensis*), Part III. Fatty acid formation in the maturing exocarp. *J. exp. Bot.*, **9**, 247.

Davis, J. B. (1978) Palm oil mill effluent: a review of methods proposed for its treatment. *Trop. Science*, **20**, 233.

Davis, J. B. and **Reilly, P. J. A.** (1980) Palm oil mill effluent – a summary of treatment methods. *Oléagineux*, **35**, 323.

Desassis, A. (1957) L'acidification de l'huile de palme. *Oléagineux*, **12**, 525.

Devendra, C., Yeong, S. W. and **Ong, H. K.** (1981) The potential value of palm oil mill effluent (POME) as a feed source for farm animals in Malaysia. In *PORIM–MOPGC workshop on oil palm by-product utilization*, PORIM, Kuala Lumpur.

Dyke, M. F-M. in **Leplae, E.** (1939) Le palmier à huile en Afrique, son exploitation au Congo-Belge et en Extrême-Orient. *Mém. Inst. r. colon. belge Sect. Sci. nat. méd.*, **7** (3), 1–108.

Eckey, E. W. (1954) *Vegetable fats and oils.* Reinhold Publishing Corp., New York.

Eggins, H. O. W. and **Coursey, D. G.** (1964) Thermophilic fungi associated with Nigerian oil palm produce. *Nature, Lond.*, **203**, 1083.

Faulkner, O. T. and **Lewin, C. J.** (1923) Native methods of preparing palm oil. II. *Second A. Bull. Dep. Agric. Nigeria*, p. 3.

Geddes, A. M. W. (1976) The low cost oil palm mill project in Sierra Leone. *Appropriate Technology*, **3**(2), 6.

Georgi, C. D. V. (1932) The centrifugal extraction of palm oil at Serdang. *Malay. agric. J.*, **20**, 446.

Georgi, C. D. V. (1933) Comparison of the press and centrifugal methods for treatment of palm oil fruit. *Malay. agric. J.*, **21**, 103.

Gray, J. E. (1922) Native methods of preparing palm oil. *First A. Bull. Dep. Agric. Nigeria*, p. 28.

Hadcock, M. (1983) It had to be simple, cheap, reliable and easy to clean. *International Agricultural Development*, Jan./Feb., 1983, 16.

Hardon, J. J. (1969) Interspecific hybrids in the genus *Elaeis.* II. Vegetative growth and yield of F_1 hybrids *E. guineensis* × *E. oleifera. Euphytica*, **18**, 380.

Hartley, C. W. S. and **Nwanze, S. C.** (1965) *Factors responsible for the production of poor quality oils.* Paper presented at the Tropical Products Institute Oil Palm Conference, London, 1965, p. 68, Ministry of Overseas Development.

Hemming, M. L. (1977) The treatment of effluents from the production of palm oil. In *International development in palm oil*, ISP, Kuala Lumpur.

IRHO (1967) *Manuel de l'huilerie de palme.* Série Sci. No. 12, Paris, new edition.

Jacobsberg, B. (1971) La Production d'une huile de palme de haute qualité. *Oléagineux*, **26**, 781.

Jacobsberg, B. (1983) Quality of palm oil. *PORIM Occ. paper*, No. 10.

Jacobsberg, B., **Deldime, P.** and **Abdul Gapor** (1978) Tocopherols and tocotrienols in palm oil. *Oléagineux*, **33**, 239.

Jasperson, H. and **Pritchard, J. L. R.** (1965) Factors influencing the refining and bleaching of palm oil. Paper presented at the Tropical Products Institute Oil Palm Conference, London, 1965, p. 96, Min. of Overseas Development.

Jorgensen, H. K. (1983) The U.P. decanter–drier system for reduction of palm oil mill effluent. *Proc. regional workshop palm oil technology and effluent treatment*, PORIM, Kuala Lumpur, p. 201.

Jorgensen, H. K. and **Gurmit Singh** (1981) An introduction of the decanter–drier system in the clarification station for crude oil and sludge treatment. In *The Oil palm in agriculture in the eighties*, ISP, Kuala Lumpur.

Kilby, P. (1967 and 1968) The Nigerian palm oil industry. *Food Res. Inst. Studies*, **7**, 2, 177; and **8**, 2, 199.

Leong, W. L. and **Berger, K. G.** (1982) Storage, handling and transport of palm oil products. *PORIM Technology*, No. 7.

Loncin, M. (1952) L'hydrolyse spontaneé autocatalitique des triglycérides. *Oléagineux*, **7**, 695.

Loncin, M. and **Jacobsberg, B.** (1963) Studies in Congo palm oil. *J. Am. Oil Chem. Soc.*, **40**, 18.

Loncin, M. and **Jacobsberg, B.** (1964) *Study on palm kernel acidification during storage.* Int. Soc. for Fat Research Congress, Hamburg.

Loncin, M. and **Jacobsberg, B.** (1965) Recherches sur l'huile de palme en Belgique et au Congo. (*Research on palm oil in Belgium and the Congo.*) Paper presented at the Tropical Products Institute Oil Palm Conference, London, 1965, p. 85, Ministry of Overseas Development.

Ma, A. N., **Chow, C. S.**, **John, C. K.**, **Ahmad Ibrahim** and **Zain Isa** (1983) Palm oil effluent survey – a survey. *Proc. regional workshop palm oil technology and effluent treatment*, PORIM, Kuala Lumpur, p. 123.

Ma, A. N. and **Ong, S. H.** (1982) Anaerobic digestion of palm oil mill. *PORIM Bull.*, No. 4, 35.

Macfarlane, N., Swetman, T. and **Cornelius, J. A.** (1975) Analysis of mesocarp and kernel oils from the American oil palm and F_1 hybrids with the West African oil palm. *J. Sci. Fd. Agric.*, **26**, 1293; and *Oil Palm News*, **19**, 12 and **20**, 1.

Manlove, D. and **Watson, W. A.** (1931) Press extraction of palm oil in Nigeria. *Tenth A. Bull. Dep. Agric. Nigeria*, p. 19.

Martinenchi, G. B. (1972) Traitements de l'huile de palme. 1. Séparation en fractions liquide et solide. *Oléagineux*, **27**, 267.

Maycock, J. H. (1975) The developments in palm oil factory design since the early 1900s. *Planter, Kuala Lumpur*, **51**, 335.

Milsum, J. N. and **Georgi, C. D. V.** (1938) Smallscale extraction of palm oil. *Malayan Agr. J.*, **26**, 53.

Naudet, M. and **Faulkner, H.** (1975) Compositions et structures glycéridiques comparées des huiles d'*Elaeis guineensis, Elaeis melanococca* et d'hybride *guineensis–melanococca. Oléagineux*, **30**, 171.

Ng, B. H., Corley, R. H. V. and **Clegg, A. J.** (1976) Variation in the fatty acid composition of palm oil. *Oléagineux*, **31**, 1.

Nwanze, S. C. (1961a) *WAIFOR ninth annual report 1960–1*, p. 96.

Nwanze, S. C. (1961b) The economics of the Pioneer mill. *J. W. Afr. Inst. Oil Palm Res.*, 3, 233.

Nwanze, S. C. (1961–4) *WAIFOR ninth, tenth and twelfth annual reports, 1960–1, 1961–2 and 1963–4*, pp. 98, 90 and 97 respectively.

Nwanze, S. C. (1962) *WAIFOR tenth annual report 1961–2*, p. 89.

Nwanze, S. C. (1963–4) *WAIFOR eleventh annual report 1962–3*, p. 88; *Twelth annual report, 1963–4*, p. 96.

Nwanze, S. C. (1965a) The hydraulic hand press. *J. Nigerian Inst. Oil Palm Res.*, 4, 290.

Nwanze, S. C. (1965b) Semi-commercial scale palm oil processing. *Paper presented at the Tropical Products Institute Oil Palm Conference, London, 1965*, p. 63, Min. of Overseas Development.

OPGL Oil palm genetics laboratory, Malaysia (1972) Progress report.

Olie, J. J. (1969) The active development of process technology in the recovery of palm oil and kernels. *Oléagineux*, **24**, 293.

Olie, J. J. (1973) The Stork twin screw press. *Oléagineux*, **28**, 33.

Olie, J. J. and **Tjeng, T. D.** (1972) Traitement et évacuation des eaux résiduaires d'une huilerie de palme. *Oléagineux*, **27**, 215.

Olie, J. J. and **Tjeng, T. D.** (1974) Kernel recovery. In *The extraction of palm oil*, pp. 63–9, Stork-Amsterdam.

Oso, B. A. (1979) Thermophylic fungi and the deterioration of Nigerian oil palm kernels – effect on oil content and quality. *Econ. Bot. 33*, 58.

Petitpierre, G. (1982) Le traitement des effluents d'huilerie de palme et la production de biogaz. *Oléagineux*, **37**, 367.

PORIM (1987) Information kindly supplied by Mr T. P. Pantzaris of PORIM Liaison Office, England.

Pritchard, J. L. R. (1969) Quality of oil palm products – user requirements, *Trop. Sci.*, **11**, 103.

Purvis, C. (1957) The colour of oil palm fruits. *J. W. Afr. Inst. Oil Palm Res.*, 2, 142.

Purvis, M. J. (1968) The Nigerian palm oil industry: a comment. *Food Res. Inst. Studies*, **8**, 2, 191.

Quah, S. K., Lim K. H., Gillies, D., Wood, B. J. and **Kanagaratnam, J.** (1983) Sime Darby POME treatment and land application systems. *Proc. regional workshop palm oil technology and effluent treatment*, PORIM, Kuala Lumpur, p. 193.

Quraishi, A. and **Macfarlane, N.** (1975) The nitrogen content and amino acid composition of palm kernel meals. *Oil Palm News*, **20**, 3.

Rossell, J. B. (1986) Classical analysis of oils and fats. In *Analysis of oils and fats,*

eds Hamilton, R. J. and Rossell, J. B., London and New York.
Siew, W. L. and **Berger, K. G.** (1981) Malaysian palm kernel oil, chemical and physical characteristics. *PORIM Technology*, 6.
Singh, Kirat and **Ng Siew Hoong** (1968) Treatment and disposal of palm oil mill effluent. *Malaysian Agric. J.*, **46**, 316.
Southworth, A. (1977) Process control. Int. Symposium on Oil Palm Processing and Marketing, Kuala Lumpur, 1976.
Stork (1960a) *Stork Palm oil-review*, **1**, 3.
Stork (1960b) *Stork Palm oil-review*, **1**, 4.
Stork (1960–1) *Stork Palm oil-review*, **1**, 5, **2**, 1.
Stork (1961a) *Stork Palm oil-review*, **2**, 2, 3.
Stork (1961b) *Stork Palm oil-review*, **2**, 2–5.
Stork (1962a) *Stork Palm oil-review*, **3**, 2.
Stork (1962b) *Stork Palm oil-review*, **3**, 3.
Stork (1963) Kernel recovery. *Stork Palm oil-review*, **3**, 4–5.
Stork (1983) *Machinery for the oil palm industry. 4. Pressing station.* Stork-Amsterdam.
Tam, T. K., Yeow, K. H. and **Poon, Y. C.** (1983) Land application of palm oil mill effluent (POME) – H & C experience. *Proc. Regional workshop palm oil technology and effluent treatment*, PORIM, Kuala Lumpur, p. 216.
Tan, B. K. and **Flingoh, C. H.** (1981) Malaysian palm oil, chemical and physical characteristics. *PORIM Technology*, 3
Thieme, W. L. and **Olie, J. J.** (1969) Discoloration of oil palm kernels in relation to processing temperature and time. In *The quality and marketing of oil palm products*, p. 144, Inc. Soc. of Planters, Kuala Lumpur.
Thomas, R. L. *et al.* (1971) Fruit ripening in the oil palm *Elaeis guineensis. Ann. Bot.*, **35**, 1219.
Turner, P. D. (1969) The importance of lipolytic microorganisms in the degradation of oil palm products in Malaysia. In *The quality and marketing of oil palm products*, p. 53, Inc. Soc. of Planters, Kuala Lumpur.
Unilever (1967) **Plantation Engineering Division**, Current Unilever developments in palm oil extraction equipment and machinery. Sabah Planters' Association. Oil Palm Technical Seminar. Mimeograph.
Van Heurn, F. C. (1921) *Considérations sur l'installation de fabriques d'huile de palme.* AVROS Com. Gen. Ser. No. 10.
Vanneck, C. and **Loncin, M.** (1951) Considération sur l'altération de l'huile de palme. *Bull. agric. Congo belge.* **42**, 57.
Velayatham, A. (1975) Palm oil process control. *Planter, Kuala Lumpur*, **51**, 386.
Weko, B. H. and **Sutiardjo,** (1969) A comparison of the hydrocyclone and clay bath separation methods for oil palm kernels. In *The quality and marketing of oil palm products*, p. 154, Incorp. Soc. of Planters, Kuala Lumpur.
Whiting, D. A. M. and **Lim, K. H.** (1983) Harrisons and Crosfield system for palm oil mill effluent treatment. *Proc. regional workshop palm oil technology and effluent treatment*, PORIM, Kuala Lumpur, p. 177.
Wohlfahrt, N. (1969) Clarification de l'huile de palme. *Oléagineux*, **24**, 699.
Wolversperges, A. (1963) The extraction of palm oil by means of screw presses. *Planter, Kuala Lumpur*, **39** (1), (2) and (3), pp. 11, 68, 111.
Wood, B. J. (1977) A review of current methods for dealing with palm oil mill effluents. *Planter, Kuala Lumpur*, **54**, 477.
Zachariassen, B. (1969) Moisture removal from palm oil. In *The quality and marketing of oil palm products*, p. 139, Incorp. Soc. of Planters, Kuala Lumpur, and *Oléagineux*, **25**, 543.

Index